Genes, Cells

A·N·D

Organisms

GREAT BOOKS IN EXPERIMENTAL BIOLOGY

EDITED BY
John A. Moore

A Garland Series

17 TITLES THAT STAND AS
MONUMENTS OF BIOLOGICAL THOUGHT

GREAT BOOKS IN EXPERIMENTAL BIOLOGY

1. *THE CELL IN DEVELOPMENT AND HEREDITY*, by Edmund B. Wilson. 1925.

2. *THE CELL THEORY: A RESTATEMENT, HISTORY, AND CRITIQUE*, by John R. Baker. 1948–1955.

3. *SPECIES AND VARIETIES, THEIR ORIGIN BY MUTATION: LECTURES DELIVERED AT THE UNIVERSITY OF CALIFORNIA*, by Hugo De Vries. 3rd edition, 1912.

4. *THE GENETICS OF DROSOPHILA*, by Thomas Hunt Morgan (with Calvin B. Bridges and A. H. Sturtevant). 1925.

5. *THE THEORY OF THE GENE*, by Thomas Hunt Morgan. 1926.

6. *AN INTRODUCTION TO GENETICS*, by A. H. Sturtevant and G. W. Beadle. 1939.

7. *RECENT ADVANCES IN CYTOLOGY*, by C. D. Darlington. 1937.

8. *PHYSIOLOGICAL GENETICS*, by Richard Goldschmidt. 1938.

9. *INVERTEBRATE EMBRYOLOGY*, edited by Matazo Kume and Katsuma Dan. 1968.

10. *EMBRYONIC DEVELOPMENT AND INDUCTION*, by Hans Spemann. 1938.

11. *THE ORGANIZATION OF THE DEVELOPING EMBRYO*, by Ross G. Harrison. 1969.

12. *PROBLEMS OF FERTILIZATION*, by Frank R. Lillie. 1919.

13. *THE BIOLOGY OF THE CELL SURFACE*, by Ernest Just. 1939.

14. *FEATURES IN THE ARCHITECTURE OF PHYSIOLOGICAL FUNCTION*, by Joseph Barcroft. 1934.

15. *THE NEUROPHYSIOLOGICAL BASIS OF MIND; THE PRINCIPLES OF NEUROPHYSIOLOGY*, by John Carew Eccles. 1953.

16. *RECOLLECTIONS OF MY LIFE*, by Santiago Ramon Y. Cajal. 1937.

17. *THE DISCOVERY AND CHARACTERIZATION OF TRANSPOSABLE ELEMENTS: THE COLLECTED PAPERS OF BARBARA McCLINTOCK*. 1987.

INVERTEBRATE EMBRYOLOGY

Matazo Kumé
Katsuma Dan

Garland Publishing, Inc.
NEW YORK & LONDON • 1988

Library of Congress Cataloging-in-Publication Data

Musekitsui dōbutsu hasseigaku. English.
Invertebrate embryology/edited by Matazō Kume and Katsuma Dan;
translated from Japanese by Jean C. Dan under NSF Grant G-8663.
p. cm.—(Genes. cells and organisms)
Reprint. Originally published: Tokyo: Bai Fu Kan Press, 1957.
Translation of: Musekitsui dōbutsu hasseigaku.
Bibliography: p.
ISBN 0-8240-1381-6
1. Embryology—Invertebrates. I. Kume, Matazō, 1899–1976.
II. Dan, Katsuma, 1904- . III. Title. IV. Series.
QL958.M9713 1988
592'.033—dc19 88-21329

All volumes in this series are printed on acid-free, 250-year-life paper.

PRINTED IN THE UNITED STATES OF AMERICA

INVERTEBRATE EMBRYOLOGY

MUSEKITSUI DOBUTSU HASSEIGAKU

INVERTEBRATE EMBRYOLOGY

Edited by

MATAZO KUMÉ

Ochanomizu University

and

KATSUMA DAN

Tokyo Metropolitan University

Translated from Japanese by

JEAN C. DAN

under NSF Grant G-8663

Published for the National Library of Medicine, Public Health Service, U.S. Department of Health, Education and Welfare, and the National Science Foundation, Washington, D. C., by the NOLIT, Publishing House, Belgrade Yugoslavia, 1968.

INVERTEBRATE EMBRYOLOGY

Originally published by
BAI FU KAN PRESS, Tokyo. 1957

Printed in Yugoslavia by
PROSVETA, Belgrade
Đure Đakovića 21.

FOREWORD

Embryology is a branch of learning which, starting with the embryo and following it through its larval stages to adulthood, inquires into the developmental processes of organs and body systems. To the extent that such inquiry takes as its aim the accurate grasping of the form of an organism at successive moments, it is a branch of morphology or anatomy. However, unlike anatomy which, strictly speaking, is concerned with the structure of the fully developed organism, the real objective of embryology is the developmental process itself. Embryology can thus be called anatomy which includes a complicated time factor. It is for this reason that so much of the work in medical anatomy runs parallel to embryological problems.

It often happens that the time factor, when it is an essential element in a particular field of study, plays an important role in determining the nature of that field. For example, in order to ensure accuracy of measurement, physiology tries to think in sections of time reduced to thousandths of a second (sigma). Genetics adjusts its time scale to the span of one generation; the theory of evolution adopts the time standard of geological eras. If we deal with time units that are too short, we transcend the capacities of our five senses and have to cling to an oscillograph as a blind man to his cane; if, however, we have recourse to over-long intervals, changes are evident only to our imagination and we lose the possibility of actually seeing them. Embryological changes, on the contrary, occur according to a time scale measured in hours, days, weeks — the units with which we are familiar in everyday life; this is no doubt the reason why embryology presents us with the most vivid images.

Embryological study often makes use of normal development charts. To unaccustomed eyes, the differences that these show between one stage and the next may appear insignificant, but if one thinks that these are the differences betwen morning and afternoon — the growth that occurred between sleeping and waking, he may be able to make a beginning at grasping the actual feeling of embryological research.

However, it is not unreasonable that embryologists, confronting objects which change thus incessantly, day and night, should sooner or later want to investigate the causes of such change. This field was cultivated chiefly by the "pure" biologists, and achieved its most brilliant flowering in the branch known as Developmental Mechanics (Entwicklungsmechanik). However, such investigation continues unabated. Researchers of the present day have gone beyond from, and are closing in upon a physiological abstraction of the concept of *Youngness*, characteristic of the cells which make up a developing embryo. Having reached this point, the embryological concept will finally embrace the phenomenon of regeneration as well as the cancer problem; at present research is being carried out on the basis of a close cooperation between embryologists on the one hand, and pathologists and physiologists on the other.

But no matter how new the field of study which is entered upon, it must have as its basis an accurate grasp of developmental processes. All about us there are examples of apparently new research attempts which have ended in utter failure because of the lack of this plain foundation. Notwithstanding the vital importance of this requirement, there are such great difficulties in the way of ppublishing a work on embryology, and particullarly a book which will cover the whole range of invertebrate embryology, that a search through the whole world will disclose only three titles: MacBride, *Textbook of Embryology* (1914); Dawydoff, *Traité d'Embryologie comparée des Invertébrés* (1928); Korschelt und Heider, *Vergleichende Entwicklungsgeschichte der Tiere* (1936).

The present book was planned at the suggestion of Professor Taku Komai; we are happy to have received contributions to it from a large number of specialists representing the highest degree of our country's knowledge. The task of bringing the many branches of the subject together into a single volume has not been easy, but we have decided to publish the result in its present form.

The editors wish to express their sincere thanks to all their colleagues who have responded to their request for contributions. Furthermore, since they are resolved to continue improving the book bit by bit with the hope of eventually bringing it somewhat closer to perfection, they will be happy if future readers will inform them of any corrections which can be made in the text.

September, 1957 *The Editors*

TABLE OF CONTENTS

INTRODUCTION

Chapter 1

1. FORMATION OF EGGS AND SPERMATOZOA

"Which came first, the chicken or the egg?" is an old and well known problem. This is indeed an embryological problem and in order to trace the source of the reproductive cells, we must necessarily go back not only to the parental generation, but to a youthful stage of it. The eggs and spermatozoa with which we will be concerned in this first section are wholly the products of the parental generation. During the youth of the parents, the gonads, although structurally complete, are not actively functioning, but even in this stage they enclose the system of cells which will later serve as the reproductive cells. These are called the *primordial germ cells.*

(1) GONIA AND AUXOCYTES

As sexual maturity approaches, the primordial germ cells begin to divide actively in order to achieve increase in number. This is called the *multiplication period* (Fig. 1.1 a); during this time these cells are simply said to be prospective reproductive cells, without there being the slightest difference, so far as their chromosomes are concerned, between them and the somatic cells. In the female, the reproductive cells of the multiplication period are called *oogonia* (sing., *oogonium*); in the male; *spermatogonia* (sing., *-nium*). When the degree of maturity of the parent animals advances still further, these gonia stop dividing, and a period of increase in cell size follows; this is called the growth period (Fig. 1.1 b). During this period, accompanying the size increase of the cell, the nucleus, called *germinal vesicle* in this stage, also undergoes a striking increase in volume, and in preparation for the reduction division which will take place shortly, the *homologous chromosomes* unite (one each from the paternal and maternal sets). The formation of this special form of chromosome, the *tetrad,* is the most important characteristic of this period (see de Robertis, Novinski and Saez 1954, Chap. 9; Kuwada (ed) 1956, Chap. 5). Because of the phenomenon of chiasma, the tetrad chromo-

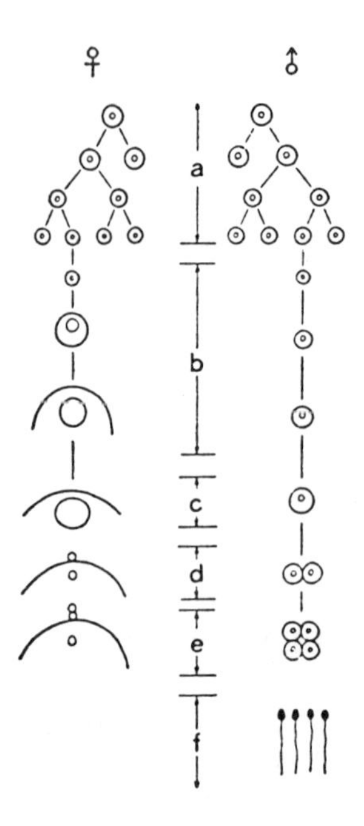

Fig. 1.1 Diagram of oogenesis and spermatogenesis

a. multiplication period; b. growth period; c. primary oocyte (left) and primary spermatocyte (right); d. secondary oocyte and secondary spermatocyte; e. ovum with polar bodies (left) and spermatids (right); f. spermatozoa.

somes often take the shape of crosses or rings at the metaphase, and it is thus possible to distinguish them from ordinary somatic chromosomes. By the time the tetrad chromosomes are fully developed, the cell also has achieved its greatest size; in the female, the cells at this stage are called *primary oocytes*, and in the male, *primary spermatocytes* (collectively *auxocytes* Fig. 1.1 a). But although in both male and female the auxocytes are larger than their respective gonia, the size increase of the oocyte is incomparably greater than that of the spermatocyte. In the case of the sea urchin, the volume ratio of oocyte to spermatocyte is 400,000 : 1.

(2) REDUCTION DIVISION

When the preliminary arrangements of the primary auxocytes are complete, they begin the *reduction division*. This process, which is also referred to as *meiosis*, is characterized by two successive cleavages. In the first meiotic division, the tetrad chromosomes are reduced to *diads*, and the two resulting cells containing such chromosomes are called *secondary oocytes* and *secondary spermatocytes*, respectively (Fig. 1.1 d). At the second meiotic division which follows immediately, the diad chromosomes finally become single unit *monads*, so that from each primary auxocyte, four daughter cells are formed. Of course in these daughter cells the maternal and paternal homologous chromosomes are randomly segregated and the number of chromosomes is reduced to one-half.

The four cells formed in the testis are called *spermatids* (Fig. 1.1 e right); each of these will later undergo a change in form and become a *spermatozoon* (*pl.* *-zoa.* Fig. 1.1 f). The corresponding process in the ovary, however, while following in principle the course described above, shows certain superficial differences. As the result of the first meiotic division which, unlike the equal cleavage of the spermatocyte, is extremely unequal, there is no perceptible diminution in the size of the larger daughter cell, while the smaller one is so minute that it is often overlooked completely (Fig. 1.1 d left). But since both these cells, regarded from the chromosome standpoint, contain diad chromosomes and are consequently identical with the secondary spermatocytes, in essence they are sister cells. However, regardless of these considerations, the smaller cell is so modified that it can take no part in development, and the term *'secondary oocyte'* is applied only to the larger cell. The small cell is not extruded from any random spot on the oocyte surface in an arbitrary fashion; since its position is fixed at a point which has a close connection with the polarity of the egg, as will be described later (p. 57), it is expressly called a *polar body*.

As the following meiotic division is again, in the same manner, extremely unequal, another polar body is formed. In order to differentiate between them, the earlier one is called the *first polar body*, and the later one, the *second polar body*.

Following the formula of spermatocyte meiosis, the first polar body ought to cleave simultaneously with the formation of the second polar body in order to complete the set of four cells. However, in actual fact the cleavage of the first polar body is ordinarily omitted so that the total number of cells resulting from meiosis is only three, and the one large cell is the mature *ovum* (pl. *ova*. Fig. 1.1 e, left). Internally the first polar body contains diad chromosomes, while the second polar body and the egg contain monads. Evidence that such production of only three daughter cells in oogenesis is definitely a secondarily imposed deficiency can be found in some molluscs. Under normal conditions oysters and mussels always produce an egg and two polar bodies, but in the absence of calcium or after treatment with trypsin which weakens the egg membrane, four cells are formed, giving a perfect parallel with the process occurring in spermatogenesis. In contrast to the egg, which possesses the capacity for development after fertilization, the polar bodies have completely lost this characteristic, with the possible exception of the polyclad order of flatworms, concerning which there is a report of polar bodies' being fertilized and developing into larvae (Korschelt u. Heider, 1936, Fig. 10). Among insects the polar bodies are often found to proliferate, without being fertilized, and take part in forming the nutritive membrane (trophamnion) which envelopes the embryo (Korschelt u. Heider 1936, Fig. 765), but this is not part of the embryo itself. Even when abnormally large polar bodies were produced experimentally in the parasitic roundworm *Ascaris* (Kautzsch 1912, 1913), and the mollusc *Crepidula* (Conklin 1916, 1917), they failed to develop. These observations lead to the interesting conclusion, that while in terms of chromosome content the polar body is a sister of the egg cell, some other factor causes it to be a sort of physiological cripple.

The spermatid has still to metamorphose into a spermatozoon, and must therefore remain in the testis for a while longer, but since the egg is mature as soon as it has given off the second polar body, there is no reason why it should not be spawned. In fact, the meiotic divisions take place within the ovary in some species, outside the body after spawning in others, and in some cases, even after sperm entrance (see pp. 18—26). For this reason, in the egg the meiotic divisions — i.e., the formation of the polar bodies — are sometimes called the *'maturation divisions'*.

(3) VARIOUS TYPES OF SPERMATOZOA

The typical spermatozoon, as shown in Figure 1.2, is equipped with three parts: a *head, a middle piece* and a *tail.* The anterior part of the head is frequently pointed, forming the *acrosome*; the remainder of it is the nucleus.

The acrosome originates in the *Golgi bodies* of the spermatocyte. The middle piece is chiefly formed from the spermatocyte *mitochondria,* and has been said to include a *centrosome.* However, the results of recent electron microscopical research do not entirely agree with the formerly held idea that the centrosome and tail originate in the middle piece. In the sea urchin spermatozoon, the centriole is ensconced in a deep depression in the nucleus located at the border between nucleus and mid-piece, and the tail pierces the mid-piece (Afzelius 1955; Rothschild 1956b). As will be described later, in the starfish spermatozoon the mid-piece and tail sometimes become separated from each other (see Chap. 9, Sect. 3, on starfish). The tail is the spermatozoan organ of locomotion; through its center

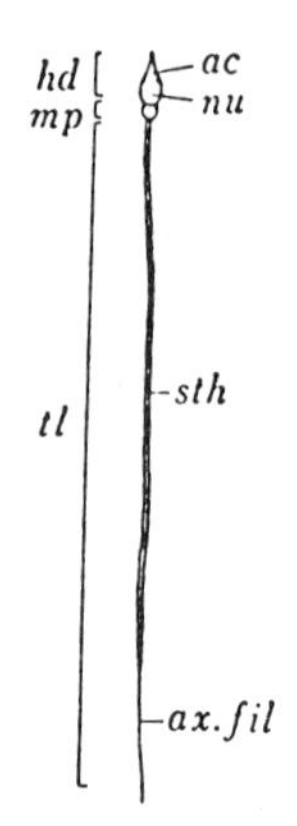

Fig. 1.2 Diagram of spermatozoon

ac: acrosome, *ax.fil:* axial filament, *hd:* head, *mp:* middle piece, *nu:* nucleus, *sth:* sheath, *tl:* tail.

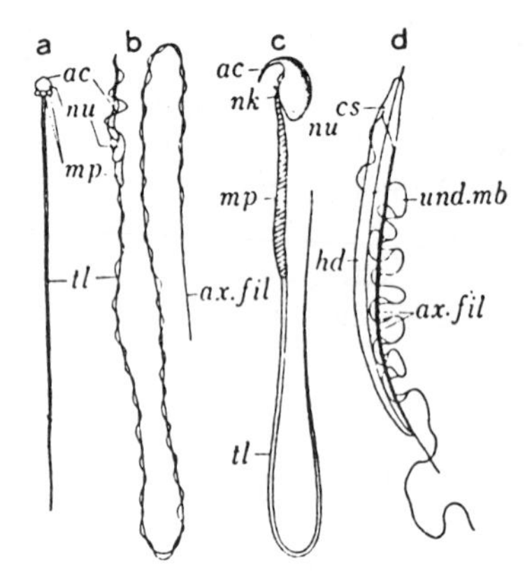

Fig. 1.3 Various types of spermatozoa (Retzius)

a. oyster, *Crassostrea*; b. bird, *Chloris*; c. mouse, *Mus*; d. toad, *Bombinator*. *ac:* acrosome, *ax. fil:* axial filament, *cs:* centrosome, *hd:* head, *mp:* middle piece, *nk:* neck, *nu:* nucleus, *tl:* tail, *und.mb:* undulatory membrane.

passes an *axial filament.* This filament is surrounded by a *sheath* which, however, is lacking at the posterior end so that the axial filament is exposed and the tail is slenderer in this region. The ordinary microscope is not well suited for observing the details of living spermatozoa — this is better done with a phase contrast microscope. The spermatozoa of man (*Homo sapiens*) sea urchins, etc., conform in general to the type described above.

However, a broad survey of animal species discloses many variations in spermatozoan morphology, such that it is sometimes even difficult to identify the head, midpiece and tail. Figure 1.3 shows some characteristic examples chosen from the drawings of Retzius (1904—1914).

As is immediately apparent from Figure 1.3, the various shapes assumed by the acrosome make it obvious that its function is to enable the spermatozoon to pierce the egg surface, and prevent it from slipping out, once it has entered. To emphasize this meaning the acrosome is also called a *perforatorium*. The most extreme variation in spermatozoan form occurs among the Arthropoda (see Fig. 10.3); parallelling their fantastic morphology, the behavior of these spermatozoa in the fertilization process also differs widely from that of other forms, as will be described below (Chap. 1, Sect. 3(2) e., f.; Chap. 10).

Sperm Dimorphism. It should be mentioned here that some animal species normally have two morphologically different kinds of spermatozoa. This is known as '*sperm dimorphism*'. As can be seen in the Prosobranchiata among molluscs (Taki 1939) and in the Lepidoptera among the insects (Toyama 1894), in addition to the difference in form, the spermatozoa of one type possess chromatin *(eupyrine spermatozoa)* while those of the other type have lost part of their chromatin *(apyrine)* and take no part in fertilization. No explanation has as yet appeared to answer the question of why such spermatozoa, lacking in fertilizing capacity, should be produced normally. Ishizaki and Kato of Tokyo University of Education are at present investigating, with the electron microscope, the apyrine sperm of the fresh water snail, *Paludina*. In animals in which sex is determined by the presence or absence in the spermatozoon of a sex chromosome, there must always be two kinds of spermatozoa so far as the chromosome situation is concerned, but these cannot be distinguished by external observation, and this case is not treated as dimorphism.

(4) FORMATION OF YOLK AND SURFACE COATS

Excepting the eggs of sponges and coelenterates, which may perform amoeboid movement (see Chap. 2; Chap. 3), animal eggs are generally spherical cells, and in morphological diversity fall far short of the spermatozoa, although in the possession of a gigantic volume and in most cases, of some sort of external covering (vitelline membrane, chorion, jelly layer, etc.), eggs clearly out-distance sperm cells. The large size of eggs is due to the fact that they contain a reserve stock of *yolk*. Since the nutriment contained in the yolk is the sole means of guaranteeing the safety of the embryo, various methods are employed in forming this yolk.

In the first place, the simplest method is for the egg to take in nutritive substances directly from the parent through the ovarian wall. An ovary of the simplest structure can be considered as a bag composed of a single layer of cells; in such a case the oogonia also form part of the ovarian wall. As shown in Figure 1.4, these cells gradually enlarge, and when they have developed sufficiently, they separate from the wall of the ovary and are set free in its lumen. The transparent body walls of coelenterates and other forms enable this process to be followed by external observation. There are also cases like that of *Urechis*, in which the eggs, after being set free from the ovary, continue their growth in the body cavity, taking nutritive substances from the body fluid.

However, it requires a long time to store up yolk by this sort of method, and other ways are used to amass a large amount of yolk in a short time. Sometimes the oogonia are accompanied by attendant nutritive cells called nurse cells. These nurse cells have a special capacity for synthesizing protein, and chiefly manufacture yolk or yolk-precursor substances and transmit them to the oocyte (Fig. 1.5).

Again, there are cases in which one oocyte annexes several of its sister cells. Since the nuclei of the amalgamated oocytes eventually degenerate, the final result is a large cell with a single nucleus; this is a method found in forms with large eggs, like the insects.

The time when amalgamation takes place varies, but the number of cells in most cases is fixed, giving the impression that they form a set. A consideration of spermatogenesis in insects makes the existence of such sets the more probable. As described above, the spermatogonia pass through the multiplication stage without amalgamating, but a certain number of divisions before the end of this stage each spermatogonium forms a membrane around itself so that all the cells resulting from the succeeding divisions make up a clump within this membrane. Such a clump is called a *cyst*, and all the cells in it divide at exactly the same pace, so that the insect testis presents a very regular and clearcut appearance. In the fruit-fly, *Drosophila melanogaster*, the spermatogonia divide four times after cyst formation, to end the multiplication period; in consequence there are invariably sixteen cells forming the group inside the cyst (Cooper 1950). Bringing the story back to the oocyte, it is extremely interesting that in the case of the water-beetle, *Dytiscus*, illustrated in Figure 1.6, fifteen nurse cells are amalgamated with the one future egg cell, giving the same total of sixteen cells as is found in the male. In other words, one completed oocyte corresponds to a cyst of the testis. Among the tipulid flies, it is reported that of the sixteen cells, one which has a special type of chromatin containing DNA undergoes meiosis, while the others which do not receive this chromatin become nurse cells (Bayreuther 1952, 1956).

Fig. 1.4 Growth of oocyte and formation of vitelline membrane in the bivalve *Cyclas* (Stauffacher)

att: point of attachment to ovarian membrane (animal pole), *g.vs:* germinal vesicle, *nuo:* nucleolus, *nut:* nutrient material flowing into oocyte, *oc:* young oocyte, *v. mb:* vitelline membrane.

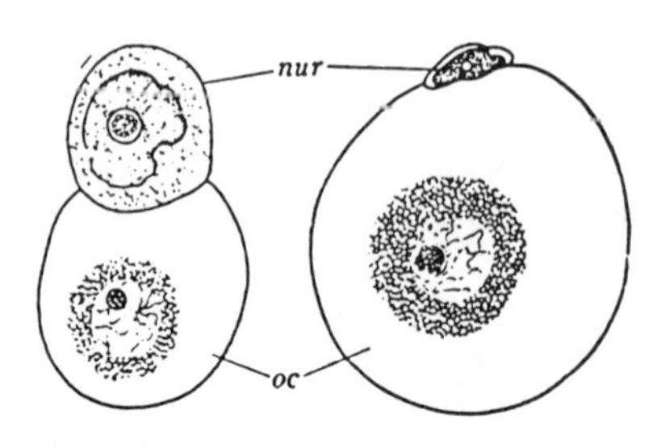

Fig. 1.5 Oocyte and nurse-cell in the annelid *Ophryotrocha* (Wilson 1925)

nur: nurse-cell, *oc:* oocyte.

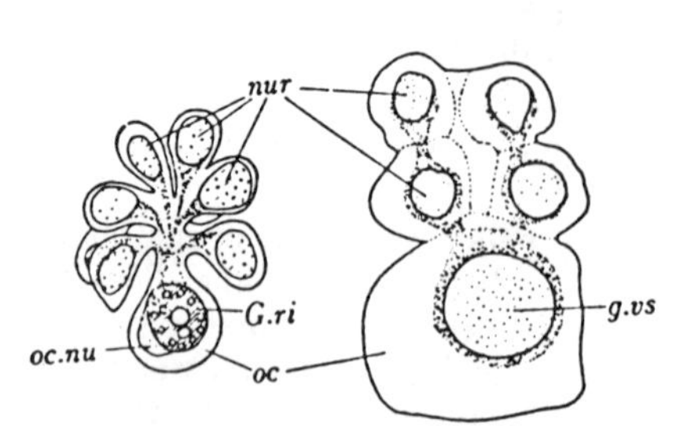

Fig. 1.6 Oocyte and nurse-cell in the beetle *Dytiscus*, showing six of the 15 nurse-cells (Wilson 1925)

G.ri: Giardina's ring, *g.vs:* germinal vesicle, *nur:* nurse-cell, *oc:* oocyte, *oc.nu:* oocyte nucleus.

Besides synthesizing yolk, these companion cells may also form protective coverings (chorion, jelly layer), and among the tunicates (see Chap. 12) they become the test cells.

In any case, in contrast to the male gonad, where every spermatogonium finally produces four spermatozoa, many sister cells are sacrificed to bring one oogonium to full development as an ovum; some are amalgamated into nurse cells, some play the secondary role of protective coverings, while still others are divested of their cytoplasm and degenerate as the polar bodies. It is thus easy to understand that the number of eggs produced by the female is less than the number of spermatozoa produced by the male. According to the calculation of Lode for humans, the number of spermatozoa produced by a man during his lifetime is of the order of 340,000,000, while the number of ova produced by a woman is about 400 at the most. If the volumes are compared, on the other hand, the sea urchin egg is 400,000 times as large as the spermatozoon, and in the much larger eggs of reptiles and birds, the ratio may as well be left to the reader's imagination.

5) CLASSIFICATION OF EGGS ACCORDING TO LOCATION OF STORED YOLK

The part of the oocyte in which the yolk formed during the growth period will be accumulated varies from species to species; and it is possible to classify eggs according to this characteristic. Moreover, these differences in the distribution of stored yolk later exert an influence on the mode of cleavage of the egg (see pp. 46—47). Since these terms are rather useful, they are briefly listed here:

(1) *alecithal* eggs: very small eggs, practically lacking in yolk (mammals)

(2) *homolecithal* eggs: yolk relatively small in amount, evenly distributed throughout cytoplasm (echinoderms, Chaetognatha)

(3) *telolecithal* eggs: most of the yolk accumulated near one pole of the egg (teleosts, amphibians, birds, cephalopods

(4) *centrolecithal* eggs: most of the yolk lying around the nucleus in the center of the egg, surrounded by a layer of cytoplasm (arthropods, sea anemones).

2. COLLECTING EGGS AND SPERM

(1) PRELIMINARY PRECAUTIONS

It goes without saying that the objects of embryological investigation are normally developing individuals. In practice, however, it is hopelessly inconvenient to try to observe animals as they grow up in their natural environments, and in the long run it is best to raise them in the laboratory. But having once decided on this procedure, the problem is to create indoors a thoroughly faithful copy of Nature. This requires in the embryologist from the outset an adequate ecological intuition. This attitude of mind is obviously necessary with respect to the larvae, but it must also extend retroactively to the care of the parents from which it is hoped to obtain offspring. We often hear about vertebrate animals which never breed in captivity. This also occurs among invertebrates; in the case of the squid *Loligo* and the sea-cucumber *Cucumaria*, it has been found (Oshima 1925) that even when the best live-cars and tanks are used, unless these animals spawn on the evening of the day they are caught, they will not spawn no matter how long they are kept.

For such reasons, the experienced embryologist pays meticulous attention to all sorts of minute details, from the lighting and ventilation of the place where he puts the containers to the food of the larvae and the way of changing the water. Since, however, these precautions are different for each animal species, the reader is referred to the book of Galtzoff et al. (1937) for particular cases, and treatment here will be limited to such common problems as those pertaining to containers, water, and the density of larval cultures.

a. Quality of water and containers

It stands to reason that the various kinds of animals develop in environments peculiarly suited to each of them; when the environmental medium is air, the embryologist has no particular problem, since the composition of air is almost uniform, and even if there are local changes, they are quickly equalized because of the high diffusibility of the gas molecules. In contrast to this, the situation is very different when water is the environmental medium. It is often noticed that when a number of containers are prepared, some organisms flourish only in certain particular ones. If these organisms are transferred to the water of different vessels, it is a usual occurrence for them to die at once, and even if they last for a while, they usually die off sooner or later. When the organism in question is an animal, this is often connected with a secondary peculiarity such as the appearance of a favorable food organism in a particular vessel; or again, with a factor like hydrogen ion concentration, which we can correct so long as we know what it is. When fresh water animals are to be kept, it often helps to place some bottom mud from their habitat in the container; the advantage in this lies in the fact that the mud acts as a buffer to keep the pH and heavy metal ion concentration at a suitable level. However, an unreasonable amount of mud is likely to cause fouling of the water and lead to failure.

On the other hand, it sometimes happens that larvae fail to grow even in an appropriate medium. An example of this is often found when the larvae of marine animals are cultured in natural sea water. In this case the difficulty lies in a defect either in the method of obtaining the water or in the quality of the containers. When the larvae are small and weak, it is difficult to change the water, and it is often necessary to keep the larvae in the same water for several days. Under these conditions, even a mildly unfavorable factor, which would produce no injury on one or two days' exposure, can be definitely poisonous when its effect is additive over a number of days.

To the list of potentially poisonous substances there is no end, but one category which must be continually guarded against is an excess of heavy metal ions. One striking example of ion effects is that if sea water is brought back to the laboratory in a metal bucket, and there transferred to a glass vessel and used as a medium for raising sea urchin larvae, practically all of them will be abnormal. This is clearly the effect of the zinc ions that have dissolved out during the 15—20 minutes that the sea water was in contact with the metal, and in fact it has been experimentally shown that zinc in a concentration of 0.00001M arrests the development of sea urchin larvae (Yanagita 1930; Rulon 1955). The bottom mud added as suggested above acts as an antidote by chelating these heavy metals. In marine biological laboratories and other places where sea water is circulated, if iron or brass is used in the pump or outlet cocks, the ions of these metals make the running sea water unfit for use in embryological experiments.

This defect can be avoided by bringing the sea water directly to the laboratory in a glass container; although this is not only a terrible nuisance but actually dangerous in bad weather or when the glass breaks, it is the most satisfactory method practicable in Japan at present. At the Woods Hole Biological Laboratory, where fertilization and later development proceed quite normally in the circulated sea water, the inside of the turbine pump is plated with lead (since this is insoluble), the pipes are lead and the outlets are made of ebonite. The Friday Harbor Laboratory pumps the sea water with a glass turbine and circulates it through glass piping.

When the larvae being grown in certain vessels develop abnormally in spite of careful scrutiny of the medium, it is well to consider that the vessels may be contaminated, and it is axiomatic that the experiment must be given up at once and all the dishes washed with soap, cleaning solution, or detergent, and thoroughly rinsed afterward to remove all traces of these cleaning agents. However, when it is suspected that the dishes have been contaminated with fixing fluid, it is probably better to throw them away; it is not far from the truth to think that anything that has once held formalin can never again be washed clean enough to use. Hence there should be a strict taboo against fixing material in containers which are used for culturing.

b. **Larval density**

One other mistake which is often made is to keep too many larvae in a given vessel. It is everyone's trouble that he wants a lot of material and there are never enough dishes, but on the other hand, if one thinks of how few larvae are found in plankton hauls and so on, he will be forced in spite of himself to recognize how low the density is under natural conditions. If larvae are cultured too densely they sometimes develop abnormally, and even when they are not abnormal their developmental pace is disorganized and it is impossible to obtain clear and accurate results. When the medium is correct and the container is clean it is too bad to lose an experiment because of crowding.

(2) DISTINGUISHING THE SEXES

The development of the offspring will not begin until a male and a female are got together to be the parents. There are of course many cases in which the sexes are clearly recognizable, but there are also many in which the males cannot be distinguished from the females. The first step in embryological research is to learn how to appraise the sexes.

The sea urchin, which has been so much used in the field of experimental embryology, is a famous example of a form in which the sexes still cannot be distinguished by the external appearance. Since I am sure there are not a few among the readers of this book who will sometime study the sea urchin, I will set down here an account of the many efforts that have, in the past, been made in this direction. When the sexes cannot be told apart, it is often necessary to kill several males in order to find the one female which is required; when this is repeated day after day it amounts to a tremendous waste, as anyone who has done even a little embryological research realizes acutely. As a result there have been some who said that the spines of male and female sea urchins are of a different shape, or that there is a sex difference in the curvature of the tests, or that members of one sex or the other come to the dark side of the tank, but these were little better than superstitions, without an objective basis. In 1939 Mrs. Harvey found that if a small amount of 1M NaCl was introduced into the mouth of a sea urchin and a little more of the same solution was injected into one of the genital pores, a small amount of gametes would be shed from that pore, so that the sex could be determined. Moreover, she discovered that if the animal was immediately returned to sea water it would not be particularly injured, and it was then easy to keep it, with its sex known, until the next time it was needed. She also said that if KCl was used in place of NaCl, the effect was even stronger, and this became the forerunner of the "KCl Method" which is now widely used in Japan to induce spawning. However, since 1M NaCl is twice the concentration of sea water, and moreover the injection procedure is rather troublesome, Harvey's method has never become very popular.

Two years later Motomura (1941a) discovered that the tube feet on the oral disc of the males of *Hemicentrotus pulcherrimus* are white, while those of the females are yellowish. This sexual dimorphism is not distinctly separable, the two characters overlapping in the center of the range, but even so it has made a great contribution to preventing the waste of experimental animals. In practice the freshly collected sea urchins are separated into three lots; those with definitely yellow tube-feet, those with definitely white tube-feet, and the intermediate ones. When gametes are needed, an animal from the intermediate lot is opened first and its sex determined; since one of the opposite sex can be chosen from the clearly differentiated lots, it is always possible to make the necessary pair without waste. Three years later, Tyler (1944) reported that it was possible to distinguish between the sexes of the American sea urchin, *Lytechinus anamesus*, by a difference in the size of the genital pores; it is regretable that both Motomura's and Tyler's characteristics are peculiar to a single species and cannot be used for a general diagnosis of sex in sea urchins.

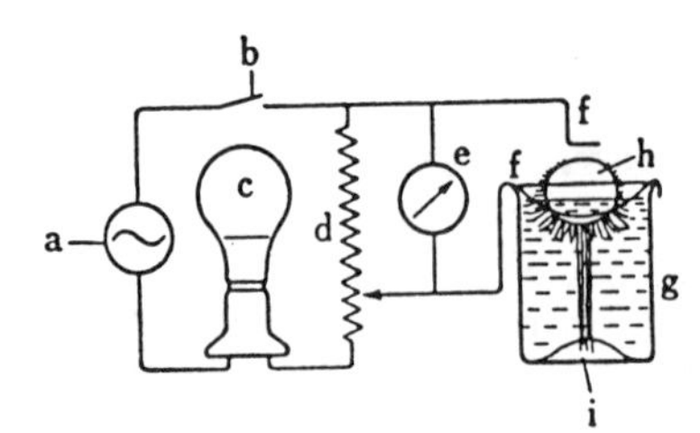

Fig. 1.7 Diagram showing apparatus for artificially inducing ovulation by electrical stimulation

a. source of electricity; b. key; c. lamp as an electric resistance; d. variable resistance; e. volt meter; f. electrode; g. beaker containing sea water; h. female sea-urchin placed oral side up; i. ovulated eggs.

After this Iwata (1950) and Harvey (1952) discovered a method of using electrical stimulus to determine sex. Iwata's apparatus is shown in Figure 1.7; this is a most convenient method and is already in wide use.

This section has been included to emphasize the fact that such modest bits of research form the necessary foundations for more brilliant discoveries.

(3) TECHNIQUES FOR OBTAINING EGGS AND SPERM

We can say that methods of obtaining eggs fall into three categories. In the first there is nothing to do but wait patiently for a chance encounter with naturally spawned eggs. In the next, it is also a matter of waiting for natural spawning, but certain measures are taken to control the time of it, even if slightly, thus enhancing the convenience of the experiment as well as the accuracy of its result. The last group of methods makes it possible to induce spawning at will. It is said to be easier to bring about fertilization in marine than in fresh water animals. To a certain extent this is true, but since there are many cases like those of *Spirocodon*, the pearl oyster or the abalone (Murayama 1953), in which fertilization does not ensue when eggs and sperm from extirpated gonads are mixed, and only those develop which the parent animals have spawned naturally, artificial induction becomes an indispensable measure. In the following, only the chief representative methods of obtaining eggs will be described, but if these are used as a bas's and adapted to particular cases, they should be rather widely useful.

a. Cases in which spawning time is naturally determined

The most obvious cases which belong to this category are those in which spawning is determined by the phase of the moon; the most universally famous of these is the swarming of the palolo worm. This is the phenomenon in which these annelids, belonging to the Lumbriconereids, come up at a definite time to cover the surface of the sea and spawn; it takes place both in the Atlantic *(Eunice fucata)* and the South Pacific *(E. viridis)*, and in this country, the swarming of the nereid, "bachi", reported by Iizuka (1903) is a comparable phenomenon. This nereid *(Tylorrhinchus heterochaetus = Ceratocephale osowai*) breeds at the mouth of the Sumida and other rivers; every year, immediately after the flood tide which occurs 1—2 hours after sundown on the four days after the dark and the full of the moon in October and November, a portion of the body *(epitoke)* filled with gametes is cut loose and swims to the surface, and shortly afterward the body wall bursts and the gametes are released. This epitoke is called a "bachi"; as bachi are used for bait in Japan, they are collected in great numbers by the fishermen; in a similar way the palolo serve as food for the natives in the islands. But an important point is that the swarming of the bachi is not confined to the Sumida River, as attested by reports from Togo-Ko in Tottori Prefecture (Inomata 1927), Hinuma in Ibaragi Prefecture (Miyoshi 1937) and the neighborhood of Tokushima (Okada 1952); if attention were paid to the matter they could probably be found widely distributed about the whole country (see Chap. 7 on the Annelida).

Taking the next examples from the echinoderms, we can bring forward the Japanese sea urchin, *Diadema setosum*, and the crinoid, *Comanthus japonica*. At Misaki, *Diadema* spawns on the full moon nights of August and September, in this agreeing exactly with the habit of the Red Sea *Diadema*, as reported by Fox (1924). The result of artificial insemination also reaches 100% fertilization on the day of the full moon, and in the natural habitat, on this day many individuals are found crowding together. Yoshida (1952) has recently cast some doubt on meaning of this habit of *Diadema*, but in general he recognizes the lunar periodicity factor in reproduction.

The spawning of *Comanthus* is connected with either the first or last quarter of the moon (or the time when the moon is at its northernmost or southernmost position) which occurs within the first twenty days of October each year. Since the proper condition of the moon occurs only once during this period, there is only one day, or at most two, in the whole year when *Comanthus* can spawn. The hour of spawning is between three and four in the afternoon, and since all the animals shed simultaneously at this time, if the date of spawning is known it is not difficult to obtain material for observation (Dan and Dan 1941).

In conclusion let us devote a few words to the land-crabs. In this connection there is the old expression "moon-lit crab". This habit of the crabs is not spawning, however — on the nights of the dark and the full of the moon they release into the sea the larvae that they have been carrying about on their abdomens. The red land crab, *Sesarma haematocheir,* which is much in evidence during warm weather, comes down to the shore in August or September at the time of high tide, which occurs about sundown the day before the full and the dark of the moon (Shimoizumi and Inamura 1951). So far as the author knows, this short report is the only reference to the behavior of Japanese crabs; but the account written by Inamura (1954) for use in teaching grade school children is an even more vivid description.

The fact that reproductive activity is connected with lunar periodicity can be widely observed throughout the animal kindgom. In addition to the annelids and echinoderms described above, various examples are known from the coelenterates to the vertebrates; the reader is referred to the work of Okuda (1939) for the detailed accounts.

b. Induction of spawning by light changes

Examples of animals in which spawning in nature takes place in the morning and evening twilight are the pulmonate snail, *Limnaea,* and the dipteran *Chironomus*; night spawning occurs in the medusan, *Spirocodon,* and the chaetognath, *Spadella.* The extent to which light is concerned in these cases has not been fully analysed, but nearly everyone must have wondered what would happen if the light were changed artificially. However, so far only complete darkness has worked, intermediate light intensities having been practically untested except on the sea cucumber. All the examples in which darkness is effective are species which normally spawn at night.

Darkening is the most markedly effective as a stimulus among the coelenterates; viz., *Gonionema, Spirocodon* (Uchida 1927; Dan and Dan 1947b; Kume 1948; Yoshida 1952), *Hydractinia* (Ballard 1942; Yoshida 1954). According to Kume (1948), the spawning reaction of *Spirocodon* is a reaction to light of the gonads themselves, as shown by the fact that extirpated gonads are induced to shed eggs and sperm by exposure to darkness. Moreover, he reports that if an adequate time has been allowed for the dark treatment, the spawning reaction will take place later even in the light, so that detailed observation of it is possible. Only it must be remembered that eggs and sperm of *Spirocodon* which have been shed separately will not unite in fertilization even if mixed together. Since the egg must apparently receive a spermatozoon the very instant it leaves the body of the female, it is necessary to put males and females in the same container when they are exposed to the dark if developmental stages are to be observed. For this purpose Kume's diagnostic characters for distinguishing the sexes are very useful.

Among the echinoderms, Ohshima (1925) discovered that the sea cucumber *Cucumaria echinata* will spawn if exposed to dim light; when the writer repeated this experiment the method was very successful, in spite of the fact that it was done in the middle of the day. It is said that the chaetognath *Spadella* and the mollusc *Chiton* both spawn at night — it would be interesting to try the effect of exposure to darkness on these forms.

I have no experience with influencing the spawning of *Limnaea* by changing the light conditions, but it is very convenient for experimental purposes that these snails are induced to lay within an hour or two after their culture medium has been renewed (Dan 1930).

c. **Spawning induced by "drying"**

Since many of the animals that live in the intertidal zone between the high and low tide lines shed eggs and sperm when the water flows back around them after ebb tide, it is possible to induce spawning by imitating this situation, removing the animals from the water for a while and then pouring sea water back over them. This method is commonly used with seaweeds, and there are also one or two beach-dwelling molluscs which are known to react to such drying and wetting. One of these is *Cumingia tellinoides*, found on the east coast of North America, and another is *Tillina juvenilis* (Miyazaki 1938). The effectiveness of the "drying period" may be greatly enhanced in summer by combining this method with the temperature effect which is discussed below — the animals are kept "dry" in the refrigerator overnight and then placed in sea water at room temperature.

d. **Method of change in temperature**

Unlike animals whose environmental medium is the air, the temperature of which normally fluctuates widely, animals living in the sea, where the temperature is relatively stable, are often found to react to a sudden temperature rise of 2—5° by spawning. In actual practice, this warm sea water method for obtaining gametes is mainly used with molluscan material, such as the edible oyster (Galtsoff 1938, 1940) and the pearl oyster, *Pinctada martensii* (Dunker) (Wada 1936, 1954); the clams *Petricola japonica* and *Mactra sulcataria* Reeve (Dan and Wada 1955) and the deep sea scallop (Kinoshita, Shibuya and Shimizu 1943; Yamamoto 1949). Particularly in the case of the pearl oyster, eggs and sperm obtained by cutting the gonads will not unite even if mixed together, and unless gametes obtained by warm water-induced shedding are used it is difficult to see the developmental process.

There are some animals which, although they potentially have a normal capacity for breeding, are much slowed down during the winter or stopped entirely by the low temperature. In these cases, if the animals are kept at a continuously high temperature, in place of the short high treatment, they can be induced to spawn. This method is successful with *Limnaea* (Dan 1930), various oysters and clams (Loosanoff and Davis 1950, 1952), as well as certain insects (the grasshoper, *Chortophaga viridifasciata)* and even with a fish such as *Oryzias latipes*.

Although it is not a matter of inducing spawning, temperature change is also used to break the diapause of insect embryos. The diapause eggs and embryos of overwintering eggs of insects stop developing at a fixed stage, and then when winter is over and spring comes, development is resumed and hatching occurs. If a diapausing embryo is left at a low temperature for several weeks and then transferred to a high temperature, the investigator can make it continue to develop at any time, without waiting for spring. Such high temperature treatment is therefore a most important resource of entomological research.

e. Sperm suspension as stimulus to spawning of eggs

The account so far has been concerned with cases in which a single treatment has induced spawning in both sexes, but there are times when it is necessary to treat males and females separately. As one of these conditions it should be borne in mind that the presence of spermatozoa in the medium is likely to induce the females to shed their eggs. In studying the pearl oyster, Wada (1936) used either a two-step method of warming to obtain spermatozoa and with this sperm suspension inducing egg spawning, or a three-step method. In this, instead of warming, he cut up ovaries in sea water and used this suspension to induce shedding of sperm. With this sperm suspension as stimulus he then obtained the naturally-shed eggs for the study of normal development (Wada 1956). He also reports that egg suspensions stimulate males of the giant clam, *Tridacna* spp., but shedding of eggs is not induced by sperm suspensions (Wada 1954).

The easiest method of obtaining eggs and sperm is that described in the next section, of simply cutting up the gonads, without any other treatment. Moreover, among the species in which this is practicable are included some, like the oysters and sea urchins, in which the females react to the presence of spermatozoa in the medium. In these cases, rather than taking advantage of this sensitivity, it is necessary to guard against it, since if one male begins to shed, all the females in the same container will be induced to spawn. During collecting trips and so on, if there are any indications that a male is beginning to shed, it should be removed promptly and the water changed or it is doubtful whether any eggs will be left inside the bodies of the mothers by the time they reach the laboratory.

f. Cases of fertilizable eggs and sperm from cut gonads

The simplest case is that in which the eggs and sperm have completed their preparation for fertilization inside the gonads, and can be cut out and used at any time during the breeding season. Among this group are the oviparous oysters *(Crassostrea = Gryphaea)*, clams (Mactridae), sea urchins, starfishes, Pomatoidae, Spirorbidae *Myzostoma*, *Ciona intestinalis*, etc.

The old method of obtaining eggs from extirpated ovaries was gradually improved in several steps, in parallel with the progress of embryological knowledge. In sea urchins, for example, the ovaries used to be cut into small pieces which were wrapped in cheesecloth to remove tissue fragments and shaken to release the eggs; the main object in this period was to obtain as many eggs as possible. This method does not give 100% fertilization.

It was mainly Just who realized and pointed out the defect in this method — that the too-violent stimulus resulted in the admixture of immature eggs which would never have been spawned under natural conditions. Since then methods of obtaining eggs have made a complete change and proceeded in the direction of collecting, so far as possible, only such eggs as are shed easily in response to a weak stimulus. When the ovaries are removed, as in the starfish, it is usual not to cut them up, but to leave them undisturbed in sea water and use only the eggs which pour out spontaneously.

However, a certain amount of contamination with body fluid is unavoidable with this method. In particular, the body fluid and epidermal secretions of the regular sea urchin *Arbacia*, the heart urchin *Clypeaster* and various sand-dollars are deleterious to fertilization (Lillie 1914; Just 1922; Pequegnat 1948; Endo and Nakajima 1953; Ohshima 1922). For that reason, washing of the eggs was added to the process. In order to do this the eggs are collected in a beaker or other deep vessel; after they have settled the water is carefully removed and fresh sea water added — it has become customary to repeat this several times. In the cases reported above, of the palolo worm and *Spirocodon*, however, as well as of vertebrate eggs in general, since the eggs must be fertilized the instant they come into contact with the sea water if fertilization is to be successful, it is obviously impossible to wash the unfertilized eggs. Inseminating eggs directly, before they are exposed to the external medium, is called the "dry method".

The sperm, however, unlike the eggs, may not be mixed with sea water at any random time. Once the spermatozoa are mature, they usually move actively in the external medium, but since they do not take in nourishment and are very small to start with, they will sooner or later use up their supply of energy and die if they are active. It is therefore necessary to keep the spermatozoa quiescent, and to accomplish this purpose the dense sperm suspension is stored, as it is shed, in a covered container. This original suspension is called "dry sperm". In order to inseminate eggs, some of this suspension is diluted each time it is needed; a dilution of about 1/10,000 is used, but if a more exact criterion is desired, the inseminated eggs can be observed with the microscope and a sperm concentration chosen which shows 2—3 spermatozoa around the egg in optical section.

g. **KCl method**

As one step milder than cutting out the gonads, it is possible to think of a method which, although it may kill the parent animal, at least induces spawning with the gonads left in place in the body. This idea points directly to the procedure of Harvey (1939) described in the section on identification of the sexes (see p. 16). Since the Harvey Method in its original form, however, is rather complicated and the yield of eggs tends to be small, as a compromise the Aristotle's Lantern of the female sea urchin is removed, the body fluid is carefully poured out and a small amount of 0.5M KCl is pipetted into the test. As soon as this is done the eggs begin to emerge from the genital pores; the animal is therefore previously inverted and rested on the rim of a glass container filled to the top so that the eggs are received directly into the sea water. This procedure, which is the most widely used in Japan at present, was devised by Kamata and reported by Iida (1942). This is a good method for obtaining a large amount of eggs, but as Kumé (1948) also found with *Spirocodon*, there is a tendency for the KCl stimulus to be too strong, and strictly speaking, to obtain best results the collection of eggs should be stopped somewhat before the very last ones are shed.

h. **Electrical stimulation method**

Finally we come to the technique by which it is possible to obtain eggs and sperm at will, without injuring the animal in the slightest. This is done simply by applying an electrical stimulus according to the method of Iwata (1950) and Harvey (1952), as was described above (p. 17). However, in the case of sea urchins, spawning takes place only while the electric current is turned on. This is adequate to give a small amount of eggs, but not practicable when large quantities are required. For this reason it is usually best to determine the sex by electrical stimulation, and then apply KCl. On the other hand, the use of Iwata's method with the mussel *Mytilus edulis* led to an unexpectedly interesting result which will be described in the next section.

3. FERTILIZATION

(1) MATURITY OF GAMETES

a. Degree of maturity of eggs

The unfertilized eggs which have been obtained by one of the methods described in the preceding section will, depending on the species, be in various stages of the meiotic divisions at the time they are laid, but the stage is fixed for each animal species. For example, fertilization in the sea urchin takes place after the completion of meiosis. The egg, therefore, when it is laid has the small egg pronucleus containing the half number (n) of chromosomes. However, early in the breeding season or after over-stimulation with KCl, cells which have not undergone reduction — that is, oocytes with the germinal vesicle intact — are shed. If these cells are inseminated, any number of spermatozoa will enter them but no development occurs. In other words, oocytes are still "immature" with respect to fertilization, and in this sense the meiotic process in the sea urchin egg is, in fact as well as in name, a "maturation division". The situation is exactly the same in *Spirocodon, Sagitta, Cerebratulus* and the crinoid *Comanthus,* among others.

On the other hand, the eggs of *Mactra, Nereis,* etc., are shed before meiosis — that is, with the large germinal vesicle intact, and are normally fertilized in this condition; while the eggs of various ascidians, some flatworms, the annelid *Chaetopterus,* the prawn *Penaeus japonicus* and others are fertilized at the metaphase of the first meiotic divison, and in *Amphioxus* and most of the vertebrates it is normal for fertilization to occur at metaphase of the second meiotic division.

Since the starfish egg is able to take in spermatozoa in the germinal vesicle stage (primary oocyte), or in the stage with one polar body (secondary oocyte) or two (ovum), the conditions are less clear-cut. However, observation of the later development shows that eggs fertilized when they have one polar body do best, and we may safely say that starfish eggs with germinal vesicles are "immature" and those with two polar bodies are "over-ripe".

In connection with the problem of maturity, the experiments on merogony are highly suggestive. As stated above, the eggs of sea urchins and *Cerebratulus* are not fertilizable while their germinal vesicles are intact, and if these cells are cut into a nucleated and an enucleate fragment, neither will be fertilized. But once the germinal vesicle has broken down, the enucleate fragment can be fertilized (Delage 1899, 1901; Wilson 1903). And not only fertilization, but also the reaction to treatments inducing artificial parthenogenesis similarly reveals this difference (Delage 1901; Yatsu 1904). Since one is led by these facts to think that the problem of maturation is not necessarily confined to the nucleus, but may also involve some sort of cytoplasmic change as well, this is known as "cytoplasmic maturation". The fact that sea urchin eggs are not fully fertilizable in the very early part of the spawning season, even though they have completed meiosis, certainly suggests that some change in the cytoplasm itself is necessary. And, indeed, there is a very great probability that such a cytoplasmic change is brought about by the substance released when the germinal vesicle breaks down.

b. Maturity of spermatozoa

Since the spermatozoa have all completed the meiotic divisions, there is no reason to expect gross differences in the condition of the nucleus, and the intermediate stages during the cytoplasmic metamorphosis from spermatid to spermatozoon are absolutely incapable of fertilization. This, however, does not by any means guarantee that every spermatozoon which has achieved morphological perfection also possesses fertilizing capacity. Fortunately this morphologically unrecognizable difference in degree of maturity in practically every case parallels the degree of sperm motility. In other words, vigorously moving spermatozoa have a high fertilizing capacity, while immobile ones are generally useless. The fact that this correlation holds good even when sperm activity is enhanced by the use of alkaline sea water, etc., is very important from the point of view of fertilization technique. When it is impossible to find a good male, alkaline sea water may be used as a last desperate expedient; it is also useful for improving the percentages of cross-fertilization in difficult crosses. Loeb's (1903) success in causing the fertilization of sea urchin (*Strongylocentrotus purpuratus*) eggs with starfish (*Asterias ochracaea*) spermatozoa by the use of alkaline sea water is a famous example of this.

c. The case of Mytilus

If Iwata's electrical stimulation method (Fig. 1.8) is applied to *Mytilus*, a striking result is observed. The electrical stimulus (30 volts) is administered for only 20 seconds, but a reaction time of about 30—40 minutes is required before the animals begin to spawn. It has been stated above that gametes cut directly from *Mytilus* ovaries are incapable of fertilization. Then what is it that happens in the gonads during these 40 minutes? In the case of the eggs, the maturation divisions are beginning (Iwata 1952). According to Iwata, following electrical stimulation of the ovary, the oocytes in the ovarian follicles begin the reduction division, and their germinal vesicles break down; at the same time the stem that connects the oocyte to the ovarian wall breaks, the egg is set free and finally shed to the outside. From the fact that oocytes in ovarian follicles removed from the body do not undergo the reduction divisions in response to direct electrical stimulation to the follicles, it appears that the stimulus is mediated by the ovary, probably by its secretion of some substance which causes the oocytes to begin meiosis. Since, in *Mytilus*, spawning appears to follow automatically as soon as the eggs begin the reduction division, the matter of whether spawning will take place or not depends entirely on this maturation process.

In connection with the above-described initiation of spawning in coelenterates by means of dark treatment, Ballard (1942) found that in *Hydractinia* and *Pennaria*, meiosis occurs on exposing the animals to about one hour of darkness followed by about 10 seconds of light, while in *Spirocodon* it is induced by darkness alone. Moreover, once the reaction has been established in *Spirocodon*, spawning will take place even in the light (Kumé 1952). The only difference between the process in *Mytilus* and that in these coelenterates is that in the latter, the eggs which are spawned have formed even the second polar body and are fully mature.

Iwata does not recognize any difference between *Mytilus* spermatozoa that are spawned in response to electrical stimulation and those that are obtained by directly cutting out the testes, but the investigators at Misaki do not necessarily agree with Iwata in this point. It is certain that spermatozoa immediately after removal from the testes are inactive, in contrast to spawned spermatozoa, which show the most vigorous activity from the moment they are shed. This indicates that something takes place in the testis as well as in the ovary. As has been said before, spermatozoa have already achieved nuclear maturity and their morphological transformation is complete — this activation may perhaps be called a sort of cytoplasmic maturation. However, since these originally immobile spermatozoa that are cut out of the testis will eventually become active if they are left standing in sea water for a while, it is apparently possible to substitute this period of suspension in sea water for the change that occurs in the testis after electrical stimulation. During the era when the greatest interest was centered in normal embryology, the problem of motility and non-motility of spermatozoa strongly attracted the attention of investigators, but it seems to have been forgotten in recent years. I think this is one of the causes underlying the conflicting inconsistencies in the results of recent research concerned with spermatozoa, but in the last few years there has begun to be an outcry that sperm motility is connected with the movement of heavy metals (Metz 1954; Fujii et al. 1955), and we may hope that the day when this will once more be taken into account is not far distant.

(2) DESCRIPTION OF THE FERTILIZATION REACTION

a. Sea urchins

When eggs of optimum maturity, which have been secured in accordance with the various considerations presented above, are mixed with maximally active spermatozoa, the fertilization reaction occurs immediately. Let us describe this process rather fully in the sea urchin. The unfertilized sea urchin egg is surrounded with transparent jelly. Since the refractive index of this jelly is the same as that of sea water, it cannot be seen by simple microscopic examination. If, however, the eggs are put into a suspension of India ink, since the ink particles cannot enter the layer of jelly, this part remains transparent (Fig. 1.8). The jelly can also be directly seen as a thick layer if it is stained with Janus Green. The eggs of

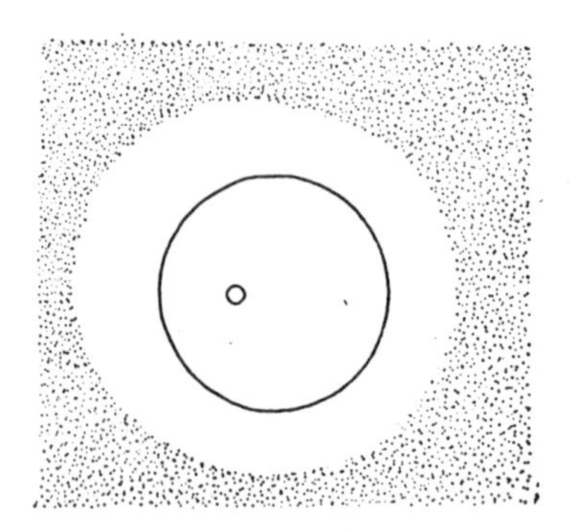

Fig. 1.8 Jelly-layer of sea-urchin egg demonstrated by adding india ink to egg suspension

such sand dollars as *Echinarachnius mirabilis* and *Astriclypeus manni* have small pigment globules scattered throughout the jelly layer so that its approximate thickness is apparent without the use of any special measures. The spermatozoa are able to pass easily through the jelly layer at any point and enter the egg surface wherever they happen to reach it. Accordingly, in the sea urchin egg the place of sperm entrance is not predetermined. On the other hand, in other eggs which are fertilized in the germinal vesicle or first polar body stage, it appears that there are many in which the spermatozoon enters the vegetal half (oyster, *Physa, Mactra,* etc.). Again, there are extreme examples like that of *Spirocodon,* in which, although there is no micropyle, the spermatozoon enters only very exactly at the animal pole (Fig. 1.11).

Since the jelly layer consists of very soft gel, the spermatozoa pass through it without even any perceptible slowing of their speed. When a spermatozoon comes up against the egg it applies the acrosome to the egg surface and rather simply penetrates it, but those which arrive later and fail to enter, without exception perform a clockwise rotational movement at the point of contact with the surface. If the fertilizing spermatozoon is carefully observed, it can be seen that although its acrosome is applied to the egg surface, its head is making small oscillations because of the violent movements of the tail. These small oscillations of the acrosome then suddenly cease, and it looks as though the acrosome had become fixed to the egg surface; this is the earliest indication of fertilization. It is of course difficult to observe these changes without using the oil immersion objective of a phase contrast microscope. Accordingly, when there are several spermatozoa in the field, the trick of catching the whole process of fertilization consists in being quick to notice the spermatozoon whose acrosome stops oscillating.

Once this binding between acrosome and egg surface is accomplished, the sperm head, in its perpendicular position, moves inside the egg as though it were gliding. In some sea urchin species the sperm tail at this time is also perpendicular to the egg surface and motionless, while in others it may be gently moving, but in either case it is impossible to think that the spermatozoon is proceding forward under its own power. In all likelihood it is the activity of the egg which draws in the spermatozoon. Since the sperm tail is very slender and moves quickly, it is impossible to see it with an ordinary microscope, and phase contrast must be used for such observation. For this purpose dark contrast objectives are most suitable (Dan 1950). The results of phase contrast obsevations by J. Dan (1950 a) will be described below.

As soon as the head of the perpendicular spermatozoon slips within the contour of the egg, the fertilization membrane is peeled off the surface at this entrance point like a blister. The inflation of the fertilization membrane is caused by the breakdown of certain granules, the cortical granules, located just below the egg surface; this will be described below (Fig. 1.20). As the fertilization membrane moves away from the egg surface, the sperm tail must be able to slip through it or it will be pulled and finally broken off. Actually, according to the hitherto accepted opinion, the sperm tail was thought to remain outside of the membrane and be carried away from the egg surface as the membrane expanded (Wilson 1895, 1925; Chambers 1933). However, phase contrast observation shows that it is usual for the tail to slip smoothly through the membrane (Fig. 1.9). Very rarely it may become caught, but this is clearly apparent because not only is the tail pulled straight by the tension, but an indentation is formed in the fertilization membrane. The error of the earlier workers was due simply to the technical predicament that the phase contrast microscope had not yet been invented. A further advantage of phase contrast is that it permits identification of the sperm head inside the egg. Since this was quite impossible with other microscopes, there was no way, except by cutting sections, to investigate the behavior of the spermatozoon between its entrance into the egg and the swelling of the sperm head to form the sperm pronucleus. For such observation with the oil immersion phase contrast objective lens, *Clypeaster japonicus* and *Mespilia globulus* are the most appropriate materials, but since these are large eggs it is necessary to compress them somewhat under the cover glass.

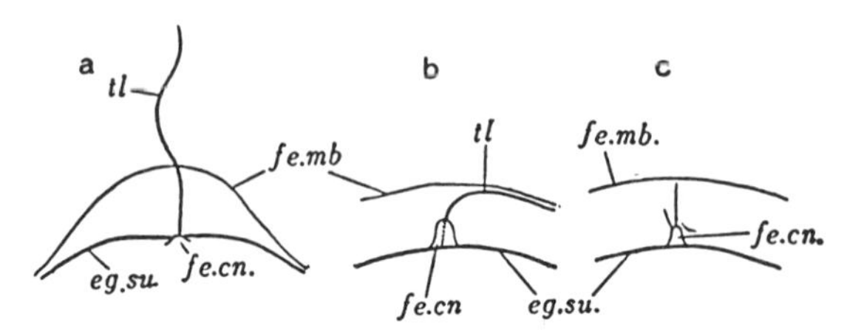

Fig. 1.9 Fertilization in the sea-urchin *Mespilia globulus* (J. C. Dan)

eg.su: egg surface, *fe.cn:* fertilization cone, *tl:* sperm tail.

When the sperm head has proceeded a few microns from the surface into the egg, the sperm tail, which has hitherto been quiescent, again begins to move. This phenomenon gives the impression that the egg, after once pulling the sperm through the surface, sets it free again inside this barrier — to such an extent does the spermatozoon move freely about. It shows no inclination whatsoever to procede in the direction of the egg pronucleus; some even turn about and head back toward the egg surface. But no matter how much the spermatozoon retraces its path, it can never go back (through the inside of the egg surface) to the outside, indicating the presence of a region several microns thick around the egg which is "off-limits" to spermatozoa. It is not possible to make satisfactorily accurate measurements, but the width of this off-limits region coincides in general with the distance covered by the spermatozoon in the quiescent state at the beginning of the entrance process, and seems to agree also with the thickness of the gel layer which develops around the egg periphery at the time of fertilization.

The formation of this gel layer is indicated by the fact that if unfertilized eggs are centrifuged, all the granules are moved by the centrifugal force, while in fertilized eggs, the granules within several microns of the surface do not move at all on centrifugation. In other words, the peripheral granules are sealed into the gel at the time of its formation. If the pigment granules of the *Arbacia* egg are used as the objectives in this experiment, the result is vividly apparent. This gel layer is called the *cortex;* it is impossible to deny the impression that sperm behavior is a result of cortical function, and that the egg plays a much more positive role in fertilization than has hitherto been imagined (see next section).

Outside the egg, in the meantime, the sperm tail is gradually becoming shorter as it is pulled inside, and its resumption of activity causes what remains of it to enter the perivitelline space inside the fertilization membrane (Fig. 1.9b).

At this time a small protuberance forms on the egg surface surrounding the sperm tail. This is called the *fertilization cone*, or the exudation cone. In the summer urchins, the whole tail is completely taken into the egg within about three minutes; after this only the fertilization cone is left (Fig. 1.9c). Just about this time, the sperm head inside the egg ceases to move, and abruptly begins to swell and change into the spherical *sperm pronucleus;* from the midpiece a centriole appears and is surrounded by a small *aster*. Once the sperm pronucleus is formed it can be seen with ordinary microscopes, and has been reported previously; the description of these observations, however, will be postponed until Section 4, and two or three other matters will be taken up here.

It has been stated above that in *Mespilia* and *Clypeaster* the time when the final tip of the sperm tail enters the egg in general coincides with the time, about three minutes after insemination, when the sperm head begins to change into the sperm pronucleus. This may give the impression that the entrance of the tail is due to the movements of the sperm head, but this is not the case. Sometimes the sperm very soon changes into the pronucleus, and in anomalous cases the tail may require as long as 15 minutes before it completely enters the egg; in either case the tail continues to be drawn in at the same rate even though the sperm head has ceased moving and begun to swell. The phase contrast microscope shows very well the part of the tail that is outside the egg and the sperm head inside, but unhappily the tail cannot be seen inside the egg, and therefore a detailed description is not possible; here also, however, it looks as though the taking in of the sperm tail is effected by the egg.

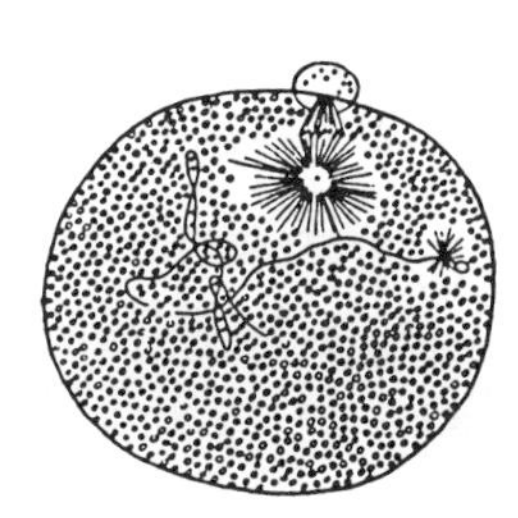

Fig. 1.10 Fertilization in the snail *Physa*, showing entire spermatozoon within the egg (Kostanecki and Wierzejski 1896)

This fact, that the sea urchin sperm tail is taken into the egg *in toto*, is opposed to the accepted opinion (see Wilson 1925), but reflection shows that this accepted opinion itself is unwarranted. If animals in general are scrutinized with respect to this problem, it is found that there are more cases in which the whole tail enters the egg; in other species there are also reports of such observations made on living material (shrimp, Hudinaga 1942); and again there are even examples like that of *Physa*, in which the total length of the sperm tail has been stained inside the egg (Fig. 1.10) (Kostanecki and Wierzejski 1896). For this reason there has been a tendency to regard sea urchins as an exceptional case. However, there have also been some workers who believed that the tail entered the egg even in the sea urchin (Selenka 1878).

With respect to the fertilization cone, there are some times and cases in which it is retracted within two or three minutes after the completion of sperm entrance, and others in which it remains a little longer, but often pointed processes are formed at its summit before it disappears (Fig. 1.9c). The sea urchin *Pseudocentrotus depressus* is exceptional in that its fertilization cone lasts for several hours; Endo (1954) recognizes its existence in the 16-cell stage.

b. *Spirocodon saltatrix*

The peculiar characteristics of fertilization in this species are that the spermatozoon never enters the egg anywhere except exactly at the animal pole, and that the 'fertilization cone' is a trumpet-shaped tube. This tube is gradually inflated and finally falls away. (J. Dan 1950b) (Fig. 1.11).

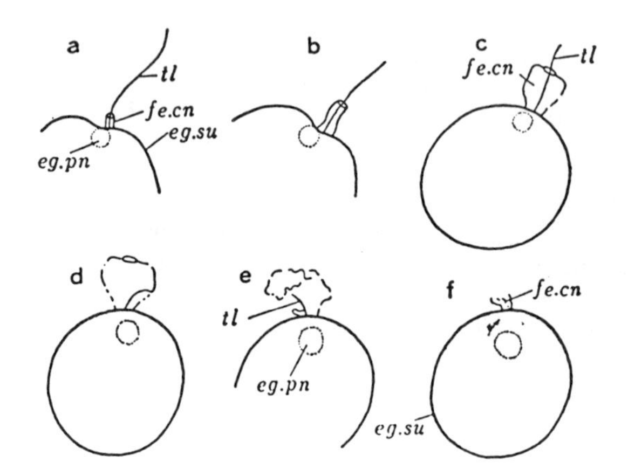

Fig. 1.11 Fertilization in the medusa *Spirocodon saltatrix* (J. C. Dan)

a. 30 seconds after oviposition; b. 3 minutes and 35 seconds; c. 6 minutes and 10 seconds; d. 10 minutes; e. 11 minutes and 20 seconds; f. 14 minutes and 20 seconds. ***eg.pn:*** egg pronucleus, ***eg su:*** egg surface, ***fe.cn:*** fertilization cone, ***tl:*** sperm tail.

c. **Nereid worms**

To study the development of the nereids it is customary to collect the epitokes by attracting them to a light at night; during May, *Perinereis cultrifera, Platynereis dumerilli,* etc., can be taken. However, although Okada (see Chap. 7) has reported the reaction of the egg cytoplasm in fertilization, there are no original descriptions of spermatozoan behavior in Japanese species; the case of *Nereis limbata,* reported by Lillie (1912) will therefore be used. The eggs of these nereid species all secrete a jelly substance instead of forming a fertilization membrane at fertilization. A spermatozoon which has attached to the egg surface remains motionless for about 15 minutes while it extends its acrosome into the cortex (Fig. 1.12c). Following it, the nucleus also becomes a thin strand, slips through the radiate structure at the cortical surface and takes a spherical shape inside; by this time 45 minutes have elapsed since the attachment of the spermatozoon. It should be noted that the mid-piece and tail, in *N. limbata,* are left outside the egg, and so end their functional life (Fig. 1.12e); it seems very likely that this was an important factor in leading Wilson to the same conclusion in sea urchins (although mistakenly in this case). The fertilization cone remains for about 20 minutes.

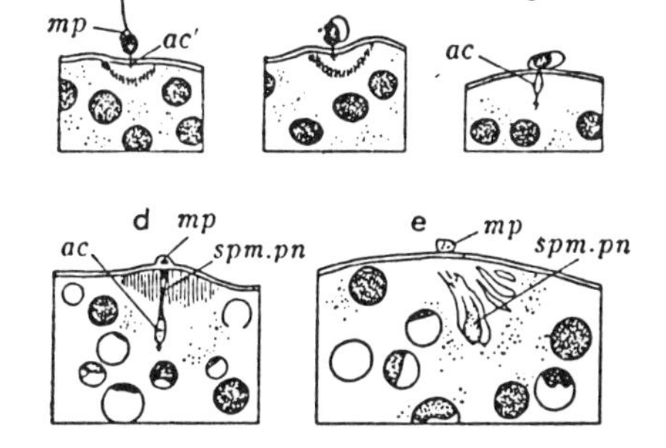

Fig. 1.12 Fertilization in the annelid *Nereis limbata* (Lillie)

a. attachment of spermatozoon to egg surface; b. beginning of acrosome penetration; c. acrosome completely within egg; d. penetration of sperm nucleus; e. formation of sperm pronucleus; middle piece left on egg surface. *ac:* acrosome, *mp:* middle piece, *spm. pn:* sperm pronucleus.

Taking advantage of this slow fertilization reaction, Lillie performed a famous experiment in which he centrifuged eggs into which the sperm head had entered in various degrees, so that the portion of the spermatozoon left outside was pulled away by the centrifugal force. The result showed that as long as a fragment, no matter how small, of the nucleus entered the egg, the fertilization reaction was able to proceed to completion (Lillie 1912).

d. **Starfish**

It was in starfish that H. Fol, in 1877, made his classical first microscopical observation of the actual phenomenon of fertilization. In addition to the fact that the jelly of starfish eggs is relatively dense, the spermatozoa have rounded heads and are unable to penetrate the jelly layer. What Fol saw, under these conditions, was that a long, slender thread coming from the egg extended through the jelly layer to one spermatozoon, which was pulled through the jelly and to the egg surface. After sperm entrance the same sort of fertilization cone as that found in sea urchins and nereids is formed (Fig. 1.13e). This thread is clearly an entirely different sort of thing form the fertilization cone, and was therefore given the special name of "reception cone". Almost fifty years later Fol's conclusion was supported by Chambers (1923, 1930); there are two facts which were very important as the basis of Fol's reasoning. These are: (i) since the sperm head is very

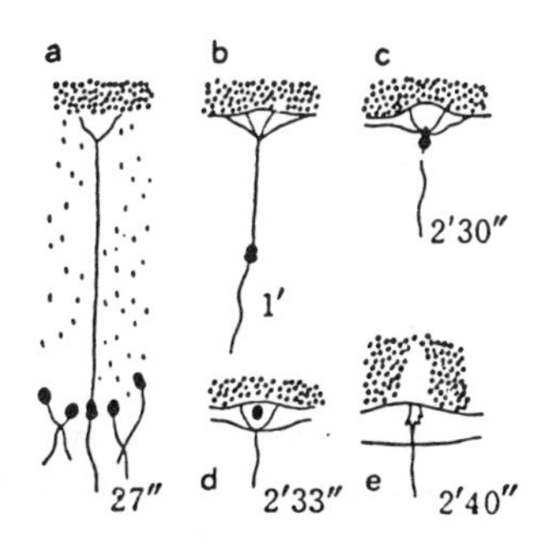

Fig. 1.13 Fertilization in the star-fish *Asterias forbesii* (Chambers)

small to begin with, there should be a marked reduction in volume after its formation if the material to form the thread originated from the spermatozoon; since Fol could not recognize such a change, he thought the thread came from the egg. (ii) Sometimes after the connection between thread and sperm head had been established the spermatozoon approached the egg with its tail in a tangential direction with respect to the egg surface. Since spermatozoa do not perform sidewise movement in their swimming, their approach to the egg must be a passive transportation effected by some force external to the sperm cell.

These two points raised by Fol are still valid, but, as will be explained in a later section (p. 55), grounds have recently been found to admit a new interpretation which will still satisfy these conditions.

e. Shrimp

The reception cone of the shrimp, *Penaeus japonicus* BATE, has certain somewhat different aspects (Hudinaga 1942). The situation at fertilization, as shown in Figure 10.4, is that when the spermatozoa reach the egg surface, the egg forms hyaline protuberances which rise temporarily under each spermatozoon (Fig. 10.4a). The egg cytoplasm then flows into the hyaline projections (b), but when among the many projections the cytoplasm reaches the tip of one and makes contact with the spermatozoon, sperm entrance takes place at this point of first contact (c, d). Inasmuch as the protoplasmic projections which formed later than this one withdraw immediately, there is no further contact whith the spermatozoa. In other words, a race takes place among the protoplasmic projections; when one has reached the goal, the others abandon the race. The protoplasmic processes are thus not only reception cones, but also act as the polyspermy-preventing mechanism.

f. Crabs and hermit crabs

It has already been stated that some crustacean spermatozoa have a very unusual structure (see p. 5). The mode of fertilizing of these spermatozoa is also very divergent from that of other types. The spermatozoa of the hermit crabs *Eupagurus* and *Galathea* have a head which is lightly twisted like a cork-screw, while the 'tail' is acorn-shaped and has three accessory branches. When a spermatozoon reaches an egg, the three accessory branches take a stand on the egg surface like a surveyor's tripod, and the nucleus is lowered at their center exactly the way the plumb is suspended from the middle of the surveyor's instrument (Fig. 1.14a). Strangely enough, at this point the acorn-shaped tail suddenly expands longitudinally, as though a tightened screw has been released (Fig. 1,14b). This explosive extension of the tail must impart a rather strong kick to the head which is directly below it; since this is conveniently twisted, it is thrust into the egg by the same principle as that of a cork-screw. Once the nucleus is completely inside the cytoplasm, the tail and accessory branches fall away from the egg surface (Koltzoff 1906).

Discoidal spermatozoa like those of the crabs, *Inachus* and *Menippe*, at first sight seem to be very different from the hermit crab spermatozoon, but the fact that, according to Binford (1913), a 'central body' inside the disc extends in an explosive fashion is highly similar. It is true that in *Galathea*, the part which

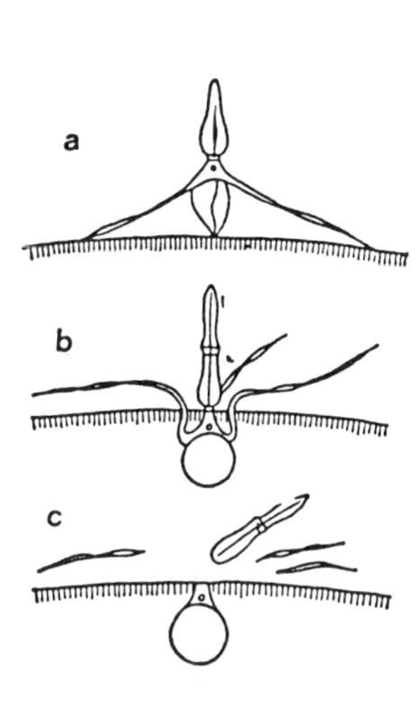

Fig. 1.14 Diagrammatic sketches showing fertilization in *Galathea* (Kortzoff 1906)

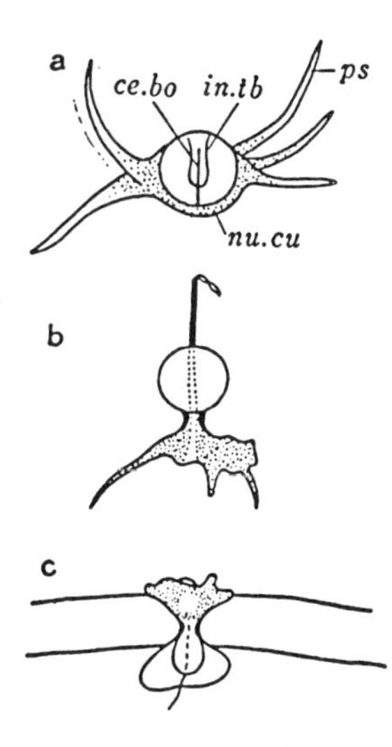

Fig. 1.15 Fertilization in *Menippe mercenaria* (Binford 1913)

a. spermatozoon; b. spermatozoon caused to burst and invert in dilute sea-water; c. spermatozoon which has reacted at egg surface. *ce.bo:* central body, *in.tb:* inner tube, *nu.cu:* nuclear cup, *ps:* pseudopodium.

expands explosively is thought to be the tail and remains outside the egg (Koltzoff, 1906), while in the crab, the exploding part works more like an acrosome, as appears in Figure 1.15, taking the lead in entering the egg cytoplasm.

g. **Sponges**

The sponges are the most complicated of all. The sponge egg is capable of amoeboid movement, and since it also takes over many vegetative cells as nurse cells during its growth, it is difficult to trace its history. However, although these

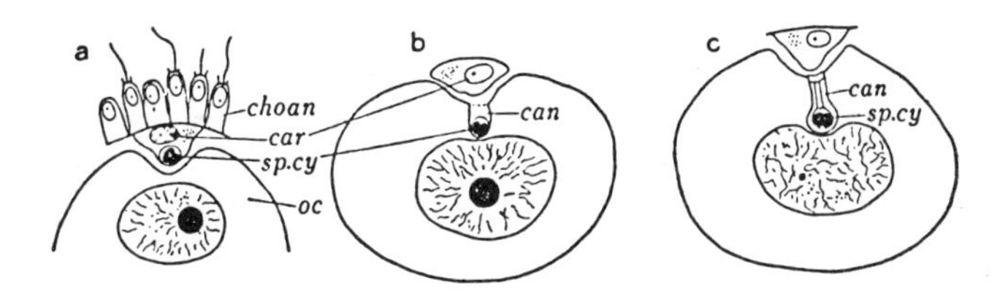

Fig. 1.16 Egg of *Sycon raphanus*, fertilized by accepting sperm cyst from carrying-cell (Duboscq and Tuzet)

a. oocyte and carrying-cell; b. reception of sperm cyst; c. fertilized egg and carrying-cell which has completed its mission. *can:* canal, *car:* carrying-cell, *choan:* choanocyte, *oc:* oocyte, *sp.cy:* sperm cyst.

are called 'nurse cells', unlike the situation in the oogenesis of other animals it is not certain specified cells which act as attendants, but choanocytes which, after taking in food, are captured one after another, so that most of the activity takes place in the canals. The most widely known fact about sponge fertilization is that the spermatozoa, which are brought from outside by the incoming currents, never fertilize the eggs directly, but first enter a choanocyte. This was first discovered by Gatenby (1927); if we take the example of the calcareous sponge *Sycon raphanus* as described by Duboscq and Tuzet (1937), we find the marked difference that sperm entrance always occurs inside the choanocyte collar, while ordinary food objects are invariably taken in outside the collar. Moreover, in view of the fact that the spermatozoa do not enter any random choanocyte, but choose those which are in contact with a full-grown oocyte, it looks as though the oocytes are also adding some effect. Once a spermatocyte has entered a choanocyte, it steadily rounds up and assumes a cyst-like structure within the choanocyte cytoplasm; at the same time the latter loses its flagellum, moves through the choanocyte canal wall to its inner side and becomes a so-called 'transport cell' (Fig. 1.16). When the transport cell encounters an oocyte, the transfer of the sperm cyst takes place. A pit is formed in the oocyte cytoplasm, and the sperm cyst is carried through this into the oocyte; this process constitutes fertilization in the sponge. The transport cell, after handing over the cyst, once again becomes a choanocyte, while the oocyte which has received the cyst begins the meiotic divisions. In the whole animal kingdom there is no case, except among the sponges, of this peculiar phenomenon in which the spermatozoon is picked up by a cell other than an egg, and only reaches the egg through the medium of this somatic cell.

(3) INTERACTION OF SPERM AND EGG

a. Agglutination reaction

It does not, however, necessarily follow that the fertilization reaction begins simply by the attachment of a spermatozoon to the egg surface. It is entirely possible that sperm and eggs exert some decisive influence on each other before they come into spatial contact.

The great discovery of Lillie, as well as the essential point of his fertilizin theory, was the phenomenon known as sperm agglutination (Lillie 1923). This is as follows. Unfertilized eggs are suspended in sea water, allowed to settle, and the supernatant removed; if this supernatant is added to a sperm suspension, the cells that were swimming entirely separate and at random in pure sea water mmediately come together to form clumps of various sizes. This is agglutination. On the other hand, the supernatant from immature or already fertilized eggs does not cause agglutination. Until the time of Lillie the study of fertilization had been limited to a simple process of description, from an onlooker's point of view; this discovery for the first time started it on a materialistic course. Lillie expressly gave the name of 'egg water' to the supernatant of unfertilized eggs, which had been endowed with the capacity to agglutinate spermatozoa, and the essential ingredient, which he believed to take part directly in agglutination, he called 'agglutinin' or 'fertilizin'.

The agglutination reaction can be observed quite clearly by mixing sperm suspension and egg water in a watch-glass. When egg water is added to an homogeneously milky sperm suspension, it takes on a granular appearance which can be seen with the naked eye. This can be studied more closely by placing a drop each of sperm suspension and egg water on a slide and observing with a microscope the region of contact between the two drops. According to Lillie, there are three stages in the agglutination process. On encountering egg water, the spermatozoa first (i) become intensely active, or show activaton; (ii) come together, which he termed aggregation; and finally (iii) undergo agglutination — of the three, the agglutination process is the most important.

If spermatozoa are carefully observed during agglutination, it can be seen that the great majority of them have their heads pointed toward the center of the clump (head-to-head agglutination), but it is not at all a case of their being bound together, each individual showing separate, small-scale vibrations, while some of them on the fringes of the clumps leave the group and swim away, exactly in the manner of honey bees during swarming (Loeb 1914). However, one other important characteristic of agglutination is that after 2—3 minutes the clumps break up and the spermatozoa resume their separate behavior. In other words, agglutination is reversible. If, however, spermatozoa are mixed with the egg water or spermatozoa of a different species, or fixing fluids or other substances are added to them, they may form clumps which bear a resemblance to those of agglutination, but since such clumps never separate, they are said to represent a processs of coagulation, which is different from agglutination. Hence Lillie believed that only the reversible type of agglutination had any meaning for embryology, and to this he attached much importance. This factor of reversibility, however, was neglected during the period which followed Lillie, and to this may be attributed much of the misunderstanding and confusion which later arose (Just 1930).

Finally, one more important point is that spermatozoa which have agglutinated and separated will not agglutinate a second time if fresh egg water is added. In other words, although agglutination is microscopically reversible, in a physiological sense it is essentially irreversible.

In summary: Lillie fully convinces us that, obscure though the details of its nature may be, since he is dealing with a substance which is specifically found in unfertilized eggs and specifically reacts with the spermatozoa of its own species, this substance must have some connection with fertilization; and we can agree to name it fertilizin. However, Lillie also tried to give to fertilizin the role of intermediary between sperm and egg, and he further wished to bestow upon it the key to the question of species specificity in fertilization — that is, why an egg is fertilizable only by the spermatozoa of its own species. However, the effort which Lillie put into this ambitious achievement was not an investigation of fertilizin itself, but a search for some other biological phenomenon which resembled fertilization. What he ran into was the phenomenon of immunity. This is a reaction occurring in vertebrate animals, by which, on introduction of an antigen into the blood serum, an antibody specifically corresponding to the antigen is formed by the serum. If the antigens are particulate objects such as bacteria, the antibodies cause these to agglutinate, after which they are removed by the action of phagocytes. Thus the immune reaction includes the phenomenon of agglutination; it is, moreover, a reaction showing an extreme degree of specificity. At that point Lillie decided to pattern his theory of fertilization after Ehrlich's theory of immunology, although it is difficult to say, after seeing the result, whether this was a fortunate or an unfortunate decision. At least, this choice of Lillie's reduced his own theory to that of an analogy with Ehrlich's and, as would be expected, gave it a highly imaginative quality.

Lillie's Fertilizin Theory is a classic known to everybody, and a repetition of it will be omitted here; its vigorous critic was Lillie's contemporary, Jacques Loeb. As was mentioned above, it was Loeb who, using alkaline sea water, succeeded in crossfertilizing sea urchin eggs with starfish sperm, and who, moreover, accomplished artificial parthenogenesis without using any spermatozoa whatsoever (1909); it is only natural that he would have much to say about Lillie's theory.

Among the various objections made by Loeb (1914, 1916), one which should be dealt with here is his claim that fertilizin, which Lillie believed to be secreted by the egg, is nothing but the jelly layer surrounding the egg. This dispute between the two was not settled at the time, but afterward the research of Tyler showed that Loeb was right. In order to measure the concentration of fertilizin, Tyler used the method of successive dilutions (Tyler 1940). In this a fertilizin solution of unknown strength is successively diluted until it just fails to cause the agglutination reaction; assuming that the last dilution in which agglutination can be recognized contains one unit of fertilizin, by calculating back the strength of the original solution can be obtained. Using this method, it is found that what Lillie believed was a steady secretion of fertilizin by the unfertilized egg is actually only a steady dissolution of the jelly substance, so that as this dissolving process advances, the agglutinating capacity of the medium increases, reaching its maximum when the jelly layer is completely dissolved. However, it was found that if Loeb's acid treatment is applied to freshly shed eggs so that the whole jelly layer is dissolved at once, the agglutinating capacity achieves the maximum level in one jump (Tyler 1941). It may be added that since Tyler found the agglutinating power of jelly substance to be lost after digestion with chymotrypsin (Tyler and Fox 1940), he argues that its active principal is a protein. The fact that later studies (Vasseur 1948; Tyler 1949; Nakano and Ohashi 1954) indicate the presence in the jelly layer of large amounts of polysaccarides does not necessarily conflict with Tyler's conclusion.

It is thus apparent that the objections to the Fertilizin Theory posed by Loeb in the first decade of its existence contained much of fact, but as the result of the efforts of Lillie and his student, Just (1914, 1915, 1930) during the 1920's, the Fertilizin Theory came to present the appearance of having been perfected, and it held full sway over this period. During the 1930's, Elster (1935) expressed dissatisfaction with it, on the basis of data gained from studies on hybridization among several genera and species of sea urchins, but soon afterward the report of a German group (Hartmann and Schartau 1939; Hartmann, Schartau, Kuhn and Wallenfels 1939; Kuhn and Wallenfels 1940) that they had chemically identified a fertilization substance (gamone) attracted attention. A short time later, however, several technical errors in this study were pointed out (Tyler 1939, Cornman 1941) and it was discredited, although it had the effect of negating the warning that Elster had sounded. At present Rybak (1955) and others can be thought of as holding an opposing point of view, while Tyler stands at the head of Lillie's followers. In his thinking that fertilizin is identical with the jelly layer substance, Tyler aligns himself with Loeb, but he is on the same track as Lillie in his firm belief that its mode of action is equivalent to an immunological phenomenon. However, since 1941 he admits two contradictions in connection with this analogy, which have an important bearing on his argument (Tyler 1941, 1948):

(1) If fertilizin is identical with the jelly substance, and if it is indispensable for fertilization, then fertilization ought not to take place if the jelly layer is completely removed. But according to Tyler's own observations, if the sperm concentration is increased, fertilization takes place even after the jelly has been removed.
(2) According to the immunological scheme, the percentage of fertilization ought to be increased if jelly solution is added to jellyless eggs. However, just the opposite is true; under these conditions less fertilization takes place. At this Tyler is compelled to recognize, in addition to the simple chemical role of fertilizin, the further, physical factor that it must be present as a jelly layer around the egg.

Before ending this section a word should be added about the situation in the starfish. Although Lillie himself made an attempt to cause agglutination of starfish spermatozoa, it happens that they do not agglutinate readily, with the result that this important material was excluded from Lillie's study. However, in recent years Tyler's school has succeeded in discovering a reliable method of agglutinating starfish sperm. These spermatozoa are non-motile when they are taken from extirpated testes, but on addition of lobster serum (Tyler and Metz 1944, 1945) or egg albumin (Metz 1945) their activity was found to increase, and when starfish egg water was further added, the sperm showed head-to-head agglutination. Since these artificially added substances play a sort of helping role, they are called "adjuvants"; as a matter of technique it should be noted that such adjuvants are required to give agglutination in Japanese starfishes also. Whether or not an adjuvant is required when eggs are spawned naturally is obviously another problem; in the author's experience the spermatozoa from fully ripe males have been observed to ag lutinate without the addition of an adjuvant.

b. **Acrosome reaction**

In the studies which have been described so far, the spermatozoon was regarded simply as a small particle. But however small it may be, it is still a living cell, and there was abound to be someone who would make its behavior the object of serious observation. That person was Popa. Since Popa's attempt was made in the pre-phase contrast microscope era, it was a difficult undertaking, and he was obliged to resort to vital staining with Janus Green in order to see the spermatozoa. According to his account, an adhesive substance is secreted from a minute pore in the very tip of the acrosome when sea urchin sperm come in contact with egg water (Popa 1927a, 1927b). However, at first glance this seemed like a fantastic conclusion; moreover, there was no way of determining to what extent the behavior of spermatozoa stained with Janus Green approached the normal. His work was therefore not given very wide credence, and so far as the author knows, this Rumanian scientist retired from the field of marine biology, leaving behind him only these two papers.

However, 25 years later, use of the phase contrast microscope made it possible to ascertain that something very like what Popa described actually does happen, and the study was immediately transferred to the electron microscope (J. C. Dan, 1952). In order to observe the spermatozoa with the electron microscope, neutralized formalin was added to the sperm suspension to give about 10% for fixation, and the fixed spermatozoa were placed directly on a collodion membrane stretched on an electron microscope mesh, and the excess fluid removed. The partially dried preparation was gently washed with distilled water to remove the sea water salts, and finally dried under vacuum.

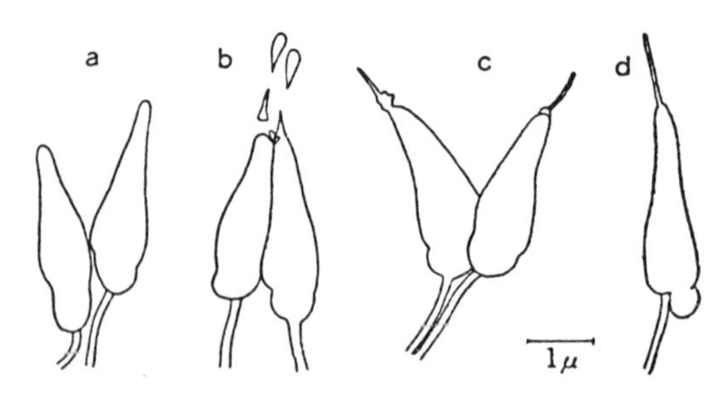

Fig. 1.17 Sketches from electron micrographs of sea-urchin spermatozoa reacting to egg-water

a. control before addition of egg-water; b. 2 seconds after addition of egg-water; c. spermatozoa during agglutination (20 seconds); d. spermatozoon fixed at 3 minutes--30 seconds after reversal of agglutination. Mid-piece is slightly shifted to the side.

Spermatozoa of the sea urchin, *Hemicentrotus pulcherrimus*, when fixed immediately after removal from the testes, appear as shown in Figure 1.17a, but when they are fixed two seconds after the addition of egg water (Fig. 1.17b), a drop-like object is found extruded at the anteriormost tip. After 20 seconds, the drop at the tip is replaced by a slender thread-like structure (c). Even after the agglutinated clumps have dispersed, this filament remains unchanged (d). However, in the course of these changes there is no sign of Popa's minute pore.

Combining the observations of living spermatozoa, made with the phase contrast oil immersion lens, and electron micrographs of spermatozoa fixed under various conditions, these changes led to the following interpretation. When the spermatozoon comes in contact with egg water, the acrosome breaks down and a sticky substance is extruded. Simultaneously the filament is formed and gradually hardens during the course of a minute or so. At two seconds, apparently the contents of the acrosome have just been released and still form a mass, and the filament is not sufficiently toughened, so that when the sperm head is dried *in vacuo* and shrinks, the connection between the two breaks and the drop-like mass is isolated. Phase contrast observation of living spermatozoa shows the mass attached to the sperm head. By 20—30 seconds the expelled mass has dissolved in the surrounding sea water and only the filament remains; this has already become tough enough so that even when it is dried, its connection with the rest of the head is not broken. The inference that an originally soft object may gradually harden is based on the example of the fertilization membrane, which is described in a later chapter. Moreover, the interpretation of the 2-second figure might be carried over to the 20-second figure and give rise to the suspicion that the filament was an artefact, pulled out by the shrinking of the sperm head. Since, however, a filament can be seen on living spermatozoa, it seems safe to say that this represents the natural shape. The spermatozoon of the red urchin, *Pseudocentrotus depressus,* undergoes an identical change, and as the heads of these spermatozoa are reatively long and slender, the decrease in length of the rest of the head when the acrosome reacts is easy to detect. Since the acrosome takes its origin from the Golgi apparatus, there is no difficulty in thinking of it as showing secretory activity.

At this point let us weigh the three following points in connection with the acrosome reaction and fertilization.

(1) Universality: Lillie's fertilizin began with Nereis and extended to sea urchins, but starfish were excluded from the argument because their agglutination is uncertain. The animals in which the acrosome reaction has so far been observed are, at Misaki, six sea urchins, three species of starfish, twelve bivalve molluscs, two annelids (Dan, J.C. and Wada 1955) and a sea-cucumber (Colwin, A.L. and L.H. Colwin 1955b). In addition, the following supplementary observations have recently been added: *Strongylocentrotus franciscanus* (Dan, J. C. unpublished); two starfish species (Colwin, L.H. and A.L. Colwin 1955a; Metz and Morrill 1955; two annelids (Metz and Morrill 1955; Colwin, A.L. and L.H. Colwin 1955a) one sea-cucumber (Colwin, L.H. and A.L. Colwin 1955b, 1956); one species of Hemicordata (Colwin, L.H. and A.L. Colwin 1954). It is thus an important point that the acrosome reaction is found to occur over a clearly wider range than the agglutination reaction.

One thing which is particularly important is concerned with a problem which was carried over from the earlier discussion: the real nature of the attraction cone in starfish fertilization. As described on page 4 (the jelly of the starfish egg is rather dense and about 20 microns in thickness; the starfish spermatozoon, with its round head, is unable to slip through this layer under its own power. This is obviously what suggested the idea of the reception cone. When adjuvant and egg water are added to these spermatozoa, a very slender and very long filament appears at the anteriormost part of the round head (Fig. 1.18) (Dan, J.C. 1954a). This filament easily exceeds 20 microns, and is therefore able to reach the egg surface through the jelly layer; and in spite of its formation, the dimensions of the sperm head do not appear reduced. Moreover, after this reaction the sperm tail comes to extend

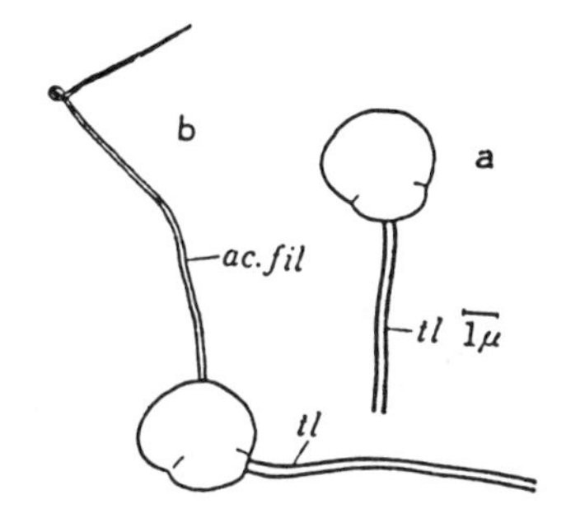

Fig. 1.18 Acrosomal reaction of starfish *Asterina pectinifera* spermatozoon

a. spermatozoon before reaction; b. spermatozoon after reaction. Acrosomal filament normally straight; notice that sperm tail is directed at right angles to filament.
ac. fil: acrosomal filament, *tl:* sperm tail.

sidewards, being inserted between the head and midpiece (Fig. 1.18b). These facts coincide perfectly with those which Fol used as the basis of his conclusion that the filament takes its origin in the reception cone (see p. 41); only this time it is as clear as possible that the filament comes from the spermatozoon. Once the filament reaches the egg surface, the egg protoplasm eventually rises around and reinforces it, so that if the first stage in the process is missed, it will be quite natural to receive an impression that the thread comes from the egg.

(2) Lysin Problem: In recent years data have bean accumulated with respect to an enzyme carried by molluscan spermatozoa, the "membrane lysin", which dissolves the egg membrane; this work brings forward an aspect of the fertilization process which could not be affirmed with sea urchin material. Tyler (1949) extracted such a lysin from the large spermatozoa of the limpet, *Megathura crenulata*, and examined with the electron microscope extracted spermatozoa fixed at several stages in the extraction process. Although the electron micrographs showed the acrosomes wholly broken down in the sample which gave the highest lytic activity, Tyler hesitated to conclude that the lysin is located in this part of the spermatozoon. Berg (1950) froze mussel testes at —30°C and then caused the breakdown of the cells by allowing the preparation to thaw at room temperature (freezing and thawing method); by this means he obtained a strong membrane-dissolving lysin. This is a convenient way of obtaining a strong lysin solution, but it has the defects of not revealing the location of the lysin within the cell and of introducing contaminating materials from the testis tissue.

However, the acrosomes of *Mytilus* spermatozoa are extremely large, so that it is even possible to determine with the 40x objective of an ordinary microscope whether or not they have reacted (Fig. 1.19). Moreover, the acrosome reaction in these spermatozoa can be induced by simply increasing the calcium content of the medium by about 10%, or by storing the sperm suspension for several hours in a refrigerator, thus avoiding the use of a complicated substance

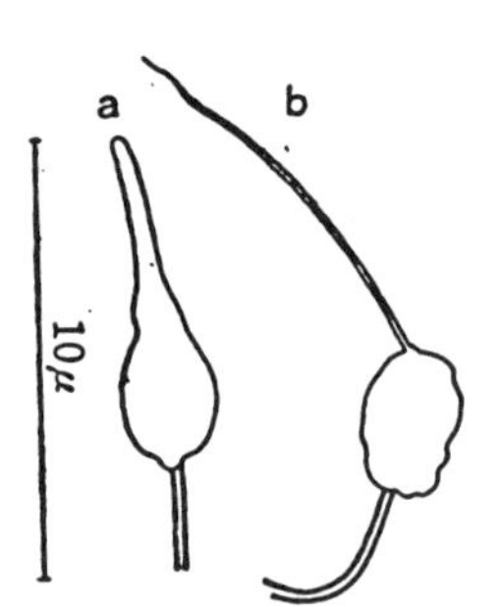

Fig. 1.19 Acrosomal reaction in *Mytilus* (J.C. Dan)
a. acrosome before reaction;
b. filamentous structure after reaction.

with unknown organic components such as egg water. In addition, these spermatozoa remain alive and intact for a long time after reaction, so that it is possible to separate by centrifugation, the suspension into a cellular fraction and a lysin sea water supernatant. By dividing the sperm of one male into several lots, inducing the acrosome reaction in these in different degrees and then testing the supernatants as well as extracts of the residual intact (unreacted) cells for lytic activity, it was found that such activity was largely in the supernatant of mostly reacted suspensions, and in the residue extract when the acrosomes were mostly unreacted. This provides strong evidence that the lysin is localized in the acrosome (Wada, Collier and J.C.Dan 1956). However, it is not yet known whether this lysin serves simply to open a path for the spermatozoon to enter the egg, or plays a more essential role in the fertilization process.

It is interesting that more than forty years ago when Meves (1915) investigated the fertilization process of *Mytilus* by means of sections, his drawings showed that, in eggs fixed two and a half minutes after insemination, the surface membrane is dissolved and a small hole left where the sperm has entered, while spermatozoa inside the vitelline membrane have lost their acrosomes. If Meves had been led by these results to undertake observation of the living process, the acrosome reaction would undoubtedly have been discovered forty years earlier.

(3) Relation between Acrosome Reaction and Agglutination: At the present time it is thought that the acrosome reaction and agglutination are two entirely distinct phenomena. Although the acrosome filament indeeed looks like a promising intermediary for bringing about agglutination, its shape does not change after it is once formed, although the clumped spermatozoa disperse. Moreover, it may be impossible to invoke this structure alone in explanation of a phenomenon which seems comparable to the swarming of insects. Furthermore, in a calcium-low medium these two phenomena are even more clearly separate (Dan, J.C. 1954b). It was Loeb's (1914, 1915) discovery that fertilization does not occur in the absence of calcium; he stated that the failure of the spermatozoon to attach to the egg surface under this condition is the cause of failure of fertilization. However that may be, it is possible to use this condition as a means of analyzing the relation between these two reactions and the phenomenon of fertilization. In the first place, with respect to the effect of lack of calcium on agglutination, Tyler (1948) states that agglutination is inhibited by the absence of calcium, tacitly referring to the Lillie concept of the correlation between agglutination and fertilization. According to the results of experiments by Dan (J.C. 1954b) on the sea urchins *Hemicentrotus pulcherrimus, Anthocidaris crassispina* and *Pseudocentrotus depressus,* contrary to Tyler's conclusion, in the lack of calcium agglutination is rather emphasized — particularly the duration of clumping is increased — but the acrosome reaction does not occur at all. Fertilization of course does not take place, but spermatozoa after agglutination in the virtual absence of calcium retain their full fertilizing capacity, and if calcium is added, will fertilize as well as the controls. However, a single exceptional case has so far been observed in which the acrosome reaction occurs in Ca-free egg water. This is *Clypeaster japonicus.* And the interesting thing is that if the egg of this species is inseminated in Ca-free sea water as the only known exception it will fertilize in the absence of calcium. These experiments do not simply demonstrate the fact that agglutination and the acrosome reaction are different phenomena — they also show that the success or failure of fertilization depends on the occurrence rather of the acrosome reaction than of agglutination.

One more marked difference between the two is that while agglutination is a highly specific reaction occurring only in response to egg water of the same species, the acrosome reaction, although egg water is the most effective stimulus, may be induced by other conditions. The *Mytilus* cases of increased calcium and reduced temperature are examples of this; in sea urchins, alkaline sea water and contact with solid surface may be cited (Dan, J.C. 1952). In short, the acrosome reaction seems to be invoked rather according to the general rule of stimulus and excitation than in response to the specificity of a gynogamone. As has already been described on page 17, the fact that alkaline sea water activates spermatozoa, which in turn increases fertilization percentages, has been amply confirmed in the experience of many workers; this again points to the close relation between the acrosome reaction and fertilization.

The acrosome reaction to the stimulus of contact touches upon two points which Tyler himself (1941) recognizes as contradictions in the immunological theory of fertilization (see p. 29). One of these is the fact that while more sperm are required per egg to fertilize eggs from which the jelly has been removed, in the end they are all fertilizable. While the contact stimulus is not very efficient it can still induce reaction, and Tyler's result is thus understandable as a matter of course. The second point is that if fertilizin solution is added to eggs from which the jelly has been removed, not only is the percentage of fertilization not increased, but it becomes lower than before addition of fertilizin. This fact compels Tyler to retreat one step in his immunological hypothesis, with the insertion of the condition that for some unknown reason fertilizin is efficacious only when it is present in its natural state, as the jelly layer surrounding the egg. The investigators at Misaki believe that, as shown in Figure 1.17, the structure of the acrosome changes moment by moment after reaction; if these changes are thought of as a sort of aging process, it becomes possible to make the assumption that this structure is effective only while it is fresh, becoming useless after a certain time. If the fertilization process is considered from this standpoint, we will find that the spermatozoon begins the reaction when it arrives at the outer edge of the jelly layer, and reaches the egg surface several seconds later. In contrast to this, if fertilizin solution is used, the reaction will begin when the spermatozoon touches the surface of the solution; it must then search for an egg, and by the time it has finally arrived at the egg it will be no surprise if the acrosome filament has already deteriorated into uselessness.

Tyler has accurately calculated the number of spermatozoa required in normal fertilization to be 4—5 per egg. Since, of course, according to Lillie's theory, one spermatozoon per egg ought to be adequate, this discrepancy is explained as due to differences in degree of maturity among the spermatozoa. This is indeed an ingenious explanation, but although differences in the degree of maturity among different males may be recognized, no experiment has yet been able to demonstrate differences in the maturity of spermatozoa from a single male. According to the idea of the Misaki group, even though the acrosome reaction may take place in an all-or-none fashion, differences will appear in the result if its effectiveness is a function of time, even though the degree of maturity is the same, providing an explanation of the necessity for 3—4 extra spermatozoa. It is significant that such considerations as these have recently induced Tyler to recognize the existence of the acrosome filament (Rothschild and Tyler 1955).

Looking back, at the conclusion of this section, one can only be amazed at the tremendous influence which Lillie's discovery of the agglutination phenomenon has had in driving forward later investigation into the problem of fertilization. But again, viewed from a slight distance, it can be seen that sperm agglutination in itself is a deleterious phenomenon which hinders the spermatozoon from reaching the egg, and even from the standpoint of Lillie's theory agglutination may be said to be an anomalous phenomenon in which a substance which was originally designed to bring about the union of egg and sperm has gone astray and united spermatozoa with other spermatozoa. May it not be that the limit to the Fertilizin Theory was set by the fact that it selected a mistake of Nature, so to speak, is its index of normal fertilizing capacity? Forty years after Lillie's achievement, the world has discovered the acrosome reaction, and progressing even further, is certainly entering the era of the chemistry of fertilization. From the standpoint of his argument, the author has been opposing Lillie's theory, but he has no intention whatsoever of denying his greatness. On the contrary, at the present time the very idea of the acrosome reaction itself is a posthumous child of Lillie's, reared on the nourishment of the agglutination phenomenon. With respect to the recent state of affairs in this field, the works of Rothschild (1956) and Rybak (1957) should be consulted.

4. INTERNAL PHENOMENA OF FERTILIZATION

(1) FERTILIZATION MEMBRANE AND THE MECHANISM OF ITS FORMATION

Among the morphological changes which accompany fertilization in the sea urchin egg, the most conspicuous is the formation of the fertilization membrane. It is almost impossible to follow sperm entrance in every case, but the appearance of the fertilization membrane, which is easily observed even with low magnification, offers an extremely simple indication as to whether or not fertilization has occurred.

The beginning of fertilization membrane formation has already been touched upon to some extent in Section 3(2), in connection with sperm entrance, and illustrated in Figure 1.9; when the sperm head penetrates the egg surface, a single-layered membrane rises as though it were being peeled off, and gradually lifts away from the surface. If this peeling-away process is observed in three dimensions, it resembles a blister-formation; this spreads around the egg to the side opposite from the sperm entrance point so that the membrane is completely separated from the egg surface and becomes a sphere one size larger than the egg and concentric with it. About one minute is required for the membrane to separate completely from the egg. This is all that can be seen in the way of morphological changes in connection with the process of membrane formation, but two or three further facts are known about its physical properties and so forth.

(1) The fertilization membrane is a soft membrane at first, but it undergoes a hardening process as it separates from the egg surface. This can be detected by a simple method. That is, some eggs are inseminated in a test tube; if the tube is shaken up and down at the time when the membranes are beginning to be lifted (after 45—60 secs.), they will be torn and the fertilized eggs become denuded. If the time between insemination and shaking is extended, at 2—3 minutes the membranes will no longer be torn (Herbst 1893; Goldfarb 1913; C.R.Moore 1916; F. R. Lillie 1921). Since the force acting on the eggs changes with the strength of shaking and the amount of sea water in the test tube, it is necessary to determine beforehand the conditions which will completely remove the membranes without breaking up the eggs. This method works especially well with *Hemicentrotus* eggs; the same result can also be achieved by drawing the eggs into a capillary pipette of appropriate bore. In experimental research the necessity often arises of removing the fertilization membrane; it is a good idea to practice thoroughly and master the knack of doing this. A.R. Moore (1930) has reported that the fertilization membranes of sand dollar *(Dendraster excentricus)* eggs will be dissolved if the eggs are transferred to 1M urea within five minutes after insemination, but not later; this result also depends on the same principle as that discussed above.

(2) The hardening of the membrane does not take place when the medium contains insufficient calcium. Since fertilization is impossible in the absence of calcium (Loeb 1915), the eggs are inseminated in sea water; if a large volume of Ca-free sea water is added after about 30 seconds to reduce the calcium concentration of the medium, the fertilization membrane will not harden for a long time (Hobson 1932). If the eggs are shaken to remove the membrane under these conditions, the surface precipitation reaction fails to occur because of the lack of calcium, and the eggs are injured or complete cytolysis takes place (see Heilbrunn 1956). It is therefore necessary to remove the membranes with a capillary pipette.

(3) It is colloid osmotic pressure which lifts the fertilization membrane away from the egg surface. If egg white, serum albumin or some other colloid is added to the sea water in which eggs with fertilization membranes are suspended, the membranes deflate until they are again in contact with the egg surface; if they are washed with pure sea water the membranes again inflate to their former dimensions. In contrast to this, the diameter is not influenced at all when the salt concentration is varied. (The diameter of the egg cell itself changes with variation in the salt concentration.) From these facts it is apparent that the fertilization membrane allows salts to pass freely but is impervious to colloid molecules; that is, it is a so-called dialyzing membrane. Accordingly, it follows that under natural conditions a colloid released into the space between fertilization membrane and egg surface (perivitelline space), acts as the cause of membrane elevation; it is accepted that the resulting colloid osmotic pressure draws sea water from the medium into the perivitelline space, causing its increase in size (Loeb 1908). This hypothesis of J. Loeb's has been fully demonstrated as fact by the work of Hiramoto (1955).

If the fertilization membrane were perfectly elastic, and colloid were present only within the perivitelline space and not in the surrounding medium, theoretically the fertilization membrane ought to expand indefinitely; however, as described above, the membrane hardens as it inflates so that an equilibrium is reached at a certain point and the expansion ceases. Hiramoto states that in the sea urchin egg there are invisible fine protoplasmic strands running between membrane and egg surface and suspending the egg in the center of the fertilization membrane; if such protoplasmic strands are present, they will also work toward quickly establishing an equilibrium condition (Hiramoto 1954). It is for these reasons that under conditions which suppress membrane hardening, the fertilization membrane becomes abnormally inflated. Sugiyama (1938) found that if all the calcium is quickly removed from the medium after fertilization, the membrane first becomes very large but soon deflates; he interpreted this as indicating that the initial over-expansion caused the membrane to leak, releasing the perivitelline colloid.

The explanation given above is probably adequate to explain the mechanism of fertilization membrane inflation, but this explanation requires that two membranes be imagined as present on the egg surface from the start, the outer forming the fertilization membrane while the inner is destined to become the surface layer of the fertilized egg. If it were possible to see two membranes on the unfertilized egg, the explanation would be perfect, but neither in the living sea urchin egg nor in sections has this attempt been successful. As a result, various people conceived the idea that the fertilization membrane of sea urchin eggs was newly formed at the instant of fertilization, and a long controversy ensued between the 'Newly Formed' and 'Pre-existing' sides.

At present, however, it appears that the latter command the field. The points contributing to this victory are: (1) if unfertilized eggs are placed in distilled water, the egg cells will swell and finally die, but after the destruction of the cells single-layered membranes remain surrounding the cytoplasmic granules (Theel 1892, Herbst 1893, Hobson 1932, Sugawara 1943a). However, if the eggs are pretreated with some reagent such as the 1M urea mentioned above (Moore, 1930), sea urchin larval hatching enzyme or trypsin (Sugawara 1943a) which dissolves the vitelline membranes, and then exploded in distilled water, the egg contents disperse completely in the medium and no outer membranes are left behind. Moreover, if eggs which have been so pretreated are inseminated, they fertilize normally and continue to develop, but no fertilization membrane is formed. From these facts it is concluded that the membrane which is left after disintegration of the unfertilized eggs in the first experiment is the precursor of the fertilization membrane, and there is no doubt that it is present on the unfertilized egg surface from the beginning.

(2) Looking over other animals besides sea urchins, we find that the unfertilized starfish egg has a surface layer which is recognizable from the beginning, and lifts off and hardens to form the fertilization membrane, so that the whole process is visible. Chambers (1921) reported that when the outermost layer of an unfertilized starfish egg was torn off with a microneedle, the egg failed to form a fertilization membrane on subsequent insemination; this result is precisely the same as that arrived at by pretreating sea urchin eggs with solvents. Since it is usual to call the outermost layer in the case of the unfertilized starfish egg the 'vitelline membrane', it is possible to make the statement with this material that the vitelline membrane is the precursor of the fertilization membrane.

In molluscs such as the oyster, the vitelline membrane surrounding the unfertilized egg is still more distinct than that of the starfish; after fertilization the perivitelline space becomes slightly wider but since the membrane itself undergoes no special change, it is still called the 'vitelline membrane' even after fertilization. The situation is the same with the chorions of such vertebrate eggs as those of teleosts and amphibia. It can therefore be said that 'fertilization membrane' is a name for a vitelline membrane which has become strikingly larger and changed its physical nature (by being hardened).

In 1939, however, Moser working with sea urchin eggs and Yamamoto (1929a, 1939b), with medaka *(Oryzias latipes)* eggs, independently discovered some closely similar new facts. These concerned the existence of cortical granules or cortical alveoli and the explosion of these at the time of fertilization. If the sea urchin egg is observed with an oil immersion lens, small granules can be seen forming a layer parallel to the egg surface; in the medaka egg there is a layer of vesicles large enough to be seen with a 40x objective. At fertilization, the breakdown of these sea urchin egg granules begins at the sperm entrance point and the wave of breakdown spreads in all directions, coming to an end at the oposite side of the egg, while the fertilization membrane rises from the part where the granules have broken down. In the medaka it is the chorion which is separated from the egg surface.

Following this, Motomura (1941b) traced the process of fertilization membrane formation in sections of sea urchin eggs, using a special Janus Green dye which stained the cortical granules. He reports that as the membrane begins to rise, the granules which hitherto have been stained become invisible, while the previously non-staining vitelline membrane begins to take up the color. This led him to think that when the cortical granules explode, this explosion product is extruded to the outside of the egg and adheres to the inner surface of the vitelline membrane, and he suggested that the addition of this substance is very likely the cause of the fertilization membrane hardening.

The work of Endo (1952, 1954a) substantiated this idea of Motomura's by means of observations on living eggs, and carried it to completion. Endo found that when he observed the higly transparent eggs of *Clypeaster japonicus* and *Mespilia globulus* with dark contrast phase contrast, the cortical granules appeared dark; when these exploded and swelled as the result of fertilization, their contrast was reversed and they appeared bright. Observations before then, with ordinary microscopes, had not succeeded in defining the granules after they exploded, but this became possible with the advent of phase contrast microscopy. These bright-appearing masses of granule-explosion product are pushed outward, becoming constricted in the process as though they were passing through small pores in the egg surface; since they are being extruded to the inner side of the vitelline membrane rather than to the true outside of the egg, they cause this membrane to rise away from the cytoplasmic surface (Fig. 1.20). As the vitelline membrane is thus stripped off and lifted away from the egg, the coacervate-like masses of exploded granule material stick to the inner surface of the vitelline membrane as flattened drops; together with the expansion of the membrane these drops spread out, and finally form a smooth inner lining of the completed fertilization membrane.

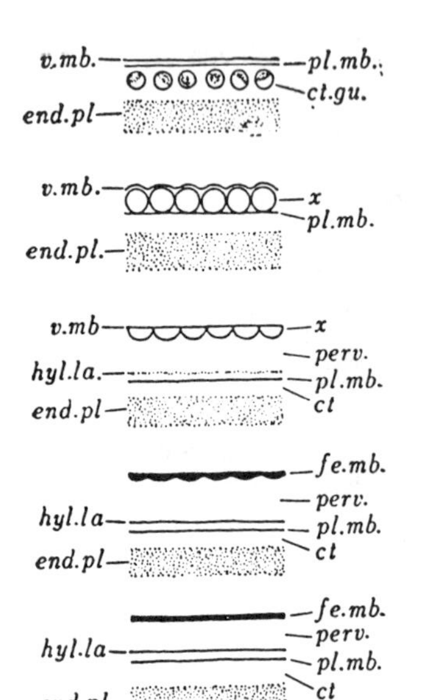

Fig. 1.20 Process of fertilization membrane formation (Endo)

ct: cortical layer — formerly "granular layer", "extra-granular zone". In egg of sea-urchin *Hemicentrotus pulcherrimus*, yellow granular layer is present just beneath cortex, *ct.gn:* cortical granule, *end.pl:* endoplasm, *fe.mb:* fertilization membrane, *hyl.la:* hyaline layer, *perv:* perivitelline space, *pl.mb:* plasma membrane (surface of fertilized egg), *x:* products of breakdown of cortical granule.

The reason why the fertilization membrane fails to become toughened in the absence of the calcium ion is that the coacervate formed by explosion of the cortical granules dissolves immediately unless calcium is present in the medium, and does not go to reinforce the vitelline membrane. In other words, it can be said that the soft fertilization membrane formed in the absence of calcium is simply the unmodified vitelline membrane. In short, the explosion product of the cortical granules is the substance which toughens the fertilization membrane.

Trypsin treatment of the unfertilized egg leads to an interesting result. It has been stated above that since this treatment dissolves the vitelline membrane, fertilization takes place normally after insemination, but no fertilization membrane is formed (Sugawara 1943a). Under these conditions, as there is no vitelline membrane for them to cling to, the explosion products are separated from the egg, but within 2—3 minutes needle-shaped or rod-shaped structures appear in the vesicles. Although such structures are never seen during the normal fertilization membrane hardening process, these rods are, in fact, thought to represent an alternate form of the fertilization-membrane hardening substance. The reason for this is that both substances originate in the cortical granules, their birefrigence is similar, and both are dissolved by sea urchin larval hatching enzyme. But why the substance forms a transparent, uniform layer against the vitelline membrane and appears as rods when isolated is not known.

The above is the process by which the fertilization membrane is formed in sea urchin eggs; it is possible to find a good many other cases in which the same mechanism seems to be operating. Moreover, cases in which the vitelline membrane separates and a perivitelline space is formed while the membrane hardens (although the presence of cortical granules has not been confirmed), as well as cases in which the existence and breakdown of cortical granules are known although the membrane change has not been fully sustantiated, may also be imagined to fall within the same category. If this is allowed, the problem becomes a wide one, including starfish and crinoids (Dan and Dan 1941), polyclads (Kato 1940), molluscs and even the gold fish (Tchou and Chen 1936). However, in contrast to practically all the other cases, in which it is possible to distinguish the vitelline membrane on the surface of the unfertilized egg, attempts to establish with certainty its presence in sea urchin eggs have been unsuccessful even with electron microscopy of ultra-thin sections (McCullock 1952a, 1952b; Endo 1956). The long dispute over the sea urchin fertilization membrane, between the 'New Formation Theory' and the 'Preexisting Theory' may be said to have been an accidental misfortune caused by this exceptional state of affairs.

(2) WAVE OF NEGATIVITY — FERTILIZATION WAVE — REFERTILIZATION

In order to understand the phenomenon of fertilization, it is of the utmost importance to know whether the wave of cortical granule breakdown described in the preceding section propagates from granule to granule, the explosion of one being the direct cause of an identical change in its neighbours, or whether there is some common mechanism behind this breakdown, of which the granule explosion is the final expression, the granules playing a role similar to that of the electric bulbs in a running news signboard.

The first person to imagine the possibility of such a common underlying mechanism was Just (1919), but since he was unaware of the existence of cortical granules, he was thinking chiefly in terms of a polyspermy block when he introduced the concept of a '*wave of negativity*'. By 'negativity' he meant the rejection by the egg of the second and all subsequent spermatozoa. From an entirely different standpoint, Yamamoto (1944), studying the cortical reaction of the medaka (teleost) egg, established the existence of a common propagation mechanism preceding the breakdown of the cortical alveoli. The evidence for this is: (1) in the medaka egg the rate of propagation of alveolar breakdown is different at the animal and vegetal poles, but practically constant for any given region; (2) even when an alveolus-free area is pricked with a needle, alveolar breakdown takes place over the whole surface; (3) if the eggs are fertilized after being centrifuged so that the cortical alveoli are displaced and accumulated in layers, the alveoli which are lying in the cortex break down, while those which have sunk into the endoplasm do not break down, even though they may be in close contact with alveoli in the cortex which do break down. Since Yamamoto's study was not concerned in particular with the mechanism of polyspermy-prevention, he calls this propagated effect the '*fertilization wave*'.

Following this work, Sugiyama (1953), using an ingenious method for applying chemical reagents to limited portions of the sea urchin egg surface, found that there are some reagents which cause cortical granule breakdown only in the treated area, and others which bring about the breakdown of the granules over the whole surface even though they are also applied locally. This led him to think that the former reagents affect the granules directly while the latter give rise to the fertilization wave. He was thus able to separate the fertilization wave and cortical granule breakdown, arriving at the conclusion expressed by the following formula:

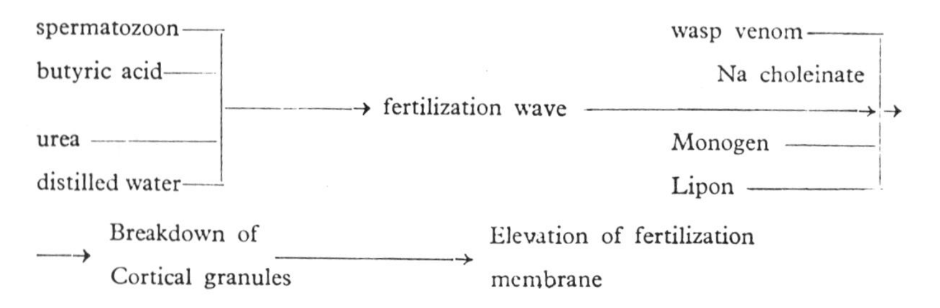

The following year Yamamoto (1954) also succeeded in separating the fertilization wave and cortical alveolar breakdown by reversibly inhibiting the latter with calcium-free Ringer.

Whether or not the fertilization wave is the so-called wave of negativity which blocks polyspermy is an interesting question to which, however, no answer can be made at the present moment. However, as one phase of the analysis of the fertilization reaction, interest in recent years has been re-directed to the fact that in sea urchin eggs it is possible to cause the entrance of spermatozoa into eggs which have already been fertilized (*refertilization*). This took its start in a report by Ishida and Nakano (1950) that, following Loeb's (1915) treatment for artificial parthenogenesis, they found a new method which used Ca-Mg- deficiency to obtain insemination of previously activated eggs. Sugiyama (1951) subsequently discovered that this method could also be applied to the refertilization of normally fertilized eggs, and unexpectedly enough, he found that it was effective even in the two-cell stage. Whether it is a matter of the fertilization wave or of such refertilization, there can be no doubt that these phenomena bear an intimate relation to the mechanism of polyspermy-prevention in normal fertilization.

(3) HYALINE LAYER

Following formation of the fertilization membrane, the next morphological change which takes place at the sea urchin egg surface is the appearance of the *hyaline layer*. This layer, as its name indicates, is translucent, a few micra in thickness, and becomes clearly evident around the fertilized egg surface 5—7 minutes after insemination. This, however, does not mean that it is suddenly formed at this time; actually, it gradually increases in thickness so that by 5—7 minutes it is easily visible to even casual observation. When Allen (1955) placed unfertilized eggs in fine capillary tubing and inseminated them from one side so that the fertilization reaction occurred over only part of the egg surface, he found that a hyaline layer was formed only where the cortical granules had broken down, indicating that these two processes must be linked in a causal relation. If this is so, formation of the hyaline layer must begin immediately after fertilization. Since the hyaline layer dissolves in calcium-free sea water, it is believed to be a kind of Ca-gel; from the classical study of Herbst (1900) it is known that this layer serves as a *Verbindungsmembran* to hold the blastomeres together during the cleavage stage.

It should be remembered that not all animal eggs have such a hyaline layer; medusan eggs, for example, are completely devoid of any sort of covering, and connection between the blastomeres is achieved by a different method (see p. 46.).

(4) SYNGAMY

a. Monaster

The external changes accompanying fertilization have been discussed in detail, but no description of the fertilization reaction could be considered complete which did not touch upon the internal events which follow sperm entrance. The account was broken off on page (38) at the moment, about three minutes after insemination, when the sperm head within the cytoplasm suddenly swells and begins to assume the characteristics of a *sperm pronucleus*. Simultaneously the midpiece breaks down and releases the centrosome into the egg cytoplasm. This statement that the midpiece 'breaks down' goes back to Meves's (1912, 1914) observation on stained sections of *Parechinus miliaris* eggs, which showed that the outer covering of the midpiece is abandoned in the cytoplasm when the centrosome is released. This old isolated observation has been confirmed in living material with phase contrast microscopy (Dan, J.C. 1950a).

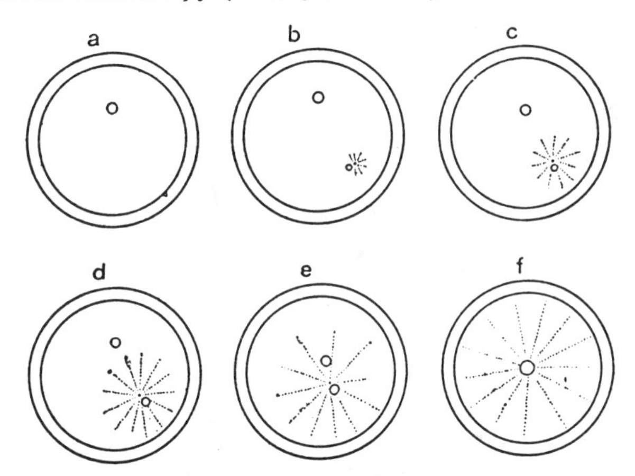

Fig. 1.21 Process of syngamy

a. entering sperm and fertilization cone; b. appearance of sperm monaster; c. growth of sperm monaster; d. migration of sperm monaster toward egg center; e. migration of egg pronucleus; f. fusion nuclei and development of monaster.

The centriole and sperm pronucleus remain close together, and around the centriole the egg cytoplasm forms the astral rays which together make up the *sperm monaster* (Fig. 1.21 b). Very early in its development, lengthening of the rays and consequent increase in diameter of the monaster have little influence on its position in the egg cytoplasm (Fig. 1.21 c), but once the rays have reached the cortex of the egg, further increase in their length causes the center of the monaster (i.e., centriole and sperm pronucleus) to move in a straight course toward the center of the egg (Fig. 1.21 d,e). The general effect is like that of pushing with a pole against the shore to send one's boat into the middle of the pond (Chambers 1939). When the tips of the astral rays which extend through the bulk of the cytoplasm reach the neighborhood of the egg pronucleus, it suddenly begins to move toward the center of the aster, and within a minute or two it reaches and meets the sperm pronucleus (Fig. 1.21 e).

Given the condition that this migration of the egg pronucleus begins when the tips of the astral rays come close to it, it follows that when the spermatozoon enters near the egg pronucleus, the monastral rays soon reach its vicinity and its migration takes place while the monaster is still small. When the sperm enters at the side opposite the egg pronucleus, on the other hand, the monaster must achieve a considerable size before its rays reach the egg pronucleus. In any case, since the monastral rays completely fill the cell sooner or later, the final position of the two pronuclei is bound to be the center of the egg. In sea urchins the two pronuclei remain thus in contact for a further period of about ten minutes, eventually uniting to form a single nucleus with the 2n number of chromosomes. This is called the *synkarion* (Fig. 1.21f). In sea urchins this synkarion proceeds to divide after a short resting stage, but there are many variations in this process. In some species the two pronuclei retain their separate identity until the beginning of mitosis (e.g., the clam, *Spisula*), and in others the first cleavage spindle is formed and the paternal and maternal chromosomes take their place as separate sets in the equatorial plate *(Cyclops, Ascaris)*.

b. Streak Stage

While the phenomenon of syngamy occurs in all animals, there is a stage following syngamy which is peculiar to sea urchins; this so-called *streak stage*

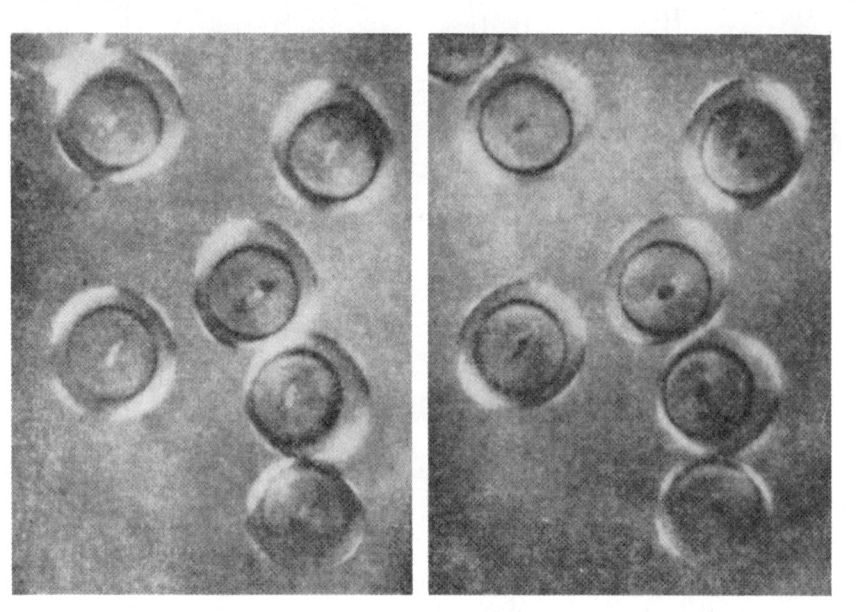

Fig. 1.22 Birefringence of streak, spindle, astral rays and fertilization membrane
Comparison of photographs shows localization of birefringence in egg as parts where black and white images are interchanged when compensator inserted between crossed Nicol prisms is turned by 90: At center of egg in middle of photograph is an early metaphase spindle, surrounded by "streak". Positive birefringence along long axis of spindle and negative birefringence of rays constituting streak produce contrasting figures. Eggs above and below this one are in late metaphase; rays of streaks very faint. Egg at upper left shows ana- or telophase figure, in which streaks have completely disappeared; two bright spots at spindle poles are small asters. Spindle tilted in egg at lower left. Since appearance of parts of fertilization membrane parallel to spindle axis changes with that of spindle, while parts vertical to spindle behave in opposite way, fertilization membrane is demonstrated to have positive birefringence in direction of tangent.

will be considered here. After about 20 minutes at temperatures ranging around 30° (or an hour or more at about 10° C), the monaster surrounding the synkarion gradually changes shape, becoming flattened in the direction of the egg axis and in turn causing the astral rays to bend (Fig. 9.1 f). As the structure takes the form of a flat plate, it appears as a conspicuously bright line when observed from the side, although seen from directly above, it looks no different from the earlier monaster. Since recognition of the streak thus depends on the angle of observation, it will not be simultaneously visible in all the eggs in any microscopic field; at any rate, its appearance provides the reason for calling this phase of sea urchin egg development the 'streak stage'. From whatever angle it is observed, the synkarion is seen to lie in the center of the flat plate.

Whether the streak stage represents no more than a change in the form of the monaster is still an open question. It has been stated that the high refringence of the plate region causes it to appear bright with the ordinary microscope; the birefringence of this region also differs from that of the monaster. In contrast to the monaster, which is composed of rays that show positive birefringence with respect to their long axes regardless of the direction from which they are observed, those of the streak show a negative birefringence with respect to their long axes (Fig. 1.22), suggesting that some radical change has taken place in the internal arrangement of their constituent molecules (Monné 1944; Inoué and Dan 1951). At this period, moreover, sea urchin eggs very often depart from the spherical shape with the appearance of indentations in their outlines. The biological meaning of these changes is quite unknown at present, but at least the direction of the streak serves as the earliest indication of the direction of the egg axis.

5. CLEAVAGE

(1) NUCLEAR DIVISION AND DIASTER

Since the nuclear division of egg cells follows the typical process through prophase, metaphase, anaphase and telophase, the details of these changes can be found described in any cytology text (see, for example, Hughes 1952); this account will be limited to the phenomena peculiar to eggs.

(1) The egg nucleus usually increases in size as the time of mitosis approaches. In fact, the synkarion is continuously growing, and suddenly in the midst of the growth process the nuclear membrane becomes invisible and metaphase begins, making it impossible to assign a definite size to the nucleus of a fertilized egg. This is sometimes inconvenient (Conklin 1902), but such increase in nuclear size has a special importance in that it constitutes, in living cells, the sole indicator of the approach of mitosis.

(2) As seen in sections or isolated mitotic apparatuses (Dan and Nakajima 1954), the prophase nucleus is already accompanied by two small asters, so that by the time the metaphase spindle is formed the diaster has attained a considerable size. It is still difficult to see the astral rays in the living condition at this stage, but since the astral centers are free of granules and appear hyaline, a bright dumbbell-shaped region in the central part of the egg is conspicuous even with low magnification. Actually, it is rather more difficult to observe this with high magnification, and a low power must be used in order to increase the contrast. This dumbbell shape is of course the result of observing the spindle at right angles to its long

axis; if it is observed along its axis, the two astral figures lie one over the other, and the mitotic apparatus appears as an even brighter circular area. In anaphase the two asters separate and the dumb-bell elongates.
(3) Not only in egg cells, but in other types of cells as well, it is rather often found that the cell becomes spherical in outline simultaneously with the disappearance of the nuclear membrane. In sea urchin species in which the surface of the egg becomes irregular during the streak stage, this change is particularly noticeable, while in many other sea urchins the hyaline layer suddenly becomes thicker at this stage, so that these two changes serve as joint indicators, two or three minutes beforehand, of the approaching cleavage.

(4) In the anaphase and telophase of mitosis there is a striking increase in the development of the diaster. In sea urchin eggs with transparent cytoplasm (viz., *Clypeaster, Mespilia, Temnopleurus, Hemicentrotus)*, it is possible to observe the astral rays in the living state with high magnification. About the time when nuclear division is complete and the daughter nuclei are formed, the astral rays reach the egg cortex and the diaster fills the whole cell; at this time the rays of the two opposing asters cross each other in the equatorial plane of the spindle. As soon as all these conditions are fulfilled, the cell begins to develop a constriction in the equatorial plane, and cleavage takes place. As has been stated above, nuclear division is already complete at this time and the two daughter nuclei have been formed; there is thus an interval of about ten minutes between the nuclear and cytoplasmic divisions.

(2) CYTOPLASMIC CLEAVAGE

a. Cleavage Process

In the case of the sea urchin egg, the time elapsing between fertilization and the first cleavage is less than 45 minutes at around 30°, about an hour at 20° and two and half hours at 10°C. The first event in the cleavage process is an elongation of the egg in the direction of the spindle axis so that its outline becomes ovoid. Next the cleavage furrow begins to appear at the middle of the long dimension — in other words, so as to bisect the spindle; finally the egg cell becomes cocoon-shaped and then divides into two cells. These daughter cells produced by the cleavage of the egg are called *blastomeres*. This whole process is completed within about 5 minutes in summer, and about 20 minutes in winter. The last remnant of the constricted portion persists for some time after the completion of cleavage as a short structure connecting the daughter blastomeres. This structure, which is commonly known as the *stalk*, is rather difficult to observe except in the absence of calcium or under some other condition which causes the blastomeres to separate (Scott 1946).

Since the egg coat (in sea urchins, the hyaline layer; in *Nereis*, the jelly) does not take any active part in the cleavage process, removal of such covering layers does not impede cleavage. On the contrary, the presence of even the normal egg coat offers some resistance to the cleavage activity, and if one of several experimental procedures is used to shrink the egg coat so that it becomes tight-fitting, cytoplasmic cleavage is suppressed, only the nucleus dividing so that a binucleate cell is formed (Gray 1931; Sugawara 1943a). It has been stated above that the chief function of these covering layers is to prevent scattering of the blastomeres after cleavage.

The first cleavage of practically all animal eggs takes place in a plane which includes the egg axis, but since in most equally cleaving eggs the two blastomeres form a practically perfect rotational figure around the spindle axis, the cleavage furrow, while it indicates the plane in which the egg axis lies, does not tell what direction the axis takes within this plane. For this reason, in order to locate the animal and vegetal poles at the first cleavage, it is necessary to resort to the use of eggs undergoing 'heartshaped cleavage', as described below, or material like the egg of the Mediterranean sea urchin, *Paracentrotus lividus,* which is provided with a special indicator in the form of a pigmented ring. It was formerly believed that the plane of the first cleavage furrow was determined by the original egg axis and the sperm entrance point, but this has been disproved (Endo 1954b).

b. **Mechanism of Cleavage**

During the last half century, two theories with respect to the mechanism of cell division have been opposing one another — the theory which locates the motive force of cell division in the cortex, against the theory which believes that it resides in the mitotic apparatus, within the cell (see Dan 1955). Only recently has the discussion reached a stage at which it is possible to hazard a conjecture connecting these two positions. According to this, the cleavage activity is initiated by the changes in the nucleus; when the mitotic apparatus is completely formed at metaphase, its influence is in some way communicated to the cortex which thereby for the first time acquires the capacity to constrict (Waddington 1952; Swann and Mitchison 1953; Hiramoto 1956). Looking back, it is apparent that the establishment of this very sensible conclusion was made possible by the fact that a method has recently been discovered by which, within a short time, the nuclear function alone can be obstructed. If the nuclear inhibitor is applied before metaphase, cell division does not take place; but if it is applied after metaphase, cell division is carried to completion even though the nuclear structure may be broken up. In consequence, research on the mechanism of cell division has now entered the stage of analyzing the effect as well as the nature of the change which is transmitted between nucleus and cortex.

c. **Unequal Cleavage**

The above description has referred to the case of equal cleavage; there are also eggs, like those of the annelids and molluscs, which undergo unequal cleavage from their first division. If the metaphase mitotic apparatus of a cell which is in the process of cleaving unequally is isolated, it is found that while the aster which will belong to the larger blastomere is spherical, that of the smaller blastomere is flat on its outer side as though it had been sliced off (Dan and Nakajima 1954). This condition is not restricted to the first cleavage of annelids and molluscs, but occurs in exactly the same way at the fourth cleavage of the Echinoidea, when the mesoderm-forming micromeres are formed. In the eggs of annelids and molluscs, also, the size difference among the blastomeres eventually shows itself to be connected with a difference in developmental fate; this strongly suggests that the largeness and smallness of the blastomeres, far from being chance characteristics, are of basic significance.

d. **Heart-shaped Cleavage**

So far, in the case of both equal and unequal cleavage, we have been concerned with the division patterns of alecithal or homolecithal eggs (see p. 7). When the yolk is concentrated at one pole of the egg (telolecithal eggs), the cleavage furrow appears earlier at the side where the yolk is relatively less (animal pole), causing the dividing cell to pass through a stage in which it is temporarily heart-shaped. One example of this among the echinoids is the egg of the sand-dollar *Astriclypeus manni*; such a cleavage pattern is caused by the fact that the spindle, instead of lying in the center of the egg, is formed closer to the animal pole (Dan and Dan 1947a). A more extreme state of heart-shaped cleavage is represented by the well-known cleavage pattern of medusan eggs (Fig. 1.23), in which the cleavage furrow advances quite one-sidedly from the animal pole, the vegetal region remaining completely inert. In this case, the spindle is bent into a V-shape, with its apex pointing toward the vegetal pole.

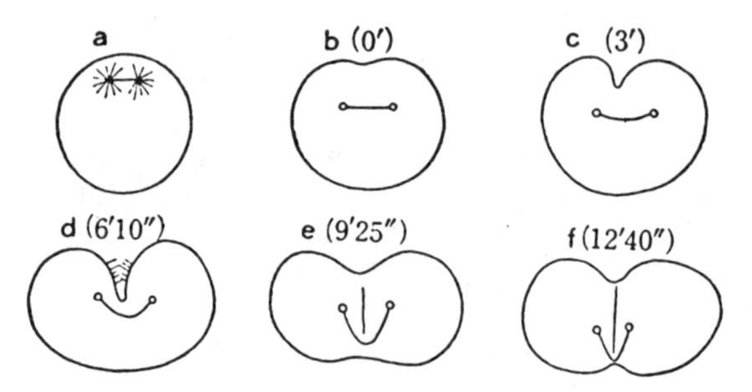

Fig. 1.23 Cleavage in the medusa *Spirocodon saltatrix*, showing one-sided advance of cleavage furrow, bending of spindle body and "spinning" between daughter blastomeres

Another aspect in which medusan cleavage differs greatly from that of other forms is that as the cleavage furrow advances, very fine cytoplasmic processes appear on the apposing walls of the furrow, unite with each other and draw the two blastomeres together (Fig. 1.23, d-f). It is by this sort of device that the medusan egg, which unlike the eggs of sea urchins and most other forms has no hyaline layer or investing membrane, prevents the separation of its blastomeres.

Dan and Dan (1947b) believe that medusan cleavage depends upon essentially the same mechanism as that of the sea urchin egg, the apparent difference being a secondary effect, resulting from the peripheral position of the spindle toward the animal pole, which causes the two asters to be rotated to some extent as the furrow passes between them. This rotation turns the two asters toward each other at the animal side and away at the vegetal, like two intermeshing gears (Fig. 1.24b); a similar effect would be achieved by turning the two sea-urchin blastomeres (Fig. 1.24d) in the directions indicated by the arrows. As a result, the spindle is bent into a V-shape, the vegetal surface is spread laterally so that the cleavage furrow there is very shallow, and the furrow at the animal side becomes still deeper. Consequently, the last connection between the two daughter blastomeres at the vegetal pole is none other than the "stalk" of sea urchin cleavage. Since cell division proceeds to completion even if the process-formation which holds the blastomeres together is inhibited, this phenomenon cannot be considered indispensable for

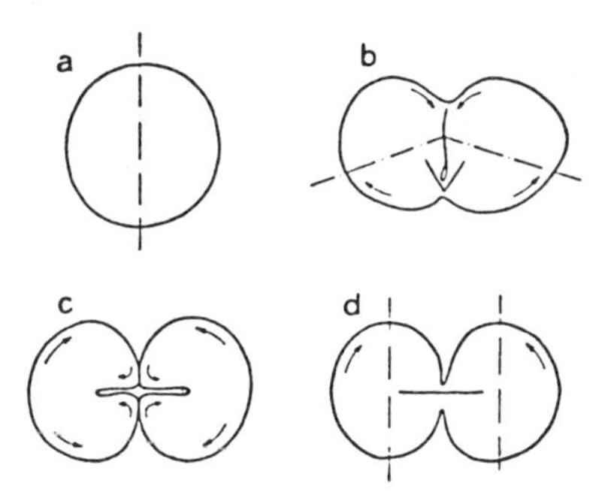

Fig. 1.24 Comparison of cleavage types in (a, b, c) medusa and (d) sea-urchin

cleavage activity; its occurrence, however, makes astral rotation take place the more readily.

What effect, then, does this rotation of the asters have on the direction of the egg axis? Since the egg contents are rotated at every division in this type of cleavage, it is to be expected that the egg axis will be disturbed, but actually the second cleavage furrow of the *Spirocodon* egg starts from the contact surfaces of the blastomeres and proceeds laterally (Fig. 1.24c), thereby following the functional egg axis established as the result of the first cleavage.

If this type of cleavage found in the medusan egg is carried to an even greater extreme, it results in a situation such that the second cleavage furrow appears at the animal pole before the first has reached the vegetal pole, so that several cleavage stages may be overlapping one another. This is called *discoidal* cleavage. The example of discoidal cleavage which is most familiar to us is that of the frog egg; among invertebrates, the squid egg cleaves in this way. Eggs which follow such a mode of cleavage are telolecithal.

e. Superficial Cleavage

Eggs like those of the arthropods, which contain a large amount of centrally located yolk (centrolecithal eggs), divide by *superficial cleavage*. This is characterized by the fact that no cell boundaries are formed in connection with the early nuclear divisions, so that a multinucleate condition results. Not until the number of nuclei has exceeded a certain level do the nuclei migrate to the egg cortex, where cell walls are suddenly and simultaneously formed around each nucleus. These cell walls, however, are formed only at the contact planes between neighboring cells, leaving the bottoms of the cells open and still directly connected with the common mass of yolk. When the time comes for the formation of a second cell layer, the bottoms of the cells constituting the outer layer are partitioned off.

f. Polar Lobe Formation

One other interesting feature of annelid and molluscan cleavage is the formation of the *polar lobe*. Although this phenomenon is not a part of the cleavage process in the strictest sense, the pushing out of the vegetal pole cytoplasm is accurately synchronized with the extrusion of the polar bodies and cleavage (Fig. 1.25). Since the cytoplasm which does not enter the polar lobe is also to be divided into two

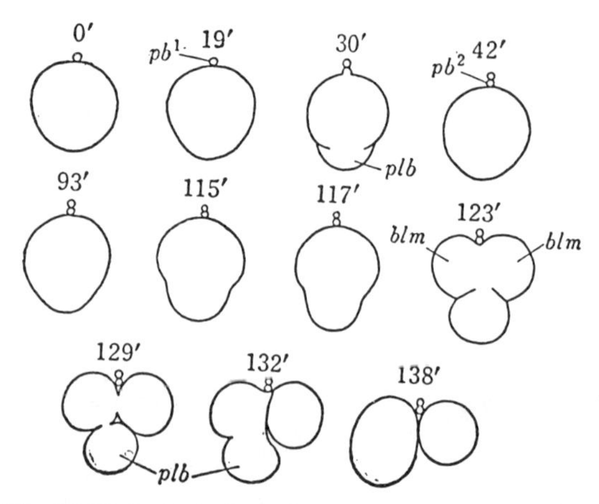

Fig. 1.25 Polar body formation and polar lobe formation in connection with first division

blm: blastomere, *pb*¹: first polar body, *pb*²: second polar body, *plb:* polar lobe.

blastomeres, a three-lobed figure, the *trefoil stage*, is formed. However, since the polar lobe is by no means an independent cell (since it is anucleate), it is never severed in the normal process no matter how slender the connecting region may become, and sooner or later the polar lobe material is combined with the cytoplasm of one of the blastomeres. Consequently, one of the blastomeres at the resting phase of the two-cell stage is small, while the other which includes the polar lobe material is large, exactly as in unequal cleavage. The polar lobe is usually formed at each cleavage until the 8-cell stage, and at no later cleavages. If the mitotic apparatus is isolated at the trefoil stage, the two asters are found to have the same shape, and it appears that the formation of the polar lobe depends on some other mechanism. In conclusion, mention should be made of the classical experiment of Wilson (1904a), in which he demonstrated that the cytoplasm contained in the polar lobe carries the burden of forming the future mesoderm.

(3) CLEAVAGE PATTERNS

The next major problem in connection with cleavage concerns the fact that the direction of each cleavage plane is predetermined — in short, this means that the location and direction of the spindle is fixed for each cleavage.. As a result, the egg of each animal species has a particular cleavage pattern. The cleavage patterns of animal eggs can be roughly divided into those of the *radial type* and those of the *spiral type*.

a. Radial Type

Of this type it can be said in general that the mitotic spindle of each cleavage takes a position at right angles to the position of the previous cleavage, while the cleavage planes appear parallel or perpendicular to the egg axis in a regular manner. The model of this can be seen in the cleavage pattern of sea cucumber eggs *(Labidoplax digitata,* Fig. 9.26). A survey of many other species shows that although they are described as undergoing radial cleavage, very few are as extremely regular as the sea cucumber egg. Sooner or later, in each case, some irregular cleavages are introduced into the pattern, but the great majority of species conform to the type at least as far as the 8-cell stage.

b. **Spiral Type**

In this type of cleavage each successive spindle lies at right angles to that of the preceding division, but since the first cleavage plane is slanted with respect to the egg axis, all the later ones also intersect it with a certain inclination. This is not particularly conspicuous in the 2-cell stage, but at the 4-cell stage, two of the diagonally opposite blastomeres line up on the upper side, while the other two lie on the lower side. In accordance with embryological custom the two upper blastomeres are labelled *A* and *C*, and the ones on the lower side are called *B* and *D*; seen from a position perpendicular to the egg axis — that is, from the side — they would appear alternately above and below: *A* *B* *C* *D*. If a 4-cell stage embryo is looked at from the animal pole, the *A* and *C* blastomeres are seen to be in direct contact with each other, while at the vegetal pole *B* and *D* are in contact. Furthermore, the vertical planes separating *A* from *C*, and *B* from *D*, meet at right angles in the center of the embryo (Fig. 11.3b).

In a great many cases the four cells divide simultaneously into eight by an unequal cleavage in which the daughter blastomeres at the animal pole side are clearly smaller than those of the vegetal side. In other words, the former are *micromeres* and the latter, *macromeres*. In this case, also, the spindles are more or less tilted with respect to the egg axis, so that the cleavage planes, unlike those in radial cleavage, are not horizontal. In *Trochus* (Robert 1902), the micromeres seen from the side are formed above and to the left of the macromeres, so that the small cells fit into the depressions between the large ones, in such a way that the boundary

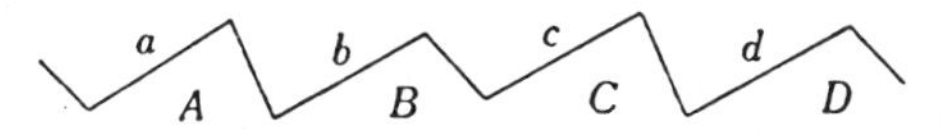

between the two layers forms a saw-like, jagged line. In later chapters it will be clear that this zigzag line is a criterion for judging whether a cleavage pattern is or is not of the spiral type. Seen from above, the four cells *A*, *B*, *C*, *D*, give off their respective micromeres in such a way that they twist in a clockwise direction around the egg axis; this type of cleavage is therefore called *right-handed, dextral, dextrotropic* or *dexiotropic*. Again following the customary system of nomenclature, the micromere which is first formed, above and to the left of the *A* blastomere, is designated as 1*a*, that of the *B* cell, as 1*b*, and so on.

This set of four micromers, 1*a*, 1*b*, 1*c*, 1*d* is called the *first quartet* (1*q*), while the macromeres, after giving off this first quartet, are labelled 1*A*, 1*B*, 1*C*, 1*D*. In short, the number 1 before the small letter indicates the first quartet, while before the large letter it indicates the macromere generation which produced the first quartet.

The eight cells again cleave simultaneously, into 16 cells. Again the cleavage plane is slanted with respect to the egg axis, but this time the spindle lies at approximately right angles to the position of the previous one; consequently, as

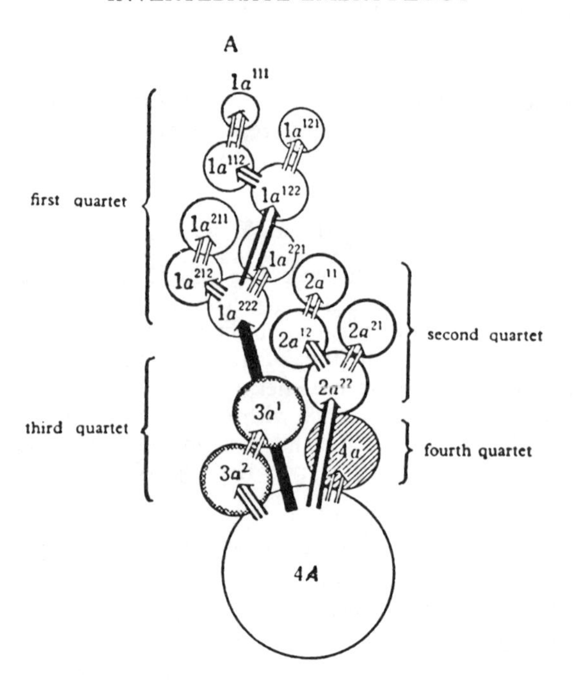

Fig. 1.26 Diagram showing change in direction of cleavage in one quadrant of *Trochus* egg (viewed from side)

, direction of first division (of macromere); , direction of second division; , direction of third division; , direction of fourth division; fine-line circles represent offspring of first quartet; thick-line circles are second quartet, shadowed circles are third quartet; hatched circle is fourth quartet; large circle is macromere.

seen from the side, each macromere is giving off its micromere above and to the right (Fig. 11.3, e.f). Observed from above, the new micromeres are formed in an anti-clockwise direction. This cleavage is therefore said to be *left-handed, sinistral, l(a)eiotropic* or *l(a)evotropic*. This cleavage gives rise to the 16-cell stage, consisting of four tiers each made up of four cells. To return to the system of nomenclature — since the quartet of smallest cells, formed in an anticlockwise direction by the next (fourth) cleavage of the micromeres (which were given off at the animal pole to form the 8-cell stage), are descendants of the first quartet, the number 1 placed before the letter naming the blastomere is retained, and an index number is added after the letter — viz., $1a^1$, $1b^1$, etc. The four blastomeres remaining in the second layer are labelled $1a^2$, $1b^2$, etc. The micromeres formed directly from the macromeres can be regarded as the second quartet; they are therefore called 2*a*, 2*b*, 2*c* and 2*d*, and the macromeres which have produced them have their labels changed to 2*A*, 2*B*, 2*C* and 2*D* (Fig. 1.26).

At each succeeding cleavage every cell proceeds to cleave, as shown in Figures 11.3, 4.2. and 1.26, in an alternately clockwise and anti-clockwise direction. In general this goes on until the fourth quartet has been formed, although some stop at the third quartet, and species are known which produce a fifth quartet. It is useful for an understanding of the embryological process to remember that as long as all the blastomeres are dividing, the number of cells belonging to the descendants of the first quartet at any particular stage is always exactly half the total number of blastomeres; when the blastula stage is reached, these occupy roughly the upper half of the embryo.

Another method of embryological description consists in dividing the embryo longitudinally into geneological lines corresponding to the descendants of the *A*, *B*, *C* and *D* blastomeres, instead of into the latitudinal quartets described above. Since separating the embryo into its *A B C D* components is equivalent to dividing it into four lengthwise (along the egg axis) segments, each of these is called a *quadrant*. It is convenient to remember that in later developmental stages the *B* quadrant forms the anterior part of the larva, the *D* quadrant forms the posterior part, and the *A* and *C* quadrants form the left and right sides, respectively. As a consequence of this fact, the anterior-posterior and lateral axes of molluscan and annelid embryos are already determined in the 4-cell stage. Watasé (1890) long ago reported that in the discoidal cleavage of the squid *Loligo*, bilaterality is established at the first cleavage; this has recently been confirmed in *Ommastrephes* (Fig. 11.40) by Soeda (1952). Other species in which bilateral symmetry is manifested very early in development are the ascidians and the nematodes (see later chapters).

In the example of *Trochus* described above, the first quartet is given off in a clockwise direction, when seen from the animal pole, and species which cleave in this way are therefore called dexiotropic. There are, however, a few cases including *Ilyanassa*, *Physa* and *Planorbis*, in which the cleavages, beginning with a left-handed separation of the first quartet, are mirror images of the more usual process; these are described as laevotropic to differentiate them from the right-handed majority. As a matter of fact, this difference between the dexiotropic and laevotropic cleavage patterns of the snails directly influences the spiraling of the adult shell; it was Crampton (1894) who made the interesting discovery that even within a single species the mirror image direction of cleavage leads to opposite rotation of the adult shell (Fig. 1.27).

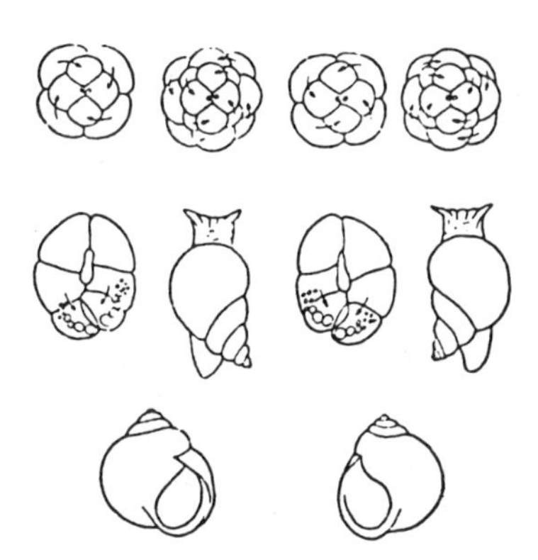

Fig. 1.27 Comparison of cleavage pattern and coiling of adult shell in dextral (left) and sinistral (right) gastropods (Conklin)

6. GERM LAYER FORMATION

(1) BLASTULA

In eggs like those of the echinoderms, which undergo more or less regular total and equal cleavage, the result is a spherical embryo composed of a single layer of cells, like a rubber ball. This embryo is called a *blastula,* and in cases resembling the echinoderms, it contains a rather spacious cavity, called the *blastocoel.* Blastulae which possess such a cavity are given the special name of *coeloblastulae*; among these are also included some forms in which the blastocoel is very small because the cells surrounding it are more columnar than usual (Fig. 1.28).

On the other hand, when the yolk is localized at the vegetal pole and the egg cleaves unequally, as in molluscs and annelids, even though a coeloblastula is formed, its blastocoel is shifted toward the animal pole. In most cases, however, the interior of such eggs is filled with the large, yolk-laden cells of the vegetal region (usually macromeres), and no space resembling a blastocoel appears. Such blastulae are called *stereoblastulae,* or *sterroblastulae.* Some stereoblastulae arise, as in the Coelenterata, from eggs which, although they cleave totally and equally, form tall blastomeres with no space left at the center; in other cases, later cleavages take place in a plane parallel to the blastular surface and the interior becomes filled with cells. Among the forms which undergo discoidal cleavage, those with especially large amounts of yolk form cells at only one side (animal pole region) of the yolk mass. Since this kind of embryo can be regarded as a modified blastula, it is called a *discoblastula.* Among the invertebrates, this type occurs in the development of the Cephalopoda. In superficial cleavage, which is similarly an example of partial cleavage, cell formation occurs only in the peripheral part of the egg;

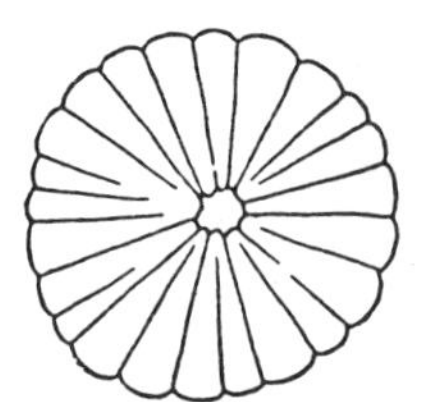

Fig. 1.28 Blastula of *Sagitta crassa* (Kumé)

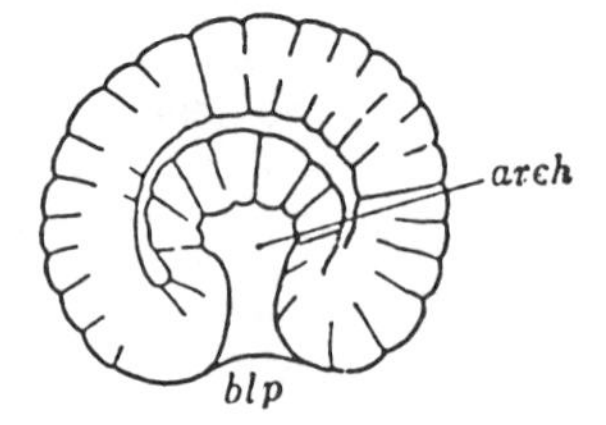

Fig. 1.29 Gastrula of *S. crassa* (Kumé)
arch: archenteron, *blp:* blastopore.

the embryo thus formed, called a *periblastula,* is typical of the insects. The cellular region in both discoblastulae and periblastulae consists of a thin layer which is generally referred to as the *blastoderm.* Because of its convenience, this term is frequently used in connection with these cellular layers even after they have reached a rather advanced stage of differentiation.

Sea urchin blastulae range from that of *Anthocidares crassispina,* with a small, sagitta-like blastocoel, to the *Clypeaster japonicus* blastula, in which the flattened blastomeres form a thin wall around a very large blastocoel; but since the differences in the apparent size of these blastulae are entirely due to differences in the amount of fluid within the blastocoel, it is obvious in each case that there has been no increase in the original amount of protoplasm contained in the egg.

(2) GASTRULATION

After reaching the blastula stage, the embryo proceeds, through the process of *gastrulation*, to become a *gastrula*. Unlike the blastula, the gastrula is composed of an inner and an outer cell layer, for the first time establishing different *germ layers*. The cell layer covering the outside of the gastrula is the *ectoderm (ectoblast* or *epiblast)*, while the inner cell layer is the *endoderm (entoblast, hypoblast)*. The endoderm often makes its appearance in the form of a cell mass.

There are various methods by which the blastula acquires two cell layers and becomes a gastrula; one of the most usual of these is *invagination*, which may also be called *emboly*. In this process the cell layer of the posterior side of the embryo — that is, the vegetal region — bends inward toward the blastocoel. The portion which thus turns inward is the endoderm, while that which remains as the outer layer of the embryo is the ectoderm. The part formed by the endoderm is called the *archenteron* or *primitive intestine*, and its opening is the *blastopore, protostomia* or *primitive mouth*. Even among the forms in which gastrulation takes place by invagination, there are some in which the wall of the archenteron is in contact with the outer body wall, leaving no blastocoel between the two *(Phoronis)*, while in others, like the echinoderms, a wide blastocoel remains around the archenteron.

There is no established theory to account for the mechanism of gastrulation. It is believed that the latter half of the sea urchin gastrulation process is accomplished through the activity of pseudopodia sent out by the cells at the blind end of the archenteron. These pseudopodia first attach their tips to the outer body wall and then contract, causing the archenteron to be drawn inward. However, we remain completely in the dark with respect to the factors underlying the first half of this process — i.e., invagination in the strict sense of the term.

In the case of such forms as the annelids and molluscs, in which there are large blastomeres containing a great deal of yolk at the vegetal side and small blastomeres distributed over the animal pole region, the cleavage of these small cells proceeds at an accelerating rate so that they spread over and invest the large blastomeres, resulting in the formation of a gastrula. Judging from later developmental results, gastrulae formed in this way preserve the same relationships as those which arise by invagination, the large cells which are enclosed being the endoderm, while the outer cell layer which surrounds them is the ectoderm. This type of gastrulation is called *epiboly*.

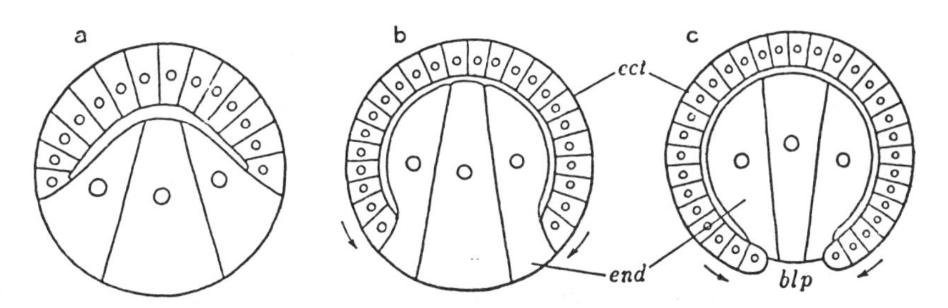

Fig. 1.30 Diagram showing gastrulation by epiboly (Boas)

blp: blastopore, *ect:* ectoderm, *end:* endoderm.

However, no archenteron cavity has yet appeared in the internal mass of endoderm cells of these gastrulae formed by epiboly, while the part that is called the blastopore is a region at the vegetal pole where the ectoderm cells fail to cover the endodermal cell mass completely, leaving part of it exposed. But as development proceeds, the ectodermal cell layer which encircles this blastopore sinks inward, forming a structure called the *stomodoeum*.

Gastrula-formation by epiboly is thus a modification of gastrulation by means of invagination, and there are in fact many cases in which the two methods occur in combination.

In discoblastulae, in which the cellular layer is formed at only one side of the embryo, one edge of the blastoderm folds inward beneath itself and elongates forward, forming the endoderm. This process can be looked upon as a modification of invagination; it also has the special name of *involution*. Among invertebrates it is seen in the cephalopods.

A method which has only a rather slim connection with those so far described is that called *delamination*. This process, which occurs mainly among the coelenterates, itself includes several variations. Delamination in the strict sense of the word is found only in the Geryoridae. In these species the cells which make up the body wall of the coeloblastula formed by cleavage all divide similarly toward the blastocoel, and the cells thus thrown into the blastocoel line up in a regular fashion and form an endodermal sac. Lankester (1877) considered this to be the most primitive form of gastrulation, but this idea is clearly in opposition to the Gastraea Theory of Haeckel (1872), which considers the gastrula formed by invagination to be at the root of the geneological tree.

There is another method of endoderm formation, called *multipolar proliferation* (or *ingrowth*), which closely resembles this type of delamination. In this, some of the cells constituting the blastular wall slip from various parts of this cell layer into the blastocoel, and eventually unite to form the endodermal sac (fresh-water *Hydra*). When this releasing of cells takes place at only one side (vegetal pole), of the embryo *(Clytia)*, it is called *unipolar proliferation* (or ingrowth). These methods are grouped together under the name of *polarization*; whichever of them occurs, the resulting sterrogastrula is solidly filled with cells, and has no archenteron or blastopore. As its development proceeds, however, an archenteron appears and a blastopore opens at the vegetal pole in the forms in which ingrowth has taken place.

(3) MESODERM FORMATION

In the past, the sponges and coelenterates have been differentiated from the other, three-layered animals *(triblastic)*, as being composed of two germ layers *(diploblastic)*. However, in these groups also, mesenchyme cells are obviously interspersed between the endodermal and ectodermal layers, showing that there is no fundamental difference between them and the so-cal'ed triblastic animals with respect to the relations among the three germ layers.

The mesenchyme cells of these animals all have their origin in the ectoderm, and it is possible to find animals other than the sponges and coelenterates whose mesenchyme comes from the same source. In the group of animals called the Protostomia, those which undergo spiral cleavage (flatworms, annelids, molluscs, etc.), certain cells from the second quartet ($2q$), or from the third ($3q$), and sometimes from both, are released into the blastocoel, where they become mesenchyme.

Mesenchymes which originate in this general manner are lumped together under the name of *ectomesoderm* or *larval mesoderm*.[1] Detailed comparative studies of the modes of origin of this kind of cells can provide extremely useful material for theories of phylogeny.

In contrast to the ectomesoderm, which always takes the form of mesenchyme, mesoderm cells of the epithelial type develop from the entoderm or at least have an intimate relation to this layer; they are therefore referred to as *entomesoderm*. What is called simply mesoderm mostly belongs to this category, and this type of tissue is sometimes also referred to as *true mesoderm*.[2]

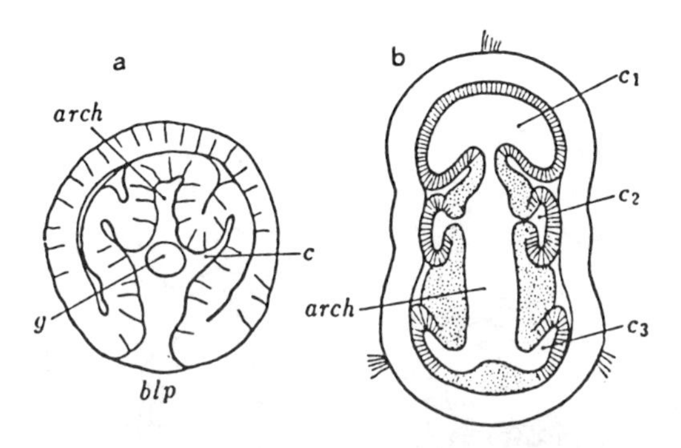

Fig. 1.31 Formation of enterocoel

a. *Sagitta crassa* (Kumé) b. *Balanoglossus kowalevskii* (Bateson) *arch*: archenteron, *blp*: blastopore, *c*: enterocoel, c_1: proboscis-coelom, c_2: neck-coelom, c_3: trunk-coelom; *g*: primordial germ cell.

There are also various ways in which this kind of mesoderm is formed. The method in which the entodermal origin is most clearly evident is that of *enterocoel formation*, found among the animals grouped together as the Deuterostomia (Chaetognatha, Echinodermata, Procordata, Brachiopoda). In this process, certain regions of the archenteron wall first bulge outward into the blastocoel, and then these bulges separate from the gut wall, forming the *coelomic pouches* (Fig. 1.31). The cell layer making up the walls of such a pouch is the mesoderm, and since the enclosed cavity originates from the archenteron, it is called an enterocoel. The number and position of these bulgings of the archenteron wall are different in different species (Fig. 1.31 a,b).

These enterocoels which are newly formed in the blastocoel are completely surrounded by mesoderm and possess different properties from those which do not have such a structure. For this reason, the enterocoel type is called *deuterocoel*, or *secondary body cavity*, in contrast with the blastocoel type, which is called *primary body cavity*, and, when it is carried over without change into the adult anatomy, *pseudocoel*.

---

[1] "Ectomesoderm" and "entomesoderm" are sometimes reffered to as "mesectoderm" and "mesentoderm"; for example, by MacBride (1914).

[2] It may be also be called "primary mesoderm (mesoblast)" in distinction to "secondary mesoderm (mesoblast)" derived from extomesoderm.

The animals belonging to the Protostomia (Platyhelminthes, Nemertini, Annelida, Mollusca, Arthropoda, etc.), follow a method different from that of enterocoel formation; their mesoderm takes its origin from a special cell, 4*d*, which is thought of as belonging to the entoderm; or in some cases, by a modification of this process. In most of the animals of this group, the egg divides by spiral cleavage, and in the forms which cleave in this way, the geneology of each blastomere — Wilson (1892) called this cell-lineage — has generally been worked out in such detail that it is clearly known what kind of germ layer, organ and even tissue each blastomere will form. As will be evident from the explanation of spiral cleavage (pp 49 ff)., the 4*d* cell in question belongs to the *D* quadrant at the posterior side of the embryo, and is one of the cells of the fourth quartet.

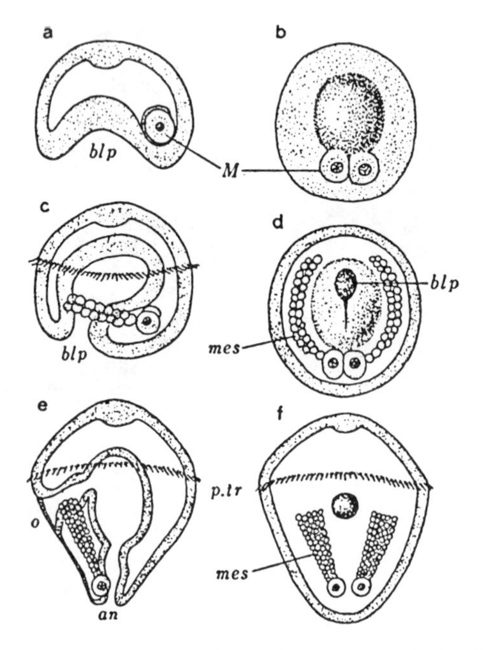

Fig. 1.32 Formation of mesodermal band from teloblast during stages from early embryo to trochophore larva (Korschelt u. Heider)

a, b. gastrula; c, d. narrowing of blastopore; e, f. trochophore. *an:* anus, *blp:* blastopore, *M:* teloblast, *mes:* mesodermal band, *p.tr:* prototroch.

In the past it was thought that the descendants of this 4*d* cell all produced mesoderm, and it was therefore called the *primary mesoblast* (or just mesoblast), abbreviated as *M,* but since more recent studies have shown that entoderm cells are also included among its descendants, it is now called the *mesentoblast,* abbreviated as *ME.* The 4*d* cell eventually slips into the blastocoel and divides into left and right cells (*Ml* and *Mr*); as the embryo continues to differentiate, these come to lie as a pair at the front side of the region which will form the future anus, and send out cells anteriorly, forming the *mesodermal bands* (Fig. 1.32). For this reason, these cells are sometimes called simply *teloblasts* or *pole cells*; with this particular meaning in view they may be called *mesoblastic teloblasts.* To describe them most accurately, these cells should be called *teloblasts of the mesodermal bands.* In cases like that of the annelids, in which a space appears within such a mesodermal band, this develops into a true coelomic cavity, and may be differentiated from the blastocoelic cavity by the name of *shizocoel.*

Even in the arthropods, which do not undergo spiral cleavage, the mesoderm makes its appearance as a pair of cellular bands. Instead of being formed by a special kind of cells, however, these arise from the cell mass surrounding the blastopore, but since their behavior is strikingly similar to that of the mesoderm bands described above, it is permissable to regard them as a sort of modification of the teloblastic method of mesoderm formation. The reasons for judging this mesoderm formed by the teloblasts to be entomesoderm are that the other cells of the fourth quartet ($4a$ — $4c$), like the macromeres, are endodermal, and that, as stated above, among the descendants of $4d$ are found endodermal as well as mesodermal cells.

In general the endomesoderm usually takes the form of epithelium, but in some cases it may also appear, like the ectomesoderm, as mesenchyme. Examples of this sort are not very numerous, but in echinoderms, for instance, such cells are released into the blastocoel as early as the blastula stage. In *Phoronis,* also, all the mesoderm is formed by this method.

7. EGG POLARITY

Before the egg begins to develop into an embryo, its only differentiation — structural or functional — is to be found in its polarity. Generally speaking, the position which the egg pronucleus takes in the cytoplasm is not, as might be expected, the geometrical center of the egg cell, but always more or less to one side. The side nearest the nucleus is the *animal pole* of the egg, while the opposite side is the *vegetal (vegetative) pole.* But in addition to this eccentric position of the pronucleus, there are various other factors, from the distribution of the contents and pigmentation of the egg to its morphology, which change from the animal to the vegetal pole along the egg axis. One example of this is the accumulation of yolk at the vegetal pole in telolecithal eggs; among the cases of localized pigmentation, there are the egg of the Mediterranean sea urchin, *Paracentrotus lividus,* which is encircled just below the equator by a band of red granules, and that of the annelid *Myzostoma,* in which the animal pole is red and the vegetal pole green. Among eggs in which the two poles have distinguishing morphological characteristics are those of most insects, and of the squids.

What, then, is the cause of such gradients in the distribution of materials, and these structural peculiarities? In general, these asymmetries are attributed to the fact that during its development the egg cell is attached to the ovarian wall by one part. In the great majority of cases this attachment point becomes the animal pole of the egg, and in addition it often happens that the micropyle, which pierces the protective membranes, is formed at this particular spot. However, the reason why this problem of the egg axis attracts so much interest is not simply because it provides an anatomical guidepost, but rather because a variety of physiological activities during development take place with the egg axis as their basis, and the body axis of the embryo is in its turn strongly influenced by the egg axis. Looking back over just the points which have been touched upon so far in this chapter, we see the formation of the polar bodies and the mode of furrow formation in heart-shaped cleavage as phenomena closely connected with the animal pole;

judging from the arrangement of the blastomeres, the direction of the cleavage spindles must be correlated with the two poles; the formation of the polar lobe appears to depend in some way on the vegetal pole. Finally, as will no doubt be described in a later chapter, the only way to account for the complicated protoplasmic movements of the ascidian egg is to think of them as occurring in conformity with the egg axis.

Considered from a physiological standpoint also, eggs and embryos are splendid examples of the axial gradient theory. As the many papers of Child have shown, the gradient of sensitivity to KCN and the oxidation-reduction potential gradient are distributed in parallel with the egg axis, and only in the gastrula stage does a second axis make its appearance, in connection with the archenteron (Child 1952). Moreover, from the point of view of 'chemical embryology', the work of Lindahl (1936) has shown that sugar metabolism is preponderant in the animal pole region, while protein metabolism dominates in the vegetal region. Furthermore, as the embryo passes through the blastula and gastrula stages and enters a free-swimming phase, the larval body axis coincides with the original egg axis, and in practically every case the larva swims with the animal pole side forward.

Thus the animal pole becomes the part, even of the future larva, which is concerned with locomotion and sensation, while the vegetal pole becomes the digestive system; and since, in the Middle Ages, motion and sensation were regarded as functions peculiar to animals, and the taking in of nourishment more characteristic of plants, these two poles were given the names 'animal' and 'vegetative'.

Finally there is the case of the ctenophores, which at first glance appear to be an exception to this grand rule of polarity, since their larvae seem to swim with the vegetal pole ahead. This circumstance has introduced a lack of unity into embryological texts. For example, in the book of Korschelt (1936, p. 213), the body axis of the swimming larva was taken as the basis for working out the direction of the egg axis in the early cleavage stages. As a result, the figure of the first (heart-shaped) cleavage shows the cleavage furrow cutting upward from the lower side of the egg, which is called the vegetal pole. While admittedly there is not much in a name but convenience, such usage, for the sake of this one example, upsets the definition of animal-vegetal polarity which is in current use for all the rest of the animal kingdom. Schleip (1929 p. 42) takes a middle course: in his illustration, the cleavage furrow cuts in from above, but the explanation describes this as the vegetal pole.

The author believes that this confusion is due to the fact that coelenterate eggs divide according to a special formula — i.e., in each of their equal divisions the rotation of the cleavage asters actually causes a turning of the egg axis (see pp. 46—47). In summary, this means that although each blastomere has its own polarity, their cleavage in various directions and their rotation in various degrees lead to the formation of an embryo which has no common body axis.

The following account explains this concept in terms of what actually happens in ctenophore cleavage. In this investigation of the development of *Beroë ovata*, Spek (1926) mainly relied on Yatsu's (1912) method, which consisted in following, in connection with cleavage, the movement of the cortical cytoplasm, which appears green with reflected light. This substance, which before cleavage is evenly distributed around the egg (Fig. 1.33a), collects at the animal pole when the cleavage furrow is being formed (Fig. 1.33b), penetrates to the inside together with the

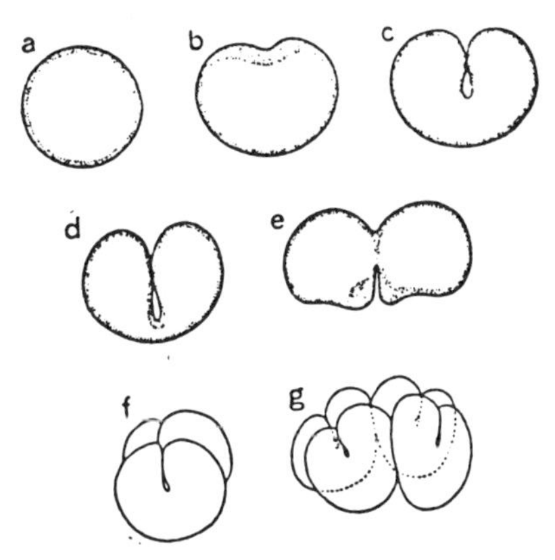

Fig. 1.33 Early cleavage and behavior of the cortical substance in the medusa *Beroë*

a ~ e. cleavage and behavior of green cortical substance (dotted peripheral area); f. second cleavage; g. third cleavage. Note that cleavage furrows in f and g advance from upper side of egg.

advance of the furrow (Fig. 1.33c, d), and at the completion of cleavage accumulates at the side opposite to its position at the beginning (Fig. 1.33e). From the facts that when the green cytoplasm moves from the animal to the vegetal pole, the nuclei make the same shift, and also that in *Spirocodon* (Fig. 1.24c), the next cleavage furrow develops from this place, it is believed that the animal pole is pulled down to the lower side by the process of cleavage furrow formation.

However, in *Spirocodon* the animal pole is fixed in the place to which it has been thus carried, so that the second cleavage takes place as shown in Figure 1.24c, while in *Beroë* both the green plasm and the nuclei move back to their original positions, so that the next cleavage furrows again begin at the upper sides of the blastomeres. Since this same process, in which the animal pole is carried down and then reverts to its original position, takes place also in the second and third cleavages, the furrows dividing the 2-cell stage into four cells and the 4-cell stage into eight cells all similarly begin at the upper side of the egg (Fig. 1.33f, g). However, when the third cleavage in *Beroë* has completed the division into eight blastomeres the green plasm as well as the nuclei become fixed at the lower side (Fig. 1.34a). The cleavage which follows this is unequal, eight micromeres containing most of the green cytoplasm being formed on the under side (Fig. 1.34b, c). Since these micromeres are responsible for forming the larval comb, this lower side takes the lead when the larva begins to swim. After resting for several hours the macromeres divide equally (Fig. 1.34d); in connection with this equal cleavage the nuclei and the remaining green plasm are shifted back to the upper side, where another set of micromeres is formed (Fig. 1.34e). These micromeres become the mesoderm.

From these facts the two following conclusions may be drawn:

(1) Although the *Beroë* larva appears at first glance to be swimming with its vegetal side ahead, the animal pole itself has been shifted to the lower side of the

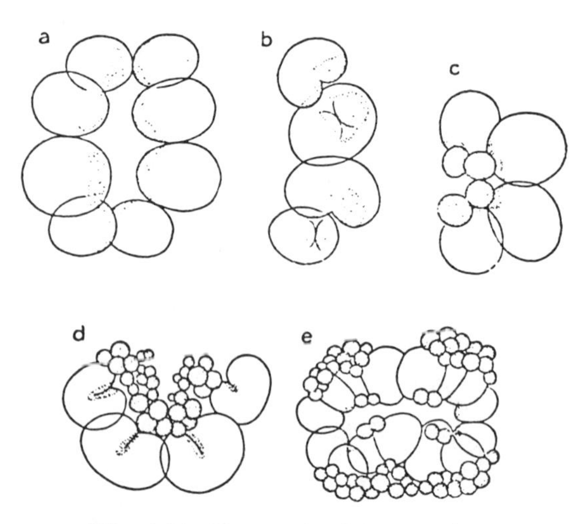

Fig. 1.34 Cleavage in *Beroë* (continued)

a. animal pole view of 8-cell stage. Green substance accumulated at lower side of each blastomere; b. vegetal pole view of four blastomeres at 8-cell stage. Cleavage furrow of next division appearing in area of green substance; c. division shown in b has resulted in unequal cleavage. Most of green substance in small blastomeres; d. during several hours following division shown in c, only blastomeres having large amounts of green substance have divided. Macromeres eventually divide; green substance remaining in them is being carried with advancing cleavage furrow to opposite side (animal pole of embryo); e. sketch of compressed embryo (animal pole side shifted inward, vegetal side turned outward).

embryo as a result of the fact that the blastomeres reversed their polarity in the third division and became fixed in that state, and it can be said that in reality the *Beroë* larva swims with its animal pole in the lead.

(2) As is apparent from Figure 1.34d, e, since the reversal of blastomere polarity is accomplished by carrying the cellular contents to the opposite side of the cell, it is possible to distribute various materials to different parts of the embryo by appropriately combining equal and unequal cleavages. As a consequence it seems not unlikely that in coelenterates it is the cleavage process itself which plays a role similar to that of the formative movements (Spemann 1938) of amphibian embryos.

8. ORGAN-FORMING SUBSTANCES

While the processes of germ layer formation described in an earlier section are processes of differentiation of developmental capacity paralleling an increase in structural complexity, there are some eggs in which certain substances which are destined to form particular tissues or even organs are recognizable (mostly in the uncleaved egg) long before these structures are formed. Such materials are called *organ forming substances,* and the field of study which follows their behavior is called *promorphology.*

The greater part of the green plasm in the cortex of the *Beroë* egg discussed in the preceding section is thus certainly a comb-forming substance; the red pigmented material on the animal-pole side of the *Myzostoma* egg largely becomes ectoderm, the green substance at the vegetal pole becomes mesoderm and the colorless material in between becomes endoderm. The polar lobe which is seen in molluscs and annelids contains the mesoderm-forming substance, so that if this is cut off at the time of the first cleavage, the larva developing from the lobeless egg is completely lacking in mesoderm (Wilson 1904a). It is also known that the hyaline pole plasm, which makes up part of the polar lobe contents, originates in the germinal vesicle. In all the forms which undergo spiral cleavage, the developmental fate not only of the polar lobe, but of each blastomere as well, is strictly predetermined, so that even if the blastomeres are separated from each other by the use of Ca free sea water, they will form only the structures which they would have formed as part of the intact larva (Wilson 1904b). This makes it likely that each blastomere contains the particular substance necessary for forming the organ for which that blastomere is responsible.

The most famous example of all is that of the ascidian egg (as fully described in the chapter on the Protochordata); in the egg of *Styela partita* the colorless substance released from the germinal vesicle becomes ectoderm, the yellow pigmented cytoplasm becomes mesoderm (especially somites), and the gray yolky cytoplasm becomes endoderm. When development has progressed somewhat further, it is possible to recognize the part which will become mesenchyme, and the parts which will form the notochord and nerve cord (see Chap. 12).

In eggs and early embryos of this kind, although the tissues and organs are still undifferentiated, the substances necessary to form these structures are contained in the various blastomeres and assembled like the pieces of a puzzle, giving advance notice of the organization to come; Conklin therefore called them '*mosaic eggs*'. These mosaic eggs are the most important objects of promorphological study. In contrast to them are *regulative eggs*, forming embryos of which the fate of each part is not determined from the start, but in which the various parts influence each other during the process of development as their final fate is gradually determined by the balance of the organism as a whole. The name refers to the fact that if a part of one of these eggs is removed at an early stage, the adjacent tissue cells are able to change their developmental course and compensate to some extent for the deficiency. The difference between mosaic and regulative eggs, however, is the difference between early and late occurrence of determination; and even in regulative eggs, regulation sooner or later becomes impossible as determination advances. In this sense, mosaic eggs may be said to be precocious.

Similar in nature to these organ-forming substances are the *germ cell determinants*. Among the copepods (*Diaptomus*, *Cyclops*) (*Amma* 1911) and in *Sagitta* (Elpatiewsky 1909; Ghirardelli 1954) there are certain special granules in the cytoplasm. During the early cleavages, these granules are not equally divided, but all go into one blastomere at each cleavage, so that among the many cells of the embryo, only certain specified ones inherit them. Since these eventually become the reproductive cells, it is apparent that the granules are serving as markers for the germ line (Keimbahn). There are also cases in which it is possible to distinguish, although not as clearly as with the granules, the part of the egg which will form the germ line. The cytoplasm near the vegetal pole of some nemertine eggs is such an example, in which a determining substance is found (*Apanteles*); the most interesting case is that of *Litomastix*, which exhibits polyembryony (formation of a number of embryos from one egg). In connection with this process, the germ-cell determinant substance is reported to break up into several units which are distributed to form the reproductive cells of each embryo (Sylvestri 1906).

9. CHROMOSOMES AND DEVELOPMENT

Genetics teaches that the characteristics of an adult animal are determined by its chromosomes. Since development can only be thought of as a process which goes forward on the basis of this grand principle, we are obliged to give full recognition to the part played by the chromosomes. Present day genetics, however, concentrates its attention too exclusively on adult characters alone, and too seldom looks back over the process of their formation. On the other hand, embryology has not advanced sufficiently to subject such phenomena as it has uncovered to genetical scrutiny, and although both sides have long recognized the obvious necessity for cooperating, a wide gap still lies between them. Since the few fragmentary facts that are known have, however, already provided valuable material for the analysis of such problems as the function of the chromosomes, two or three of them will be considered here.

The mechanism of sex determination by means of the sex chromosomes can surely be called a fully established fact. Even with respect to the sex chromosomes, however there is a tendency to think of them, in connection with heredity, only in terms of the mature male and female parent animals. But the example of the free-martin in cattle makes it very clear that even the embryo, of which the sex has just been determined by fertilization, is already asserting this sex. The free-martin is a sexual abnormality that is formed when two calves of unlike sex happen to be attached to the same placenta so that the blood plasms mix (Newman and Patterson 1910). Although these fetuses are obviously still far too young to possess any sort of capacity for reproduction, they nevertheless have enough sexual differentiation to obstruct the development of their partner's reproductive system. Transplantation experiments also show that there is an incompatability between the tissues of the two sexes. On the other hand, there are cases in which such mixture of the sexes does not present any obstacle, as in insect gynandromorphs, which result from the fact that the sex chromosomes are not distributed uniformly throughout the whole body. In the moth *Lymantria,* however, disorders in the genes controlling sex lead to the production of intersexual forms Goldschmidt 1927).

The direction of spiralling in spiral cleavage, as described in the section on cleavage, is one known example of *maternal inheritance* (Boycott and Diver 1923). It is impossible to think that a character appearing so soon after fertilization as the mode of cleavage could be affected by the spermatozoon which has just entered the egg. Changing the subject to a consideration of organ-forming substances, it is obvious that the nature of such substances must be controlled by the genes derived from the two parents of the female producing the eggs, and there is no way in which the spermatozoon can be imagined to affect egg cytoplasm which was already completely equipped before fertilization. Consequently the paternal influence on such characters is only expressed after a delay of one generation, and the phenomenon of maternal inheritance is a natural result.

The results of mixing the chromosomes of rather distantly related animals have been demonstrated by intergenus and interspecies crosses in echinoderms and teleosts. It is well known that in the great majority of these cases the hybrid larvae cease to develop as they approach the gastrula stage. As the result of recent enzymological (Gustafsson and Hasselberg 1951) and immunological (Cooper 1948) research, it is now known that the larva begins to synthesize its own particular proteins in the gastrula stage, and it is believed that the combinations of distantly related chromosome sets reveal their disharmonies at this time. The fact that lethal genes cause the larvae bearing them to die at particular developmental stages specific for each gene is also tied in with particular physiological defects.

If one thinks in terms of heredity only, the haploid number of chromosomes, comprising one full set, ought to be enough to answer the purpose. For testing this point we have the method of artificial parthenogenesis. By its use the larva, certainly starting with only the haploid number of chromosomes, can be made to advance past gastrulation, but if those which have developed well and approached the adult form (late pluteus stage in the sea urchin; metamorphosis stage in the frog) are examined, it will usually be found that they have the double chromosome number, or even three or four times the haploid number (Dalcq 1928; Kawamura 1943). In the sea urchin the doubling of the chromosome number is brought about by the formation of a monaster instead of a diaster, so that the chromosomes which have divided into two are all brought together into one nucleus — that is, one cytoplasmic cleavage is omitted while the nuclear contents double (Herlant 1914a, 1914b). In *Urechis* the chromosomes which are supposed to be thrown out as polar bodies unite with the egg nucleus (Tyler and Bauer 1937). Thus to a certain extent definite facts are known about the methods by which the doubling of the chromosome number is effected, but the reason why one set of chromosomes is insufficient to let development proceed is quite unknown at present.

Finally, an example of the relation between the chromosomes and differentiation is the chromosome diminution often found occurring in the Nemertini and Arthropoda (see Chaps. 6 and 10). The cells which retain all their chromatin become the reproductive cells, while those which have thrown out part of their chromatin become somatic cells. Even from the standpoint of pure logic it is clear that the reproductive cells need to preserve all the hereditary material for the sake of the next generation, while the body cells may not necessarily require all this genetic stuff in order to carry out the special task allotted to them. In Section 8 above it was stated that the matter of which cells will become those of the germ line and which will become somatic cells is apparently settled by the intrinsic differentiation of the egg cytoplasm; assuming that this is true, it still does not contradict this line of logic.

Cytology, until now, has strongly advocated the concept of a fixed number of chromosomes, and this gives the appearance of being a fundamental cellular principle. But unfortunately, this conclusion was chiefly derived from studies made on the reproductive cells, which have been widely investigated on the basis of a different set of requirements. However, as other than reproductive cells have begun to be investigated in recent years, it is found that the chromosome numbers of somatic cells are surprisingly variable (Beatty 1954), providing reason to think that somewhere in this area there may be hidden another key to the locked door of the tissue differentiation problem. If this attempt is successful, it may even pave the way to the long and eagerly awaited coalition between genetics and embryology.

REFERENCES

Allen, R. D. & B. Hagström. 1955: Interruption of the cortical reaction by heat. Exptl. Cell Research, **9**: 157—167.

Amma, K. 1911: Ueber die Differenzierung der Keimbahnzellen bei den Copepoden. Arch. Zellforsch., **6** : 495—576.

Ballard, W. W. 1942: The mechanism for synchronous spawning in *Hydractinia* and *Pennaria*. Biol. Bull., **83** : 329—339.

Bayreuther, K. 1952: Extrachromosomale feulgen positive Körper (Nukleinkörper) in der Oogenese der Tipuliden. Naturwiss., **39** : 71.

———— 1956: Die Oogenese der Tipuliden. Chromosoma, **7** : 508—557.

Beatty, R. A. 1954: How many chromosomes in mammalian somatic cells? International Rev. Cytol., **3** : 177—197.

Berg, W. E. 1950: Lytic effects of sperm extracts on the eggs of *Mytilus edulis*. Biol.Bull., **98** : 128—138.

Binford, R. 1913: The germ-cells and the process of fertilization in the crab, *Menippe mercenaria*. J. Morph., **24** : 147—201.

Boycott, A. E. & C. Diver. 1923: On the inheritance of sinistrality in *Lymnaea peregra*. Proc. Roy. Soc. London ser. B., **95** : 207—213.

Chambers, E. L. 1939: The movement of the egg nucleus in relation to the sperm aster in the echinoderm egg. J. Exptl. Biol., **16** : 409—424.

Chambers, R. 1921: Microdissection studies. III. Some problems in the maturation and fertilization of the echinoderm eggs. Biol. Bull., **41** : 318—350.

Chambers, R. 1923: The mechanism of the entrance of sperm into the starfish egg. J. Gen. Physiol., **5** : 821—829.

———— 1933: The manner of sperm entry in various marine ova. J. Exptl. Biol., **10** : 130—141.

Child, C. M. 1952: Indicator gradient patterns in öocytes and early developmental stages of echinoderms: A reexamination. Biol. Bull., **104** : 12—27.

Colwin, A. L. & L. H. Colwin 1955a: Concerning the spermatozoon and fertilization in the egg of *Sabellaria vulgaris*. Biol. Bull., **109** : 357.

——— & ——— 1955b: Sperm entry and the acrosome filament (*Holothuria atra* and *Asterias amurensis*). J. Morph., **97** : 543—568.

——— & ——— 1955b: The spermatozoon and sperm entry in the egg of the holothurian *Thyone briareus*. Biol. Bull., **109** : 357—358.

——— & ——— 1956: The acrosome filament and sperm entry in *Thyone briareus* (Holothuria) and *Asterias*. Biol. Bull., **110** : 243—257.

Colwin, L. H. & A. L. Colwin 1954: Sperm penetration and the fertilization cone in the egg of *Saccoglossus kowalewskii* (Enteropneusta). J. Morph., **95** : 351—372.

——— & ———1955a: Some factors related to sperm entry in two species of *Asterias*. Biol. Bull., **109** : 359.

Conklin, E. G. 1902: Karyokinesis and cytokinesis in the maturation, fertilization and cleavage of *Crepidula* and other gastropoda. J. Acad. Nat. Sci. Philadelphia, **12** : 1—122.

———— 1916: Effects of centrifugal force on the polarity of the eggs of *Crepidula*. Proc. Natl. Acad. Sci., **2** : 87—90.

———— 1917: Effects of centrifugal force on the structure and development of the eggs of *Crepidula*. J. Exptl. Zool., **22** : 311—419.

Cooper, K. W. 1950: Normal spermatogenesis in *Drosophila*. Biology of *Drosophila* (edited by M. Demerec) Chapter **1** : 1—61.

Cooper, R. S. 1948: A study of frog egg antigens with serum-like reaction groups. J. Exptl. Zool., **107** : 397—438.
Cornman, I. 1941: Sperm activation by *Arbacia* egg extracts, with special relation to echinochrome. Biol. Bull., **80** : 202—207.
Crampton, H. E. 1894: Reversal of cleavage in a sinistral gastropod. Ann. N. Y. Acad. Sci., 8 : 167—170.
Dalcq, A. 1928: Les bases physiologiques de la fécondation et de la parthenogénèse. Univ. France Press.
Dan, J. C. 1950a: Sperm entrance in echinoderms, observed with the phase contrast microscope. Biol. Bull., **99** : 399—411.
——————— 1950b: Fertilization in the medusan *Spirocodon saltatrix*. Biol. Bull., **99** : 412—415.
——————— 1952: Studies on the acrosome. I. Reaction to egg-water and other stimuli. Biol. Bull., **103** : 54—66.
——————— 1954a: Studies on the acrosome. II. Acrosome reaction in starfish spermatozoa. Biol. Bull., **107** : 203—218.
Dan, J. C. 1954b: Studies on the acrosome. III. Effect of calcium deficiency. Biol. Bull., **107** : 335—349.
Dan, J. C. & K. Dan. 1941: Early development of *Comanthus japonica*. Jap. J. Zool., **9** : 565—575.
Dan, J.C. & K. Wada. 1955: Studies on the acrosome. IV. The acrosome reaction in some bivalve spermatozoa. Biol. Bull., **109** : 40—55.
Dan, K. 1930: Collecting eggs of *Lymnaea japonica* in winter. Zool. Mag. **42** : 156—158. (in Japanese)
——————— 1950: The value of phase contrast microscopy for cell physiology. Kagaku, **10** : 441—448. (in Japanese)
——————— 1955: Mechanism of cell division. Recent Biology, 5 : 1—28. Baifu-kan (in Japanese) Tokyo.
Dan, K. & T. Nakajima 1954: Isolated 'Mitotic Apparatus'. Kagaku, **24** : 354—358. (in Japanese)
Dan, K. & J. C. Dan 1941: Spawning habit of the crinoid, *Comanthus japonica*. Jap. J. Zool., **9** : 555—564.
——— & ——— 1947a: Behavior of the cell surface during cleavage. VII. On the division mechanism of cells with excentric nuclei. Biol. Bull., **93** : 139—162.
——— & ——— 1947b: Behavior of the cell surface during cleavage. VIII. On the cleavage of medusan eggs. Biol. Bull., **93** : 163—188.
Dawydoff, C. 1928: Traité d'Embryologie comparée des Invertébrés. Paris
Delage, Y. 1899: Sur l'interprétation de la fécondation merogonique et sur une théorie nouvelle de la fécondation normale. Arch. Zool. Exptl. Gén. ser. 3, **7** : 511—527.
——————— 1901: Etudes éxperimentales sur la maturation cytoplasmique et sur la parthénogénèse artificielle chez les échinodermes. Arch. Zool. Exptl. Gén. ser. 3, **9** : 285—326.
de Robertis, E. D. P., W. W. Nowinski & F. A. Saez. 1954: General Cytology. Philadelphia
Duboscq, O. & O. Tuzet. 1937: L'ovogénèse, la fécondation et les premiers stades du dévelopment des éponges calcaires. Arch. Zool. Exptl. Gèn., **79** : 157—316.
Elpatiewsky, W. 1909: Die Urgeschlechtzellenbildung bei *Sagitta*. Anat. Anz., **35** : 226—239.
Elster, H. J. 1935: Experimentelle Beiträge zur Kenntnis der Physiologie der Befruchtung bei Echinodermen. Arch. Entw. -Mech., **133** : 1—87.
Endo, Y. 1952: The role of cortical granules in the formation of the fertilization membrane in eggs from Japanese sea urchins. Exptl. Cell Res., **3** : 406—418.
——————— 1954a: Fertilization membrane. Cyto-chemical Symposium, **2** : 37—63. (in Japanese)
——————— 1954b: Relation of sperm entrance point and first cleavage plane to polarity in sea urchin eggs. Zool. Mag., **63** : 164—165. (in Japanese)
Endo, Y. & T. Nakajima. 1953: Inhibiting action of body fluid on fertilization in sea urchins. Zool. Mag., **62** : 106. (in Japanese)
Fol, H. 1877: Sur le commencement de l'hénogenie chez divers animaux. Arch. Zool Exptl. Gén., **6** : 145—169.
Fox, H. M. 1924: Lunar periodicity in reproduction. Proc. Roy. Soc. ser. B., **95** : 523—550.
Fujii, T., S. Utida, T. Mizuno & S. Nanao. 1955: Effects of amino acids and some chelating substances on the motility and oxygen uptake of starfish spermatozoa. J. Fac. Sci. Tokyo Univ., **7** : 335—345.
Galtsoff, P. S. 1938: Physiology of reproduction of *Ostrea virginica*. II. Stimulation of spawning in the female oyster. Biol. Bull., **75**: 286—307.
——————— 1940: Physiology of reproduction of *Ostrea virginica*. III. Stimulation of spawning in the male oyster. Biol. Bull., **78**: 117—135.

Galtsoff, P. S., F. E. Lutz, P. S. Wilch & J. G. Needham. 1937: Culture Methods for Invertebrate Animals. Cornell Univ. Press, Ithaca, N. Y.
Gatenby, J. B. 1927: Further notes on the gametogenesis and fertilization of sponges. Q. J. Mi. Sci., **71:** 173—188.
Ghirardelli, E. 1954: Determinante germinale e nucleo nelle uovo dei Chetognati. Boll. d. Zool., **21:** 241—247.
Goldfarb, A. J. 1913: Studies in the production of grafted embryo. Biol. Bull., **24:** 73—101.
Goldschmidt, R. 1927: Physiologische Theorie der Vererbung. Berlin
Gray, J. 1931: Experimental Cytology. Cambridge
Gustafson, T. & I. Hasselberg. 1951: Studies on enzymes in the developing sea urchin egg. Exptl. Cell Res., **2:** 642—672.
Haeckel, E. 1872: Die Kalkschwämme. Berlin
Hartmann, M., O. Schartau, R. Kuhn & K. Wallenfels. 1939: Ueber die Sexualstoffe der Seeigel. Naturwiss., **27:** 433.
Hartmann, M. & O. Schartau. 1939: Untersuchungen über die Befruchtungsstoffe der Seeigel. Biol. Zentlbl., **59:** 571.
Harvey, E. B. 1939: A method of determining the sex of *Arbacia* and a new method of producing twins, triplets and quadruplets. Biol. Bull., **77:** 312.
——————— 1952: Electrical method of "sexing" *Arbacia* and obtaining small quantities of eggs. Biol. Bull., **103:** 284.
Heilbrunn, L. V. 1956: The dynamics of living protoplasm. New York
Herbst, C. 1893: Ueber die künstliche Hervorrufung von Dottermembran an unbefruchteten Seeigeleiern nebst einige Bemerkungen über die Dotterhautbildung überhaupt. Biol. Zentlbl., **13:** 12—22.
——————— 1900: Ueber des Auseinandergehen von Furchungs- und Gewebezellen in Kalkfreien Medium. Arch. Entw.-Mech., **9:** 424—463.
Herlant, M. 1914a: Sur l'éxistence d'un rhythme périodique dans le déterminisme des premiers phènomenes du développement parthénogénétique expérimental chez l'oursin. Comp. rend. Acad. Sci., **158.**
——————— 1914b: Sur le mécanisme de la premier segmentation. Comp. rend. Acad. Sci., **159.**
Hiramoto, Y. 1954: Nature of the perivitelline space in sea urchin eggs. Jap. J. Zool., **11:** 227—243.
Hiramoto, Y. 1955: Nature of the perivitelline space in sea urchin eggs. III. On the mechanism of membrane elevation. Annot. Zool. Jap., **28:** 183—193.
——————— 1956: Cleavage of sea urchin eggs after removal of spindle. Zool. Mag., **63:** 89. (in Japanese)
Hobson, A. D. 1932: On the vitelline membrane of the egg of *Psammechinus miliaris* and of *Teredo norvegica*. J. Exptl. Biol., **9:** 93—106.
Hudinaga, M. 1942: Reproduction, development and rearing of *Penaeus japonicus* Bate. Jap. J. Zool., **10:** 305—393.
Hughes, A. 1952: The Mitotic Cycle. New York & London
Iida, T. 1942: A method of collecting sea urchin eggs without removal of ovaries. Zool. Mag. **54:** 280. (in Japanese)
Iizuka, A. 1903: Observations on the Japanese Palolo, *Ceratocephale osawai*. J. Coll. Sci. Tokyo Imp. Univ. **17**, Art. 11: 1—37.
Inomata, S. 1927: Swarming season of *Tylorrhynchus heterochaetus* in Togo Lake, Tottori Prefecture. Zool. Mag., **39:** 153. (in Japanese)
Inoué, S. & K. Dan. 1951: Birefringence of the dividing cell. J. Morph., **89:** 423—455.
Ishida, J. & E. Nakano. 1950: Fertilization of activated sea urchin eggs deprived of fertilization membrane by washing with Ca-Mg-free media. Annot. Zool. Jap., **23:** 43—48.
Iwata, K. S. 1950: A method of determining the sex of sea urchins and of obtaining eggs by electric stimulation. Annot. Zool. Jap., **23:** 39—42.
Iwata, K. S. 1952: Mechanism of egg maturation in *Mytilus edulis*. Biol. J. Okayama Univ., **1:** 1—11.
Just, E. E. 1914: Breeding habits of the heteronereis form of *Platyneres megalops* at Woods Hole, Mass. Biol. Bull., **27:** 201—212.
——————— 1915: An experimental analysis of fertilization in *Platynereis megalops*. Biol. Bull., **28:** 73—114.
——————— 1919: The fertilization reaction in *Echinarachnius parma*. I. Cortical response of the egg to insemination. Biol. Bull., **36:** 1—10.
——————— 1922: The effect of *Arbacia* blood on the fertilization-reaction. Biol. Bull., **44:** 10—16.
——————— 1930: The present status of the fertilizin theory of fertilization. Protoplasma, **10:** 300—342.

Kato, K. 1940: On the development of some Japanese polyclads. Jap. J. Zool., **8:** 537—572.

Kautzsch, G. 1912: Studien über Entwicklungsanomalien bei *Ascaris*. I. Ueber Teilungen des zweiten Richtungskörper. Arch. Zellforsch., **8:** 217—251.

———— 1913: Studien über Entwicklungsanomalien bei *Ascaris*. II. Arch. Entw.-Mech., **35:** 641—691.

Kawamura, T. 1943: Problem of the nucleus in amphibian fertilization. Jap. J. Exptl. Morph., **1:** 117—140. (in Japanese)

Kinoshita, T., S. Shibutani, & J. Shimizu 1943: A method of inducing spawning in *Pecten (Patinopecten) yessoensis* Jay. Bull. Jap. Soc. Sci. Fisheries, **11:** 168—170 (in Japanese)

Koltzoff, N. K. 1906: Studien über die Gestalt der Zelle. I. Untersuchungen über die Spermien der Decapoden, als Einleitung in das Problem der Zellgestalt. Arch. Mikr. Anat., **67:** 364—572.

Korschelt, E. u. K. Heider. 1936: Vergleichende Entwicklungsgeschichte der Tiere. Jena

Kostanecki, E. v. & A. Wierzejski. 1896: Ueber das Verhalten der sog. Achromatischen Substanzen in befruchteten Ei nach Beobachtungen am *Physa fontinalis*. Arch. Mik. Anat., **47:** 309—386.

Kuhn, R. & K. Wallenfels. 1940: Echinochrom als prosthetische Gruppen hochmolekularer Symplexe in der Eiern von *Arbacia pustulosa*. Ber. Deutsch. Chem. Ges., **73:** 458—464.

Kumé, M. 1948: Spawning and related phenomena in *Spirocodon saltatrix*. Collecting and Breeding, **10:** 270—271. (in Japanese)

Kuwata, Y. 1956: Cytology. Baifu-kan (in Japanese)

Lankester, R. 1877: Notes on the embryology and classification of the animal kingdom. Q. J. M. S. (n.s), **17:** 399—454.

Lillie, F. R. 1912: Studies of fertilization in *Nereis*. III. The morphology of the normal fertilization of *Nereis*. IV. The fertilizing power of portions of the spermatozoon. J. Exptl. Zool., **12:** 413—477.

———— 1914: Studies on fertilization. VI. The mechanism of fertilization in Arbacia. J. Exptl. Zool., **16:** 523—588.

———— 1921: Studies on fertilization. IX. On the question of superposition of fertilization on parthenogenesis in *Strongylocentrotus purpuratus*. Biol. Bull., **40:** 23—31.

———— 1923: Problems of Fertilization. Chicago University Press, Chicago.

Lindahl, P. E. 1936: Zur Kenntnis der physiologischen Grundlagen der Determination im Seeigelkeim. Acta Zool., **17:** 179—365.

Loeb, J. 1903: Ueber die osmotischen Eigenschaften und die Entstehung der Befruchtungsmembran bei Seeigeln. Arch. Entw.-Mech., **26:** 82—88.

———— 1908: Ueber die Befruchtung von Seeigeleiern durch Seesternsamen. Pflüger's Arch. Ges. Physiol., **99:** 323—356.

———— 1909: Die chemische Entwicklungserregung des tierischen Eies. Berlin.

———— 1914: Cluster formation of spermatozoa caused by specific substances from eggs. J. Exp. Zool., **17:** 123—140.

———— 1915: On the nature of the conditions which determine or prevent the entrance of the spermatozoon into the egg. Amer. Naturalist, **49:** 257—285.

———— 1916: The Organism as a Whole. New York.

Loosanoff, V. L. & H. C. Davis. 1950: Conditioning *V. mercenaria* for spawning in winter and breeding its larva in the laboratory. Biol. Bull., **98:** 60—65.

——— & ——— 1952: Temperature requirements for maturation of gonads of northern oysters. Biol. Bull., **103:** 80—96.

MacBride, E. W. 1914: Text-book of Embryology. London.

McCullock, D. 1952a: Fibrous structures in the ground cytoplasm of the *Arbacia* egg. J. Exp. Zool., **119:** 47—63.

McCullock, D. 1952b: Note on the origin of the cortical granules in *Arbacia punctulata* eggs. Exptl. Cell Res., **3:** 605—607.

Metz, C. B. 1945: The agglutination of starfish sperm by fertilizin. Biol. Bull., **89:** 84—94.

———— 1954: The adjuvant action of chelating agents in agglutination of starfish sperm. Biol. Bull., **109:** 317.

Metz, C. B. & J. B. Morrill Jr. 1955: Formation of acrosome filaments in response to treatment of sperm with fertilizin in *Asterias* and *Nereis*. Biol. Bull., **109:** 349.

Meves, F. 1912: Verfolgung der sogenannten Mittelstückes des Echinidenspermiums im befruchteten Ei bis zum Ende der ersten Furchungsteilung. Arch. Mikr. Anat., **80:** 81—123.

———— 1914: Verfolgung des Mittelstückes des Echiniden-spermiums durch die ersten Zellgenerationen des befruchteten Eies. Arch. Mikr. Anat., **85:** 1—8.

——— 1915: Ueber den Befruchtungsvorgang bei der Miesmuschel *(Mytilus edulis L)*. Arch. Mikr. Anat., **87:** Abt. 2 47—62.

Miyazaki, K. 1938: On the development of *Tellina juvenilis* Hanley. Bull. of the J. S. S. F., **7:** 179. (in Japanese)

Miyoshi, S. 1937: Swarming of *Ceratocephale osawai* at Konuma. Botany and Zoology, **7:** 2017—2024. (in Japanese)

Monné, L. 1944: Cytoplasmic structure and cleavage pattern of the sea urchin egg. Ark. f. Zool., 35A Heft 3 No. 13.

Moore, A. R. 1930: Fertilization and development without the fertilization membrane in the egg of *Dendraster eccentricus*. Protoplasma, **9:** 18—24.

Moore, C. R. 1916: On the superposition of fertilization on parthenogenesis. Biol. Bull., **31:** 137—180.

Moser, F. 1939: Studies on a cortical layer response to stimulating agents in the *Arbacia* egg. J. Exp. Zool., **80:** 423—445.

Motomura, I. 1914a: The sexual character of the sea urchin, *Strongylocentrotus pulcherrimus* A. Agassi. J. Exp. Zool., **16:** 431.

——— 1941b: Materials of the fertilization membrane in the eggs of echinoderms. J. Exp. Zool., **16:** 345—363.

Murayama, S. 1935: On the development of the Japanese abalone, *Haliotis gigantea*. J. Coll. Agr. Tokyo Imp. Univ., **13:** 227—233.

Nakano, E. & S. Ohashi. 1954: On the carbohydrate component of the jelly coat and related substances of eggs from Japanese sea urchins. Embryologia, **2:** 81—86.

Newman, H. H., and J. T. Patterson. 1910: Development of the nine-banded armadillo from the primitive streak stage to birth; with special reference to the question of specific polyembryony. Jour. Morph., **21:**

Oshima, H. 1922: Inhibiting action of dermal secretion on fertilization in sea urchins. Zool. Mag., **43:** 59—64. (in Japanese)

——— 1925: On the maturation and fertilization of sea cucumber eggs. Science Bull. Fac. Agric. Kyushu Univ., **1:** 70—102. (in Japanese)

Okada, K. 1952: In the swarming of Japanese Palolo. Bull. exptl. Biol., **2:** 181—185. (in Japanese)

Okuda, S. 1939: Influence of phase of moon on reproduction phenomena in some animals. Botany and Zoology, **7:** 911—918, 1087—1096. (in Japanese)

Pequegnat, W. E. 1948: Inhibition of fertilization in *Arbacia* by blood extracts. Biol. Bull., **95:** 69—82.

Popa, G. T. 1927a: A lipo-gel reaction exerted by follicular fluid. Biol. Bull., **52:** 223—237.

——— 1927b: The distribution of substances in the spermatozoa *(Arbacia* and *Nereis)*. Biol. Bull., **52:** 238—257.

Retzius, G. 1904—1914: Biologische Untersuchungen I—XV. Jena

Robert, O. 1902: Recherches sur la développement des troques. Arch. Zool. Exp. Gén. ser. 3, **10:** 269—538.

Rothschild, Lord 1956a: Fertilization. Methuen Co. London

——— 1956b: The fertilizing spermatozoon. Discovery, **18.**

Rothschild, Lord & A. Tyler. 1955: Acrosomal filaments in spermatozoa. Exp. Cell. Res., Suppl. **3:** 304—311.

Rulon, O. 1955: Developmental modifications in the sand dollar caused by zinc chloride and prevented by glutathione. Biol. Bull. **109:** 316—327.

Rybak, B. 1957: Recherches sur la biologie des spermatozoides d'oursin. Bull. biol. d. France e. Belgique, Suppl.: 1—177.

Schleip, W. 1929: Die Determination der Primitiventwicklung. Leipzig.

Scott, A. 1946: The effect of low temperature and of hypotonicity on the morphology of the cleavage furrow in *Arbacia* egg. Biol. Bull., **91:** 272—287.

Selenka, E. 1878: Befruchtung des Eies von *Toxopneustes variegatus*. Leipzig

Shimoizumi, J. & K. Tanemura 1951: On the release of Zoëa in *Sesarma haematocheir*. Zool. Mag., **60:** 51—52. (in Japanese)

Soeda, J. 1952: Artificial fertilization and early cleavage of *Ommastrephes sloani pacificus*. Bull. of the Hokkaido Regional Fish. Res. Lab., F. A. **5:** 1—15. (in Japanese)

Spek, J. 1926: Ueber gesetzmässige Substanzverteilung bei der Furchung des Ctenophoreneies und ihre Beziehung zu den Determinationsproblemen. Arch. Entw.-Mech., **107:** 54—73.

Spemann, H. 1938: Embryonic development and induction. Yale University Press.

Stauffacher, H. 1894: Eibildung und Furchung bei *Cyclas cornea*. Jen. Zeits. f. Naturwiss., **28:** 196—246.

Sugawara, H. 1943a: Hatching enzyme of the sea urchin, *Strongylocentrotus pulcherrimus*. J. Facult. Sci. Tokyo Imp. Univ. ser. 4, **6:** 109—127.

———— 1943b: The formation of multinucleated eggs of the sea urchin by treatment with proteolytic enzymes. J. Facult. Sci. Tokyo Imp. Univ. ser. 4, **6:** 129—139.

Sugiyama, M. 1938: Effect of some divalent cations upon the membrane development of sea urchin egg. J. Facult. Sci. Tokyo Imp. Univ. ser. 4, **4:** 501—508.

———— 1951: Re-fertilization of the fertilized eggs of the sea urchin. Biol. Bull., **101:** 335—344.

———— 1953: Physiological analysis of the cortical response of the sea urchin eggs to stimulating reagents. II. The propagating or non-propagating nature of the cortical changes induced by various reagents. Biol. Bull., **104:** 216—223.

Swann, M. M. & J. M. Mitchison. 1953: Cleavage of sea-urchin eggs in colchicine. J. Exp. Biol., **30:** 506—514.

Sylvestri, F. 1906: Contribuzioni alla conoscenza biologica degli Imenotteri parassiti. I. Biologia del *Litmastix truncatellus* Dalm. Ann. R. Scuola Sup. Agric. Portici **6.**

Taki, Y. 1939: Mollusca from Attuka Province. First Report of the Academic survey of Manchuria and Mongolia. 29—38. (in Japanese)

Tanemura, K. 1954: Observations of release of Zoëa in *Sesarma haematocheir*. Children's Science, **19:** 20—23. (in Japanese)

Tchou, S. & C. H. Chen. 1935: Fertilization in goldfish. Contr. Inst. Zool. Nat. Acad. Peiping, **3:** 35.

Theel, H. 1892: On the development of *Echinocyanus pusillus*. Nova Acta p. Soc. Uppsala

Toyama, K. 1894: On the spermatogenesis of the silkworm. Bull. Agric. Coll., **2.**

Tyler, A. 1939: Crystalline echinochrome and spinochrome. Their failure to stimulate the respiration of eggs and of sperm of *Strongylocentrotus*. Proc. Nat. Acad. Sci., **25:** 523—528.

———— 1940: Agglutination of sea urchin eggs by means of a substance from the eggs. Proc. Nat. Acad. Sci., **26:** 249—256.

———— 1941: The role of fertilizin in the fertilization of eggs of the sea urchin and other animals. Biol. Bull., **81:** 190—204.

———— 1944: Sexual dimorphism in sea-urchins. Anat. Rec., **89:** 573.

———— 1948: Fertilization and immunity. Physiol. Rev., **28:** 180—219.

———— 1949: Properties of fertilizin and related substances of eggs and sperm of marine animals. Amer. Naturalist, **83:** 195—219.

Tyler, A. & H. Bauer. 1937: Polar body extrusion and cleavage in artificially activated eggs of *Urechis caupo*. Biol. Bull., **73:** 164—180.

Tyler, A. & S. W. Fox. 1940: Evidence for the protein nature of the sperm agglutinins of the keyhole limpet and the sea urchin. Biol. Bull., **79:** 153—165.

Tyler, A. & C. B. Metz. 1944: Natural heteroagglutinins in lobster serum. Anat. Rec., **89:** 568.

——— & ——— 1945: Natural heteroagglutinins in the serum of the spiny lobster, *Panulirus interruptus*. J. Exp. Zool., **100:** 387—406.

Uchida, T. 1927: Studies on Japanese hydromedusae. I. Anthomedusae. J. Facul. Sci. Tokyo Imp. Univ. ser. 4, **1:** 145—242.

Vasseur, E. 1948: Chemical studies on the jelly coat of the sea-urchin egg. Acta Chem. Scand., **2:** 900—913.

Wada, S. K. 1936: Thermo-stimulus to spawning eggs and spermatozoa in *Pteria martensii*, Proc. Sci. Fisheries Ass. **7:** 131—133. (in Japanese)

———— 1954: Spawning in the tridacnid clams. Jap. J. Zool., **11:** 273—285.

———— 1956: Some considerations on "forced disharge" and "forced retention" of eggs in the pearl-oyster, *Pteria martensii*. Fisheries Production (in press) (in Japanese)

Wada, S. K., J. R. Collier & J. C. Dan 1956: Studies on the acrosome. V. An egg membrane lysin from the acrosome of *Mytilus edulis* spermatozoa. Exp. Cell Res., **10:** 168—180.

Waddington, C. H. 1952: Preliminary observations on the mechanism of cleavage in the amphibian egg. J. Exp. Biol., **29:** 484—489.

Watasé, S. 1890: Studies on Cephalopods. I. Cleavage of the ovum. J. Morph., **4:** 247—302.

Wilson. E. B. 1892: The cell-lineage of *Nereis*. J. Morph., **6:** 361—480.

———— 1895: An Atlas of the Fertilization, Karyokinesis of the Ovum. Columbia University Press. New York.

———— 1903: Experiments on cleavage and localization in the nemertine egg. Arch. Entw.-Mech., **16:** 411—460.

———— 1904a: Experimental studies in germinal localization. I. The germ-regions in the egg of *Dentalium*. J. Exp. Zool., **1:** 1—72.

———— 1904b: Experimental studies in germinal localization. II. Experiments on the cleavage-mosaic in *Patella* and *Dentalium*. J. Exp. Zool., **1:** 197—268.

———— 1925: The Cell in Development and Heredity. New York
Yamamoto, G. 1949: Concerning larval culture and the spat of *Pecten*. Zool. Mag., **58:** 111. (in Japanese)
Yamamoto, T. 1939a: Physiological problems of fertilization in fish eggs. Kagaku, **9:** 450—453. (in Japanese)
———— 1939b: Changes of the cortical layer of the egg of *Oryzias latipes* at the time of fertilization. Proc. Imp. Acad. Tokyo, **15:** 269—271.
———— 1944: Physiological studies on fertilization and activation of fish egg. I & II. Annot. Zool. Jap., **22:** 105—136.
———— 1954: Physiological studies on fertilization and activation of fish eggs. Exptl. Cell Res., **6:** 56—68.
Yatsu, N. 1904: Experiments on the development of egg fragments in *Cerebratulus*. Biol. Bull., **6:** 123—136.
———— 1912: Observations and experiments in the Ctenophore egg. J. Coll. Sci. Tokyo Imp. Univ., **32:** 1—21.
Yanagita, T. 1930: Oligodynamic action between fertilization and zinc ion in sea urchin eggs. Zool. Mag., **42:** 344—345. (in Japanese)
Yoshida, M. 1952: Spawning in *Spirocodon saltatrix*. Zool. Mag., **61:** 358—366. (in Japanese)
———— 1952: Some observations on the maturation of the sea urchin, *Diadema setosum*. Annot. Zool. Jap., **25:** 265—271.
———— 1954: Spawning habit of *Hydractinia epiconcha*, a hydroid. J. Fac. Sci. Tokyo Univ., **7:** 67—78.

PORIFERA

Chapter 2

INTRODUCTION

Since sponges are animals of the simplest organization above protozoa, it might be imagined that their developmental processes are also very simple. While the organization of sponge larvae is indeed of a low order, with respect to the developmental mechanism of each part we are confronted with many intricate problems which are not involved in the development of other animals.

For instance, the gemmule formation which occurs in asexual reproduction of some sponges is believed to be a special adaptation for surviving adverse environmental conditions. It is very difficult, however, to recognize radical differences between the processes of gemmule formation and ordinary bud formation, because various intermediate processes are found as a result of the low degree of differentiation of the parent animals.

As regards the processes involved in the development of the fertilized egg, it is hardly possible to establish, as in other animal groups, definite demarcations between the germ layers, owing to the nearly complete lack of differences among the blastomeres. The egg itself has no definite shape or characteristic capsular membrane, and wanders among the somatic tissues by amoeboid movement, only distinguishable from them by its larger size.

Moreover, in the adult mesoglea, there exist amoeboid, undifferentiated cells called *archeocytes* which retain the full capacity to differentiate into any type of cell. Other, higher animals are also postulated to have some kind of primitive cells with a high capacity for differentiation, which function in regeneration, but this activity occurs only as the result of injury or loss of some part of the body, and under no other circumstances. In the sponges, on the contrary, the archeocytes very readily and very frequently become active, making it sometimes impossible to draw a definite line between archeocytes and fertilized eggs. In other words, it can be said that an adult sponge always stores in some part of its body embryonic cells which have a high potency to differentiate.

As early as 1907, H. V. Wilson demonstrated that if sponges are squeezed through bolting silk so that they are completely separated into a suspension of individual cells, they will reaggregate and regenerate a complete sponge. This result has been fully confirmed, by Galtsoff (1925) and other workers. Wilson found that masses consisting of only epidermal cells or of only collar cells (choanocytes) do not form a perfect animal, while such masses as include some archeocytes are able to complete regeneration. This conclusion can best be understood in the light of the above-described totipotency of the archeocytes.

1. REPRODUCTION OF SPONGES

Reproduction in the sponges is effected either sexually or asexually. In asexual reproduction, a bud is formed externally or internally; the former process is called *budding* and the latter, *gemmule formation*. Sexual reproduction is accomplished by means of eggs and spermatozoa as in other animals.

(1) ASEXUAL REPRODUCTION

Formation of asexual gemmules occurs in all fresh-water sponges and some marine forms. Although the process of gemmule formation has long been studied, many points still remain to be clarified.

Carter (1849) claimed that gemmules are formed from single cells, which might be called "ovum-bearing cells", while Goette (1886) held the opinion that they arise from aggregates of a number of cells. Marshall (1884), Wiezejski (1886), Zykoff (1892) and Weltner (1892) believed that a gemmule is simply a collection of the various kinds of cells that are concerned in the developmental process. Evans (1900), through a detailed study of the composition of gemmules, reached the conclusion that the gemmule is very probably not formed from a single cell, since no mitotic divisions are observed during its formation, although in its earliest stage it may be composed of a single cell or a mass of peripherally disposed cells.

According to more recent studies, the gemmules of fresh-water sponges begin as an aggregation of archeocytes, to which nurse cells (*trophoblasts*) are added. These cells contain granules of reserve food substances such as glycoproteins and lipoproteins. Other amoebocytes surround the cell mass, forming a layer of cylindrical cells (Fig. 2.1c). Meanwhile *scleroblasts* secrete *amphidisk spicules* (Fig. 2.1b); these are arranged radially within the cylindrical layer or between the inner and outer membranes surrounding the gemmule, while a *micropyle* forms an opening through all these enveloping layers, completing the formation of the gemmule (Fig. 2.1c).

Fresh-water sponges belonging to the family Spongillidae produce, in autumn, a large number of gemmules within the body. The animals then disintegrate, and some of the gemmules remain within the remnant of the parent sponge, while others fall to the bottom. These structures usually show a yellow tint, and are able withstand adverse natural conditions such as cold and desiccation. They germinate when the temperature rises in spring; according to Zeuthen (1939), the gemmules will hatch in about three days at a water temperature of 13° to 21° C. The cells contained in the gemmule escape from the micropyle, and with some differentiation and rearrangement develop into a young sponge. More precisely, the large multinucleate archeocytes divide into uninucleate ones, some of which

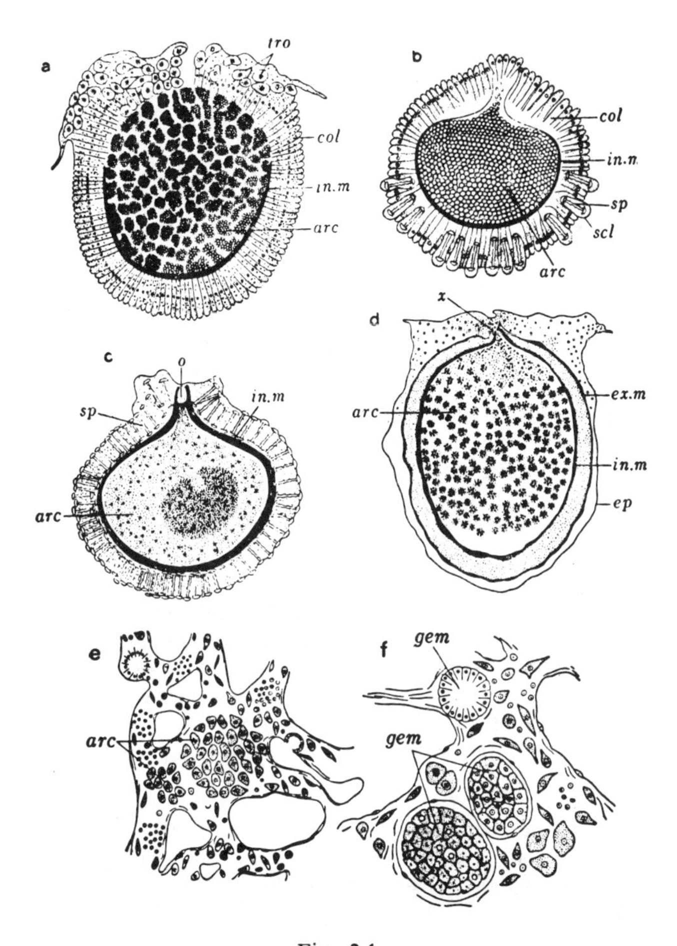

Fig. 2.1

Young gemmule of fresh-water sponge *Ephydatia*.
Trophocyte, columnar layer, inner membrane and archeocyte beginning to differentiate; b. Formation of gemmule and columnar layer complete; amphidisc spicules forming in scleroblast; c. Completed gemmule with micropyle and surface layer of amphidisc spicules. (Evans); d. Gemmule of fresh-water sponge *Spongilla* in process of hatching. Gemmule is covered with outer membrane; archaeocytes pour out of micropyle. Epidermis of young sponge formed on outside of gemmule. (Brien); e, f. Early stages of gemmule formation in marine sponge *Esperella*. (Mycale) Gemmule formed as aggregation of archaeocytes. (Wilson). *arc:* archaeocyte, *col:* columnar layer, *ep:* epidermis of young sponge, *ex.m:* outer membrane of gemmule, *gem:* gemmule, *in.m:* inner membrane, *o:* micropyle, *scl:* scleroblast, *sp:* amphidisc spicule, *tro:* trophocyte, *x:* archeocyte plasm streaming from micropyle

are distributed to form an epidermis, while others differentiate into the choanocytes and *porocytes* which surround the flat epidermal cells. The spicules are secreted by scleroblasts, which are specialized archeocytes. Other uninucleated archeocytes remain undifferentiated, serving as phagocytic amoebocytes or later forming germ cells. Within about a week, this larva will have developed into a complete young sponge.

Such asexual gemmule formation among the marine Monaxonidae has been studied by Wilson (1890), in a species belonging to *Tedania*, and by Kadota (1922) in *Reniera okadai* and *Reniera japonica*. The gemmules begin as an aggregation of archaeocytes enclosed by a thin envelope of flat cells. The surface cells of the mass become flagellated epidermal cells, except at the posterior pole, and the gemmule then escapes as a flagellated larva. On settling, the larvae lose their flagella, their body cells differentiate along various lines, and they develop into animals with body construction like that of the young arising from fertilized eggs.

(2) SEXUAL REPRODUCTION

Sponges very often reproduce sexually, but the details of this process are not clearly established. Regarding the origin of the germ cells, several opinions have been put forward: some authors claim that they are derived from archaeocytes, while others maintain that they originate from choanocytes (Fig. 2.2) or from the amoebocytes within the mesogloea. The oogonia are generally believed to differentiate from amoebocytes. The oocyte has a large nucleus with a prominent nucleolus. During the growth period, it acquires reserve material directly or through specialized nurse cells (Fig. 2.2d). Upon attaining a certain size, it undergoes the meiotic divisions.

Spermatogenesis has been observed in only a few sponges; the spermatogonium is an enlarged amoebocyte surrounded by one or more layers of flattened cells which also are modified amoebocytes. This mass of cells is called a *spermatocyst*; the enclosed spermatogonium undergoes several divisions, finally producing spermatozoa. According to Gatenby (1919), the spermatogonia are transformed choanocytes; and the cells of an entire flagellated chamber may change into sperm cells. These spermatozoa enter other sponges by way of the water current, and

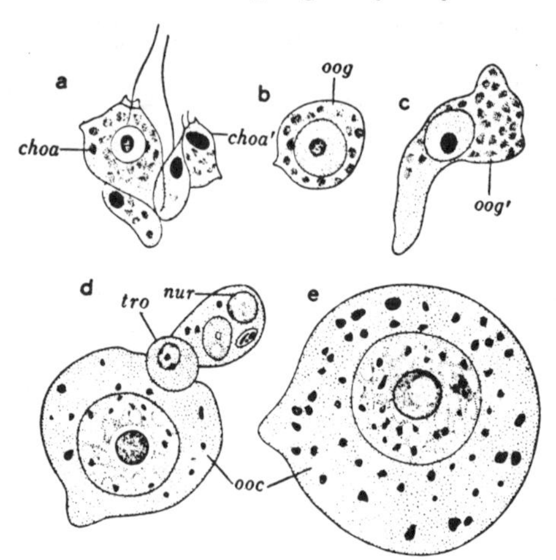

Fig. 2.2 Formation of egg from collar-cell in the sponge *Grantia compressa* (Dendy)

a. Transformation of collar-cell into oogonium: b. oogonium; c. amoeboid state of oogonium; d. oocyte, developed from oogonium, receiving trophocyte caught by nurse cell; e. mature oocyte. *choa:* collar-cell (choanocyte), *choa'*: collar-cell beginning to change into oogonium, *nur:* nurse cell, *ooc:* oocyte, *oog:* oogonium, *oog'*: amoeboid oogonium, *tro:* trophocyte.

fertilize the eggs present there. In the Calcarea, according to Gatenby, the spermatozoon first enters a choanocyte and is transferred by it to the egg. Choanocytes of this type are also known to serve as nurse cells. Tuzet (1937) also reports that in *Cliona* and *Reniera,* the spermatozoon enters an amoebocyte which then transfers it to an egg.

The fertilized egg undergoes equal or unequal holoblastic cleavage and develops into a blastula. Using the cilia which develop on its epidermal cells, the blastula works its way through the parent tissues into the excurrent system, and emerges from the osculum. After swimming for a time, the larva attaches to some object and in several hours develops into a small sponge.

According to Marshall (1884), *Spongilla lacustris,* a fresh-water species, undergoes an alteration of generations like that of the coelenterates (Cnidaria). In other words, the animals which develop from gemmules produce reproductive cells; these unite and give rise to sponges which in turn form gemmules.

Although some sponges are dioecious, the majority are hermaphroditic. In the latter case, eggs and sperm mature at different times; sometimes they are produced in different parts of the animal — the eggs in the basal region, the sperm apically. The shallow-water sponges usually breed at definite seasons of the year, while the deep-water forms probably breed the year round (Hyman 1940).

2. DEVELOPMENT OF REPRESENTATIVE SPECIES

(1) CALCAREA

The sponges belonging to this group reproduce both sexually and asexually. Asexual reproduction is by budding; in some forms the offspring remain attached to the parent animal (*continuous budding*), while in others the buds eventually become detached and develop into new individuals. This type is called *discontinuous budding.*

The origin of the reproductive cells has not been clearly established. Several authors have described their origin in wandering amoeboid cells, although others maintain that they originate as choanocytes.

Poljaeff (1882) studied the spermatogenesis of a species of *Sycon.* He found that the nucleus of the spermatogonium divides into two, one of which moves to the periphery of the cell, while the other remains near its center. The peripheral cytoplasm accompanies the former nucleus and separates from the main part of the spermatogonium, forming a covering cell; the spermatogonium proper divides to form a mass of spermatocytes. While the cover cell remains undivided, the spermatocytes undergo the mitotic divisions, each forming 4 to 6 spermatids which later transform into spermatozoa. These spermatozoa as a mass are covered with an envelope formed by the covering cell, constituting a *sperm ball.*

In the process of oogenesis, an egg mother cell, or oocyte, which is believed to arise from an amoebocyte, increases in size by taking in trophocytes, and becomes an ovum. During the maturation period, the ovum extrudes two polar bodies.

According to the investigation of Maas (1900) on the development of the two genera, *Sycon* and *Grantia,* the oogonia are transformed amoebocytes resembling archaeocytes, with very large nuclei and conspicuous nucleoli. These pass into a radial canal where they undergo two mitoses, each producing four oocytes.

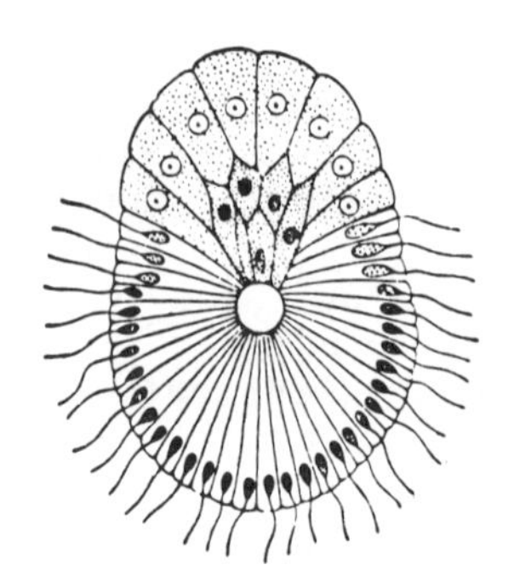

Fig. 2.3 Amphiblastula of *Leucosolenia variabilis* (Minchin)

The oocytes migrate by amoeboid movement back into the mesogloea, where they increase greatly in size by the aid of special trophocytes.

These trophocytes are transformed choanocytes; during the transformation, the choanocytes lose their collar and flagellum to become amoeboid, and one attaches to the surface of each oocyte. In fertilization, other choanocytes entrap spermatozoa which are brought in from outside by the water currents. These choanocytes also become amoeboid, containing the captured spermatozoon in a cyst-like condition. Attaching itself to the surface of an oocyte, one of these cells transfers its sperm cyst to the oocyte, and fertilization is thereby accomplished.

The fertilized egg of *Leucosolenia variabilis* undergoes the meiotic divisions and cleaves holoblastically. At the 16-cell stage, the flat disk-shaped embryo lies below the parental choanocyte layer; the eight blastomeres forming a tier apposed to these choanocytes are destined to be the future epidermis, while the other eight will give rise to choanocytes. The latter divide rapidly, and acquire flagella on their ends facing the blastocoel. The future epidermal cells remain undivided for some time, forming a group of eight large round cells which resemble the macromeres in other metazoan embryos. In the center of this group is formed a cavity which opens to the outside and serves the same function as the adult *osculum*. This stage is designated *stomoblastula* by Duboscq and Tuzet (1937). It is followed by a process of *inversion*, in which the embryo turns inside out through the osculum-like opening, so that the flagellated ends of the small blastomeres (micromeres) are on the outside. The embryo is now the typical hollow larva of the Calcarea, called an *amphiblastula* (Fig. 2.3). The surface of this oval larva consists of a layer of columnar flagelated cells; the blastocoel is rather extensive and filled with a colloidal fluid. The larvae of *Leucandra blanca* have large, round cells at the future posterior pole (Fig. 2.4A). These cells multiply until they come to constitute the greater part of the larval body. The larva is surrounded by a layer of trophoblasts which supply it with nutrients. Finally, with its flagellated end ahead, it works its way into the radial canal system and swims free of the parent animal.

Gastrulation takes place during the ensuing swimming phase, the flagellated cell layer invaginating and forming an inner mass of cells. Some of the flagellated cells lose their flagella, become amoeboid and fill the blastocoel. This solid type of larva is called a *parenchymula*. The flagellated cells become spherical, with inconspicuous flagella, and crowd together in the blastocoel; dispersed among these cells are found amoebocytes which later change into porocytes. These form cell masses here and there inside the larva and an irregular cavity appears between them (Fig. 2.4D), which later develops into the *spongocoel (paragastral cavity)*. Some of the epidermal cells migrate to the interior and become scleroblasts. Except for its apical part the entire paragastral cavity is lined with choanocytes having collar and flagellum. The apical part is the future osculum of the adult. The larva constructed from these tissues gradually becomes tubular and develops into a small sponge. Each of the masses of porocytes which lie near the surface forms a canal, the animal thus acquiring numerous small incurrent pores.

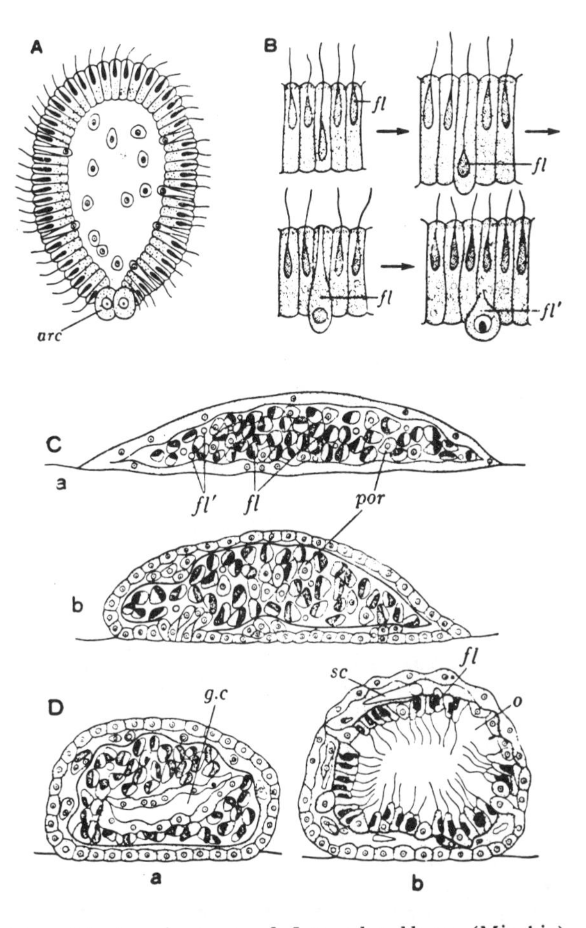

Fig. 2.4 Development of *Leucandra blanca* (Minchin)
A. swimming larva, showing archaeocytes (*arc*); B. flagellated cells from larval wall become amoeboid and move inward; C (a, b). flattened amoeboid larva just attaching. Cells inside larva differentiated into amoeboid and flagellated cells and porocytes; D (a, b). as development proceeds, mouth, gastric cavity and spicules are formed. *arc:* archaeocyte, *fl:* flagellated cell, *fl':* amoeboid flagellated cell, *g.c:* gastric cavity, *o:* mouth, *por:* porocyte *sc:* spicule.

The osculum opens to the outside at the apical end, the choanocytes lining the paragastral cavity become functional and ingest food, and the yound sponge grows into an adult.

The developmental processes of *Sycon raphanus* are somewhat different from these of *Leucosolenia* which have just been described, although both genera belong to the Calcarea. According to Minchin (1878), in *Sycon* the oocytes are fertilized within the maternal tissues but the embryos pass their early stages in the flagellated chamber. The fertilized egg undergoes regular, holoblastic cleavage; since the first three cleavage planes are all meridional, the blastomeres of the 8-cell stage lie in a single layer (Fig. 2.5a). The next two cleavages are both equatorial, and the embryo of the 32-cell stage acquires a small central cavity.

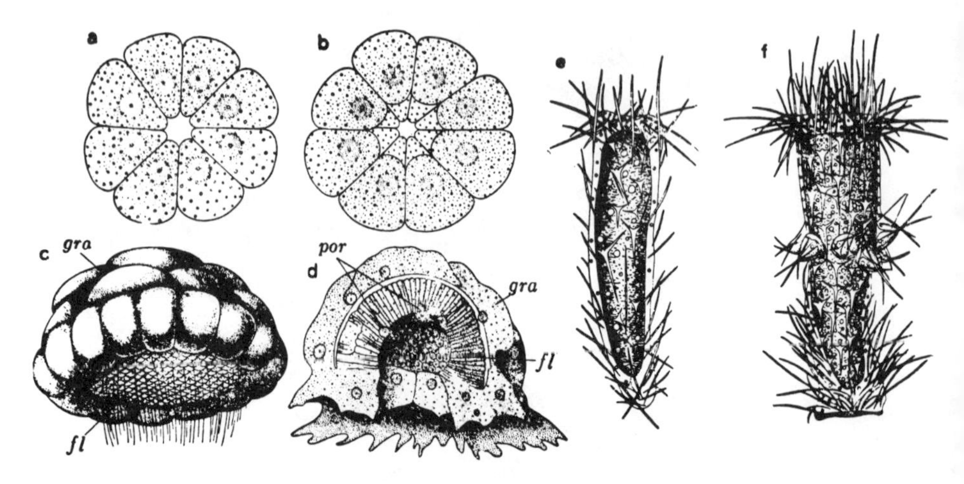

Fig. 2.5 Development of *Sycandra raphanus* (*Sycon raphanus*)

a. 8-cell stage; b. 16-cell stage; c. blastula; surface of flagellated cells covered by granular cells; d. cross-section of attached larva, showing central cavity and porocytes (Schulze); e. ascon type of *Sycandra*; f. larva of *Sycandra* (Maas).

The cleavage pattern after this is more or less irregular, leading to the formation of a blastula composed of one layer of cells, of which the eight at the posterior pole are large and filled with granules. The inside is hollow, forming the blastocoel or cleavage cavity. In this type of embryo the blastocoel has no opening to the outside, although a temporary aperture makes a fleeting appearance between the granule-rich cells of the flattened posterior side. These granular cells are archaeocytes; while they are cleaving to form a total of 32, the other cells of the blastula elongate greatly, becoming columnar and each acquiring a flagellum. The archaeocytes continue to divide rapidly, and push their way as a solid mass into the blastocoel *(pseudogastrula)*. At this stage the larva moves, by means of its flagella, through the maternal tissues and into one of the flagellated chambers, and finally by way of the osculum to the outside. It then becomes hollow, acquires bright red pigmentation and swims about for 1—2 days, finally settling on some substratum where it undergoes a striking transformation. The flagellated cells which have made up the anterior part of the larva shift to the inside of the posterior part, hence the anterior region becomes flattened and the blastocoel is compressed to a narrow, slit-like space (Fig. 2.5c, d).

After becoming attached to some object, the animal elongates into a closed cylinder, composed of an outer layer of flattened cells and an inner layer of flagellated cells. The gelatinous *mesoglea* develops between these two layers. Some of the flattened cells of the outer layer migrate into the space between the outer and gastral layers and become porocytes as each cell is pierced by a central canal, while others become scleroblasts within the mesogloea. In this manner the young sponge with a simple canal system of the asconoid type develops a highly complicated system of flagellated chambers (Fig. 25e, f).

(2) HEXACTINELLIDA

These "glass sponges" reproduce sexually as well as asexually; the latter method involves budding, as in the other groups of sponges. Ijima (1901) suggested that certain aggregations of archaeocytes *(archaeocyte congeries)* which are present in the body of *Euplectella marshalli* correspond to the gemmules in other kinds of sponges. The occurrence of eggs and sperm balls was observed in *Euplectella aspergillum* by F. E. Schulze (1880, 1887). Okada (1928), studying the reproduction of *Farrea sollasii*, a deep-water species from Sagami Bay, found clear evidence of sexual reproduction. In this glass sponge, the eggs are formed within a mass of archaeocytes. The egg cell is larger than the somatic cells, and has an alveolar nucleus. The mature egg is 0.7 mm in diameter, with numerous yolk granules packed around its large alveolar nucleus, and lipid-containing vacuoles in the peripheral cytoplasm (Fig. 2.6a). The spermatogonia are also differentiated archaeocytes. The spermatozoon has a spherical head and a slender tail. Fertilization appears to take place within the maternal tissues; cleavage is total and more or less unequal (Fig. 2.6b - d), and the resulting blastula is of the parenchymula

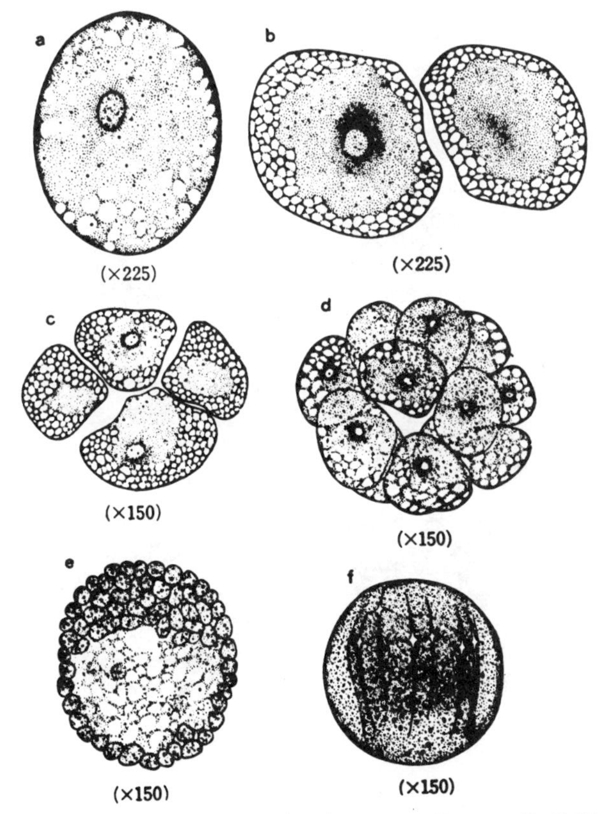

Fig. 2.6 Development of hexactinellid sponge *Farrea sollasii* (1) (Okada)

a. egg in body of adult sponge; b. 2-cell stage; c. 4-cell stage; d. 16-cell stage; e. section of larva at blastula stage; f. larva with earliest-formed stauractine spicules.

type. This embryo, still developing within the parental tissues, is at first planula-like and spherical (Fig. 2.6e - f), later becoming ovoid. It is composed of an outer layer of cells, rich in yolk granules, which surround a mass of mesogloea containing amoeboid cells. These arrange themselves in a layer surrounding a central cavity and change their form to become the future choanocytes. The flagellated chambers have not yet appeared, but six cross-shaped spicules *(oxystauractines)* are formed near the body surface (Figs. 2.6f; 2.7g). Larvae of the later stages are oval, and become bluntly pointed at both anterior and posterior ends (Fig. 2.7h), where the tips of the stauractines come together. The number of these spicules has now increased to 12, and their axes have become thicker. The cells comprising the posterior two-thirds of the larva are loaded with fatty material which will be utilized as energy source during its pelagic life. The flagellated chambers are already beginning to be formed at this stage.

The most advanced larva which was found within the maternal tissues is slightly pointed at the anterior pole and rounded posteriorly (Fig. 2.7i). It has developed two kinds of spicules — small *discohexasters* in the interior of the body

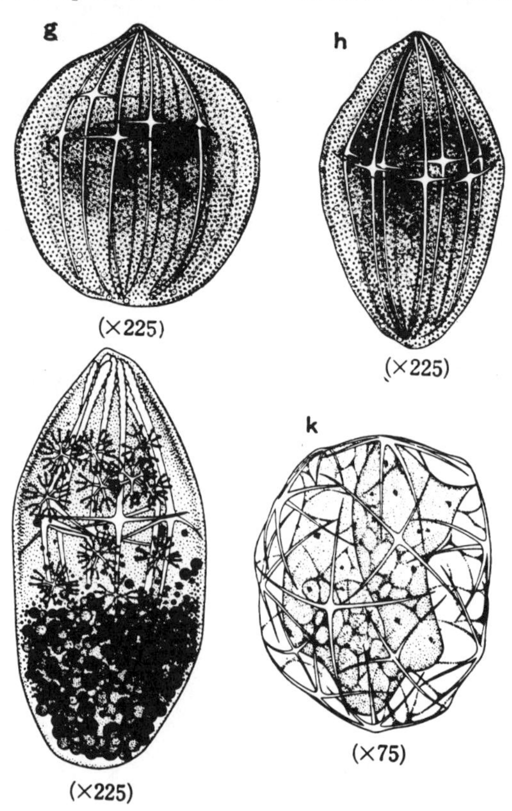

Fig. 2.7 Development of hexactinellid sponge *Farrea sollasii* (2) (Okada)

g. fairly well-developed larva; h. elliptical larva with oxy-stauractines; i. most advanced larva found within maternal body. Oxystauractines, fine spicules and hexactines are formed; k. young sponge has escaped to outside, is about to attach. Spongocoel, gastric cavity and mouth are formed.

in addition to the large stauractines at the surface. The stauractines are thickened anteriorly, and bear small processes. The posterior, anterior and lateral rays of the stauractines measure 105—120 μ, 110—120 μ and 60 μ, respectively. It is probably at this stage that the larva acquires flagella, escapes from the maternal body and settles onto some object; the smallest specimen which has been found in the attached state is barrel-shaped, 0.54 mm long 0.36 mm wide, somewhat flattened anteriorly and bluntly pointed posteriorly (Fig. 2.7k). The flagelated chambers appear as several protuberances of the spongocoel toward the dermal layer. At the surface, small spicules such as *pentactines* and stauractines develop, and the *dictyonine net* begins to be formed within the body near the point of attachment to the substratum. Spicules are not yet formed in the gastral layer.

In a young sponge which has developed still further (0.68 mm long, 0.57 mm wide), the shape is ovoid, and the openings of some of the flagellated chambers have broken through although the adult osculum is not yet formed. Spiculation becomes more composite with the addition of the *clavules,* which appear near the posterior rays of the surface pentactines. The paragastral cavity is formed, with choanocytes lining its central portion, and the osculum breaks through to the outside. Sexual reproduction takes place throughout the year in deep-sea sponges such as *Farrea.*

(3) TETRACTINELLIDA (TETRAXONIDA)

The sponges of this group reproduce sexually as well as asexually by external budding, as is widely found in other sponge groups, and also by internal budding. The former type of budding occurs in the genus *Donatia,* in which the bud first appears as an external protuberance, develops spicules and becomes detached from the parent animal (Fig. 2.8). The bud eventually attaches to some object and develops into an adult. Internal budding, which occurs in the genera *Ficulina* and *Cliona,* is a kind of gemmule-formation. Although the buds in these groups differ from the gemmules of freshwater sponges in certain respects, such as lacking spicules, a micropyle and an inner membrane, they are similarly composed of aggregated yolky archaeocytes surrounded by a layer of columnar cells (Fig. 2.1e,f).

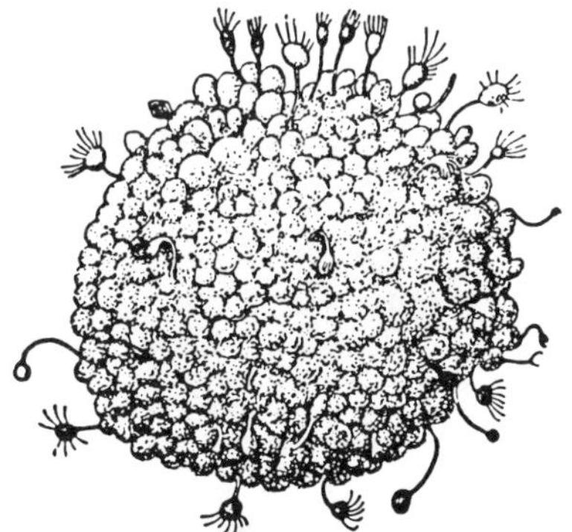

Fig. 2.8 External budding in tetractinellid sponge *Danatia* (Hentschel)

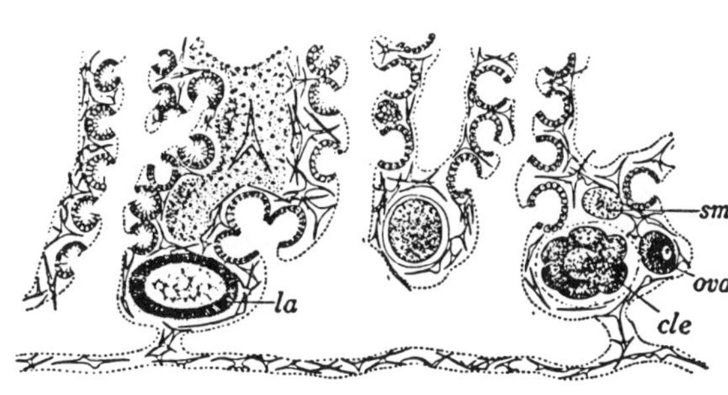

Fig. 2.9 Part of tetractinellid sponge *Plakina* (Schulze)

cle: embryo at cleavage stage, *la:* larva, *ova:* ovum, *sm:* sperm ball.

The sexual reproduction of this group was studied in detail in *Plakina* (Fig. 2.9) by Schulze (1880) and Maas (1909). The egg undergoes equal holoblastic cleavage; the blastula is ovoid to pear-shaped, its outer surface composed of flagellated cells, while a very few granule-filled archaeocytes are found in the blastocoel. At a later stage the epidermal cells making up the posterior half of the body wall lose their flagella and become filled with granules, and the mesoglea cells are pushed into the blastocoel (Fig. 2.10a, b.) The larva attaches by its anterior, flagellated half, which flattens and invaginates inward so that the flagellated cell layer becomes

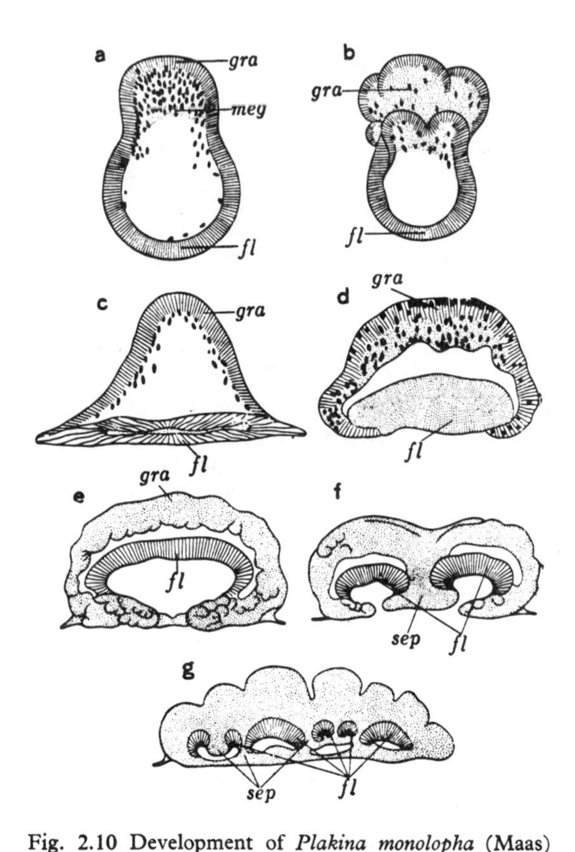

Fig. 2.10 Development of *Plakina monolopha* (Maas)

a. amphiblastula, granular cells budding into interior. These granular cells are present in anterior part of outer layer of larva; flagellated cells in posterior part; b. larva about to attach to substrate; c. attaching larva begins form change; d. larva in late attachment stage; flagellated cells beginning to invaginate; e. "rhagon" stage; f, g. two stages in subdivision of rhagon cavity by downgrowth of septa. *fl:* flagellated cells, *gra:* granular cells, *meg:* mesoglea cell, *sep:* septum.

surrounded by the granule-containing columnar epithelial layer making up the posterior half of the body wall (Fig. 2.10c, d). A blastopore is formed during this process, but it is closed by the surrounding columnar epithelial cells which collect there and eventually form the basal part of the sponge. The larva at this stage is called a *rhagon* (Fig. 2.10e). The *rhagon, cavity* which was formed by the invagination of the flagellated layer, is divided into separate cavities, together with the corresponding parts of the flagellated layer, by partitions originating in the dermal layer. These cavities become filled with flagellated cells but later reappear, forming spherical flagellated chambers (Fig. 2.10f, g).

A detailed study has been made by Meewis (1939) on the development of *Halisarca*, a primitive sponge belonging to the Tetractinellida. According to Meewis's study, which follows the developmental processes to stages more advanced than those studied in *Plakina*, the osculum breaks through the body wall only after completion of the rhagon stage; excurrent canals grow inward from the osculum, making connection with the flagellated chambers through the paragastral cavity. Meewis also confirmed the observation that the dermal cells and amoebocytes of the adult sponge are derived from the original posterior half of the larva, and found that the inversion of the embryonic cell layers occurring in the Monaxonida does not take place in this form.

(4) KERATOSA (HORNY SPONGES)

The sponges of this group also reproduce both sexually and asexually, asexual gemmule formation occurring in all the freshwater species and in such marine forms as *Mycale* (Fig. 2.1e, f), *Chalina, Cladorhiza* and *Reniera*. The processes of gemmule formation have been briefly described in the preceding section. Among

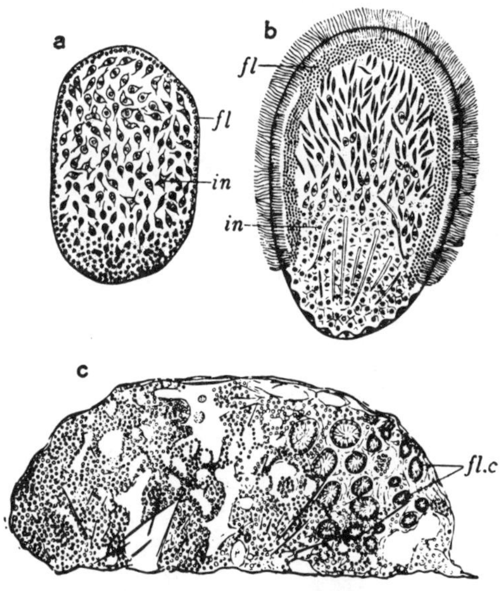

Fig. 2.11 Development of marine monaxonid sponge *Esperella* (*Mycale*) (Wilson)

a. parenchymula larva, containing cell mass in its interior; b. swimming larva; c. cross-section of young sponge; flagellar chambers developing. *fl:* flagellated cell layer, *fl.c:* flagellar chamber, *in:* inner cell mass.

the marine forms, the larva developing from the gemmule is identical with that arising from a fertilized egg.

Most of the animals of this group are dioecious, while the hermaphrodites are *protandrous* — the spermatozoa are produced earlier than the eggs. Both kinds of gametes are derived from archaeocytes; the egg has a finely granular cytoplasm and a large nucleus, and is surrounded by a cellular envelope which, originating from parenchyme tissue close to the egg, probably functions as a trophic membrane. The spermatogonium divides into two cells; one of these differentiates into an envelope cell, while the other becomes a spermatocyte. This latter divides several times to form spermatozoa. The details of oögenesis are not clear, but the fertilized egg is known to undergo unequal holoblastic cleavage, forming a stereogastrula, or parenchymula larva (Fig. 2.11a). The anterior surface of this pear-shaped or oval larva is occupied by a layer of cylindrical flagellated cells, and the scleroblasts and main spicules develop near the posterior pole (Fig. 2.11b). The parenchymula corresponds to the gastrula of other animals; it swims about by means of its flagella for approximately 24 hours, and then becomes attached by its anterior surface to some object. It then undergoes the process of inversion of the cellular layers constituting the body wall. The outer flagellated cells become situated inside the body, where they differentiate into the paragastral layer cells lining the flagellated chambers. At the same time, the amoebocytes from the inside emerge to the surface, where they arrange themselves to form the dermal layer and the scleroblasts (Fig. 2.11c). By further increases in size and complexity this young sponge develops into an adult.

In freshwater sponges, Brien and Meewis (1938) have found that the fertilized egg cleaves to form a stereoblastula, composed of three sizes of cells: micromeres, mesomeres and macromeres. The micromeres, at first situated at the anterior pole, spread over the embryo and give rise to flagellated surface cells. While the embryo is still enclosed in the maternal body, the cells of the inner mass transform into scleroblasts, choanocytes and amoebocytes. The first two kinds of cells are derived from the macromeres. When the embryo escapes from the maternal tissues, it is a flagellated larva, already provided with spicules and flagellated chambers. After swimming for several days, it becomes attached by its anterior pole. The developmental process of the fresh-water sponges thus differs from that of the other monaxonid species in that there is no inversion of the embryonic cell layers.

REFERENCES

Brien, P. & H. Meewis, 1938: Contribution à l'etude de l'embryogénese des spongillidas. Arch. Biol., **49**.

Duboscq. O. & O. Tuzet, 1937: L'ovogénese, la fécondation, et les premiers stades du développement des éponges calcaires. Arch. Zool. Expt. Gen., **79**.

Evans, R. 1899: Structure and metamorphosis of the larva of *Spongilla*. Quart. Jour. Micros. Sci., **42**.

—————— 1912: *Ephydatia blembingia*, with an account of the formation and structure of the gemmule. Quart. Jour. Micros. Sci., **44**.

Gatenby, J. B. 1919: Germ-cells, fertilization and early development of *Grantia*. Linn. Soc. London Jour. Zool., **34**.

Hammer, E. 1908: Histologie und Entwicklung von *Sycon*. Arch. Biontologie, **2**.

Hyman, L. H. 1940: The Invertebrates, vol. **1**, Protozoa through Ctenophora. McGraw-Hill, New York and London.

Ijima, I. 1901: Studies on the Hexactinellida. Contrib. 1, Euplectellidae. Jour. Coll. Sci. Univ. Tokyo, vol. **15**.

Katoda, J. 1923: Monaxonida *Reniera ikadai* Katoda: Several observations on a new species. Collected Works of J. Katoda. (in Japanese)

Maas, Q. 1892: Metamorphose von *Esperia*. Mitt. Zoöl. Stat. Neapel., **10**.

—————— 1898: Entwicklung der Spongien. Zoöl. Zentbl., **5**.

—————— 1893: Embryonal-Entwicklung und Metamorphose der Cornacuspongien. Zoöl. Jahrb. Abt. Ontog. Tiere, **7**.

—————— 1898: Metamorphose von *Oscarella*. Ztschr. Wiss. Zoöl., **63**.

—————— 1900: Die weitere Entwicklung der Syconen nach der metamorphose. Ztschr. wiss. Zoöl., **67**.

—————— 1909: Entwicklung der Tetractinelliden. Verh. Deut. Zool. Gesell., **19**.

Meewis, H. 1939: Embryogénèse des Myxospongidae: *Halisarca*. Arch. Biol., **50**.

—————— 1878: Metamorphose von *Sycandra*. Ztschr. wiss. Zoöl., **31**.

Minchin, E. 1896: Larva and postlarval development of *Leucosolenia*. Proc. Roy. Soc. London, **60**.

Okada, Yaichiro, 1928: On the development of a hexactinellid sponge, *Farrea sollasii*. Tokyo Univ. Facult. Sci. Jour., Sect. IV vol. **2**. no. 1.

Oshima, H. & Yaichiro Okada, 1943: Animal Genealogy. vol. 1. Porifera. Tokyo, Yokendo (in Japanese)

Polejaeff, J. 1882: Sperma und Spermatogenesis bei *Sycandra*. Akad. Wis. Wien. Math. Nat. Kl. Sitzber., **86**.

Schulze, F. E. 1875: Ueber den Bau und die Entwicklung von *Sycandra raphanus* Haeckel. Ztschr. wiss. Zoöl., **25**.

—————— 1878: Die Metamorphosis von *Sycandra raphanus*. Ztschr. wiss. Zoöl., **31**.

Wilson, H. V. 1894: Observations on the gemmule and egg development of marine sponge. Jour. Morph., **9**.

—————— 1907: On some phenomena of coalescence and regeneration in sponges. J. Exp. Zool., **5**: 245.

Woodland, W. 1908: Scleroblastic development of hexactinellid and other siliceous sponge spicules. Quart. Jour. Micros. Sci., **52**.

Zeuthen, E. 1939: Hibernation of Spongilla. Ztschr. Vergleich. Physiol., **26**.

COELENTERATA

Chapter 3

I. CNIDARIA

1. GENERAL DEVELOPMENT

While asexual reproduction is continuously occurring among the coelenterates, sexual reproduction usually takes place only once a year. Most of these animals breed sexually sometime between early spring and autumn, but there are also species which reproduce during the winter, and tropical species may breed sexually several times a year. Among such forms as the free-living medusae, the eggs break through the body wall in conjunction with the contracting movements of the medusa, and are thrown out, to be fertilized either suspended in the sea water or after sinking to the bottom. Among the species belonging to the sessile polyp type, the eggs are fertilized while they are still attached to the maternal body. In case fertilization takes place in suspension, the egg usually passes through a free-swimming *planula* stage before settling to the bottom and assuming the sessile *polyp* form. *Medusae* are given off from the polyp by a process of budding, and these develop reproductive organs and reproduce sexually. Even when the polyp produces eggs, these may in some cases develop into free-swimming planulae and then attach, to give rise to sessile polyps by much the same process. While these may bud off medusae in some species, in others no medusa stage appears, reproductive cells being formed in one part of the polyp, and growing directly into a new polyp. In the Actinozoa, no medusa stage ever appears; the polyp stage alone gives rise to planulae which attach and become polyps, producing reproductive cells which grow into planulae. In the Hydrozoa the reproductive cells are ectodermal in nature, while in all the other Coelenterata, including the Ctenophora, they arise from the endoderm.

It is usual for coelenterates to pass through a planula stage, but ecological conditions may result in the omission of this phase. For example, in the sea anemone *Actinia equina*, the embryo completes its development within the body of the mother animal, and after assuming the form of a small sea anemone, it is released to the outside by contraction of the adult body wall. In another sea anemone, *Epiactis japonica,* the planulae never become free-swimming; the developing embryos, after being expelled to the outside, attach to the body wall of the mother and there grow into small sea anemones.

Among coelenterate eggs, those of the Cnidaria may be said to cleave with radial symmetry, while ctenophore cleavage is bilaterally symmetrical. Of course cleavage in most of the Cnidaria is radial, but there is much variation among the Hydromedusae. In some species the blastomeres shift their positions markedly in the 4-cell stage *(Spirocodon, Köllikeira),* while in others (e.g., *Turritopsis, Ectopleura),* bilateral symmetry emerges in a later cleavage stage. The first cleavage furrow cuts inward from the animal pole, but there is practically no cleavage activity at the vegetal pole (Fig. 3.4a-c). As cleavage advances, a blastula is formed. This may be either a coeloblastula or a sterroblastula (see Fig. 1.28). In many cases formation of a sterroblastula results from the fact that the egg contains a large amount of yolk, but the free-swimming Hydromedusae form such blastulae even though their eggs have very little yolk. The blastula advances to the gastrula stage, and at this point the endoderm is formed, in the great majority of cases by invagination. However, even among the eggs with little yolk, gastrulation takes place by polarization and by delamination, and there are some cases of epiboly among the ctenophores. Moreover, the process of invagination as it occurs in these species is colored to some extent by variations suggestive of polarization or delamination. Invagination begins with a vigorous proliferation of the cells at the vegetal pole of the blastula, these cells accumulating on the inner side of the blastular wall. In polarization, the cells of a limited area at the vegetal pole proliferate into the blastocoel. Delamination is a special mode of gastrulation peculiar to the coelenterates, in which all the cells of the blastula divide into the blastocoel. In this way the outer and inner body layers, the ectoderm and endoderm, are formed: later the *mesogloea* is secreted between these two layers.

The cases so far described have been concerned with eggs having a small amount of yolk, but among the Actinozoa and Hydrozoa, there are species with opaque eggs containing large amounts of yolk. It is difficult to make accurate observations of the mode of cleavage in such eggs, but they appear to divide by a sort of superficial cleavage. In some cases the blastomeres separate into small ectodermal and large endodermal cells, while in other forms a gastrula is formed without any such differentiation between the two types of cells, the outer layer simply becoming the ectoderm, and the inner layer, the endoderm. In such cases, yolk remains within the blastocoel, and is gradually absorbed. These types of cleavage are specialized modes of dealing with unusually large amounts of yolk, and should not be systematically differentiated from typical radial cleavage.

As germ layer formation proceeds, the ectodermal cells develop cilia, sensory and gland cells are differentiated, and a free-swimming planula is produced. The group of Stauromedusae form a single exception in that they lack cilia, and are only able to creep about on the bottom. Most forms have a blastocoel during this planula stage, but the Trachymedusae, Narcomedusae and Siphonophora among the Hydromedusae have solid planulae without any blastocoel.

In many cases the planula becomes attached and turns into a polyp, but there are also some which transform, while they are still in the swimming stage, into a temporary swimming polyp form. This phenomenon occurs in the Zoantharia and Ceriantharia; each of these types has its own particular morphology and is known by a special larval name. *Porpita* and *Velella* among the Siphonophora also develop into a special type of larva. These various larvae all undergo metamorphosis to acquire the adult form. In the Scyphomedusae, after the planula has developed into a polyp, this in turn produces a special *strobila stage*, which gives rise asexually to the young larval medusa (*ephyra*). This ephyra passes through a metamorphosis to become the adult medusa.

2. HYDROZOA

Among the Hydrozoa, each species has a polyp stage and a medusa stage, the two stages appearing alternately to give a regular *alternation of generations*. The extent of development of each stage, however, differs greatly from one species to another, so that the polyp alone may be well developed, with the degenerate medusa never separating off from it and reduced to the form of a *sporosac*. Or, at the other extreme, the whole life cycle may be pelagic, only the medusa being well developed while the polyp is present in a modified or rudimentary state, or makes only a very brief appearance. The Hydrozoa are usually divided into the orders Hydroida, Trachylina and Siphonophora; the Hydroida is the only one of these groups which has a conspicuous polyp stage.

The eggs of the Hydrozoa are of ectodermal origin, and the course of their early development, particularly cleavage and the process of endoderm formation, is exceptionally rich in variations. Since most medusan eggs do not contain large amounts of yolk, their cleavage is equal and easy to observe. In addition to the usual radial cleavage, they may show bilateral and even temporary spiral cleavage. In the process of germ layer formation, also, they may follow the method of polarization as well as the more usual methods of invagination and epiboly. This situation suggests that the low order of differentiation characterizing the Hydrozoa as a class results in the inclusion within the group of this array of different forms. Among the sessile species of Hydrozoa which do not give rise to a medusa stage, there are some in which the eggs contain so much yolk that their cleavage cannot be clearly observed, and it is also impossible to follow their germ layer formation satisfactorily.

(1) DEVELOPMENT OF *TUBULARIA MESENBRYANTHEMUM*

The development of *Tubularia* will be described as representative of a group which lacks the medusa stage and retains only the polyp form. *Tubularia*, living in shallow water on rocks and loose stones along the sea shore, or attached to eel grass or other sea weeds, is one of the rather commonly seen species of our sea coasts. The polyp reaches a height of 3—8 cm, and can be called one of the larger hydropolyps. It is also generally found on all parts of the European and American shores, where its embryology has been known for a very long time.

Tubularia belongs to the category mentioned above, which lacks a free-swimming medusan phase; the reproductive cells are continuously produced asexually in *gonophores* formed on the polyp by budding. Such a gonophore, moreover, does not separate from the polyp, but remains as a sporosac, which can be thought of as a sort of degenerate medusa. The gametes mature within these structures, and the eggs are fertilized within the female gonophores. The hydranth of *Tubularia* has two circlets of tentacles, an *oral circlet* around the mouth and an *aboral* (or *proximal*) *circlet* at the base of the hydranth; the gonophores are always formed in the space between these. At the base of the hydranth appear a number of processes which elongate and branch. On the end of each is then

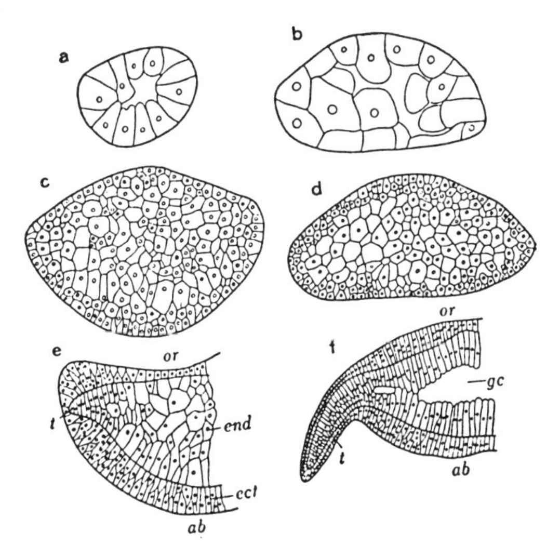

Fig. 3.1 Development of *Tubularia mesembryanthemum* (Brauer)

a. blastula; b. budding of cells from blastular wall; c~f. formation of endoderm and tentacle. *ab:* aboral pole, *ect:* ectoderm, *end:* endoderm, *gc:* gastric cavity, *or:* oral pole, *t:* tentacle rudiment.

formed a gonophore, so that the whole structure resembles a bunch of grapes. The species of *Tubularia* are dioecious, the gonophores of a given colony being either all male or all female, and very often exhibiting a certain degree of morphological sex differentiation. In any case, a *manubrium* is present in the center of the gonophore, and this gives rise to sperm sacs in the male, or several eggs in the female. The eggs are fertilized *in situ* by spermatozoa which swim out from the male gonophores. In our country, breeding generally takes place during the spring and early summer.

The fertilized eggs begin to cleave within the female gonophore. *Tubularia* eggs contain a large amount of yolk, and as a result their cleavage is rather irregular. Furthermore, the presence of a number of eggs in one gonophore causes them to crowd each other and increases the irregularity to some extent. These cleavages are likely to be somewhat polynucleate, and the cytoplasm tends to accumulate at the surface, while the yolk is left in the center, giving the effect of superficial cleavage. The ectoderm cells arrange themselves over the surface; the endoderm cells at first lie irregularly on the inner side of the ectoderm, but eventually they form a definite layer. In other words, cleavage and germ layer formation take place simultaneously. Examining these processes one at a time, we find that at about the 16-cell stage of the irregular cleavage, a small space is formed at the center of the mass of blastomeres, and the embryo enters the blastula stage. As cleavage proceeds still further, this small blastocoel disappears and the embryo becomes a solid *morula*. To form this morula, the cells making up part of the blastular wall divide rapidly many times in succession, and the cells produced in this way proliferate into the blastular cavity. The part of this embryo which originally made up the blastular wall later becomes the ectoderm, while the cells

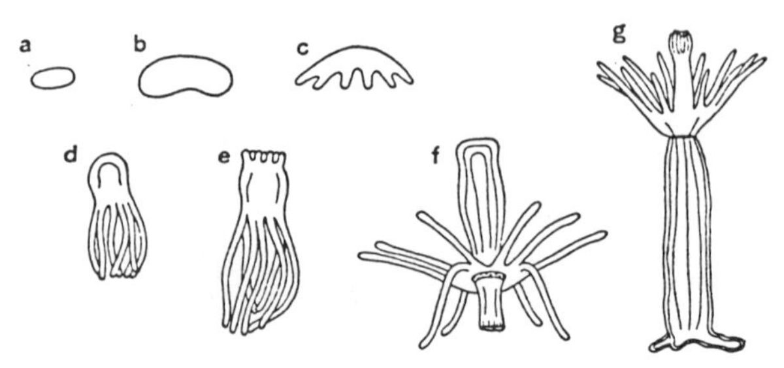

Fig. 3.2 Development of *Tubularia indivisa* (Allman)

a~c. process of gastrulation; d~f. actinula; g. attached actinula.

which moved inward to fill the blastocoel become the endoderm. In other words, a gastrula is formed by a process similar to delamination. In the center of such an embryo, small empty spaces gradually make their appearance. These eventually collect in the central part of the embryo, giving rise there to a single cavity which later becomes the part of the adult called the *gastral cavity* or *coelenteron*. At this stage the embryo is somewhat compressed, becoming roughly plate-shaped. A number of small processes develop around the edge of this plate; these are the rudiments of the aboral tentacles which will appear later. As development proceeds, these begin to bend toward the concave side of the embryo and increase in length. Meanwhile a region appears on the opposite (convex) side where both ectoderm and endoderm become thin, and eventually a mouth opening forms here. About this time the tendency to bend toward the aboral side becomes more pronounced, and gradually the larval shape changes from plate-like to cylindrical (Fig. 3.2a-d), while around the mouth opening at its upper end a number of small processes are formed. These are the rudiments of the oral tentacles. In the meantime the aboral tentacles have already become fairly long and slender, and the aboral end of the body itself grows to reach about the same length as these tentacles. The larva at this stage is called an *actinula*.

All the development to this point has taken place inside the gonophores, but in the actinula stage the larvae escape to the outside through the mouths of their respective gonophores. Once free, they immediately sink toward the bottom, landing on rocks, stones or the fronds of sea weeds, where they spend a short time moving about by means of their oral and aboral tentacles before finally becoming attached to these objects by their aboral ends. After attachment the actinula increases in height, and the part bearing the oral and aboral tentacles forms a *hydranth*, while the *stalk* becomes covered with a chitinous *perisarc* which it secretes around itself, completing the organization of a new small hydropolyp of *Tubularia*. From the attached base a number of *stolons* are put out, and these in turn give rise to new polyps, forming a *colony*. The hydranths of this new *Tubularia* do not yet have gonophores, being still sexually immature; these will appear for the first time in the breeding season of the following spring.

The development of *Tubularia* is thus completed without including the free-swimming planula stage seen in the life-histories of most other coelenterates, and with the larva well on its way to the adult state when it first escapes from the gonophore as an actinula.

While *Tubularia*, moreover, is thus seen to lack a medusan phase, there is a hydroid of which the polyp presents an outward appearance identical with that of the *Tubularia* polyp, but in which the gonophore becomes a separate medusa. Although this form is closely related to *Tubularia,* it is nevertheless treated as a separate genus, *Ectopleura.*

(2) DEVELOPMENT OF *GONIONEMA DEPRESSUM*

This species is common along the Pacific Coast of Honshu from the central part southward, and can be collected throughout June by feeling for it among the eel-grass. The gonads are somewhat different in color, the testes being yellowish brown while the ovaries are dark brown. Since this species naturally spawns at night, in the laboratory it will begin to spawn within 15 minutes after being placed in a dark-room. Previously exposing the animals to strong light improves the spawning reaction. The eggs are brownish and opaque, spherical, and ca. 0.07 mm in diameter. They are released to the outside through breaks in the ovarian walls, and fertilized as they fall to the bottom. The cleavage furrow is formed at the animal pole, and cuts through to the vegetal pole, dividing the egg into two blastomeres in about an hour. Cleavage is equal and total. The second cleavage also begins at the animal pole and proceeds to the vegetal pole, at right angles to the first cleavage plane. This division requires about 50 minutes for completion. The third cleavage, furrow cuts equatorially, producing four upper and four lower blastomeres, of which the four upper ones are turned by 45° and wedged between the four lower ones. This resembles spiral cleavage, but the embryo fails to show the later characteristics of this mode of cleavage. After the third cleavage, the division rate shows a gradual slight acceleration, and eventually the blastula stage is reached. The next event is endoderm formation, which is accomplished by the method of delamination. While this is going on, the ectoderm cells acquire cilia, and the larva takes on the planula form and rotates within the chorion by means of its cilia. After it has been rotating for about an hour, the chorion is torn and the larva is set free. Once outside the chorion, the larva elongates, swimming with its broader end forward. About twelve hours are required to arrive at this free-swimming stage of development, and the planula has achieved a length of more than 0.1 mm. It swims about for several days, and then sinks to the bottom, where it loses its cilia and creeps about for a while before changing into a polyp. This polyp first develops a mouth opening, and then two tentacles are formed on opposite sides of this opening. Somewhat later, two more tentacles appear

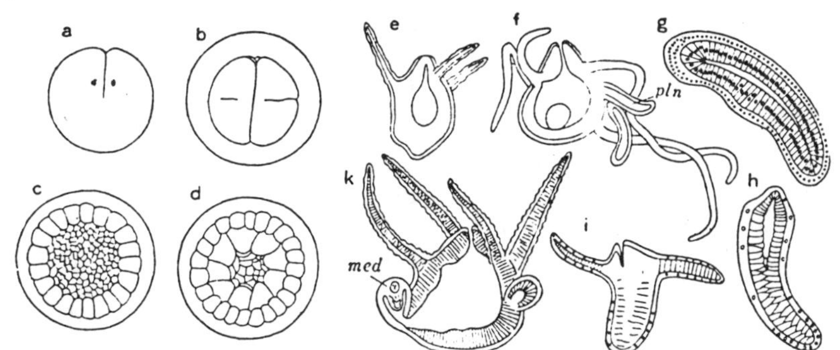

Fig. 3.3 Development of Adriatic marine medusa *G. vindobonensis* (Joseph)

a. first cleavage; b. beginning of third cleavage; c. blastula; d. formation of endoderm; e. polyp; f. budding of planulae from polyp; g~i. change in shape of freed planula; k. budding of medusa from polyp. *med:* bud of medusoid form, *pln:* planula.

beside the mouth, opposite each other at right angles to the first pair, making a total of four tentacles. Nothing more than this is known about the polyp phase of this species, but the development of an Adriatic species, *G. vindobonensis*, (Fig. 3.3), has been investigated, and it is likely that the local species follows about the same course. The number of tentacles of the polyp may increase to five or six, but during the 4-tentacle stage, bodies like planulae but lacking cillia *(frustules)* are budded off. This budding is repeated several times, and the frustules are believed to grow into polyps. The polyps then produce from one to several medusa buds, each of which develops *radial canals*, a manubrium, tentacles, *statocysts*, a *velum* and other structures, and is finally set free as an independent medusa. At this time it has twelve tentacles and eight statocysts; these gradually increase in number. In this country the polyps develop from the fertilized eggs in June, while the young medusae are believed to appear during May. This means that the polyps are alive for at least ten or eleven months, but even in the European species, nothing is known of them during this period.

The development of *Craspedacusta sowerbyi*, a freshwater medusa belonging to the same order as *Gonionema*, follows a closely similar course. In the latter species, the polyp lacks a cuticular layer, and is a much reduced form, while the polyp of *C. sowerbyi* is still more degenerate, never forming tentacles, and having only clustered nematocysts around the mouth. However, it is very similar to *Gonionema* in such characters as the mode of medusa-budding. The young *Craspedacusta* medusa has eight tentacles but no statocysts when it is released; these develop during its free-swimming life. The medusa makes its appearance for at most a few months, but the polyp is believed to live practically a full year.

(3) DEVELOPMENT OF *SPIROCODON SALTATRIX*

In all parts of this country *Spirocodon saltatrix* breeds during the winter. The eggs of this medusa are about 0.06—0.07 mm in diameter, without an external membrane, and their cytoplasm is colorless and highly transparent. In nature they spawn at night; if placed in a dark-room, a nature animal will spawn within 30—60 minutes. The meiotic divisions are completed within the ovaries; the eggs are fertilized as soon as they are shed, and the first cleavage furrow divides the egg into two, starting at the animal pole and proceeding towards the vegetal pole (Fig. 3.4a-c). The furrow does not cut through at the vegetal pole, the mitotic spindle becomes V-shaped, and the daughter nuclei are drawn downward and towards the furrow (p. 46). After about 30 minutes, the second cleavage furrow appears,

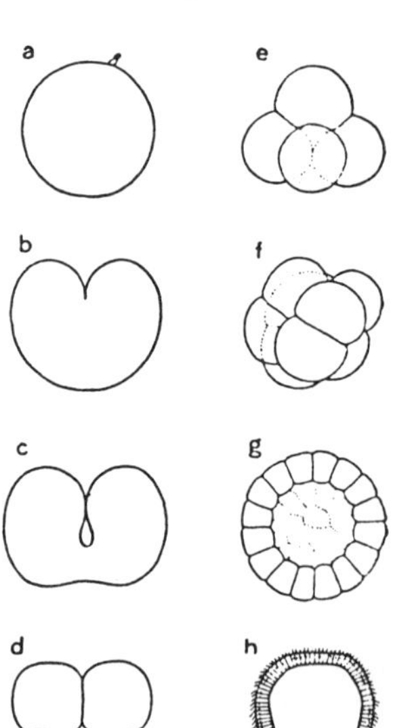

Fig. 3.4 Cleavage in the medusa *Spirocodon saltatrix* (Uchida)

a. fertilized egg; b. c. successive stages of first cleavage; d, e. 4-cell stage; f. 8-cell stage; h, g. successive stages of blastula.

perpendicular to the first, producing four blastomeres. Although the third cleavage furrow cuts across equatorially, the 8-cell stage blastomeres are tilted at an angle of 45° so that the upper and lower layers are wedged against each other in an alternating way, resembling the arrangement in spiral cleavage (Fig. 3.4f). With the fourth cleavage, 16 blastomeres are formed, and the embryo gradually increases the number of its cells and becomes a blastula. At first this is elliptical, but it gradually becomes egg-shaped. By about ten hours after fertilization, the cells of the blastula are covered with cilia, the anterior end has become rounded and the posterior end pointed, and compared with the cells of the anterior end, those of the posterior end are very long and slender. The larva at this stage is already a swimming planula, and the endoderm is apparently formed from the posterior ectoderm by the method of polarization. The planula spends about a week in this swimming stage, but it is not known when it changes to the polyp form. However since small medusae with four tentacles and four tentacle swellings make their appearance every January, medusa-budding is believed to take place at that season. The smallest medusae are 1 mm in diameter with short tentacles; the ring canal is without diverticula, and the manubrium is rounded and without lips. As this medusa grows and its tentacles elongate, a pair of new tentacles arise on each side of the old ones, so that the perradial and interradial clusters of tentacles become connected with each other. The newly arising tentacles tend to push the older ones upward, making the bell margin eight-lobed. The radial canals send out on either side horizontal arms which later branch and form a complicated system. The upper parts of the radial canals, however, do not give off such branches, but bend down into the region that supports the manubrium, causing it to grow downward and carry the manubrium with it. The radial canals of this region increase in length and become coiled, and the reproductive organs develop here. The manubrium acquires four lips, which eventually become folded and frilled. Following such a course of development, this medusa may reach a height of 50 mm, and is one of the largest and most highly differentiated and morphologically complicated of the Anthomedusae.

Turritopsis nutricula is a common species in this country, but the record of its developmental history is incomplete. However, the fact that its cleavage is bilaterally symmetrical (Fig. 3.5) and resembles that of the ctenophores deserves special mention. It forms a sterroblastula, and its polyp is known to develop from a planula.

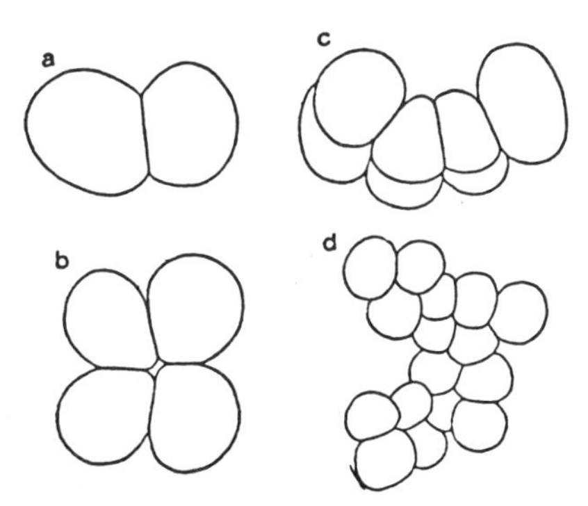

Fig. 3.5 Cleavage in the medusa *Turritopsis nutricula* (Brooks)
a. 2-cell stage; b. 4-cell stage; c. 8-cell stage; d. 16-cell stage.

Among the Leptomedusae, the mode of cleavage of *Aequorea* is known; this is in general similar to the cleavage of such standard forms as *Gonionema,* although in the Leptomedusae the polyps are sheathed, and statocysts are formed in the medusae.

(4) GENERAL DEVELOPMENT OF HYDROIDA

As described above, the members of the order Hydroida have both a polypoid and a medusoid generation. In principle these appear alternately, giving rise to an alteration of generations, although actually there are many cases in which the medusa fails to separate from the polyp, and the gonads stop with the formation of a sporosac, as was seen in *Tubularia.* Various degrees in this reduction of the medusoid generation can be recognized; these are divided into *eumedusoid, cryptomedusoid, heteromedusoid, styloid,* etc. (Fig. 3.6). Moreover, in some species the degree of reduction is different between the sexes, the female, for instance, separating as a free medusa while the male is represented by only a sporosac. This complicated situation leads to a good deal of taxonomic as well as embryological confusion.

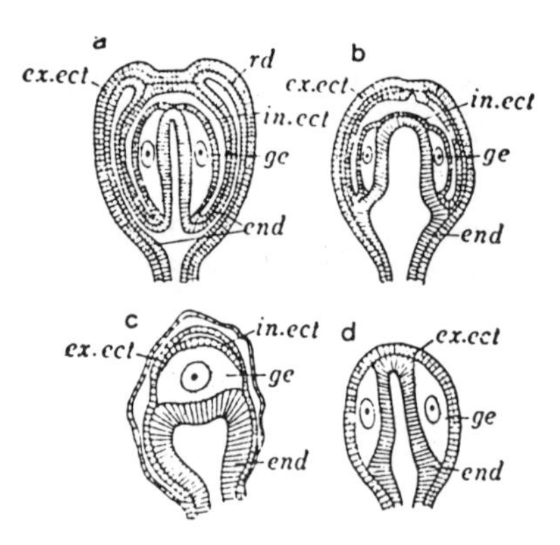

Fig. 3.6 Various types of sporosacs (Kühn)
a. eumedusoid; b. cryptomedusoid; c. heteromedusoid; d. styloid. *end:* endoderm, *ex. ect:* external ectoderm, *ge:* germ cell, *in.ect:* internal ectoderm, *rd:* radial canal

The reproductive cells of the Hydrozoa are ectodermal in origin, and usually penetrate into the endoderm after completing maturation. The spermatozoa are flagellate; the eggs are usually devoid of a covering membrane, and those produced by a medusa tend to be small and have little yolk, while eggs that are formed in a sporosac are likely to be large and yolky. Depending on the amount of this yolk, the later development varies considerably. The eggs with little yolk are shed into the sea water and fertilized there; these forms undergo comparatively regular cleavage. As described for *Gonionema,* cleavage is total and practically equal, the first two cleavage planes being perpendicular to each other while the third is horizontal. From that point on the cleavage conforms to the radial type, with micromeres formed at the animal pole and macromeres at the vegetal side. The simply organized Hydromedusae are also known to cleave totally but irregularly, and, like *Turritopsis nutricula* described above, show bilateral symmetry. Even eggs which undergo regular radial cleavage tend to form blastulae with small blastocoels or even with none at all, and in many cases the blastocoel is later completely filled with endoderm. While the alecithal medusan eggs form coeloblastulae, the sporosac eggs of such genera as *Clava, Gonothyrea* and *Plumularia* are likely to develop into sterroblastulae. Endoderm is formed in the coeloblastula type by invagination or polarization, while it is likely to appear irregularly in sterroblastulae. As was described above, yolky eggs like those of *Tubularia* and *Eudendrium* cleave irregularly, giving rise to multinucleate cells, with the cytoplasm accumulated at the surface and the yolk left in the center of the embryo. This produces an effect like that of superficial cleavage, with ectoderm cells arranged around the surface, and few nuclei in the interior, where endoderm cells are formed in an irregular manner.

As the coeloblastula proceeds to the gastrula stage, it usually swims free as a planula. During this free-swimming stage the planula grows, taking on an elongate oval shape, with gland cells and sensory cells developed at its anterior end. In most cases the planula, after leading this free-swimming existence for a while, attaches to some substrate by its anterior end. At the beginning of its metamorphosis it loses its cilia, and *pedal discs* or *stolons* are formed. At the opposite end tentacles develop, a mouth opens and a hydranth is produced. The middle part becomes the stalk, consisting of the *coenosarc,* which secretes the chitinous perisarc around itself. The stalk gives off branches, while the stolon may form a *hydrorhiza,* leading to the development of a hydroid colony. In the colonial hydroids, there is specialization among the individuals making up the colony, nutritive polyps being formed first, and later those with reproductive and other specialized functions.

In some species the embryo is retained within the gonophore while it develops to the planula stage. In other cases the brooding may last even longer, resulting in the complete elimination of the planula stage from the developmental process. *Tubularia* is such an example — no planula is formed, the embryo within the gonophore developing directly into an actinula. This larva has a mouth, surrounded by tentacles, and constitutes a tiny polyp. Furthermore, it is known that in such genera as *Orthophrix* and *Sertularia,* the embryos are reared in special brood sacs *(marsupia)* formed by gonothecae, which contain masses of a gelatinous substance.

(5) GENERAL DEVELOPMENT OF TRACHYLINA

The order Trachylina includes the two suborders, Trachomedusae *(Aglantha, Liriope)* and Narcomedusae *(Solmaris, Cunoctantha, Solmundella)*. Both of these groups include species which do not form polyp colonies, and in most of them there is no polyp generation at all, the whole life cycle consisting solely of swimming medusae, although among the Narcomedusae there are some species which pass a larval stage as parasites of other medusae. In a few cases the polyp form is represented during the developmental process by a briefly appearing actinula larval stage, but with these minor exceptions there is nothing which can be called alternation of generations.

The reproductive cells of the Trachylina are also ectodermal in origin; in the Trachomedusae the gonads develop on the upper part of the radial canals, while in the Narcomedusae they are formed on the floor of the stomach. The

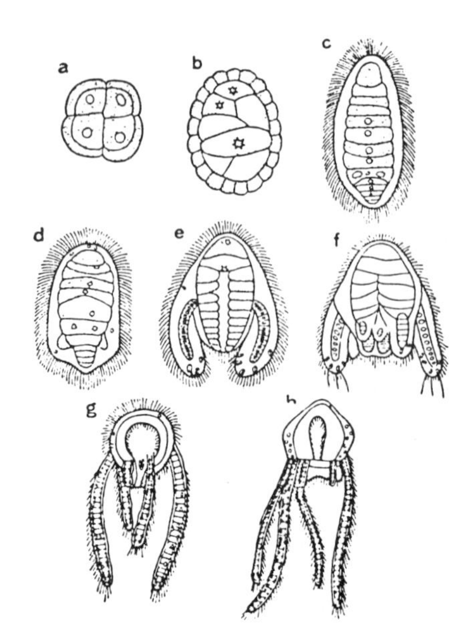

Fig. 3.7 Development of *Aglaura hemistoma* (From Uchida, after Metschnikoff)

a. 8-cell stage; b. formation of endoderm and ectoderm; c. planula; d. early stage of tentacle development; e. formation of gastric cavity; f. increase in number of tentacles; g, h. formation of young medusa.

eggs of the Trachomedusae are generally small, but in the Narcomedusae they are large. As in the Hydroida, the cleavage pattern varies considerably according to the species. The blastula may also be either a coelo- or sterroblastula; in the coeloblastulae, endoderm formation takes place by multipolar proliferation, while the sterroblastulae form endoderm by the delamination method. A planula is formed at the time of gastrulation, and this is followed by the secretion of a large amount of mesogloea. A mouth opens at one end of the planula, tentacles are formed there, and after passing through an actinula phase, the larva gradually metamorphoses into a medusa.

Among the Trachymedusae, the mode of development of *Aglaura* and *Liriope* is known; in *Aglaura* (Fig. 3.7) cleavage is total and unequal or equal, leading to the formation of a morula, which in turn becomes a ciliated planula. Endodermal cells at the posterior end of this larva protrude outward (Fig. 3.7d), forming the cores of the tentacles. The number of tentacles gradually increases, and the cilia are lost as the bell of the medusa takes shape. The development of *Liriope* shows certain differences, cleavage being total and practically equal, producing a coeloblastula which forms its endoderm by a process of delamination. The planula and actinula stages do not occur, however, and mesogloea is abruptly laid down between the ectoderm and endoderm. The large size of the endoderm cells causes the small ectoderm cells to be stretched into a thin layer covering the outer surface. This embryo eventually forms the tentacles, velum, manubrium and other organs characteristic of the adult medusa.

Of the Narcomedusae, the developmental process has been followed in *Solmundella, Cunoctantha* and others. *Solmundella* is not parasitic, and its life

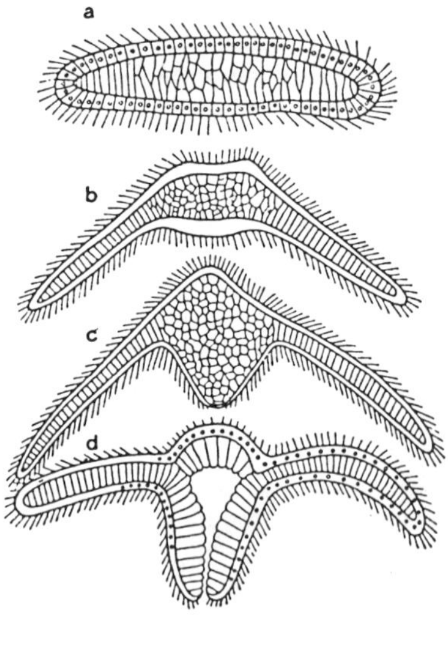

Fig. 3.8 Development of *Solmundella* (Metschnikoff)

history does not include alteration of generations (Fig. 3.8). Its cleavage is total, and the sterroblastula forms its endoderm by a process of multipolar proliferation. The embryo then develops cilia and becomes a planula, both extremities of which lengthen and form tentacles, while the central region thickens. This later becomes the stomach, with a mouth opening at its lower side (Fig. 3.8d). This larva gradually acquires mesogloea, sense organs are formed, and it becomes an adult medusa.

However, even among the Narcomedusae, the genus *Cunoctantha* has a rather complicated developmental history. It forms a planula, which in turn develops into an actinula-type larva with tentacles and a mouth but lacking sensory organs and a bell. This larva is parasitic in the stomach cavity or water vascular system of medusae of its own or other species, where it produces more actinulae like itself by asexual budding. All these actinulae transform finally into medusae — in other words, this animal combines a parasitic way of life with alteration of generations.

(6) GENERAL DEVELOPMENT OF SIPHONOPHORA

The Siphonophora are medusae, adapted for a pelagic life, which form highly polymorphic colonies by budding. Among the various individuals making up the colony, those which develop reproductive organs constitute the gonophores. Most of these reach a degree of development similar to the sporosacs of the Hydrozoa, the gonads forming from the part corresponding to the manubrium. The gonads of a single gonophore are either male or female, but the colony as a whole may have either gonophores of all one sex, or of both sexes, depending on the species.

In general the eggs cleave totally and equally to form a morula; gastrulation usually takes place by delamination. The sterrogastrula develops cilia on its outer surface and forms a planula, the shape of which may be spherical, oval, or some modification characteristic of its species. Development after the planula stage varies to some extent from one species to another. Among the suborder Calyconecta, including the genera *Muggiaea, Diphyes, Hippopodius* and others, the planula is solid and ovoid, with a ciliated surface. This ovoid planula buds from one side of its upper part a *bell nucleus,* which will become the under surface of the *primary swimming bell.* Tentacles form on the lower part of this, and at the opposite (lower) end of the planula, the *primary gastrozooid* is formed. When the primary swimming bell is completely developed, a stem grows out between bell and gastrozooid and buds extensively to produce the various other individuals, eventually forming a large colony.

In the suborder Physonectae *(Agalma, Physophora),* the ectoderm cells on the upper side of the ciliated planula increase in size and form a projecting structure like an umbrella; this becomes the *pneumatophore,* while tentacles arise from the side of the larva. A mouth opens at the lower side and the primary gastrozooid is formed. However, in *Agalma* a *protective bract* is formed, before the pneumatophore has developed, and this performs the function of keeping the larva afloat. Eventually the pneumatophore is formed beneath the bract, which then degenerates.

In the suborder Cystonectae *(Rhizophysa, Physalea,* etc.), the development seems to be generally similar to that of the Physonectae; the pneumatophore develops to a still larger size and is separated from the gastrozooid by a septum, and no swimming bells are formed. The early development of *Rhizophysa* and also of the Physonectae appears to take place in deep water, and its details are unknown.

Finally, the development of such forms as *Velella* and *Porpita*, which belong to the suborder Disconectae, shows rather wide divergence from the foregoing. The youngest known larva of these species is called a *conaria*. This is a hollow, nearly spherical larva with a slightly thickened area on its upper side, which later becomes the pneumatophore, while the nearly spherical portion forms the primary gastrozooid. The conaria is said to sink into deep water where it continues to develop, and only reappears at the surface after reaching the next stage, which is called a *rataria*. The rataria larva metamorphoses and gradually builds up the complex structure of the adult siphonophore.

In several species belonging to the Calyconecta, some of the units making up the colony, grouped on a single stem as *cormidia*, may separate from the main colony and take up an independent existence, during which their gonads reach sexual maturity. These units, called *eudoxids*, carry out sexual reproduction, the fertilized eggs giving rise to colonies like the original one. In other words, in these species an ordinary phase which carries on asexual reproduction is succeeded by the eudoxid phase which reproduces sexually, thus accomplishing a sort of alternation of generations.

3. SCYPHOZOA

Among the Scyphozoa, as in the Hydrozoa, in general there are both polypoid and medusoid types, but the polypoid generation is insignificant in compa-

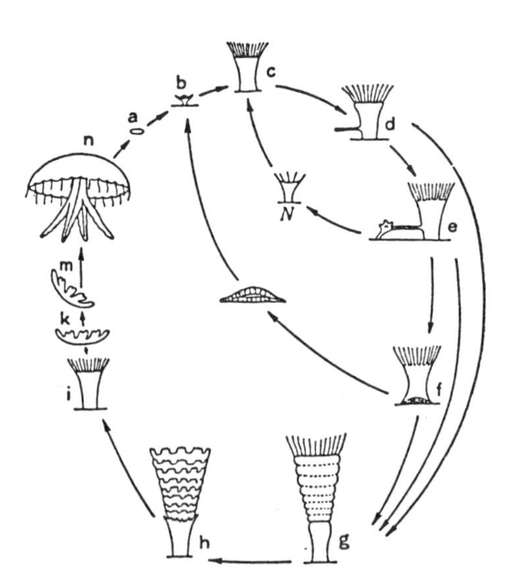

Fig. 3.9 Life-history of *Aurelia aurita* (Hêrouard)

a. planula; b~f. scyphopolyp, budding of scyphopolyp; g~m. liberation of ephyra by strobilation of scyphopolyp; n. medusa.

rison with the large and conspicuously well developed medusa. It is well known that these two types appear in turn, to establish a general alternation of generations. The Scyphozoa are classified into five orders: 1) Stauromedusae, 2) Cubomedusae, 3) Coronatae, 4) Semaeostomae, 5) Rhizostomae. Of these, the Stautomedusae are sessile, in contrast to the Cubomedusae which are the most highly adapted for a pelagic life, but these two orders exhibit many morphological as well as embryological differences from the other three orders, and at the same time have a good many characteristics in common. The developmental history of the Cubomedusae has still not been thoroughly investigated, but it is known that no ephyra is formed, and therefore alternation of generations cannot be taking place. The Coronatae, Semaeostomae and Rhizostomae, on the other hand, are alike in having an ephyra stage, and generally exhibiting alternate sexual and asexual generations. With respect to the embryology of these three orders, there are some doubtful points in the case of the deep-sea Coronatae, but the Semaeostomae and Rhizostomae have many aspects in common, although the metamorphosis of the rhizostome medusa is especially complex.

Morphologically speaking, the Stauromedusae are medusoid in their upper part and polyploid in their lower part; since they do not give off a free-living medusa, they have no phase of asexual reproduction, and therefore, as suggested above, no alternation of generations. Cubomedusan development is known to the point at which a polyp is produced, but the manner of medusa formation still remains undiscovered. In the other three orders the usual alternation of generations can be observed; as this takes place regularly, the fertilized egg gives rise to a planula, which becomes attached, as a *scyphopolyp*. This polyp develops into a strobila, which produces, by an asexual process called *strobilization*, a series of ephyra larvae. The adult form is achieved by metamorphosis of this ephyra. Some wholly pelagic species like the semaeostome *Pelagia*, however, make an exception to this general rule by omitting the scyphopolyp stage; in these forms the planula develops directly into an ephyra larva.

The reproductive cells of the Scyphozoa are endodermic in origin. Except for the Stauromedusae, the course of development followed by the fertilized egg is in general the same for all the orders. Cleavage is total and radial, the 8-cell stage showing some spiraling tendency. The first two cleavages are usually equal or nearly equal; as cleavage proceeds a morula is formed, and then a coeloblastula. In some cases a certain amount of yolk may be left in the blastocoel. Endoderm formation usually takes place by invagination, but polarization and epiboly also occur in some species. The gastrula eventually becomes a planula, which swims about for a while and then attaches and becomes a scyphopolyp which develops a pedal disc, a mouth opening, and two tentacles that progressively become four, eight and finally sixteen. At a later stage, septa and subumbrellar funnels are formed. As the scyphopolyp develops, one or more lateral grooves are formed; above each groove eight pairs of *lappets* arise, the tentacles retrogress, and eight sensory organs *(rhopalia)*, one between each member of the paired lappets, complete the characteristic form of the ephyra. The grooves gradually deepen until the ephyra is completely separated as an independent unit. During this ephyra larva's free-swimming existence it acquires tentacles on each body radius, the gastrovascular system gradually takes on a complex structure, and finally transforms, with a further increase in number of tentacles and other lobes and appendages, into the adult animal. An ephyra which has growth to some extent and is about to metamorphose is known as a *metephyra*.

(1) DEVELOPMENT OF *THAUMATOSCYPHUS DISTINCTUS*

As the Stauromedusae are largely cold water species, in this country most of them are to be found in the waters off Northeast Honshu and Hokkaido, except for such forms as *Stenocyphus, Haliclystus* and *Kishinoueya,* which occur along the shores of Central Honshu. *Thaumatoscyphus,* which is also a cold water medusa, is very common off Hokkaido at Muroran, Akkeshi, Abashiri, etc., from June to September, living attached to *Zostera* and other marine algae. The development of this species has been studied in this country only; in summary it is as follows.

Thaumatoscyphus is dioecious, one pair of gonads lying along each interradius so that there is a total of eight. The animals are sexually mature during July and August, both males and females shedding their gametes into the sea, where fertilization takes place. If several sexually mature male and female specimens are left together overnight in a jar of sea water, during the night the gametes are expelled, by the contractions of the umbrella, through the gastrovascular cavity to the outside by of way the mouth; fertilization takes place as soon as they come together. The spermatozoa are small, with triangular heads and tails 70—80μ in length. The pale yellow eggs are also very small, about 50μ in diameter, and contain microscopically visible yolk granules. The egg pronucleus always lies close to one pole of the egg. After fertilization the egg is surrounded by a single, rather inconspicuous membrane. Two or three hours after fertilization the egg begins to cleave, totally and about equally. The second cleavage always occurs perpendicularly to the first. While both the blastomeres formed at the first cleavage divide at the second cleavage, they usually do not do so simultaneously; in most cases there is a considerable difference between their cleavage times. The third cleavage occurs horizontally, a little above the equator of the embryo. The four smaller blastomeres of the upper half turn horizontally to a slight extent (Fig. 3.10a), so that they tend to slip between the four lower blastomeres. Until this stage the cleavage has proceeded with radial symmetry; after this, while it is still practically equal and total, it gradually loses its regularity (Fig. 3.10c) and the outline of the blastomeres becomes indistinct. The embryo becomes a morula, but instead of going on to form a blastula, the interior is filled with cells which are about the same as those of the outer layer (Fig. 3.10d). Endoderm formation begins, at about the 16—32-cell stage, by a process of unipolar proliferation; i.e., the outer layer of one part of the morula gradually moves toward the interior, giving rise to two cell layers (Fig. 3.10e). The cells which have moved inward also divide in their new location. The endoderm is formed in this fashion, but since this embryo has

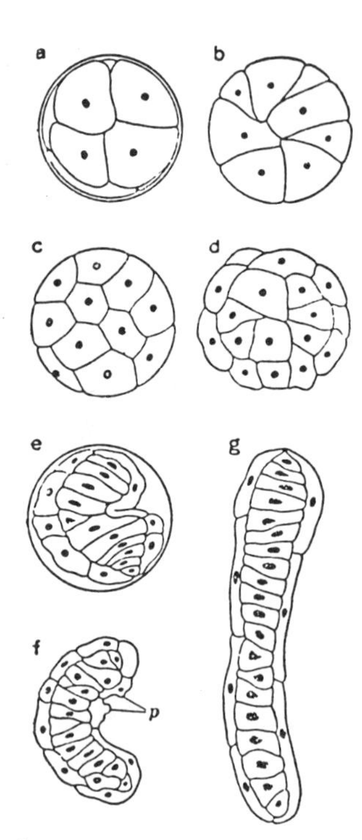

Fig. 3.10 Development of the scyphomedusa *Thaumatoscyphus distinctus* (Hanaoka)

a. 8-cell stage, micromeres showing slight horizontal rotation; b. 16-cell stage; c, d. more advanced stages; e. young planula within fertilization membrane; f. planula with ectodermal process (*p*); g. extended planula.

no blastocoel, no blastopore is formed — it is more as though the ectodermal layer is enclosing the mass of endoderm cells. There are about ten cells in this mass at first; the number increases somewhat as development proceeds.

After endoderm differentiation has taken place, the ectoderm cells flatten and become arranged in a layer around the outside of the embryo, while the endoderm cells, now rather large and vacuolated, line up longitudinally in the interior. At this stage the embryo breaks out of the thin fertilization membrane and becomes a long, slender, worm-like planula (Fig. 3.10f,g). This stage begins about 30—40 hours after fertilization. The planula of *Thaumatoscyphus* thus has an ectodermal layer composed of a few flattened cells; its endoderm cells are also more or less flattened, with the nuclei in the centers of the cells surrounded by a thicker mass of protoplasm than is found in the other parts of the cells. This larva differs from other planulae in having no blastocoel and also no trace of ciliation, so that it cannot swim, and creeps about the bottom with a vermiform type of locomotion. This creeping movement is performed by alternating contraction and expansion of both ectoderm and endoderm; when the ectoderm contracts, small protoplasmic processes can be seen to protrude from the body surface (Fig. 3.10f). Such planulae are often found collected into a mass, which is usually surrounded by a sort of shell made of sand grains and bits of miscellaneous debris.

The development of *Thaumatoscyphus* has been observed only to this extent, but information is available about later stages in other Stauromedusae such as *Haliclystus, Sasakiella,* etc. When the planula finally attaches it takes on a hemispherical shape, and before transforming reproduces asexually by budding. The planulae produced in this way grow and form other hemispherical cell masses which finally transform into polyps. Ordinarily the Stauromendusae do not reproduce asexually, but this one case is an exception. The polyp eventually forms a *hypostome, pedal disc,* tentacles, *anchors, infundibulum, gastral filaments,* etc.; new tentacles are subsequently formed in a regular arrangement at each perradius. In some species (i.e., *Kishinouyea*), the rhopaloid bodies (anchors) are not present in the adult, while in others (*Sasakiella*), they remain as tentacle-like projections.

The foregoing is a general description of the course of development in the order Stauromedusae, but actual observations are limited to a rather small number of cases. For example, only the early development of *Haliclystus* and *Thaumatoscyphus,* and the later stages of *Haliclystus* and *Sasakiella* have been studied. To summarize the points in which the Stauromedusae differ radically from the four other orders: they do not form a blastocoel, the planula is non-ciliate and therefore non-swimming, resorting instead to a vermiform creeping type of locomotion. The planula transforms into a scyphopolyp, which in turn becomes a single young medusa. No alternation of generations, therefore, occurs in the stauromedusan life cycle.

It has already been stated that these Stauromedusae have a number of characteristics in common with the Cubomedusae; the development of the latter, however, is not very well known. In the case of the cubomedusan *Charybdea,* the planula and scyphopolyp have been observed, but the interval between this scyphopolyp and the medusa remains unaccounted for, and it is consequently not known whether asexual reproduction occurs — in other words, whether or not there is an alternation of generations, although judging from the morphology of the young medusa, it is believed likely that no ephyra stage exists.

Observations made on *Charybdea* show that fertilization takes place in the gastrovascular cavity, and cleavage is total and equal and follows about the same sequence as that found in the other scyphomedusae, as far as the fourth cleavage. A morula stage is succeeded by the formation of a coeloblastula, the blastocoel of which is presently filled with endoderm cells by means of multipolar proliferation or delamination, and the embryo becomes a gastrula. This gastrula develops cilia and changes into a small, pear-shaped planula with pigment granules in its central part (Fig. 3.11a,b). After swimming for two or three days it becomes sessile. It then loses its cilia and becomes a plate-like mass of cells; from this mass a pair of tentacles arise (Fig. 3.11c,d). As the tentacles elongate the body also

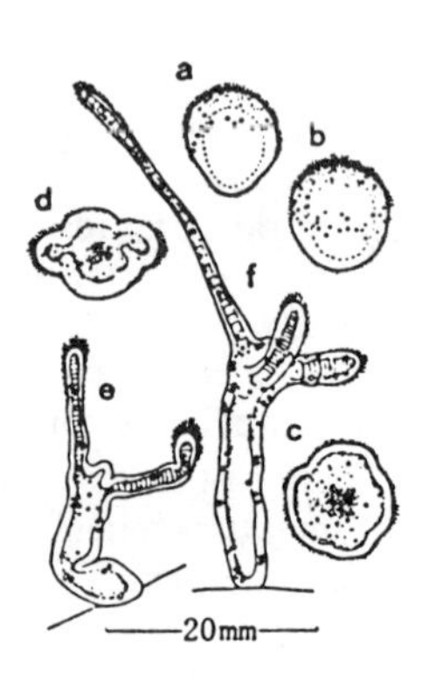

Fig. 3.11 Early development of the scyphomedusa *Charybdea* (Okada)

a, b. successive stages of planula; c. attached planula; d. formation of two primordial tentacles; e, f. formation of scyphopolyp.

Fig. 3.12 Ephyra of *Nausithoë punctata* resulting from strobilation of the giant polyp "*Stephanoscyphus*" (Komai)

gf: gastric filament, *lb:* marginal lappet.

increases in length, a small hypostome is formed, and two new tentacles make their appearance (Fig. 3.11e, f). This is about all that is known at present of the development of this species.

(2) DEVELOPMENT OF *NAUSITHOË PUNCTATA*

The medusae belonging to the order Coronatae, which includes *Nausithoë*, are morphologically closest to the ephyra larva of the Scyphomedusae. Many of these medusa are deep-water species and their embryology is almost completely unknown, but a rather detailed study has been made of the development of *Nausithoë*, which is a coastal form. The eggs are generally small, and their cleavage is typically radial, equal and total, passing through regular 2-, 4-, 8-, 16- and 32-cell stages to form a morula and then a blastula. There are two types of gastrulation: endoderm formation may take place either by invagination of the vegetal pole or by irregular, monopolar proliferation at this side of the embryo. The gastrula develops cilia and becomes a planula, and begins to swim after escaping from its membrane. The planula of *Nausithoë* attaches by a plate-like cell mass on its under side, loses its cilia, and separates into an inner and outer cell layer. A hard transparent layer is secreted over the surface, and later a mouth opens and tentacles are formed. Unlike the other Scyphomedusae, this species has an extremely well developed polyp, which is known by the name of *Stephanoscyphus*. This polyp reaches a height of about 10 centimeters, and the transverse constrictions of strobilation begin to appear simultaneously in a number of places, finally cutting loose the ephyrae, which begin a free-swimming life. The ephyra of *Nausithoë* differs from those of the Semaeostomae and the Rhizostomae in the symmetry of its lappets (cf. Fig.

3.12 with Figs. 3.15 and 3.16). At first there are no tentacles on the adradii, but these appear later; gastral filaments first occur singly on the interradii and gradually increase in number; and finally one gonad is formed on each adradius. There are slight variations in the rhodalia and gastrovascular filaments, but nothing especially striking.

(3) DEVELOPMENT OF *AURELIA AURITA*

This Scyphomedusa is widely distributed throughout the temperate zone waters of the Pacific and Atlantic Oceans and the Mediterranean Sea, and has been the object of much morphological and embryological study in both Europe and America. In this country also it is found in large numbers along both the Pacific and Japan Sea coasts of Hokkaido in the neighborhood of Oshoro, and is one of the most commonly encountered Scyphomedusae.

Aurelia is dioecious; the gonads are purplish in the young medusae, but turn brown in older animals. They develop on the four interradii, each being much folded, and forming the shape of a horseshoe pointing inward. Under the gonads are the *subgenital pits,* the lower sides of which are thin and rupture during the breeding season, making a connection with the outside. The spermatozoa are shed by the males through these apertures into the sea water; some of them find their way into the gastral cavities of the females where the eggs are fertilized, and where they develop to the planula stage. They then leave through the mouth opening, although in some cases they remain for a while longer attached to the peripheries of the oral arms, lappet tentacles and other appendages.

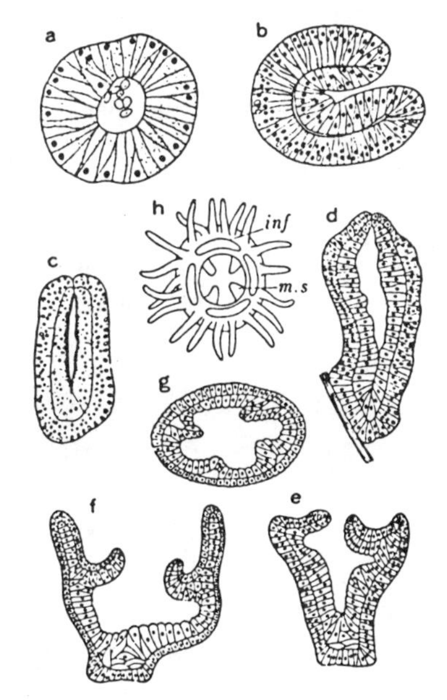

Fig. 3.13 Development of *Aurelia aurita* (Hein, Friedmann)

a. blastula; b. germ layer formation; c, d. stages in transformation of attached polyp; e, f. formation of gastric cavity and oral cone; g. formation of mesenteries; h. scyphopolyp. *inf:* infundibular duct, *m.s:* mesentery.

The fertilized egg begins to cleave within the gastral cavity. The cleavage of *Aurelia* takes place in a rather regular fashion: the cleavage furrow starts from the animal pole, and the first two cleavage planes are vertical, while the third is horizontal, all three cleavages being total and practically equal, although the 8-cell stage is more or less spiral. Cleavage leads to the formation of a spherical blastula; this contains a blastocoel which is small at first, but becomes rather spacious as cleavage proceeds and the blastula increases in size. A considerable number of cells migrate from the blastular wall into the interior so that the blastocoel becomes almost full of them, and at first glance the embryo looks like a solid blastula. These cells which have moved into the blastocoel, however, disintegrate before long and are eventually resorbed by the cells of the blastular wall.

Germ layer formation in *Aurelia* is accomplished by a typical invagination of the vegetal pole cells. As a result, the hitherto single-layered blastula becomes a hollow gastrula with a two-layered body wall made up of ectoderm and endoderm (Fig. 3.13a,b). A small pore remains open where the vegetal wall invaginated; this is the blastopore, which persists without being completely closed, although it becomes very narrow and is finally reduced to a slender tube (Fig. 3.13c). During this stage the yolk granules contained in the cells begin to be absorbed, and the spherical gastrula gradually becomes oval, with the somewhat wider anterior end distinguishable from the more pointed posterior end. The ectoderm acquires cilia at this time, and the gastrula becomes a planula and swims away from the body of the adult.

The planula has a small blastocoel; its ectoderm gives rise to nematocysts, sensory cells and gland cells, while its endoderm cells contain large vacuoles. The planula swims about for 4—5 days, and then loses its cilia, attaches at the bottom by its broad end, and forms a pedal disc, the ectoderm cells of which secrete an adhesive substance that holds it to the substratum. This attached planula becomes somewhat flattened, so that its originally cylindrical form becomes compressed to that of a cup (Fig. 3.13d,e). The endoderm cells around the mouth proliferate at a particularly rapid rate and form a hypostome, and the mouth opens in the center of this structure (Fig. 3.13f). Eventually two small projections form at opposite points on two sides of the mouth; these are the first two tentacles. Two more tentacles are then formed perpendicular to the first pair. These tentacles are solid, composed of ectoderm cells including many nematocysts. At this time, also, endodermal projections, formed at positions alternating with the four tentacles, become mesenteries *(septa)* which extend into the gastral cavity (Fig. 3.13g,h). From the upper parts of these mesenteries, the ectoderm gives rise to muscular bands which form the longitudinal *septal muscles,* extending to the base of the larva. The animal which develops in this manner is called a *scyphopolyp* or *scyphistoma* (Fig. 3.14a).

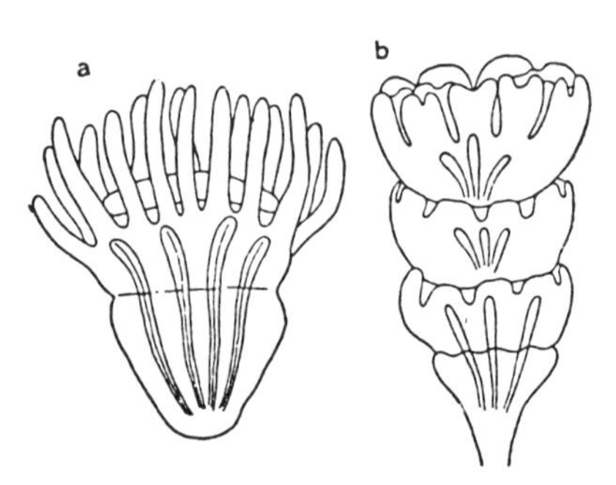

Fig. 3.14 Strobilization of *Aurelia aurita* (Claus)

a. scyphopolyp; b. strobila.

Alternating with each of its four tentacles, four new tentacles are formed, making a total of eight. These later increase to 16, and in some cases to 20 or 24. The nematocysts borne on the tentacles proliferate and become clumped into many small, wart-like projections.

The scyphistoma of this stage feeds and increases in size, reproducing asexually by budding, and also forming new polyps by sending out hollow stolons from its base. In other words, the scyphistoma asexually produces more scyphistomae. Other changes also take place in the interior of the scyphistoma. The mouth, after changing to a hypostome, further develops into a manubrium; the margins of the septa thicken and give rise in the gastral cavity to the rudiments of the *gastral filaments.* On the perradii and interradii of the oral end, paired marginal lappets are formed, with the rudiments of the modified sense organs (rhopalia) lying in the cleft between the members of each pair. At this time a number of transverse constrictions become apparent in the long, slender, tubular polyp. These constrictions gradually become deeper, giving the polyp the appearance of a stack of inverted plates as it enters the strobila phase (Fig. 3.14b).

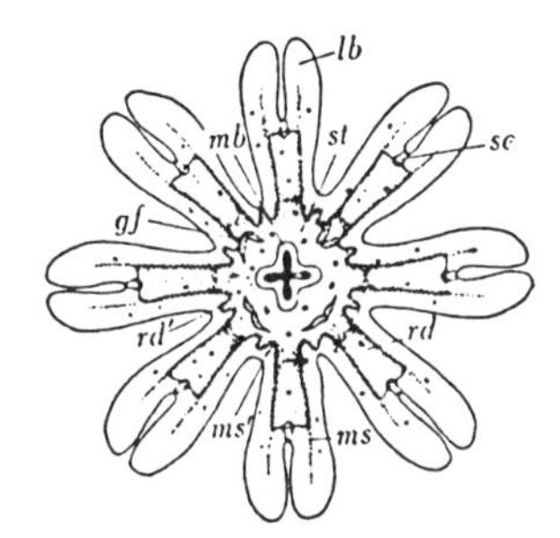

Fig. 3.15 Ephyra larva of *Aurelia aurita* shortly after liberation

gf: gastral filament, *lb*: marginal lappet, *mb*: manubrium, *ms*: radial muscle, *ms'*: circular muscle, *rd*: radial sac, *rd'*: perradial sac, *se*: sensory organ, *st*: stomach.

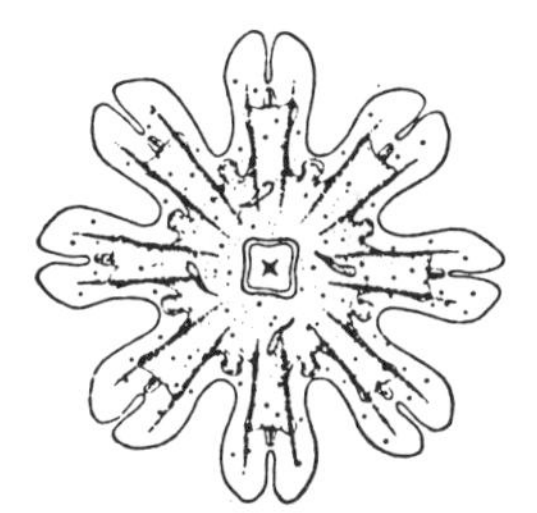

Fig. 3.16 Ephyra larva of *Mastigias papua* immediately after liberation (Uchida)

Of this pile of plate-like structures, the topmost one has undergone the changes in its oral surface described above, but in a descending order the others become smaller and their degree of modification less. The uppermost plate, provided with marginal lappets and rhopalia, is finally separated from the strobila by the process of strobilation. This separated unit is an ephyra, and as soon as one has been cut loose, the next in line increases in size, becomes modified structurally, and is cut off in its turn. (Fig. 3.15, 3.16).

The newly released ephyra is a pale reddish-brown, flower-shaped animal, 0.3—0.4 mm in diameter, with eight pairs of marginal lappets, and between each member of the lappet pairs, a rhopalium which becomes a statocyst. In the center of the subumbrella there is a short manubrium, ending in a mouth opening with four lips. A circular (coronal) muscle and radial muscle bands are developed. The *gastrovascular tract,* extending inward from the *mouth,* is formed of a *gullet, gastral cavity* and eight radial sacs. The number of gastral filaments within the gastral cavity also increases. This ephyra swims free in the sea, and is regularly found in spring plankton hauls.

The ephyra finally transforms into an adult *Aurelia.* In the process, new tentacles are formed on each adradius, and their number continues to increase; additional marginal lappets are also formed in the adradial regions. The manubrium develops long *oral arms*, and the subgenital pits, gastrovascular system, gonads and other structures increase in size. The gastrovascular system expands by the formation of adradial canals, which arise as projections from eight perradial sacs. At the time of its release the ephyra is flower-shaped, but it loses its color and becomes almost circular with the development of adradial tissue, and the great increase in the number of lappets makes individual ones less conspicuous. The young medusa which has completed this transformation is still only about 1 cm in diameter; from this point on, it increases steadily in size, while the radial canals become more complicated, the number of gastral filaments increases, and the oral arms develop elaborate frills and folds.

In addition to *Aurelia*, the semaeostomes *Dactylometra, Cyanea, Chrysaora* and *Pelagia* are among the Scyphozoa in which the embryology has been most thoroughly investigated. In general, their development is closely similar to that of *Aurelia*. *Pelagia* alone differs in that its planula develops directly into an ephyra, without alternation of generations. This planula of *Pelagia* is ciliated, but a small mouth forms on its lower side and this region gives rise to endoderm tissue which invaginates into the interior and forms a large gastral cavity. Eight lappets are first formed on the lower side, these soon become eight pairs, and rhopalia develop in the spaces between them. By this process the larva changes directly into an ephyra without passing through a scyphistoma stage.

As has been stated above, there are many points of similarity in the developmental histories of the Semaeostomae and the Rhizostomae; among the members of the latter order, the species *Mastigias papua*, which is found in Japan, has been particularly well investigated. The early development of this species closely resembles that of *Aurelia*: its cleavage is total and practically equal, and it forms a coeloblastula with, however, a small amount of yolk left in the blastocoel. This blastula develops into a gastrula by invagination. The gastrula eventually becomes a planula, which spends a few days swimming about and then attaches and changes into a scyphistoma. This polyp gradually increases the number of its tentacles from an original two to four and then eight. It becomes a strobila, and although the process by which the ephyra is produced has not been observed in this species, it is known that in *Cassiopea* and other rhizostomes, only one ephyra forms at a time (Fig. 3.17), and there is no stack of discs like that found in *Aurelia*. Like

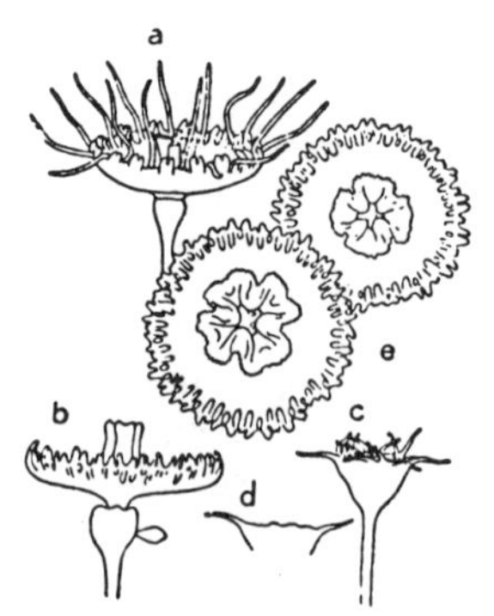

Fig. 3.17 Stages in the development of the scyphomedusa *Cassiopea* (Bigelow)
a. scyphopolyp; b. budding of planula from scyphopolyp; c, d. further development of tentacle after liberation of ephyra; e. ephyra.

the ephyrae of the semaeostomes, the rhizostome ephyra has a distinct mouth (Fig. 3.16 and 3.17). The subsequent processes by which the ephyra transforms into the adult medusa, however, are highly characteristic, and quite different from those occurring in the semaeostomes. The most outstanding of these differences are the failure to form tentacles and the complicated development and transformation of the gastrovascular system and oral arms. In this process, the edges of the four lips around the mouth opening spread horizontally, developing a fringe of small tentacles (Fig. 3.18 a,b). Each adradial portion becomes deeply indented, while

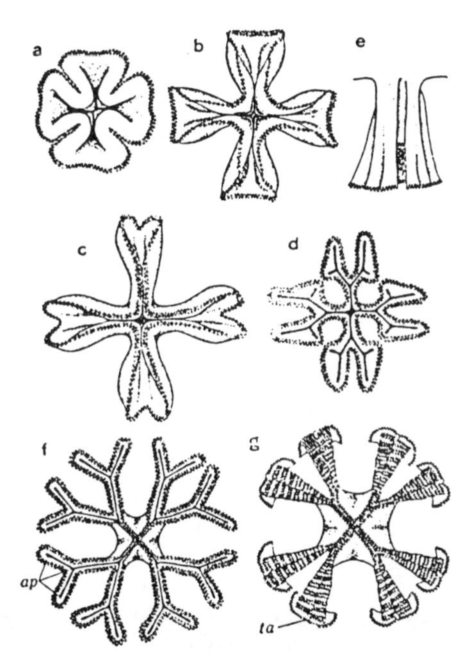

Fig. 3.18 Process of manubrium formation in *Mastigias papua* (Uchida)
ap: opening of oral arm, *ta*: rhopalium.

shallow constructions appear in the perraddii (Fig. 3.18c). These arms later bifurcate again (Fig. 3.18f), acquiring an extremely complex structure as the bifurcation process is repeated many times. During this elaboration of the manubrium, the central mouth is closed off and many small *suctorial mouths* appear among the fringes of the oral arms, connected by *brachial canals* which are subdivisions of the gastrovascular cavity formed by the repeated bifurcations of the oral arms. In *Mastigias* and similar species, the manubrium consists of four *stomach pillars* supporting eight oral arms; each oral arm has two *wings* and bears an elongated, club-shaped *appendage*. Later in development the gastrovascular canals branch to form a complicated network.

4. ANTHOZOA

GENERAL DEVELOPMENT

Since the Anthozoa have no medusan phase, their life history, consisting solely of a polypoid generation, is relatively simpler than those of the other Coelenterata. The fertilized egg develops into a planula, which attaches and becomes a polyp. Tentacles are formed, septa become prominent, and asexually reproducing species in most cases form branching, plate-like or clumped colonies. The reproductive cells of the Anthozoa are of endodermal origin, and there are many species in which the eggs contain large amounts of yolk and undergo superficial cleavage so that it is difficult to observe their mode of germ layer formation. In species with only a small amount of yolk, cleavage is total and radial, although there are some species which show a temporary appearance of spiral cleavage in the 8-cell stage. In most cases germ layer formation takes place by invagination, some yolk usually remaining in the blastocoel. However, this yolk is mostly absorbed by the time the larva becomes a planula and begins its swimming life. This lasts for several days; there are some species which develop into colonies by budding or fission. In solitary forms, the septa become highly developed. Motile species, whether solitary or colonial, form chiefly muscular bodies, while the sessile species produce a calcareous or keratinoid *skeleton*. These skeletons are secreted by mesogloeal cells which arise as modified ectodermal cells and gradually migrate into the interior.

Among the Anthozoa, the order Actiniaria, comprising the sea anemones, has been most thoroughly studied, but since no detailed observations on Japanese species are available, the development of *Sagartia troglodytes*, a European species, will be described.

This sea anemone breeds every year in the beginning of August, when the temperature of the sea water reaches 15°. The sperm are shed into the sea water through pores in the tips of the tentacles or in the pedal disc of the males, and reach the females, which respond to this stimulus by moving toward the source of the spermatozoa until they are in contact with the male animals. The mature eggs, 0.08—0.09 mm in diameter, are fertilized as soon as they are spawned, and form a fertilization membrane. About an hour after fertilization the first cleavage takes place; this is equal and total, and the cleavage furrow begins at the animal pole and cuts through toward the vegetal pole. The second cleavage furrow is perpendicular to this, forming four blastomeres. The next furrow arises in the equatorial plane, dividing the embryo into four upper and four lower blastomeres; at this 8-cell stage a blastocoel has already formed, which becomes larger as cleavage proceeds. The embryo continues cleaving regularly with radial symmetry until it finally reaches the 128-cell stage, about four hours after fertilization. The conditions of cleavage and the mode of endoderm formation of this species are similar to those of other sea anemones. The blastula has a blastocoel which is not, however, extensive, and a rather large number of yolk cells fill one end of its narrow lumen. At this stage it still has no cilia, and like the cleaving eggs, lies motionless on the bottom. The endoderm is formed by invagination. When this process is practically complete, the ectoderm invaginates and forms the future *stomodaeum*. While the endoderm is invaginating, cilia develop on the body surface, and the planula begins to swim, about 15 hours after fertilization. After the beginning of the swimming stage, the body becomes longer and thinner, and a bunch of especially long cilia develops at the end opposite the mouth (Fig. 3.19d,e). The planula swims with this end forward, and later attaches with this same end. The possession of this bundle of long cilia is a characteristic feature of the sea anemone. In this stage, the so-called *Edwardsia stage,* the septa also become conspicuous. After six or seven days of life as a planula, the larva sinks to the bottom and becomes attached. The size of the body increases, and tentacles appear in a regular order; internally the number of septa increases and the gonads approach sexual maturity. In this species, two eggs sometimes fuse after being spawned, and develop as a single large embryo, with certain slight differences in the developmental process (Fig. 3.20b).

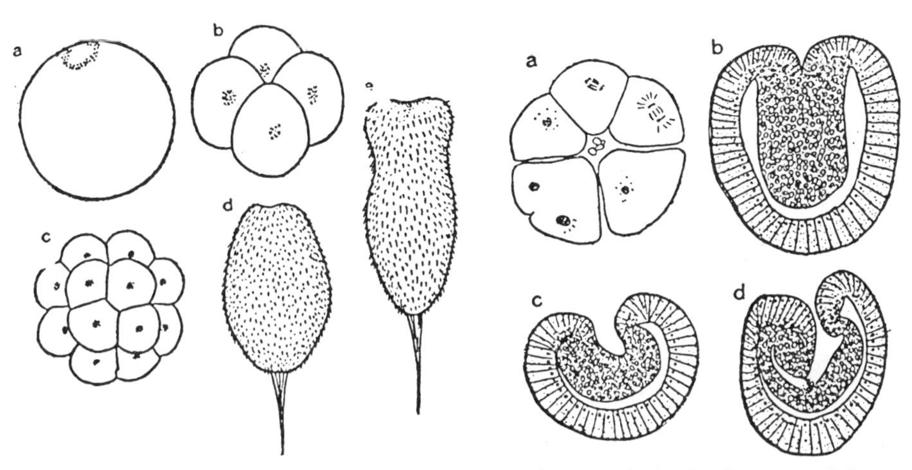

Fig. 3.19 Early development of the sea anemone *Sagartia troglodytes* (1) (Nyholm)

a. fertilized egg; fusion nucleus at animal pole; b. 4-cell stage; c. 16-cell stage; d, e. successive stages in formation of swimming planula

Fig. 3.20 Early development of *S. troglodytes* (2) (Nyholm)

a. cross-section of embryo at 16-cell stage; b. embryo derived from two fused eggs; c. embryo 8 hours after fertilization; endoderm beginning to form; d. free-swimming larva with completely formed oral canal.

In *Halcampa duodecimcirrata,* an elongated sea anemone which buries its body in the sand, the egg contains a large amount of yolk, and undergoes superficial cleavage (Fig. 3.12a,b). This species breeds in October and November on the coast of Sweden. The males first come out of the sand and shed their sperm; in response to this stimulus, the females also come to the surface of the sand and spawn. The eggs are shed from the mouth, and traveling along the tentacles, are dis-

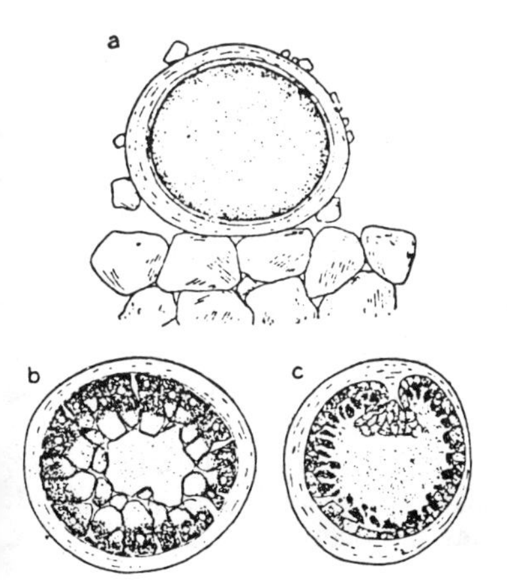

Fig. 3.21 Development of the sea anemone *Halcampa duodecimcirrata* (Nyholm)

a. fertilized egg with thick jelly layer; b. embryo at 25 hours; cleavage occurs superficially and yolk remains in center of egg; c. embryo at 35 hours; endoderm forming.

persed into the sea water. They are about 0.3 mm in diameter and opaque because of the large amount of yolk. In many cases, they have already been fertilized within the coelenteron. Such eggs have extruded the polar bodies, and are surrounded by a rather thick jelly layer to which adhere grains of sand and other particles (Fig. 3.21a). Under the fertilization membrane there is a thin peripheral cytoplasmic layer surrounding the endoplasm. The nucleus lies in this cortical layer and undergoes mitosis there, and the daughter nuclei collect in the peripheral

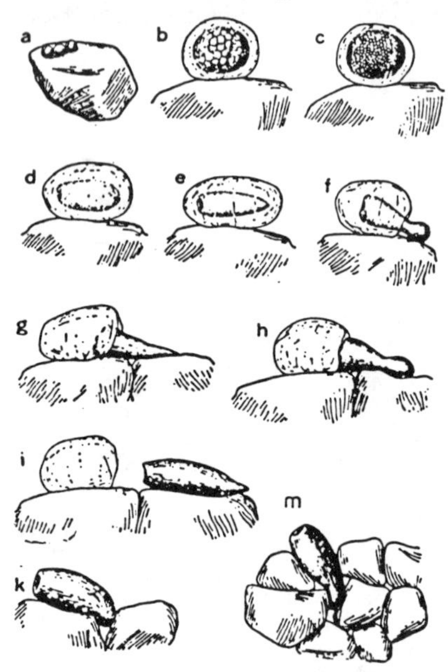

Fig. 3.22 Several sketches of *Halcampa duodecimcirrata*, showing the course of changes from the freshly laid eggs to the adult about to burrow into the sand (Nyholm)

layer, presenting an appearance like that of the superficial cleavage pattern o. arthropod eggs, while the interior cytoplasm is completely filled with yolk granules Endoderm formation takes place, about 35 hours after fertilization, by invagination of one pole into the interior, solidly filled though it is with yolk; the endoderm tissue spreads from the point of invagination into the interior. The embryo is still solid even after the endoderm has been completely formed (Fig. 3.21c), but a stomodaeum develops and the yolk is gradually absorbed. The septa are then formed, and as their outlines become distinct, the last of the yolk cells are disappearing. The shape of the larva at this stage is an elongated oval. The part which will become the oral plate flattens (Fig. 3.22d,e) and a mouth opens in the center of it, while the opposite end becomes pointed. The development to this point has taken place within the jelly case, but the larva now begins to move inside the case, which is finally torn as the larva emerges, pointed end foremost (Fig. 3.22f,g,h). Once free of the case, the larva immediately digs its way into the sand, and there continues its development. In this species the planula has no swimming stage whatsoever.

In some sea anemones the young develop attached to the outside of the adult, or in brood pockets or within the coelenteron. A species which thus broods its young, *Epiactis japonica*, is found in the northern part of this country. The adults of this species very often have small sea anemones attached to the outer surface of the body. The breeding season is from October to December; the species is hermaphroditic, producing eggs on the younger septa and sperm on the older ones. The sperm are shed first, and several hours later the eggs are shed. Sperm

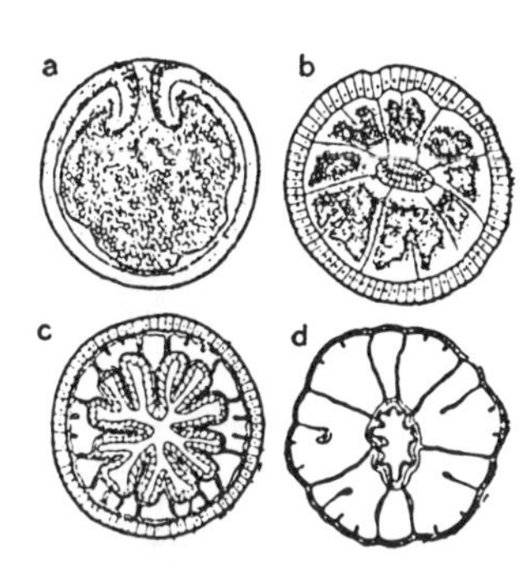

Fig. 3.23 Development of *Epiactis prolifera* (Uchida, Iwata)

a. gastrula stage; gullet is about to form, although yolk remains in interior of egg; b. septa have already developed, but yolk is not yet absorbed; c. oral canal and septa are formed and yolk has been absorbed; d. more advanced stage (late Edwardsia stage) (a. longitudinal section; b~d. cross-sections)

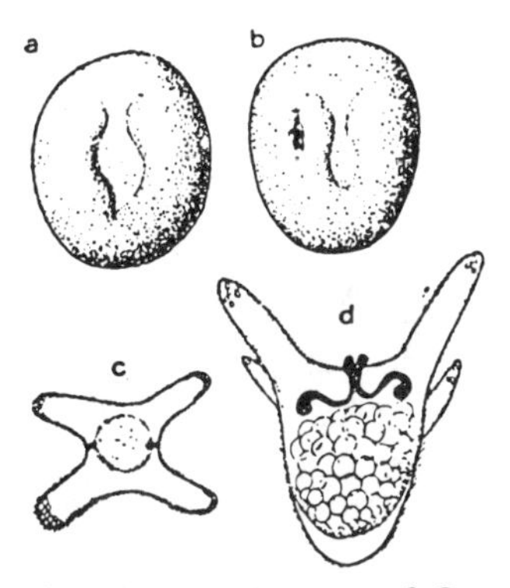

Fig. 3.24 Development of *Cerianthus lloydii* (Nyholm)

a, b. successive stage in gastrulation; c. arachnactis stage (from above); d. same stage (lateral view).

and eggs are expelled from the pharynx together with a whitish mucous secretion, as a result of rhythmical muscular contractions of the body wall. The spermatozoa have a head, midpiece and long tail; the length of the tail is 45μ, and of the other parts, 2.5μ. The spermatozoa will live for five days in sea water at a temperature of 7.5°. The eggs are reddish-brown in color, and are among the largest coelenterate eggs, with a diameter of about 1 mm. In general, species which develop without a free-swimming phase tend to have a small number of large eggs with a generous supply of yolk. The eggs which have been expelled from the pharynx along with the viscous secretion adhere to the body of the parent. Of course there are some which fail to stick and fall to the bottom. Where the eggs have adhered, the body wall tissue, containing many mucous glands, indents downward to form a pouch which envelops the eggs. Among the eggs developing under these conditions, the youngest which have been found are already in the gastrula stage, with a small blastocoel visible at one pole of the emrbyo (Fig. 3.23a). The eggs are apparently fertilized before being spawned, and they are surrounded by a thick gelatinous membrane. In the gastrula of this species, although the blastocoel is solidly filled with yolk, endoderm is formed and the septa develop, while a pharynx appears in the upper part. As long as the outlines of the septa remain vague, there is still yolk packed into one end of the blastocoel (Fig. 3.23a,b), but by the time the septa become distinctly defined, the yolk is completely absorbed (Fig. 3.23c,d). At first eight complete septa are formed; later 24 incomplete septa are added, making a total of 12 pairs. At the time when there are 12 septa, 12 tentacles are formed, and the young sea anemone has 24 tentacles when it leaves the parent and begins an independent existence.

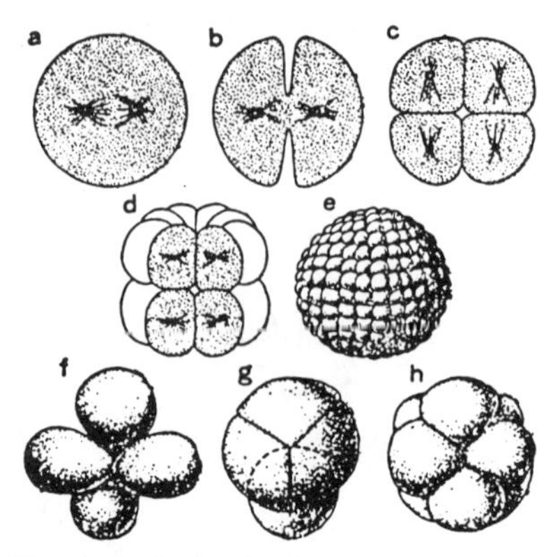

Fig. 3.25 Early development of *Pachycerianthus multiplicatus* (Nyholm)

a ~ c. from first cleavage to 4-cell stage; d. 16-cell stage; e. blastula; f, g. migration of one blastomere in 4-cell stage; h. embryo at 8-cell stage, showing spiral form.

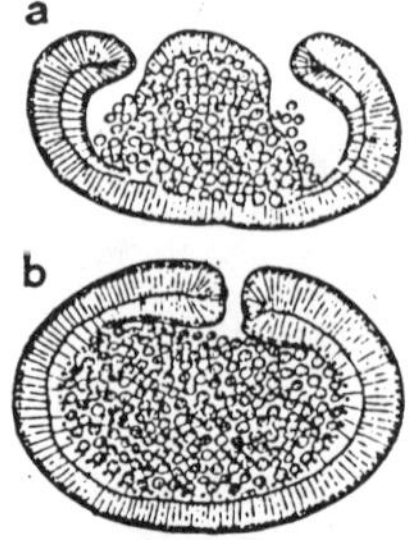

Fig.3.26 Successive stages in gastrulation of *P. multiplicatus* (Nyholm)

a. formation of endoderm by gastrulation; b. segmentation cavity filled with yolk.

We thus find that the development of sea anemones varies, depending on the amount of yolk in the eggs: cleavage may be equal or unequal, total or superficial, leading to the formation of a coelo- or sterroblastula; and endoderm formation, although most often accomplished by invagination, may sometimes occur as the result of epiboly, with some admixture of polarization. In eggs which cleave totally, equal cleavage is usual, although there are some which cleave unequally; the symmetry is radial, with a temporary tendency to resemble spiral cleavage appearing in some cases. (Eggs with relatively little yolk form a blastocoel, but very often some yolk cells remain in the the blastocoel until they are absorbed. Yolky eggs undergo superficial cleavage, forming a sterroblastula.) The planulae sometimes have a cluster of long cilia on the aboral end. In other cases they lack cilia entirely, transforming immediately into a polyp. After passing through an *Edwardsia* stage, characterized by the possession of eight complete septa, the larva enters the *Halcampula* stage, characterized by six pairs of septa. The numbers of tentacles and septa vary from one species to another.

The members of the order Ceriantharia outwardly resemble the sea anemones, but they differ considerably from them in internal structure and other characteristics. Actually the sea anemones are more closely related to the stony corals, while the Ceriantharia are closer to the black corals. It is only recently that the embryology of the Ceriantharia has been elucidated.

One northern European species of this order, *Cerianthus lloydii*, an hermaphrodite, breeds in February. The eggs have already extruded their polar bodies before being spawned; they are shed about an hour after the shedding of the sperm, and fertilization takes place externally. There is no visible change in the eggs after fertilization, so that they might be thought to be dead until invagination begins after 24 hours; by the third day they have become gastrulae. If these eggs are fixed during their early developmental stages, it can be seen that they undergo superficial cleavage and form sterroblastulae. A blastocoel appears and gradually becomes larger after the endoderm is formed, and a pharynx develops. Next the ectoderm and endoderm together extend out in four directions with the pharynx as center, forming a rectangular figure (Fig. 3.24c,d); these extensions are the rudiments of the *(marginal)* tentacles. After this two more *(oral)* tentacles appear at the sides of the mouth, making a total of six tentacles; internally septa are formed, and the so-called *Arachnactis* larva begins its pelagic life. The number and length of the tentacles increase, the body lengthens and the larva sinks to the bottom and burrows into the sand, where it constructs around itself a sheath of sand and slime.

Pachycerianthus multiplicatus, another sea anemone, has eggs which are yolky, but, unlike those of *Cerianthus,* undergo total, somewhat unequal cleavage. This species also is hermaphroditic, the sperm being shed first, and then the eggs. The polar bodies are already extruded before fertilization, and the egg does not form a special fertilization membrane. Cleavage is total and equal; since the cleavage furrow advances from both animal and vegetal poles, and the polar bodies have already been lost, it is imposible to differentiate between the two poles. The first cleavage occurs 15 or 20 minutes after fertilization; about 10 minutes later the second follows at right angles to it, forming four blastomeres. At this stage, a shifting in the relation of the blastomeres, like that seen in *Spirocodon,* takes place (Fig. 3.25f,g), so that one blastomere moves to a different level. This later returns to its original place and the next cleavage takes place, giving rise to an 8-cell stage which rather resembles that of spiral cleavage (Fig. 3.25h). Radial symmetry is restored, however, at the 16-cell stage, and cleavage continues regularly, forming a blastula (Fig. 3.25d,e). The center of this blastula is solidly filled with yolk, but it nevertheless gastrulates by invagination, one pole of the embryo becoming flattened and invaginated to form the endoderm (Fig. 3.26). The yolk which remains in the blastocoel is gradually absorbed, the body surface becomes ciliated; internally a pharynx and septa are formed and the larva begins to swim. The body lengthens, the oral end becoming flattened and even slightly concave (Fig. 3.27e,f,g) while the aboral end narrows. The larva swims with the aboral end forward, and in this position burrows into the sand with a spiral movement. The body continues to increase in length (Fig. 3.27i,k,m), tentacles are formed and their number increases gradually, the number of septa increases, and labial tentacles arise. This species is an example of the type which develops directly, without passing through an *Arachnactis* stage.

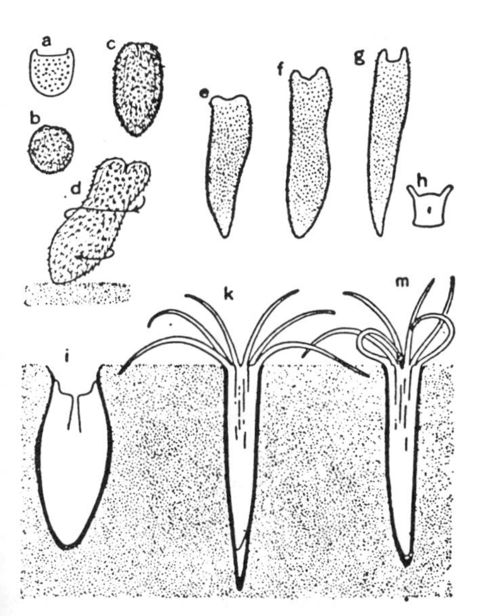

Fig. 3.27 Later development of *P. multiplicatus* (Nyholm)

So far nothing is known concerning the development of the black corals

(Antipatharia), probably because they are mainly found in the seas of the Far East, and in very deep water. Very little, also, is known about the embryonic stages of the order Zoantharia, except that they have two special types of larvae, the *Zoanthina* larva and the *Zoanthella* larva; there is a report that these have been reared to the adult stage.

Since the stony, or true corals (order Madreporaria) mostly live in the paths of the warm ocean currents, there are many which breed the year round, although some have fixed and limited breeding seasons. A good many of the species have been studied: since the eggs are heavily yolk-laden in many cases, superficial cleavage is very common, and those which undergo total cleavage often do so unequally. Gastrulation is generally achieved by invagination. The resulting planula is pear-shaped or elongated ovoid, this early part of the developmental process on the whole resembling the corresponding phases in the sea anemone, although these planulae are often brown because of the symbiotic algae (zooxanthellae) which they harbor.

The planulae fix by their aboral ends and change to the polypoid form, developing septa in the same way as do the sea anemones, and also passing through the *Edwardsia* and *Halcampula* stages. The main special characteristics of these species are the slight development of muscle tissue and the formation of an exoskeleton, together with extensive asexual reproduction by which colonies are built up. Exoskeleton formation proceeds hand-in-hand with the elaboration of the septa and eventually becomes highly complex; at the same time, increase in the number of tentacles accompanies the internal development. The young polyps have a strong tendency to aggregate, which leads to their forming very extensive colonies.

One fact which should be specially recorded with respect to the stone corals is that the solitary corals making up the genus *Fungia* exhibit an alternation of generations. In this genus the fertilized egg cleaves and forms a planula, which attaches and develops into a branched *trophozoid*; this gives rise to *anthoblasts* by budding. The anthoblast serves a function like that of the strobila in scyphozoan development, its periphery expanding and undergoing repeated lateral fission as in the formation of ephyrae, the corresponding larva being called an *anthocyathus*. This larva develops into the adult *Fungia*, which reproduces sexually as well as by budding.

In the foregoing account of the development of the subclass Hexacorallia, it appears that the peculiar characteristics of each order pertain less to the early stages than to the later processes such as metamorphosis and development of the septa. In the subclass Octocorallia, the morphology of the various individual polyps is generally rather uniform; consequently their development follows the same general course, and is, moreover, quite similar to the early stages of the Hexacorallia. The subclass includes the three orders: Alcyonacea, Gorgonacea and Pennatulacea; in none of these orders are there species of which the developmental history has been thoroughly investigated. The available observations indicate that the eggs are yolky, cleaving superficially in many cases, and cleavage being unequal when total. A sterroblastula is formed, and the yolk which fills the blastocoel at first is gradually absorbed as development proceeds. The endoderm is formed by delamination; the planula is ovoid or pear-shaped and free-swimming. Eventually it loses its cilia and attaches by its aboral end. After this the pharynx forms by invagination, tentacles appear around the periphery of the oral disc and septa develop internally. Next the calcite of the spicules is deposited, and finally asexual reproduction occurs and a colony is formed. Although the members of the Pennatulacea are colonial, they retain the capacity to move about. For this reason the structure of the colony is highly organized, and during its later development, especially during the budding process, the order of appearance of the new polyps is extremely regular. Among the Pennatulacea, the developmental history of such genera as *Renilla, Veretilium* and *Pteroides* has been followed. The egg of *Veretilium* cleaves totally, and almost equally or in some cases unequally; a sterroblastula is formed, and endoderm formation takes place by delamination. The planula is pearshaped; at first it is solid, but supporting tissue soon appears between the ecto- and endodermal layers. As the planula grows a mouth opens, and the pharynx invaginates; at the anterior end the *dorsal (asulcal) septa,* and at the posterior end, the *stalk septa* are formed. The planula finally becomes elongate and slender, and attaches to some substratum by its aboral end. Tentacles develop, and these become pinnulate. In the stalk region, a row of cells arises in the central part of the septa, marking the place where the axial skeleton will later be formed. From the center of the polyp body wall, secondary polyps are budded off in a bilaterally symmetrical fashion, and growth takes place by a regular repetition of this process. Tubular individuals are also formed and a well organized colony is produced, variations in the arrangement of the buds giving rise to the individual features of the different species.

The chief characteristics by which the development of the members of the class Anthozoa differs from that of the other coelenterates are: a free-swimming phase is lacking, and the eggs are therefore likely to be large and yolky; this in turn influences cleavage and germ layer formation. A free-swimming larval stage rarely appears, the planula in most cases developing directly into a polyp. The polyp forms a pharynx and system of septa; in some groups a muscular structure is predominant, while other are characterized by a calcareous or horny skeleton.

REFERENCES

Hydrozoa

Brauer, A. 1891: Ueber die Entstehung der Geschlechtsprodukte und die Entwicklung von *Tubularia mesembryanthemum*. Allm. Zeitschr. wiss. Zoöl. **52:** 551—579.

Brooks, W. K. 1907: On *Turritopsis nutricula* (McCrady). Proc. Boston Soc. Nat. Hist., **33:** 429—460.

Dan, K. & J. C. Dan 1947: Behavior of the cell surface during cleavage. VIII. On the cleavage of medusan eggs. Biol. Bull., **93:** 163—188.

Götte, A. 1907: Vergleichende Entwicklungsgeschichte der Geschlechtsindividuen der Hydropolypen. Zeitschr. wiss. Zoöl., **87:** 1—335.

Hargitt, G. T. 1909: Maturation, fertilization and segmentation of *Pennaria tiarella* (Tyres) and of *Tubularia crocea* (Ag.) Bull. Mus. Comp. Zool. Harvard Coll., **53:** 161—212.

Joseph, H. 1925: Zur Morphologie und Entwicklungsgeschichte von *Haleremita* und *Gonionemus*. Ein Beitrag zur systematischen Beurteilung der Trachymedusen. Zeitschr. wiss. Zoöl., **125:** 374—434.

Kühn, A. 1913: Entwicklungsgeschichte und Verwandschaftsbeziehungen der Hydrozoen. 1. Die Hydroiden. Ergeb. Fortsch. Zool., **4:** 1—284.

Lowe, E. 1926: The embryology of *Tubularia larynx*. Quart. Journ. Micr. Sci., **70:** 599—627.

Metschnikoff, E. 1886: Embryologische Studien an Medusen. Wien. Okada, Y. K. 1932: Dévelopment post-embryonnaire de la Physalie pacifique. Mem. Coll. Sci. Kyoto Imp. Univ., **8:** 1—26.

Perkins, H. F. 1902: The development of *Gonionema murbachii*. Proc. Acad. Nat. Sci. Philadelphia, **54:** 750—790.

Uchida, T. 1927: Studies on Japanese Hydromedusae. 1. Anthomedusae. Jour. Fac. Sci. Imp. Univ. Tokyo, Sec. IV. Zool. **1:** 145—241.

———— 1943: Animal Geneology. vol. 2 Coelenterata Edited by H. Oshima & Y. Okada, Tokyo, Yokendo (in Japanese)

Scyphozoa

Bigelow, R. P. 1900: Anatomy and development of *Cassiopea*. Mem. Boston Soc. Nat. Hist., **5:** 191—236.

Friedemann, O. 1902: Untersuchungen über die postembryonale Entwicklung von *Aurelia aurita*. Zeitschr. wiss. Zoöl., **71:** 227—267.

Hanaoka, K. 1934: Notes on the early development of a stalked medusa. Proc. Imp. Acad., **10:** 117—120.

Hargitt, C. W. & G. T. 1910: Studies in the development of Scyphomedusae. Jour Morph., **21:** 217—262.

Hein, W. 1900: Untersuchungen über die Entwicklung von *Aurelia aurita*. Zeitschr. wiss. Zoöl., **76:** 401—438.

Komai, T. 1935: On *Stephanoscyphus* and *Nausithoë*. Mem. Coll. Sci. Kyoto Imp. Univ., Ser. B., **10:** 289—339.

———— 1936: On *Stephanoscyphus*. Zool. Mag., **48:** 535—544. (in Japanese)

Okada, Y. K. 1927: Note sur l'ontogenie de *Charybdea rastonii* Haacke. Bull. Biol. d. France et Belgique, **61:** 241—248.

Uchida, T. 1926: The anatomy and development of rhizostome medusa, *Mastigias papua* L. Agassiz, with observations on the phylogeny of Rhizostomae. Jour. Fac. Sci., Imp. Univ. Tokyo, Sec. Zool., **1:** 45—95.

———— 1929: Studies on the Stauromedusae and Cubomedusae, with special reference to their metamorphosis. Jap. Jour. Zool., **2:** 103—193.

———— 1934: Metamorphosis of a Scyphomedusa *(Pelagia panopyra)*. Proc. Imp. Acad., **10:** 428—430.

———— 1943: Animal Geneology. vol. 2 Coelenterata Edited by H. Oshima & Y. Okada, Tokyo, Yokendo. (in Japanese)

Wietrzykowskj, W. 1912: Recherches sur le développement des Lucernaires. Arch Zool. exp. et gen. ser. 5, **10:** 1—95.

Anthozoa

Nyholm, K. -G. 1943: Zur Entwicklung und Entwicklungsbiologie der Ceriantharien und Aktinien. Zoöl. Bid. f. Uppsala, **22:** 87—248.

———— 1949: On the development and dispersal of Athenaria actinia with special reference to *Halcampa duodemicirrata* M. Sars. Zoöl. Bid. f. Uppsala, **27:** 465—506.

Uchida, T. 1943: Animal Geneology vol. 2 Coelenterata Edited by H. Oshima & Y. Okada, Tokyo, Yokendo. (in Japanese)

Uchida, T. & F. Iwata 1954: On the development of a broodcaring actinian. Jour. Fac. Sci., Hokkaido Univ., Ser. 6, Zoology, **12:** 220—224.

II. CTENOPHORA

INTRODUCTION

Within the phylum Coelenterata, the Ctenophora are clearly differentiated from the other classes. For this reason the coelenterates are often broadly divided into the two main categories of Cnidaria and Ctenaria. The Ctenaria consists of only the Ctenophora, which lack nematocysts and are thereby definitely distinguished from the Cnidaria. In addition, the ctenophores always possess *comb-plates,* which are comb-shaped structures composed of bands of more or less modified cilia. These plates are arranged in eight meridional rows on the body surface, and constitute the locomotory apparatus of these species.

The bodies of the Ctenophora are biradially symmetrical; one of their planes of symmetry usually includes a pair of tentacles, while the other, perpendicular to it, coincides with the long axis of the flattened pharynx. The former is therefore referred to as the *tentacular plane,* and the latter, as the *pharyngeal plane.* At one end of the main body axis is a *sense organ,* and the *mouth* is located at the other end. Within the mouth, the *pharynx* of ectodermal origin connects with an endodermal *infundibulum.* This lies directly under the sense organ, and sends off canals, which run alogside the comb rows, tentacles and pharynx. The paired *tentacles* consist of freely elongating and contracting muscular strands with numerous branches; each can be retracted into a depression in the body surface known as a *tentacular sheath.* Attached to the surface of the tentacles are many *colloblasts.* These are considered to take the place of nematocysts, and are similarly formed of single cells, although their structure is different from that of nematocysts. The members of the order Beroidea have no tentacles at any time in their life-span. Among the Lobatea and Cestidea, the main tentacles have degenerated in the adults, leaving *secondary tentacles* derived from their branches.

1. GONADS

All the ctenophores are hermaphroditic, and the gonads develop within the body wall on the outer sides of the eight meridional canals which parallel the comb rows. Ovaries and testes arise along the whole length of these canals, the ovaries on the sides facing the principal (tentacular and pharyngeal) planes and the testes on the interradial sides (Fig. 3.35, *ov, te*). Although certain nonreproductive cells are also found in both ovaries and testes, nothing equivalent to the yolk cells characterizing some of the Platyhelminthes occurs in these animals.

2. FERTILIZATION

When eggs and sperm mature, they are released into the canals and carried out of the body through the mouth. The eggs, which are covered with a thin gelatinous layer, float in the sea water and are fertilized externally. In *Coeloplana,* a creeping ctenophore, there are several epidermal invaginations along the meridional canals; the tips of these tubular depressions widen into sacs adjoining the ovaries. These sacs seem to serve the function of *seminal receptacles,* the spermatozoa entering them from the outside by way of the tubular portion, and being stored there. It is probable that the sperm cells later pass through the tissues and reach the ovaries, where they fertilize the mature eggs within them. In the sessile ctenophores *Lyrocteis* and *Tjalfiella,* fertilization takes place inside the water canals.

3. CLEAVAGE (Fig. 3.28a-i)

Since the great majority of the larvae in these species are planktonic, development takes place as they float in the sea. In *Coeloplana* the eggs are ejected from the mouth, and develop in the space between the underside of the flattened body and the substratum. In *Lyrocteis* and *Tjalfiella,* lateral chambers are formed by the distension of the meridional canals; the eggs are fertilized and develop to larvae within these brood pouches, as in viviparous animals. These modifications are believed to have arisen in response to the various special habits of the species.

The eggs of these animals (Fig. 3.28a) are surrounded by a thin, structureless membrane (*m*); the contents consist of a surface *ectoplasm (ec. pl),* and an interior *endoplasm (en. pl)* which makes up the greater part of the egg. The ectoplasm contains many small granules, and the entire layer appears more or less opaque, while the endoplasm, especially in its central part, has a conspicuously foamy appearance.

Cleavage is regular, and proceeds according to a strict biradial symmetry. The cleavage furrow appears first at one pole of the egg, and proceeds toward the other pole (Fig. 3.28b). The plane of this first cleavage coincides with the future pharyngeal plane: the second plane is perpendicular to this, and corresponds to the tentacular plane. These two cleavage planes divide the egg into the equal-sized blastomeres of the four-cell stage. The third cleavage is peculiar to the Ctenophora, the cleavage plane forming diagonally between the horizontal and vertical axes and dividing the blastomeres unequally (Fig. 3.28c,d). Since the resulting smaller blastomeres (E) line up along the second cleavage plane, the embryo acquires its biradial symmetry as early as the eight-cell stage.

The following cleavage proceeds unequally, each of the eight cells budding off a small blastomere. The succeeding processes, at least during the early developmental stages, are also regular, resulting in a faithful adherence to the pattern of biradial symmetry. At the fourth cleavage, all the blastomeres give off micromeres nearly simultaneously to form the 16-cell stage (Fig. 3.28e). Among the macromeres, those labelled E in the figure produce micromeres (e) which are slightly smaller than those (m) given off by the macromeres designated by M.

The e. 1 cells divide unequally into e. 1.1 and e. 1.2, and almost simultaneously another set of micromeres, e. 2, is formed from the E blastomeres, giving rise to the 24-cell stage. Shortly afterward the m. 1 cells divide into m. 1.1 and m. 1.2, giving the 28-cell stage (Fig. 3.28f), and the m. 2 micromeres are formed from the M macromeres to complete the 32-cell stage. This stage lasts comparatively long, and then e. 1.1 and e. 1.2 divide to produce e. 1.1.1, e. 1.1.2, e. 1.2.1 and e. 1.2.2, and the e. 2 blastomeres divide into e. 2.1 and e. 2.2 to give the 44-cell stage. Next m. 1.1 and m. 1.2 cleave into m. 1.1.1, m. 1.1.2, m. 1.2.1 and m. 1.2.2 and at almost the same time the E macromeres produce e. 3 micromeres. These e. 3 cells are larger than the other micromeres, having diameters equal to about half those of the E macromeres (56-cell stage). The m. 2 cells then divide into m.2.1 and m.2.2, to give the 60-cell stage. Finally the M macromeres cleave

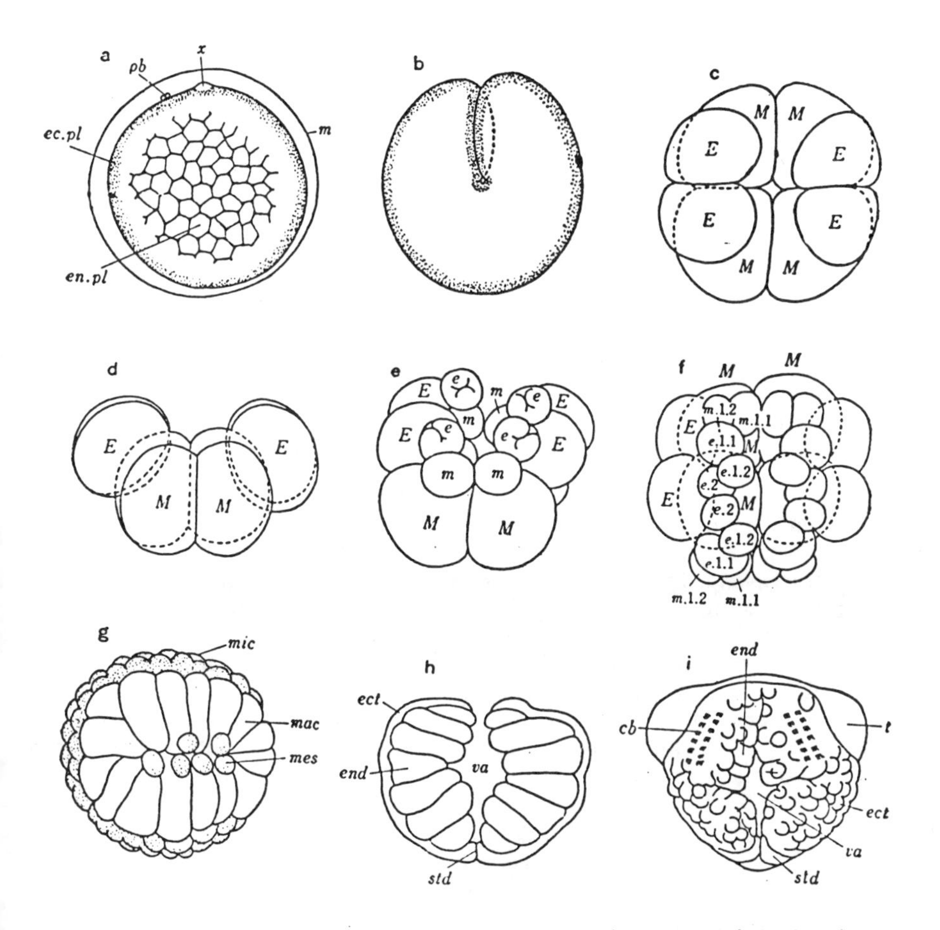

Fig. 3.28 Early development of *Coeloplana bocki* (about 100 times natural size) (Komai 1922)

a. egg before first cleavage; gelatinous substance is contained in space between egg surface and structureless membrane surrounding it. Eggs of most other comb-jellies have large amounts of this substance separating structureless membrane from egg surface; b. mid-cleavage; c. 8-cell stage (from animal pole); d. 8-cell stage (from side); e. transition from 16- to 20-cell stage; f. 28-cell stage (from animal pole); g. gastrulating embryo (from vegetal pole); h. optical cross-section of embryo at completion of gastrulation; i. more advanced stage (interior shown through surface). *cb:* comb plate, *E:* small terminal macromere, *ec.pl:* ectoplasm, *ect:* ectoderm, *end:* endoderm, *en.pl:* endoplasm, *M:* large central macromere, *m:* structureless membrane surrounding egg, *mac:* macromere, *mes:* so-called 'mesodermal cell', *mic:* layer of micromeres, *pb:* polar body, *std:* stomodaeal rudiment, *t:* tentacular rudiment, *va:* gastro-vascular system, *x:* point of sperm entrance.

into M1 and M2, and the embryo reaches the 64-cell stage (Table 1.3). Cleavage thus proceeds with the eight macromeres giving rise to successive generations of cells in an orderly fashion, although there is a tendency for the micromeres

Table 3.1 **Cleavage in Coeloplana bocki, showing order of cleavage in each quartet to 64-cell stage.**

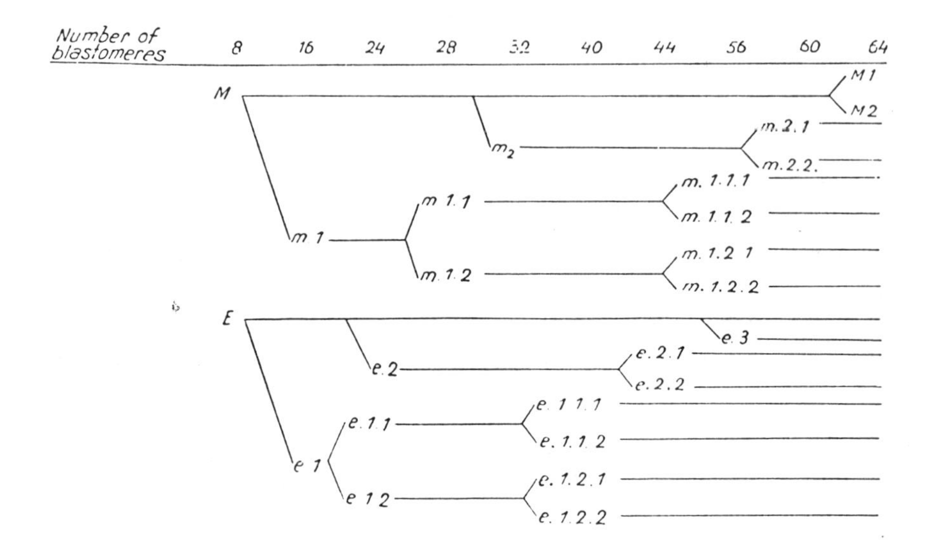

derived from the smaller E macromeres to divide earlier than those from the larger M macromeres. The order of cleavage given above accords with observations made on *Coeloplana*, but the process is practically the same in the other ctenophore species. While slight irregularities may occassionally be found, in general cleavage proceeds to this point in a regular fashion. After this time also, the micromeres continue to divide, and their number is increased by further budding of micromeres from the macromeres. During this process there may be some differences in cleavage rate among the four quadrants, so that the pace of each is not necessarily identical with the others. In the meantime, the E cells cleave, to give a total of 16 macromeres.

4. GASTRULATION (Fig. 3.28g-i)

As the number of micromeres increases, they form a layer which spreads over the macromeres and finally covers them. About this time, the macromeres are arranged in the shape of a bowl, with the micromere layer as a lid over its upper surface, while its lower surface is exposed. On the lower surface in this stage, a small cell is budded off from each macromere (Fig. 3.28g, *mes*). These small cells are believed to correspond to the mesodermal cells found in the Platyhelminthes and other groups.

During the subsequent development, the number of micromeres continues to increase, until they finally enclose the macromeres. In other words, *epibolic gastrulation* takes place. When this change is complete, the opening at the vegetal pole of the gastrula is closed, while a small pore makes a temporary appearance at the opposite, animal pole (Fig. 3.28h), and then closes. This marks the completion of the gastrula, the layer of micromeres forming the ectoderm and the macromeres making up the endoderm.

Cydippid larva

At one end of the gastrula a *stomodaeum* is formed (Fig. 3.28 *std*). It is very probable that the "mesodermal cells" mentioned above lie at the inner end of this invagination. At about the same time the rudiments of the tentacles *(t)* are formed as a pair of raised ectodermal thickenings. At the opposite end from the mouth the rudiment of the sense organ (Fig. 3.29 *ot. vs*) is formed. Prior to this, meridional rows of comb plates, at first not the full number, appear on the body surface in the interradial positions (Fig. 3.28 *cb*). These are a modification of the surface cilia.

The sense organ consists of tall columnar ectodermal cells provided with cilia; this part invaginates and forms a vesicle. Some of these cells secrete a calcareous *statolith* (Fig. 3.29 *ot*), which is suspended within the cavity supported by special cilia. In this manner a typical balancing organ or *statocyst* is formed.

After this each organ continues to develop; the stomodaeum deepens and becomes the *pharynx (ph)*, and its tip extends to the center of the larva. The endodermal cavity is roughly divided into four regions which later separate into the central *infundibulum* and the various *canals*. The rudiments of the tentacles *(t)* gradually become prominent, while the adjacent regions invaginate to form pockets, in the bottoms of which the bases of the tentacles become attached like roots. The tentacles gradually elongate into cord-like structures as the larva attains the fully formed *cydippid* stage.

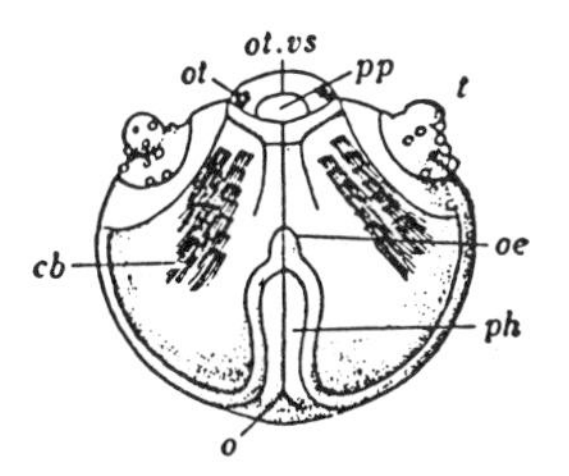

Fig. 3.29 Cydippid-type larva of *C. bocki* (× 100) (Komai)

cb: comb plate, *o:* mouth, *oe:* oesophagus, *ot:* statolith, *ot.vs:* statocyst, *ph:* pharynx, *pp:* polar plate (thought to be a kind of sensory organ), *t:* tentacular rudiment.

This cydippid larval stage (Fig. 3.29) occurs in the development of all the ctenophores, being common to all the orders, except that in the Beroidea no tentacle rudiments are formed. The cydippid is in general spherical, the tentacular and pharyngeal axes are of approximately the same length, and the pharynx is circular in cross-section. In other words, if the tentacles are disregarded, the larva is nearly radially symmetrical.

5. METAMORPHOSIS

When it reaches the condition described above, the larva breaks out of the egg membrane and swims in the sea by moving its comb plates. In the Cydippida there are few changes after this stage. The number of comb plates increases, the cilia grow longer, the tentacles lengthen and branch and become freely movable in and out of their sheaths. The water-vascular system develops completely and the general shape of the body gradually changes toward that of the adult.

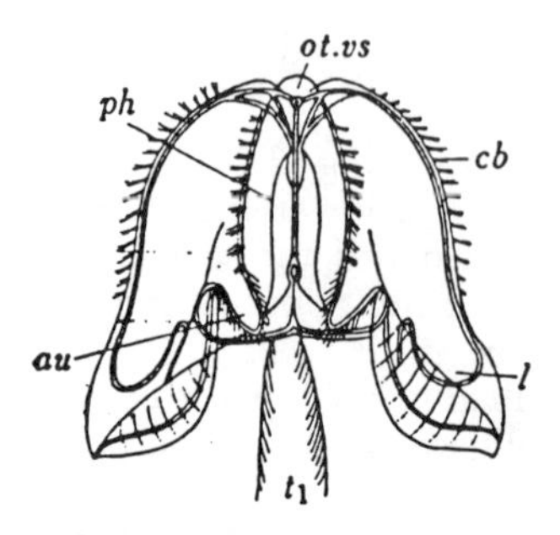

Fig. 3.30 Young comb-jelly of *Bolinopsis vitera* (×ca. 4.5) (Chun 1892; Krumbach 1925)

au: auricular lappet, *cb:* comb plate, *l:* oral lobe, *ot.vs:* statocyst, *ph:* pharynx, t_1: main tentacle.

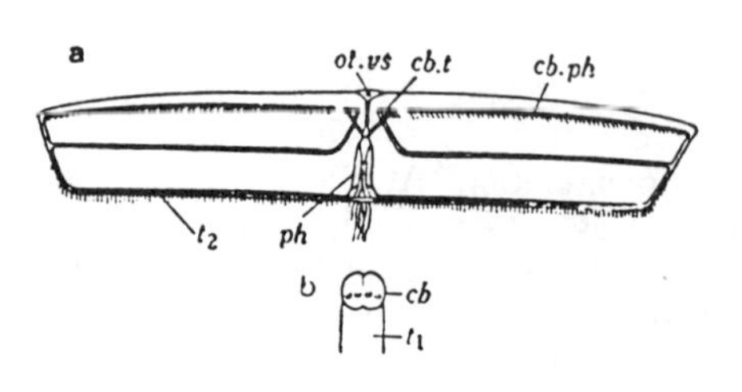

Fig. 3.31 a. *Cestum veneris* (adult). (×0.5) b. Cydippid-larva of same species. (×ca. 5) (Chun 1880)

cb: comb plate, *cb.ph:* subsagittal comb row, *cb.t:* subtentacular comb row, *ot.vs:* statocyst, *ph:* pharynx, t_1: main tentacle, t_2: sub-tentacle.

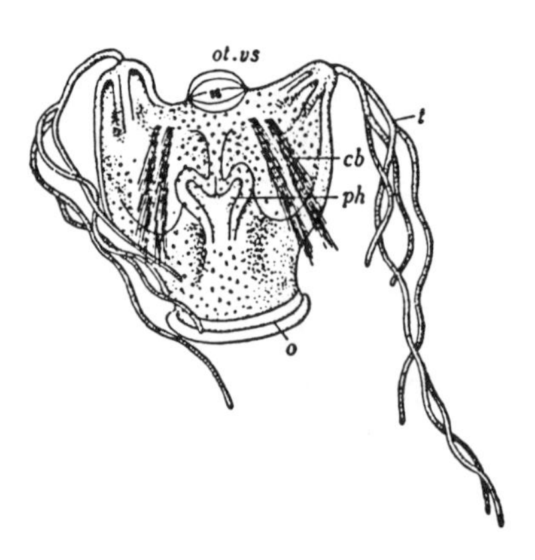

Fig. 3.32 Fully developed larva of *C. bocki* (×100) (Komai 1922)

cb: comb-plate line, *o:* mouth, *ot.vs:* statocyst, *ph:* innermost part of pharynx *t:* tentacle.

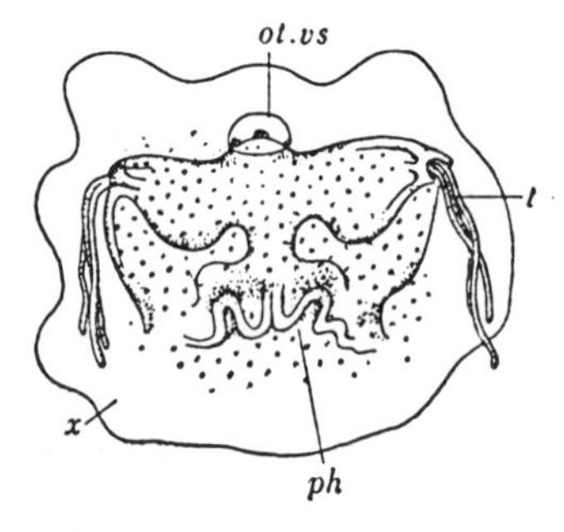

Fig. 3.33 Larva of *C. bocki* in metamorphosis (×ca. 100) (Komai 1922)

ot.vs: statocyst, *ph:* innermost part of pharynx, *t:* tentacle, *x:* skirt-like extension of mouth.

In the Lobatea (Fig. 3.30), the whole body of the cydippid larva shortens in the tentacular axis and lengthens in the pharyngeal direction. The lower parts of the body in the pharyngeal plane are prolonged into conspicuous sleeve-like structures *(l)*, and the water canals grow into this region and make complicated zigzag revolutions there. At the bases of the sleeves are formed ear-shaped projections *(au)*. The main cores of the tentacles retrogress, and their branches elongate to form the so-called secondary tentacles.

Among the Cestida (Fig. 3.31), the difference between the pharyngeal and tentacular axis is still more pronounced, so that the body takes the shape of a ribbon. The four rows of comb plates adjacent to the pharynx (*cb. ph*) grow very long, forming a border along the upper edge of the ribbon, while the plate-rows next to the tentacles *(cb. t)* become vestigial, consisting of a few comb plates situated near the center of the upper margin. The secondary tentacles (t_2) hang from the lower margin of the ribbon. In spite of this exaggerated shape shown by the adult, the larva (Fig. 3.31b) is a cydippid, like those of the other ctenophore orders.

In *Coeloplana* (Order Platyctenea), the mouth of the cydippid larva (Fig. 3.32) widens, and the metamorphosing larva moves over the substratum by means of the cilia on the stomodaeal surface. Shortly afterward, the comb plates are shed, pigment cells develop, and the whole body becomes flattened so that it resembles a turbellarian (Fig. 3.33). In *Lyrocteis* and *Tjalfiella,* the mouth expands and the animal attaches to some object by the inner stomodaeal surface. The body elongates in the tentacular axis and the tentacle sheath regions extend upward like chimneys, so that the general shape of the body is somewhat like a harp. In most of the species of the order, the water canals form a highly ramified network.

The form of the adult in the order Beroida is not very different from that of the cydippid larva, except for the complete absence of tentacles in all stages, and the markedly reticulate water vascular system.

REMARKS

As described above, there are certain unique features in ctenophore development; the most striking of these is the strict adherence to the principle of biradial symmetry. The quadrants of the developing egg which are separated from each other by the early cleavage planes follow their own courses of differentiation, so that if they are artificially isolated, they develops as parts of the whole, forming only imperfect partial larvae.

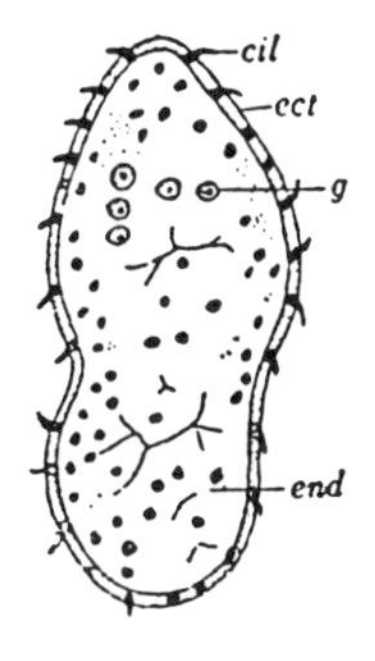

Fig. 3.34 Longitudinal section of *Gastrodes parasiticum* planula (×200) (Komai 1922)

cil: cilia, *ect:* ectoderm, *end:* endoderm, *g:* early germ cell.

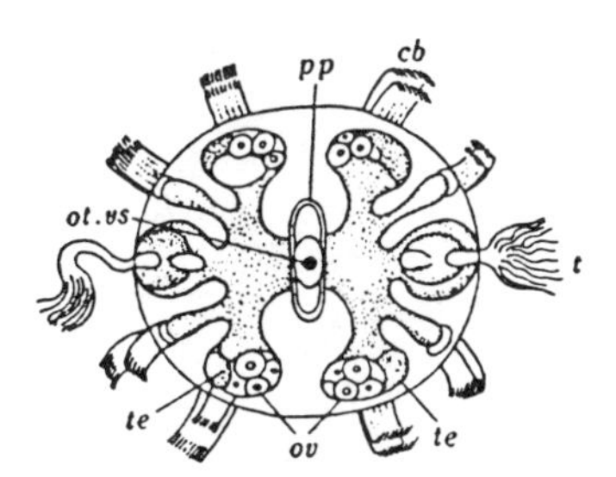

Fig. 3.35 Sexually mature larva of *Bolinopsis vitrea* (Chun 1892; Krumbach 1925)

cb: comb plate, *ot.vs:* statocyst, *ov:* ovary, *pp:* polar plate, *t:* tentacle, *te:* testis.

The blastula and gastrula stages of these species present a rather different appearance from those of the other coelenterates, but this is probably not a fundamental difference. The parasitic platyctenean *Gastrodes*, for example, forms a planula larva (Fig. 3.34).

One further point which should be noted in connection with ctenophore development concerns the occurrence of a peculiar reproductive phenomenon known as *dissogony*. It has been observed that reproductive cells develop in the larvae of the Lobata before metamorphosis. These are shed, and then another period of sexual maturity occurs following metamorphosis (Fig. 3.35).

REFERENCES

Ctenophora

Chun, C. 1880: Die Ctenophoren des Golfes von Neapel. Fauna u. Flora. Golf. Neapel. Monogr. 1. Leipzig.

——— 1892: Die Dissogonie. Eine neue Form der geschlechtliche Zeugung. Festschr. R. Leuckarts, Leipzig; 77—108.

Hyman, L. H. 1940: The Invertebrates. 1. Protozoa through Ctenophora, McGraw-Hill New York.

Komai, T. 1922: Studies on two aberrant Ctenophores, *Coeloplana* and *Gastrodes*. Kyoto.

——— 1942: The structure and development of the sessile ctenophore *Lyrocteis imperatoris* Komai. Mem. Coll. Sci., Kyoto Imp. Univ. Ser., **17:** 1—36.

Korschelt, E. u. K. Heider 1936: Vergleichende Entwicklungsgeschichte der Tiere, **1**. Jena.

Krumbach, Th. 1925: Ctenophora. in Kükenthal-Krumbach Handbuch der Zoölogie. **1:** 905—995, Berlin and Leipzig.

Yatsu, N. 1911: Observations and experiments on the ctenophore egg. II. Annot. Zool. Japon. **7:** 333—346.

——— 1912: Observations and experiments on the ctenophore egg. I. Jour. Coll. Sci. Univ. Tokyo, **32:** 1—21.

PLATYHELMINTHES

(Class Turbellaria)

Chapter 4

INTRODUCTION

This chapter deals only with the development of the single class Turbellaria among the phylum of the Platyhelminthes; in order to make clear the relation between the Turbellaria and the other members of the phylum, a brief systematic description will be given here.

Since the flatworms hold the lowest position among the Bilateria, they are correspondingly objects of great embryological interest. The phylum is divided into three classes: Turbellaria, Trematoda and Cestoidea, although some people add the Nemertini as a fourth class. The Turbellaria are further divided into five orders: Acoela, Rhabdocoela, Allocoela, Tricladida and Polycladida. In the past a sixth order, the Temnocephala, was recognized, but this has recently been included among the Allocoela. The Trematoda are further divided into Monogenea and Digenea, and the Cestoidea into Cestodaria and Cestoda. Very many of the species are hermaphroditic; the only free-living groups are the Turbellaria and the Nemertini, the Trematoda and Cestoidea all living as parasites, and practically all parasitizing vertebrates in their adult stage.

There are two theories concerning the origin of the Platyhelminthes: the Ctenophore Theory (A. Lang) and the Planula Theory (Ludwig v. Graff). The former theory is based on the fact that there are certain structural similarities between the Polycladida and the ctenophore *Coeloplana*. The latter theory, on the other hand, holds that the turbellarian order Acoela resembles the coelenterate planula stage, and suggests that the present Acoela had their origin in some acoeloid ancestor which evolved from a planuloid form. According to this idea, all the platyhelminth worms originate from the Acoela, which gave rise to the Allocoela and Rhabdocoela on the one hand, and the Tricladida and Nemertini on the other; finally, the Allocoela are believed to have produced the Polycladida, while the Rhabdocoela gave rise to the parasitic Tremadoda and Cestoidea.

Of these groups, the Polycladida and Nemertini are like the Acoela in forming simple eggs, while the Rhabdocoela and Allocoela, like the Trematoda and

Cestoidea, have composite eggs. A detailed explanation of these types of eggs will be given below: at any rate, it is obvious that the reproductive systems of species forming composite eggs are definitely more complicated than those of species laying simple eggs.

Comparing the developmental processes of these types of eggs, it appears that while the eggs of the Acoela and Polycladida, together with those of the Nemertini, undergo typical spiral cleavage, the cleavage patterns of the composite eggs are quite different, and show an extreme degree of modification. Furthermore, while the members of the former group give rise in their later development to the free-swimming *Müller's*, *Götte's*, and *pilidium larvae*, and each undergoes a clear-cut process of metamorphosis, this bears no comparison with the extremely striking metamorphosis and conspicuous variety of the forms exhibited by the members of the latter group, in particular by the parasitic species.

Among the Monogenea, which resemble the Rhabdocoela and are comparatively simple trematodes, there are genera like *Gyrodactylus* which exibit *polyembryony*, while in the Digenea the life-history becomes extremely complicated — besides exhibiting *heterogony*, or the alternation of an hermaphroditic with a parthenogenetic generation, a concurrent exchange of host animals also takes place. That is, the adults *(maritae)* of the sexual generation *(marital generation)*, which lay fertilized eggs, usually parasitize vertebrate animals as their *definitive hosts;* from the fertilized eggs laid by these adults develop *parthenitae*, the individuals of the *parthenitic generation* which reproduce by parthenogenesis and which select only molluscs as their *intermediate hosts*. The larva which first develops from the fertilized eggs is called a *miracidium;* its structure is extremely similar to that of the rhabdocoel larva. The miracidium metamorphoses, becoming a *sporocyst;* next the sporocyst gives rise to *rediae;* and the rediae to *cercariae*, both by parthenogenetic reproduction, and the adult form is finally achieved by the metamorphosis of the cercariae.

The cestodes, on the other hand, while they undergo striking metamorphosis, usually do not reproduce during the larval stages. The first embryo which develops from the fertilized egg, in the Cestodaria, is the *lycophora*, having ten hooks, and in the Cestoda, the *onchosphaera* with six hooks, also called a *hexacanth*. These embryos develop into various different larvae, depending upon their species. In species like those making up the family Diphyllobothriidae which require two intermediate hosts, the onchosphaera passes through a *procercoid* stage to become a *plerocercoid*. However, in forms like *Taeniarhynchus*, *Taenia* and *Echinococus*, included in the order Cyclophyllidea, which require only one intermediate host, the development proceeds from an onchosphaera embryo to a *cysticercus* ("bladderworm"), *cysticercoid*, or *plerocercus larva*, depending on the species. Among the cystocerci are other larval types such as the *coenurus* and *eschinococus*, in which a large number of head segments proliferates within the bladder, constituting a beautiful case of polyembryony.

The phylogenetic relationship of the Platyhelminthes and the Mesozoa also poses an interesting problem. On one side it is held that the Mesozoa are truly primitive animals, not forms which have degenerated as the result of a parasitic habit (Hartmann '25, Hyman '40). The view which maintains, however, on the basis of the strong structural resemblance between these animals and the trematode miracidium as well as the cestode onchosphaera, that they originated in the parasitic flatworms (Nouvel '47, Stunkard '54), seems to be a more reasonable one.

1. REPRODUCTION

In general the flatworms have a strong capacity for regeneration, and asexual reproduction by *transverse fission* is widely practised. This form of asexual reproduction takes place repeatedly over a rather long period, before the body attains full growth and the gonads and copulatory apparatus are developed so that sexual reproduction can occur. In ordinary planarians, after the body has been divided posterior to the pharynx, a tail region develops on the isolated anterior part, and a head region regenerates on the posterior part, forming two complete new animals. In this sort of fission, the body divides without going through any previous process of regeneration; this is called *architomy*. In the Rhabdocoela, however, the organs which are due to be lost in fission are first formed, and division takes place afterward. This kind of fission is called *paratomy* (Fig. 4.1).

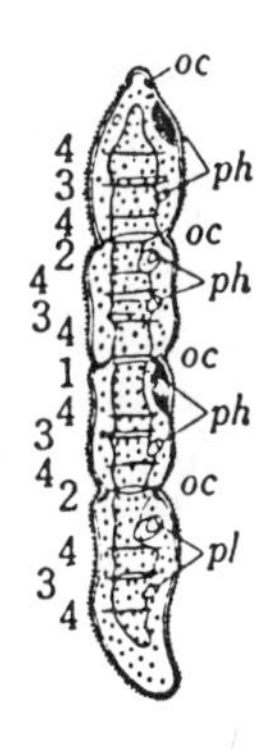

Fig. 4.1 Paratomy of *Microstomum lineare* (Graff)
Eyes, brains, mouths, pharynxes, etc. are formed before division; one individual subdivides into about 18 sub-individuals. Numbers show order of division. *oc:* eye-spot, *ph:* pharynx.

Sexual reproduction begins when the reproductive organs are fully developed; although these species are almost without exception hermaphroditic, self-fertilization does not take place as a rule (see p.[142]). There is a well developed *copulatory pouch* or *seminal receptacle,* and a true copulation takes place between any two mature individuals by the union of the respective male and female reproductive organs. In the rhabdocoels, some of the fresh-water triclads and the polyclads, the spermatozoa are bundled into *spermatophores,* which are placed in the copulatory pouch or seminal receptacle of the partner. The most simplified mode of copulation, *hypodermic injection,* occurs widely, in many species among the Acoela and Polycladida, and in several of the Rhabdocoela, Allocoela and marine Tricladida. In this method of impregnation, one animal punctures the epidermis of the other by means of the copulatory apparatus, and injects its spermatozoa into the parenchyma. The sperm travel through the tissue spaces and enter the uterus, where fertilization takes place. Since this occurs just before spawning, it has been reported (Selenka 1881) that in exceptional cases in *Thysanozoon brocchi* fertilization may take place after spawning, but Kato ('40) believes this to be an error.

2. METHOD OF OBTAINING EGGS

In order to keep polyclad worms and make them spawn, a number of animals with mature gonads should be placed together in a glass container of running sea water which is covered with a piece of gauze or cloth. Small species kept in Petri dishes will spawn if the sea water is changed occasionally. As food, the carnivorous species should be given bits of molluscan or crustacean tissue, and herbivorous species may be fed with the various simple algae which attach to rocks at the low tide line. The mating reactions of the animals are also easy to observe. Since they ordinarily spawn at any time of the day or night, with care eggs can be secured immediately after deposition. Various turbellarians attach their egg masses to the under-surface of stones at the low tide line; if the characteristics of these egg masses are learned, they can also serve as a source of material for observation.

The eggs are surrounded by an *egg shell,* and large numbers of the shells are held in round masses or irregular thin plates by a covering gelatinous substance. If the eggs masse are loosened from the substratum with a scalpel or spatula and transferred to Petri dishes, they can be cultured until hatching with occasional changes of the sea water. The general course of development may be seen by bringing the egg mass under a microscope, but for detailed observation the shell should be removed with a needle and the embryo released into the sea water. Practically all the spawned eggs will be found to be fertilized.

For artificial fertilization an animal should be selected in which the uterus is full of eggs. It is necessary to apply mechanical stimulation to the body to cause the release of excess mucous secretion from the epidermis; this should be removed, and then the body wall in the region of the uterus torn with a needle and the eggs collected in sea water. Sperm may be taken in the same way from the spermatheca or the sperm duct. The embryos obtained by artificial fertilization develop to a certain point in about the same way as those within the egg shells.

In order to observe cleavage, it is convenient to stain the eggs vitally with Nile blue sulfate. The egg mass may be fixed *in toto* with Gilson's or Bouin's fixative and stained with hematoxylin, and the embryos removed from their shells in clove oil for detailed observation. Whole larvae may be mounted by smearing the egg mass on a slide; for cutting paraffin sections, the usual method of attaching the larvae to a leaf of *Ulva* should be followed. If the egg mass as a whole is to be paraffin-sectioned, it is advisable to use dioxane or the celloidin-paraffin double embedding method.

If triclads with well-developed reproductive organs can be found, they are rather certain to spawn in the laboratory, laying their eggs on the under-sides of decayed leaves or other objects. It is probably useless to expect that animals which are actively reproducing asexually by fission will become sexually mature, even if they are kept for a long time. Specimens of the Acoela, Allocoela and Rhabdocoela can be kept easily, and will spawn in small laboratory containers.

3. OOGENESIS

One very conspicuous characteristic of the Platyhelminthes is their possession of a special gland, called the *yolk gland,* which forms *yolk cells.* These are combined with the egg cells to make up *composite eggs.* This characteristic originated with the Turbellaria, but even in this class, no yolk gland develops in the Acoela, Polycladida and some species among the Rhabdocoela and Allocoela, and consequently they do not form composite eggs. The eggs laid by these species are no different from the eggs of most animals, but by contrast with the composite eggs they are called *simple eggs*. The turbellarian species in which yolk glands develop and composite eggs are formed are those making up the order Tricladida, as well as some of the Rhabdocoela and Allocoela. Since the yolk is accumulated outside the egg cell in composite eggs, these are sometimes called *ectolecithal eggs*, while the simple eggs are called *endolecithal* since their yolk is located inside the egg.

In the composite egg, around the ovum which is formed in the ovary are arranged the yolk cells, and outside of these is the egg shell (Fig. 4.10). This shell is believed to consist mainly of material secreted by a *shell gland*. Within one shell there may be included as many as several thousand yolk cells, and from one to ten egg cells. Such "eggs" are large in general, usually being several millimeters in diameter, while the specially large eggs of some of the terrestrial triclads may reach a diameter of 20 mm.

Among the simple eggs, from one to several are surrounded by a thin shell[1], which is in turn enclosed in a gelatinous substance; the eggs are deposited in a single layer making up *egg-plates* of various shapes. Lang and others have thought that the egg shell was secreted by a shell gland, but this is actually formed by a shell-forming substance secreted to the outside by the egg cell itself (Hofsten '12, Kato '40). If the ovarian eggs of the simple-egg type are observed at the time of germinal vesicle breakdown, certain eosinophylic granules can be found to appear in the cytoplasm. As the egg enters the uterus, these granules move to the egg surface and form a layer there, and about the time the eggs enter the vagina, the granules are beginning to fuse with each other. As the egg is laid these granules are expelled from the egg to the outside, where they take up water, inflate and change into the egg shell. From the so-called shell glands a sticky gelatinous material is secreted; together with more of the same substance secreted from the ventral epidermis of the animals, this holds the eggs together and contributes to the formation of the egg plate.

There are conspicuous differences in mode of development between simple and composite eggs. The processes in simple eggs are closer to the primitive state, while those of the composite eggs are subjected to extreme modifications because of the existence of the yolk cells.

4. DEVELOPMENT OF SIMPLE EGGS (POLYCLADIDA)

The Polycladida are characterized by highly typical spiral cleavage. Their development is the object of the historically significant discovery that homology of blastomere determination, which had been found among the annelids and molluscs (phyla which also undergo spiral cleavage), exists even in such a phylogenetically low group of animals as the Turbellaria (Lang 1884, Wilson 1898). Since then, extremely detailed investigation into the embryology of this group has been continued by Surface ('07) and Kato ('40).

[1] In *Notoplana, Thysanozon,* and others, one egg; in *Planocera,* usually seven; *Prosthiostomum siphunculus,* twelve eggs as observed by Lang.

As was described in the previous section, the egg shell is rapidly inflated and its formation completed immediately after spawning; the egg cell which it surrounds may take on a somewhat irregular shape at that time. The size of the ovum varies with the species: in *Planocera reticulata* the diameter is 310—330 μ, but the 130—150 μ diameter of *Notoplana humilis* is about average, and there are some eggs as small as that of *Stylochus uniporus* (88—95 μ). The larger and more yolky the egg, the more slowly does it develop; the time required until hatching is three weeks in *Planocera,* two weeks in *Notoplana* and one week in *Stylochus*. The meiotic divisions begin about an hour after the egg is laid, when a depression appears at the animal pole, and the first polar body is formed there. After this the egg seems to shrivel and its outline becomes extremely irregular for a time; it then regains its spherical shape and gives off the second polar body.

(1) **CLEAVAGE** (Fig. 4.2)

The egg divides totally and unequally, and as stated in the foregoing section, in the typical pattern of spiral cleavage (see Introduction). The first cleavage plane cuts vertically through the animal and vegetal poles and divides the egg into two nearly equal-sized blastomeres (*AB*, *CD*). Lang states that one of these blastomeres *(CD)* is always larger than the other *(AB)*, but this cannot be confirmed in a large number of Japanese species (Kato '40). The following cleavage is also vertical, but unequal, dividing the blastomeres into two fairly large *(B, D)*, and two smaller ones *(A, C)*. At this time the *D* blastomere is the largest, and its position indicates the future posterior side of the embryo. As a consequence, the *B* blastomere marks the anterior side; viewed from the animal pole, the *A* blastomere lies on the left, and the *C* blastomere on the right side. Seen from the side at the time of the second cleavage, the positions of the two spindles intersect each other practically at right angles. As a result, two of the four blastomeres lie above the other two: a pair of smaller blastomeres is formed on the animal pole side, and a pair of larger ones at the vegetal side. This cleavage is clearly laevotropic. The blastomers eventually shift their positions and line up in a single plane, but at this time *A* and *C* are in contact at the animal pole, while *B* and *D* make contact at the vegetal pole, forming the so-called *cross furrow*.

The third cleavage is horizontal and unequal, dividing the blastomeres in a strongly dexiotropic direction. In general, the larger a blastomere is, the more rapidly it cleaves; in this case *D* thus cleaves first, followed by *B*, *C* and *A*. The blastomeres cut off toward the animal pole side (micromeres) are smaller than those left on the vegetal side (macromeres). However, this size difference between the micromeres and macromeres varies with the species: in *Pseudostylochus obscurus* and *Notoplana dilicata* there is a large difference (Kato '40), but in such forms as *Discocoelis* it is hard to recognize any disparity (Lang 1884).

The fourth cleavage, which follows next (8 — 16-cell stage), is laevotropic; first the macromeres at the vegetal side divide, forming the second quartet of micromeres ($2a$ — $2d$), but before these are completely separated, the first micromere quartet divides into $1a^2$ — $1d^2$ on the vegetal side and $1a^1$ — $1d^1$ on the animal pole side. $1a^2$ — $1d^2$ correspond to the trochoblasts of annelids and molluscs.

The fifth cleavage (16 — 32-cell stage) is dexiotropic, leading to the formation of a typical spiral cleavage 32-cell stage. First of all $2D$ and $2B$ give off $3d$ and $3b$ dexiotropically; there is a considerable lag before $2A$ and $2C$ separate off

$3a$ and $3c$, and in the meantime $2d$ and $2b$ cleave dexiotropically to form $2d^1$, $2b^1$ and $2d^2$, $2b^2$. This time, $2d^1$, $2b^1$ at the animal pole side are smaller than $2d^2$, $2b^2$ on the vegetal side. While the cleavage of $2d$, $2b$ is proceeding, the four blastomeres $1a^1$ — $1d^1$ also divide dexiotropically, producing $1a^{11}$ — $1d^{11}$. At the same time the remaining $2a$, $2c$ micromeres divide and finally the other two macromeres, $2A$ and $2C$, give off $3a$ and $3c$.

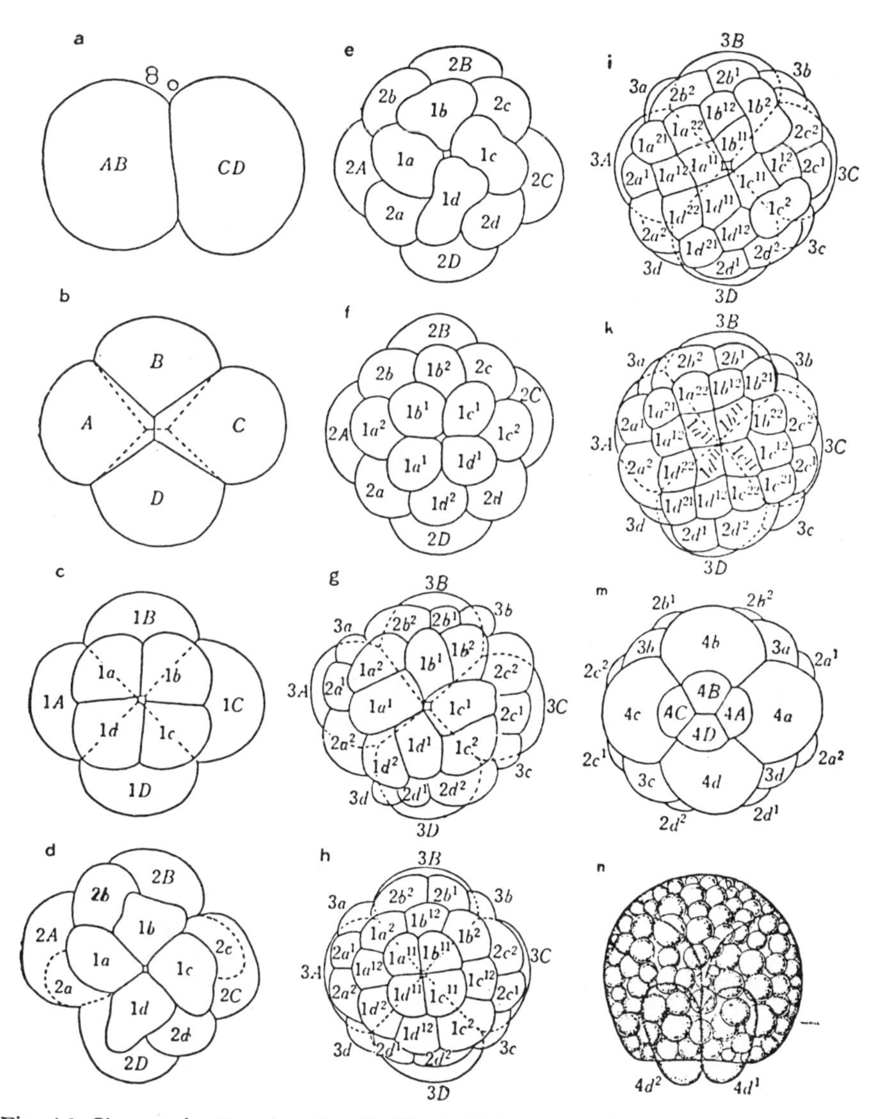

Fig. 4.2 Cleavage in *Notoplana humilis* (Kato) (Unless otherwise stated, sketches show embryo from animal pole side)

a. 2-cell stage (lateral view); b. 4-cell stage; c. 8-cell stage; d. transition from 8- to 12-cell stage; e. 12-cell stage; f. 16-cell stage; g. 24-cell stage; h. 28-cell stage; i. transition from 28- to 32-cell stage; k. 32-cell stage; m. 36-cell stage (vegetal pole); n. cleavage of mesentoblasts

Thus the 32-cell stage embryo, which has just completed its fifth cleavage, consists of 16 cells derived from the first micromere quartet (1*a*—1*d*), eight cells derived from the second micromeres (2*a*—2*d*), four cells comprising the third set of micromeres (3*a*—3*d*) and the four macromeres (3*A* — 3*D*). Among these the third micromere quartet (3a—3*d*) are the smallest and the macromeres (3*A*—3*D*) are the largest. In the subsequent divisions the isochrony of cleavage is increasingly lost, so that the 32-cell stage is succeeded by a 36-cell, and then 40-, 44-, 45, 53-, 61-cell stages.

After the embryo has reached the 32-cell stage, the next event is the cleavage of the macromeres (3*A*—3*D*) to form the 36-cell stage. This division is highly peculiar in that the 'macromeres', which are left at the vegetal pole (4*A*—4*D*), in spite of their name are minute, while the four 'micromeres' given off at the animal pole side contain the major part of the yolk and are many times larger. The four little macromeres (4*A*—4*D*) do not divide after this; when the embryo gastrulates they move to the interior where they later degenerate and are absorbed. The three cells 4*a*, 4*b* and 4*c* also fail to cleave further, and in the same role of providing yolk for the embryo they later degenerate and are absorbed. On the other hand, the 4*d* cell at the posterior side of the embryo proceeds to divide actively, becoming the source of the future endodermal as well as the mesodermal tissue. This cell is the counterpart of the one which Conklin (1897) named the 'mesentoblast'.

(2) GASTRULATION (Fig. 4.3)

Gastrulation takes place chiefly by a process of epiboly. The cells derived from the first to the third quartet of micromeres proliferate by repeated divisions and gradually extend toward the vegetal pole, enveloping the cells of this region

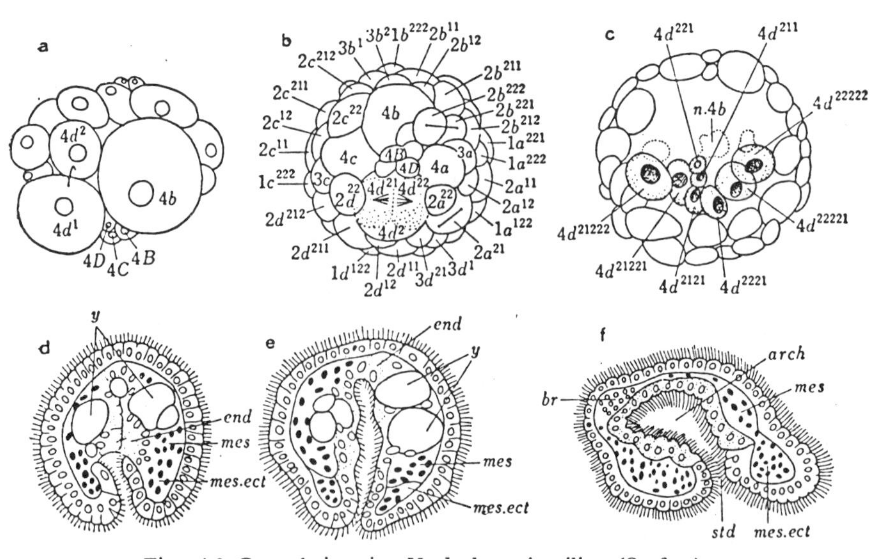

Fig. 4.3 Gastrulation in *Hoploplana inquilina* (Surface)

a. cross-section of embryo at 45-cell stage; $4d^2$-blastomere moves into segmentation cavity; b. vegetal polar view of embryo at 66-cell stage; $4d^2$ divides into right and left cells; c. subsequent division of $4d^2$ (cross-section near vegetal pole); d. invagination of stomodaeum; e. opening of archenteron after stomodaeal invagination; f. early stage of Müller's larva. *arch.* archenteron, *br:* cerebral ganglion, *end:* endoderm, *mes:* mesoderm, *mes.ect:* ectomesoderm, *std:* stomodaeum, *y:* yolk.

and finally forming a stereogastrula, without a blastocoel left in its center. The cells which fill the interior of the gastrula are chiefly the three yolk-laden blastomeres (4*a*—4*c*) of the fourth micromere quartet and the descendants resulting from the divisions of the 4*d* cell; in addition there are the brain cells derived from the first micromere quartet and the ectomesoderm cells which arise from the second micromere quartet.

The origin of each of these is as follows. Before the division of the fourth micromere quartet is complete, the first micromere quartet ($1a^{11}$—$1d^{11}$) undergoes a distinctly laevotropic cleavage, forming the small $1a^{111}$—$1d^{111}$ and the fairly large $1a^{112}$—$1d^{112}$, making the 40-cell stage. These $1a^{111}$—$1d^{111}$ cells were designated the *apical cells* by Selenka (1881), and they are believed to invaginate and proliferate to give rise to the brain and the organs of the forebrain (Kato '40). According to the detailed investigations of Surface ('07) on *Hoploplana* (Table 4.1), all the cells of the first micromere quartet lead the other blastomeres in repeated

Table 4.1 **Cell lineage in *H.* inquilina (Surface) Division of *B* and *C* blastomeres same as that of *A*. (end: endoderm, mes: mesoderm)**

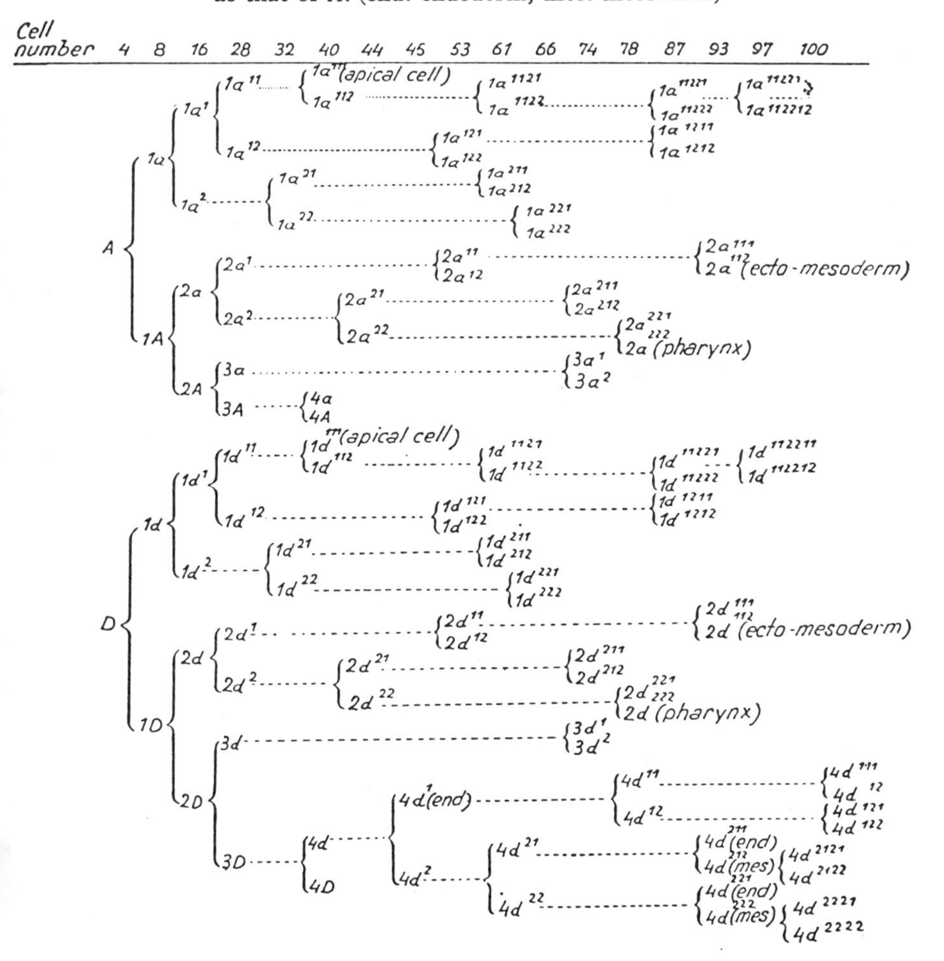

cleavages, and finally their descendants are arranged in a spherical formation around the apical cells at the animal pole, thus making up the main part of the anterior dorsal ectoderm of the larva.

The descendants of the second micromere quartet, on the other hand, chiefly form the ectoderm of the posterior ventral part of the embryo. Of these, the four cells $2a^{222}$—$2d^{222}$ in particular are clearly located at the vegetal pole, in close contact with the macromere quartet, and when the embryo gastrulates and finally forms a pharynx, these cells are believed to take part in its development. When $2a^{11}$—$2d^{11}$ divide, the rather large blastomeres $2a^{112}$—$2d^{112}$ derived from them move into the interior of the embryo and develop into ectomesoderm. The descendants of these cells become located around the mouth, and later form the musculature around the pharynx.

The cell-lineage of the $4d$ blastomere in the polyclads is complicated, and specialized to a certain extent (Table 4.1). About the time when the apical cells appear and the second micromere quartet is beginning to cleave (44-cell stage), the $4d$ blastomere forms a mitotic figure at right angles to the surface of the embryo and cleaves into a large and a small cell. The large cell ($4d^1$) remains at the surface of the embryo while the small one ($4d^2$) moves into the blastocoel (Fig. 4.3a). According to Surface, $4d^1$ which remains at the surface is the *endoderm mother cell (entoblast)*, while the $4d^2$ cell which migrates inward is the *mesentoblast*, which also gives rise to endoderm[2]. Next each of these cells divides to form the left-right pairs, $4d^{21}$ and $4d^{22}$, as well as $4d^{11}$, $4d^{12}$ (Fig. 4.3b). The $4b^2$ daughter cells divide again into large and small cells; the smaller ones become *endoderm*, while the larger ones ($4d^{212}$, $4d^{222}$) continue to divide repeatedly and become the source of the *mesoderm* (Fig. 4.3c).

The daughter blastomeres of $4d$, which remain on the surface of the embryo, continue to divide and send more endoderm cells into the interior, but at about this time, these blastomeres together with the other micromeres of the fourth quartet ($4a$, $4b$, $4c$) become covered by the *ectoderm* cells as the result of the epiboly which is taking place. The endoderm cells, which have been multiplying during this time inside the embryo, form a clump located at the lower side of the embryo near the $4b$ cell, but as yet no cavity suggestive of an archenteron has appeared in it. At the vegetal pole, where the little macromeres are located, an inconspicuous blastopore appears, but this is eventually closed and obliterated. The embryo acquires cilia over its whole surface, and begins to turn slowly within the shell.

The larva which has arrived at this stage may continue its process of development indirectly by passing, like *Planocera*, through a *Müller's larva* stage, or like *Stylochus*, through a stage of *Götte's larva;* or instead, like *Notoplana*, it may develop directly without taking on a special larval form.

(3) DIRECT DEVELOPMENT

Direct development takes place in such genera as *Leptoplana, Notoplana, Pseudostylochus, Stylochus, Stylochoplana, Discocoelis* and *Hoploplana*, all of which belong to the polyclad suborder, Acotylea.

---

[2] According to Surface ('07), before $4d$ divides into left and right paired cells it first undergoes this cleavage parallel to the surface of the embryo, but Kato ('40), agreeing with Wilson (1898), does not confirm this. It is not clear whether this is due to an error in observation on the part of Surface, or to a species difference.

Taking mainly *Notoplana humilia* as an example, we find (Fig. 4.4) that as the ciliated gastrula begins to rotate within its shell, its main animal-vegetal (or aboral-oral) axis little by little tends to incline toward the anterior side. Between the ectoderm and the endoderm cells which can be seen in the interior of the embryo, a pair of *mesodermal stands,* formed by the proliferation of cells arising from the mesentoblast, is expanding laterally; these structures make it possible to define clearly the bilateral symmetry of the embryo. The epithelial cilia increase considerably in length, the rotation becomes faster, and rudiments of *rhabdites* begin to appear among the ectoderm cells.

Internally, the yolk-laden cells of the fourth micromere quartet fuse with each other, their cell boundaries disappear, and presently the single mass begins to divide into a large number of cells (Fig. 4.4a, b). The endoderm cells proliferate, spreading in a network around this mass of yolk cells and repeatedly extending their processes into the yolk, so that it is eventually absorbed. The embryo bends anteriorly, so that the aboral pole shifts to its front end and the oral pole finally takes up a position near its posterior end. At this stage the body of the embryo assumes an oval shape, and a pair of *eye-spots* appear in the ectoderm near the aboral pole (two weeks after spawning); these eventually sink from the surface into the mesodermal layer. At this stage, also, the rudiments of the *brain* and *frontal gland* make their first appearance at the aboral pole.

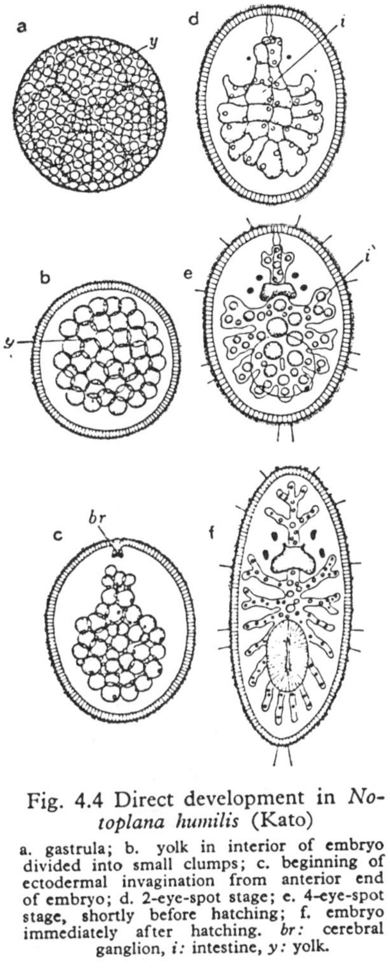

Fig. 4.4 Direct development in *Notoplana humilis* (Kato)

a. gastrula; b. yolk in interior of embryo divided into small clumps; c. beginning of ectodermal invagination from anterior end of embryo; d. 2-eye-spot stage; e. 4-eye-spot stage, shortly before hatching; f. embryo immediately after hatching. *br:* cerebral ganglion, *i:* intestine, *y:* yolk.

The endoderm continues to develop, becoming sac-shaped, and beginning to undergo extensive ramification. The mesoderm is especially well developed in the anterior part of the embryo, although it also forms a large cell mass posteriorly, which will become the *pharynx rudiment.* During this stage the ectoderm in the former blastopore region invaginates in the form of a slender cord which passes through the pharynx rudiment and reaches the lumen of the intestine. A split next appears in this cord, extending through it and connecting with the intestinal space. The whole structure becomes tubular, and develops into the epidermis of the *pharyngeal sheath.*

As development proceeds, annular folds arise in the pharynx rudiment, and the pharyngeal sheath is formed. The mode of formation of the pharynx in *Hoploplana villosa,* however, is rather different, although this species[3] follows

[3] This species, unlike the Japanese *Hoploplana,* undergoes indirect development.

the same direct course of development. As in the American species *H. inquilina*, the pharynx of *H. villosa* has a wide opening from the beginning of its invagination process, and at that stage establishes a connection with the endodermal cell mass in which an intestinal cavity has begun to appear (Fig. 4.3d-f).

Within another week the eye-spots are divided into two pairs, the frontal gland is clearly defined and the musculature of the pharyngeal wall is seen to have a radial arrangement, although the mouth is not yet open to the outside. A small number of immobile *sensory filaments* appear around the edge of the body. Hatching takes place at about 21 days; it is believed that the breaking through of the egg shell is the function of the frontal gland. The larva has already nearly attained the shape of the adult when the opening of the mouth takes place five days after hatching.

(4) INDIRECT DEVELOPMENT

The species which undergo indirect development, forming a Müller's or Götte's larva, are restricted to the polyclad suborder Cotylea and, as will be described below, a very few acotylean species. These larvae resemble the trocho-

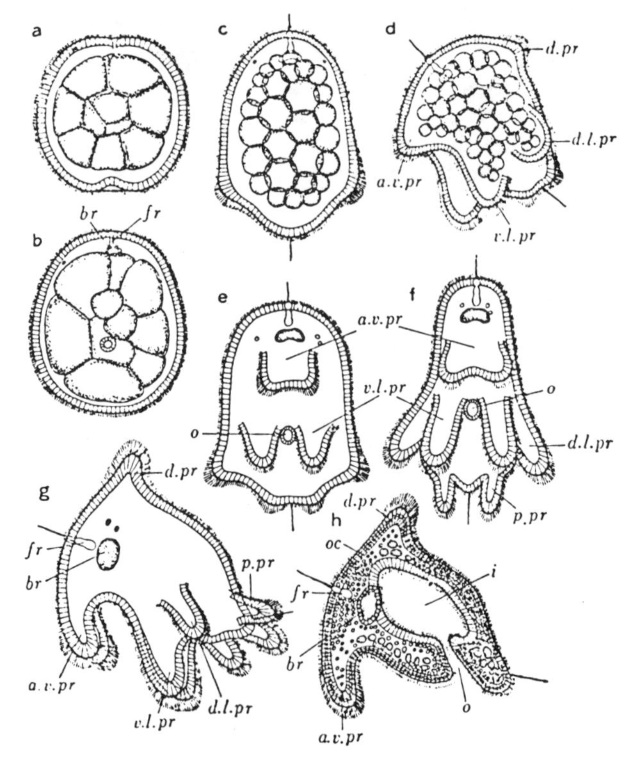

Fig. 4.5 Indirect development in *Planocera multitentaculata* (Kato)

a. gastrula; b. formation of procerebral organ rudiment; c. 2-eye-spot stage; d. side view of embryo at 3-eye-spot stage; e. ventral view of same; f. Müller's larva (ventral view); g. same larva (side view); h. longitudinal section of same. *a.v.pr:* antero-ventral process, *br:* brain, *d.l.pr:* dorsolateral process, *d.pr:* dorsal process, *fr:* procerebral organ, *i:* intestine, *o:* mouth, *oc:* eye-spot, *p.pr:* posterior process, *v.l.pr:* ventrolateral process.

phore larva, but since they are like the pilidium and Desor larvae of the Nemertini in having a simple ciliated band and lacking an anus, these larvae are all included in the category of *protrochula*.

It was Johannes Müller (1850) who first discovered that the Müller larva represents a developmental stage of the Turbellaria. The species forming this larva nearly all belong to the Cotylea, although among the Acotylea, Müller's larvae have been found by Surface ('07) in *Hoploplana inquilina* and by Kato ('40) in *Planocera multitentaculata*. In the latter species, the general course of development is similar to that of the directly developing forms. Five days after spawning, epiboly is complete and the blastopore is closed over; on the sixth day, as in *Hoploplana*, an ectodermal invagination takes place in the vicinity of the original blastopore and a broad, short pharynx, which connects with the intestinal lumen, is formed. Cilia develop on the inner surface of the pharynx and intestine, and about this time an invagination at the aboral pole gives rise to the rudiments of the brain and frontal gland. Only a single eye-spot appears, however, rather far to the left of the body midline. As in direct development, rhabdite rudiments are formed in the ectodermal cells, the embryo bends anteriorly and a bundle of sensory filaments, the *apical sensory tuft*, appears at the anterior end. Internally, as in the former case, the yolk mass is subdivided into small cells (Fig. 4.5c).

Next the ventral side of the embryo becomes rather flattened, and longitudinal depressions appear at the sides of the mouth. The epidermis at the anterior end of these depressions protrudes somewhat, giving rise to the rudiment of an *antero-ventral process*. Following this a *ventro-lateral process* is formed on each side of the mouth, and posterior to these a pair of *dorso-lateral processes*, and finally a single small *dorsal process* is formed. The cells making up these processes are tall columnar, with long cilia. It is considered most probable that these cells arise from the $2a^2$—$2d^2$ quartet (corresponding to the annelidan trochoblasts).

At about the tenth day, another eye-spot is added on the right of the midline, and a second sensory tuft appears at the posterior end. About this time the rotational speed increases, and also the left eye-spot divides, producing a larva with three eye-spots. Each of the processes grows in size, the antero-ventral one especially increasing markedly in width, while a new pair, the *posterior processes*, are added at the posterior end, making a total of eight such processes or *lobes*, and marking the full development to the Müller larva.

Such a larva escapes from its shell on the 14th or 15th day and begins a free-swimming life. It is about 0.2—0.3 mm in length; seen from the side it forms a rounded triangle, with the antero-ventral process as the front angle, the dorsal process forming the top angle and the posterior processes completing the triangle at the rear. Besides the unpaired antero-ventral and dorsal processes there are one pair each of ventro-lateral, dorso-lateral and posterior processes; each process bears long cilia. There is a circular *mouth* located behind the antero-ventral process; from this a narrow *stomodaeum* leads into a spacious *intestine*. Around the stomodaeum is a mass of mesoderm cells, the rudiment of the pharynx. There are sensory tufts on both anterior and posterior ends, and the brain and frontal gland are well developed, although the latter will degenerate a short time after the larva hatches.

According to Lang, who found Müller's larvae of *Thysanozoon* and *Yungia* in his plankton hauls and studied their metamorphosis (Fig. 4.6), the larva flattens as it increases in length, the various processes diminish in size and finally disappear, and for the first time the larval shape begins to resemble that of an adult polyclad. The most striking internal change which accompanies this metamorphosis is the formation of the pharynx. The innermost part of the stomodaeum swells to a spherical shape and a mass of mesoderm cells collects to form a protuberance which will become the pharynx. The pharynx sheath forms in the empty space around the protuberance; the connection between the pharynx sheath and the stomodaeum consists of an extremely narrow slit. At the time of metamorphosis the stomodaeum is pulled outward and its wall forms part of the body wall.

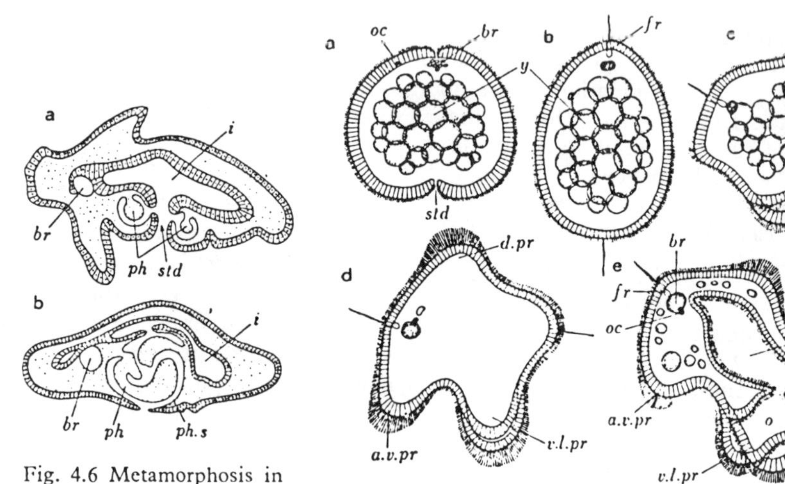

Fig. 4.6 Metamorphosis in *Yungia aurantica* (Lang)

a. longitudinal section of Müller's larva; b. longitudinal section after metamorphosis. *br.* brain, *i:* intestine, *ph:* pharynx of adult, *ph.s:* pharyngeal sheath, *std:* stomodaeum.

Fig. 4.7 Götte's larva of *Stylochus uniporus* (Kato)

a. young embryo; b. 1-eye-spot stage; c. 2-eye-spot stage; beginning of process formation; d. Götte's larva; e. longitudinal section of larva. *a.v.pr:* antero-ventral process, *br:* brain, *d. pr:* dorsal process, *fr:* frontal organ, *i:* intestine, *o:* mouth, *oc:* eye-spot, *std:* stomodaeum, *v.l.pr:* ventro-lateral process, y: yolk.

The Götte larva (Fig. 4.7) was so named by Lang (1884) in honor of its first discoverer; in place of the Müller larva's eight processes, it has only four: an antero-ventral, a dorsal, and a pair of latero-ventrals. The acolydean family of Stylochidae have this type of larva, which can be seen in the Japanese species *Stylochus unipones, S. aomori,* etc.

The developmental process of this larva is essentially like that of *Planocera.* The embryo which reaches this stage is 0.1 mm in length; as stated above it has four processes, the brain is differentiated near the anterior end of the body, and the frontal gland is not very highly developed. It metamorphoses several days after hatching. Lang believed that this larval form was an intermediate stage in the development of a Müller larva, but the results of Kato's ('40) rearing experiments show that metamorphosis occurs without any increase in the number of processes, and it is thus clear that this is an independant larval form.

In the Japanese species, *Planocera reticulata,* development proceeds through the Müller larva stage inside the egg shell, and the young animal emerges only after metamorphosis. Kato ('40) has called this an *intracapsular Müller's larva,* and this mode of development, *intermediate development*. The larva of this type is flattened, its dorsal process is lacking and instead, the tip of the antero-ventral process is bilobed; the intestine is restricted in size.

Another interesting point in connection with the Müller larva is the existence of a small (less than 1 mm in length) turbellarian found in California, *Graffizoon lobata* (Cotylea), which, although it closely resembles the Müller larva, becomes sexually mature. This is regarded as a case of *neoteny* (Heath '28).

5. DEVELOPMENT OF SIMPLE EGGS (ACOELA)

In this group, *Polychoerus caudatus* has been studied by Gardiner (1895) and *Convoluta roscoffensis*, by Breslau (1909).

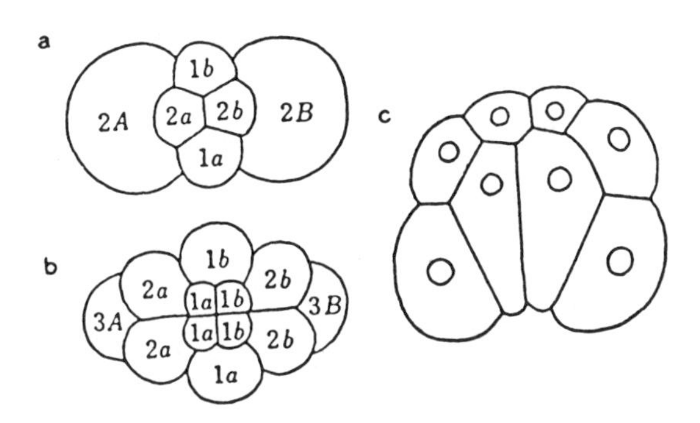

Fig. 4.8 Cleavage in the Acoela

a. *Polychoerus caudatus*, 6-cell stage; b. *P. caudatus*, 16-cell stage (Gardiner); c. *Convoluta roskoffensis*, 12-cell stage (Bresslau); macromeres have already been surrounded by micromeres.

The general mode of development is similar to that of the polyclads, with the exception that only the first cleavage plane includes the egg axis, so that only two macromeres *(A, B)* are formed (Fig. 4.8a, b). Thereafter cleavage proceeds in a spiral fashion, producing the first to the fourth micromere sets as in polyclad cleavage, but with only two micromeres in each set. In *Convoluta,* according to Breslau, the macromeres 3*A*, 3*B* are already in the interior, surrounded by the overlapping micromeres, by the time they form the fourth set of micromeres (Fig. 4.8c). This process occurs somewhat later in *Polychoerus,* but in either case a stereogastrula is formed. The cells filling the interior of this gastrula are chiefly 4*A*, 4*B* and the descendants of 4*a* and 4*b*; it is believed likely that there are also ectomesodermal cells derived from the third set of micromeres. When epiboly is complete and the blastopore is closed, the cells on the surface remain there as the ectodermal layer; internally, a clear distinction has been established between the cells of the endoderm and mesoderm.

The mouth is formed by an invagination of the ectoderm in the region of the former blastopore, but no intestinal cavity appears. There is no free-swimming larval stage, and the definitive adult carries over the structure of the stereogastrula. The nervous system is formed by the first set of micromeres.

6. DEVELOPMENT OF COMPOSITE EGGS (TRICLADIDA)

The development of the Tricladida was early studied by Metschnikoff (1883), Ijima (1884) and Hallez (1887), and later by Mattiesen (1904) and Fulinsky (1914, 1916).

In these species it is usual to find, within one egg shell (capsule), a large number of yolk cells — sometimes as many as several thousand — surrounding a few egg cells. In the ordinary planaria there are usually four to six eggs, while in *Dendrocoelum lacteum* there are said to be 20—40. The cleavage of these eggs is extremely peculiar, in that the blastomeres become isolated from each other and distribute themselves among the yolk cells. Under these conditions it is difficult to find evidence of spiral cleavage, and in addition, regular germ layer formation does not take place. The yolk cells surrounding the blastomeres proceed to fuse together, forming a syncytium, while some of the blastomeres transform into wandering amoeboid cells and, separating from the other blastomere masses, migrate to the vicinity of the yolk syncytium. There they form a thin membrane, which acts as a boundary between the ordinary yolk cell masses outside and the yolk syncytium inside. This membrane is called the *outer membrane*, or *provisional ectoderm* (Fig. 4.9).

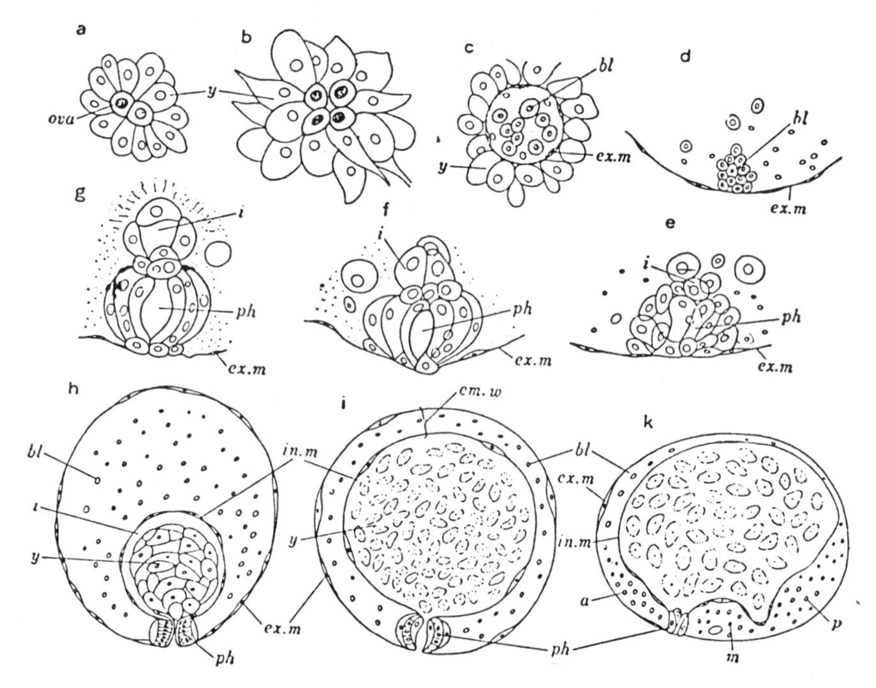

Fig. 4.9 Development of the Tricladida (Ijima, Mattiesen, Fuliński)

a, b. early cleavage stages; c. formation of external membrane (temporary ectoderm); d. aggregation of blastomeres toward ventral side; e ~ g. differentiation of embryonic pharynx and internal membrane (temporary intestine); h. absorption of yolk into temporary intestine; i. formation of embryonic wall by extension of temporary intestine; k. germinal cord formed at ventral side and anterior, mid- and posterior cell masses appear. *a:* anterior cell mass, *bl:* blastomere, *em.w:* embryonic wall, *ex.m:* external membrane, *i:* temporary intestine, *in.m:* internal membrane, *m:* mid-cell mass, *ova:* ovocyte, *p:* posterior cell mass, *ph:* temporary pharynx, *y:* yolk cell.

At the same time another group of wandering cells assembles at one point on this membrane; eventually a cavity appears in this cell mass and the whole thing takes the shape of a small sac. Next a mouth opens to the outside of the membrane and an *embryonic pharynx* is formed, at the inner end of which is a sac that functions as a *temporary intestine.* The wall of this structure is also called the *internal membrane*; as the embryonic pharynx sucks in the yolk from the outside, the temporary intestine swells enormously and its wall, pushing away the blastomeres and yolk which lie outside of it, gradually approaches the outer membrane, and together with it, forms the *embryonic wall* (Fig. 4.9 h, i).

Within the embryonic wall the blastomeres, suspended in yolk, proliferate actively, and the resulting cells eventually gather at the ventral side, which is determined by the embryonic pharynx, and form the *germinal cord* along the mid-ventral line. This embryonic cord consists of three cell masses, the middle one near the embryonic pharynx, and the others lying anterior and posterior to it.

The embryo eventually becomes flattened, and the embryonic pharynx as well as the inner and outer membranes begin to degenerate. However, the cells of the part facing the inner surface of the germinal cord are destined to become the endoderm of the adult; they therefore spread out in the region where the inner membrane is degenerating, and there form the wall of the intestine. The other parts of the germinal cord give rise to the ectoderm and endoderm; in particular the anterior cell mass forms chiefly the ectoderm of the anterior part of the body, taking the place of the degenerating outer membrane to become the *adult epithelium.* Furthermore, two groups of cells in close contact with this epithelium differentiate into *cerebral ganglia.* These two ganglia eventually fuse into a single unit from which the other nerves of the body grow out. In the central part of the embryo, the embryonic pharynx degenerates; shortly afterward spaces appear in the cellular mass and the definitive pharynx takes shape around them. In this species there is no clear-cut stomodaeal invagination.

7. DEVELOPMENT OF COMPOSITE EGGS (RHABDOCOELA AND ALLOCOELA)

The development of these species has been investigated by Hallez (1909), Bresslau (1904), Ball (1916) and others. Usually each egg capsule contains one egg surrounded by several hundred yolk cells, although in *Plagiostoma girardi* there are 10—12 eggs in a capsule. The number of yolk cells varies widely with the type of the egg. In the rhabdocoel family Typhloplanae, the genera *Mesostoma, Typhloplana, Bothromesostoma,* etc. lay two types of eggs: small *summer* (or *subitaneous) eggs* with thin shells, and large, thick-shelled *winter eggs* (also called *dauerei, latent* or *dormant eggs).* Summer eggs have few yolk cells, and winter eggs many; the latter resemble ordinary rhabdocoel eggs. In the species *Mesostoma Ehrenbergi,*

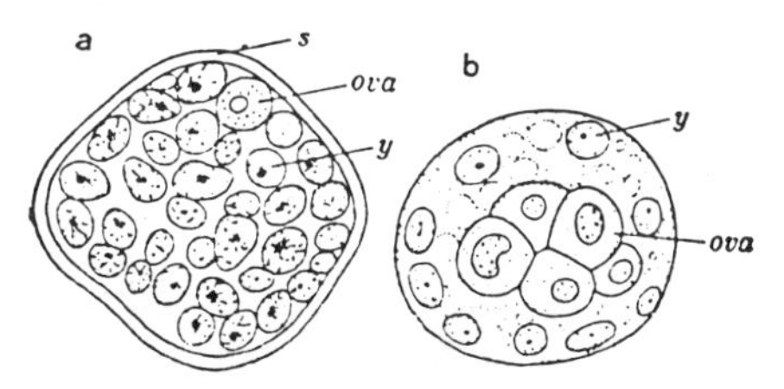

Fig. 4.10 Ova of *Mesostoma ehrenbergi* (Bresslau)

a. latent or dormant egg, within thick shell; b. subitaneous egg, cleaving, *ova:* egg, *s:* egg shell, *y:* yolk cell.

which is famous because of its large size, the worms which hatch in the spring from winter eggs form summer eggs before the copulatory organs are fully developed; these are self-fertilized and hatch and grow to maturity within the parental body, finally rupturing it to emerge. This occurs repeatedly during the summer, until the reproductive organs finally reach maturity in the autumn and the animals copulate and form winter eggs. Only at the death of the parent are these released into the water, where they pass the winter.

As in the eggs of the Tricladida, the development of these species has also undergone extreme modification. Although a faint trace of spiral cleavage persists, it is difficult to follow the cell lineage, and no process resembling germ layer formation through gastrulation has been seen.

As the egg, surrounded by yolk cells, proceeds to cleave, the blastomeres form into clusters which gather at one side of the yolk. During this time the yolk cells fuse together and form a syncytium. The part where the blastomere masses are collected will be the ventral side of the future embryo, while the yolk region will be the dorsal side. In the triclads, the embryonic pharynx and other temporary structures form at this stage, but in these species no such performance takes place. However, the masses of blastomeres eventually take the form of a germinal cord, exactly as this appears in the triclads, with anterior, central and posterior masses lined up along the antero-posterior axis. The cells of the germinal cord near the surface of the embryo become arranged in a layer and form the epithelium of the ventral surface; this gradually spreads until it extends to the dorsal surface, enclosing the yolk mass. The cells on the opposite side of the germinal cord — i.e., those in contact with the yolk — become endodermal, and as they digest the yolk, form the intestinal sac. As in the triclads, a pair of cerebral ganglia appear in the anterior cell mass and later become connected; the general nervous system proliferates from these ganglia. The central cell mass forms a pharynx, the inner surface of which is covered with ectoderm by a cord-like invagination from the surface layer. This invagination constitutes the stomodaeum, which does not appear in the triclads. However, as in the triclads, the cells of the posterior cell mass form the posterior portion of the body.

REFERENCES

Ball, S. 1916: Development of *Paravortex gemellipara*. Jour. Morph., **27**.

Bresslau, E. 1904: Die Entwicklung der Rhabdocölen. Zeitschr. wiss. Zoöl., **76**.

——— 1909: Die Entwicklung der Acölen. Ver. d. Zoöl. Ges., **19**.

——— 1930: Turbellarien. Kükenthal's Handb. Zoöl., **2**.

Fulinski, B. 1914: Entwicklungsgeschichte von *Dendrocoelum*. Bull. Acad. Sc. Krakau.

Gardiner, E. G. 1895: Early development of *Polychoerus caudatus*. Jour. Morph., **11**.

Götte, A. 1882: Entwicklungsgeschichte der Würmer. Hamburger, Leipzig.

Hallez, P. 1889: Embryogenie des Dendrocoeles d'eau douce. Paris.

Hartman, M. 1925: Mesozoa. Handb. d. Zool. Kükenthal u. Krumbach, **1**.

Heath, H. 1928: A sexually mature turbellarian resembling Müller's larva. Jour. Morph. Physiol., **45**.

Hofsten, N. 1912: Eischale und Dotterzellen bei Turbellarian und Trematoden. Zool. Anz., **39**.

Hyman, L. H. 1951: The Invertebrates. 2. McGraw-Hill, New York.

Ijima, J. 1884: Bau und Entwicklungsgeschichte der Tricladen. Zeitschr. wiss. Zoöl., **40**.

Kato, K. 1940: On the development of some Japanese polyclads. Jap. Jour. Zool., **8**.

Lang, A. 1884: Polycladen, Fauna u. Flora Golf v. Neapel, **11**.

Mattiesen, E. 1904: Embryologie der Süsswasser Dendrocölen. Zeitschr. wiss. Zoöl., **77**.

Nouvel, H. 1947—48: Les Dicyémides. Arch. Biol., **58, 59**.

Stunkard, H. W. 1954: The life-history and systematic relations of the mesozoa. Quart. Rev. Biol., **29**.

Surface, F. M. 1907: Early development of a polyclad, *Planocera inquilina*. Proc. Acad Nat. Sci. Philadelphia, **59**.

Wilson, E. B. 1898: Considerations on cell lineage and ancestral reminiscence. Ann. N. Y. Acad. Sci., **11**.

NEMERTINI

Chapter 5

INTRODUCTION

The Nemertini are divided into two main groups: the Anopla, which lack stylets on the proboscis and have the mouth located posterior to the brain; and the Enopla, in which stylets are present on the proboscis and the mouth lies in front of or below the brain. A summary of the classification is as follows:

Nemertini	Anopla	Paleonemertea —	*Tubulanus, Cephalotrix*
		Heteronemertea —	*Micrura, Lineus, Cerebratulus*
	Enopla	Hoplonemertea —	*Prosorhochmus, Emplectonema, Prostoma*
			Stichostemma, Tetrastemma, Drepanophorus
		Bdellonemertea —	*Malacobdella*

Development among the Nemertini may be either direct or indirect; in general the former category includes the Paleonemertea and the Enopla, while indirect development is a peculiarity of the Heteronemertea. However it must be admitted that the number of species which has been adequately investigated is very small. Historically speaking the Heteronemertea have been studied for the longest time; after Johannes Müller (1847) had given the name of *Pilidium gyrans* to a little swimming animal which he found in plankton taken off the island of Heligoland, he observed (1854) that this passed through a metamorphosis to become a young nemertinean worm. Since that time there have been reports concerning the development of this helmet-shaped or pilidium larva and its metamorphosis published by Metschnikoff (1869) Bütschli (1873) Burger (1894), Coe (1899) C.B.Wilson (1900) Salensky (1886, 1912) and others.

On the other hand, Desor (1850) observed that the *Lineus ruber* egg (O.F. Müller) does not form a swimming larva, but instead, following the cleavage period, rapidly acquires a ciliated epithelium and revolves inside the egg membrane; after completing the internal formation of an adult epithelium, it finally bursts out of the membrane. Investigations concerning the development of the larva of this genus, which is called "*Desor's larva*", have been published by Barrois (1877), Hubrecht (1886), Arnold (1898), Nusbaum and Oxner (1913), Schmidt (1934) and others. Studies on the forms showing direct development, besides that of Ikeda (1915) on *Stichostemma grandis*, have been published by Dieck (1873), Barrois (1877), Salensky (1884, 1909, 1914) Lebedinsky (1897), Coe (1904), Hammersten (1918), Reisinger (1926), Dawydoff (1928) and others.

The sexes are separate in almost all the nemertinean worms, except for a few hermaphroditic genera among the Hoplonemertea. In the Paleonemertea and Heteronemertea, the small eggs are shed almost simultaneously from the ovaries, which are aligned segmentally along the body wall; while in most of the Holonemertea, spawning takes place continuously, at intervals of a day or a week, one large egg maturing and being released from each ovary.

Fertilization is usually external, although there are a few viviparous species (*Prosorhochmus viviparus, Lineus viviparus, L. biliniatus, Geonemertes australiensis, G. agricola, Prostoma lacustre, Stichostemma graecense),* in which the young larva is formed inside the body of the mother. In *Cephalothrix galathea, C. rufifrons* and *Carcinonemertes carcinophila* the eggs are fertilized internally and then spawned.

For observation of development and experimentation in the laboratory, the heteronemerteans *Lineus, Cerebratulus* and *Micrura* are the most suitable,

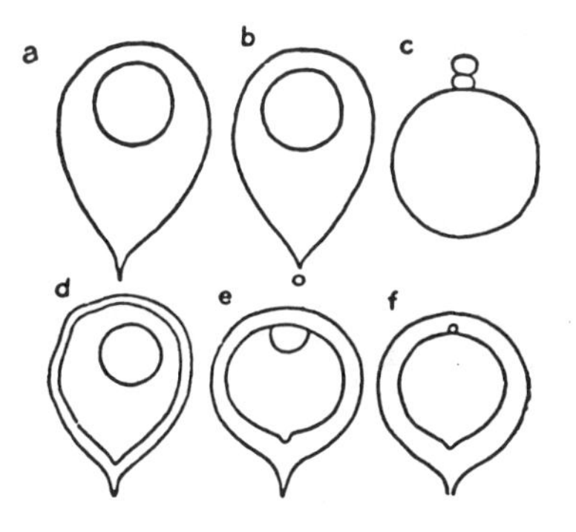

Fig. 5.1 Changes in shape of conical protrusion in egg after oviposition (E.B. Wilson)

a ~ c. egg of *Lineus torquatus* (egg membrane removed. x140) a. just after oviposition; b. protrusion separates as vacuole; c. formation of first and second polar bodies after fertilization (Iwata); d ~ f. *Cerebratulus lacteus* d. just after oviposition; e. at 1 hour; formation of first polar spindle; f. formation of first polar body after fertilization.

and among these three, *Cerebratulus* is the best. Females collected during the breeding season and kept in sea water tanks will spawn, and if males and females are placed together it is very easy to secure material for observation of developmental stages. The breeding season differs according to species and locality; the eight species studied at Akkeshi (Hokkaido) all breed during the summer, particularly in the latter part of July and the first ten days of August. At Asamushi (northern Honshu) *Emplectonema gracile* breeds in April. *Lineus alborostratus* breeds during August at Akkeshi and in April at Onomichi (near Hiroshima).

To obtain unfertilized eggs, a piece of the body is cut off, and longitudinal cuts are made along the sides; the muscular movements of the animal will then cause the eggs to be expelled. These should be transferred with a pipette to fresh sea water. In such eggs the germinal vesicle is still intact and clearly visible, and normal fertilization does not take place. Immediate insemination usually results in polyspermy, but if the eggs are left uninseminated, the germinal vesicle soon breaks down in response to the sea water stimulus, and the first polar spindle is formed. Preparation for polar body formation is completed between 10 and 30 minutes after the eggs come in contact with the sea water, the time depending on the temperature and the degree of maturation of the eggs. The eggs are then ready to be fertilized. If a small incision is made in the body wall of a male worm, sperm exudes as a white, viscous thread-like mass. A very small amount of this stirred into the egg suspension with a pipette will bring about fertilization of the eggs. In the case of the pilidium larva, it is necessary to provide food for the swimming stage that precedes metamorphosis, in the form of small diatoms, ciliates, etc., care being taken to keep from fouling the culture medium. By this method the early cleavage stages have been followed and the developmental capacity of each blastomere determined in the studies of E.B.Wilson (1903), Yatsu (1904, 1910), Zeleny (1904) and Hörstadius (1937). According to their results, each of the two blastomeres resulting from the first cleavage is able to give rise to a complete larva of half size, while defects appear in larvae reared from separated 4-cell stage blastomeres.

1. EARLY DEVELOPMENT

(1) CLEAVAGE

The eggs which contain a large amount of yolk are spherical when spawned, but those of *Cerebratulus* and *Lineus*, which have little yolk, are rather irregular in shape (Fig. 5,1), with a conical protuberance at one side, and within the cytoplasm a germinal vesicle of which the outline is clearly visible. Within about 10—20 minutes after shedding, whether fertilization has taken place or not, the germinal vesicle breaks down and its material, distinguishable as a clear mass, begins to move toward the side of the egg opposite the conical protuberance. By the time the first polar body is formed, the clear mass has reached the opposite pole. According to E.B. Wilson, in *Cerebratulus lacteus* the cone-shaped protuberance disappears at this time in most cases, but in *Lineus torquatus* (unpublished observation) it pinches off from the egg as a small vesicle before the egg becomes spherical (Fig. 5.1a, b). In these eggs the egg axis is believed to be already determined before fertilization: the side with the conical protuberance is the vegetal pole, at which the egg is attached to the ovarian wall, while the side to which the germinal vesicle substance moves, and where the first and second polar bodies are given off after fertilization, is the animal pole.

Cleavage is of the typical spiral type. After expulsion of the polar bodies, the egg is divided into two equal-sized blastomeres by the first cleavage plane (Fig. 5.2a), which passes close to the polar bodies and parallel to the egg axis. At the second cleavage, a longitudinal plane forming a right angle with the first cleavage plane produces four blastomeres of equal size (Fig. 5.2 b,c). The third cleavage divides the egg along the equatorial plane, forming four equal-sized "macromeres" (1*A*—1*D*) at the vegetal side, and dextrally rotated with respect to these, four equal-sized "micromeres" (1*a*—1*d*). At this point it is important to note that the so-called "micromeres" are larger than the "macromeres", a characteristic which is not found in any of the other spirally cleaving groups (Fig. 5.2d). At the fourth cleavage the blastomeres are divided by furrows parallel to the equatorial plane, this time with a sinistral rotation as opposed to the dextral

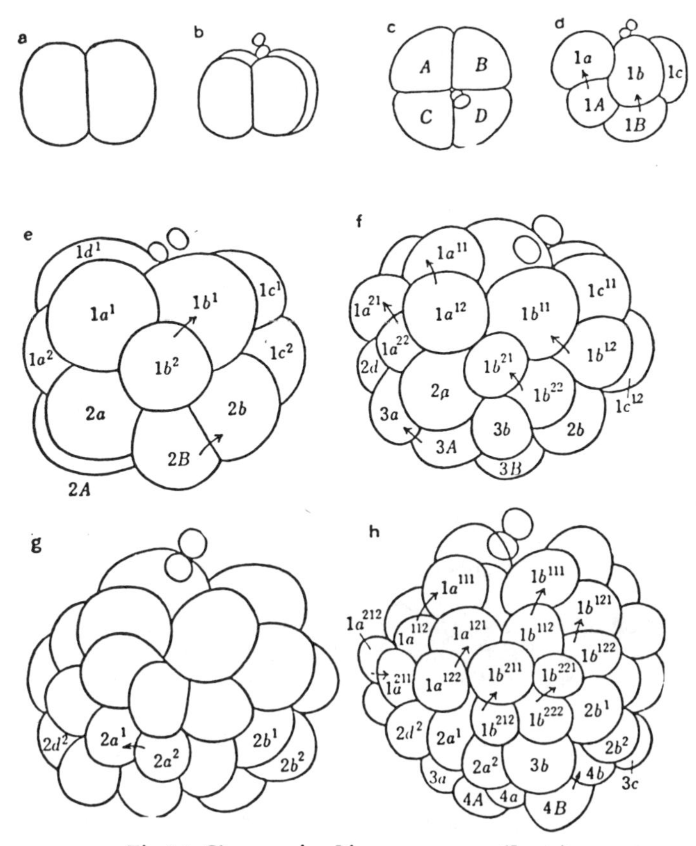

Fig.5.2 Cleavage in *Lineus torquatus* (Iwata)

a. 2-cell stage; b. 4-cell stage (side view); c. 4-cell stage (from animal pole); d. 8-cell stage; e. 16-cell stage; f. 28-cell stage; g. 32-cell stage; h. 52-cell stage. (a ~ d. x160; e ~ h. x300)

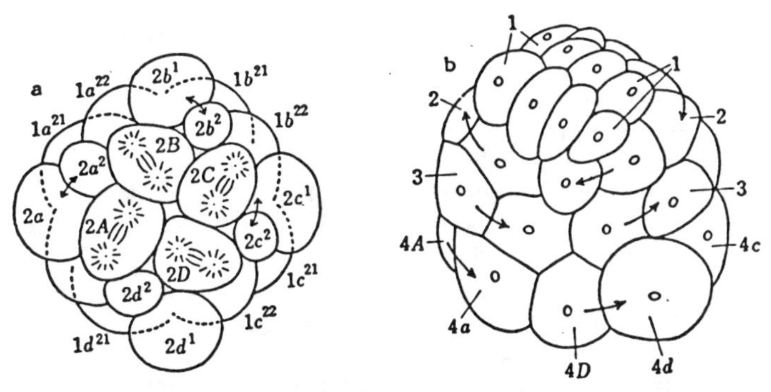

Fig. 5.3

a. Embryo of *Malacobdella grossa* at 28-cell stage, seen from vegetal side. (Hammarsten) b. Embryo of *Lineus ruber* at 64-cell stage in side view. (Nusbaum u. Oxner)

rotation of the third cleavage (Fig. 5.2e). This produces eight blastomeres in the animal hemisphere; an upper ring, $1a^1$—$1d^1$ and a lower ring $1a^2$—$1d^2$; and eight in the vegetal half: four secondary micromeres, *2a*—*2d*, above, and four macromeres, *2A*—*2D*, below, making up the 16-cell stage. The fifth cleavage is accomplished with a second dextral twist, the 8 animal hemisphere blastomeres dividing almost simultaneously, forming the four sets of blastomeres: $1a^{11}$—$1d^{11}$, $1a^{12}$—$1d^{12}$, $1a^{21}$—$1d^{21}$ and $1a^{22}$—$1d^{22}$. The cleavage order of the 8 vegetal blastomeres varies according to the species; in *Lineus* and *Cerebratulus* the lowest group (*2A*—*2D*) divides first to form the third quartet of micromeres *3a*—*3d* (Fig. 5.2 f) and the third quartet of macromeres *3A*—*3D*, resulting in a 28-cell stage (Fig. 5.2 g). The order is reversed in *Emplectonema* and *Malacobdella*, the upper cells dividing before the lowermost. In the 32-cell stage of *Malacobdella*, the blastomeres of the $2a^1$—$2d^1$ quartet project outward, giving the embryo a square outline when seen from the vegetal side (Fig. 5.3a). Reports of later cleavages are available to the sixth cleavage in *Malacobdella* and *Lineus*, and to the eighth cleavage in *Emplectonema* and *Cerebratulus*; in all cases right and left torsion continue to alternate.

(2) GASTRULATION

When the embryo reaches the *blastula stage*, its surface becomes thickly covered with delicate cilia, and it begins to rotate slowly within the egg membrane. The speed of rotation increases with time, and eventually what looks like one long flagellum can be detected at the animal pole. This is the *apical tuft*, composed of a number of long, slender cilia which form a bundle. At this stage the embryo makes its way out of the egg membrane and swims free in the sea water, while the animal pole region which bears the apical tuft becomes pointed and the vegetal side flattens like a flask shortened in its vertical axis (Fig. 5.4a,c). By the time a circular *blastopore* can be detected in the center of the flattened vegetal side, the embryo has entered the *gastrula stage*, and swims rapidly forward in a spiral fashion with the animal pole in the lead. In the following *pilidium larval stage*, the ectoderm of the vegetal side extends down around the blastopore, forming the *lateral lappets*, and the larva acquires complete bilateral symmetry and assumes the shape characteristic of this type (Fig. 5.4e). The animal pole side will become the dorsal side of the adult form, while the vegetal region will form the ventral side. The blind end of the *archenteron* turns toward one side of the embryo as it elongates; this side will be the future posterior end, and the opposite side will form the anterior, or head end.

In directly developing larvae, the ectoderm in the vegetal pole region expands one-sidedly, giving the embryo an oval shape and gradually shifting the blastopore toward the apical tuft. The part bearing the apical tuft becomes the head region, and the vegetal pole forms the posterior end of the body. The side traversed by the shifting blastopore becomes the ventral side, and the opposite side forms the dorsal side of the worm (Fig. 5.9).

The blastocoel appears early in the cleavage period, and during the blastula stage may be large or small, depending on the species. Those with little yolk, which will eventually become pilidium larvae, have large blastocoels, while the yolky eggs have small ones. During the gastrula stage, this blastocoel becomes smaller as the result of the invagination of the endoderm and the formation of mesenchyme. In the blastula stage of pilidium larvae, a dish-shaped indentation

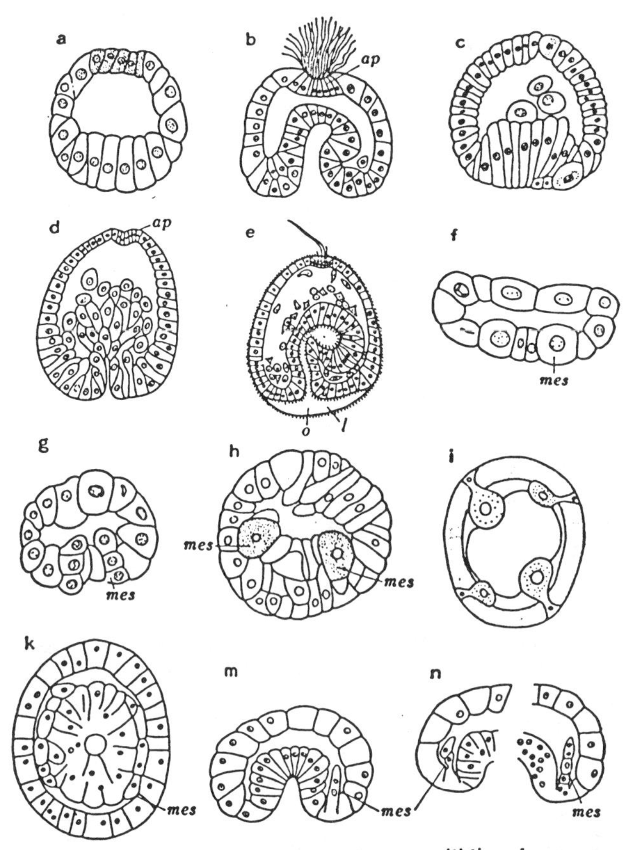

Fig. 5.4 Blastula, gastrula and young pilidium larva

a. b. *Cerebratulus lacteus* (C. B. Wilson); c ~ e. *C. marginatus* (Coe); f, g. *Prosorochmus viviparus* (Salensky); h, i. *Malacobdella grossa* (Hammarsten); k ~ n. *Lineus ruber* (Arnold). *ap:* apical plate, *l:* lateral lobe, *mes:* mesoderm, *o:* mouth.

appears at the animal pole. This part is formed of closely packed, columnar cells which are smaller than those of the surrounding ectoderm; a little later this will become the *apical plate* and give rise to the long cilia described above. With the appearance at this time of a pair of mesoderm mother cells at two sides of the vegetal pole, the embryonic symmetry changes from radial to bilateral. In the gastrula stage, the endoderm cells invaginate to form a pouch-shaped archenteron; since the blastopore moves inward together with the neighboring ectoderm cells, the larval mouth does not coincide with the blastopore. In the pilidium larva, the blastopore remains open during the larval period.

The reports with respect to mesoderm formation are nearly all fragmentary, although full accounts have been published by Nusbaum and Oxner on *Lineus ruber* and by Hammarsten on *Malacobdella grossa*. However, no consistent conclusion treating mesoderm formation conjointly with the development of the coelom has been put forward. At the 64-cell stage of *Lineus ruber*, one blastomere of the fourth micromere quartet, the 4*d* blastomere, which is especially large and further distinguishable by its color, is recognized to be the *mesoderm mother cell* (Fig. 5.3b);

this blastomere later becomes two cells situated on opposite sides of the blastopore. Like their counterparts in annelids and molluscs, these proliferate inward to form *strands*, extending into the blastocoel, in which the *body cavity* later develops. According to E.B. Wilson, in *Cerebratulus marginatus* a pair of large cells can be distinguished in the late blastula stage just before gastrulation begins, lying on opposite sides of the flattened vegetal surface. These later proliferate inward and become *mesenchyme*, instead of forming strands; moreover, they do not arise by division of one mesoderm mother cell, but are mesoderm cells which originate from widely separated ectoderm cells. On the other hand, Coe (1899) recognized as mesoderm one large cell located at the vegetal pole, and besides this certain cells which separate from the endoderm, but he expressed the opinion that these latter were pathological cells resulting from an abnormality of the egg, and that the mesoderm probably takes its origin, as in the annelids, from one blastomere.

The reports cited above hold the mesoderm mother cells to be a pair of blastomeres lying symmetrically on two sides of the blastopore, but Lebedinsky maintains that in *Tetrastemma* and *Drepanophorus*, the mesoderm mother cells are four cells which form two pairs lying anteriorly and posteriorly to the blastopore, and later proliferating inward as the strands which give rise to the coelom. Moreover, in *Malacobdella* Hammarsten recognizes as mesoderm the four blastomeres $2a^{111}$—$2d^{111}$, which arise from the second micromere quartet $2a$—$2d$ and lie around the blastopore, and states that two of these blastomeres lying diagonally opposite each other are larger than the other two (Fig. 5.4 i). These two blastomeres proliferate into the blastocoel, but do not form strands such as are found in annelid larvae. Adhering to the body wall, the descendants of these cells give rise to *muscle*; or they may be suspended in the body cavity or become attached to the nervous system, proboscis rudiment, gut, etc.. In this same species, in addition to this kind of mesoderm, mesenchyme cells originating from the ectoderm are formed during later development. Dawydoff distinguishes two types of origin of the mesoderm in *Tubulanus*. In one of these, the $3D$ blastomere, which does not continue cleaving after the 32-cell stage but is distinguished by its structure rather than its size, later becomes two cells which lie symmetrically at the two sides of the early blastopore, and continue to divide in the blastocoel into a small number of cells. In the other type of origin, from the ectoderm, he states that four cells, arising from the $2a$—$2d$ quartet of blastomeres and surrounding the blastopore in radial formation, divide to become mesenchyme.

2. LARVA AND METAMORPHOSIS

(1) INDIRECT DEVELOPMENT

The forms which have been found to pass through the pilidium stage are two species of *Cerebratulus*, two species of *Micrura* and one species of *Lineus*. In addition, four species of this type of larva discovered in plankton hauls have been named. Although the developmental processes of the pilidium and Desor larvae are somewhat different, their mode of metamorphosis is the same. The former, assuming its characteristic shape, develops as it swims free in the the sea and feeds, and moulting after metamorphosis, takes on the adult form; while the latter grows by digesting and absorbing the yolk of its own and sometimes of other eggs (see below), differing from the pilidium in that it does not swim or feed. However, the author has observed that the larva of *Micrura akkeshiensis* swims although it does not have the pilidium morphology, digests and absorbs the egg yolk, forms the various organ systems and moults. It thus shows a mode of development just intermediate between the pilidium and Desor larvae (unpublished observation).

a. **Pilidium (Helmet Larva)**

One day after the larva has escaped from the egg membrane and begun to swim, the blastopore sinks inward and the *mouth* is formed external to it. This region between the larval mouth and the blastopore forms the *foregut,* which will become the *esophagus* of the adult. The part extending crookedly posterior is the *gut,* which will be the *intestine* in the future adult. The cells of the foregut are flat like those of the ectoderm, and carry short cilia. According to Coe (1899), a valve develops near the blastopore and regulates the entrance and exit of food particles. He states that this region is formed of a small number of non-ciliated amoeboid cells which send out processes to make the valve aperture narrower, or widen it by retracting them.

The characteristic pilidium shape is achieved within 1—2 weeks after shedding the egg membrane. That is, the ectoderm on the two sides of the oval mouth opening extends downward to form the lateral lappets. The cells along the lower edges of these are large and crowded with long cilia. In some species the lateral lappets are extremely conspicuous, while in others they are less so; the larvae of *Cerebratulus lacteus* (Fig. 5.5a) and *Pilidium gyrans* belong to the former group, while that of *Micrura caeca* belongs to the latter (Fig. 5.5c). In *M. caeca* the dorsal

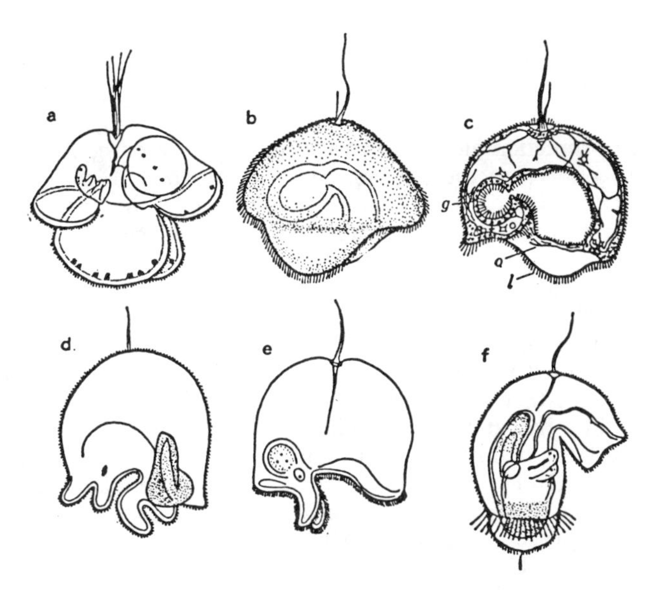

Fig. 5.5 Pilidium larvae of various species
a. *Cerebratulus lateus*; b. *C. marginatus*; c. *Micrura caeca* (Coe); d. *Pilidium brachiatum* (E. B. Wilson); e. *P. auriculatum* (Leuckart u. Pagenstecher); f. *P. recuvatum* (Fewkes).
g: gut, *l:* lateral lobe, *o:* mouth.

ectoderm is umbrella-shaped, causing the back of the larva to be broader than the ventral side. On the posterior wall of the mouth a protruding *lip* bearing especially long cilia extends forward and downward. The ingested food enters the mouth through the anterior part, while the undigested material is expelled to the outside, along the back wall in the region of the lip.

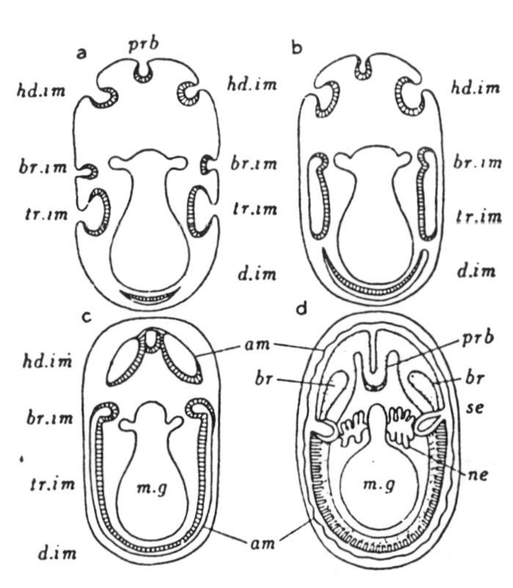

Fig. 5.6 Diagrammatic sketches of pilidium larva during metamorphosis (Korschelt u. (Heider)

am: amnion, *br:* brain, *br.im:* imaginal disc of brain, *d.im:* dorsal imaginal disc, *hd.im:* imaginal disc of head, *m.g:* mid-gut, *ne:* nephridium, *prb:* proboscis rudiment, *se:* head sensory organ, *tr.im:* imaginal disc of trunk.

At an early stage, mesenchyme cells can be recognized suspended in the transparent body fluid filling the coelom between the body wall and the wall of the gut; by the pilidium stage these have differentiated into amoeboid mesenchyme and muscle cells bearing long *muscle fibers*. These long, slender fibers of the muscle cells join together the various parts of the larval body and direct its movements. The mesenchyme forms the muscular layers of the future adult body. According to observations on *Micrura caeca* reported by Coe (1899), the muscle cells are densely aggregated on the underside of the apical plate and the lower edges of the lateral lappets; they are also attached to the body wall and the wall of the gut. The most conspicuous among the groups of muscle fibers are two strong bands extending from the apical plate to the lower borders of the left and right lateral lappets; the next most prominent are strong bundles running in an antero-posterior direction along the two sides of the mouth. A small number of muscle fibers connecting the anterior part of the foregut with the apical plate can also be distinguished, as well as a few bundles in the anterior and posterior parts of the larva which send out many branches, and certain groups connecting the gut with the body wall.

Among the four species of pilidium larvae discovered as plankton which have been named although the adult species remains unidentified, *Pilidium brachiatum* is helmet-shaped with an umbrella-like ectoderm and a pair of small lateral lappets divided into two parts (Fig. 5.5d); in *P. recuvatum* (Fig. 5.5f) the body is retort-shaped, lacking lateral lappets and having a ring of long cilia around its lower part like the annelidan trochophore larva.

Metamorphosis is initiated by invaginations of the ectoderm at several places around the circumference of the lower part of the umbrella; these invaginations become connected to each other, secondarily forming a new body wall surrounding the gut, and the process is completed when cilia appear on this new body wall and the larval skin is burst and cast off. The invaginations occur at eight places, six of these being paired and two unpaired. The paired ones are the *cephalic discs*, lying at the two sides of the head end (Fig. 5.6, *hd. im*), the *trunk discs*, beside the gut (Fig. 5.6, *tr. im*) and the *cerebral discs*, located between the other two pairs, beside the foregut (Fig. 5.6, *br. im*); the unpaired ones are the *proboscis disc,* at the anteriormost part of the head end (Fig. 5.6, *prb*), and the *dorsal disc,* at the posterior end (Fig. 5.6, *d. im*). The proboscis disc has been recognized only in the Desor larva, not having been confirmed in pilidium larvae.

The invaginations first appear as hollows in the larval body wall, which gradually become deeper and at the same time spread out parallel to the body wall, while their openings eventually close. The inner walls of the pouches formed in this way are made up of deeply columnar cells; later they become many-layered and form the adult epithelium. This part is called the *imaginal disc* or *blastodisc.* The outer wall, composed of a single layer of flattened cells, is very thin and is called the *amnion.* The *amniotic cavity* lies between these two layers. The dorsal disc alone is not formed by invagination; in this case a few cells of the posterior body wall divide, bringing about the formation of two layers (delamination), which later proliferate, forming blastodisc and amnion.

Early in the blastodisc formation stage, mesenchyme cells arrange themselves in a single layer over the inner face of the blastodisc which faces the body cavity and the outer surface of the gut; as blastodisc development proceeds, these form a single-layered membrane which eventually becomes multi-layered, giving rise to the musculature of the body wall, the proboscis muscles and the mesenchyme of the coelomic fluid.

The development of the several blastodiscs and the order of their fusion is as follows (Figs. 5.6, 5.7). First the left and right cephalic discs grow anteriorly and fuse together, simultaneously uniting with the proboscis disc. The trunk discs gradually extend posteriorly as well as ventrally, eventually joining each other at the mid-ventral line; during this time they also fuse anteriorly with the cerebral discs. The dorsal blastodisc appears late, and unites with the trunk disc which advances upward from the ventral side. As this process continues, the cephalic and cerebral discs unite at the sides of the body in the anterior part, while posteriorly the dorsal disc spreads forward and sidewise, finally joining the cephalic disc.

Independently of this body wall formation process, the proboscis disc forms a pouch in the upper part of the body cavity and gradually grows posteriorly. Salensky does not recognize the formation of a proboscis disc, maintaining that the proboscis takes its origin in an invagination of the upper layer of the blastodisc formed by the union of the left and right cephalic discs. The *brain,* formed on the left and right sides by the proliferation of ectoderm cells which appear in the cephalic discs, sinks into the body cavity. It gives off the ventral *ganglia* below and at the same time the left and right ganglia are connected by the *dorsal* and *ventral commissures.* Since the lateral nerves are formed well before the union of the dorsal with the cephalic blastodiscs, they grow out of the ventral ganglia and gradually extend posteriorly, running between the blastodisc and the mesenchyme. The cerebral discs do not participate in the formation of the body wall. Later they become tubular and extend into the body cavity; the cells at their tips proliferate, forming the *cerebral sense organs* and making connection with the dorsal ganglia. The proboscis disc becomes the single-layered *proboscis epithelium*; the mesodermal mesenchyme attached to its inner side forms a many-layered *proboscis sheath.* This undergoes delamination and differentiates inward as *proboscis muscle,* while the space between the layers develops into the *rhynchocoel.*

The regions of the adult epithelium formed by each of the respective blastodiscs are: anterior tip of head to posterior end of lateral cephalic grooves — cephalic discs; ventral side — trunk discs; dorsal and lateral body walls — dorsal disc. *Nephridia* appear as a pair of ectodermal invaginations between the larval mouth and the cerebral discs, and these openings persist until a late developmental stage. After this, however, they are closed off and *efferent ducts* are secondarily formed.

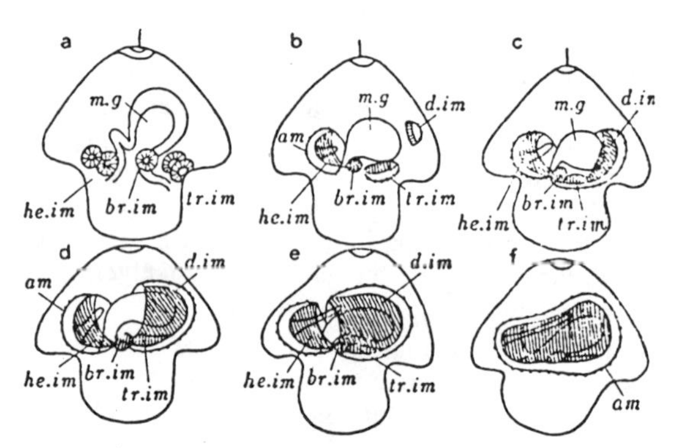

Fig. 5.7 Diagrammatic sketches of pilidium larva during metamorphosis (Salensky)

am: amnion, *br.im:* imaginal disc of brain, *he.im:* imaginal disc of head, *m.g:* mid-gut, *tr.im:* imaginal disc of trunk.

There is no doubt that the mesenchyme which is massed against the blastodiscs becomes multi-layered and forms the circular and inner longitudinal muscle layers of the body wall as well as the mesenchyme of the coelomic fluid, but opinions differ about the origin of the outer longitudinal muscle layer and the cutis which lies between this and the epithelium. According to Salensky, they have an ectodermal origin in the pilidium larva, while Nusbaum and Oxner state that they arise from the mesoderm in these larvae.

It can be thought that the coelom of the adult takes its origin from the blastular cavity, but there is no general conclusion which ties together the relation between the coelom and the rhynchocoel or the circulatory system. Salensky maintains that the central part of the foregut, between the larval mouth and the blastopore which marks the border between the foregut and the intestine, is constricted by the body wall muscles, and after moulting the adult mouth is formed there. Considerably after the completion of adult epithelium formation, the intestine unites with the body wall and an opening forms, which is the *anus*.

b. **Desor Larva**

Unlike the pilidium larva, the Desor larva does not form a complete amnion at the time of blastodisc formation; after the several amniotic membranes separate from the larval body wall, they disappear without uniting into a common membrane (Fig. 5.8). The archenteron which is formed by invagination of the endoderm at gastrulation closes temporarily. Following this, the position of the blastopore is moved inward by the invagination of the ectoderm, and the foregut develops between the mouth region and the blastopore. The esophagus and intestine of the adult are formed secondarily, at the expense of the yolk contained in the cells which make up the tubular wall, and the blastopore reopens between esophagus and intestine. Schmidt (1934) has reported a different type of yolk-absorption. According to his account, egg clumps consisting of a few eggs held together by an adhesive substance secreted by the ovary are surrounded by mucus from the epithelium and laid in two long strings. In each clump it becomes possible to recognize eggs which will continue to develop and others which will stop; the latter are soon absorbed by the former.

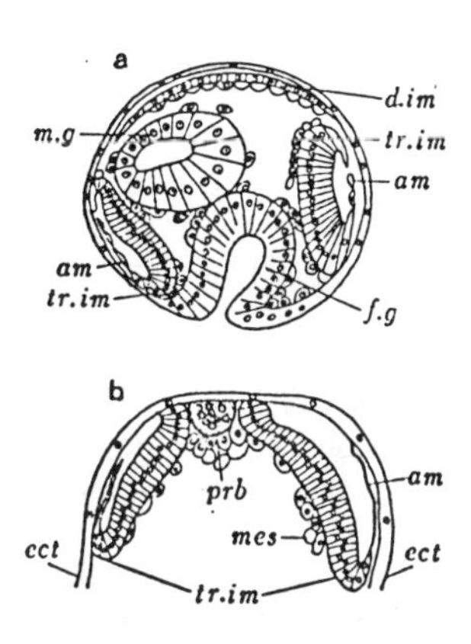

Fig. 5.8 Formation of blastodisc in Desor's larva (Arnold)
am: amnion, *d.im:* dorsal imaginal disc, *ect:* ectoderm, *f.g:* foregut, *m.g:* mid-gut, *mes:* mesoderm, *prb:* proboscis rudiment, *tr.im:* imaginal disc of trunk.

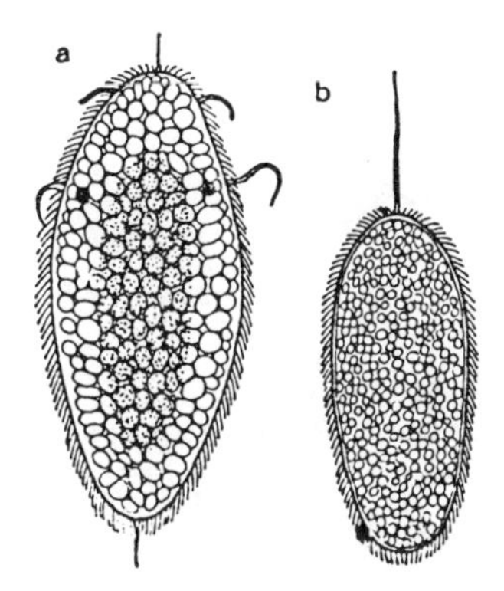

Fig. 5.9 Development in Paleonemertea (Iwata)
a. *Procephalothrix simulus* (x190);
b. *Tubulanus punctatus* (x190).

(2) DIRECT DEVELOPMENT

In general the directly developing species do not exhibit the sort of unique morphology found in the later pilidium stages. The larvae become oval in shape and thickly covered with cilia; a tuft of long cilia develops at the anterior end and they enter a free-swimming stage. In *Procephalothrix simulus,* besides tufts of long cilia formed at the head and tail ends, two additional pairs develop (Fig. 5.9a). The larval epithelium generally changes directly into that of the adult, although it is known that in *Cephalothrix* and *Emplectonema* a ciliated young worm bursts out of the larval epithelium. The proboscis develops from an ectodermal invagination at the anterior end, exactly as in the pilidium larva.

The blastopore of the gastrula stage is moved inward by an invagination of the ectoderm, and the region between the blastopore and the mouth of the future adult becomes the foregut. In *Cephalothrix,* according to Smith, the blastopore closes for a time, later reopening and uniting with the intestine. In the Enopla, after the larval mouth closes, the foregut shifts toward the ciliated anterior end, joining the proboscis sheath and opening into the rhynchocoel. Together with the proboscis it forms a common passage leading into the anteriormost part of the head (Fig. 5,10). The mouth, esophagus, stomach and pylorus are formed from the foregut. The intestinal caecum, lying below the esophagus, is evaginated from the lower side of the intestinal wall at the border between the foregut and the intestine.

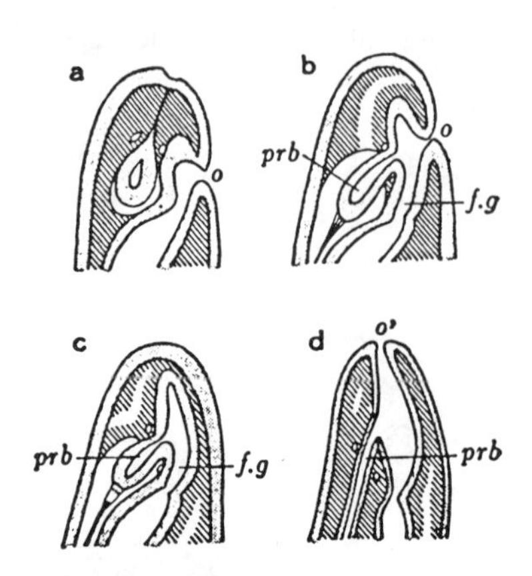

Fig. 5.10 Diagrams showing development of mouth and proboscis in *Malacobdella grossa* (Hammarsten)
f.g: fore-gut, *o:* mouth in early stage, *o':* opening of mouth and proboscis in final state, *prb:* proboscis.

The brain develops from a pair of ectodermal proliferations in the head region, which grow and sink inward. The cerebral sense organs are formed from a pair of invaginations on the ventral side of the head, while the *cephalic gland* develops from an invagination in its anteriormost part. In the Hoplonemertea the anus appears early, as an ectodermal invagination of the posterior body wall, but in the paleonemertean *Cephalothrix,* it opens after the intestine has been completely formed, even later than in the Heteronemertea, the opening appearing after the intestine and the ectoderm have united.

According to Salensky, in *Prosorhochmus* the mesenchyme of mesodermal origin, within the body cavity, adheres to the inner side of the body wall and to the intestine, where it forms a layer, later developing into the body wall musculature and the mesenchyme which lies between this layer and the intestine. The circulatory system first appears as tubules surrounded by mesenchyme cells, which lie on the inner surface of the body wall and become the *lateral blood vessels,* while the *dorsal blood vessel* is formed from proboscis sheath mesenchyme. In *Malacobdella,* Hammarsten found that the ectoderm of the posterior body wall grows inward and forms a *sucker,* and at the same time proliferates inside the body and forms mesenchyme of ectodermal origin.

REFERENCES

Arnold, G. 1898: Entwicklungsgeschichte des *Lineus gesserensis*. Trav. Soc. Nat. Pétersbourg, **28:** 21—27.

Barrois, J. 1877: Mémoire sur l'embryologie des Nemertes. Annales des sciences nat. ser 6, Zoologie, **6:** 1—232.

Bürger, O. 1894: Studien zu einer Revision der Entwicklungsgeschichte der Nemertinen. Ber. Nat. ges. Freiburg, **8:** 111—141.

Bütschli, O. 1873: Einige Bemerkungen zur Metamorphose des Pilidium. Arch. f. Naturg., Jahrg. **39:** 276—282.

Coe, W. R. 1899: On the development of the Pilidium of certain nemerteans. Trans. Connecticut Acad., **10:** 235—262.

——————— 1904: The anatomy and development of the terrestrial nemertean *(Geonemertes agricola)* of Bermuda. Proc. Boston Soc. Nat. Hist., **31:** 531—570.

——————— 1943: Biology of the nemerteans of the Atlantic coast of North America. Trans. Connecticut Acad., **35:** 129—328.

Dawydoff, C. 1928: Sur l'embryologie des Protonemertes. C. r. Ac. Sc. Paris, **186:** 531—533.

——————— 1928: Reversibilite des processus du developpement. C. r. Ac. Paris, **186:** 911—913.

Delsman, H. C. 1915: Eifürchung und Gastrulation bei *Emplectonema gracile*. Tijdschr. Ned. Dierk. Ver., **14:** 68—109.

Desor, E. 1848: On the embryology of Nemertes. Boston Journ. Nat. Hist., **6:** 1—18.

Dieck, G. 1874: Beiträge zur Entwicklungsgeschichte der Nemertinen. Jen. Z. f. Naturwiss., **8:** 500—520.

Fewkes, J. W. 1883—85: On the development of certain worm larvae. Bull. Of Museum Compar. Zool. Harvard College, **11:** 167—208.

Gegenbaur, C. 1854: Bemerkungen über Pilidium, Actinotrocha und Appendicularia. Zeitschr. wiss. Zoöl., **5:** 345—352.

Hammarsten, O. 1919: Embryonalentwicklung der *Malacobdella grossa*. Arb. Zool. Inst. Stockholm, **1:** 1—89.

Hoffman, C. K. 1867—77: Beiträge zur Kenntnis der Nemertinen. I. Zur Entwicklungsgeschichte von *Tetrastemma Varicolor* Oerst. Niederl. Archiv f. Zool., **3:** 205—215.

Hörstadius, S. 1937: Experiments on determination in the early development of *Cerebratulus lacteus*. Biol. Bull., **73:** 317—342.

Hubrecht, A. A. W. 1886: Contribution to the embryology of the Nemertes. Quart. Journ. of Micr. Sc., **26:** 417—448.

Ikeda, I. 1915: A new fresh water Nemertine from Japan. Annot. Zool. Jap. **8** : 239—255.

Korschelt, E. u. K. Heider. 1936: Vergleichende Entwicklungsgechichte der Tiere. Jena.

Lebedinsky, J. 1896/97: Entwicklungsgeschichte der Nemertinen. Arch. f. mikr. Anat., **49** : 503—556.

Leuckart, R. und A. Pagenstecher. 1858: Pilidium, die Larve einer Nemertine. Arch. Anat. Phys. Jahrg. **19** : 569—588.

Metschnikoff, E. 1869: Studien über die Entwicklung der Echinodermin und Nemertinen. Mem. de l'Acad. de St. Pétersbourg, ser. 7, **14** : 1—65.

Müller, J. 1847: Fortsetzung des Berichts über einige Tierformin der Nordsee. Arch. f. Anat. u. Physiol. : 157—179.

———— 1854: Ueber vershiedene Formin von Seethierne. Arch. Anat. Phys. Jahrg. 69—97.

Nusbaum, J. u. M. Oxner. 1913: Die Embryonalentwicklung des *Lineus ruber*. Zeitschr. wiss. Zoöl., **107** : 78—191.

Reisinger, E. 1926: Nemertini. Biologie der Tiere Deutschlands, **17** : 1—24.

Salensky, W. 1884: Recherches sur le développement de *Monopora* (*Borlasia*) *vivipara*. Arch. de Biologie, **5** : 517—571.

———— 1886: Bau und Metamorphose des Pilidiums. Zeitschr. wiss. Zoöl., **43** : 481—509.

———— 1909: Embryonalentwicklung der *Prosorhochmus viviparus*. Bull. Ac. Imp. St. Pétersburg, **3** : 325—340.

———— 1912: Entwicklung der Nemertine im Innern des Pilidiums. Mem. Acad. Imp. St. Pétersburg, ser 8, **30** : 1—71.

———— 1914: Ueber die Entwicklungsgeschichte der *Prosorochmus viviparus*. Mem. Acad. Imp. St. Pétersburg. ser. 8, **33** : 1—36.

Schmidt, G. A. 1934: Ein zweiter Entwicklungstypus von *Lineus gesserensis-ruber*. Zoöl. Jb. Anat., **58** : 107—659.

Smith, J. E. 1935: Early development of *Cephalothrix rufifrons*. Quart. Journ. Micr. Sc., **77** : 335—378.

Wilson, C. B. 1900: The habits and early development of *Cerebratulus lacteus*. Quart. Journ. Micr. Sc. (2), **43** : 99—191.

Wilson, E. B. 1882: On a new form of Pilidium. Stud. of the Biol. Lab., Johns Hopkins University, **2** : 341—345.

———— 1903: Experiments on cleavage and localization in the Nemertine-egg. Arch. Entw.-mech., **16** : 411—458.

Yatsu, N. 1904: Experiments on the development of egg fragments in *Cerebratulus*. Biol. Bull. **6** : 123—136.

———— 1910: Experiments on cleavage and germinal localization in the egg of *Cerebratulus*. Journ. Coll. Sc. Tokyo. **27** : 1—17.

Zeleny, C. 1904: Experiments on the localization of developmental factors in the nemertine egg. Journ. Exp. Zool., **1** : 293—329.

NEMATHELMINTHES

Chapter 6

INTRODUCTION

The normal development of the phylum Nemathelminthes, and particularly of its order, Nematoda, has been studied by A. Schneider, O. Bütschli, van Beneden, Zur Strassen, Zoja, Erlanger, Martini, Fauré-Fremiet, Meves and H. Müller, among many others; since their time the reproductive cells of these animals have also been the object of investigations in cytology and physiological embryology.

Table 6.1 **A summary of the Eunematoda.** (Korschelt & Heider 1936)

Family		Outstanding genera
1. Enoplidae	free-living, particularly marine, nematodes	*Enoplus, Trilobus*
2. Anguillulidae	nematodes free-living in fresh water, soil and decaying organic mater; parasitic in plants and a few animals	*Anguillula, Tylenchus, Rhabditis, Heterodera Diplogaster*
3. Angiostomidae	parasitic nematodes, famous for their alternation of generations	*Rhabdomena, Angiostoma, Allantonema, Atractonema, Phaerularia*
4. Gnathostomidae	spirurine nematodes parasitic in the digestive tracts of vertebrates	*Gnathosoma (Cheiracanthus), Cucullanus*
5. Filariidae	generally elongated parasites of the circulatory systems, digestive tracts and other organs of vertebrates	*Filaria, Spiroptera, Dracunculus*
6. Trichotrachelidae	nematodes parasitic in various organs of a wide range of animals	*Trichocephalus, Trichosoma Trichinella*
7. Strongylidae	parasitic in the digestive tracts and other organs of vertebrates	*Eustrongylus, Strongylus, Dochmius, Ancylostoma*
8. Ascaridae	intestinal parasites of various vertebrates	*Ascaris, Parascaris Crossophorus*
9. Oxyuridae	intestinal parasites of various vertebrates	*Heteracis, Oxyuris*
10. Mermitidae	long, thread-like nematodes, lacking an anus in most species; free-living or parasitic in the coelomic cavities of insects	*Mermis, Hexamermis, Hydromermis*

In particular, the uniqueness of nematode reproductive processes and living conditions early attracted the attention of embryologists, leading to the publication of many important papers by a great number of workers between 1850 and 1920. To the classical studies listed above, one might add those of Leuckart, Goette, Hallez, Carnoy, and then of Conté, Neuhaus, Krüger, Walton, Bonfig, Vogel, Pai, Kreis. Most of these men dealt with either the parasitic nematodes *(Ascaris, Oxyuris, Strongylus, Ancylostoma, Cucullanus, Spiroptera)* or the free-living forms *(Anguillula, Rhabditis, Diplogaster)*. Among these genera, the results of the most thorough investigations, made by Boveri, Zoja, Zur Strassen, Müller and Bonfig on *Parascaris equorum,*[1] agree in essence with those obtained by H. Spemann, Martini, Ziegler and S. Pai on *Strongylus, Cucullanus, Rhabditis, Anguillula.* In this country little embryological investigation of a physiological sort has been done, but many important results have been achieved in the field of clinical parasitology by such workers as Koizumi, Yoshida, Kobayashi, Asada, Morishita, Shimamura and Itagake. One reason for the paucity of physiological studies on the early developmental stages is that the impenetrable membranes have had a dsastrous effect on efforts at experimental manipulation. Now that a method has ibeen devised for removing these membranes at will, it has become possible to carry out various experimental procedures, and incidentally experimentation in parasitology should be benefited (Tadano and Tadano 1955).

As representing the most widely studied nemathelminth class, Nematoda, the group of Eunematoda will be chosen as the chief object of the following description — in particular, the early developmental stages of the horse parasite *(Parascaris), Rhabditis ikedai* and *Rh.* sp., supplemented by information gained from other nematode species.

1. EARLY DEVELOPMENT

(1) SPERM AND EGG CELLS

a. Spermatozoa

The nematodes have amoeboid spermatozoa. The spermatozoon of *Parascaris* consists of the sperm cell and an acorn-shaped body *(Glanzkörper)*; this cytoplasmic portion has a greater volume than that of the nucleus, and contains small bodies known as *Plastosomen* (Meves 1911, Romeis 1913, Held 1916). These are mitochondrial in nature, and are usually found around the nucleus. The spermatozoa of *Rhabditis ikedai* and *Rh.* sp.[2] are oval or trianguloid, and within the female seminal receptacle, most of them lie with the pointed end toward the oviduct. In Ringer's solution the spermatozoa exhibit active amoeboid movement, putting out a pseudopod of clear cytoplasm at one end, while the nucleus and mitochondria accumulate at the other. The fingershaped or branching pseudopods are formed continuously at any point so that the spermatozoa rotate frequently; in a quiescent or moribund state they become spherical.

---

[1] *Parascaris equorum* (Goeze) = *Ascaris megalocephala* Cloquet

[2] A species parasitizing *Armadillidium vulgare* (Latreille)

b. **Oocytes**

The mature oocytes of *Parascaris* are triangularly ovoid, with one pointed and one rounded pole; the latter faces the oviduct wall so that in cross-section the oviduct presents the appearance of a wheel, each egg having a pronounced morphological polarity (Schleip 1929). As the oocytes pass into the uterus they become spherical. The outermost layer of such oocytes is a thin semi-transparent membrane, and the cytoplasm contains many granules. Schleip (1929) described the irregular masses of granules as yolk granules, while Fauré-Fremiet considered them to be lipoid bodies. Other small brown granules, held by Meves (1911) and Carnoy (1887) to be Plastosomen (= Chondriosomen), were called "hyaline granules" by Schleip. In addition to these, Fauré-Fremiet stated that the oocytes contain granules of glycogen.[3]

A polarity is definitely fixed in the immature oocyte, but early in the spherical stage of the egg, the relation between the oocyte axis and the egg axis is not established (Schleip 1929). In *Rhabditis ikedai* and *Rh.* sp., the oocyte takes on a spherical form as it descends the oviduct, at the same time undergoing an increase in volume.

As a result of peristaltic movements of the lower part of the oviduct, the contents of the oocyte are mixed and transformed as it is passed to the *spermatheca.*

(2) FERTILIZATION

Fertilization in the Nematoda is internal; sperm entrance takes place in the spermatheca preceding the meiotic divisions, and syngamy occurs simultaneously with the formation of a *perivitelline space.* The prospective point of sperm entrance is not predetermined. The eggs of *Parascaris equorum, Ascaris suum, As. lumbricoides,* etc., may begin to develop as the result of fertilization and the provision of oxygen, at the proper temperature of 37°C, although there are considerable differences in this respect, depending on the species. The egg of *Rhabditis* changes from a spherical to an ovoid shape as the result of sperm entrance. In *Parascaris,* granules collect in the center of the egg, the vacuoles begin to grow larger, and a thick egg membrane is formed. The spot where the spermatozoon is attached becomes hyaline, and sperm entrance takes place there. At this time the cytoplasmic granules in the vicinity of the sperm entrance point begin to move about energetically and swell, and the sperm pronucleus is immediately surrounded by these granules. The plasma membrane in the region where the sperm entered becomes indistinct, but a few vacuoles collect there and dissolve, to repair the membrane. Boveri (1888) maintained that a particular cytoplasmic region is formed around the sperm in which minute granules collect, and this he believed to be a special type of cytoplasm, *Archoplasma,* which would later form the sperm aster; while Held (1912, 1916) considered it to be a mass which envelopes only the sperm pronucleus. Following sperm entrance, the cytoplasmic granules and vacuoles begin centrifugal and centripetal movements. This significant phenomenon, which continues to make its appearance during succeeding developmental stages,

[3] Histological studies on ova and spermatozoa have been performed by Panijel (1947), Pasteels (1951), Nigon and Delavoult (1952), Izumi (1952), Yanagisawa and Ishii (1954).

is one aspect of the mutual relation between the cytoplasm and the dynamic state of the nucleus.

a. **Formation of Egg Membrane and Expulsion of First Polar Body**[4]

The eggs of the oviparous species have a thick and resistant egg membrane. In the egg membrane of *Oxyuris vermicularis,* four layers, labelled from the outside A, B, C and D, can be distinguished. Egg membranes of *Ascaris,* depending on the mode of formation and properties of the membrane, show five layers; the outermost is counted as the first layer (Zawadowsky 1926; Tadano and Tadano 1950).

This first layer is formed by a secretion product of the uterine wall. The second to the fifth layers are produced by the egg cytoplasm, each layer being composed of a *basal membrane* and an *additional membrane.* The fifth layer in particular has properties different from those of the other layers, and resembling the D layer of *Oxyuris.* In viviparous species like *Rhabditis* or in ovoviviparous species, the egg membrane is generally thin.

i) *Formation of First Layer (Protein Membrane)*

The wavy-profiled first layer is formed of granules secreted by the wall of the uterus, which swell and adhere to the egg surface, and then fuse together. This egg membrane has no connection with fertilization, appearing also on unfertilized eggs. When the granules are large, the surface waves are large-scaled, as in *Ascaris suum,* and the free sides from processes. This sticky layer contains a large amount of albumin and is easily peeled off the egg. It is usually lacking in viviparous species like *Rhabditis,* and in ovoviviparous species.

ii) *Formation of Second Layer*

The formation of the second layer of the egg membrane begins after sperm entrance. The large granules migrate centrifugally, from the center of the egg to the periphery, swelling into vacuoles as they go. At this time the medium-sized granules adhere arround the vacuoles and join together to form a lipoid layer, which attaches to the inner side of the already existing plasma membrane (this corresponds to the basal membrane), thus forming the additional membrane of the second layer. Next the granules migrate centripetally. When it is first formed, the membrane is irregular, but during the formation of the third layer it becomes smooth (Fig. 6.1, a1).

iii) *Formation of Third Layer*

As soon as the second layer of the egg membrane has been formed, small granules move from the central region to the periphery, swell, and together with a hyaline substance become attached to the inner surface of the second layer, forming a thin basal membrane. To produce the additional membrane, large

[4] Further reports concerning the mode of formation and structure of the egg membrane can be found in Fauré-Fremiet (1913), Wottge (1937), Zawadowsky (1927, 1928, 1929).

granules swell to give rise to vacuoles which adhere to the inner side of the basement membrane, resulting in a great increase in the thickness of this layer. The formation of the fourth layer causes the third layer to be compressed, reducing its thickness by about one-third and smoothing its irregularities. Compared with the second and fourth layers, however, the third layer is definitely thick (Fig. 6,1, b2).

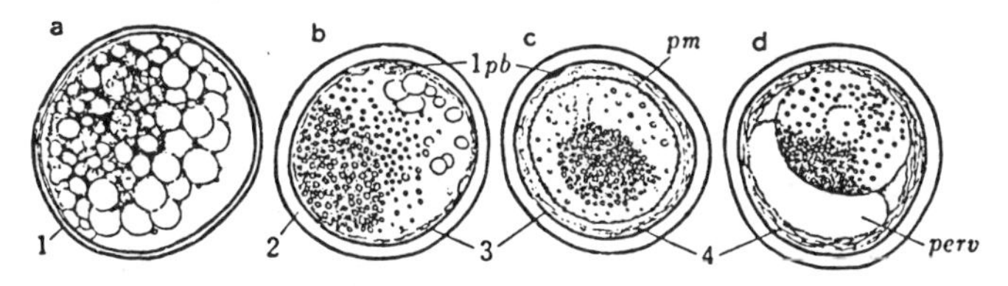

Fig. 6.1 Egg membrane formation in *Ascaris megalocephala* (Tadano)

a. formation of second layer in centrifuged egg; b. formation of first polar body and fifth layer in centrifuged egg; c. formation of plasma membrane; d. formation of perivitelline space in centrifuged egg. *l*: second layer, 1*pb*: first polar body, 2: third layer, 3: fourth layer, 4: fifth layer, *perv*: perivitelline space, *pm*: plasma membrane.

iv) *Formation of Fourth Layer*

In the same way as in the third layer, a basal membrane arises; then by the movement to the periphery and vacuolization of large and small granules, the additional membrane of the fourth layer is formed and attached to the basal membrane. As the vacuoles reach the periphery they become extremely large and lie one on top of another, giving the effect of soapbubbles. Around or between the bubbles minute granules can be seen. When the large vacuoles dissolve, these granules also break down and merge their contents with those of the vacuoles. This layer also acquires a regular, smoothed outline when the fifth layer is formed (Fig. 6.1, c3). It resembles the second layer except that it is lustrous and strong. In this stage the diminution in number of cytoplasmic granules becomes conspicuous, and the interior of the cell appears brighter, while the color of the cytoplasmic granules changes from yellowish-black to brownish-yellow.

v) *Extrusion of First Polar Body*

The oocyte nucleus, accompanied by granules, moves to the periphery of the cell and there gives off the first polar body. This takes place just as the fifth layer of the egg membrane is being formed; the polar body acquires a flattened shape as it is pressed against the inner wall of the fourth layer (Fig. 6.1, b, c).

vi) *Formation of Fifth Layer*

After extrusion of the first polar body, the egg nucleus moves back to the center of the oocyte. Around the periphery a homogeneous region appears; swollen vacuoles and small granules move into this region, fusing and dissolving together to give rise to very delicate lamellae. In this process the small granules break down less readily than the vacuoles. The result of this activity is a gradual increase in amount of a fibrous material on the inner side of the fourth layer; at its end the fifth layer is complete, although it is still rough and thick. Its thickness is reduced, however, by the formation of the perivitelline space, and it becomes more dense.

Especially in the egg of *Parascaris equorum*, the inner side of the fifth layer is denser than other regions and differently affected by treatment with various reagents (Fig. 6.1 d). In *Ascaris suum* also, the fifth layer is closely attached to the fourth layer, and its inner surface is undulatory.

According to Zawadowsky, Schmidt, Wottge and Fauré-Fremiet, the second to the fourth layers of the *P. equorum* egg are composed of chitin. However, each layer from the second to the fifth shows a different reaction to histochemical tests.

Table 6.2 **Histochemical properties of the egg-shell of** ***Ascaris.***

Tadano & Tadano (1955 b, e)

Histochemical reaction	Fixation	Layer of egg shell					Method of
		1st	2nd	3rd	4th	5th	
Xanthoproteic	Formaldehyde Alcohol	+	+	+ +	+	—	
Ninhydrine	Formaldehyde	+	+ + +	—	+ + +	—	Berg Romieu
Berlin blue	Freezing	+	+ +	—	+ +	+ +	
Millon's	Formaldehyde	+	+ +	±	+ +	±	
Alloxan	Formaldehyde	—	+ +	—	+ +	—	
Lecithin	Formaldehyde	+ +	+	+	+	+ +	Fischer
Lipoid	Potassium dichromate 8 40% formaldehyde 2 Glacial acetic acid 1	+	+	+	+	+ +	Ciaccio
Fatty acid	Formaldehyde Freezing	—	±	—	±	+ +	Fischer
Sudan III	Formaldehyde Freezing	+	+	—	+	+ +	
Nadi (oxidase) Labile oxidase	Formaldehyde Freezing	+	+ +	+ +	+ +	±	
Chrome	Zenker 3.5% potassium dichromate	+	+ +	+ +	+ +	±	
Silver	Bouin Formaldehyde Freezing	+	+ +	+	+ + +	—	Masson Herpel Ogata Törö

— = negative + = weak positive +++ = very strong positive
± = dubious ++ = strong positive

Protein predominates in the second to the fourth layers, while the fifth layer is lipoid (Table 6.2). The fifth and fourth layers exhibit strong birefringence, the third and second layers, a certain amount, and the first layer none.

If these various layers are treated with strong acid or alkali, each of them shows delicate fibrous structural elements (Tadano and Tadano 1955b, e).

b. Formation of Hyaline Membrane, Plasma Membrane and Perivitelline Space

Following the formation of the fifth layer, there appears in the egg cortex, in contact with its inner surface, a zone of clear protoplasm containing a small amount of granules, which can be distinguished from the rest of the egg contents. Small granules scatter outward from the interior, swell, and become arranged in this clear zone, separated only very slightly from the fifth layer. These granules fuse to form the *plasma membrane* (Fig. 6.1c *pm*). This membrane is irregular at first but becomes smooth when the perivitelline space and external cytoplasmic zone are formed. In addition to the granules, the substance of the clear zone plays an important part in the formation of the plasma membrane. In a parallel process, a thin *hyaline membrane* appears outside the plasma membrane, formed by substances included in the granules which gave rise to the plasma membrane. Certain swollen vacuoles exude through the hyaline membrane and spread out between it and the inner surface of the fifth layer to form the perivitelline space. It is at this time that the hyaline and plasma membranes become thinner and denser.

As is shown in Figure 6.1d, the transparent colloidal contents of the vacuoles which formed the perivitelline space are shifted to the heavy side under centrifugal force. Wottge (1937) states that the perivitelline fluid of the *P. equorum* egg appears with the migration of vacuoles. Perivitelline space formation in *Rhabditis ikedai* and *Rh. sp.* begins in the vicinity of the animal pole immediately before the second polar body extrusion, and the perivitelline fluid is given out from the egg. According to Leuckart (1866), the perivitelline space of nematode eggs usually has its origin in a granular substance which fills the interior of the egg. Bütschli (1875), Zeigler (1895) and Auerbach (1874) believed that the vesicles of the *Rhabditis* egg are pushed out, forming the perivitelline space, as the egg moves into the uterus. After this space has been formed, the egg becomes exceedingly transparent.

c. Second Polar Body Formation[5] and Fusion of Pronuclei

Paralleling the formation of the perivitelline space, the second polar body is extruded, the egg pronucleus returns to the center of the egg, the sperm and egg pronuclei unite and granules gather around the fusion nucleus. The same process takes place in *Rhabditis* and *Dipolagaster*. The volume of the mature *Parascaris* egg is reduced to about 70% of that of the unfertilized egg.

The mature eggs of most nematodes are ovoid, but occassionally they are spindle-shaped like those of *Trichocephalus dispar*, while the eggs of *Spiroptera microstoma* are almost cylindrical. Usually the surface of the egg membrane is smooth, but some have processes or pits. In addition there are forms with lid-shaped structures or plugs at the two poles of the shell *(Trichocephalus trichuris)*, and others with filose or racemose arrangements of fibers *(Mermis nigresens, Tetrameres Nouval)*.

---

[5] In some cases the second polar body may be very large. Schleip (1924) called these "Nebenzellen".

(3) CLEAVAGE AND GERM LAYER FORMATION

a. 1 — 2-cell stage

The cleavage of *Parascaris* is practically equal and total. The first cleavag plane is perpendicular to the polar axis, dividing the egg into animal pole sid S_1 *(AB)* and vegetable pole side P_1 blastomeres. The separation into somati and germ-cell lines takes place from the two-cell stage, as is made most clearl evident by the phenomenon of *chromatin diminution*. One cleavage cycle ma be divided into two phases. The earlier covers the time from the appearance o the mitotic figure to the completion of nuclear division and the formation of th cleavage furrow, so that the position of the cleavage plane is determined and it basic area established; the later phase includes the time covered by the new for mation of the cleavage plane and the completion of the daughter nuclei. Th cleavage surfaces are newly formed with hyaline protoplasm and mitochondria granules as the minimum in the way of essential elements; the migration an swelling of the granules can be seen during this process. The apposed surface of the blastomeres constitute the new surfaces; these are flattened against each other

i) *Early phase of the first cleavage: establishment of external cytoplasmic zone behavior of pronuclei and formation of daughter nuclei*

Immediately after the maturation divisions, the *Parascaris* egg is fairl oval, with a rather thick peripherial zone of pale yellowish-green hyaline sub stance (Fig. 6.2a), distinguishable from the interior cytoplasm which is crowde with a large number of variously sized granules, and has large vacuoles scattere here and there throughout it. Granules of all sizes change to vacuoles and the dissolve into this hyaline zone, causing an increase in its refringence and height ening the contrast between the external and interior cytoplasm, and the eg takes on a spherical shape (Fig. 6.2b).

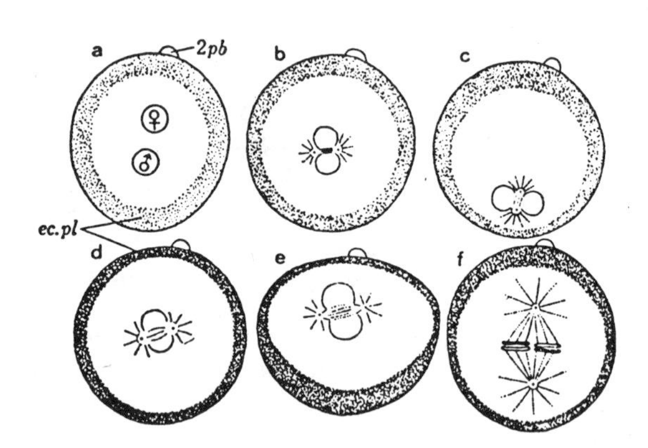

Fig. 6.2 Localization of mitotic apparatus and behavior of ectoplasm (Tadano)

2pb: second polar body, *ec.pl:* ectoplasm.

The egg and sperm pronuclei make contact in the center of the egg. Next the contact surfaces of the pronuclei disappear and the two pronuclei are joined side by side, forming the transparent fusion nucleus in which the chromosomes are clearly visible. As the sphere of the fusion nucleus gradually expands into a spindle-shape, the spindle fibers become visible, and asters develop at the two ends. A little way outside of the spindle, a spindle-shaped boundary profile can be differentiated from the endoplasm; the region inside this boundary will tentatively be given the name of nuclear region (Fig. 6.3e, *nu*). This membrane is the preserved nuclear membrane of the fusion nucleus, and its two ends connect with the asters.

The mitotic figure moves with the internal cytoplasm to the vegetal pole, and there is less peripheral cytoplasm at the vegetal than at the animal pole (Fig. 6.2c). When the mitotic figure moves back to the central region, there is a return of peripheral plasm to the vegetal pole and the formation of new plasm takes place so that the peripheral layer becomes uniform around the egg. Unlike the previous peripheral cytoplasm, this plasm, yellow in color and containing many small granules, forms a firm layer of definitive *peripheral cytoplasm* (Fig. 6.2d). This layer is less easily damaged than the internal cytoplasm, and harder to move by centrifugal force. If the plasma membrane is torn, it flows out in a viscous state.

As the mitotic figure develops it moves toward the periphery at the animal pole, and the egg takes an asymmetrically oval shape (Fig. 6.2e). On this occasion the peripheral plasm appears in greater amount at the vegetal side, and most of the remainder is near the animal pole. Turning through 90° the mitotic spindle elongates further, and as it returns to the center of the egg the mitotic process reaches the metaphase. At this time there is more of the peripheral cytoplasm at the animal pole and an equally smaller amount in all the other regions, while the spindle axis is parallel to the polar axis (Fig. 6.2f).

This rotation of the mitotic figure within the uncleaved egg can be observed also in other species (Auerbach 1874; Ziegler 1895; Erlanger 1897; Boveri 1910a; Spek 1918). After expulsion of the polar bodies in the egg of *Rhabditis* sp., the egg pronucleus moves to the sperm pronucleus at the vegetal pole, the two pronuclei unite and then return to the center of the egg where they rotate through 90°. There are some forms *(Rh. nigrovenosa, Diplogaster, Rh. dolichura, Tylenchus)* which show amoeboid movement of the cytoplasm during the period from syngamy through the elongation of the spindle[6], while in other species an active protoplasmic streaming takes place *(Rh. ikedai, Rh.* sp.). In the egg of *Parascaris equorum,* after rotation of the mitotic figure, the distance between the center of the asters and the egg surface is greater at the animal pole side. The astral rays reach the cortical layer at the poles. With the elongation of the spindle the whole egg elongates and the cleavage furrow is formed. This furrow formation consists in an expansion toward the equatorial zone of the plasma membrane, the hyaline membrane which surrounds it and the peripheral cytoplasm (ectoplasm) in contact with its inner side. At this time there is a conspicuous migration of granules toward the furrow. As the furrow expands, the egg surface at the two poles undergoes an especially striking expansion. The asters penetrate deeply into the peripheral cytoplasm, and the astral rays reach the cortex. Paralleling the expansion

[6] In some cases the amoeboid movement is so violent that a cleavage furrow-like appearance results, but since this later disappears, it is called a *false furrow* (Erlanger 1897).

of the furrow, the nuclear region also divides into two. That is, as the cleavage furrow advances centrally from the outside of the equatorial zone, a furrow forms cutting the nuclear region also (Fig. 6.3f), and finally the two blastomeres are connected with each other at only one point (Fig. 6.3a). This region of last connection consists of nuclear region (mainly spindle fibers), plasma membrane and hyaline membrane. As shown in Figure 6.3g, h, these are cut through, and the egg is divided into a larger S_1 blastomere, containing more of the peripheral cytoplasm, and a smaller P_1 blastomere, which has more endoplasm. If blastomeres which have thus completed the first part of the cleavage process are subjected to centrifugal force, there is no exchange of their contents between the two cells. The blastomeres which derive from S_1 are the Somazellen: i.e., S_1 is the *primary ectoderm cell;* while P_1 *(Propagationszelle* 1) gives rise to the *primary germ cell (Urgeschlechtszelle)*. This situation is the same in *Cucullanus elegans* (Martini 1903), *Anguillula aceti* (Pai 1928), *Rh. ikedai* and *Rh. nigrovenosa* (Ziegler 1895) and *Strongylus paradoxus* (Spemann 1895), and can be thought of as applicable to the Nematoda in general. S_1 is usually larger than P_1, but occasionally there are cases in which equal cleavage occurs normally. In *Rh. teres* the relative size of the first two blastomeres depends upon whether the union of the pronuclei takes place in the animal or vegetal half of the egg (Zeigler 1895). According to Tadano and Tadano (1955f), the activity of the mitotic figure in *Rh.* sp., however, increases together with the elongation of the spindle. Moreover, they consider that the activity of the aster at the vegetal side is more vigorous than that of the animal side, so that the latter is simply a reflection of the former. S_1 is larger than P_1 because the mitotic figure is located asymmetrically toward the vegetal pole when it stops its migratory activity and brings about the beginning of the first cleavage.

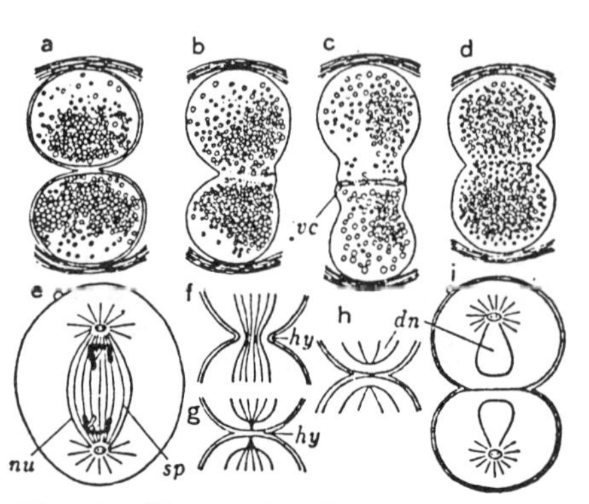

Fig. 6.3 Cleavage in *Ascaris megalocephala* (Tadano)

a, e ~ g. "First process" of cleavage; b ~ d, h ~ i. "Second process". *dn:* daughter nucleus, *hy:* hyaline membrane, *nu:* nuclear region, *sp:* spindle fiber, *vc:* vacuole.

ii) *Late phase of the first cleavage*

The peripheral cytoplasm which expanded into the cleavage furrow in the early phase becomes opaque, and the apposed faces of the furrow lose their curvature and become flat (Fig. 6.3h, i); collecting small granules on the way, cytoplasmic vacuoles accumulate in this area and dissolve. The vicinity of the equatorial zone swells and its periphery bulges out, and both blastomeres begin to elongate in the direction of the egg axis, this elongation appearing alternately at the two sides. During this process the blastomere surfaces at the two poles are restrained by the egg membrane, so that the stalk connecting the blastomeres is bent to one side or the other (Fig. 6.3b). If the egg membrane is removed, this bending fails to occur, and the blastomeres become elongate and slender in the direction of the egg axis. At this time vacuoles with attached small granules line up along the furrow wall, and fuse together to form the plasma membrane (Fig. 6.3c). As the bulging of the surface in the equatorial zone disappears, the plasma membranes along the furrow become thicker. At this stage the endoplasm disperses and a layer of homogeneous peripheral cytoplasm appears around all the surfaces of the two blastomeres except for the apposed furrow faces, where no such zone is seen (Fig. 6.3d). The nuclear region, which was divided between the two blastomeres, forms at first two pear-shaped masses (Fig. 6.3i), but small mitochondria-like granules from the endoplasm crowd around these and break down, resulting in spherical daughter nuclei with thick nuclear membranes.

In the eggs of *Rh. ikedai* and *Rh.* sp., the furrow of the 'early phase' cuts simultaneously into the equatorial zone from all sides, but stops, leaving one-third of the diameter uncleaved. This roughly corresponds to the nuclear region: it is thus the spindle fibers which remain uncleaved. With the disappearance of these spindle fibers, granules appear in the equatorial zone, and the nuclear region is divided into two parts. These unite with swollen granules which have migrated into the equatorial zone, and form the walls of the furrow (Tadano and Tadano 1951, 1954d). These eggs do not show the bulging of the periphery in the equatorial zone, or the polar elongation that was seen in the *Parascaris* egg during the 'later phase'; instead there is a wave-like slackening of the newly formed furrow walls, followed by the appearance of a uniform surface. Consequently the early and later phase of cleavage in the *Rhabditis* egg apparently differ from those of *Parascaris*.[7]

b. 2—4-cell Stage

The second cleavage plane is parallel to the egg axis in the S_1 *(AB)* blastomere, and perpendicular in the P_1 blastomere, producing a T-shaped 4-cell stage which is later made rhombic-shaped by a shift in the position of the P_2 blastomere (Fig. 6.4). Both of these arrangements can be seen off and on during later development, and constitute the fundamental patterns of the blastomeres. At the second cleavage the ends of the chromosomes of the S_1 blastomere are cut off and thrown out into the cytoplasm. With this process of chromatin diminution, the S_1 blastomere divides into *A* and *B* blastomeres. The P_1 cell cleaves into P_2 and S_2 *(EMSt)* without undergoing chromatin diminution. When the mitotic figure rotates at the time of S_1 cleavage, peripheral zone cytoplasm (ectoplasm) appears at the surface nearest the spindle poles. In most cases the P_1 blastomere begins to cleave slightly after S_1, but this order may be reversed. The mitotic figure of the P_1 blastomere completes its rotation in an early stage. The characteristic shift from the T-shape to the rhombic, which appears in early nematode development, parallels a movement of the nuclei and a change (migration, new formation) in the condition

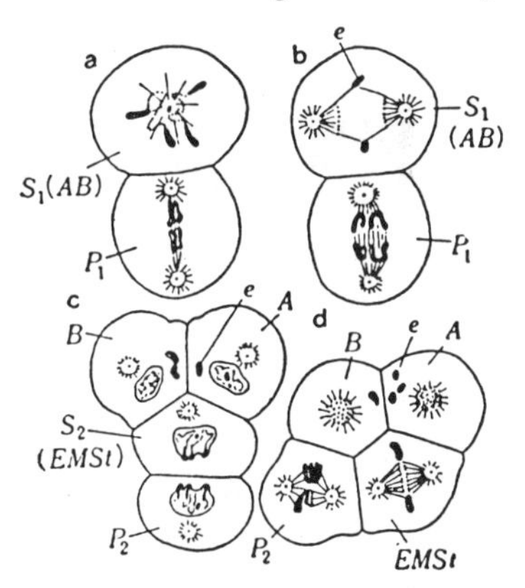

Fig. 6.4 Chromatin diminution in egg of *Ascaris megalocephala* (Boveri)

a. metaphase of second cleavage; *b.* anaphase of second cleavage; *c.* telophase of second cleavage: *d.* metaphase of third cleavage. *e:* eliminated chromatin.

[7] In accordance with these superficial differences in mode of cleavage, it is possible to distinguish an '*Ascaris form*' and '*Rhabditis form*'.

of the peripheral cytoplasm in the 'later phase' of the second cleavage. At this time the contact surface between the S_2 *(EMSt)* and the daughter blastomeres of S_1 *(A* and *B)* is expanded, a large amount of peripheral zone cytoplasm appears at the side in contact with the *A* cell and this region becomes spherical. At the opposite, *B*-cell side, in the 'later phase' of the P_1 blastomere cleavage, P_2 bends around and aproaches the *B* blastomere, and finally *B* and P_2 unite side by side, giving the embryo a rhomboidal configuration (Fig. 6.4c, d). In this process there occurs a pronounced change in the state of the ecto- and endoplasm.

Table 6.3 **Cell-lineage in Parascaris equorum.**

(based on Boveri 1899, 1910)

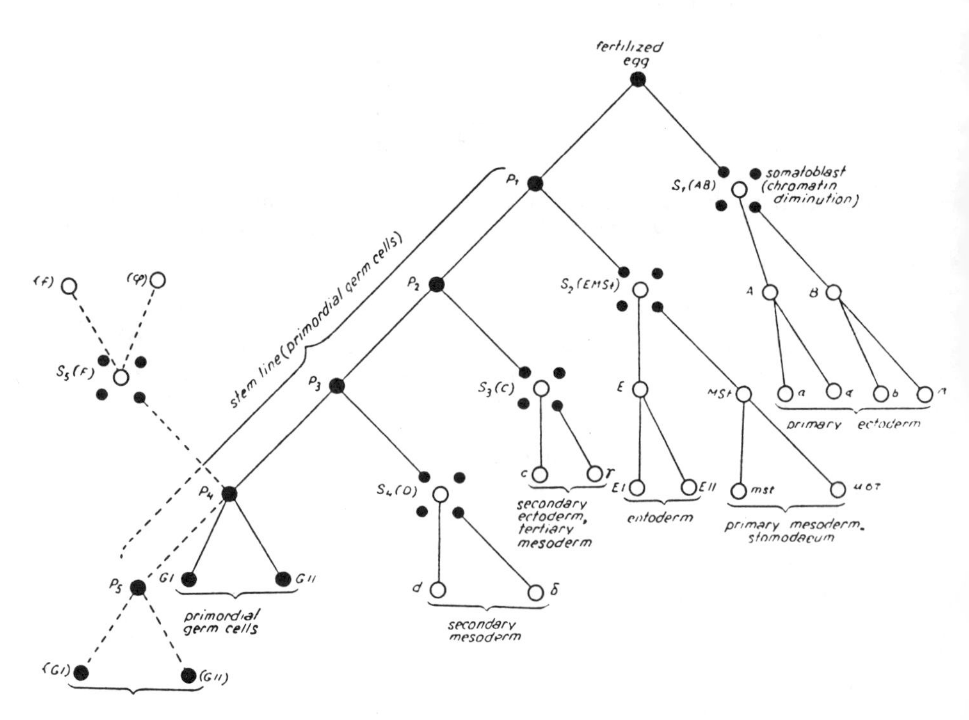

As is shown in Table 6.3, the *A* and *B* blastomeres are the *primary somatoblasts; A* and *B* may be the same size, or *A* may be slightly larger than *B*, while *B* contains slightly more endoplasm. S_2 in the *secondary somatoblast;* i. e., the *endodermal-mesodermal-stomodaeal stem cell;* while P_2 is the prospective *secondary propagative cell.* S_2 is larger than P_2, has a larger amount of peripheral cytoplasm than P_2 but less than the *A* and *B* blastomeres, and the largest amount of endoplasm. In the early phase of the second cleavage of *Rh.* sp., the mitotic spindles of the two blastomeres elongate parallel to each other but obliquely with respect to the egg axis, and give rise to the cleavage furrows. At the end of this process, the blastomeres take a rhomboidal configuration, so that the T-shape is not well defined. The amount of peripheral cytoplasm is less than that of *Ascaris* eggs. In *Rh. ikedai,* on the other hand, the T-shaped larva makes its appearance.

c. 4-cell Stage — Gastrula

The somatic cells produced by repeated cleavages are designated by S_1, S_2, etc.; furthermore, $S_1=AB$, $S_2=EMSt$, $S_3=C$, $S_4=D$, $S_5=F$. *A, B, C, D, E, F, M,* and *St* are all located in the median plane of the embryo; when these divide to produce cells lying symmetrically to the left and right of the median line, those on the right half are designated by Roman letters, and those on the left by Greek: e. g., $MSt \rightarrow mst$, $\mu\sigma\tau$. Successive generations are indicated by appropriate exponent and index symbols: $a \rightarrow a\mathrm{I}$, $a\mathrm{II}$; $a\mathrm{I} \rightarrow a\mathrm{I}_1$, $a\mathrm{I}_2$; $a\mathrm{I}_1 \rightarrow a\mathrm{I}_1^{1}$, $a\mathrm{I}_1^{11}$, etc. Following the 4-cell stage, the *A* and *B* blastomeres divide, simultaneously in most cases, although *A* always cleaves first in *Strongylus paradoxus* (Spemann 1895). There are also some species in which P_2 divides before either *A* or *B*.

The mitotic spindles of *A* and *B* lie perpendicular to the embryonic axis The daughter cells derived from the *A* and *B* blastomeres in this third division. are arranged in right-left pairs, *a*—α and *b*—β, respectively (Fig. 6.5a). These are asymmetrical when first formed, but become more or less symmetrical during the 'second process'. The α, β blastomeres lie somewhat forward and below *a* and *b*, and are more or less divergent. The *a* and α blastomeres lie slightly to the left of the midline, and since *a* and β come to lie in contact with each other, an asymmetry makes its appearance between the lefr and right halves of the embryo. As is shown in Figure 6.5b, the P_2 blastomere which lies above the midline in the posterior part of the embryo divides into P_3 and S_3 *(C)* = *tertiary somatoblast*. When S_2 *(EMSt)* divides, chromatin diminution again takes place. Goette (1882) named S_3 the 'Schwanzzelle', maintaining that it gives rise to the epidermis covering the posterior half of the embryonic body (i.e.,= secondary ectoblast). However, Müller (1903) and Boveri (1910) established the fact that it also gives off part of the mesoderm.

Among the primary ectoblasts, the ones on the right side (*a* and *b*) incline downward and posteriorly, while those on the left (α and β) slant toward the lower anterior end of the embryo, and the left-hand ones protrude more than those on the right. The *b* and β blastomeres are thus of about the same height. S_2 *(EMSt)* divides (with chromatin diminution) in the same direction as P_2, forming *MSt* anteriorly and *E* posteriorly. The embryo now has eight cells, with a 'Ventralfamilie' arranged in an arc along the ventral midline, consisting of *MSt, E,* P_3 and S_3 (Fig. 6.5b, c). As Table 3 shows, the four blastomeres resulting from the division of *S*I (*a*, α, *b*, β) all give rise to ectoderm only. P_3 cleaves to produce the *fourth propagative cell* and the *secondary mesoblast*. The S_3 *(C)* blastomere lies dorsally at the posterior end of the embryo. The axis passing through S_3 and P_3 is perpendicular, while that joining *E* and *MSt* is horizontal; both lie in the midplane and at right angles to each other. After the division of S_2, α and β are pushed out in a downward direction, *a* and S_3 *(C)* come into contact with each other, *MSt* lies at the front end and S_3 *(C)* at the back end, and *a*, α, *E* and P_3 are contiguous to these, lying in an annular arrangement along the midline (Fig. 6.5c, d). The space between these sets of blastomeres will later develop into the blastocoel.

Next, some of the primary ectoblast cells begin to cleave; in this stage the embryo becomes rather asymmetrical. The *b*, β blastomeres move to the middorsal region, and away from each other; *a*II fits into the upper part of the space thus formed, while S_3 *(C)* moves around from the posterior end into contact with aII. At the fourth cleavage, the primary ectodermal cells a, α give rise to aI, aII and αI, αII, which spread out over the anterior dorsal part of the embryo, bringing the total number of blastomeres to ten. The six cells, aI, aII, *MSt*, *E*, *C*, and P_3 lie in an annular formation on the midline of the embryo (Fig. 6.5d), and the space surrounded by these blastomeres is closed at the sides by b and β. Next aI and αI take up bilaterally symmetrical positions, equalizing the numbers of cells on the two sides of the embryo (Fig. 6.5e). The total number of blastomeres is increased to 12 when b divides dorso-ventrally into bI and bII, and β into βI and βII. The right-side blastomeres aI, aII, bI, bII assume a T-formation, while the left-side αI, αII, βI and βII line up in a rhombic shape. *MSt* S_3 *(C)* and *E* begin to cleave. S_3 *(C)* and *MSt* divide into left and right cells: S_3, undergoing chromatin diminution, forms c and γ, while *MSt* divides into *mst* and $\mu\sigma\tau$.

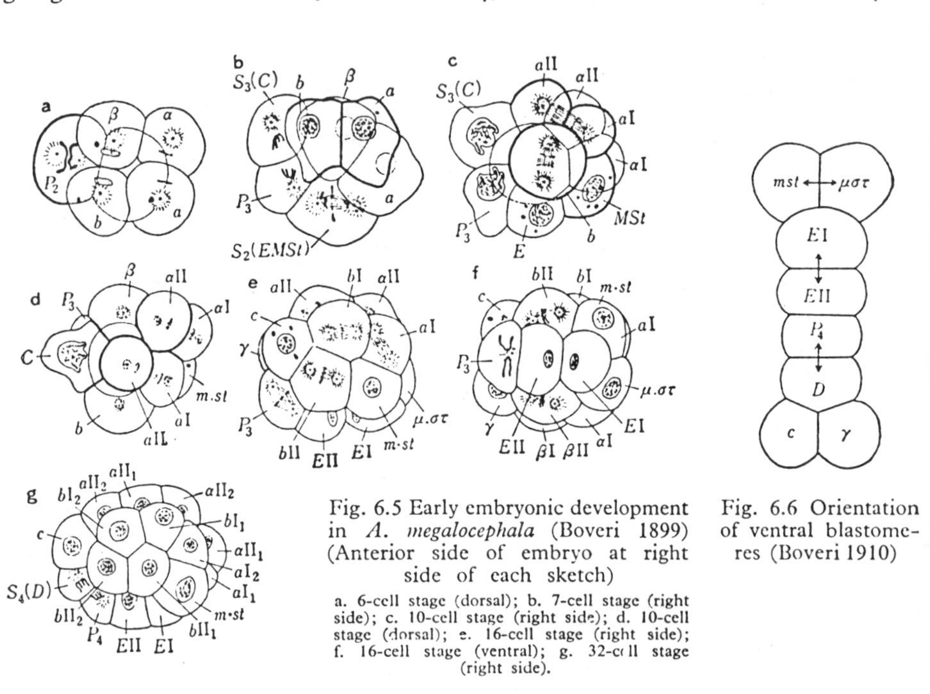

Fig. 6.5 Early embryonic development in *A. megalocephala* (Boveri 1899) (Anterior side of embryo at right side of each sketch)

a. 6-cell stage (dorsal); b. 7-cell stage (right side); c. 10-cell stage (right side); d. 10-cell stage (dorsal); e. 16-cell stage (right side); f. 16-cell stage (ventral); g. 32-cell stage (right side).

Fig. 6.6 Orientation of ventral blastomeres (Boveri 1910)

E, the endoblast, divides transversely in the ventral mid-line to form *E*I and *E*II, but the division of P_3, which also lies in this posterior region, lags behind the others. Blastomeres c and γ take positions on the dorsal side of P_3 which lies in the mid-ventral line at the posterior end of the embryo, and bII and βII face each other (Fig. 6.5f). The aII blastomere moves to a position above them. In the midline *mst* and $\mu\sigma\tau$ are in contact, and then αI pushes between them. At this time the embryo as a whole shows a considerable degree of symmetry, although on the right side *mst* and bI are in contact, while $\mu\sigma\tau$ and βI are some distance apart on the left (Fig. 6.5f). Before P_3 divides, the primary ectoderm-producing blastomeres begin to undergo the fifth cleavage: i.e., the aI and αI blastomeres divide dorso-ventrally into aI_1, aI_2 and αI_1, αI_2, bI divides transversely to give bI_1 and bI_2, and bII divides into bII_1 and bII_2. These line up in a rhombic configuration, and on the other side βI and βII give rise to βI_1, βI_2, βII_1 and βII_2. On the dorsal side *a*II and αII also divide transversely to produce aII_1, aII_2, αII_1, and αII_2, and the daughter cells of *a*I, *a*II, αI, αII, *b*I, *b*II, βI and βII assume a symmetrical arrangement. In this fifth cleavage, the primary ectoderm-producing blastomeres thus give rise to 16 cells (Fig. 6.5g).

As the two anterior ventral blastomeres *mst* and μστ migrate from their position in front of the daughter cells of *a*I and αI to the ventral side of these blastomeres, they come to lie to the right and left of *E*I (Fig. 6.5g). Next the midventrally located P_3 cleaves, forming S_4 *(D)* at the posterior extremity of the embryo and P_4 on the ventral side (Fig. 6.7a). This S_4, which will produce the *quartenary somatic cell line*, gives rise to the *secondary mesoblast*. As Figure 6.6 shows, *E*I, *E*II, P_4 and D_1, among the descendants of the P_1 blastomere, are now located on the midventral line of the embryo, while *mst* and μστ lie side-by-side at the anterior end, and *c* and γ are similarly arranged posteriorly. These ventral blastomeres form an arc which curves dorsally. Next the primary ectoderm-forming line of blastomeres carry out the sixth cleavage, and with their ventral migration, *mst* and μστ move into a position between *E*I, *E*II and the daughter cells formed by *b*II and βII. As the result of this shift, *mst* and μστ become separated from aI_1 and αI_1 (Fig. 6.7a).

After this the *mst* blastomere divides transversely into an anterior *st* and a posterior *m* blastomere, while on the left side μστ simultaneously gives rise to στ and μ, and these assume symmetrical positions. Strassen held that all these blastomeres are primordial mesoderm cells, while Boveri believed that only *m*, μ are mesodermal, the *st*, στ blastomeres being ectodermal in nature and destined to form the stomodaeum. For this reason he called them *stomatoblasts*.

At this stage the embryo has become a blastula with a small blastocoel (Fig. 6.7a-d). The endoblast cells *E*I and *E*II in the mid-ventral line divide laterally into *e*I, *e*II, εI and εII; both blastomeres at the right side move forward, giving a rhombic configuration to these four cells and the embryo as a whole loses its spherical form and begins to elongate anteroposteriorly. The *c* blastomere undergoes chromatin diminution and divides obliquely upward into *c*I, *c*II, while γ divides into γI, γII. These blastomeres are *secondary ectoblast* and *tertiary mesoblast cells*. From this stage, the endoblasts are gradually enveloped dorsally and laterally in primary ectoblast cells, while the descendants of P_4 cover the posterior ventral side. From the two sides of the embryo the primary mesoblast cells, particularly the stomatoblasts, approach the midventral line (Fig. 6.7c), and gastrulation begins. Next, as shown in Figure 6.7c, e, f, the stomatoblasts divide transversely, forming *st*I, *st*II, στI and στII, and *m* and μ also divide in the same direction. As *m*, μ begin to cleave, the secondary mesoblast S_4 *(D)*, which lies posterior to P_4, undergoes chromatin diminution and divides, giving rise to the bilateral pair of blastomeres *d* and δ. The stomatoblasts also divide, slightly ahead of the mesoblasts, and all move toward the midline, gradually causing the endodermal blastomeres *e*II, *e*I, εI and εII to slip into the blastocoel. Eventually the mesodermal cells follow them (Fig. 6.7f), and the increase in number of blastomeres causes a reduction in the size of the blastopore. On the ventral side of the embryo, a ring is formed around the blastopore, consisting anteriorly and laterally of the blastomeres *m*I, *m*II, *st*I, stII, μI, μII, στI, and σrII, and posteriorly of *d*, δ and P_4. This latter blastomere then moves to a position below the endoderm cells, while the ectodermal cells shift toward the ventral side and the stomatoblasts *st*I, *st*II, στI, στII approach still more closely to the midline. The ectodermal cells then surround the stomatoblasts from the lower anterior part of the embryo. The primary and secondary mesoblasts (*m*I, *m*II, μI, μII) lying alongside of and behind the blastopore become surrounded by ectoderm, making the blastopore opening still smaller. In this process, the stomatoblasts *st*I, *st*II, στI, στII do not invaginate together with the mesodermal blastomeres *m*I, *m*II, μI, μII. The descendants of S_3 (*C*) increase to eight cells and occupy the posterior part of the embryo.

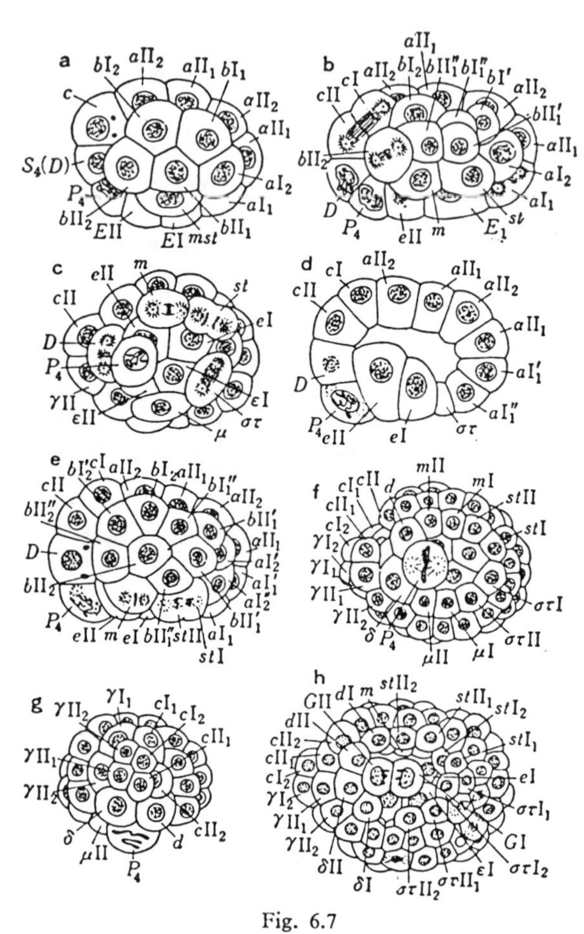

Fig. 6.7

a. 24-cell stage (right side). *mst*, μστ blastomeres shift posteriorly; b. More advanced stage (right side). c. Gastrula (ventral). d. Gastrula (optical cross-section). e. Gastrula (right side). f. Gastrula (ventral). g. Gastrula (posterior). h. Gastrula (ventral); appearance of *G*I- and *G*II-blastomeres. (Boveri 1899,

As shown in Figure 6.7g, the four blastomeres $c\mathrm{II}_1$, $c\mathrm{II}_2$, $\gamma\mathrm{II}_1$, and $\gamma\mathrm{II}_2$ form a bilaterally symmetrical line. These blastomeres are also called *ventral cells* (*Bauchzellen* — Strassen). In the posterior dorsal region, *c*I and γI cleave transversely to produce $c\mathrm{I}_1$, $c\mathrm{I}_2$, $\gamma\mathrm{I}_1$, $\gamma\mathrm{I}_2$, and these cells envelope the posterior end of the embryo. As these mesodermal blastomeres invaginate, the mesoblasts and stomatoblasts derived from the S_2 *(MESt)* blastomere are undergoing division.

Next the secondary mesoblast cells *d*, δ on the ventral posterior side divide into *d*I, *d*II, δI, δII; these surround the P_4 blastomere and become arranged in a wide arc in the anterior part of the embryo, while P_4 shifts to a position directly below the endoderm cells in the midventral line.

The next event is the transverse cleavage of P_4 to produce the *primordial germ cells* *G*I and *G*II. Accompanying the advancing endodermal invagination, a connection is established between the *st*, στ and *d*, σ blastomeres, and the primary mesoblasts *m*, μ invaginate. *G*I and *G*II are carried inside the blastopore by the migration of the blastomeres at the posterior end of embryo. The ectoderm cells move to the sides of the embryo and the secondary mesoderm cells invaginate. The number of stomatoblasts increases to eight (stI_1, stI_2, $stII_1$, $stII_2$, $\sigma\tau I_1$, $\sigma\tau I_2$, $\sigma\tau II_1$, $\sigma\tau II_2$,), and stI_2 and $\sigma\tau I_2$ come into contact with each other. Next the posterior stomatoblasts establish contact with the secondary ectoblast cells on the posterior ventral side of the embryo. (Fig. 6.7h), and GI and *G*II, lying at the blastopore, are enveloped in ectoderm. At this time the mesoderm cells separate from the ventral cell layer and become completely invaginated. The four endoblasts inside the embryo and these primary mesoblasts proceed to divide, and the latter form the lateral embryonic body wall.

After this the stomatoblasts produce a depression, and the formation of the stomodaeum begins. While this process is going on, the secondary ectoblast cells spread over the posterior ventral surface of the embryo, and the *G*I and *G*II blastomeres slip out of the ventral cell layer and move to the interior. The stomodaeal cells in the blastopore region bulge inward, forming a *stomodaeal pocket* anterior to the germ cells *G*I and *G*II (Fig. 6.8 a—c). *G*I and *G*II aie pushed still closer to the endoblasts; ectoderm cells proliferate from the sides and the posterior part of the embryo, and spreading over *G*I and *G*II, come together in the midventral line. The germ cells separate from the ectoderm and move inward, the anterior cell (*G*I) penetrating farther into the gut rudiment than the other (*G*II), and the stomatoblasts establish connection with the ectoblasts. In this process the blastopore is completely closed off, and posterior to it the two germ cells are surrounded by mesoderm and endoderm. The ventral ectoderm gradually acquires a smooth contour (Fig. 6.8 d—f). In addition to the stomatoblasts, some ectoderm cells also take part in forming the stomodaeum (Boveri 1899). By this process the stomodaeal invagination pushes inward, leaving a small opening and slanting somewhat ventrally; this will later move forward and become the larval mouth. Seen from the side, the stomodaeum becomes shortened, and elliptical in cross-section (Fig. 6.8g). In the interior of the embryo the endoderm develops into a cell mass, in contact anteriorly with the stomodaeum, and is flanked by bands

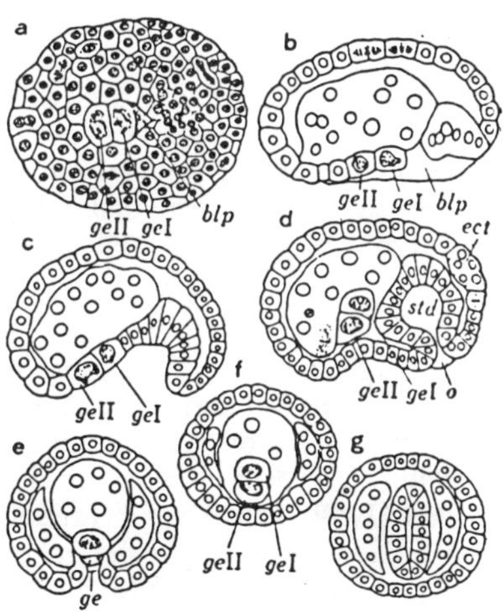

Fig. 6.8

a. Formation of stomodaeum (ventral view). b. Formation of stomodaeum (mid-plane). c. Appearance of stomodaeal invagination (mid-plane). d. More advanced stage: invagination of germ cells and closure of blastopore. e. Cross-section through germ cells of embryo shown in c. f. Cross-section through germ cells of embryo shown in d. g. Cross-section at stomodaeum. (Boveri) *blp:* blastopore, *ect:* thick part of ectoderm, *ge:* primodial germ cell, *o:* mouth, *std:* stomodaeal invagination.

of mesoderm. At the ventral side of the endodermal cell mass are situated the two germ cells.

The various problems associated with the above processes have received somewhat varied interpretations from workers investigating such nematodes as *Ascaris, Cucullanus, Bradynema, Rhabditis* and *Anguillula,* and accordingly different names have been used to designate the blastomeres. In *Anguillula aceti* and *Parascaris equorum,* the first two blastomeres have been called 'anterior' and 'posterior': 'vordere Zelle S_1', 'hintere Zelle P_1' (Pai 1928, Martini 1903), while in *Parascaris* they have also been described as 'dorsale Zelle S_1', 'ventrale Zelle P_1' (Strassen 1906); and Boveri (1899) called them 'animale Zelle S_1' and 'vegetale Zelle P_1'. In *Ascaris* and *Rhabditis,* the dorsoventral axis of the embryo is not yet fixed in the 2-cell stage; the terms 'animal' and 'vegetal' are therefore the appropriate ones. According to Schleip (1924), the primary egg axis is determined before sperm entrance; Strassen (1906) held that it becomes fixed between fertilization and the first cleavage; Boveri (1910) maintained that it becomes evident when the polar bodies are extruded. Pai (1928) held that the place of sperm entrance becomes the anterior part of the larva. In *Parascaris* and *Rhabditis,* however, the morphological difference in the cytoplasm at the animal and vegetal poles becomes progressively apparent during the period between polar body extrusion and syngamy, and is quite clear by the time syngamy is complete. The dorso-ventral and antero-posterior axis of the larva, as well as the median plane, are roughly indicated after the 2-cell stage, when the embryo shifts from the T- to the rhombic-shape; the median plane is the plane which includes the diameters of all four blastomeres (plane of page in Fig. 6.4d).

With respect to the fate of the blastomeres; in *Parascaris,* S_1, forms most of the epidermis (ectoderm). The primary mesoblast *M* (*Mes.* 1), which separates from S_2, forms most of the mesoderm; the *E* blastomere gives rise to the endoderm; *St* forms the stomodaeal rudiment; S_3 (*C*) supplies the primary endoblast cells and forms part of the mesoderm. S_4 (*d*) also forms mesoderm. There are some still unresolved questions in connection with the division of the P_4 blastomere. Boveri (1899) held that P_4 divides into P_5 and S_5 (*F*), P_5 then dividing into *GI* and *GII*, while S_4 forms the *f* and φ blastomeres at the posterior end of the embryo. Müller (1903), Zoja (1895), and Strassen (1896), on the other hand, considered that *GI* and *GII* are derived from P_4. The same sort of thing is also seen in other nematode species — *Cucullanus elegans* (Martini 1903), *Strongylus paradoxus* (Spemann 1895), *Rh. nigrovenosa* (Ziegler 1896). Boveri (1910) later maintained that the production of P_4 is normal in *Parascaris,* but that P_5 and S_5 (*F*) may be formed as an exception[8]. According to Pai, the P_5 blastomere always appears in *Anguillula aceti,* as well as the five groups of somatic cells, the S_1 blastomere giving rise to the primary ectoderm cells, S_2 (*EMSt*) to primary and secondary endoderm and mesoderm, S_3 and S_4 to the secondary and tertiary ectoderm, S_5 to tertiary endoderm, P_5 to the primordial germ cells. The S_5 blastomere produces the gonad tissue. In *Parascaris,* however, it is not clear whether S_4 (or possibly Boveri's S_5) is responsible for forming the gonad. Strassen discovered two cells (*Terminalzelle*) lying anterior and posterior to *GI* and *GII* after gastrulation; he believed that these cells participate in the formation of the gonad. These terminal cells can also be

[8] The embryological texts of Korschelt u. Heider (1936) and Schleip (1929) do not refer to the problem of Boveri's P_5, S_5 blastomeres.

Table 6.4 **Summary of the early development of Rhabditis nigrovenosa and Rh. ikedai.** Ziegler (1896) and Tadano (1952)

Cleavages of S_1 (= primary somatoblast	Time between cleavages	Total no. S_1 line cells	Activities of other blastomeres	Time for each process	Total elapsed time	Total no. cells
				at 20—23°C		
			oocyte enters uterus	60′	60′	1
			first polar body given off	45′	105′	
			second polar body given off	30—60′	135—165′	
			syngamy	15—25′	150—190′	
			cleavage spindle appears	15—45′	165—235′	2
	45′	1	egg cleaves into S_1 (AB) (= primary somatoblast = primary ectoblast) and P_1 (= stem cell = primary reproductive cell)			
first		1→2	S_1 (AB) divides into A and B		2.5 to 4.75 hrs	3
	60′		P_1 divides into S_2 (EMSt) (= entomeso-stomatoblast) and P_2 (= stem cell)	47—48′		4
second		2→4	(A — a, α; B — b, β)		3.5 to 5.75 hrs	6
	60—70′		S_2 $(EMSt)$ divides into MSt (= meso-stomatoblast) and E (= entoblast)	10—15′		7
			P_2 divides unequally into C (= entomesoblast) and P_3 (= stem cell)	15—25′		8
third		4→8			4,5 to 6.5 hrs	12
			MSt divides into mst and $\mu\sigma\tau$; E divides into EI and EII	ca. 30′		14
	60—80′		C divides into c and γ	10—20′		16
fourth			P_3 divides into $S_4(D)$ (= = mesoblast) and P_4 (= = stem cell)		5.5 to 8 hrs	(24)*
		8 ↘ (16)*	mesoblasts mst and $\mu\sigma\tau$ divide	30′		(26)
	ca. 80′		entoblasts EI and EII divide	ca. 30		(28)
fifth		16 ↘	ectomesoblast c and γ divide	ca. 30′		(30)
		(32)	($S_4(D)$ divides into d and δ); 4 entoblasts of S_2 line begin to invaginate		6.75 to 9,5 hrs	<60
	ca. 90′		P_4 (= primordial germ cell) divides into GI and GII			
sixth		32 ↘	ectomesoblasts derived from c & γ divide			
		(64)	GI and GII invaginate; blastopore closes		8.3 to 11 hrs	<96
	70—90′		4 invaginated entoblasts divide, form two rows along embryonic long axis			
		64 ↘ (128)	mesoblasts of $S_4(D)$ line divide; mesoblast of MSt line begin to divide and invaginate		9.5 to 12.5 hrs	<192
seventh						

* Numbers of blastomeres in parentheses represent theoretical numbers; actual numbers smaller becaus of progressive loss of synchrony in cleavage.

recognized in *Rhabditis,* but it is not known whether they are derived from S_4 or S_5. According to Boveri, the P_5 blastomere is produced at the 82-cell stage; Strassen places its formation at the 51-cell stage, and Martini states that it is formed at the 52-cell stage in *Cucullanus elegans.*

Various opinions have also been expressed with respect to the mode of gastrula formation in *Parascaris.* Strassen believed it to take place by invagination, while Boveri held that the absence of an archenteron points to epiboly. The observations of Hallez (1885, 1887) and other workers also led to results agreeing with Boveri's conclusion. However, Seurat (1920) and Martini (1907) formulated the following summary of the modes of gastrulation in various nematodes. a) conjunctive emboly: invagination by emboly in which an archenteron is formed to some extent, *Rh. bufonis;* b) disjunctive epiboly: cells released into the blastocoel as the result of epiboly, *Parascaris equorum;* c) conjunctive epiboly: *Camallanus lacustris, Pseudalius minor. Cosmocerca ornata* shows an intermediate mode of gastrulation.

In the early development of *Parascaris equorum,* there are differences among the types of cytoplasm of the cell groups which form each of the primary germ layers: the endodermic cells contain more endoplasm, and the cells destined to form ectoderm have relatively more ectoplasm. In *Rh. ikedai* and *Rh.* sp., however, these differences are less pronounced.

Table 4 shows that in *Rh. nigrovenosa* and *Rh. ikedai* the beginning of endoderm invagination occurs when the descendants of the S_1 blastomere reach the number of 32 cells. In this stage the total number of blastomeres is about 60; when there are approximately 180 cells the blastopore closes. In *Anguillula,* gastrula formation ends at the 141-cell stage; at this time S_1 has divided into 115 cells (Pai). Before the gastrula stage, the arrangement of the blastomeres is sometimes asymmetrical, but as the embryo approaches this stage it becomes symmetrical. The time elapsing between syngamy and gastrulation varies with the species as well as with temperature. At 30° *Parascaris* requires about 18 hours; *Rhabditis ikedai, Rh. nigrovenosa* require about 7 hours at 20—23°. The *Parascaris* egg undergoes the first cleavage at about 2 hours at 30—35°; in this same period the *Rhabditis* egg, at 20—23°, goes from first polar body extrusion to the 8-cell stage (Table 4), and completes one generation in 4—5 days.

Although nematode development as described above differs in certain respects from that of *Parascaris,* it agrees in essence with the reports of Boveri (1887, 1899, 1910) and Strassen (1896), and the research of these two men forms the foundation of nematode embryology.

(4) CHROMATIN DIMINUTION

a. Phenomenon

The phenomenon in which part of the chromatin is discarded into the cytoplasm was first discovered in *Parascaris equorum* by Boveri (1887c), and it was also he who made the later detailed observations of the process (1890, 1892). His work was repeated and confirmed by Dostoiewsky (1888), Schneider (1891), Herla (1893). Boveri (1899) adapted the term 'diminution', as used by Herla, to describe this phenomenon, calling it 'chromatin diminution'. It was furthermore shown that this process occurs in *Ascaris lumbricoides* (Mayer 1895; Bonnevie 1901), *As. rubicunda, As. labiata* (Meyer 1895), and *Rh. nigrovenosa* (Schleip 1911). The details of the process differ to some extent among these species; it has not yet been recognized as occurring in any other nematodes.

As has been stated above, nematode cleavage proceeds in a determinative fashion, and the differentiation of the somatic cell line from the germ cell line begins as early as the 2-cell stage. The germ line cells preserve their chromosomes intact through the division process, but the somatic cells cast off material from the ends of their chromosomes into the cytoplasm between metaphase and anaphase of each nuclear division, reducing the amount of chromatin left in the nucleus. This behavior makes it possible to distinguish between the germ cells and the somatic cells making up all the other organs of the body. If we consider the two blastomeres, P_1 and S_1, resulting from the first cleavage, we see that P_1 possesses the capacity to form nearly the whole embryo. It can therefore be described as 'universal', while the S_1 blastomere, by contrast, may be called 'partial'; the nuclear division leading to this situation is of profound significance from the embryological point of view. In *Parascaris equorum,* the ends of the S_1 (AB) chromosomes thicken into a flask-shape by the metaphase (Fig. 6.4a, b), and during the period between metaphase and telophase they are cast out of the nuclear region into the cytoplasm (Fig. 6.4b—d) where they lie scattered, in a row or chain, or all clumped at one side.

The central part of the chromosomal rod becomes a series of particles connected by a hyaline matrix. The particles then separate into granules, the hyaline matrix disappears, and the granules are distributed to the two new blastomeres to form the daughter nuclei. According to Boveri, there are about 60 of these granules. In the newly formed blastomere the daughter nucleus appears, with the granules clustered in its center, and the discarded chromatin lies scattered in the cytoplasm around it (Fig. 6.4d, *A* and *B* blastomeres). One opinion holds that these flask-shaped portions of the S_1 chromosomes become recognizable only after syngamy, and do not contain any genes. While these end portions are strongly stainable with gentian violet, they closely resemble the P_1 chromosomes, but they appear to lose their stainability somewhat when the central part breaks up into separate particles. According to T. P. Lin (1954), the discarded material is *heterochromatin,* and contains a large amount of DNA; Tadano and Tadano (1956) find that it gives a very strongly positive Feulgen reaction. Van Beneden et Neyt (1887) reported that at the mitotic metaphase these portions of the chromosomes which are to be discarded either fail to develop spindle fibers , or develop them poorly. *Parascaris equorum* includes the two varieties, *univalens,* with a chrot mosome number n=1, and *bivalens,* in which n=2. However, each of these unis may also be thought of as a *compound chromosome* composed of a large number of small granular chromosomes. According to Geinitz (1915), *P. equorum* var. *univalens* has a single large compound chromosome composed of an X-*complex* made up of 8 components, and an *autosomic complex* containing 22 components.[9] Since *P. equorum.* var. *bivalens* discards a larger amount of chromatin than *univalens,* the somatic resting nuclei of the two varieties after diminution are about the same size. On the other hand, the resting nuclei of the germ line cells (*P* 1—4) and of the primordial germ cells (*GI* and *GII*) are clearly larger in var. *bivalens* (Bonfig 1925). Moreover, the vesicular resting nuclei of the somatic cells after chromosome

[9] Results of research concerned with ascarid chromosomes, and particularly the sex chromosomes, have been published by Walton (1921, 1924), Edwards (1910, 1911), Jeffrey and Haertl (1938).

diminution are oval or spherical, while the resting nuclei of the germ line cells characteristically have a number of pouch-shaped processes (Fig. 6.4c) equal to the number of ends of the chromosomes contained in them (Boveri 1887, van Beneden et Neyt 1887, 1888b).

This chromatin diminution phenomenon takes place during each cleavage of the somatic line blastomeres from S_1 to S_4 (or S_5), so that only the germ cells descended from the P_4 (or P_5) blastomere retain a set of complete chromosomes.

b. Mechanism

Strassen (1895, 1896) believed the cause of chromatin diminution to depend on some essential factor located in the nucleus. Boveri (1910a, b), who found that in dispermic and centrifuged eggs diminution takes place only in the animal-pole side blastomere, considered the cause of the diminution phenomenon to lie in some peculiarity of the cytoplasm (Heteropolie) at either the animal or the vegetal pole. Beams (1937, 1938) also found that when suppression of cytoplasmic cleavage by ultracentrifugation caused the formation of multinucleate egg cells, all these nuclei carried out chromatin diminution. This led him to agree with Boveri in searching for a cytoplasmic cause, and he concluded that chromatin diminution is due to a 'chemical diminisher' originating in the cytoplasm, with a distribution gradient along the uncleaved egg axis, decreasing toward the animal pole.

Furthermore, it is becoming apparent that it may be possible by means of chemical reagents to bring about a reversible conversion of germ line blastomeres into somatic blastomeres and vice versa. In eggs treated with LiCl, the germ line blastomeres are induced to undergo chromatin diminution, and in other respects such as size, rate of cleavage, and endo-ectoplasmic relation, these blastomeres resemble normal somatic line blastomeres. Moreover, chromatin diminution is inhibited in somatic line blastomeres which have been treated with NaSCN, and again, the resemblance to normal germ line cells extends to their size, division rate and endo-ectoplasmic distribution. If the ammonia treatment (Yamada 1950) which induces dorsalization in isolated ventral ectoderm of amphibian gastrulae is applied to *P. equorum* eggs, chromatin diminution is inhibited in blastomeres of the mesoderm-forming line. The secondary mesoblast S_4 (*D*) and its daughter cells are particularly affected, and the other characteristics of these cells are again like those of normal germ line cells (Tadano and Tadano 1955, 1956). The factor which brings about this phenomenon of chromatin diminution, however, has not yet been definitely established.

2. LATER DEVELOPMENT

(1) EXTERNAL FORM

a. Cell Constancy

In many nematodes the various organs are characterized by *cell constancy*, as shown by the studies of Martini (1906, 1907) on *Oxyuris*, Goldschmidt (1908, 1909, 1910) on *Parascaris* and Pai (1928) on *Anguillula*. In a word, this means that every cell in the nematode body has a definite shape, size, and position, and that the number of cells is determined by the time molting is complete, with no changes occurring thereafter, except that the boundaries between cells may disappear to some extent after ecdysis, making it difficult to confirm the cell constancy of such tissues as the epithelium in any but the small nematodes.

b. Body Shape

After the gastrula stage the embryo loses its oval form; the anterior end becomes thicker while the posterior end becomes thinner and bends anteroventrally, performing localized movements. As the result of extremely accelerated growth of the dorsal ectoderm, the head makes contact with the anterior part of the ventral surface, and the ectoderm of the posterior end also develops and thickens. Next the anterior end curves ventrally, bringing the dorsal part into contact with the inside of the egg membrane (Fig. 6.9). As it grows inside the membrane, the embryo takes on a spiral or S-shape. When the membrane is thin, as in *Metastrongylus elongatus* or *Rh. ikedai*, its shape changes as the embryo elongates.

c. Epithelium

The cells making up the epidermis in the central part of the curved embryo are arranged in longitudinal rows, forming regular layers and sections. In *Anguillula*, according to Pai, the S_1 (AB) blastomere gives rise to about 4/5 of the epithelium, and S_3 (C), to about 1/5. In *Cucullanus* the cells form two rows each on the ventral and lateral surfaces and one row dorsally, consisting of 11, 12 and 24 cells. These will form the whole epidermis, from the anterior end to the posterior; three-fourths of the cells are descended from the S_1 (AB), and one-fourth from the S_3 (C) blastomere, the latter covering the posterior dorsal surface of the embryo (Fig. 6.9c) (Martini 1903). According to Boveri, at least 12—14 cells derived from the S_3 (C) blastomere take part in forming the epidermis in *P. equorum*; the remainder is composed of descendants of the S_1 (AB) blastomere. The cells making up the epidermis in the anterior and posterior parts of the embryo become flattened and lose their regular alignment as growth proceeds, and a tough surface layer, the *cuticula*, is soon formed, as well as an underlying *subcuticular layer*. Even in the juvenile stage the cell boundaries in the cuticula are indistinct, and they gradually disappear, giving rise to a *syncytium*. The subcuticular layer cells are arranged in a linear fashion along the lateral, dorsal and ventral surfaces[10]; in some places they retain the regular distribution of the early stages.

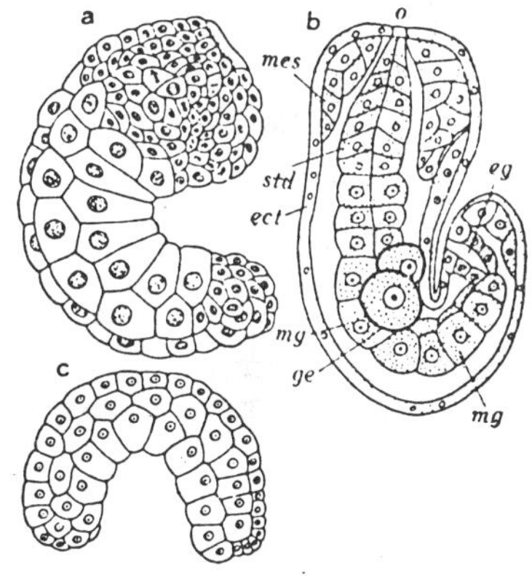

Fig. 6.9

a. Embryo of *Parascaris equorum*, to show orientation of ectodermal cells. (Müller 1903) b. Optical cross-section of embryo of *Rhabditis nigrovenosa*. (Neuhaus 1903) c. *Embryo* of *Cucullanus elegans*, showing orientation of ectodermal cells. (Martini 1903) *ect:* ectodermal cell, *eg:* end-gut, *ge:* primordial germ cell, *mes:* mesodermal cell, *mg:* mid-gut, *o:* mouth, *std:* stomodaeum.

10 The lines of subcuticular cells are called 'dorsal and ventral median line' and 'lateral line' according to their situation; in *Anguillula* they are composed of 56 cells.

(2) ORGANOGENESIS

a. Digestive Tract

i) *Foregut*

Strassen described the stomodaeum as formed by the cells derived from the stomatoblast *St*; Boveri added to these the blastomere S_1 (*AB*); Martini held that *St* forms the muscular part of the stomodaeum and the other parts are formed by the S_1 (*AB*) blastomere. The *foregut* consists of a *pharynx* and an *oesophagus*. In *Anguillula*, 59 cells make up the pharynx, while the oesophagus, forming the junction between pharynx and midgut, consists of five cells (Pai 1928).

ii) *Midgut and Hindgut*

The rudiment of the midgut arises from 12 large endoderm cells derived from the *E* blastomere which separated from S_2 (*EMSt*). These are arranged in two rows, and consist of not more than 18 cells: e.g., the midguts of *Cucullanus elegans* and *Rh. nigrovenosa* are made up of 16 cells, and that of *Anguillula*, of 18. In general the small nematodes have smaller numbers of cells.

As the embryo grows, these longitudinally arranged endodermal cells form a simple tube with a lumen which is at first restricted, but later widens. The tube becomes connected anteriorly with the foregut and at its posterior end with the hind gut. Next the lumina of foregut and midgut join, and a constriction is formed in this region. Since the midgut of *Anguillula* has no glands, Pai considered it to have an absorbing function. According to Martini and Pai, the hindgut is derived from the S_4 (*D*) blastomere. In *Anguillula*, it is formed of 20 cells which are said to make contact with an infolding of the ectoderm in the posterior part of the embryo when they form the hindgut, but this is claimed not to be so in other nematodes. *Rh. ikedai*, however, resembles *Anguillula* in this respect. The hindgut of the male connects with the ejaculatory duct and opens into the cloaca (Fig. 6.11c—f).

b. Central Nervous System

The central nervous system in the ascarids is derived from the S_1 (*AB*) blastomere. When the stomodaeum is formed, the surrounding S_1 line blastomeres enlarge, proliferate toward the interior and form the *nerve ring* (Fig. 6.8d, *ect*) (Boveri 1899). In *Anguillula* the cells descended from S_1 give rise to the pharyngeal ring as an ectodermal thickening in the region of the pharynx; these proliferate internally and form eight *ganglia* (Pai, Ganin, Goette, Strassen 1892, Neuhaus). According to Pai, this thickening is easily confused with another in this region composed of mesoderm and ectoderm which will form the salivary gland. The formation of the *longitudinal nerve trunk* appears to follow immediately after the ectodermal thickening takes place, but the details of the process are not known. The nervous system of *Anguillula* is made up of 185 cells.

c. Excretory System

According to Jammes (1892) and Pai, the *excretory duct* or *lateral duct* is derived from the ectoderm; in *Anguillula* cells descended from the S_1 (*AB*) blastomere push into the body cavity and form this duct, with an opening at the side of the pharyngeal bulb. Hamann, Strassen and Conte state that it arises from mesodermal cells which appear near the pharyngeal bulb and extend in the form of a cylinder along the body wall. In more highly differentiated species, *lateral canals* starting at the posterior end run forward along the lateral lines at the two sides, to join in the oesophageal region, where the common duct empties into an *excretory ampulla* in the ventral midline. This canal system usually consists of four ducts, but there are many species differences, and the details are not clear.

d. Musculature, Connective Tissue, Coelom

The mesoderm derived from the S_2 (*EMSt*) blastomere grows in the form of bands which extend anteriorly and posteriorly between the endodermal and ectodermal cell walls. Later the band structure loosens and the mesodermal cells disperse. As the epidermis and gut wall take shape, the blastocoel, or *primary body cavity,* appears between them. Next, the cells which lie scattered about in these regions form *connective tissue,* and others in contact with the epidermis form the body *musculature.* Both types of tissue extend as fibers under the cuticula. The body wall musculature of *Anguillula* consists of 64 cells, and the connective tissue of 61 (Pai). The space between the gut and the muscle layer disappears with the development of the reproductive organs.

e. Reproductive Organs

Immediately after gastrulation in *P. equorum* and *Rhabditis,* there can be seen two 'terminal cells' lying at the anterior and posterior sides of the two primordial germ cells *GI* and *GII* (Fig. 6.10a, b). The former give rise to the somatic parts of the reproductive system, and the latter become the reproductive cells. Pai maintained that in *Anguillula* the S_5 (*F*) blastomere produces the reproductive system.[11] In the female, the anterior cell derived from S_5 (*F*) forms the somatic part of the ovary, while the posterior cell forms the *oviduct, uterus* and *seminal receptacle* (Fig. 6.10e). In the male, the anterior cell forms the *vas deferens, seminal vesicle* and *ejaculatory duct,* and the posterior cell gives rise to the somatic tissues of the testis (Fig. 6.11c, d, f).

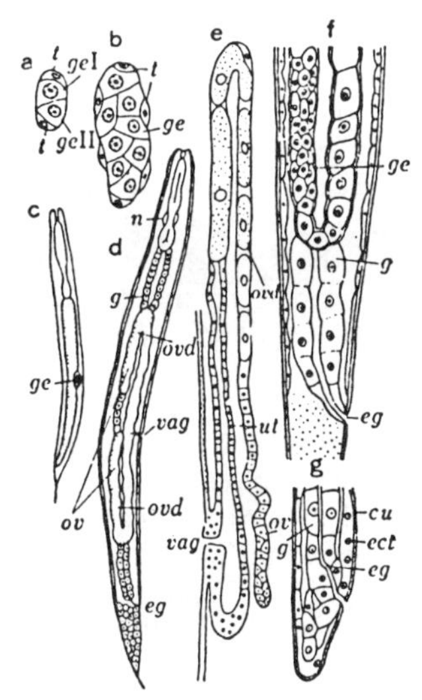

Fig. 6.10 Female reproductive system

a, b. *Rhabditis aberrans* (Krüger); c, d, f. *Rh. ikedai* (Tadano); e, g. *Anguillula aceti* (Pai). *cu:* cuticula, *ect:* ectoderm, *g:* gut, *ge:* germ cell, *n:* circular nerve, *ov:* ovary, *ovd:* oviduct, *t:* terminal cell, *ut:* uterus, *vag:* vagina.

The primordial germ cells *GI* and *GII*, which in the young worm lie midventrally in the central or somewhat posterior part of the body, are larger than the other cells (Fig. 6.9b; 6.10c). For some time after their appearance these cells remain in a resting state, and only when the young worm grows and molts do the two germ cells and the two terminal cells begin to divide. The latter surround *GI* and *GII*, and the cell

[11] The blastomere which Pai called S_5 (*F*) in *Anguillula* corresponds to Strassen's 'terminal cells'.

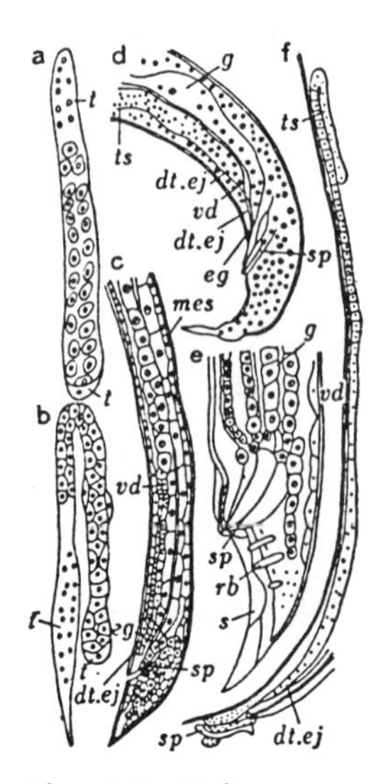

Fig. 6.11 Male reproductive system

a, b, f. *Anguillula aceti* (Pai); c, d. *Rh. nigrovenosa* (Neuhaus); e. *Rh. ikedai* (Tadano). *dt.ej:* ejaculatory duct, *eg:* end-gut, *g:* gut, *mes:* mesoderm, *rb:* ribs, *s:* genital alae, *sp:* spiculum, *t:* terminal cell, *ts:* testis, *vd:* vas deferens.

mass as a whole elongates in a cylindrical fashion. Eventually the anterior end bends backward (Figs. 6.10 d, e; 6.11b), and the posterior part connects in the female with the *gonopore,* and in the male with the *cloaca* (Fig. 6.11c — f).

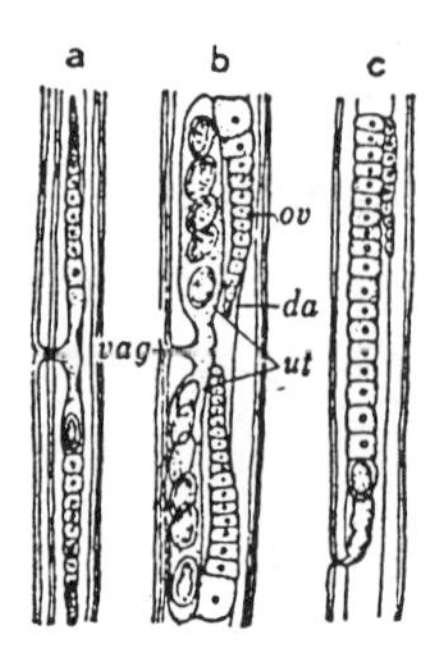

Fig. 6.12 Various types of female reproductive organs

AI-type; b. AII-type; c. C-type.

In *Rh. ikedai* the gonopore opening is located at the center of the gonad rudiment, slightly posterior to the mid-point of the body; in other species the position may be different. The gonopore opens to the outside, and leads inward to the *vagina* (Fig. 6.10 d, e). In *Allantonema* a mesodermal protuberance from the ventral body wall makes contact with the uterus to form the vagina (Leuckart). According to Strubell (1888), the muscle cells of this region form the *sphincter* in *Heterodera.* After the first molt in the males, the caudal end bends and the *copulatory spicules* and *genital papillae* appear, although these are still small and non-functioning. In the anterior part of the cloaca is the rudiment of the *rectum,* and posteriorly there is a pouch-shaped mass of cells of the S_3 (C) line which will form the *copulatory bursa*; the spicules develop in the space between these two rudiments (Fig. 6.11c—f). With this the gonadal rudiment undergoes a striking development: a pair of chitinous spicules are formed, located in the dorsal side of the rectum. In *Rh. ikedai,* chitinous *ribs* are also formed at the sides in the posterior end; (Fig. 6.11e *b*); the development of these structures elevates the epidermis and gives rise to the copulatory bursa. The formation of the caudal region is completed by successive moltings. After four or five molts, conspicuous structural differences between the males and females make their appearance. In most cases, the male reproductive system is a simple tubular one, while that of the female is paired and rather complicated, with characteristic species differences which may be divided into four types.

Type A-I: with the gonopore in the center, the ovary extends anteriorly and posteriorly; found in the free-living nematodes *Diplogaster, Chromadora,* etc. (Fig. 6.12a). Type A-II: like A-I, but with the ends of the ovaries more elongated and turned back toward the center; *Actinolaimus, Rh. ikedai*, etc. (Fig. 6.12b). Type B: a pair of parallel ovaries; *As. lumbricoides, Parascaris equorum*, etc.[12] Type C: a single ovary with the gonopore at one end; *Monhystera, Cephalobus, Rh. Dujardin*, etc. (Fig. 6.12c). A-II and B are the most usual types.

(3) LARVA

In general, the embryo in the oviparous species *(Ascaris)* develops outside the maternal body to a stage resembling the adult worm, while in the ovoviviparous and viviparous forms *(Rhabditis)*, this growth takes place within the uterus. The number of eggs spawned is small in the free-living species, and large in the parasites; it is particularly small in species such as *Rhabditis, Enterobius, Strongylus*, etc., which are considered to be primitive forms. Growth generally takes place by direct development, and except for the specialization of organs of parasitic adaptation, there is no metamorphosis. During its growth, the embryo undergoes two or three changes as the result of hatching and molting. When a young worm, after hatching, can be morphologically distinguished from sexually mature individuals, it is called a *larva*. The young of oviparous species undergo marked growth after hatching, the reproductive organs, in particular, reaching complete development in a comparatively brief time.

a. Hatching

Studies of the hatching process have been made by Stewart (1921), Asada (1923), Ida (1930), MacRae (1935), Yoshida (1939), O'Connor (1951) and Takahashi (1953), but the mechanism is not yet thoroughly understood. Hatching in the ascarids occurs about the time of the first molt. The young of the horse and pig ascaris will hatch in Ringer's solution, and the egg membrane also dissolves in the host's intestinal fluid, setting the young worm free. If this larva has attained sufficient growth it will undergo a second molt *in vitro*. On the basis of these observations it is considered that in addition to the state of the larva itself, external factors such as the intestinal fluid, temperature, pH, and the mechanical action of the digestive tract play a decisive role in the control of hatching. It is clear, moreover, that there are different degrees of internal and external influence as well as species differences in the membranes.

b. Molting

The larva usually molts four times before it becomes an adult, although some free-living species such as *Actinolaimus* molt five times (Maupas 1899; Kreis 1930). The lifetime of a worm which molts four times can be divided into five stages; the stages to the fourth molting are then called 'larval', and the fifth is the 'adult stage'. There are few changes in morphology during the larval stages, since practically all the organs except those of the reproductive system are already

[12] In *P. equorum*, individuals are sometimes found with three complete ovaries, as well as others with one ovary, or two, of which only one contains eggs. Such irregularities occur with a frequency of about 1%.

present in the first stage. The reproductive organs develop conspicuously in the third stage and are nearly complete in the fourth. When ecdysis proceeds rapidly, it may require only a few hours; when it goes slowly, it may take several months. At the time of ecdysis the larva becomes quiescent and rigid, and spaces appear under the cuticula, first in the head and then in the tail region; the cuticle thus becomes separated and finally torn by the movements of the larva. In some forms it is shed all at once, in others bit by bit, and some species shed even the lining of the mouth cavity and oesophagus. In the males of *Actinolaimus* the posterior end is renewed at the third molt. The larvae of some species form cysts under unfavorable environmental conditions and pass through a dormant stage as *encysted* or *ensheathed larvae,* molting when conditions become favorable. *Heterodera* is one form which passes through a long dormant period (Strubell 1888, Nagakura 1918). In the species which undergo successive molts, the total length increases by about three times within a few days after the third molt. Among the parasitic species, ecdysis bears a special significance with respect to infection: the larva gives up its free-living habit at the time of the second ecdysis and waits for a chance to penetrate a host. If, however, it does not find a host, it may fail to undergo the second molt and become ensheathed; in some species *(Ascaris)* the larva waits within the egg membrane for a chance to infect a host. Forms which parasitize a single host undergo the second and succeeding molts within the host, while in forms with an intermediate host, the second-stage larva encysts within its tissues and molts, waiting as a third-stage larva for the chance to parasitize its final host. In other forms the larva does not encyst, but finds and penetrates its definitive host by its own exertions.

3. METHOD OF REPRODUCTION AND SEX

Nematodes are usually dioecious, but there are also some hermaphroditic forms. The former carry on sexual reproduction by copulation, the latter by self-fertilization; there are also forms which reproduce parthenogenetically. Again, in the genera *Rhabdias* and *Strongyloides* there are cases like that of *Rhabdias bufonis,* parasitizing the lungs of frogs, in which the parasitic generation is hermaphroditic and produces fertilized eggs by self-fertilization; these develop into the *Rhabditis-type larvae* of the free-living generation which become dioecious adults. The offspring of these worms become *Filaria-type larvae* and enter upon a parasitic habit, completing the alternation of generations.

The hermaphroditic individual in most such cases has undergone the change from a female, but sometimes males may transform into hermaphrodites. In *Rhabdias bufonis,* both females and hermaphrodites have two sex chromosomes (XX) while the males have one (X). When spermatozoa are formed by the parasitic hermaphrodite individuals, the spermatogonia, which are really female cells, have XX-chromosomes. However, one X-chromosome is discarded during meiosis, so that spermatozoa with one X-chromosome and others with none are formed. In consequence, both males and females appear in the free-living generation, in spite of the fact that they are produced by self-fertilization. During the course of spermatogenesis in the free-living generation, the spermatozoa lacking an X-chromosome degenerate; as a result all the offspring are XX females, which enter the lungs of frogs and become the hermaphrodites of the parasitic generation (Boveri 1911; Schleip 1911). This process furnishes the basis for regarding these animals as 'female-type hermaphrodites'. Additional evidence comes from some species in which males and hermaphrodites are found existing together; these hermaphrodites produce sperm during their juvenile stages and, after reaching maturity, eggs which develop into hermaphrodites following self-fertilization. This results in a gradual reduction in the number of males. In some species these males completely lose the capacity to copulate, and disappear. The *Rhabditis* hermaphrodite forms spermatozoa in the part of the body between the sperm receptacle and the oviduct. In *Rh. sechellensis* the spermatozoa, which are formed in a limited region in the lower part of the oviduct, mature first, while *Rh. gurneyi* forms eggs and sperm together at first, and alternately later. The *Rh. ikedai* hermaphrodite resembles *Rh. sechellensis* in an early stage, but afterwards shows similarities to *Rh. gurneyi*. This large hermaphrodite has a much convoluted oviduct, and reproduces prolifically. On the other hand, *Rh. elegans* and *Bradynema rigidum* have hermaphrodites of the male type. In *B. rigidum,* foı example, there are both male and female larvae, but the male acquires an ovary in addition to its well developed testis, and grows to the adult stage, while the female larvae disappear (Strassen 1892). Since the spermatozoa mature first in hermaphrodites of this type, the supply may be inadequate to fertilize the later-ripening eggs, which then develop parathenogenetically (Maupas 1900, 1901). In *Rh. schneideri,* moreover, reproduction takes place solely by parthenogenesis. In *Rh. aberrans,* the sperm pronucleus within the egg fails to unite with the egg pronucleus, and finally disappears, and the egg develops parthenogenetically (Krüger 1913). Among the nematodes in general there tend to be fewer males than females, and the sex ratio varies with such factors as species and strain, place and time of collection and conditionsthf culture (Maupas 1900; Hertwig 1922; Bělař 1924; Dotterweich 938; Tadano ond Tadano 1955). The sex ratio of *Rh. ikedai* has been found to vary with the strain (Motomura 1948).

REFERENCES

Asada, J. 1923—24: Supplemental notes to a study of the migration pathways of round-worms within the animal body. Tokyo Med. Mag. **2324** : 695—699, **2327** : 857—862, **2357** : 346—356. (in Japanese)

Baer, G. J. 1951: Ecology of Animal Parasites. Urbana.

Beneden, E. van et Y. Neyt 1887: Nouvelles recherches sur la fécondation et la division mitosique chez *l'Ascaride megalocephale*. Bull. Acad. Roy. Belg., **14** : 214—295.

——————— 1888a: Sur la fécondation chez *l'Ascaride megalocephale*. Anat. Anz., **3** : 104.

——————— 1888b: Präparate über Copulation der Geschlechtsprodukte, Reifung des Eies, Befruchtungsvorgang und Mitose bei *As. megalocephala*. Anat. Anz., **3** : 707—709.

Bělăr, K. 1923: Über den Chromosomenzyklus von parthenogenetischen Erdnematoden. Biol. Zentralbl., **43** : 513—519.

Bonfig, R. 1925: Die Determination der Hauptrichtungen des Embryos von *As. megalocephala*. Zeitschr. wiss. Zoöl., **124** : 407—465.

Bonnevie, K. 1901: Über Chromatindiminution bei Nematoden. Jena. Zeitschr. Naturwiss., **36** : 275—288.

Boveri, Th. 1887a: Über die Befruchtung der Eier von *As. megalocephala*. Sitz. — Ber. d. Ges. f. Morph. und Phys., **3** : 71—80.

——————— 1887b: Zellenstudien. I. Jena. Zeitschr. Naturwiss., **21** : 424—515.

Boveri, Th. 1887c: Über den Anteil des Spermatozoon ander Teilung des Eies. Sitz.-Ber. d. Ges. f. Morph. und Phys., **3** : 151—163.

——— 1887e: Über Differenzierung der Zellkerne während der Furchung des Eies von *As. megalocephala.* Anat. Anz., **2** : 688—693.

——— 1888: Zellenstudien. II. Jena. Zeitschr. Naturw., **22** : 685—882.

——— 1892: Über die Entstehung des Gegensatzes zwischen den Geschlechtszellen und den somatischen Zellen bei *As. megalocephala.* Ber. d. Ges. f. Morph. und Phys., **8** : 114—125.

——— 1899: Die Entwicklung von *As. megalocephala* mit besonderer Rucksicht auf die Kernverhältnisse. Festschr. f. Kupffer Jena, 383—430.

——— 1910a: Die Potenzen der Ascaris-Blastomeren bei abgeänderter Furchung. Festschr. R. Hertwig, **3** : 131—214.

——— 1910b: Über die Teilung zentrifugierter Eier von *As. megalocephala.* Arch. f. Entw.-Mech., **30** : 101—125.

——— 1911: Über das Verhalten der Geschlechtschromosomen bei Hermaphroditismus. Verh. Phys. med. Ges. Würzburg, **41** : 83—79.

——— & M. N. Stevens. 1904: Über die Entwicklung dispermer Ascariseier. Zoöl. Anz., **27** : 408—417.

Brumpt, E. et M. Neveu-Lemaire 1951: Travaux pratiques de parasitologie. Paris.

Bütschli, O. 1876: Zur Entwicklungsgeschichte von *Cucullanus elegans* Zid. Zeitschr. wiss. Zoöl., **26** : 103—111.

Carnoy, J. B. et H. Lebrun 1897; La fecondation chez *l'As. megalocephala.* Verh. Anat. Ges., **11** : 65—68.

Cobb, N. A. 1918: Free-living Nematodes, in Ward and Whipple's Fresh-water Biology. 459—505. New York.

Conte, A. 1902: Contributions a l'Embryologie des Nematodes. Ann. Univ. Lyon **1**. Fas. 8.

Dostoiewsky, A. 1888: Eine Bemerkung zur Furchung der Eier von *As. megalocephala.* Anat. Anz., **3** : 646—648.

Erlanger, R. von, 1897: Beobachtungen über die Befruchtung und die ersten zwei Theilungen an den lebenden Eiern kleiner Nematoden. Biol. Zbl. **17** : 152—160, 339—346.

Faure-Fremiet, E. 1912: Sur la maturation et fécondation chez l'*As. megalocephala.* Bull. de la Soc. zool., **37** : 83—84.

——— 1912: Graisse et glycogène dans l'oeuf de l'*As. megalocephala.* Bull. de la Soc. zool., **37** : 233—234.

——— 1913: La formation de la membrane interne de l'oeuf d'*As. megalocephala.* Compt. rend. de la Soc. biol., **74** : 1183—1184.

Fülleborn, F. 1928: Über den Infektionsweg bei *Rhabdias bufonis* (*Rhabdonema nigrovenosum*) des Frosches. Centralbl. Bakt. Paras. **109** : 444—462.

Geinitz, B. 1915: Über Abweichungen bei der Eireifung von *As. canis.* Arch. Zellforsch. **13** : 588—633.

Goldschmidt, R. 1903: Histologische Untersuchungen an Nematoden. Zoöl. Jahrb. Abt. Anat. u. Ont. **18** : 1—57.

——— 1909—1910: Das Nervensystem von *As. lumbricoides* und *megalocephala.* Zeitschr. wiss. Zoöl. **92** : 306—357, Festschr. Hertwig, **2** : 253—254.

Goette, A. 1882: Untersuchungen zur Entwicklungsgeschichte der Würmer, *Rhabditis nigrovenosa* Leipzig.

Grassi, B. 1887: Trichocephalus-und Ascarisentwicklung. Zbl. f. Bakt. Paras. kde., **1** : 131—132.

Hallez, P. 1883: Sur la spermatogénèse et les phénomènes de la fécondation chez l'*As. megalocephala.* Compt. rend. Acad. Sci., **98** : 695—697.

——— 1885: Sur le développement des Nématodes. Compt. rend. Acad. Sci., **101** : 170—172.

——— 1887: Nouvelles études sur l'embryogénie des Nematodes. Compt. rend. Acad. Sci., **104** : 517—720.

Hamann, O. 1892: Zur Entstehung des Exkretionsorgans, der Seitenlinien und der Leibeshöhle der Nematoden. Centralbl. f. Bakt. Paras., **11** : 501—503.

Held, H. 1912: Über den Vorgang der Befruchtung bei *As. megalocephala.* Verh. Anat. Ges. 26 : 242—248.

——— 1916: Untersuchungen über den Vorgang der Befruchtung. I. Arch. f. mikr. Anat., **89** : 59—224.

Hertwig, P. 1922: Beobachtungen über die Fortpflanzungsweise und die systematische Einteilung der Regenwurmnematoden. Zeitschr. wiss. Zoöl., **119** : 539—558.

Hogue, M. J. 1910: Über die Wirkung der Zentrifugalkraft auf die Eier von *As. megalocephala*. Arch. f. Entw.-Mech., **29** : 109—145.

Ida, S. 1930: On the egg shell constitution in round-worm eggs. J. Keio Med. Soc. **10** : 6. (in Japanese)

———— 1930: On the mechanism of ecdysis of the mature larva in round-worms. J. Keio Med. Soc. **9** : 1. (in Japanese)

Itagaki, S. 1941: Pathology of round worm infections in domestic animals. Koseido. (in Japanese)

Izumi, S. 1952: Biological Studies on Ascaris Eggs. III. Jap. Med. Jour. **5** : 45—51.

Jeffrey, E. C. & E. J. Haertl 1938: The nature of certain so-called sex-chromosomes in Ascaris. La Cellule. **47** : 239—244.

King, R. L. & H. W. Beams 1937: Effect of ultracentrifuging on the egg of *As. megalocephala*. Nature. 139 : 369—370.

——— & ——— 1938: An experimental study of chromatin in Ascaris. Jour. Exptl. Zool. **77** : 425—438.

Kobayashi, H. 1921—1922: Growth and habits in the Nematoda. Zool. Mag. **33** : 342—348, **33** : 393—405. **33** : 469—472, **34** : 41—51. (in Japanese)

Koizumi, T. 1938: Introduction to Human Parasitosis. Iwanami. (in Japanese)

Korschelt, E. & K. Heider. 1936: Vergleichende Entwicklungs-geschichte der Tiere, Bd. **1** : 300—318.

Kreis, H. S. 1930: Die Entwicklung von *Actinolaimus tripapillatus* (v. Daday). Zeitschr. Morph. Okolog., **18** : 322—346.

Krüger, E. 1912: Die Phylogenetischen Entwicklung der Keimzellenbildung einer freilebenden Rhabditis. Zoöl. Anz., **40** : 233—237.

Krüger, E. 1913: Fortpflanzung und Keimzellenbildung von *Rhabditis aberrans*. n. sp. Zeitschr. wiss. Zoöl., **105** : 87—123.

Leuckart, R. 1866: Zur Entwicklungsgeschichte der Nematoden. Arch. f. Heil., **2** : 195—235.

———— 1883: Über die Lebensgeschichte der sogenannten Anguillula sterocoralis und deren Beziehungen zu der sogenannten Anguillula intestinalis. Abh. Sächs. Ges. wiss., 85—107.

Martini, E. 1903: Über Furchung und Gastrulation bei *Cucullanus elegans*. Zeitschr. wiss. Zoöl., **74** : 501—556.

———— 1906—1909: Über Subcuticula und Seitenfelder einiger Nematoden. Zeitschr. wiss. Zoöl., **81** : 699—766, **86** : 1—54. **93** : 535—624.

———— 1907: Die Konstanz der histologischer Elemente bei Nematoden. Verh. Anat. Ges., **22** : 132—134.

Maupas, E. 1899: La mue et l'enkystement chez les nématodes. Arch. Zool. Exptl., **7** : 563—628.

———— 1900—1901: Modes et formes de reproduction des Nématodes. Arch. Zool. Exptl., **7** : 463—496.

McRae, A. 1935: The extra-corporeal hatching of Ascaris eggs. Jour. Paras., **21** : 222.

Meves, F. 1911: Über die Beteilung der Plastochondrien an der Befruchtung des Eies von *As. megalocephala*. Arch. mikr. Anat., **76** : 683—713.

Meyer, O. 1895: Celluläre Untersuchungen an Nematodeneiern. Jena. Zeitschr. Naturwiss., **29** : 331—410.

Müller, H. 1903: Beiträge zur Embryonalentwicklung der *As. megalocephala*. Zoologica. **41** : 30

Morishita, K. 1953: Ascarids and ascaridiasis. Nagai Shoten (in Japanese)

———— 1930: Strongylidae. Iwanami Biological Course, Iwanami Shoten. Tokyo (in Japanese)

———— 1943: Nematomorpha. Animal Geneology **1** : 816—872. Yokendo, Tokyo (in Japanese)

Motomura, I. 1948: On the sex of *Rhabditis ikedai* Tadano. Ecological Studies, **11** : 113—116. (in Japanese)

Nagakura, K. 1918: Life history and anatomy of *Heterodera radicicola*. Zool. Mag., **30** : 199—204, 253—255, 296—300, 331—339, 413—421. (in Japanese)

Nigon, V. et R. Delavault 1952: L'evolution des acides nucléiques dans les cellules reproductives d'un Nématode pseudogame. Arch. de Biol., **63** : 393—410.

Neuhaus, C. 1908: Die postembryonale Entwicklung der *Rhabditis nigrovenosa.* Jen. Zeitschr. f. Naturw., **37** : 653—690.

O'Conner, G. R. 1951: Morphological and environmental studies on the hatching of ascarid egg *in vitro.* Jour. Paras., **37** : 179.

Pai, S. 1928: Die Pasen des Lebenscyclus der *Anguillula aceti* Ehrbg. Zeitschr. wiss. Zoöl., **131** : 293—344.

Panijel, J. 1947: Contribution a l'étude biochimique de la fécondation chez *As. megalocephala.* Bull. de la Soc. de Chim. Biol., **29** : 1098—1106.

Panijel, J. et J. Pasteels 1951: Analyse cytochimique de certains phénomènes de recharge en ribonucléoproteines. Arch. de Biol., **62** : 353—370.

Ping, T. Lin 1954: The chromosomal cycle in *Parascaris equorum.* Chromosoma, **6**(**3**): 175—198

Potts, F. A. 1910: Notes on the free-living nematodes. I. The hermaphrodite species. Quart. J. Micr. Sci., **55** : 443—484.

Romeis, B. 1913: Beobachtungen über die Plastosomen von *As. megalocephala.* Arch. mikr. Anat., **81** : 128—172.

Sala, L. 1895: Experimentelle Untersuchungen über die Reifung und Befruchtung der Eier bei *As. megalocephala.* Arch. mikr. Anat., **44** : 657—674.

Schleip, W. 1911: Das Verhalten des Chromatins bei *Angiostomum* (*Rhabdonema*) *nigrovenosum.* Arch. Zellforsch., **7** : 87—138.

———— 1924: Die Herkunft der Polarität des Eies von *As. megalocephala.* Arch, f. mikr. Anat. u. Entw.-Mech., **100** : 573—598.

———— 1929: Die Determination der Primitiventwicklung. Leipzig.

Schneider, A. 1866: Monographie der Nematoden. Berlin.

Seurat, L. G. 1920: Histoire naturelle des Nématodes de la Berbèrie. Alger.

Shimamura, T. 1916: Studies on visceral parasites with special reference to Askaron, a toxic substance of round-worms, and its physiological action. Zool. Mag. **28**(**334**) : 334. (in Japanese)

Spek, J. 1918: Die amöboiden Bewegungen und Stiömungen in den Eizellen einiger Nematoden. Arch. Entw.-Mech., **44** : 5.

Spemann, H. 1895: Zur Entwicklung des *Strongylus paradoxus.* Zool. Jb. Morph., **3** : 301—317.

Stewart, F. H. 1912: On the life-history of *As. lumbricoides.* Paras., **13** : 37—47.

Strassen, O. Zur 1892: *Bradynema rigidum* v. Sieb. Zeitschr. wiss. Zoöl., **54** : 655—747.

———— 1895: Entwicklungsmechanische Beobachtungen an Ascaris. Verh. D. Zoöl. Ges., **5** : 83—95.

———— 1896: Embryonalentwicklung der *As. megalocephala.* Arch. Entw.-Mech., **3** : 27—195.

———— 1906: Die Geschichte der T-Riesen von *As. megalocephala.* Zoölogica, **40** : 39—342.

Strubell, A. 1888: Untersuchungen über den Bau die Entwicklung des Rubennematoden *Heterodera schachtii.* Bibliotheca zoöl., **2** : 1—49.

Tadano, M. 1950: Notes on a new Nematode species, *Rhabditis ikedai* n. sp., from the slug, *Incillaria confusa* Cockarell. Zool. Mag., **59** : 289—291. (in Japanese)

———— 1952: Early development of the Nematode, *Rhabditis ikedai.* Zool. Mag., **61**: 289—291. (in Japanese)

Tadano, M. & Y. Tadano 1953: Embryological studies on Nematodes. II Mechanism of stratum formation in round-worm egg membrane (abstract). Zool. Mag., **62** : 32—33 (in Japanese)

—— &—— 1954a: Concerning some granules which have a close connection with the asters and cleavage furrow in centrifuged eggs of *Rhabditis* sp. Zool. Mag., **63** : 164. (in Japanese)

Tadano, M. & Y. Tadano 1954b: Removal of the round-worm egg-membrane and the mechanism of hatching. Zool.Mag., **63** : 42—43. (in Japanese)

—— & —— 1954c: Behavior of the cortical cytoplasm in early cleavage of the round-worm. Zool. Mag., **63** : 406. (in Japanese)

——— & ——— 1954d: Vacuolation (swelling) and diffusion of granules during ascarid cleavage, and new formation of the cleavage furrow. Zool. Mag., **63** : 407. (in Japanese)

——— & ——— 1955a: New formation of the cleavage furrow surface in the *Ascaris* egg. Sci. Rep. Fac. Lib. Arts Ed. Gifu Univ., **3** : 267—269. (in Japanese)

——— & ——— 1955b: Structure of the ascarid egg membrane and its removal. Kagaku **25** : 309—310. (in Japanese)

——— & ——— 1955c: Removal of the ascarid egg membrane by a two-step method using Antiformin and an enzyme. Zool. Mag. **64** : 14. (in Japanese)

——— & ——— 1955d: Culture of *Rhabditis ikedai* Tadano. Sci. Rep. Fac. Lib. Arts Ed. Gifu Univ., **3** : 271—279. (in Japanese)

Tadano, Y. 1951: Studies of cleavage in the eggs of nematode. I. Sci. Rep. Tohoku Univ. (Biol.), **19** : 100—103.

Tadano, Y. & M. Tadano 1955e: Embryological studies on the Nematodes. I. The structure and the properties of the eggshell of *Ascaris*. Sci. Rep. Fac. Nat. Sci. Gifu Univ., **3** : 237—244.

——— & ——— 1955f: On the elongation and the rotation of the spindle, and the different behavior in each of the amphiasters, in relation to the cleavage-plane Sci. Rep. Fac. Nat. Sci. Gifu Univ., **3** : 281—287.

——— & ——— 1955g: Embryological studies on the Nematodes. II. New formation of the plasma membrane and the egg-membranes of the *Ascaris* egg. Sci. Rep. Fac. Nat. Sci. Gifu Univ., **3** : 245—266.

Takahashi, K. 1953: Experimentally induced hatching of round-worm eggs. Jap. Jour. Parasitology, **2** : 102. (in Japanese)

Vogel, R. 1925: Zur Kenntnis der Fortpflanzung, Eireifung, Befruchtung und Furchung von Oxyuris. Zoöl., **42** : 243—271.

Walton, A. C. 1918: The oogenesis and early embryology of *As. canis* Werner. Jour. Morph., 30 : 527—603.

——— 1924: Studies on Nematode gametogenesis. Zietschr. f. Zellen-u. Gewebslehre, **1** : 167—239.

Wottge, K. 1937: Die stofflichen in der Eizelle von As. *megalocephala*. Protoplasma, **26** : 31—59.

Wülker, E. 1923: Über Fortpflanzung und Entwicklung von *Allantonema* und verwandten Nematoden. Ergebn. Zoöl., **5** : 389—507.

Yamada, T. 1950: Dorsalization of the ventral marginal zone of the Triturus gastrula. Biol. Bull., **98** : 98—121.

Yanagisawa, T. & K. Ishii 1954: Relation of protoplasmic granules to formation of egg shell in round worms. Jap. Jour. Med. Sci. & Biol., **7** : 215. (in Japanese)

Yoshida, S. 1917—1918: On growth in round worms. Zool. Mag., **29** : 301—317, **30** : 1—13, (in Japanese)

——— 1919: On the development of *As. lumbricoides*. Jour. Paras., **5** : 105—114.

——— 1923: Zoological aspects of ascaridiasis. Tokyo Med. Mag. 2342, 2344. (in Japanese)

Yoshida, S. & Toyoda, K. 1939: Artificial hatching of ascarid eggs. Livro Jubilare Prof. Travassos. 569.

Zacharias, O. 1913: Die Chromatin-Diminution in den Furchungszellen von *As. megalocephala*. Anat. Anz., **43** : 33—53.

Zawadowsky, M. M. 1927B: Aussere Entwicklungsbedingungen der Eier von *As. megalocephala*. Arch. f. Entw.-Mech., **109** : 14—23.

——— 1929: Th: nature of the egg-shells of various species of Ascaris eggs. Trans. Lab. Exper. Biol., **4** : 201—205.

Zawadowsky, M. M. & L. G. Schalimov 1928: Die Eier von *Oxyuris vermicularis* und ihre Entwicklungsbedingungen. Zeitschr. Paras., **2** : 14—43.

Ziegler, H. E. 1895: Untersuchungen über die ersten Entwicklungsvorgänge der Nematoden. Zeitschr. wiss. Zoöl., 60 : 351—410.

Zoja, R. 1896: Untersuchungen über die Entwicklung der *As. megalocephala*. Arch. mikr. Anat., **47** : 218—260.

ANNELIDA

Chapter 7

INTRODUCTION

(1) CLASSIFICATION

The taxonomic positions of the annelid species which will be treated in this chapter can be summarized in the following way:

- Archiannelida — *Polygordius, Dinophilus.*
- Chaetopoda
  - Polychaeta
    - Errantia — *Amphione, Polynöe, Lepidonotus. Podarke, Scoloplos. Syllis, Autolytus, Trypanosyllis, Sphaerosyllis, Pionosyllis, Grubea, Exogene. Tylorrhynchus (=Ceratocephale), Nereis, Perinereis. Alciopa. Eunice, Lumbriconereis, Marphysa, Diopatra, Ophryotrocha. Spio, Polydora. Chaetosphaera.*
    - Sedentaria — *Chaetopterus, Telepsavus. Arenicola. Asychis. Owenia. Sternaspis. Polyophthamus. Audouina. Ctenodrilus. Amphitrite, Terebella. Capitella. Sabellaria. Branchiomma, Sabella. Hydroides (=Eupomatus), Serpula, Protula, Pomatoceros, Salmacina, Spirorbis.*
  - Myzostomida — *Myzostoma.*
  - Oligochaeta — *Lumbriculus. Tubifex, Peloscolex. Limbodrilus. Bimastus. Pachydrilus. Criodrilus. Bdellodrilus. Lumbricus. Rhynchelmis. Allolobophora (=Eisenia), Pheretima. Drawida*
- Hirudinea
  - Rhynchobdellae — *Clepsine, Piscicola.*
  - Gnathobdellae — *Hirudo, Nephelis, Aulastoma.*
- Echiuroidae — *Urechis. Echiurus. Thalassema. Bonellia.*

(2) ASEXUAL REPRODUCTION

The annelids generally have a great capacity for regeneration, in many cases reforming a new head or tail region on a cut-off fragment to produce a complete individual. Perhaps this strong regenerative capacity developed into a method of reproducing the individual — at any rate there are some annelids which depend solely upon reproduction by asexual fission, and others in which sexual and asexual generations alternate. For example, in such forms as *Ctenodrilus*, one of the tube-dwelling polychaets (Sedentaria), and the oligochaet *Lumbriculus*, a wholly asexual process of *monogony* takes place, and there is said to be no indication of a sexual reproductive process. Again, in the syllids, the males and females separately undergo a process of *schizogony* by means of unequal fission *(paratomy)* or *budding* (Fig. 7.1), after which the large numbers of male and female schizogonts carry on sexual reproduction (Y. K. Okada 1933). A case of this sort may well be validly considered to fall within the generally accepted definition of alternation of generations. Such phenomena, however, as alternation of generations and asexual reproduction are on the whole confined to a few special species, and among the phylum Annelida in general, sexual reproduction is clearly the principal method.

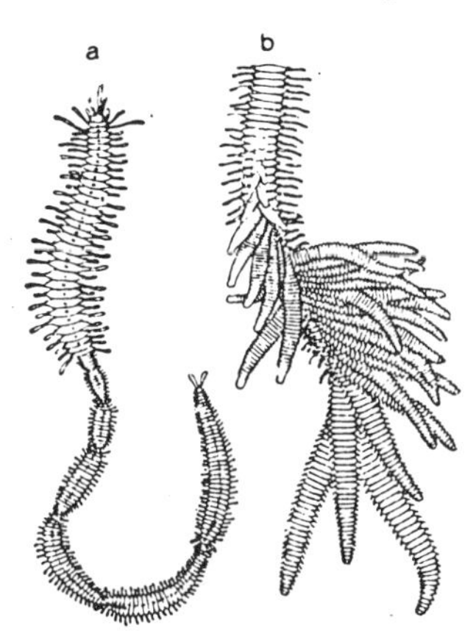

Fig. 7.1 Schizogamy in Polycheata (Okada)
a. paratomy of *Autolytus purpureimaculata;* b. budding in *Trypanosyllis asterobia.*

(3) SEXUAL REPRODUCTION

In the developmental process which follows sexual reproduction, most annelids form a *trochophore* larval stage which metamorphoses into the typical vermiform adult. It can probably be said that the so-called *indirect development* is the general rule. Of course there are interspecies differences with respect to the characteristics of the trochophore, as well the extent of metamorphosis, but the process of indirect development is widely found among such marine annelids as the Archiannelida, Polychaeta and Echiuroidea. On the other hand, the developmental stages of the Oligochaeta and Hirudinea take place in a protective *cocoon.* In this case, the larval period is secondarily reduced, in accordance with the change in environmental conditions caused by the development of the cocoon, and the larva grows directly into a vermiform adult within the cocoon, and only then hatches. In other words, it follows the process known as *direct development.*

(4) SEXUALITY

In connection with the annelidan reproductive system, the geneological relationships are coupled with various habits of spawning, fertilization and brooding, so that the final picture presents an extremely intricate aspect (Nomura 1930). Among the Archiannelida, Polychaeta and Echiuroidea, dioecism is the rule, but some hermaphroditic species are known to exist. The gonads are formed in the wall of the coelomic cavity; the reproductive cells leave the gonads as soon as they ripen, and fill the body cavity, finally passing through the nephridial ducts or special genital pores to the outside at the time of spawning. Gonadal tissue is seldom formed by all the body segments from the head end to the tail — in most cases it appears in a rather large number of central segments, excluding a few at either end. In the hermaphroditic species, half of the gonad-forming body region produces testis and the other half forms ovary; the demarcation between the two is always distinct.

Even in the dioecious species, *sexual dimorphism* does not usually appear, but with the arrival of the breeding season the color of the ripe gametes filling the coelom can be seen through the body wall, making it possible to distinguish between the sexes in a good many cases. For instance, in the sexual form of *Tylorrhynchus,* the body wall of the female is green, while that of the male is a pale pinkish-white. When *Arenicola cristata* arrives at the breeding season, the body color of the female is purplish while the male is white.

On the other hand, some species are known to show a strikingly conspicuous sexual dimorphism. The archiannelidan genus *Dinophilus* and the echiuroidean *Bonellia* are good examples of this (Fig. 7.2 and 7.36). The males and females of these genera are so unlike that they were thought to belong to entirely different species; moreover, the females are 20—50 times larger than the so-called *dwarf males,* which have undergone a regression in body organization so that they are like parasitic forms in comparison with the females.

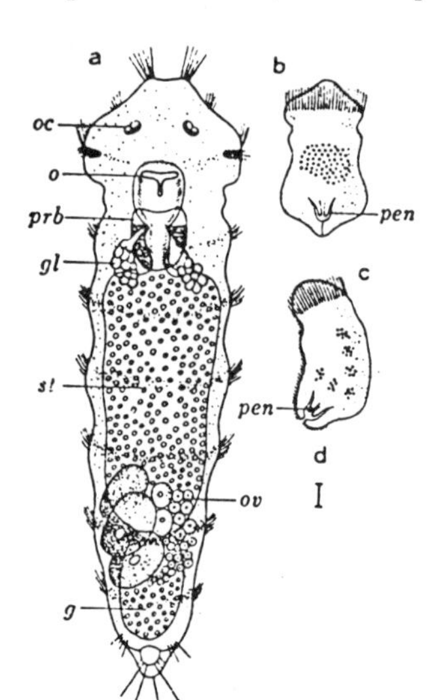

Fig. 7.2 Sexual dimorphism in *Dinophilus apatris* (Korschelt)

a. dorsal view of female; b. dorsal view of male; c. lateral view of male; d. line shows length of male on same scale as female in a. *g:* gut, *gl:* salivary gland, *o:* mouth; *oc:* eye-spot, *ov:* ovary, *pen:* penis, *prb:* proboscis, *st:* stomach.

Among the Oligochaeta and Hirudinea, hermaphroditism is the rule. The gonads develop so that they overlap in the coelom, and are confined to a small number of body segments. In the adjacent segments, the female accessory reproductive organs consist of *oviduct* and *seminal receptacle;* the male accessory reproductive organs include a *sperm reservoir, sperm duct* and *prostate gland;* the male and female *genital pores* each have separate openings on the ventral sides of special segments. Among the many oligochaet species, however, a few cases have been found in which the sexes are separate.

(5) REPRODUCTIVE SWARMING

In the Archiannelida, Polychaeta and Echiuroidea, the ripe gametes are cast out by one means or another into the water, where *external fertilization* takes place and development begins. In a good many species among the pelagic polychaets (Errantia), in which the annual reproductive season is limited to a very brief period, a conspicuous *swarming* phenomenon takes place in conjunction with a *lunar periodicity.* This reproductive swarming is a phenomenon in which mature sexual individuals *(epitocae = heteronereid form)* appear simultaneously at the surface of the sea in tremendous numbers and swim about shedding their eggs and sperm. Such swarming as found among the polychaets has been the object of attention because it occurs periodically every year, as reliably as the days follow each other, at a particular phase of the moon. Remarkable breeding cycles dependent on the lunar cycle are frequently reported in other than annelidan species, but there is probably no group in which such a large number of species show such beautiful lunar periodicity in their spawning as do the Polychaeta (Okuda 1939).

The first scientific report of an animal reproductive cycle exhibiting lunar periodicity concerned the so-called Pacific palolo worm, the sexual form of a large polychaet, *Eunice viridis.* This palolo worm is an extremely long, slender worm (90 × 0.3 cm), with a great number of segments bearing parapodia and setae, and its body filled with mature gametes. It used to be regarded as a curious and wonderful creature because it possessed no slightest trace of a head, but in 1875, Whitmee resolved the mystery of the palolo worm when he described the breeding habit of *Eunice viridis:* the posterior three-fourths of this polychaet abandons the head and anterior segments at the time of swarming, and swims off as an epitoca (Fig. 7.3a). *Eunice viridis* lives in the coral reefs of Samoa, Fiji and the islands of the Gilbert Archipelago in the South Pacific. Every year at about the last quarter moons of October and November — i. e., on the eighth and ninth days after the full moon, during the twilight that precedes the dawn, its massive assemblies can be seen. These swarming palolo worms, once they have shed their gametes, promptly become the food of fishes, or sink to the bottom and die. The inhabitants of the islands also value the palolo as a gastronomic blessing sent from Heaven. Its flavor is said to excell that of fresh fish roe, and the islander's feelings of frustrated anticipation as they wait for its appearance become linked with the mysteriousness of the lunar cycle to give rise to a variety of beliefs and legends.

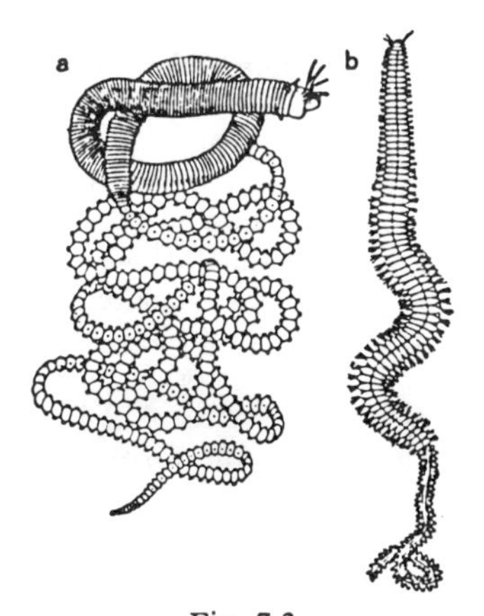

Fig. 7.3
a. Pacific palolo (Woodworth) b. Japanese palolo (Izuka)

Other species which exibit reproductive swarming equal to that of the Pacific palolo are the Atlantic palolo and the Japanese palolo. The "Atlantic palolo" refers to the reproductive phase of the polychaet *Eunice fucata,* which occurs in great numbers around the Florida peninsula of the United States and near Puerto Rico in the West Indies. The Japanese polychaet *Tylorrhynchus heterochaetus*

(= *Ceratocephale osawai)* also performs in a spectacular manner, so that its sexual form has been given the separate name of "bachi", or Japanese palolo. The pronounced lunar periodicity of this bachi swarming has been described by Izuka (1903), Oinuma (1926), Miyoshi (1939), Yo K. Okada (1950), K. Okada (1952), Kagawa (1954). In this case, unlike that of the Pacific palolo, the epitoca is formed from the head and anterior segments (Fig. 7.3b).

This worm lives in the sandy mud of estuaries where the water is practically fresh; every year myriads of them swarm after sunset for three days at the high water of the spring tides occurring during the tenth and eleventh lunar months. As they are carried by the out-going tide toward the lower reaches of the river where the salt content is high, the eggs and sperm are shed from the posterior part of the body. The bachi thus swarm twice in each lunar month, but the swarming which occurs at the dark of the moon is especially marked, while that at the full moon is clearly on a smaller scale.

Again, the Japanese brackish-water nereid, *Nereis japonica,* swarms in the same way as the bachi during the twelfth and first months of the lunar calendar (Izuka 1908). In this case, however, the reproductive segments do not separate from the rest of the body; the whole animal, keeping the form of the sexually immature *atoca,* becomes filled with reproductive cells and swims to the surface to spawn. The gametes are shed through the nephridial ducts of each segment.

In addition to these examples, a considerable number of species among the lumbriconereids, nereids, perinereids, syllids and other groups are known to swarm, according to reports from Woods Hole (North America), Napoli (Italy) Concarneau (France), Madras (India) and the Malay Peninsula. The relationships between such swarming habits and lunar periodity differ somewhat according to the species and localities concerned, but in every case swarming is connected with a special phase of the moon. The structural characteristics of the swarmers also show as many kinds as there are species, from the mode of the nereids, in which the complete atoca swims to the surface, to that of the bachi, which casts off the posterior half of its body, or the palolo which comes swimming up in a headless state. Although it is obvious that no proliferation of the individual is involved in the formation of the epitocae, a thread of connection may perhaps be found between their origin and the asexual proliferation by schizogony found in the syllids.

(6) EGG-LAYING AND BROODING HABITS

Fertilization takes place externally also in most of the sedentary, burrowing polychaets, but the phenomenon of swarming which was found among the Errantia is not seen in this group. Instead, in many cases large numbers of fertilized eggs are formed into an *egg-mass* which is attached to the sea-bottom or some object such as sea-weed, near the burrow. This spawning habit is commonly seen in such genera as *Arenicola, Capitella, Sabella,* etc. In the case of *Arenicola cristata,* the ripe gametes filling the body cavity are shed, through six pairs of nephridia which serve as genital ducts, into the burrow where the animal lives. The eggs and sperm are caught in a jelly-like substance which is simultaneously secreted from the body surface; it is probable that the spermatozoa make their way through this jelly medium and gradually reach the eggs in the vicinity to fertilize them. A few hours later great numbers of fertilized eggs make their appearance on the sandy bottom near the burrows of the females in the form of balloon-like egg-masses (Okada 1940, 1941). The basal part of the egg-mass forms a slender stalk which extends into the sand as far as the burrow (Fig. 7.4a). In *Arenicola claparedii*, the fertilized eggs are held together by the jelly into an *egg-tube* (Okuda 1938), and go through their early development attached to the surface of the mother (Fig. 7.4b).

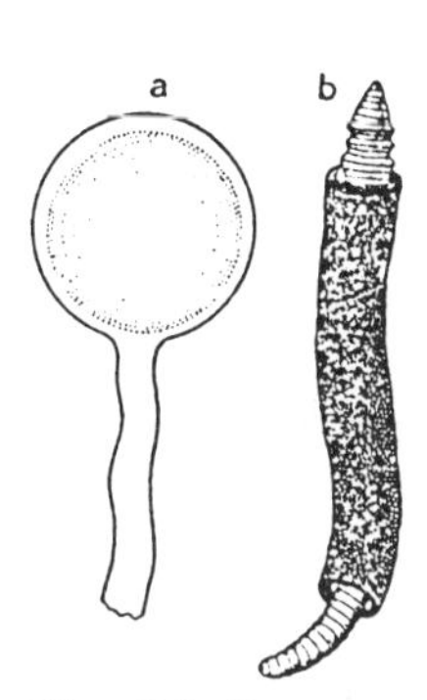

Fig. 7.4 Egg masses of two *Arenicola* species

a. egg-mass of *Arenicola cristata* (Okada); b. egg-tube of *Arenicola claparedii* (Okuda).

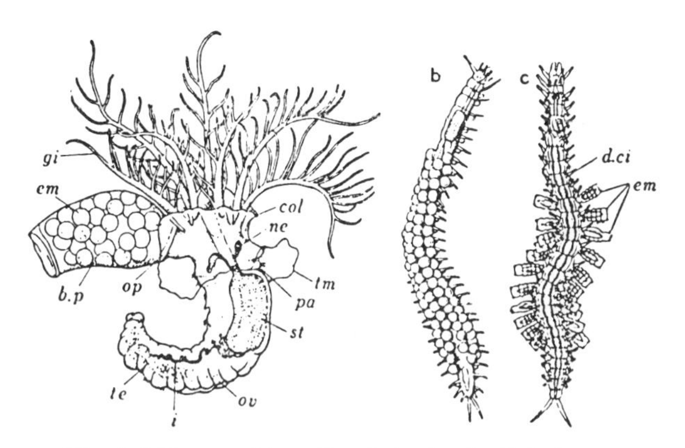

Fig. 7.5 Types of brooding behavior in Polychaeta

a. brood pouch of *Spirorbis argutus* (Abe); b. egg mass of *Grubea limbata* (Korschelt); c. egg mass of *Pionosyllis elegans* (Korschelt). *b.p*: brood pouch, *col*: collar, *d.ci*: dorsal cirrus, *em*: embryo within brood pouch, *gi*: gill, *i*: intestine, *ne*: nephridium, *op*: operculum, *ov*: ovary, *st*: stomach, *te*: testis, *tm*: thoracic membrane

When the egg mass acquires an even closer connection with the parental body, so that the eggs are carried by some part of the latter, a brooding habit may be considered to have developed. In most species of *Spirorbis* (Abe 1943), the early embryonic stages are passed in a *broodchamber* formed by the *operculum* which is itself a transformed tentacle (Fig. 7.5a). This sort of brooding habit, moreover, is not necessarily limited to the Sedentaria — it is also found from time to time among the Errantia. In *Grubea limbata* and *Pionosyllis elegans*, the females carry the egg masses attached to their dorsal surface (Fig. 7.5b, c). In *Polynoë cirrata* they are borne on the dorsal cirri, and in *Autolytus cornutus*, *Exogene gemmifera* and *Sphaerosyllis pirifera*, on the ventral side. None of these, however, can be thought of as having yet developed into a full-fledged *brood-cavity*. As the brooding habit advances still further, such a brood-cavity comes to be developed inside the body. Among the Polychaeta, such highly developed brood-cavities are very seldom found, having been reported only in three species of Errantia, *Syllis vivipara*, *Eunice sanguinea* and *Polydora ciliata*, and a single Sedentaria species, *Pomatoceros triqueter*. In these species the males and females are said to come together at the time of reproduction in so-called *pseudo-copulation*.

In the spawning habit of the tube-dwelling polychaets also, cases are occasionally found in which the spawning periodicity is connected with the phases of the moon and the tides. According to the results of the author's investigation of the spawning of *Arenicola cristata* at Asamushi, Aomori Prefecture, the spawning of the animals in any one habitat occurs at four-day intervals, at the same hour of the same day, and this periodicity coincides with one of the component curves which can be derived from the tidal periodicity curve for Mutsu Bay. Moreover, the quantity of the spawning — i.e., the number of egg masses — fluctuates with a periodicity that follows the spring and neap tides, showing peaks when the moon is near the first and last quarters (Fig. 7.6). In *Polyophthalmus pictus* and *Amphitrite ornata,* the spawning is similarly correlated with a component tidal periodicity, showing a peak at a particular phase of the moon (Okuda 1939).

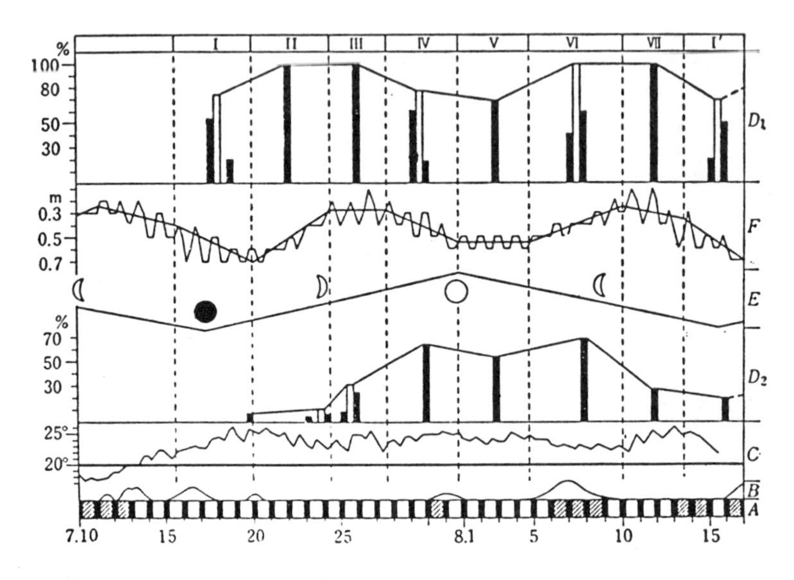

Fig. 7.6 Spawning cycle in *Arenicola cristata* (Okada)

Abscissae: dates from 10 July to 17 August. Note that spawning curves D_1 and D_2 run closely parallel to tide-curve (*F*). (*A*): Diurnal cycle (■: night. ▨: rainy day. □: clear day). (*B*): Wave level. (*C*): Temperature in sand. (D_1): Spawning at Station 1. (D_2): Spawning at Station 2. Curves show time of appearance of egg-masses and amount of spawning expressed by number of egg-masses divided by half-number of animals in habitat x100%. (▌: actual values for time and amount of spawning. ▯: total of values in one tide-level cycle). (*E*): Lunar cycle (white circle: full moon; black circles: new moon.). (*F*): Tide-level curve obtained from tide-time curve: values of ordinate increase downward. Maxima: spring tides; minima: neap tides. (I), (II), ... , (VII): seven sections of tide-level curve in one lunar month.

Among the oligochaets and leeches, this spawning habit of the sedentary polychaets can be considered to have undergone further ecological changes to develop a unique pattern of reproduction. These species are hermaphroditic, but usually self-sterile. According to the report of Oishi (1930) concerning pseudo-copulation in a common earthworm, *Pheretima communissima,* two individuals come into close contact by means of an adhesive substance secreted from the *clitellum,* lying with their long axis reversed, and the genital pore of each applied to the seminal receptacle of the other. The seminal fluid then flows from the respective sperm ducts into the seminal receptacles. By means of this pseudo-copulation, the two animals thus exchange their sperm and store it in the seminal receptacle. It is thus apparent that pseudo-copulation has no direct connection with fertilization.

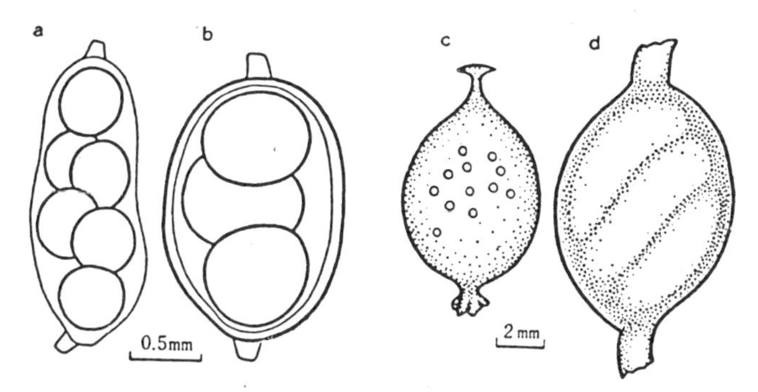

Fig. 7.7 Egg sacs of several oligochaetes

a. fresh-water oligochaete *Tubifex rivulorum*, 6 eggs (Penners); b. fresh-water oligochaete *Limnodrilus udekemianus*, 3 eggs (Penners); c. earthworm *Eisenia foetida*, 12 eggs (Hirai); d. earthworm *Drawida hattamimizu*, 3 late-stage embryos (Ohfuchi)

The ripe eggs are expelled through the female genital pore into mucus secreted by the clitellum. Practically as soon as the eggs are shed, the worm backs out of the encircling clitellum, simultaneously expelling into it from the seminal receptacle the sperm received during pseudo-copulation. The eggs are thus externally fertilized, immediately after being deposited in the clitellum, in the medium provided by the mucus secretion of the clitellum, with sperm which had been stored in the seminal receptacle. The mucus secreted by the clitellum forms the cocoon; internally it consists of an albuminous slime which supports the developing embryos and is used to some extent by them as a source of nutriment. Oligochaet cocoons usually contain only a few eggs, and their characteristics are fixed and peculiar to each species (Fig. 7.7), but the number of eggs contained in each cocon and its size may vary considerably within a single species (Penners 1933, Ohfuchi 1938).

The reproductive organs of the leeches are even more complex than those of the oligochaets; these animals are hermaphrodites, provided with a so-called *penis* and *vagina*. In contrast to the oligochaet pseudo-copulation, fertilization is accomplished internally, the leeches using the penis to introduce *spermatophores* into the partner's vagina. The fertilized eggs are laid in cocoons containing albuminous slime. The form and size of the cocoon among the leeches are largely influenced by the number of included eggs. In the fish-leech *Pisciola*, the slender cocoon 1 mm in length contains only one egg, while the genus *Hirudo* of the Gnathobdellae has a barrel-shaped cocoon of 2 cm, containing more than 20 eggs, and the cocoon of *Clepsine* among the Rhynchobdellae is said to contain 200—300 eggs. Although leech cocoons are attached at the time of spawning to aquatic plants or other solid underwater objects, they often lie scattered and buried in the bottom mud, but the cocoons of *Clepsine* are carried on the ventral surface of the parent until the time of hatching.

1 EARLY DEVELOPMENT OF ARCHIANNELIDA AND POLYCHAETA

(1) FERTILIZATION

Polychaet eggs usually have a homolecithal structure. Among the various species are some having small eggs with clear cytoplasm and very little yolk *(Lepidonotus* sp., diameter 65 μ; *Podarke obscura,* 63 μ; *Hydroides uncinatus,* 55 μ; *Polygordius,* about the same size*)*, and others with rather large eggs having opaque yolk *(Nereis dümerilii,* 310 × 390 μ; *Scoloplos armiger,* 250 μ), sometimes in an almost telolecithal distribution *(Diopatra amboinensis)*. In most species, however, the egg diameter lies between 100 and 200 μ; each species shows a beautiful color tint of its own and finely granular homogeneous cytoplasm, but many eggs also contain so-called oil globules in addition to the yolk granules (Fig. 7.9a). A well-defined membrane can always be recognized surrounding the egg, and in many cases a jelly layer is formed outside of this when the egg is shed. This layer of jelly does not simply serve as a protective covering, but is also believed to have a close connection with the physiology of fertilization (Yamamoto 1947, 1952; Okada 1955). In most species the egg is practically a perfect sphere, but it may occasionally be flattened in the direction of the egg axis, or form a rather elongated oval. In any case the egg nucleus, in the form of a germinal vesicle, lies more or less acentrically toward the animal pole (Fig 7.9a). The formation of the polar bodies takes place after sperm entrance, in parallel with the course of fertilization.

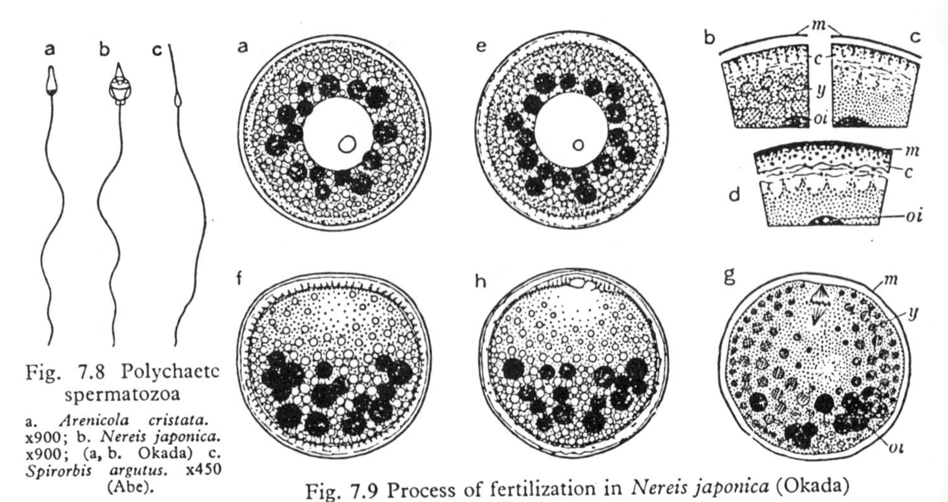

Fig. 7.8 Polychaete spermatozoa

a. *Arenicola cristata.* x900; b. *Nereis japonica.* x900; (a, b. Okada) c. *Spirorbis argutus.* x450 (Abe).

Fig. 7.9 Process of fertilization in *Nereis japonica* (Okada)

a. unfertilized egg; b. immature state of ooplasm (fixed); c. ooplasm of mature egg (fixed) d. section of egg undergoing cortical change after insemination; e. fertilized egg (30 min after insemination); f. fertilized egg, showing bipolar differentiation (60 min. after insemination); g. fertilized egg, after breakdown of germinal vesicle (fixed); h. completion of polar body formation (2.5 hrs. after fertilization) c: cortical ooplasm, *m:* egg membrane or fertilization membrane, *oil:* oil drop, *y:* yolk granule.

The spermatozoa are formed into *spermatophores* within the body cavity when these clumps of ripe sperm are shed into the sea water they break up into active individual spermatozoa (Okada 1941, 1953). The spermatozoa in all the species are flagellate, with a small head and a long tail, but individual characteristics can be recognized in the structure and size of the head region (Fig 7.8)

Among the Polychaeta, fertilization usually takes place externally. It would thus seem possible, using sea water as the medium, to secure artificial fertilization experimentally at any time, but in actual fact there are subtle conditions which underlie the activation of the egg cells as well as the spermatozoa. There is a surprising number of cases in which artificial fertilization proves difficult to obtain. In *Tylorrhynchus,* for example, fertilization is successful only if a special method of insemination is followed (Yamamoto 1947; Okada 1952; Kagawa 1952). On the other hand, in species fertilizing easily on simple insemination, the process of polychaet fertilization, which is rather different from that of sea urchins, has been clearly elucidated. In particular, the investigation of the phenomenon of fertilization in *Nereis* attracted early attention, and the egg of this worm, studied as a homologue of the sea urchin egg, can be said to have made an invaluable contribution toward explaining the principles of animal fertilization in general (Lillie 1911a, 1911b, 1912a, 1912b; Spek 1930; Costello 1949). In this section the results of recent observations on the Japanese nereid *Nereis japonica* will be brought together to describe the process of fertilization in the polychaets.

In the living state the unfertilized egg of this species is rather flattened in the direction of the egg axis, the diameters averaging $160 \times 170\,\mu$. The cytoplasm immediately under the membrane is differentiated into a *cortical layer* 6—7 μ in thickness. At first glance this layer appears to be practically transparent, but close examination reveals that it is densely filled with *cortical granules* in the form of short rods and particles about one micron in size[1] (Fig 7.9a, b, c). The *endoplasm* contains opaque yolk granules not found in the cortex, and is pale green; around the germinal vesicle can be seen 30 or more *oil droplets* (Fig. 7.9a). The germinal vesicle lies somewhat eccentrically, toward the animal pole; within it can be seen a nucleolus. If the unfertilized egg is fixed with Champy's fixing mixture and stained according to Kull's method, the structure not only of the cortex, but also of the endoplasm, shows a different appearance depending on the state of maturity of the egg. In cytoplasm of the immature egg, the fine granules of yolk are distinguished from the granular protoplasm, since they are fixed as small globules (Fig. 7.9b), while the cytoplasm of a fertilizable egg is fixed in such a way that the small yolk globules cannot be recognized (Fig. 7.9c).

If the eggs of *Nereis japonica* are inseminated in sea water of which the salinity lies between 17.71 and 13.28‰, they will fertilize readily, and a *cortical reaction* takes place. Insemination is immediately followed by several subtle changes such as formation of the jelly layer, formation of the fertilization membrane, appearance of the fertilization cone. As the material of the cortical granules is expelled from the egg, the cortical layer disappears, and finally the germinal vesicle breaks down. Simultaneously, oöplasmic segregation is seen to take place, an astral figure appears and very shortly afterward the polar bodies are extruded and the egg and sperm pronuclei unite.

When a spermatozoon makes successful contact with an egg, the contour of the egg undergoes a very slight shrinkage and the formation of the jelly layer begins. The thickness of this layer increases with time, so that it finally equals

---

[1] If these are observed with phase contrast, their distribution is found to be fairly irregular, but it is possible to distinguish between *outer cortical granules* which lie peripherally and *inner cortical granules* which are scattered along the inner side of the cortex.

more than 1.5 times the egg diameter.[2] In the nereid egg, no jelly layer is formed unless fertilization takes place. It is therefore possible to judge whether or not an egg is fertilized by the presence or absence of such a layer.[3]

At the point of sperm contact, the fertilization cone appears, and with it as the starting point, the cytoplasmic cortex displays a fine structure of wavy lines while the cortical granules of the outermost layer begin to dissolve toward the egg membrane (Fig. 7.9d). At the same time a peri-vitelline space begins to appear and the vitelline membrane is strengthened to form the fertilization membrane, while the remaining cortical granules show a new finely radial structure and continue to dissolve into the peri-vitelline space. These various cortical changes are complete within 20 minutes after insemination, and the surface of the membrane regains its stretched spherical condition (Fig. 7.9e). If the cytoplasmic structure at this stage is examined in fixed material, it is seen that the membrane of the germinal vesicle is still present, surrounded by a granular cytoplasm, and the small yolk globules again present a globular appearance. On the other hand, the thick, intricately formed cortex of the unfertilized egg has entirely disappeared. These results indicate that the cortical granules of the unfertilized egg consist of *precursor substances* which contribute to the formation of the jelly layer and the strengthening of the fertilization membranes.

Thirty to 60 minutes after insemination, the germinal vesicle breaks down and its (nuclear) membrane disappears. At about the same time ooplasmic segregation begins, the oil droplets and yolk granules move to the vegetal hemisphere, and the animal hemisphere becomes filled with the characteristic granular protoplasm (Fig. 7.9f). By about two hours after insemination, a distinct astral figure is seen in about the center of the egg. In this period the nuclear contents are acentrically located at the animal pole, indicating the metaphase of the first polar body division (Fig. 7.9g). The aster presently disappears; 2.5—3 hours after insemination, polar body formation is complete (Fig. 7.9h) and the union of the egg and sperm pronuclei takes place.

(2) CLEAVAGE PROCESS

A survey of the numerous papers which have been published about polychaet cleavage indicates that a common mode of cleavage is clearly recognizable. This is called 'annelidan cleavage'; it is total, unequal and spiral. The mode of cleavage in general is always strongly influenced by the amount of yolk contained in the egg cell. In this group, however, no species is known in which the egg has so much yolk as to cause it to undergo superficial or partial cleavage. When the amount of yolk is small, cleavage tends to be proportionally equal, but never loses its essentially spiral character.

---

[2] The increase in thickness of the jelly layer takes place rapidly on insemination. If the thickness of this layer at 10—15°C is followed quantitatively, it is found to measure 0.5 times the egg diameter at 7 minutes after insemination, 0.8 times at 15 minutes, 1.1 times at 30 minutes, and practically 1.5 times after 2 hours.

[3] The unfertilized bachi egg will form a jelly layer if it is simply transferred to sea water ("water-activation"). For this reason it is impossible to use the presence or absence of a jelly layer as a criterion of the fertilized condition in dealing with bachi eggs. Nereid eggs do not respond to this simple water-activation.

A considerable number of studies have been made of cell-lineage in the polychaets; in particular may be cited the contribution to the fundamental knowledge of annelidan cleavage made by the study of E. B. Wilson (1892) on *Nereis limbata,* and the splendid achievement of Woltereck's (1903) investigation of *Polygordius.* In this article, the course of early cleavage will be described as it occurs in the nereids, combining the author's observations made on *Nereis japonica* with the results of Wilson's cell lineage studies on *Nereis limbata* (see Table 7.1).

Table 7.1 **Cell-lineage of Nereis limbata.**
(adapted from Wilson 1892)

First Cleavage	Second Cleavage	Third Cleavage	Fourth Cleavage	Fifth Cleavage	Sixth Cleavage	
No. of cells: 2	4	8	16	23 · 29 · 32	36 · 38 · 42 · 58	
Fertilized egg → AB	A	1a	$1a^1$	$1a^{11}$	$1a^{111}$	apical cell
					$1a^{112}$	annelidan cross cell
				$1a^{12}$		molluscan cross cell
			$1a^2$	$1a^{21}$	$1a^{211}$, $1a^{212}$	trochoblasts
				$1a^{22}$	$1a^{221}$, $1a^{222}$	trochoblasts
		1A	2a	$2a^1$		
				$2a^2$		left somatoblast
			2A	3a		
				3A		endomere
	B	1b	$1b^1$	$1b^{11}$	$1b^{111}$	apical cell
					$1b^{112}$	annelidan cross cell
				$1b^{12}$		molluscan cross cell
			$1b^2$	$1b^{21}$	$1b^{211}$, $1b^{212}$	trochoblasts
				$1b^{22}$	$1b^{221}$, $1b^{222}$	trochoblasts
		1B	2b	$2b^1$		
				$2b^2$		median somatoblast
			2B	3b		
				3B		endomere
Fertilized egg → CD	C	1c	$1c^1$	$1c^{11}$	$1c^{111}$	apical cell
					$1c^{112}$	annelidan cross cell
				$1c^{12}$		molluscan cross cell
			$1c^2$	$1c^{21}$	$1c^{211}$, $1c^{212}$	trochoblasts
				$1c^{22}$	$1c^{221}$, $1c^{222}$	trochoblasts
		1C	2c	$2c^1$		
				$2c^2$		right somatoblast
			2C	3c		
				3C		endomere
	D	1d	$1d^1$	$1d^{11}$	$1d^{111}$	apical cell
					$1d^{112}$ → $1d^{1121}$, $1d^{1122}$	annelidan cross cell
				$1d^{12}$		molluscan cross cell
			$1d^2$	$1d^{21}$	$1d^{211}$, $1d^{212}$	trochoblasts
				$1d^{22}$	$1d^{221}$, $1d^{222}$	trochoblasts
		1D	2d(=X)	X, 1x	X, 2x; X, 3x; X^l, X^r	teloblasts of first somatoblast
			2D	3d		(second somatoblast)
				3D	4d(=M) → M^l, M^r	teloblasts of mesodermal bands
					4D	endomere

A short time after expulsion of the polar bodies from the fertilized egg, the first cleavage furrow appears at the animal pole, and advances in a diagonally vertical direction,[4] dividing the egg into an *AB* and *CD* blastomere (Fig. 7.10a).

[4] The words 'vertical' and 'horizontal' are used with reference to an egg having its main axis in a vertical position. (vertical = meridional, and horizontal = equatorial)

The *AB* blastomere is clearly smaller than *CD*. After the 2-cell stage has lasted for 60 minutes, the second cleavage furrows appear near the animal pole sides of the blastomeres, this time intersecting the first cleavage furrow at right angles. This cleavage divides *AB* into the equal-sized blastomeres *A* and *B*, while *CD* divides into a smaller *C* and a larger *D*, to give rise to the 4-cell stage (Fig. 7.10b).[5] *C* is about the same size as *A* and *B*. The several blastomeres of the 4-cell stage represent the four cleavage quadrants, *A* lying on what will be the left side of the future embryo, *B* lying anteriorly, *C*, at the right side and *D*, posteriorly. For the present, the chief egg axis, which connects the animal and vegetal poles, may be considered to coincide with the dorsoventral axis.

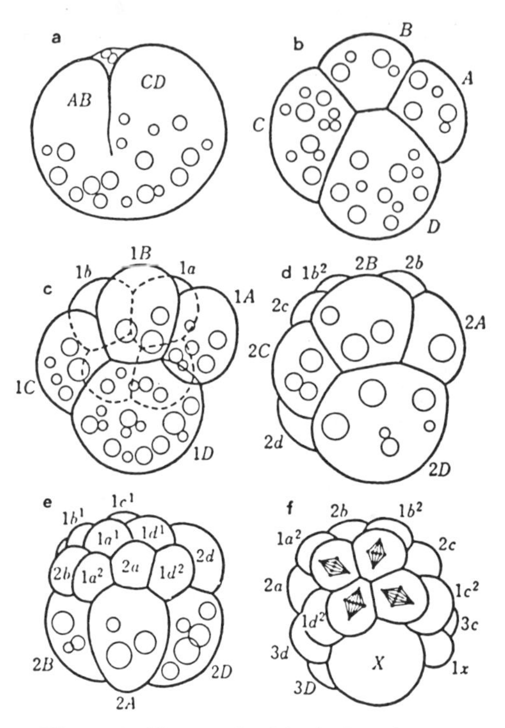

Fig. 7.10 Cleavage in *Nereis japonica* (Okada)
a. 2-cell stage (left side); b. 4-cell stage (vegetal pole); c. 8-cell stage (vegetal pole); d. 16-cell stage (vegetal pole); e. 16-cell stage (left side); f. 23-cell stage (animal pole).

The 4-cell stage again lasts about 60 minutes before the third cleavage furrows appear in each blastomere, this time advancing in a diagonally horizontal direction and separating off the equal-sized micromeres 1*a*, 1*b*, 1*c* and 1*d* diagonally upward to form the 8-cell stage. These micromeres, which contain no yolk, are interposed between the dorsal sides of the macromeres 1*A*, 1*B*, 1*C* and 1*D*, forming the so-called *first quartette* of micromeres which surround the animal pole. The macromeres 1*A*, 1*B* and 1*C* are all of about the same size, while 1*D* is especially large (Fig. 7.10 c). From about this time the material of the oil droplets which are included in the macromeres begins to be consumed, causing the droplets to fuse together, so that they gradually become larger while their number shows a sharp decrease.

In this third cleavage, which produces the 8-cell stage, the micromeres are given off toward the right as seen from the animal pole. From this point is is possible to calculate back and formulate the hypothesis that the second cleavage is laeotropic and the first, dexiotropic, and that the cleavage process advances with the successive divisions alternating their direction to form a spiral with the egg axis as its center. In spiral cleavage, it is generally the rule that the odd-numbered divisions are dexiotropic, the even-numbered ones laeotropic; cleavage among the nereids conforms to this rule.

[5] The eggs of *Polygordius*, *Podarke* and *Lepidonotus* can practically be described as alecithal. As a result, both first and second cleavages are equal, and the *A*, *B*, *C* and *D* blastomeres of the 4-cell stage are all of the same size. Unequal cleavage appears for the first time at the third cleavage.

At the fourth cleavage, the macromeres 1*A*, 1*B*, 1*C* and 1*D* divide obliquely toward the left and unequally, giving rise to the second micromere quartette, 2*a*, 2*b*, 2*c* and 2*d*. At almost the same time the micromeres of the first quartette divide equally and laeotropically, becoming eight micromeres ($1a^1$ and $1a^2$ - - - - $1d^1$ and $1d^2$) as the embryo reaches the 16-cell stage (Fig. 7.10d, e). When the second micromere quartette is formed, the *D* quadrant undergoes an atypical cleavage in producing the 2*d* blastomere. Although this 2*d* cell does not contain any yolk, it is equal in size to the macromeres; it is destined to produce the ectodermal cells which later form the embryonic body wall of a wide area in the central posterior and ventral regions. In other words, since this cell is separated off as the rudiment of the trunk ectoderm, it is known as the *first somatoblast,* and is usually indicated by the symbol *X,* or sometimes *x*. During the three following cleavages *X* divides unequally, giving off three micromeres (1*x*, 2*x*, 3*x*) obliquely downward, and then cleaves equally into X^l and X^r which lie symmetrically to left and right of the midline and constitute the *teloblasts* of the *first somatoblast* (Table 7.1).

At the fifth cleavage, the third micromere quartette is formed by dexiotropic divisions. Theoretically the other micromere groups should simultaneously be undergoing dexiotropic divisions, conjointly with the special unequal cleavage of the first somatoblast, to give rise to the 32-cell stage. In fact, however, from this stage the strict synchrony of cleavage begins to slacken, and discrepancies in the speed of division appear between one quandrant and another, or between quartettes. In general, the micromeres of the first quartette and the first somatoblast cells divide more rapidly than the others. Also there is a tendency among the micromeres of the third quartette for the *D* and *C* quadrants to precede *A* and *B* in cleaving. As a result, the fifth cleavage can most accurately be described as first producing a 23-cell stage (Fig. 7.10f), and then taking the embryo through a stage with 29 cells before it finally reaches the 32-cell stage (Table 7.1).

Since the macromeres 3*A*, 3*B* and 3*C* are the mother cells which will give rise to the future endoderm, they are called the *endomeres*. 3*D*, on the other hand, only becomes an endomere after giving off the micromere 4*d* at the succeeding division. This 4*d* cell, as the mother cell of the future mesoderm, is known as the *mesoblast* or *second somatoblast,* and usually designated by *M*. Moreover, the micromeres $2a^2$, $2b^2$ and $2c^2$, which are formed in the *A*, *B* and *C* quadrants at the fifth cleavage, are known respectively as the *left somatoblast, median somatoblast* and *right somatoblast* (Table 7.1).

The sixth cleavage eventually gives rise to a 64-cell stage, but by this time the synchrony of cleavage is more than ever confused, so that quickly cleaving blastomeres are already dividing for the seventh time before the slow ones carry out their sixth division. As synchrony breaks down, the typical spiral cleavage configuration which has been visible until this time begins to be lost, the special cleavage modes of *X* and *M* are interposed, the arrangement of the blastomeres changes from quadrilateral to bilateral symmetry, the number of invididual cells increases and the embryo arrives at the blastula stage.

The amount of yolk contained in the egg exerts a strong influence, as cleavage proceeeds, on the process of blastula formation. *Nereis* and *Arenicola,* with generous supplies of yolk, form *sterroblastulae* with practically no blastocoels. In the almost yolk-free *Polygordius* embryo, however, the regularity of spiral cleavage is well preserved, and the transparent *coeloblastula* has a spacious blastocoel. Wilson (1892) has elucidated the process of formation of the *apical cells,* the *tro-*

choblasts, the *teloblasts of the mesodermal bands* and of the *ventral plate,* etc., even in the sterroblastula of *Nereis,* while Woltereck (1903) has described the cell-lineage of the coeloblastula of *Polygordius*; the latter work is a masterpiece of precise detail. The ensuing account of the development of the blastula will follow the fates of the blastomeres as Woltereck observed them in *Polygordius.*

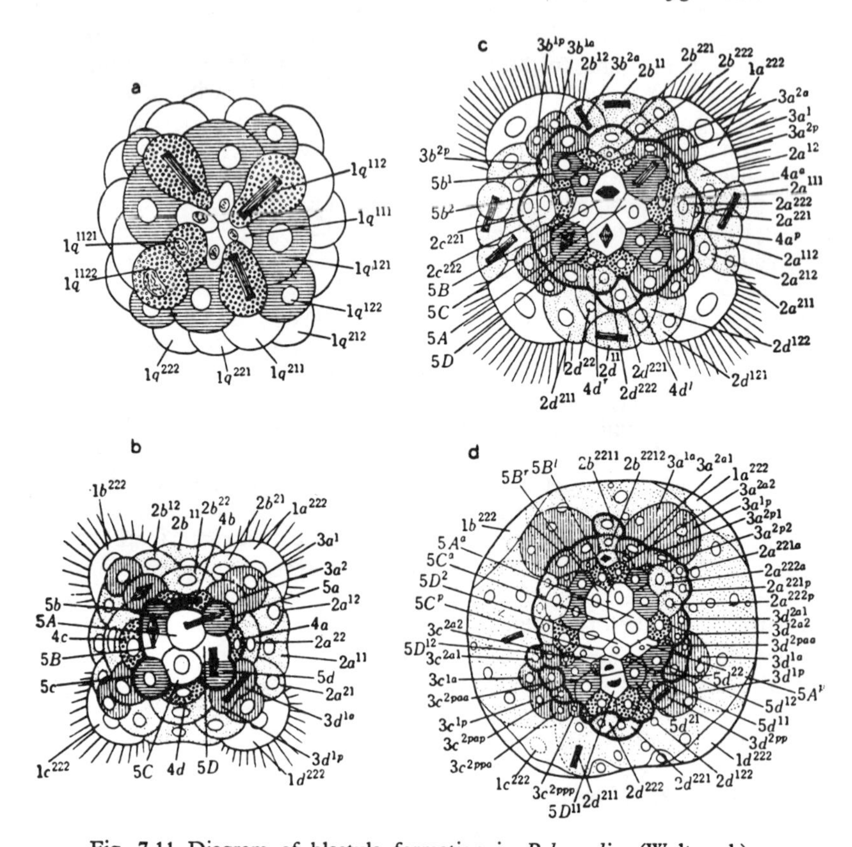

Fig. 7.11 Diagram of blastula formation in *Polygordius* (Woltereck)
a. 76-cell stage (animal pole); b. 76-cell stage (vegetal pole); c. 112-cell stage (vegetal pole); d. late blastula stage (vegetal pole). Heavy line encircling vegetal polar region shows boundary of future blastopore; encircled cells will invaginate to form archenteron.

The sixth cleavage of this embryo, which gives rise to the 64-cell stage, produces the fourth quartette by laeotropic division from the macromeres; the first quartette micromeres divide for the third time, those of the second quartette for the second time, and the third quartette cells cleave for the first time, all toward the left. In the resulting embryo the animal hemisphere is made up of the 32 cells derived from the first micromere quartette (the cell groups $1q^{111}$ and $1q^{112}$, $1q^{121}$ and $1q^{122}$, $1q^{211}$ and $1q^{212}$, $1q^{221}$ and $1q^{222}$[6]. At the vegetal pole lie the four macromeres (*4A, 4B, 4C* and *4D*) as well as the four blastomeres of the fourth quartette (*4a, 4b, 4c, 4d*), and forming a band around the equator are the 16 blastomeres

[6] "q" stands for "quartette", i.e., it is used as a substitute for the symbols "a, b, c and d."

derived from the second quartette (the cell groups $2q^{11}$ and $2q^{12}$, $2q^{21}$ and $2q^{22}$) and the eight blastomeres derived from the third quartette ($3q^1$ and $3q^2$).

As the embryo enters upon the seventh cleavage, in theory each quartette should be proceeeding to divide, more or less together with the dexiotropic formation of the fifth set of micromeres; but in practice retardation in cleavage time tends to appear in one quartette or another, possibly because the differentiation of the larval organs is about to take place. As seen from the animal pole at this time, the centrally located $1q^{111}$ group of first micromeres forms a *rosette* consisting of *apical cells*, while the conspicuous cells of quartette $1q^{112}$ are interposed between them.[7] Surrounding these are the blastomeres[8] of $1q^{121}$ and $1q^{122}$, and in the equatorial band lie the blastomeres of $1q^{211}$, $1q^{112}$, $1q^{221}$ and $1q^{222}$, which will become the *trochoblasts* (Fig. 7.11a). Among these cell groups, retardation of cleavage begins to appear in the apical cells and the trochoblasts. The apical cells will eventually form a bundle of long cilia *(apical tuft)* which will develop into the *apical organ,* believed to be the sensory organ of the trochophore larva. The trochoblasts are cells which will give rise in the trochophore to the *prototroch,* the larval organ of locomotion; already in this stage cilia are beginning to appear on the blastomeres of the $1q^{222}$ and $1q^{221}$ groups. The activity of these cilia causes the embryo to rotate within its membrane at a gradually increasing rate, until it presently breaks out of the membrane.[9] At this time the cilia simply occur isolated in the four quadrants (Fig. 7.11b), but as development proceeds they become connected with each other and eventually form a complete prototroch which encircles the equatorial region of the embryo.

In the vegetal hemisphere, four *radial zones* and four *inter-radial zones,* one derived from each quartette, are distinguished. The micromere groups of the 64-cell stage derived from the second quartette ($2q^{11}$, $2q^{12}$, $2q^{21}$, $2q^{22}$) occupy the radial zones of the equatorial region (Fig. 7.11b). Of these, the groups $2q^{11}$, $2q^{12}$ and $2q^{21}$ presently divide to form $2q^{111}$ and $2q^{112}$, $2q^{121}$ and $2q^{122}$, $2q^{211}$ and $2q^{212}$; they then stop dividing for a while and make posterior lateral contact with the prototroch cells (Fig. 7.11c). However, the members of the $2q^{22}$ group, which is to give rise to the first somatoblast, pursue rather different courses of cleavage depending upon the quadrant in which they lie: the *A* and *C* members of $2q^{22}$ carry out two successive divisions, the first horizontal and the second transverse, to form the blastomere groups $2q^{221a}$ and $2q^{221p}$, $2q^{222a}$ and $2q^{222p}$.[10] The *B* quadrant member ($2b^{22}$) divides horizontally and then longitudinally, forming the four blastomeres $2b^{2211}$ and $2b^{2212}$, $2b^{222r}$ and $2b^{222l}$. The *D* quadrant blastomere $2d^{22}$ divides once into $2d^{221}$ and $2d^{222}$ (Fig. 7.11d).

The cell groups $3q^1$ and $3q^2$ of the third micromere quartette lie in the inter-radial zones. The $3q^1$ cells of each quadrant cleave transversely to form $3q^{1a}$ and $3q^{1p}$ and thereafter stop dividing and become associated with the ectoderm cells bordering the equatorial prototroch cells (Fig. 7.11c). The $3q^2$ blastomeres

---

[7] The conspicuous development of these four blastomeres is a characteristic feature of annelidan cleavage; together they are called the "*annelidan cross.*"

[8] These blastomeres fail to become prominent in the annelidan embryo, but attain a conspicuous development among the Mollusca, and are therefore referred to as the "*molluscan cross*".

[9] The time of hatching varies greatly from species to species, tending to be later in proportion to the amount of yolk. In many species hatching occurs only after attainment of the complete trochophore form.

[10] Index symbol *a*=anterior, *b*=posterior

of the *A* and *B* quadrants undergo a transverse and then a horizontal cleavage, giving rise to $3q^{2a1}$ and $3q^{2a2}$, $3q^{2p1}$ and $3q^{2p2}$; those of the C and *D* quadrants divide once transversely, and then one of these daughter cells divides horizontally and the other transversely, to produce the quartettes $3q^{2a1}$ and $3q^{2a2}$, $3q^{2pa}$ and $3q^{2pp}$. These quartettes derived from $3q^2$ are interposed between those arising from $2q^2$ to form an area of ectoderm surrounding the endoderm cells (Fig. 7.11d).

The blastomeres of the fourth quartette occupy the radial zones. When the fifth quartette is being formed, the *A* and *C* members of $4q$ undergo unequal transverse cleavage to form large $4q^a$ and small $4q^p$ blastomeres, while the cells of the *B* and *D* quadrants cleave equally and longitudinally into $4q^r$ and $4q^l$ (Fig. 7.11c). Among these cells, $4d$, as in *Nereis*, becomes the mother cell of the future mesodermal bands, and is given the name of mesoblast or second somatoblast, and represented by the symbol *M*. $4d$ soon divides equally and longitudinally; since the daughter blastomeres are the cells which will be located at the posterior ends of the left and right mesodermal bands, they are called the teloblasts of the mesodermal bands. The rest of the fourth and the members of the fifth quartette, as well as the macromeres, are destined to be endoderm cells, and all will take part in forming the wall of the *archenteron* (Fig. 7.11c, d). By the time this fifth quartette of micromeres is being produced in the vegetal hemisphere, the blastula as a whole can be said to have attained its full development.

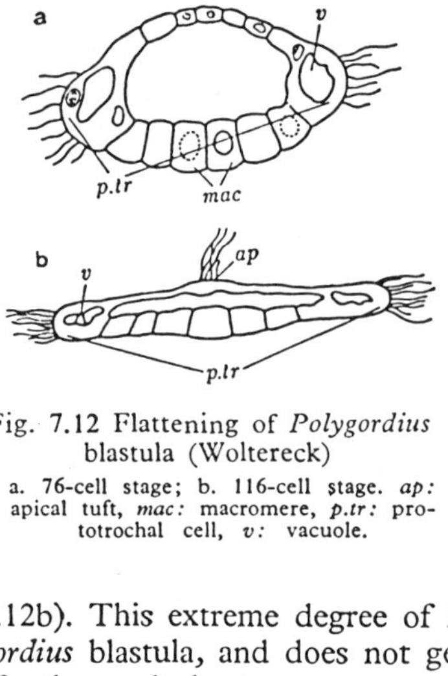

Fig. 7.12 Flattening of *Polygordius* blastula (Woltereck)

a. 76-cell stage; b. 116-cell stage. *ap:* apical tuft, *mac:* macromere, *p.tr:* prototrochal cell, *v:* vacuole.

If the internal structure of the *Polygordius* blastula is investigated in sectioned material, it is found that the size of the macromeres has gradually diminished as the result of their successive cleavages, so that it is difficult to distinguish them from the other blastomeres on the basis of size alone. In the 64-cell stage the blastula is a flattened sphere with a large blastocoel, but as cleavage advances the vegetal side becomes much more strongly compressed and the larval shape changes to that of a plate. The prototroch cells, which have ceased dividing, develop vesicles, and can be recognized as large ciliated cells lying at the equator of the embryo (Fig. 7.12a); the apical cells also eventually produce a bundle of long cilia (Fig. 7.12b). This extreme degree of flattening, however, is a peculiarity of the *Polygordius* blastula, and does not generally occur to such an extent in the blastulae of other polychaets.

(3) ARCHENTERON FORMATION

Once the blastula is fully grown, the center of cleavage activity for the most part shifts to the endodermal cell group and the ectodermal cells surrounding and interposed among them. In this region cell division continues actively, forming a pouch-shaped or solid cellular rudiment which protrudes from the vegetal pole into the blastocoel as the embryo makes progress toward a gastrular organization.

Archenteron formation among the Archiannelida and Polychaeta may take place either by invagination or by epiboly. It is hardly necessary to point out that the basic factor in causing this difference is to be found in the type of the blastula. In the coeloblastula, the cleavage of the endodermal cell groups always precedes that of the ectoderm, and the endodermal cells move into the blastocoel and form the archenteron. In the sterroblastula, which is unable to form a blastocoel because of the large amount of yolk contained in its endoderm, these cells are likely to lag behind the others in dividing, while the vigorous cleavage activity of the surrounding ectoderm cells causes them to over-spread and envelop the endoderm cells, thereby forming the archenteron. In actuality, most species produce the archenteron by a combination of invagination and epiboly; i.e., there is no clear line of distinction between the two processes, the mode of gastrulation being secondarily influenced by the amount of yolk. The manner in which the successive cleavages lead to the formation of the archenteron has been followed in detail in the schematic coeloblastula of *Polygordius*; this account will serve as basis of the following description.

In the *Polygordius* blastula which has formed the micromeres of the fifth quartette, the cells of the endodermal group lying in the plate of the flattened vegetal pole region continue to divide and invaginate into the blastocoel, giving rise to a sac-shaped archenteron. The surrounding ectoderm cells next extend over this region and close the blastopore. The earliest indication of invagination appears in the cleavage of the macromeres after they have released the fifth micromere quartette. For the first time the cleavage of $5A$, $5B$, $5C$ and $5D$ emerges from the confines of the spiral pattern and shows a bilaterally symmetrical relation to the chief embryonic axis. $5A$ and $5C$ cleave transversely and unequally into $5A^a$ and $5A^p$, $5C^a$ and $5C^p$; $5B$ divides equally and longitudinally to form $5B^l$ and $5B^r$. $5D$ alone cleaves in a rather special fashion. First it divides into $5D^1$ and $5D^2$; $5D^1$ next forms $5D^{11}$ and $5D^{12}$. Moreover, the direction in which $5D$ cleaves in less transverse than directed obliquely outward (Fig. 7.11d). The nine cells formed by these cleavages show a tendency to invaginate into the blastocoel at the central position occupied by $5D^2$ and $5D^{12}$. The micromeres of the fourth and fifth quartettes also successively divide transversely or longitudinally, and consequently push against the endoderm cells from the periphery and gradually crowd them into the blastocoel. As invagination proceeds in this manner, the margin of the *blastopore* becomes distinctly visible. The blastopore at this stage is laterally compressed into an antero-posteriorly elongated opening, with $4b^r$ and $4b^l$ forming its anteriormost margin and $4d^r$ and $4d^l$ at its posterior side. The left margin is occupied by $5a^1$, $5a^2$, $4a^a$, $5d^{12}$ and $5d^{11}$, while the right margin consists of $5b^1$, $5b^2$, $4c^a$, $5c^{12}$ and $5c^{11}$ (Fig. 7.13a).

Next, as invagination advances, the two cells $4a^a$ and $4c^a$, which lie at the centers of the two sides of the blastopore, draw together. At the same time the ectodermal cells grow from the two sides toward the mid-line so as to cover the blastopore. In this stage, its outline still has the form of a figure 8, but as the overgrowth of the ectodermal cells continues, it closes as though the sides were being stitched together, except for small openings left at its anterior and posterior ends (Fig. 7.11a, b). The anterior opening is not completely closed even later, and when the ectoderm cells immediately adjacent to it invaginate, they form the *stomodaeum* there. The anterior wall of the stomodaeum is occupied by the blastomeres $2b^{2211}$, $2b^{2212}$, $2b^{222r}$ and $2b^{222l}$; the left side wall by $3a^{2a1}$, $3a^{2p1}$, $2a^{221a}$ and $2a^{222a}$; the right side wall by $3b^{2a1}$, $3b^{2p1}$, $2c^{221a}$ and $2c^{222a}$; and the posterior wall by $3c^{2a1}$, $3c^{2a2}$, $3d^{2a1}$ and $3d^{2a2}$. At the time of stomodaeum formation, the ectodermal blastomeres $2a^{221p}$, $2a^{222p}$, $2a^{2a2}$, $3a^{2p2}$, $2b^{2a2}$, $3b^{2p2}$ move into the blastocoel and become the *larval mesenchyme.*

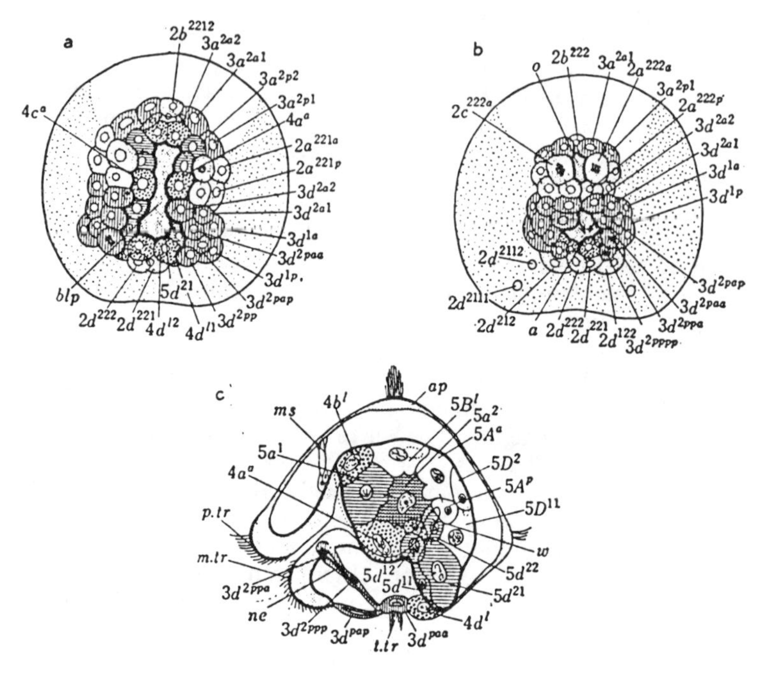

Fig. 7.13 Gastrulation and trochophore of *Polygordius* (Woltereck)

a. gastrula, after invagination of archenteron (vegetal pole); b. gastrula at closure of archenteron (vegetal pole); c. longitudinal section of trochophore larva of Archiannelida. *a:* anus, *ap:* apical plate, *blp:* blastopore, *ms:* retractor muscle of oesophagus, *m.tr:* metatroch, *ne:* protonephridium, *o:* mouth, *p.tr:* prototroch, *t.tr:* telotroch, *w:* boundary wall between stomach and intestine.

The posterior opening of the blastopore is very soon covered over by the ectoderm, but reopens when $2d^{222}$ and $2d^{221}$ give rise to the *proctodaeum.* In connection with these epibolic activities, the teloblasts of the mesodermal bands M^{l} and M^{r}, as well as $3c^{2pa}$ and $3c^{2pp}$, which will develop into the left protonephridium and $3d^{2pa}$ and $3d^{2pp}$, which will form the right protonephridium, all begin to show signs of moving into the blastocoel. The process of invagination completed, the endodermal cells give indications of their approaching differentiation into stomach and intestine (Fig. 7.13c).

It is of course not possible to apply this description of archenteron formation in *Polygordius* to all the Polychaeta without some modification. In the genera in which there is practically no invagination of the endodermal cells and the archenteron is formed almost entirely by epiboly *(Nereis, Arenicola, Amphitrite, Serpula, Protula,* etc.), each genus shows certain secondary changes in this process correlated with the amount of yolk included in its eggs. In *Arenicola,* for example, while the endodermal cell mass maintains a wide-open blastopore, the growth of the ectodermal cells has already begun, causing an elongation of the larva in the ventro-posterior direction. Only later is the blastopore closed by the overgrowth of the lateral and posterior ectoderm (Fig. 7.14).

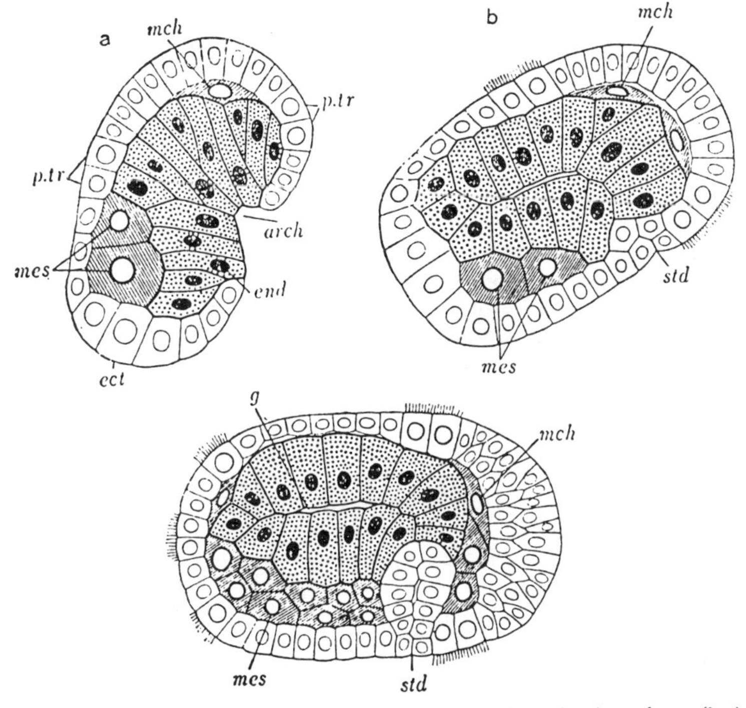

Fig. 7.14 Longitudinally sectioned gastrula (a) and trochophore larva (b,c) of *Arenicola cristata* × 300 (Okada)

arch: archenteron, *ect:* ectoderm, *g:* gut, *mch:* mesenchyme, *mes:* mesodermal band, *p.tr:* prototrochal cell, *std:* stomodaeum.

As gastrulation proceeds and the vegetal pole region moves into the blastocoel, a new dorso-ventral embryonic axis is established. The digestive tract, which produces a *mouth* and an *anus,* elongates in the antero-posterior direction along the ventral side, and according to the new axis, the animal pole side of the egg is converted into the antero-dorsal region (Fig. 7.14).

(4) ARCHIANNELIDAN TROCHOPHORE

While gastrulation is going on, cilia arise on the surface of the gastrula to serve as its organs of locomotion, and shortly afterward it becomes a free swimming larva, begins to secure food for itself and grows. In most species, the apical cells of the gastrula gradually become thickened and form an apical organ equipped with a bundle of long cilia; from the prototroch cells there develops the *prototrochal girdle,* which encircles the equatorial region. In connection with the overgrowth which resulted in closing the blastopore, the descendants of the first somatoblast develop into the *ventral plate,* which comes to have the stomodaeum at its anterior end and the proctodaeum at its posterior end. In such a manner the larval organization progresses toward the structure characteristic of the *trochophore,* which is considered to be the prototype of Annelida.

The young trochophore, or *protrochophore,* usually shows the following characteristics: (1) it has an apical organ and a proto-trochal girdle; (2) it has a digestive tract, with the mouth and anus already determined. That is, a "*trochosphere*" (Fig. 7.19a) is imagined as the basic form of the external appearance of the protrochophore, with an apical organ bearing a tuft of long cilia[11] at the side corresponding to the animal pole of the egg, and a band of powerful cilia differentiated into the prototrochal girdle corresponding to the trochoblasts located around the equator. The upper part of the larva between the prototrochal girdle and the apical organ is called the *episphere,* or *head blastema.* The *hyposphere,* below the prototrochal girdle, arises as the result of the complicated transformation brought about by the invagination or epiboly of gastrulation; in brief, the mouth becomes located just below the ventral edge of the prototrochal girdle, and the anus at the opposite side from the apical organ, while the digestive tract connecting them is bent in a right angle (Figs. 7.15d, 7.16b). Cilia eventually develop around the anus, forming the *telotroch.*

As the protrochophore continues to develop, it enters the *metatrochophore* stage. In the episphere of this larva *eye-spots* are differentiated, as well as *radial nerves* which grow out from the apical organ. In the lower hemisphere, the rudiment of the trunk develops together with the growth of the ventral plate, and the swimming capacity of the larva is increased by the formation of a *metatroch*[12]. Other characteristics of this stage include the development of mesenchyme cells and the *protonephridia,* and the clear differentiation of the digestive tract into an *oesophagus, stomach* and *intestine.* The most typical case of such annelidan-larval trochophore organization is shown by the *Polygordius* larva. Since this larva was first discovered by Lovén in 1842, it is also sometimes called *Lovén's larva* (Figs. 7.13c, 7.19a). Among the Polychatea as a whole, however, considerable species differences are to be found with respect to the time of appearance of the above-mentioned larval organs and their degree of development.

(5) POLYCHAET TROCHOPHORE

The time of appearance of the larval organs is strongly influenced by the amount of yolk, while the significance of the mode of development is probably less a matter of geneological affinites than of ecological adaptation to the free-swimming habit. The less the amount of yolk in the eggs of any species, the earlier is the stage in which the larval organs appear and hatching takes place; furthermore, the longer the larval swimming period, the more schematic is the organization of the trochophore. In the opposite sense, if the eggs contain much yolk and the larva begins its bottom-dwelling life almost immediately after hatching, it will form a degenerate trochophore.

---

[11] In some species the apical tuft secondarily shows a tendency to degenerate.

[12] When a complete ciliated band is formed parallel to the prototrochal girdle and telotroch, this may be called a *paratroch.*

It is possible to classify the protrochophores of the Polychaeta into the *monotrochal, atrochal* and *mesotrochal* types of larvae, according to the mode of formation of the ciliated bands. The monotrochal larva is provided with the standard basic organization described for the archiannelidan larva, with a single prototrochal girdle encircling the central part of the body (Fig. 7.15a). In most cases, the formation of the prototrochal girdle is followed by the appearance of the apical tuft and the telotroch (Fig. 7.8b). In species which brood the young in a cocoon, not all these organs are fully formed, but at least a weak ciliary band corresponding to the prototrochal girdle can always be recognized. In this sense, practically all the polychaet species have a monotrochal protrochophore stage.

In contrast to this, the atrochal larva can be thought of as a trochophore in which the ciliated bands have degenerated completely. In such larvae the apical cilia are distinctly present, but no cilia constituting anything that can be called a ciliary band, although the embryo hatches with

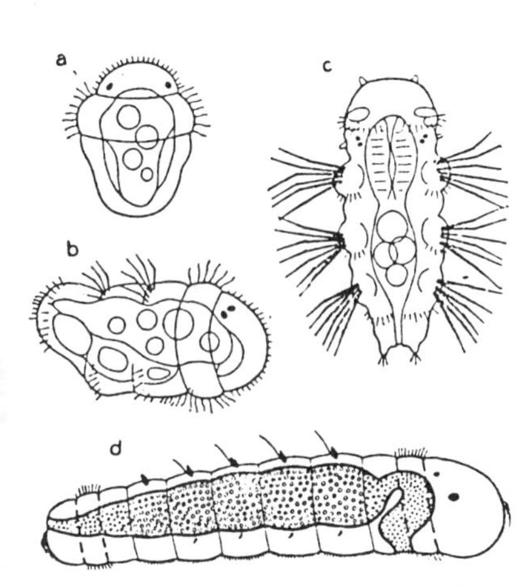

Fig. 7.15 Trochophore larvae of Polychaeta (1)

a. monotrochal larva of *Tylorrhynchus heterochaetus* x120; b. nectochaetal larva of *T. heterochaetus*; 7 days after fertilization. x120 (a, b. Okada); c. nectochaetal larva of *T. heterochaetus*; 20 days after fertilization. x120 (Yamamoto); d. nectochaetal larva of *Arenicola cristata*; 10 days after fertilization. x100 (Okada).

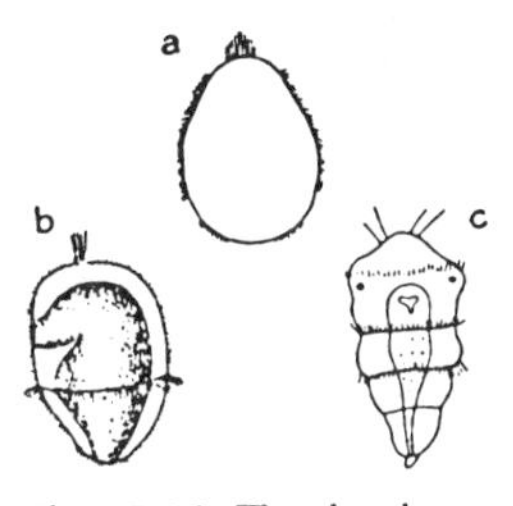

Fig. 7.16 Trochophore larvae of Polychaeta (2)

a. atrochal larva of *Marphysa* (Aiyar); b. mesotrochal larva of *Chaetopterus pergamentaceus* (E. B. Wilson); c. polytrochal larva of *Ophryotrocha puerilis* (Korschelt).

its whole surface thickly covered with very delicate cilia. Parts of the larval surface formed immediately after hatching lack cilia, however, and the ciliated portions are considered to be the vestiges of degenerated prototrochal girdle and telotroch, although at first glance the external appearance of such a larva suggests the planula larva of the coelenterates (Fig. 7.16a). Atrochal larvae are known in species belonging to the genera *Sternaspis, Lumbriconereis* and *Marphysa.*

In the mesotrochal larva, both prototroch and telotroch have degenerated, and a single band of cilia which correponds to the metatroch develops in an exaggerated form, as though to compensate for the missing prototroch (Fig. 7.16b). Mesotrochal larvae are found in species belonging to *Chaetopterus, Telepsanus* and related genera.

The monotrochal type of protrochophore adds to its lower hemisphere body segments produced by the growth of the ventral plate and the mesodermal bands, and thereby advances to the *metatrochophore* stage. At this time the metatrochal bands also increase, and the larva develops into the standard type of this stage, the *polytrochal larva.* Depending on the extent of development of the metatrochal bands, polytrochal larvae may be classified into three groups. The *amphitrochal larva* has on each segment a metatroch which forms a complete ciliated band encircling the segment parallel to the prototroch. In the *gastrotrochal larva,* the metatroch fails to extend around the dorsal sides of the segments, forming a band of weak cilia on the ventral side only; while in the *nototrochal larva,* the opposite is the case, the cilia forming only dorsally.

The basic form of the polytrochal larva is most faithfully represented by the amphitrochal larva (Fig. 7.16c). Actually, however, there are relatively few cases in which a complete ciliated band is found on each of the segments formed during the metatrochophore stage; more often vestiges of degenerated metatrochs appear as rows of weak cilia on the ventral surfaces of each segment (Fig. 7.15b, d). It is difficult to trace a geneological significance in these degrees of development of the metatrochs. Cases are known in which closely related species, of the same family and genus, produce entirely different types of polytrochal larvae when their ecological habits are different. For example, among the Terebellidae, *Terebella conchilega,* in which the larva is free-swimming, has a typical nototrochal stage while the larva of *T. meckelii,* which almost completely lacks a free-swimming stage, is said not to produce any metatrochal cilia.

Except in a few species in which the free-swimming stage is especially pronounced, a tendency is soon shown in the polytrochal larvae of most polychaets for the metatrochal cilia to degenerate. At the two sides of each body segment the rudiments of *parapodia* are formed, and numerous paddle-shaped *swimming setae,* or *nectochaetae,* appear (Fig. 7.15b, c, d). On arriving at this stage, the larva is gradually beginning to exhibit some peculiarly polychaetan characteristics, although in most species the free-swimming habit continues. The larva in this period is known as the *nectochaetal larva* or *polychaet larva.*

The tufts of swimming setae which characterize this larva not only serve as powerful swimming organs, but also take care of the movements involved in maintaining equilibrium and in rising and sinking. The shapes of these larvae, moreover, show an extraordinary diversity, which is probably the result of a high degree of ecological adaptation originating in the mode of development of the setae. This variety is so great that one is in fact at a loss in attempting to define the standard basic type of nectochaetal larva. If one were forced to make a decision in this respect, the choice would probably fall on the *Nereis* larva. In the nereid genera *Nereis* and *Tylorrhynchus,* three paired tufts of setae begin to be formed as the larva proceeds toward the polytrochal stage, and it finally comes to possess the characteristics of a full-fledged nectochaeta (Figs. 7.15c, 7.21a). As development continues, new segments are added to its posterior end, and the tufts of setae also increase accordingly.

Among the numerous polychaet species, there are not a few in which the free-swimming habit is very feebly developed. In *Arenicola,* for example, the growth of the swimming setae is poor, and the larva spends part of its time swimming, and part creeping (Fig. 7.15d). The larva of *Marphysa,* also, which passes through an atrochal stage as a protrochophore, fails to form the metatroch characteristic of the polytrochal larva, and develops directly into a peculiar nectochaeta (Fig. 7.17a).

On the other hand, there are also many nectochaetal larvae in which the free-swimming habit is highly evolved. In these larvae the swimming setae are strongly developed so that they present a spectacular appearance; examples of this type of specialization can be found in the *Polynöe-larva* and *Spio-larva*. (Fig. 7.17b, c, d). In the *Spio*-larvae of *Polydora* and *Sabellaria*, for instance, numerous powerful swimming setae which are longer than the whole body length are formed dorsally on both sides of the anterior segments close to the prototroch, so that they completely enwrap the larval body. The larva of the deep-sea polychaet *Chaetosphaera* is well-known as the most conspicuous example of such structure (Fig. 7.17e).

A type of nectochaetal larva which is believed to have developed in the opposite direction from the *Spio*-larva by reducing the number of its swimming setae is het larva of the polychaet *Myzostoma*, which parasitizes certain echinoderms (Fig· 7.17f). Other peculiar nectochaetal larvae include that of *Ophryotrocha*, and the types represented by the *Mitraria-larva* and the *Rostraria-larva*. The nectochaeta of *Ophryotrocha* forms swimming setae while continuing to keep the strongly developed metatroch of its polytrochal stage (Fig. 7.17g). The *Mitraria*-larva can be seen in the development of *Owenis* and *Asychis*; these larvae resemble the nemertine pilidium in external appearance, but have a band of delicate cilia as the prototroch and close to this, extending posteriorly on the dorsal surface, long needle-like swimming setae (Fig. 7.17h). The Rostraria-larva has a *prostomium* which gives rise to a *rostrum*; on this are formed long processes similar to tentacles (Fig. 7.17i). Larvae of this type are frequently found among the genera *Amphione*, *Audouinia* and *Branchiomma*. The variation producing these thousand and one shapes of the nectochaetal larvae can be found arising within a single species in response to differences in ecological conditions. *Nereis dümerilii*, for example, is divided into pelagic and sedentary individuals; the former are said to produce a pelagic type of larva with well developed swimming setae, while the larvae of the latter are non-pelagic, with poorly developed setae. The same is reported to be the case in *Spio filicornis*.

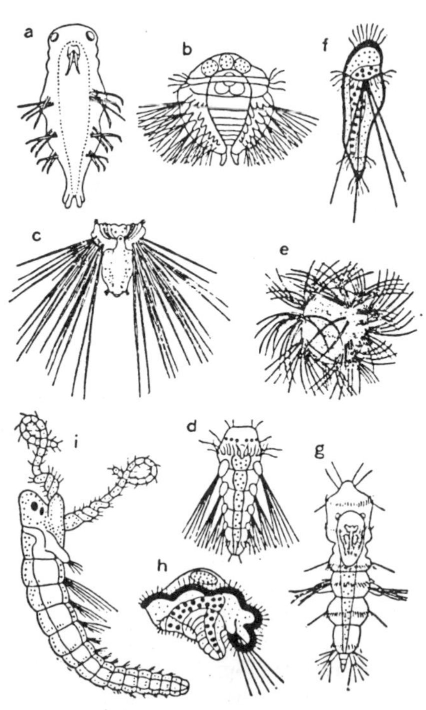

Fig. 7.17 Nectochaetal larvae of Polychaeta

a. *Marphysa* (Aiyar); b. *Polynöe* (Okuda); c. *Spio* (Haecker); *d. Polydora* (Okuda); e. *Chaetosphaera* (Haecker); f. *Myzostoma* (Beard); g. *Ophryotrocha* (Korschelt); h. *Mitraria*-type larva; i. *Rostraria*-type larva (Okada).

(6) MESODERM FORMATION

The most noteworthy progress made by the larva during the trochophore stage is the formation of the mesoderm, which consists of the *larval mesoblast* and the *mesodermal bands*. As described above, the mesoblast arises chiefly from the blastomeres of the third quartette. These cells, which slipped into the blastocoel at the time of gastrulation, take up scattered positions within this cavity and eventually develop into the *retractor muscle cells* (Figs. 7.13c, 7.18). Throughout the trochophore stage, these muscles enable the larval body, and especially the digestive tract, to expand and contract.

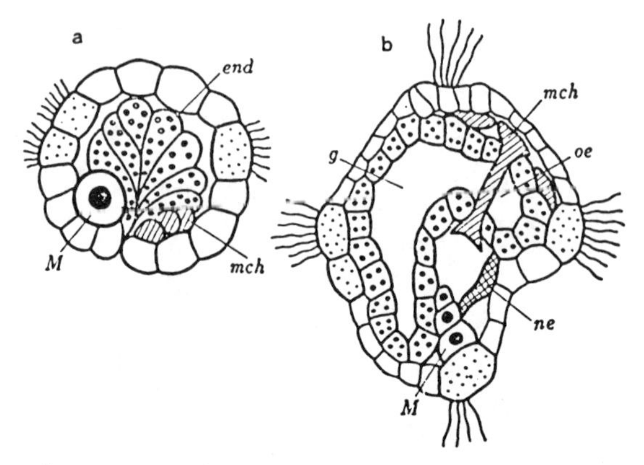

Fig. 7.18 (a) gastrula and (b) trochophore larva of *Hydroides* (Shearer)

end: endoderm, *g:* gut, *M:* mesoblast, *mch:* mesenchyme cell, *oe:* oesophagus, *ne:* nephridium.

The mesodermal bands, which have their origin in the mesoblast (4*d*, *M*), will develop into the mesoderm of the future adult. As has already been described, this cell, lying at the posterior end of the endodermal region at the time of gastrulation, divides equally to form the teloblasts of the mesodermal bands. These teloblasts undergo two unequal divisions, and the two pairs of micromeres thus formed are left in the endodermal region while the teloblasts themselves gradually move away[13] (Fig. 7.18a, b). Before the larva enters the protrochophore stage, these *M* cells have become completely transposed into the blastocoel (Fig. 7.18a). There they divide repeatedly to form the mesodermal bands — the paired lines of cells which proliferate ventro-laterally along the digestive tract (Fig. 7.18b). At first the mesodermal bands grow out in a direction roughly parallel to the prototroch, but after the turning of the larval axis which accompanies the development of the trochophore, they take an antero-posterior position intersecting the prototroch (Fig. 7.18). In the polytrochal larva, the proliferation of mesoderm cells increases still more in response to the growth of the ventral plate. By the time the larva acquires swimming setae and reaches the nectochaetal stage, the bands are segmenting into mesodermal somites which show the beginnings of coelom formation.

13

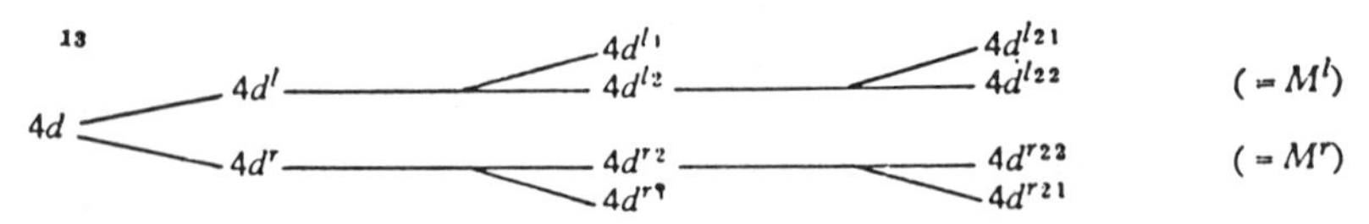

The micromeres $4d^{l1}$, $4d^{r1}$, $4d^{l21}$, $4d^{r21}$ remain in the endodermal region.

(7) METAMORPHOSIS

The free-swimming protrochophore larva presently forms on its hyposphere the rudiment of the *trunk* and becomes a metatrochophore, adding new segments to its posterior end and preparing for the process of metamorphosis. In most cases the larva largely gives up its swimming habit before it has acquired as many as ten segments, and turns to a new way of life as either a creeping or fixed organism. Together with this change in its habits, rapid shape transformations bring the larva close to the vermiform organization of the adult. In the process of metamorphosis, the swimming setae and other larval organs retrogress, the trochophore episphere contracts and the *tentacular cirri* (sing. *us*) and *branchial filaments* develop; in the meantime, the trunk segments which have begun to extend posteriorly undergo rapid enlargement and growth.

a. Metamorphosis of archiannelidan trochophore

Two modes of metamorphosis can be recognized within this group: the *exo-larval* and *endo-larval* types. The former has been described by Hatschek and other workers, as it is found in *Polygordius neopolitanus*. In the trochophore of this species, the posterior end of the episphere projects as a trunk with a seg-

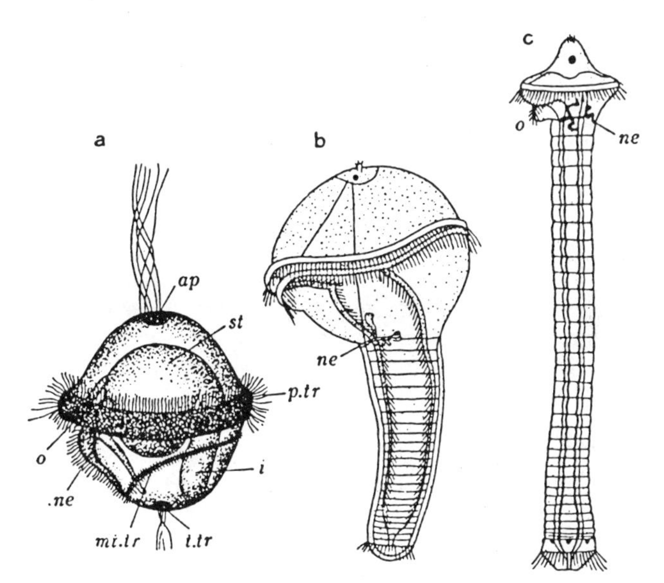

Fig. 7.19 Exo-larval type of metamorphosis in *Polygordius neapolitanus* (MacBride, Korschelt)

a. trochophore larva of Archiannelida; b. trochophore larva beginning to form trunk segments; c. episphere of metamorphosing larva beginning to shrink. *ap:* apical plate, *i:* intestine, *mt.tr:* metatroch, *ne:* protonephridium, *o:* mouth, *p.tr:* prototroch, *st:* stomach, *t.tr:* telotroch.

mented mesodermal band, the growth of which brings the larva to the beginning of metamorphosis (Fig. 7.19a, b). However, the fully developed prototroch still functions actively, enabling the larva to continue its free-swimming life for some time. During this period it assumes a vertical posture, with its steadily elongating trunk dangling downward (Fig. 7.19b). When there are about 30 segments, the wall of the episphere begins to thicken as it changes into a shrunken, cone-shaped pre-oral lobe, and the prototroch gradually degenerates; as growth of the trunk rapidly increases the number of segments, metamorphosis reaches completion and the young worm is prepared for a bottom-creeping habit (Fig. 7.19c). This type of metamorphosis agrees in principle with that found in most polychaets.

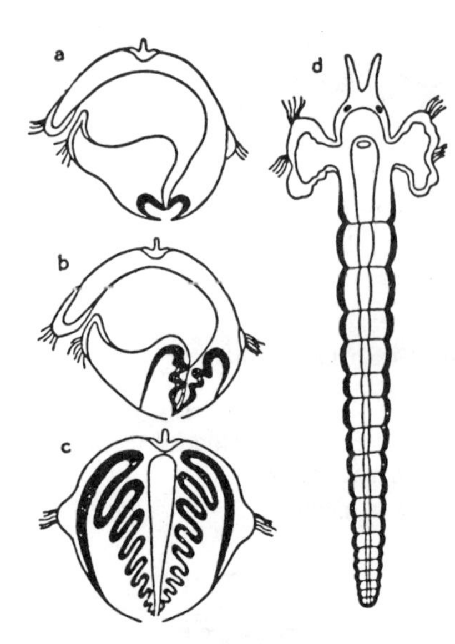

Fig. 7.20 Endo-larval type of metamorphosis in *Polygordius appendiculatus* (Korschelt)

Endo-larval metamorphosis follows a peculiar mode discovered by Woltereck, Salensky and Söderström in the North Sea *Polygordius appendiculatus* and the Black Sea form, *P. ponticus*. In this case, from the outset the extension of the trunk takes place not toward the outside as in the exo-larval type, but folded inward from a point at the posterior side of the episphere, until ten trunk segments have been formed (Fig. 7.20a, b, c). Eventually, however, these segments are everted to the outside as in the exo-larva, and metamorphosis is completed in the same manner (Fig. 7.20d). Metamorphosis among the Mitraria-type larvae is believed to follow this mode almost exactly (Fig. 7.17h).

b. **Metamorphosis of the Nectochaetal Larva**

The nectochaetal larvae of the Polychaeta undergo metamorphosis by various processes which are peculiar to each species. On the whole, in nectochaetal larvae which show a high grade of adaptation to the free-swimming habit, the morphological changes accompanying metamorphosis are likely to take place gradually, because of the strongly developed ciliary bands and powerful swimming setae. On the other hand, the adult form tends to be assumed rapidly in species which will take up a sessile habit after metamorphosis. Using *Nereis pelagica* as an example of the former type, and the tubiculous worm *Branchiomma vesiculosum* to represent the latter the concrete changes involved in these modes of metamorphosis will be compared.

The larva of *Nereis pelagica* reaches a typical nectochaetal stage several days after fertilization. This larva has a pair of eye spots and an *acrotroch* in its head region, and behind the prototroch are three segments provided with tufts and a telotroch. In the third week after fertilization, these ciliary bands begin to retrogress in regular sequence, and in their place a pair of *first tentacles* are formed at the anterior tip of the head and paired *second tentacles* at the sides of the mouth. The *caudal cirri* (sing. *-rus*) are a pair oft entacles located at the posterior end of the trunk (Fig. 7.21a). From this time a progressive enlargement and extension of the trunk sets in, and the larval course of development deliberately turns in a direction appropriate to a creeping existence. By the fourth week, a fourth segment having parapodia and setae appears at the posterior end of the trunk; in the meantime, the setae of the first segment have degenerated and its parapodia have developed into the third pair of *tentacles*. This constitutes the process of metamorphosis; with further development segments are added to the posterior end in regular order, and the growing worm acquires the vermiform shape of the adult (Fig. 7.12b).

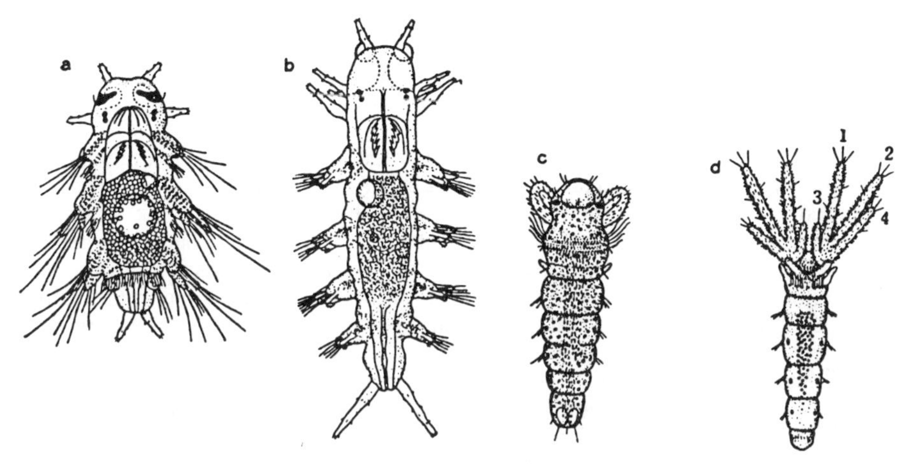

Fig. 7.21 Metamorphosis of nectochaetal larva (D. P. Wilson)

a. larva of *Neris pelagica* 18 days after fertilization. x90; b. larva of *N. pelagica* 23 days after fertilization. x90; c. larva of *Branchiomma vesiculosum* 9 days after fertilization. x90; d. larva of *B. vesiculosum* 20 days after fertilization. x90 (numbers indicate order of formation of branchial filaments).

The nectochaetal larva of *Branchiomma vesiculosum*, on the other hand, has poorly developed ciliary bands and swimming setae. In the second week after fertilization, it forms a pair of eye spots and bulging rudiments of the branchial filaments in the head region, produces three or four trunk segment and becomes a larva closely similar to the *Rostraria* type (Fig. 7.12c). This soon arrives at the stage of metamorphosis, and while it is secreting a mucous tube and becoming attached to the bottom, it completes the process of metamorphosis within a few days. By the third week, the head region, which has lost its eye spots, ciliary bands and other larval organs, shrinks to a proboscis-like prostomium; the branchial filament rudiments elongate anteriorly and divide into feathery branches, and the trunk extends its growth posteriorward to give rise to the final form of the tubicolous polychaet (Fig. 7.21d).

2. EARLY DEVELOPMENT OF THE OLIGOCHAETA

Among the oligochaets, the whole course of embryonic development from fertilization to the time of hatching takes place within the protective cocoon. In its basic aspects the process closely resembles that of the polychaets in many points, but there is no trace whatsoever in oligochaet development of a free-swimming stage with a prototroch. After the gastrula stage, the proliferative growth of the ectoderm cells produces the trunk rudiment, and growth proceeds step-by-step in a direct fashion toward the adult vermiform morphology, without the complicated process of metamorphosis seen in the Polychaeta. A voluminous literature concerned with oligochaet embryology has been accumulated; these studies indicate that the direct mode of development is characteristic not only of terrestrial and fresh-water oligochaets, but of the marine forms as well.

(1) CLEAVAGE

The amount of yolk contained in the eggs of oligochaets varies with the species. Generally speaking, the eggs of aquatic worms are large and yolky, while the terrestrial species have small eggs with little yolk. There are not a few examples, however, of striking size differences among the eggs of even closely related species. In very many cases, the special characteristics of each species which appear as development proceeds find their chief cause in the quantity of yolk contained in the egg cell. One particularly striking feature of many oligochaet egg cells is their inclusion of a so-called *pole-plasm*. The egg cytoplasm consists of an opaque, yolky portion, and a clearly different, active protoplasm. This material is about evenly distributed throughout the immature oocyte, but when polarization of the egg substance takes place at the time of polar body extrusion, it collects in separate masses near the animal and vegetal poles. For this reason these masses of active protoplasm located at the two poles of the mature egg are called 'pole plasm' (Fig. 7.22a).

As in the Polychaeta, the mode of cleavage in the Oligochaeta is basically total, unequal and spiral. A strict comparison, however, will show that some of the conspicuous blastomeres in polychaet development play a less striking role in the oligochaets, and vice versa. The degree of size inequality among oligochaet blastomeres is so marked from the earliest divisions that at first glance the cleavage

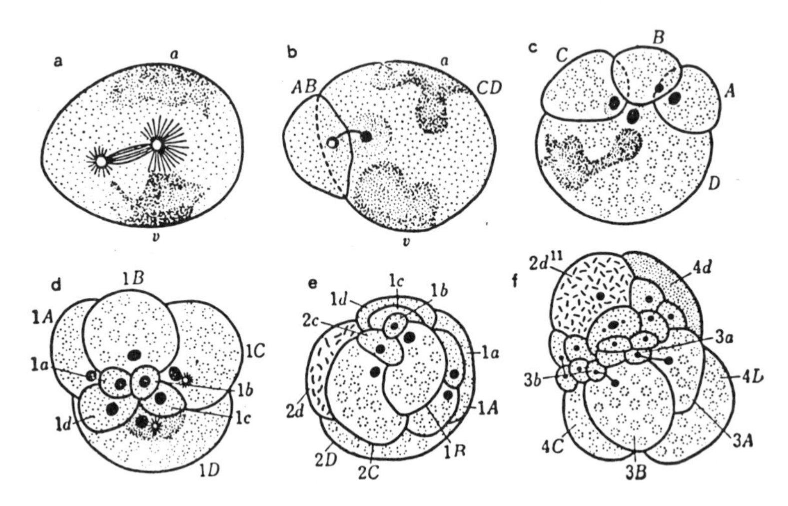

Fig. 7.22 Cleavage in *Tubifex rivulorum* (Penners)

a. mitotic apparatus at first division; b. 2-cell stage (left side); c. 4-cell stage (vegetal pole); d. 8-cell stage (animal pole); e. frontal view of embryo at 10-cell stage; f. frontal view of embryo at 22-cell stage. a: animal pole, *v:* vegetal pole.

appears to be less spiral than irregular. The following summary of the cleavage process as it occurs in the oligochaets is based on Penners' (1922) description of *Tubifex rivulorum.*

The pronouncedly unequal first cleavage of *T. rivulorum* divides the egg so that both masses of pole-plasm are contained in the large *CD* blastomere (Fig. 7.22b). At the second cleavage, *CD* divides unequally into a small *C* and a large *D* blastomere, while *AB* cleaves equally into *A* and *B* to produce the 4-cell stage (Fig. 7.22c). Equatorial division occurs for the first time at the third cleavage, giving rise to an 8-cell stage which includes the first quartette of micromeres; 1*D* is especially large, and still monopolizes both pole-plasms (Fig. 7.22d). In principle, the cleavage of these blastomere-groups is spiral: the third cleavage is dexiotropic, as in the polychaets, and the fourth cleavage is laeotropic. But perhaps because of the marked inequality, the synchrony of cleavage is already disturbed at this fourth division, so that 1*D* and 1*C* separate off their second quartette micromeres before 1*A* and 1*B*, giving rise to a 10-cell stage (Fig. 7.22e). At this cleavage, the animal and vegetal pole-plasms for the first time separate into different blastomeres. The 2*d* blastomere is much the larger, and includes the animal pole-plasm; as in the Polychaeta, this cell is destined to be the first somatoblast. The 2*D* blastomere contains the vegetal pole-plasm and yolk.

After the 10-cell stage, the synchrony of cleavage of the various quartettes becomes still more disrupted, giving rise to a divergence in the time of micromere formation of the several quadrants. As a consequence, the division process fails to follow the typical spiral course with 16-cell and 32-cell stages. In general, the *D* quadrant divides earliest, followed by the *C* quadrant. The 22-cell stage, which follows the stage with 10 cells, consists of the four yolky macromeres (3*A*, 3*B*, 4*C* and 4*D*) in the vegetal hemisphere, and in the upper half, the two large blastomeres $2d^{11}$ (first somatoblast) and 4*d* (second somatoblast) as well as 16 micromeres. The position of 4*d* is perhaps better described as posterior than as belonging to the upper hemisphere; while it contains no yolk, it includes most of the vegetal pole-plasm, and as in the polychaets, is destined to be the mesoderm mother cell. The fact that the 'active' protoplasm of the two pole-plasm masses is distributed between the first and second somatoblasts is connected with the important role which these blastomeres play in the ensuing development of the embryo.

The 16 micromeres which occupy the rest of the upper hemisphere lie interposed between the first somatoblast and the four macromeres of the lower hemisphere, forming a micromere band which fills this space (Fig. 7.22f). Shortly afterward, an extremely limited blastocoel develops in the center of the blastula. Among the terrestrial oligochaets, the eggs of which have less yolk, the inequality between the blastomeres is somewhat less extreme, and the embryo formed as the result of cleavage is correspondingly closer to a coeloblastula.

(2) GERM LAYER FORMATION

The process of archenteron formation also conforms to that in the Polychaeta: in the yolk-laden aquatic oligochaet eggs, gastrulation takes place chiefly by epiboly, while the blastulae of the less yolky terrestrial worm eggs mostly invaginate to form a gastrula. The eggs of the marine oligochaet *Pachydrilus lineatus* contain a large amount of yolk. In the blastula of this species, *ectodermal bands* formed by the proliferation of the first somatoblast descendants develop prominently on each side before the endodermal cell group has completed formation of the archenteron; these cellular bands undergo marked growth and envelop the endodermal cells from both sides, bringing about closure of the blastopore (Fig. 7.23a). This relationship is the same in principle as the growth of the ventral plate which leads to blastopore closure in the Polychaeta.

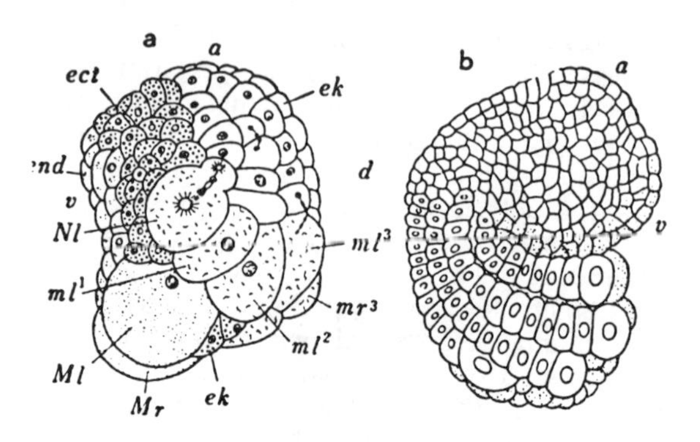

Fig. 7.23 Oligochaete gastrulae

a. lateral view of gastrula of *Pachydrilus* (Penners); b. lateral view of late-stage embryo of *Bdellodrilus* (Tannereuther). a: anterior end, *d:* dorsal side, *ect:* ectoderm, *ek:* ectodermal band, *end:* endoderm, *Ml:* telocyte of left mesodermal band, *Mr:* telocyte of right mesodermal band, ml^1: : ml^2 : ml^3 : mr^3 : *Nl:* telocytes of ectodermal band, *v:* ventral side.

When, however, the eggs have relatively little yolk, as in *Lumbricus*, the embryo has a rather well-developed blastocoel, and the archenteron is formed by the invagination of the blastomeres located in the vegetal hemisphere. In this case also, the ectodermal bands attain a conspicuous growth, so that the external appearance is not very different from that of the gastrula formed by epiboly. In either case the ciliary bands which characterize the trochophore larva have completely degenerated and fail to appear. Moreover, the striking development of the ectodermal and mesodermal bands makes it easy to distinguish among the three differentiating germ layers. The ectodermal bands arise from the first somatoblast; by the gastrula stage it has increased to become four conspicuous rows of cells on each side of the embryo (Fig. 7.23b). By the blastula stage this first somatoblast 2d has divided repeatedly, $2d^{11}$ being formed from $2d^1$, and then $2d^{111}$; this $2d^{111}$ for the first time undergoes equal cleavage, giving rise to the left and right teloblast T^l and T^r.[14] During the progress of the embryo from blastula to gastrula, these teloblasts divide in a special fashion,[15] eventually producing on each side the four teloblasts N, m^1, m^2 and m^3. These form the starting points from which the cells proliferate on each side to give rise to the cellular bands. The latter occupy the animal hemisphere, and together with the other micromere quartettes comprise the ectoderm.

---

[14] Penners (1930) called T^r and T^l the *anterior teloblast*.

[15] According to Penners (1930), T divides into m and N, m into m^{1+2} and m^3, and finally m^{1+2} forms m^1 and m^2. In other words, T gives rise to the four teloblasts N, m^1, m^2, m^3, in a ventro-dorsal alignment.

The mesodermal bands take their origin from the second somatoblast, *4d*. This blastomere divides equally into right and left cells to form on each side the mesoderm mother cells M^r and M^l;[16] in connection with the epibolic growth that closes the blastopore, M^r and M^l are crowded into the posterior end of the blastocoel, where they proceed to give rise to the mesodermal bands. In short, it is possible to distinguish in this stage an endodermal cell group forming the archenteron in the vegetal hemisphere, an ectodermal group consisting of the micromere quartettes and the cellular bands originating from the four teloblasts on each side of the animal hemisphere, and mesoderm in the form of the two bands proliferating bilaterally into the blastocoel from the two teloblast cells.

The ensuing proliferation of these cellular bands causes the originally spherical embryo to become first oval and then elongated. The accompanying shift in the embryonic axis which was observed in the trochophore also takes place at this time in the oligochaet embryo on the same principle, the primary egg axis being abandoned in favor of the elongating antero-posterior axis as determiner of embryonic polarity. Accompanying the subsequent vigorous development of the mesoderm, body segments proliferate posteriorly, and the embryo gradually acquires a vermiform morphology. Presently it begins to show feeble peristaltic movements within the cocoon, and hatches by breaking through the cocoon wall.

3. EARLY DEVELOPMENT OF HIRUDINEA

At first glance, the leeches, with their morphological peculiarities such as the formation of suckers, flattening of the trunk and development of a *sinus system* and *botryoidal tissue* to replace their lost body cavity, appear to be quite different from the Chaetopoda, but their developmental process is actually very similar to that of the Oligochaeta in the embryonic stages. Among the leeches, for example, the whole course of early development also takes place directly, and similarly within the protective covering of the cocoon, and especially during these early stages there are more points of similarity between the leeches and oligochaets than between the latter and the polychaets. This fact has aroused much interest in leech development from early times, and led to the accumulation of an extensive literature covering their embryology. The following account will present a description of leech development, using the rhynchobdellid *Clepsine* and the gnathobdellid *Nephelia* as examples.

Like the oligochaet egg, the leech egg contains animal and vegetal poleplasm. Its mode of cleavage can be described as spiral in principle, although it does not follow the regular course seen in the Polychaeta; the extreme degree of size inequality among the blastomeres completely upsets the synchrony of the division cycles, producing an irregularity even greater than that found in oligochaet cleavage.

In *Clepsine*, the process goes on in typical spiral fashion until the formation of the first micromere quartette (Fig. 7.24a). In the ensuing division leading to the 16-cell stage, the *D* and *C* quadrants clearly show a tendency to precede the others in cleaving to produce the second and third micromere quartettes (Fig. 7.24b). The *2d* blastomere, as in the oligochaets, is separated off as a large cell containing the vegetal poleplasm, and destined to play the role of first somato-

[16] Penners (1930) named M^r and M^l the posterior teloblasts, emphasizing their relation to the anterior teloblasts.

blast. The cleavage synchrony, however, is completely lost after the 16-cell stage, only the micromeres and the blastomeres of the *D* quadrant continuing to divide at a rapid pace (Fig. 7.24c, d).

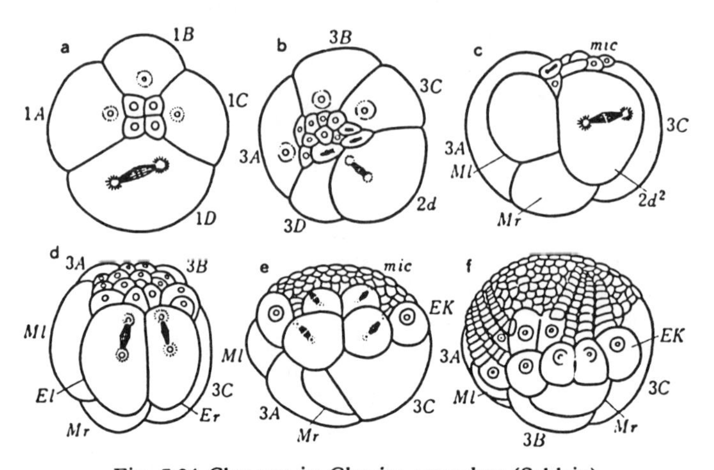

Fig. 7.24 Cleavage in *Clepsine sexoculata* (Schleip)
EK: telocyte of ectodermal band, *El:* telocyte of left ectodermal band, *Er:* telocyte of right ectodermal band, *mic:* micromere, *Ml:* left mesoblast, *Mr:* right mesoblast.

On the other hand, the *D* quadrant has no connection with the endodermal layer, which is made up of cells derived from 3*A*, 3*B* and 3*C* only. This peculiarity has been noted as a characteristic unique to the genus *Clepsine*. The 3*D* blastomere, which includes the vegetal pole-plasm, precociously becomes the second somatoblast, and divides obliquely into equal-sized blastomeres which form the mesoderm mother cells M^r and M^l on the two sides of the embryo (Fig. 7.24c). The first somatoblast 2*d* at once divides unequally, giving rise to a small $2d^1$ and a large $2d^2$; $2d^2$ again divides, equally and longitudinally this time, to form E^r and E^l, the end cells of the ectodermal bands on the right and left sides (Fig. 7.24d). These eventually, as in the Oligochaeta, develop into the ectodermal teloblasts *(EK)* on each side (Fig. 7.24e, f). The mesoderm mother cells M^1 and M^r also lie diagonally at first, but gradually take on a bilaterally symmetrical arrangement (Fig. 7.24e, f).

In the gnathobdellid *Nephelis*, the uniquely hirudinean characteristics are even more marked than in *Clepsine*. After 1*A* and 1*B* have separated off the first micromeres, and 2*C* has given off the second micromere, these macromeres become wholly yolk cells, and fail to divide again during the entire cleavage period. The division of the *D* quadrant blastomeres thus provides most of the cells to form the embryonic body. The first somatoblast *(2d)*, which separates off from 1*D*, continues to cleave, and develops into the bilateral teloblasts of the ectodermal bands. The 2*D* macromere divides twice more, forming 3*d* and 4*d*. 2*c*, 3*d* and 4*d* eventually make their way into the interior of the embryo, where they are destined to form the endoderm, together with the yolk cells 1*A*, 1*B* and 2*C*. It is at this stage that 4*D*, which holds the vegetal pole-plasm, for the first time undergoes equal cleavage, to produce the two mother-cells of the mesoderm (Fig. 7.25a).

Among the leeches, then, the cleavage pattern is fixed at an early stage, and each blastomere follows a direct course toward its final destiny in an even more pronounced fashion than was seen in the oligochaets. In the leech blastula,

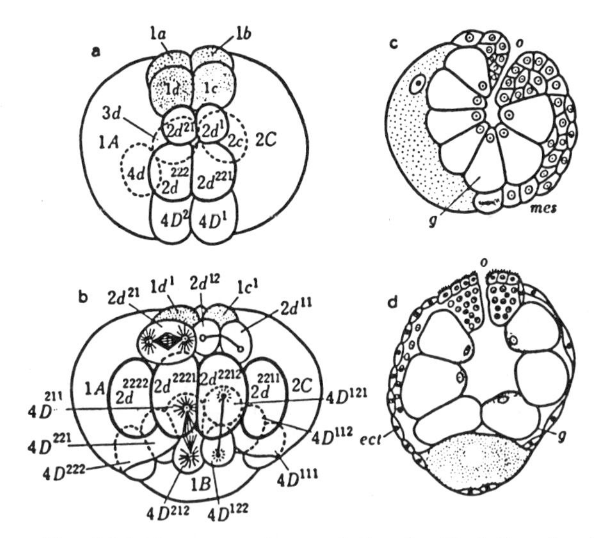

Fig. 7.25 Cleavage and gastrulation in *Nephelis vulgaris* (Dimpker, Sukatschoff)

a. 16-cell stage, from side of *D*-quadrant; b. 27-cell stage, from side of *D*-quadrant; c. longitudinal section of embryo at stage of stomodaeum formation; d. more advanced stage. *ect:* ectoderm, *g:* gut, *mes:* mesodermal band, *o:* mouth.

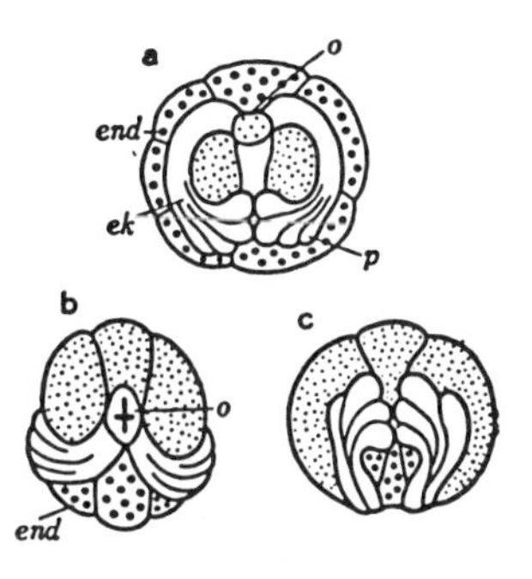

Fig. 7.26 Process of archenteron formation in *Clepsine* (Whitman)

a. dorsal view of embryo showing growth of ectodermal bands; b. frontal view; endodermal cells being overgrown by ectoderm (epiboly); c. posterior side of embryo at same stage. *ek:* ectodermal bands, *end:* endoderm, *o:* mouth, *p:* telocyte of ectodermal band.

the few yolk-containing macromeres in the vegetal hemisphere constitute the endodermal rudiment; the numerous micromeres which will form the ectoderm of the head region lie in the animal hemisphere. Moreover, the teloblasts of the ectodermal bands and the mesoderm mother cells, located in this stage on the two sides at the posterior end of the embryo, are already beginning to bud off daughter cells toward the anterior end to form the ectodermal and mesodermal bands (Fig. 7.25b).

The archenteron of the *Clepsine* gastrula is formed almost wholly by epiboly. As the result of the growth of the ectodermal bands and the vigorous proliferation of the micromeres, the endodermal cells are enveloped into the interior of the embryo as though a cap were being pulled down over them. The growth of the ectodermal bands first carries them around the lateral edges of the micromere area until they reach the anterior side, where they fuse into a girdle encircling the equatorial region (Fig. 7.26a). However, the continued cell proliferation causes the growth in the peripheral parts of the ectodermal girdle and micromere area to turn toward the vegetal pole. The resultant enveloping of the vegetal cell group then proceeds gradually from the posterior toward the anterior side of the embryo, and thereby gives rise to the archenteron (Fig. 7.26b, c). In the meantime, small cells with endodermal characteristics are budded off from the yolk cells which are in the process of being wrapped up in this manner, and these develop into the cellular gut wall surrounding the yolk cells.

In the process of archenteron formation as it occurs in *Nephelis*, the directness with which each of these cleavage patterns follows a predestined course is even more clearly evident. For example, 2*c*, 3*d*, 4*d* and the other cells which were thrown off into the blastocoel give rise to the endodermal layer while they are being enveloped by the overgrowth of the micromeres and the ectodermal bands. Moreover, the macromeres 1*A*, 1*B* and 2*C*, which transformed into yolk cells early in the cleavage process, become a yolk mass which undergoes no further cleavage, and is left behind at the ventroposterior side of the gastrula (Fig. 7.25c, d).

In either case, soon after the endodermal cell layer has enveloped the yolk cells in the blind sac of the archenteron, a small ectodermal invagination, the stomodaeum, forms at the place where the anterior tips of the ectodermal bands have met and fused. This then connects with the archenteron. Once the embryo has reached the gastrula stage, the proliferation of the cells making up the ectodermal and mesodermal bands causes the trunk region to grow, stretching the external form of the embryo into an oval; this continues to elongate and develops toward the vermiform shape of the adult. In many of the species, the embryo begins to shed its egg membrane about this time, and carries on the processes of organogenesis within the cocoon, nourished by the albuminous material with which it s filled; almost the entire later development takes place inside the cocoon.

4. EARLY DEVELOPMENT OF THE ECHIUROIDEA

The Echiuroidea are short, thick worms of an elongate ovoid shape, with a bifurcate *proboscis* at their anterior end. The trunk has neither the parapodia found among the Polychaeta nor a similar metameric organization. Since the peculiar external appearance of these animals resembles that of the Sipunculoidea, the two groups are sometimes lumped together as the Gephyrea. So far as their developmental processes are concerned, however, the annelidan characteristics of the Sipunculoidea are decidedly feeble, while the Echiuroidea are in many points exceedingly close to the Polychaeta. Echiuroidean development, for instance, includes a typical trochophore stage, and body segments appear in that stage, although they disapper after metamorphosis; the proboscis can be considered to be a modification of the polychaet prostomium; the nephridia form metamerically; setae survive on the anterior end of the body. There is, consequently, hardly room for doubt at present that the Echiuroidea are different from the Sipunculoïdea, and they are usually treated as belonging to the Annelida.

(1) FERTILIZATION

Among the Echiuroidea in general, fertilization takes place externally; as in the Polychaeta, eggs and sperm cast out into the water meet and unite. However, in *Bonellia*, which exhibits a peculiar sexual dimorphism, the dwarf male lives like a parasite on the female, and fertilization is consequently also carried on inside the body, the fertilized eggs being laid in an egg mass surrounded by jelly. The egg mass of *B. viridis* is a long, slender ribbon, while that of *B. fuliginosa* is said to be spherical. In *Urechis*, on the other hand, where fertilization is external, the unfertilized eggs can be made to develop by simply adding spermatozoa to them in their sea water medium. As a consequence these eggs have frequently served as a valuable material for experimental embryological studies. The following account will give a summary of the fertilization process in the Japanese species, *Urechis unicinctus.*

The spermatozoon of this species is a typical flagellate cell with a round head portion and a long tail; its total length is 70 μ (Fig. 7.27a). The unfertilized egg (diameter, 105 μ) has usually one, but sometimes two or three large indentations, like a partly deflated rubber ball (Fig. 7.27b, c). The surface is covered with a distinct egg membrane; while the cytoplasm contains some granules, on the whole it is practically transparent and has a faintly yellowish color. In about the center of the egg there is a germinal vesicle 60 μ in diameter in which a nucleolus is clearly recognizable.

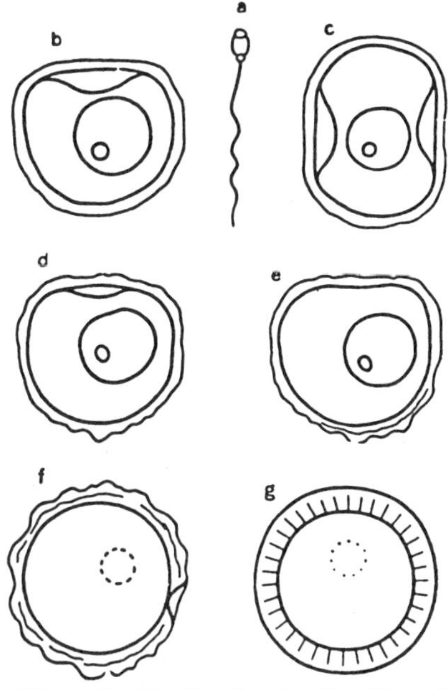

Fig. 7.27 Fertilization in *Urechis unicinctus* (Okawa)

a. spermatozoon; b. unfertilized egg; c. unfertilized egg with two concavities; d. fertilized egg; (5 min. after insemination); e. fertilized egg (10 min.); f. fertilized egg (20 min.); g. fertilized egg (30 min.).

When a spermatozoon attaches to the egg membrane, a small, hyaline fertilization cone forms on the surface of the egg and extends to the membrane. The spermatozoon passes through this process and penetrates the egg surface. As it enters, a sort of deformation appears in the cortical layer of the egg: the surface indentation immediately begins to fill out, and within 10—15 minutes after insemination the egg is completely spherical. This deformation is the first visible change resulting from fertilization in the *Urechis* egg. In the eggs of *Urechis caupo,* Tyler (1932) distinguished three different types in the process by which the fertilized egg becomes spherical. Ohkawa (1952) reports that these three types are found also in *U. unicinctus,* all occurring among the eggs formed by a single animal. In the commonest type, the indentation completely disappears within several minutes after insemination, and the egg maintains a spherical form thereafter (Fig. 7.27d, e). In a few eggs, however, the rounding out of the egg immediately after insemination is incomplete; a few minutes later it forms a more marked indentation, and then becomes spherical again. Or the identation may disappear once completely, reappear several minutes later and then disappear again finally.

Whichever the case may be, the germinal vesicle breaks down by the time the egg has arrived at a spherical condition, and the nucleolus disappears. As the surface of the egg is changing toward a spherical condition, the egg membrane becomes wavy and elevates; only after the egg is completely spherical is this membrane reinforced to form a fertilization membrane (Fig. 7.27f, g). Within the perivitelline space appear numerous fine radiating filaments. After the fertilized egg has attained a stable spherical shape, its diameter is definitely less than that of the unfertilized egg.

Shortly after the completion of the fertilization membrane, or 40—50 minutes after insemination, a slight polarization of the egg cytoplasm takes place, and the first polar body is given off from the somewhat flattened animal pole. The second polar body is formed about 20 minutes later, and the sperm pronucleus which entered earlier unites with the egg pronucleus to complete the process of fertilization some 70—80 minutes after insemination.

(2) CLEAVAGE AND GERM LAYER FORMATION

Cleavage and the formation of the germ layers in the echiuroids closely resemble those of the polychaets. The eggs of *Urechis* and *Thalassema* cleave almost equally in a spiral fashion (Fig. 7.28), while the yolky egg of *Bonellia* undergoes typical spiral cleavage, characterized by an extreme size difference between the macromeres and micromeres. In either case, the first, second and third quartettes are formed, and the fate of each blastomere corresponds exactly to those of the Polychaeta. The group of macromeres which remain after giving rise to the third micromere quartette similarly separate off the mesoderm mother cell in the D quadrant and all become endodermal (Fig. 7.28).

Further development leads in *Urechis* and *Thalassema* to the formation of a coeloblastula; cell proliferation soon produces a thickening at its vegetal pole, which presently invaginates to form the archenteron (Fig. 7.29a). The mesoderm mother cell is located at the posterior side of the blastopore, and the mesodermal bands begin to develop *pari passu* with the process of blastopore closure as it progresses from posterior to anterior (Fig. 7.29b). In contrast to this, the vegetal macromere group in *Bonellia* is overgrown by the ectodermal layer and forms the archenteron with the blastopore still wide open. The gastrulae of both types develop apical tufts of cilia and prototrochal girdles, and hatch as trochophore larvae.

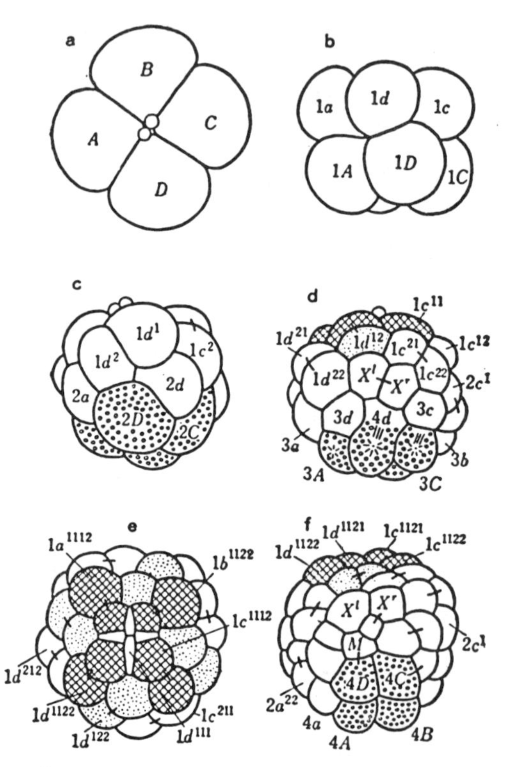

Fig. 7.28 Cleavage in *Urechis caupo* (Newby)
a. 4-cell stage (animal pole); b. 8-cell stage (posterior side); c. 16.-cell stage (posterior side); d. 32-cell stage (posterior side); e. 64-cell stage (animal pole); f. 64-cell stage (posterior side).

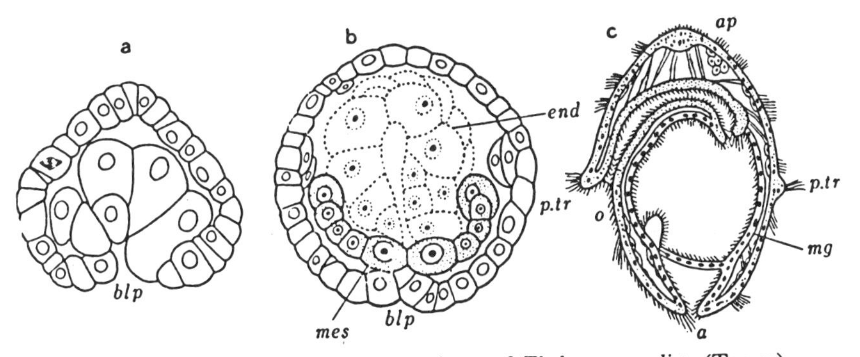

Fig. 7.29 Gastrula and trochophore larva of *Thalassema melitta* (Torrey)

a. median longitudinal section of gastrula at invagination of archenteron; b. horizontal section of gastrula at closure of blastopore; c. median longitudinal section of trochophore larva. *a:* anus, *ap:* apical plate, *blp:* blastopore, *end:* endoderm, *mes:* mesoderm, *mg:* mid-gut, *o:* mouth, *p.tr:* prototroch.

(3) LARVAL STAGE AND METAMORPHOSIS

While the young trochophore which has just begun its free-swimming life is ciliated over practically its whole body surface, the prototroch stands out as a band of particularly well developed cilia. As in the polychaet larva, there is a prominent ventral plate, and conspicuous rows of cilia parallel the mid-line on the ventral side. A long oesophagus extends from the mouth to a spacious mid-gut. This organ is constricted into anterior and posterior chambers, the posterior one leading through a short hind-gut to the anus. The larval mesenchyme cells, already scattered throughout the body, are forming the retractor muscles which connect the digestive tract with the body wall (Figs. 7.29c; 7.30a, b).

The mesodermal bands do not develop very actively at first, but by the time the larva arrives at the height of the free-swimming stage, they show vigorous growth, forming the trunk rudiment and dividing it into mesodermal segments. The rudiments of the protonephridia can be recognized in the young larva as paired cell masses lying ventrally, anterior to the anus; these are pushed forward by the subsequent growth of the mesodermal bands, and develop into conspicuous branching protonephridia (Fig. 7.30a).

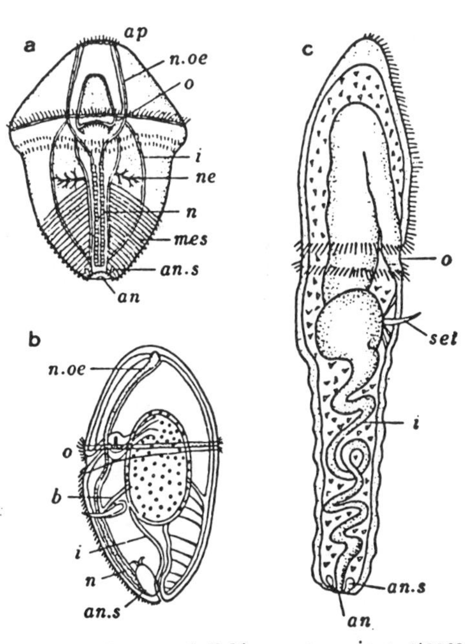

Fig. 7.30 Larva of *Echiurus* at various stages

a. ventral view of trochophore larva (Hatschek); b. median longitudinal section of trochophore larva (Baltzer); c. larva in metamorphosis (right side) (Conn). *an:* anus, *an.s:* anal sac, *ap:* apical plate, *b:* blood vessel, *i:* intestinal duct, *mes:* mesodermal segment, *n:* ventral nerve cord, *ne:* archinephridium, *n.oe:* circumpharyngeal connection, *o:* mouth, *set:* setae.

According to Baltzer's account, the urechid larva spends one to two months as a trochophore well adapted to a free-swimming habit, but eventually it undergoes a striking metamorphosis in both external form and internal structure as it turns to bottom-dwelling. The metamorphosing larvae are about 1.5 mm long, and can often be found in swarms among the bottom layer of plankton.

In the course of metamorphosis, the larva loses its prototroch and increases in length. The winding digestive tract also elongates proportionally. On the surface of the trunk the characteristic urechid *papillae* are beginning to appear, and *ventral setae* are formed on the anterior end of the ventral surface, as well as *caudal setae* around the anus. After metamorphosis the prostomium grows anteriorly to form the cone-shaped *proboscis;* the trunk elongates markedly to produce a vermiform body, and the metameric organization which appeared during the free-swimming period is gradually lost (Fig. 7.30b, c).

5. ORGANOGENESIS

The following account will describe the processes of organogenesis as they occur in the later developmental stages of the Annelida, with the Polychaeta as the center of attention.

(1) SEGMENTS

The body of an adult annelid consists of three principal parts: a prostomium, a trunk and an anal segment or *pygidium.* The term 'prostomium' refers to the front end of the body, anterior to the mouth; this arises from the episphere of the trochophore stage, which persists in a vestigial form even after metamorphosis. The prostomium gives rise to the tentacular cirri, eye spots, and *superpharyngeal* (or *cerebral*) *ganglion.* Since the mesodermal bands do not extend as far forward as this, the space inside the prostomium is filled with mesenchyme cells, and no true coelom is formed.

The main part of the trunk is made up of a succession of metameric segments. After metamorphosis, these segments are gradually extended posteriorly by the growth of the ectoderm cells. In the fully differentiated segments there is a *coelom* surrounded by a *peritonium,* while the posterior part of the trunk consists of an undifferentiated region which will produce new segments.

The pygidium, lying contiguous to this undifferentiated-segment region, constitutes the last segment, which surrounds the anus. Its central cavity consists of the unexpanded blastocoel remnant, and like the prostomium, the pygidium cannot be called a true segment.

The undifferentiated-segment which makes up the posterior end of the trunk gradually forms new segments anteriorly as the growth of the mesoderm brings about cell proliferation. The segment-producing capacity of this undifferentiated region is, in a sense, used up during the course of later development in forming the number of segments highly characteristic of each species, but even in the fully grown adult this part of the body retains a strong, although concealed, capacity for segment formation.

In connection with the process of metamorphosis, the two or three anteriormost segments of the trunk, which characterized the larval stage, lose their setae, change their parapodia into tentacular cirri or branchial filaments, form a *peristomium,* and together with the prostomium make up the *head* of the adult worm.

(2) SETAE, SETAL SAC, PARAPODIA

The possession of setae is a conspicuous characteristic of the phylum Annelida, and serves as an important taxonomical criterion. As the body segments are formed, regularly aligned, bud-like *setal sac* rudiments appear on both sides of the body wall. These develop into setal sacs by an inward proliferation of their cells and invagination into the mesodermal layer. The setae are formed by a secretion of the cells making up the walls of the setal sacs; as they increase in length and thickness they extend outward, developing into the various shapes characteristic of each species. The number of setae found among the Polychaeta is extremely large, and since they are accumulated into tufts on the parapodia, they present a rather complicated appearance, but their basic principle is actually the same as that of the Oligochaeta. The polychaet parapodium is a stumpy swelling of the body wall at the sides of each segment; in most cases it branches into a dorsal *notopodium* and a ventral *neuropodium*. The swimming setae of the larval period, as described above, show nearly as many shapes as there are species and modes of life. These are lost after metamorphosis, however, and the adult setae make their appearance anew in conjunction with the development of the parapodia, forming dorsal and ventral setal tufts on the notopodia and neuropodia, respectively.

(3) SENSE ORGANS AND NERVOUS SYSTEM

There are few highly developed sense organs among the Polychaeta. In many cases the eye spots which were formed during the free-swimming larval stage persist in a vestigial form on the sides of the prostomium after metamorphosis, but the sensory tuft of cilia disappears. In many species, tentacular cirri are newly formed on the head after metamorphosis. These various sensory organs are no more than simple groups of sensory cells differentiated from the ectoderm. *Nereis* and *Alciopa* have rather highly specialized cup-shaped eye spots, and *Hydroides*, *Terebella* and *Sabella*, among others, also form *statocysts*.

The process by which the nervous system develops in the Archiannelida is applicable without change to the Polychaeta. That is, the superpharyngeal ganglion is formed by a thickening of the apical plate ectoderm in the episphere of the trochophore larva; the *ventral nerve cord* rudiment is derived from the *neural plate*, which develops from the so-called *ventral plate*, formed where the two sides of the trunk ectoderm unite ventrally. The anterior end of the neural plate forms the *subpharyngeal ganglion*, and also gives off a pair of processes antero-dorsally, which connect with the superpharyngeal ganglion and complete the formation of the *circumpharyngeal connective* (Fig. 7.31). At least part of the elements making

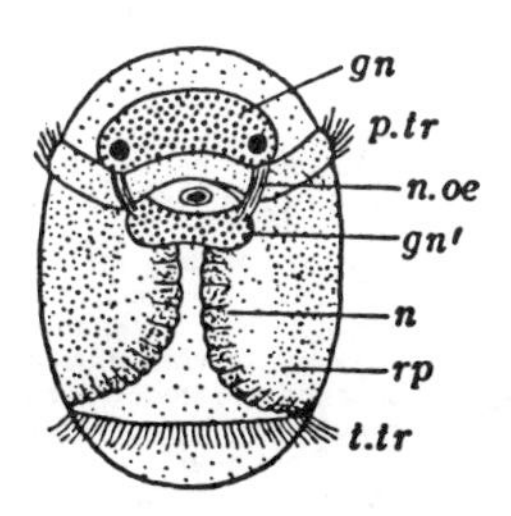

Fig. 7.31 Nerve primordia in trochophore larva of *Capitella* (Eisig)

gn: superpharyngeal, *gn':* subpharyngeal ganglion, *n:* primordium of ventral nerve cord, *n.oe:* circumpharyngeal connection, *p.tr:* prototroch, *t.tr:* telotroch, *rp:* trunk blastema.

up this nerve ring are said to arise from the superpharyngeal ganglion. As development proceeds, the neural plate successively divides — starting at the front end and proceeding backward — into segments corresponding to the body segments, to produce the ventral nerve cord.

Among the urechids also, the cerebral ganglion rudiment appears in the same way, as an ectodermal thickening of the trochophore apical plate, gives rise to a nerve ring encircling the oesophagus, and connects with the ventral nerve cord. This latter arises as a segmented structure along the midline of the ventral plate, but secondarily loses its segmentation (Fig. 7.30b).

The development of the nervous system in the Oligochaeta and Hirudinea is rather different from that of the Polychaeta. The embryos of these groups have no larval organ corresponding to the apical plate, but the four cell rows of the ectodermal bands are conspicuously developed. The most ventrally located of these cells correspond to the polychaet neural plate; these are called the *neuroblasts.* The ventral nerve cord connections, which are precisely the same as those in the Polychaets, are formed by the growth of these neuroblasts, except that, at least in *Tubifex* and *Lumbricus,* the superpharyngeal (cerebral) ganglion consists of only ventral nerve cord elements. In other words, it is believed that the subpharyngeal ganglion sends out such neural elements in the dorsal direction, and these give rise to the superpharyngeal ganglion. On the other hand, there is also a claim that the superpharyngeal ganglion has no connection with the ventral nerve cord in the Oligochaets, rather taking its origin, on the same principle as that governing this process in the Polychaeta, from the ectoderm cells of the prostomium.

(4) EPIDERMAL LAYER

The *epidermis* which covers the head region is made up for the most part of cells derived from the micromeres. The ectodermal bands give rise to the rows of neuroblasts, as well as rows of *myoblasts* which will later become the *circular muscles*; the remaining cells form the epidermis of the trunk. The ectodermal bands are prominent in oligochaet and leech embryos, and the fates of their component cells are well established. Among the four rows of cells which arise at the mid-ventral line on each side of the embryonic body and extend dorsally, the most ventrally located ones become the neuroblast rows (N^l and N^r) and, as described above, constitute the rudiment of the ventral nervous system. Of the six remaining rows, the most dorsally located two, m^{13} and m^{r3}, are believed to form the setal

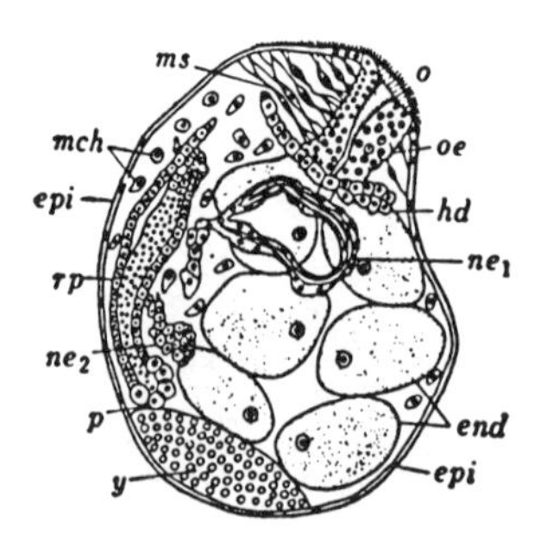

Fig. 7.32 Longitudinal section of *Nephelis* larva (Bergh)

end: endoderm, *epi:* epithelium, *hd:* head blastema, *mch:* mesenchyme cell, *ms:* muscle, ne_1*:* primary archinephridium, ne_2*:* secondary archinephridium, *o:* mouth, *oe:* oesophagus, *p:* telocyte of ectodermal band, *rp:* trunk blastema, *y:* yolk cell.

sacs, and the median rows m^{l1}, m^{r1} and m^{l2}, m^{r2} mainly develop into the circular muscles of the body wall (Fig. 7.23). There are especially prominent circular muscles around the mouth in the leeches (Fig. 7.32).

(5) DIGESTIVE TRACT

The ectoderm continuous with the mouth invaginates to form the *stomodaeum,* which connects with the endoderm of the trochophore archenteron to give rise to the *oesophagus*; the endodermal layer itself differentiates into a large sac-like stomach and an intestine which elongates posteriorly as development proceeds. The hindmost end of the intestine, in turn, connects with a small ectodermal invagination, the *proctodaeum,* and the *anus* opens there. Before the mass of endoderm cells thus bundled in together by the process of epiboly finally differentiates into a digestive tract, however, a complicated process of cellular fusion and separation takes place. In general the large yolk-laden cells lie in the center of the mass and tend to unite with one another while the small peripheral cells without yolk multiply and draw away from each other, becoming organized into an intestinal wall, which gradually digests and absorbs the yolk cells and finally forms the intestinal tube. The wall of the digestive tract is later surrounded by muscle cells and a peritoneal layer differentiated from the mesoderm, and develops into an alimentary tract capable of making chewing and swallowing movements.

(6) MESODERMAL SEGMENTS

The mesoblasts are clearly evident from a very early developmental stage, as they become the teloblasts of the mesodermal bands on the two sides of the embryo, and produce a succession of mesoderm cells. The growth of these bands is a conspicuous characteristic of the annelidan developmental process. As the formation of the trunk progresses, the mesodermal bands begin to form segments one by one from the anterior toward the posterior, separating off a cell mass for

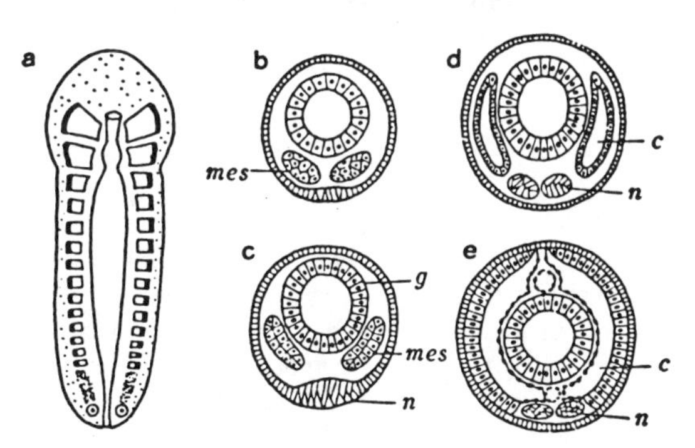

Fig. 7.33 Differentiation of mesodermal band (Korschelt)

a. frontal section of embryo showing differentiation of somites and teloblasts; b. cross-section, showing undifferentiated mesodermal band; c. differentiation of somatic and splanchnic layers; d. appearance of coelom; e. cross-section through well-differentiated somites. *c:* coelom, *g:* gut, *mes:* mesoderm, *n:* ventral nerve cord.

each segment, like boxes in a row (Fig. 7.33a). These mesodermal cell masses can probably be called *somites* (Fig. 7.33b). Close to the teloblasts, moreover, are segments consisting of still undifferentiated mesodermal band tissue, which continue to grow as the trunk develops.

(7) COELOM AND BLOOD VESSELS

After a short time a space appears in the center of each somite, and it differentiates into an inner *splanchnic layer* and an outer *somatic layer* (Fig. 7.33c). Since the spaces of the segments expand around the digestive tract from both sides toward the dorsal surface, the originally small slit develops into a spacious coelom (Fig. 7.33d). The splanchnic layer forms a flat tissue which gives rise to the *peritoneum*, a membranous epithelium surrounding the digestive tract and blood vessels, and the muscles of the digestive tract. The somatic layer develops into cuboidal and columnar tissue, which mostly goes to form the muscular layers making up the body wall.

The cellular layers constituting the anterior and posterior walls of each segment unite with those adjacent to them, to form *septa* (sing. *septum)* between each two segments. The splanchnic layer of each of the paired somites expands to the dorsal and ventral sides of the digestive tract where the two layers meet and fuse, giving rise to a *mesentery* (Fig. 7.33e). This kind of segmented mesoderm is not found in the prostomium. The first somite begins with the peristomium, but the septa between the segments and the mesenteries are likely to undergo secondary modifications in connection with such processes as the formation of the head, and develop in various ways characteristic of the different species.

The blood vessels develop in the space left between the wall of the gut and the splanchnic layer of the somites. This space represents a remnant of the blastocoel, which has been compressed by the expansion of the coelom; endothelial cells appear here, and become organized locally into *primitive vessels*. These structures eventually become united end-to-end above and below the digestive tract to form the *dorsal* and *ventral* blood-vessels. In the same way *lateral blood-vessels*, connecting the dorsal and ventral vessels, also develop in each segment. These major vessels later divide into fine branches to form a completely closed circulatory system.

In the Hirudinea also, the somites develop rapidly, at first forming a large coelomic cavity in each segment, as in the Oligochaeta. By the time the characteristic leech *sinus system*, which corresponds to a circulatory system, is being formed, the coelomic cavities become filled with *parenchyme cells*, and secondarily retrogress. The left and right somites first unite above the ventral nerve cord, leaving a small space which becomes the *ventral sinus*. They then extend around the digestive tract to meet at its dorsal side; the narrow space left here forms the *dorsal sinus*. The septa between the segments disappear in the vicinity of the dorsal and ventral sinuses, but thicken laterally to form masses of mesoderm from which cells separate off and become mesenchyme; this tissue fills in the cavities of the coeloms, at the same time giving rise to other spaces which form the *lateral sinuses* and *botryoidal tissue*.

(8) NEPHRIDIA

Among the Annelida, a distinction is made between the larval and adult nephridia. The larval form is the *protonephridium* found in the trochophore stage; the development of this organ is particularly conspicuous in the leech embryo. In the embryo of *Nephelis* there are two pairs of protonephridia (Fig. 7.32); three pairs occur in *Hirudo* and four pairs in *Aulastoma*. In each case these are reported to arise from a part of the ectodermal band. The adult nephridia, on the other hand, are formed in each segment during the late embryonic period as *segmental organs*. A large number of attempts have been made, particularly in oligochaet embryos, to elucidate the origin of the cells which produce these organs, the *nephroblasts*. The results of these observations, however, have not always agreed very well, with the result that there are at present two opposing views of the matter. One of these maintains that the nephridia have an ectodermal origin: that at least the nephroblasts arise from the ectodermal cells comprising the ectodermal cell line *m'* (Wilson 1899, *Lumbricus*; Vejdowsky 1892, *Rhynchelmis*; Staff 1910, *Criodrilus)*. The other side insists that the nephridia have a mesodermal origin. Bergh (1888, 1899) early expressed this opinion with respect to *Criodrilus*, and

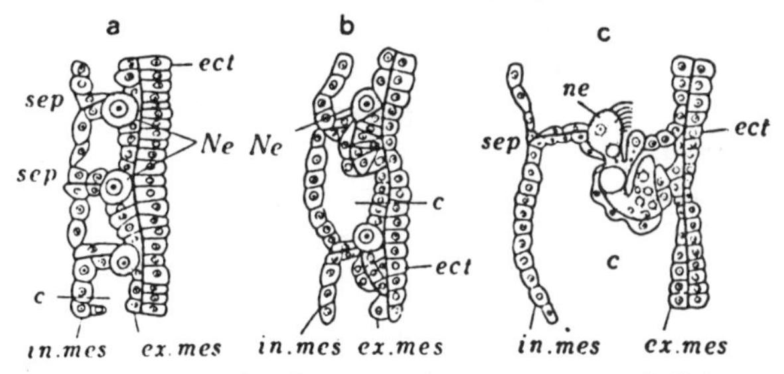

Fig. 7.34 Longitudinally sectioned embryo of *Criodrilus*, showing process of formation of nephridial duct (Bergh)

c; coelom, *ect:* ectoderm, *ex.mes:* somatic layer of mesoderm, *in.mes:* splanchnic layer of mesoderm, *Ne:* nephroblast, *ne:* nephridial duct, *sep:* septum.

it has received support from many later workers (Penners 1923, *Tubifex*; Iwanoff 1928, *Tubifex*, Vanderbrock 1932, *Allolobophora)*. The results of Lillie's (1905) observations on the polychaet *Arenicola* also practically prove that the nephroblasts have a mesodermal origin. In the leeches, moreover, the appearance of the nephroblasts in the cellular layer forming the wall of the somites has been established. At any rate, although the nephroblasts show some special characteristics attributable to species differences, they always appear on the anterior part of the somatic layer in each of the regularly developed segments (Fig. 7.34a). Next they produce a small, winding line of cells which extends backward between the somatic layer of the mesoderm and the ectoderm; this row of cells finally develops into a fine tubule which connects with the ectoderm to form a *nephridium* (Fig. 7.34b, c).

In the Echiuroidea, the protonephridia degenerate at metamorphosis; in their place there appears in the posterior part of the ventral side an *anal sac* which has three pairs of nephridial rudiments, and will later form the genital ducts and nephridia (Fig. 7.30b).

(9) REPRODUCTIVE SYSTEM

The formation of the reproductive system takes place last among the processes of organogenesis, only attaining a conspicuous degree of development after the vegetative organs are nearly completed. To be precise, however, it must be recognized that the formation of this system actually begins with the appearance of the *primordial germ cells.* These cells are separated from the line of body cells at a very early developmental stage, but fail to cleave further for some time thereafter, lying among the mesoderm cells of the splanchnic layer so that it is often difficult to trace their whereabouts. About the time the peritoneum reaches its full development, the primordial germ cells make their appearance in special regions of the septa, for the first time becoming clearly evident as the gonadal rudiments. Since these cells have a characteristic vesicular nucleus, it is possible (with considerable effort) to distinguish them from the mesoderm cells.

Malaquin (1925) reports that the first appearance of the primordial germ cells in the polychaet *Salmacina* can be recognized in the gastrula stage: i.e., the pair of mesodermal teloblasts (M^r and M^l), before they begin to produce the mesodermal bands, first separate off the primordial germ cells.[17] These paired cells lie in a bilaterally symmetrical position near the blastopore, but the rapid multiplication of the mesoderm cells derived from the teloblasts makes it difficult to distinguish them. However, they take no part in the mesodermal proliferation, dividing for the first time during the metatrochophore stage, and dispersing into the somites, which are just developing at this time.

Tracing the course of the primordial germ cells in the oligochaets gives essentially the same result. According to the observations of Penners and Stablien (1930) on *Tubifex* and *Limnodrilus,* the primordial germ cells, which have separated from the mesodermal teloblasts, first undergo two divisions and then come to lie between the mesodermal bands and the endoderm cells (Fig. 7.35a). As development proceeds and the segments are being formed one by one, the primordial germ cells become buried in the septal tissue between the segments from the eight to the twelfth (Fig.7.35b).

These primordial germ cells, which are lodged in certain regions of the septa, form the gonads in these positions. Meyer (1929) states that in *Tubifex,* the difference between the testis and ovary rudiments can first be seen about the time of the 27-segment stage. In this stage the testis rudiment consist of one germ cell in the septum, while the ovarian rudiment is already made up of two germ cells (Fig. 7.35c). As development progresses still further, these cells continue to divide *in situ,* producing a number of germ cells, while at the same time the clearly different cells of the septum surround them to form a small lump of immature gonad (Fig. 7.35d-f).

17

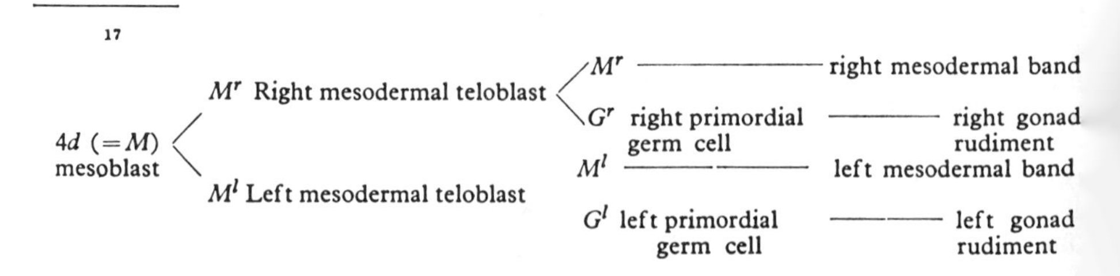

Among the Polychaeta, the gonads are formed on a larger scale, appearing in each segment; as the breeding season approaches, the germ cells undergo a rapid proliferation, in many cases being released into the body cavity and maturing there.

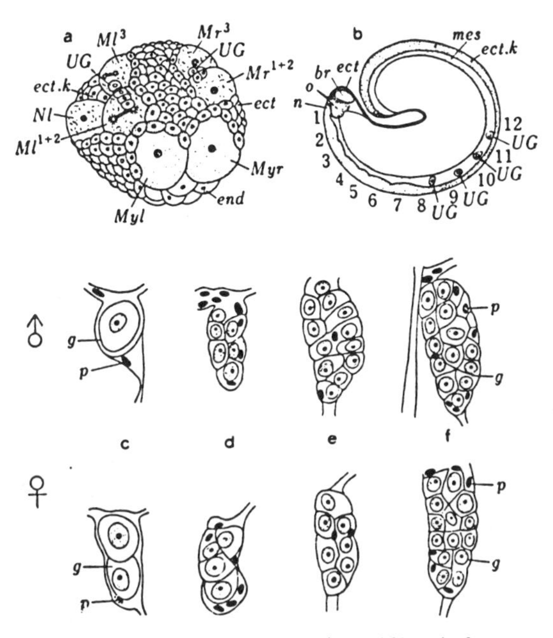

Fig. 7.35 Formation of gonad in *Tubifex rivolorum*

a. appearance of primordial germ cells at gastrula stage; b. orientation of primordial germ cells in late embryonic stage; (a, b. Penners) c~f. process of gonad formation. (Meyer) *br*: cerebral ganglion *ect*: ectoderm, *ect.k*: ectodermal band, *end*: endoderm, *g*: germ cell, *mes*: mesoderm, *Ml,Mr.Myl,Myr*: myoblasts, *n*: ventral nerve cord, *Nl*: left neuroblast, *o*: mouth, *p*: septal cell. *UG*: primordial germ cell, (1~12: somite numbers)

The genital ducts and other accessory organs are formed last in the course of organogenesis, usually arising as modifications of some part of the peritoneum. In not a few cases the nephridia serve also as genital ducts during the breeding season.

(10) SEX DIFFERENTIATION IN BONELLIA

The echiuroid *Bonellia* is famous for its exaggerated sexual dimorphism. The male is extremely small (1—3 mm in length), and all its internal structure except the reproductive system is degenerate like that of a parasite (Fig. 7.36♂). The female, on the other hand, has a large oval trunk, to the anterior end of which is attached a slender forked proboscis about twice as long as the trunk; its total body length of 15 cm is no less than magnificent when compared with that of the male (Fig. 7.36♀). According to Baltzer, the female requires two years to reach maturity, while the male is mature in one to two weeks. In other words, this is a most extreme example of a dwarf male.

The rather yolky egg of *Bonellia* develops to a gastrula with the archenteron formed by epiboly. Hatching takes place relatively late, and no typical protrochophore stage appears during the course of its development. From the gastrula stage it goes directly to an elongated oval metatrochophore stage and hatches, to begin its free-swimming life. Such a free-swimming larva has a pair of eye spots and a mass of endoderm cells for an archenteron, but its mouth and anus have not yet opened. There is of course still no difference between the sexes in this stage (Fig. 7.36a).

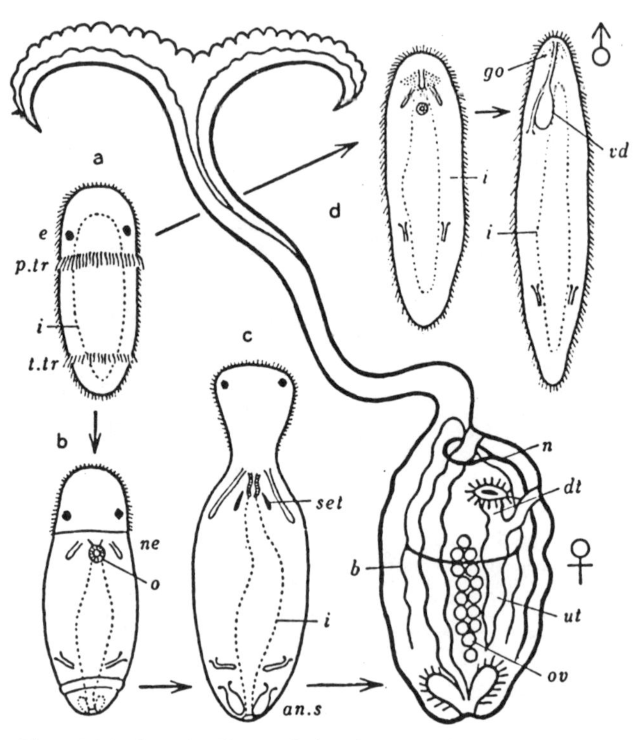

Fig. 7.36 Sexual differentiation in *Bonellia viridis* (Baltzer)
a. sexually immature larva; b,c. larva destined to become female; d. larva destined to become male. (♂) mature male, x20. (♀) mature female, natural size.
an. s: anal vesicle, *b:* blood vessel, *dt:* oviduct, *e:* eye spot, *go:* gonal, *i:* intestine, *n:* nerve, *o:* mouth, *ov:* oocyte, *p. tr:* prototroch, *set:* setal sac, *t. tr:* telotroch, *ut:* uterus, *vd:* seminal vesicle.

Some of these sexualy undifferentiated larvae become caught on the proboscis of the adult females, and continue to exist there as symbionts, differentiating into males. This development toward maleness is determined within several days after they are caught by the female, and these larvae then move along the proboscis until they reach the mouth cavity, oesophagus, or some other region in the anterior end of the digestive tract. Here they develop exactly as though they were parasitic on the female, and reach maturity; they then leave the female digestive tract, and this time invade the uterus. Several mature males are usually found in the uterus of a mature female, as well as a large number of developing males in the anterior part of the digestive tract. Those larvae, on the other hand, which undergo the latter part of their development as free-living organisms unconnected with an adult female differentiate into females.

Although the eye spots and ciliary bands of these male-determined larvae begin to degenerate, both their external morphology and internal structure show evidence of their metamorphosis into males (Fig. 7.36b). After about two days, gonadal tissue which includes spermatocytes accumulates in the anterior part of the mesodermal layer surrounding the body cavity. Very shortly thereafter, mature spermatozoa can be found in the body cavity. The mature males within the uterus exhibit a fully mature organization, including a so-called *seminal vesicle* (Fig. 7.36♂), which is said to be formed as a modification of the (ectodermal) stomodaeum when this structure degenerates, and is therefore considered to have a unique embryonic origin, observable only in the male of *Bonellia*. The *sperm duct*, which collects the mature spermatozoa within the body cavity into the seminal vesicle, is probably of mesodermal origin.

The undifferentiated larvae destined to be females remain unchanged for one or two days after hatching, but eventually their metamorphosis into females begins. In the first place, the enlargement of the body cavity brought about by the growth of the trunk causes the mesodermal tissue to begin to disperse, and a distinct demarcation gradually arises between the trunk proper and the anteriormost region formed by the prostomium. At the same time, the rudiments of the various organs begin to become obvious, and the cilia on the body surface suddenly degenerate (Fig. 7.36c). This larva which has completely lost its cilia takes up a bottom-dwelling habit, and the trunk lengthens as its development continues, while the prostomium elongates even more markedly to produce the proboscis (Fig. 7.36c). In this stage the digestive tract is completed so that it can take in nutrition, and three pairs of nephridia as well as almost all the other organs are formed. Setal sacs arise on the ventral side of the trunk at the anterior end, and the proboscis begins to bifurcate in the region where the eye spots are located. The time when the ovary, oviduct, uterus and other organs can be seen differentiating from the mesoderm is definitely the final stage in the developmental process.

REFERENCES

Abe, N. 1943: The ecological observation on *Spirorbis* especially on the post-larval development of *Spirorbis argutus* Bush. Sci. Rep. Tohoku Imp. Univ., Biol., **17**.

Aiyar, R. G. 1931: Development and breeding-habits of a Polychaet (*Marphysa*). Jour. Limm. Soc. London, 37.

Berg, R. S. 1888: Zur Bildungsgeschichte der Exkretionsorgane bei *Criodrilus*. Arb. Zoöl. Inst. Wurzburg, **7**.

——— 1890: Neue Beiträge zur Embryologie der Anneliden. I. Zur Entwicklung und Differenzierung des Keimstreifens von *Lumbricus*. Zietschr. wiss. Zoöl., 50.

——— 1891: Neue Beiträge zur Embryologie der Anneliden. II. Die Schichtenbildung im Keimstreifen der Hirudineen. Zeits. wiss. Zoöl., **52**.

——— 1899: Nochmals über die Entwicklung der Segmental-organe. Zeitschr. wiss. Zoöl., **66**.

Child, C. M. 1900: The early development of *Arenicola* and *Stermaspis*. Arch. Entw. -Mech., **9**.

Costello, D .P. 1949: The relation of the plasma membrane, vitelline membrane and jelly in the egg of *Nereis limbata*. Jour. Gen. Physiol. **32**.

Delsman, H. C. 1916: Eifurchung und Keimblattbildung bei *Scoloplos armiger*. Tijschr. Ned. Dierk. Vereen, **2**.

Eisig, H. 1898: Zur Entwicklungsgeshichte der Capitellined. Mitth. Zoöl. Stat. Neapel, **13.**

Iwanoff, P. P. 1928: Die Entwicklung der Larvalsegmente bei den Anneliden. Zeitschr. wiss. Biol Abt. A, **10.**

Izuka, A. 1903: Observations on the Japanese Palolo, *Ceratocephale osawai.* Jour. Coll. Sci. Tokyo, **17.**

——————— 1908: On the breeding habit and development of *Nereis japonica* n. sp. Annot. Zool. Jap., **6.**

Kawamura, T. 1951: Activation of *Ůrechis unicinctus* egg. Jap. J. Exptl. Morphology, **7.** (in Japanese)

Kagawa, Y. 1952a: Refertilization phenomenon in acid sea water-treated *Ůrechis unicinctus* eggs. Nat. Sci. Rep., Col. Lib. Arts. Tokushima Univ., **2.** (in Japanese)

——————— 1952b: Effect of changes in salt concentration of sea water on fertilization of coelomic eggs of *Ceratocephale.* Bull. Exptl. Biology, **3.** (in Japanese)

——————— 1954: Histological observation of process of heteronereid formation in *Tylorrhynchus.* Nat. Sci. Rep. Col. Lib. Arts, Tokushima Univ., **4.** (in Japanese)

Kleinberg, N. 1880: The development of the earth-worm, *Lumbricus trapezoides.* Quart. J. Mic. Sci., **19.**

Kowalevsky, A. 1871: Embryologische Studien an Würmern und Arthropoden. Mem. Acad. St. Pétersbourg, **16.**

Lefévre, P. G. 1945: Certain chemical factors influencing artificial activation of *Nereis* eggs Biol. Bull., **89.**

Lieber, A. A. 1931: Zur Oogenese einige Diopatra-Arten. Inaugural-Dissertation, Eberhard-Karls-Univ., Tübingen.

Lillie, F. R. 1911a, b: Studies of fertilization in *Nereis,* I. & II. Jour. Morph., **22.**

——————— 1912a, b: Studies of fertilization in *Nereis,* III. & IV. Jour. Exptl. Zool., **12.**

Lillie, R. S. 1905: The structure and development of the Nephridia of *Arenicola cristata* Stimpson. Mitt. Zoöl. Stat. Neapel., **17.**

Malaquin, M. A. 1925: La ségrégation, au cours de la ontogénèse, de deux cellules sexuelles primordiales, souches de la lignée germinale, chez *Salmacina dysteri* (Huxley). C. R. Acc. Sci. Paris, **180.**

Mead, A. D. 1897: Early development of marine annelids. Jour. Moiph., **13.**

Meyer, A. 1929: Die Entwicklung der Nephridien und Gonoblasten bei *Tubifex rivulorum* Lam., nebst Bemerkungen zum natürlichen System der Oligochäten. Zeitschr. wiss. Zoöl., **133.**

Miyoshi, W. 1939: Swarming of *Ceratocephale osawai* at Konuma. Bot. and Zool. **7.** (in Japanese)

Monroy, A. 1948: A preliminary approach to the physiology of fertilization in *Pomatoceros triqueter* L. Arkiv Zoöl., **40.**

Newby, W. W. 1932: The early embryology of the echiuroid, *Ůrechis.* Biol. Bull., **63.**

Nomura, M. 1930: Annelida Iwanami Biology Course. Iwanami Shoten, Tokyo (in Japanese)

Ohfuchi, S. 1938: On the cocoon of *Drawida hattamimizu* Hatai. Zool. Mag., **50.**

Oinuma, S. 1926: Regularly occurring swarming in *Ceratocephale osawai.* Jour. Okayama Med. Assoc., **432.** (in Jananese)

Oishi, M. 1930: On the reproductive processes of the earthworm, *Pheretima communissima* (Goto et Hatai). Sci. Rep. Tohoku Imp. Univ., Biol., **5.**

Okada, K. 1940: A study on the spawning cycle in *Arenicola cristata* Stimpson. Ecological Review, **6.** (in Japanese)

——————— 1941: The gametogenesis, the breeding habits, and the early development of *Arenicola cristata* Stimpson, a tubicolous polychaete. Sci. Rep. Tohoku Imp. Univ., Biol., **16.**

——————— 1952: Swarming of Japanese palolo. Bull. Exptl. Biol., **3.** (in Japanese)

——————— 1953: Activation of mature spermatozoa of *Arenicola* by means of vital staining Bull. Exptl. Biol., **3.** (in Japanese)

——————— 1955: On the fine structure of the jelly and egg membrane in *Ceratocephale.* Proc. Embryol. Symp., Sugashima M.M.B.S. (in Japanese)

Okada, Yô K. 1933: Two interesting syllids, with remarks on their asexual reproduction. Memoirs Coll. Sci., Kyoto Univ., Ser. **8.**

——————— 1950: A note on the so-called Japanese palolo. Annot. Zool. Jap., **23.**

Okuda, S. 1938: Notes on the spawning habit of *Arenicola claparedii* Levinsen. Annot. Zoöl Jap., **17.**

——————— 1939: Lunar periodicity in reproduction. Bot. & Zool., **7.**

Penners, A. 1922: Die Furchung von *Tubifex rivulorum* Lam. Zoöl. Jahrb., Anat, Ontog., **43**

——————— 1923: Die Entwicklung des Keimstreifs und die Organbildung bei *Tubifex rivulorum* Lam. Zoöl. Jahrb., Anat, Ontog., **45.**

———— 1929: Entwicklungsgeschichtliche Untersuchungen an marinen Oligochäten. I. Furchung, Keimstreif, Vorderdarm und Urkeimzellen von *Pelescolex benedeni* Undkem. Zeitschr. wiss. Zoöl., **134.**

———— 1930: Entwicklungsgeschichtliche Untersuchungen an marinen Oligochäten. II. Furchung, Keimstreif und Keimbahn von *Pachydrilus* (*Lumbricillus*) *lineatus* Müll. Zeitschr. wiss. Zoöl., **137.**

———— 1933: Über Unterscheide der Kokons einiger Tubificiden. Zoöl. Anz., **103.**

Penners, A. & A. Stablein 1930: Über die Urkeimzellen bei Tubificiden (*Tubifex rivulorum* Lam. und *Limnodrilus udekemianus* Claparede). Zeitschr. wiss. Zoöl., **137.**

Sato, H. & M. Tanaka 1932: Early development of *Urechis unicinctus* (von Drasche). Zool. Mag.‘ **44.** (in Japanese)

Schleip, W. 1914: Die Furchung des Eies der Russelegel. Zoöl. Jahrb., Anat., **37.**

Segrove, F. 1941: The development of the serpulid *Pomatoceros triqueter* L. Quart. Jour. Mic. Sci., **82.**

Shearer, C. 1911: Development and structure of trochophore of *Hydroides uncinarus* (*Eupomatus*). Quart. Jour. Mic. Sci., **56.**

Spek, J. 1930: Zustandänderungen der Plasmakolloide bei Befruchtung und Entwicklung des *Nereis*- Eies. Protoplasma, **9.**

Staff, F. 1910: Organogenetische Untersuchungen über *Criodrilus laccuum* Hoffen. Arb. Zoöl. Inst. Wien, **18.**

Sugiyama, M. 1955: Embryological Experiments on Invertebrata. Nakayama Shoten, Tokyo (in Japanese)

Sukatschoff, B. 1900: Beiträge zur Entwicklungsgeschichte der Hirudineen. I. Zur Kenntnis der Urnieren von *Nephelis vulgaris* und *Aulostomus galo*. Zeitschr. wiss. Zoöl., **67.**

———— 1903: Beiträge zur Entwicklungsgeschichte der Hirudineen. II. Über Furchung und Bildung der embryonalen Anlagen bei *Nephelis vulgaris*. Zeitschr. wiss. Zoöl. **73.**

Svetlov, P. G. 1923: Sur la segmentation de l'oeuf chez *Bimastus constrictus*. Bull. de l'Inst. recherch. biol. Univ. Perm., **1.**

———— 1924: Sur la segmentation de l'oeuf chez *Rhynchelinis limosella*. Bull. de l'Inst. recherch. biol. Univ. Perm., **2.**

———— 1928: Untersuchungen über die Entwicklungsgeschichte der Regenwürmer. Tran. Laborat. Zoöl. Stat. biol., Sebastopol. Ser. **2.**

Tannreuther, G. W. 1915; The embryology of *Bdellodrilus philadelphicus*. Jour. Morph., **26.**

Torrey, J. C. 1903: The early embryology of *Thalassema mellita* (Conn) Ann. N.Y. Acad. Sci., **14.**

Treadwell, A. L. 1901: The cytogeny of *Podarke obscura*. Jour. Morph., **17.**

Tyler, A. 1932: Changes in volume and surface of *Urechis* eggs upon fertilization. Jour. Exptl. Zool., **63.**

Tyler, A. & H. Bauer 1937: Polar body extrusion and cleavage in artificially activated eggs of *Urechis caupo*. Biol. Bull., **73.**

Tyler, A. & J. Schultz 1932: Inhibition and reversal of fertilization in the eggs of the echiuroid worm, *Urechis caupo*. Jour. Exptl. Zool., **63.**

Whitman, C. O. 1878: Embryology of *Clepsine*. Quart. Jour. Mic. Sci., **18.**

Wilson, D. P. 1932: The development of *Nereis pelagica* Linnaeus. J. Mar. Biol. Assoc. U. K., **18.**

———— 1935a: The development of the Sabellid, *Branchiomma vesiculosum*. Quart. Jour. Mic. Sci., **78.**

———— 1936b: The development of *Audouina tentaculata* (Montagu). J. Mar. Biol. Assoc. U. K., **20.**

Wilson, E. B. 1889: Embryology of the earthworm. Jour. Morph., **3.**

———— 1892: The cell-lineage of *Nereis*. Jour. Morph., **6.**

von Wistinghausen, C. 1891: Untersuchungen über die Entwicklung von *Nereis dümerilii*. Mitth. Zoöl. Stat. Neapel, **10.**

Woltereck, R. 1903: Beiträge zur praktischen Analyse der *Polygordius*-entwicklung. Arch. Ent-mech., **18.**

Yamamoto, T. 1947: Optimal salt concentration for fertilization and development of the brackish water polychaete, *Ceratocephale osawai*. Ecological Review, **1.** (in Japanese)

———— 1952: On the cortical changes, especially in the cortical granules, during fertilization and artificial activation in *Ceratocephale*. Bull. of Exptl. Biol. **2.** (in Japanese)

PROSOPYGII

Chapter 8

I. PHORONIDEA

INTRODUCTION

When J. Müller (1846) first discovered the *actinotrocha,* which is the larval form of this group, he thought it represented an independent species, and named it *Actinotrocha branchiata.* Müller regarded the *ventral pouch* in the abdomen of this creature as its reproductive organ. Later, Schneider (1862) was of the opinion that this animal was a larval form belonging to the Asteroidea. It was A. Kowalevsky (1867) who first showed it to be the larva of a phoronidean. At the same time, however, he mistook the preoral lobe for the posterior portion of the body, and thought that the blastopore as such changed over to become the anus.

Studies concerned with the development of the Phoronidea have been published by Caldwell (1882—1885), Masterman (1898), de Selys Longchamps (1902) and others. Recently Rattenbury (1954) has published especially noteworthy results on this problem, particularly regarding the cell-lineage. With respect to Japanese species, there is the famous investigation of Iwaji Ikeda (1901) on *Phoronis ijimai.* Some points remain obscure, however, concerning the origin of the mesoderm and the type of cleavage, which is higly important in connection with determining the phylogenetical position of this group.

1. FERTILIZATION

Although there are some dioecious species such as the Californean *Phoronopsis viridis* among this group, most of the species are hermaphroditic. The gonads develop on the coelomic epithelium near the ventral blood vessels. The mature gametes are first released into the coelomic cavity and then projected outside the body through the nephridia. The problem of the place where fertilization occurs has been the subject of some discussion, leading to a general belief that the eggs are fertilized as they are spawned through the nephridia. However, Rattenbury's work (1953) on *Ph. viridis* and that of Kume (1953) on *Phoronis australis* from Misaki have clearly shown, supporting a similar opinion once expressed by Kowalevsky (1867), that fertilization takes place in the coelomic cavity.

This fertilization in the coelomic cavity occurs late in the course of the first maturation division and as long as the eggs stay in the coelom, the process of maturation remains at a standstill. As the eggs are discharged from the body, the polar bodies are formed and the eggs begin to cleave. In order to obtain the eggs of the phoronids and observe the process of development, it is only necessary to prick the body wall with a needle, so that the eggs suspended in the coelom can exude to the outside. 50—100 eggs can be obtained from one individual. According to Ikeda, the spawning season of *Ph. ijimai* lasts from November to the following June or July. In *Ph. australis* it is possible to collect the eggs at least in spring and summer.

2. CLEAVAGE

After the fertilized eggs are liberated from the body cavity through the nephridia, in many species they remain attached with mucus to the tentacular ring for a certain definite period; in *Ph. ijimai* and *Ph. australis* they develop here into larvae with two pairs of tentacles. This is, however, not the case with *Ph. viridis* (Rattenbury 1954), *Phoronis mulleri* (de Selys Longchamps 1902) and *Phoronis architecta* (Brooks and Cowles 1905); these species spawn their eggs freely into the sea water. In general, the eggs of species in the latter group are smaller than those of the former (0.06 mm in diameter in *Ph. viridis*, 0.13 mm in *Ph. australis*).

Soon after being released from the adult body, the fertilized egg forms the first polar body and then the second, and in most cases the first polar body also divides at this time. During this period there can be seen, encircling the polar bodies on the surface of the egg of *Ph. australis*, a ring of cilium-like processes which keep up a flickering movement. In some extreme cases these processes may even appear on the polar bodies (Kume 1953); although their significance is not known, they remain until at least the 4-cell stage (Fig. 8.1a).

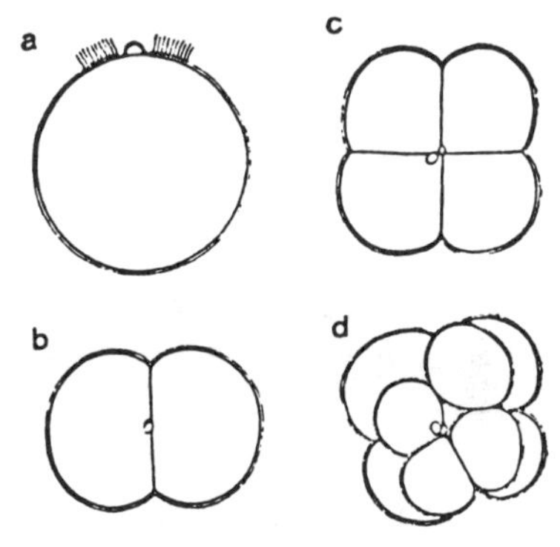

Fig. 8.1 Cleavage in *Phoronis australis* (Kume)

The blastomeres are equal in size through the second cleavage, the first unequal division taking place at the third cleavage in such a way that the four cells of the vegetal half are somewhat larger than those of the animal half (Fig. 8.1d). Since the dividing nucleus lies in the animal hemisphere in *Ph. australis*, cleavage begins at the animal pole, as in the medusan *Spirocodon* and some others, and advances toward the vegetal pole, while fine protoplasmic processes arise at the animal pole side and bind the two blastomeres together.

Although the matter has been investigated in a large number of species, no extensive report has been made with respect to the course of cleavage after this stage. However, according to the recent study of Rattenbury (1954) on *Ph. viridis*, this species clearly undergoes spiral cleavage (Fig. 8.2). While this result is highly important, further studies should be made to determine whether all phoronidean cleavage conforms to this type.

In *Ph. viridis,* equal cleavage continues to the third division, resulting in eight blastomeres of approximately the same size. But this cleavage takes place in a clockwise direction, so that the four cells of the animal side (first micromeres *1a, 1b, 1c, 1d)* are interposed between the four cells of the vegetal side (first macromeres *1A, 1B, 1C, 1D).* Similar cleavages follow each other according to the general plan of spiral cleavage, alternately clockwise and counter-clockwise, the divisions of all the blastomeres taking place almost synchronously. The sixth cleavage, however, which results in the 64-cell stage, goes irregularly to some extent.

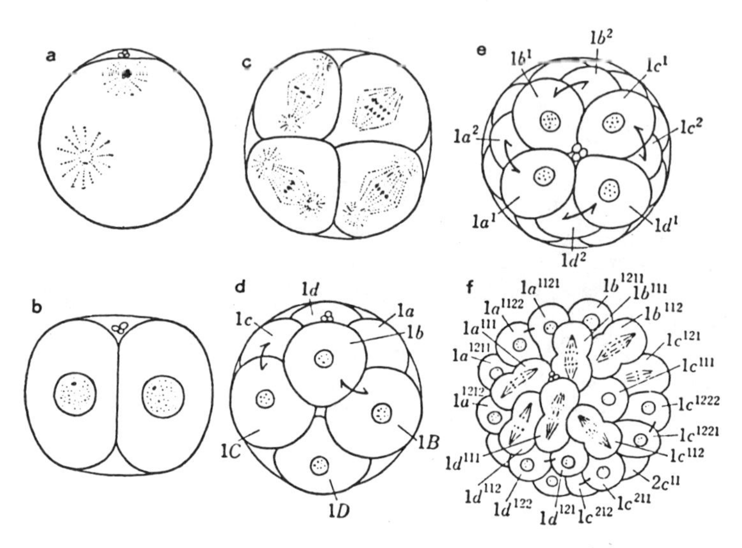

Fig. 8.2 Cleavage in *Phoronopsis viridis* (Rattenbury)

a. fertilized egg; b. 2-cell stage; c. 4-cell stage; d. 8-cell stage; e. 16-cell stage; f. seventh cleavage.

Thus the fourth macromeres, which have just given off the fourth micromeres, are larger than the other blastomeres. At this stage the blastocoel is already extensive, and contains suspended fragments of cytoplasm, a phenomenon frequently encountered among phoronid species.

From the seventh cleavage onward, the daughter blastomeres derived from the first and second micromere groups divide more rapidly than the others, and the macromeres divide at a correspondingly slower rate, so that it becomes difficult to follow the cell lineage in these stages. The embryo of about 13 hours, undergoing the eighth cleavage, bears a cilium on each of its cells. A particularly rapid rate of cleavage is shown by the blastomeres descended from the first and second micromere groups, which are located in the region that will form the dorsal side of the embryo; these consequently become the very smallest of the blastomeres.

3. GASTRULATION

Before long, the ventral surface of the embryo flattens. This portion, which is called the *ventral plate*, is composed of descendants of the relatively large fourth macromeres and the relatively small fourth micromeres. Later on, this region will differentiate into mesodermal and endodermal cells. On this account Ikeda (1901) called it the *meso-entoblast*. Soon several cells at the anterior edge of the ventral plate sink into the blastocoel; these are mesoderm cells, most of which are derived from the descendants of the *A* blastomere (Fig. 8.3a,b).

Next the ventral plate begins to invaginate. The archenteron formed by this invagination, the tip of which consists mainly of the large cells descended from the fourth macromeres, stretches diagonally toward the ventro-posterior part of the embryo. The blastopore is large and circular when it is first formed, but it gradually becomes ovoid. This may be due to the marked proliferation undergone by the ectodermal cells making up the dorsal side of the embryo. (Fig. 8.3c.d).

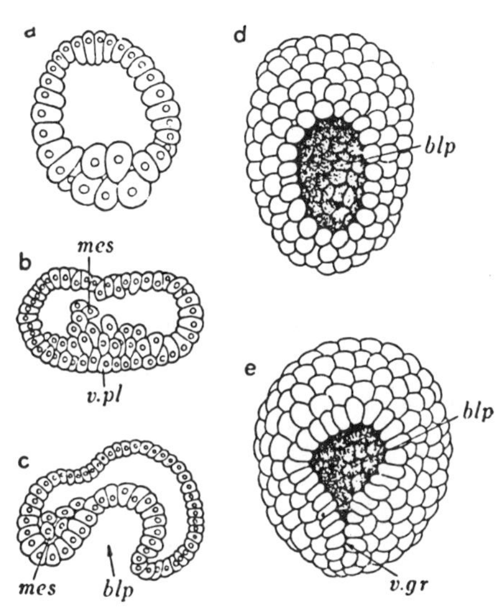

Fig. 8.3 Gastrulation in *Phoronopsis viridis* (Rattenbury)

blp: blastopore, *mes*: mesodermal cell, *v.gr*: ventral groove, *v.pl*: ventral plate.

There are various conflicting theories with regard to the origin of the mesoderm in the Phoronidea. At one time Caldwell (1885) and Masterman (1900) strongly advocated an enterocoelic theory, whereas Roule (1900), Ikeda (1901), de Selys Longchamps (1902) and Rattenbury (1954) fail to agree with them.

The blastopore, after first changing its shape from round to ovoid, next gradually closes from the posterior side. Where the blastopore has closed, its trace can be seen as the *ventral groove*, which finally has a small triangular pore remaining open at its anterior tip; this later changes into a lateral slit. While these changes in the shape of the blastopore are going on, the mesodermal cells are proliferating inside the embryo, at the same time becoming clearly independent from the endodermal cells, and spreading throughout the blastocoel in the form of mesenchyme, part of which encloses the archenteron wall. Among these mesodermal cells, there are none which can be specially identified as teloblasts, nor any which have changed over from ectoderm in the manner of ectomesoblasts.

4. FORMATION OF THE ACTINOTROCHA LARVA

With the completion of gastrulation, the embryo gradually takes on the *actinotrocha* form peculiar to this group, as the result of unequal growth and differentiation in various of its parts. In the first place, eight ectodermal cells ($1q^{111}$ and $1q^{112}$) located at the top of the embryo, just below the polar bodies, begin to elongate and produce long cilia, and an *apical plate* is formed with these cells as the chief components. The ectoderm between this apical plate and the neighborhood of the anterior tip of the blastopore gradually spreads, until it extends over the blastopore like an umbrella, and thus forms the *preoral lobe* characteristic of the larvae in this group. (Fig. 8.4 a-e). As the preoral lobe elongates above the blastopore, the entrance to the latter gradually shifts forward, and a new larval *mouth* is formed at this point. The region between mouth and blastopore differentiates into the ectodermal *oesophagus*.

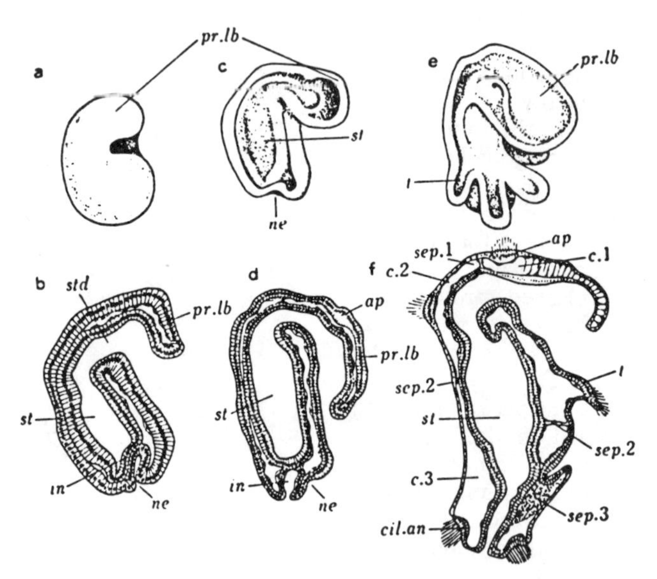

Fig. 8.4 Formation of actinotrocha larva (Ikeda)

a. larva beginning to expand preoral lobe; b. longitudinal section of embryo at more advanced stage; c. larva at stage in which preoral lobe protrudes and stomodaeum is formed; d. longitudinal section of embryo at above stage; e. embryo with 3 pairs of tentacles; f. longitudinal section of embryo at above stage. *ap:* apical plate, *c.1:* preoral body-cavity, *c.2:* collar-body-cavity, *c.3:* trunk body-cavity, *cil.an:* perianal ciliated belt, *in:* intestine, *ne:* nephridial pit, *pr.lb:* preoral lobe, *sep.1:* preoral septum, *sep.2:* postoral septum, *sep.3:* ventral mesentery , *st:* stomach, *std:* stomodaeum, *t:* larval tentacle.

A ciliary band densely crowded with especially long cilia is formed around the preoral lobe; this corresponds to the prototroch of the trochophore. A similar ciliary band is also formed below the mouth and obliquely around the posterior part of the embryo. This is thought to correspond to the metatroch. Tentacles appear along this metatroch; at first a pair of them, on either side of the mid-ventral line, and later two more pairs, one on the dorsal part of each side.

As growth proceeds, the embryo elongates downward, causing a stretching of the archenteron, at the blind tip of the lower part of which a cell-cord made up of several cells is formed. This cord transforms into a tubular *intestine*, and the archenteron itself will become the larval *stomach* (Fig. 8.4 b, d). The end of the intestine soon reaches the lower tip of the embryo, where an opening forms, to become the *anus*.

Well before the opening of the anus, a small pit has already appeared in the ectoderm of the lowest part of the embryo, at the ventral side of the intestine. This is the *nephridial pit,* which deepens as development proceeds, and later elongates in a tubular form, its tip eventually bifurcating into right and left branches. When this occurs, the opening, which had been single at first, divides into two.

Ganglionic cells differentiate from the base of the apical plate, and mesenchyme cells derived from the mesoderm proliferate, especially in the cavity included within the preoral lobe; it is not yet possible, however, to establish clearly the formation of a body cavity. When the larva remains in the parental tentacular ring, it begins its free-living stage in this condition. According to Hiraiwa (1925), the larva has a strong capacity for regeneration at this stage.

5. FREE-LIVING STAGE LARVA

The trunk of the free-swimming larva continues to elongate downward, and the number of tentacles gradually increases. The full number, which depends upon the species, ranges from 8 to 24 pairs. As the trunk elongates, a new ciliary ring is formed around the anus. This *perianal ciliated belt,* which probably corresponds to the telotroch of the trochophore, plays an important role in larval locomotion at this stage.

It is not known how many days are required before the larva can metamorphose; at any rate, during this free-swimming period it forms the various organs necessary for beginning its benthic existence. The formation of the *larval tentacles* is first completed, and later the ectoderm at their bases begins to thicken, and differentiates into the primordia of the *adult tentacles.* Although these primordia are generally formed at the bases of the larval tentacles, there are some species in which they form on the tentacular walls. A more important structure is the *ventral pouch,* also called the *metasoma,* which begins to be formed in a relatively early embryonic stage, as a thickening of the ventral ectoderm below the tentacular ring. This invaginates into the interior of the body to form a sac which grows around the digestive tract as development proceeds Mesodermal cells become attached around the sac; these later differentiate into muscle tissue.

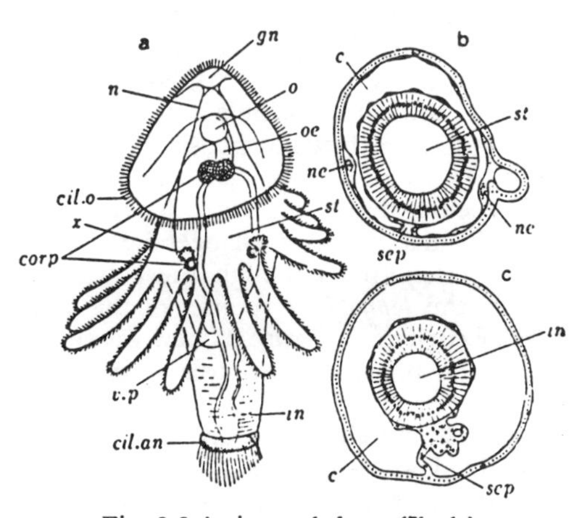

Fig. 8.5 Actinotroch larva (Ikeda)

a. larva with 14 tentacles; b. cross-section through stomach of above larva; c. cross-section of same larva through intestine. *c:* trunk body-cavity, *cil.an:* perianal ciliated belt, *cil.o:* preoral ciliated belt, *corp:* blood corpuscle, *gn:* ganglion *in:* intestine *n:* nerve fiber *ne:* nephridial duct, *o:* mouth, *oe:* oesophagus, *sep:* ventral mesentery. *st:* stomach, *v.p:* ventral pouch, *x:* rosette cells of nephridial duct.

The nervous system of the larva is poorly developed, as is also that of the adult, and no extensive differentiation of nervous tissue takes place, except for *nerve fibers* which grow out around the single *ganglion* formed below the apical plate. The branches of the excretory organ elongate gradually, and at their coelomic ends, at the two sides of the stomach, flower-like groups of cells are formed. Muscular development is also poor; the *retractor muscles* at the sides of the oesophagus arise to the left and right of the ganglion and attach to the body wall near the tentacular ring. These muscles are used to move the preoral lobe.

Various complicated problems concerning the formation of the body cavity and the blood-vascular system remain unsolved. The mesoderm cells proliferate in the blastocoel, some of them spreading throughout the cavity in the form of mesenchyme cells, while others become epidermal and adhere to the body wall and the wall of the digestive tract, especially in the trunk region, where the body cavity becomes completely covered with this coelomic epithelium. Anterior and posterior septa are later formed across this coelomic cavity, dividing it into three regions: the *preoral, collar,* and *trunk coeloms*. The preoral coelom, which is the cavity of the preoral lobe, is filled with mesenchyme cells, and the *preoral septum*, which divides it from the collar coelom, is incomplete. The collar coelom is separated from the trunk coelom by the *postoral septum*, a complete membrane formed posterior to the tentacular ring. The trunk coelom is thus an independent body-cavity. Within this coelom, a *ventral mesentery* runs longitudinally along the ventral side of the intestine, further dividing the ventral portion of the coelom into right and left cavities. The ventral pouch is formed along this mesentery.

Masterman (1898, 1900), who considered that the larva of *Phoronis* consisted of three body regions, preoral, collar and trunk, each of them including an independent body cavity, postulated that the Phoronidae, like *Balanoglossus*, belong to the Hemicorda. Ikeda (1901), however, has suggested that it is difficult to regard the preoral and collar coeloms as independent, and maintained that the body cavities before and after metamorphosis have different origins. According to Ikeda, it is only the trunk coelom that remains as such after metamorphosis, since the preoral coelom disappears with the resorption of the preoral lobe, and the collar coelom shrinks and then changes into the *ring-vessel* of the adult. Further, he was of the opinion that although the collar coelom sends branchesi nto the larval tentacles, the *lophophore cavities* of the adult are not formed by the retention of the collar coelom, but arise as new cavities from the bases of the tentacles.

In the pre-metamorphosis larva, the only blood vessel is the rudiment of the *dorsal vessel*, which is situated in the center of the dorsal wall of the stomach, between the muscle tissue and the endoderm. However, some cells which can be identified as precursors of the blood corpuscles are already found scattered among the mesoderm cells of the early larva; these proliferate and eventually differentiate into two pairs of *corpuscle masses* at the sides of the stomach (Fig. 8.5).

6. METAMORPHOSIS

The metamorphosis of the actinotrocha larva takes place rapidly, requiring, according to Ikeda, only 15 to 20 minutes. As MacBride also records, "on one occasion we left an advanced Actinotrocha in a watch-glass, left the room for a short time, and on coming back found a young *Phoronis*". (1914, p. 383)

The enlarged ventral pouch, folded within the trunk, is showing signs of evaginating; contractions of the trunk muscles cause it to evert so that the whole length protrudes to the outside (Fig. 8.6). Since the tip of the pouch is connected to the digestive tract, all of the latter posterior to the stomach is drawn into the everted pouch. This causes the intestine to become U-shaped and the anus, which until now was located at the posterior end of the body, is shifted to a position near the tentacular ring. At this time the preoral lobe with the larval tentacles is absorbed into the larval stomach, and the perianal ciliated belt also proceeds to degenerate.

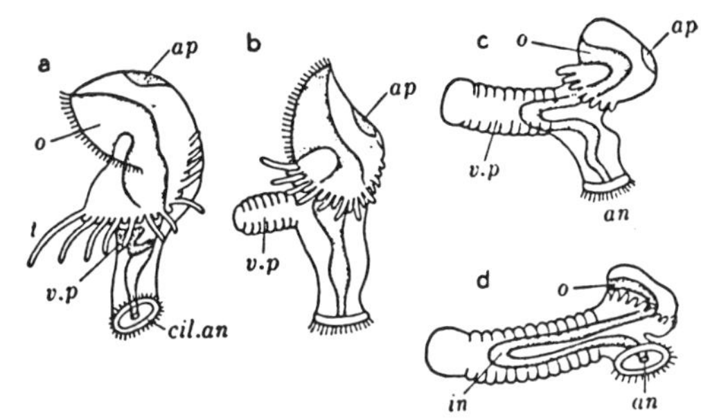

Fig. 8.6 Metamorphosis of actinotrocha larva (Korschelt u. Heider)

ap: apical plate, *an:* anus, *cil.an:* perianal ciliary belt, *in:* intestine, *o:* mouth, *t:* tentacle, *v.p:* ventral pouch.

Within the body, in the meantime, the masses of blood corpuscles at the sides of the stomach disperse rapidly and begin to circulate in the dorsal vessel which is now pulsating. The collar coelom shrinks and changes into the *ring vessel,* to complete the formation of the closed *blood-vascular system* of the adult *Phoronis.* The flower-shaped cell masses at the ends of the excretory organs disappear, and the excretory tract comes to open directly into the body cavity.

REFERENCES

I. Phoronidae

Hiraiwa, K. 1925: Regeneration in Actinotrocha. Zool. Mag., **37.** (in Japanese)

Ikeda, I. 1901: Development, structure and metamorphosis of Actinotrocha. Jour. Coll. Sci. Tokyo Univ., **13.**

——— 1902: On the occurrence of *Phoronis australis* Haswell near Misaki. Annot. Zool. Jap., **4.**

———1903: On the development of sexual organs and of their products in *Phoronis.* Annot. Zool. Jap. **4.**

Korschelt u. Heider 1936; Vergleichende Entwicklungsgeschichte d. Tiere. Jena.

Kume, M. 1953: Some observations on the fertilization and the early development of *Phoronis australis.* Nat. Sci. Rep. Ochanomizu Univ., **4.**

MacBride, E. W. 1914: Text-book of Embryology, London.

Rattenbury, J. C. 1953: Reproduction in *Phoronopsis viridis.* The annual cycle in the gonads, maturation and fertilization of the ovum. Biol. Bull., **104.**

——— 1954: The embryology of *Phoronopsis viridis.* J. Morph., **95.**

II. BRYOZOA

INTRODUCTION

All the Bryozoa[1] (or Polyzoa) form colonies, each member *zooid* of which is very small, consisting of a *polypide* having a tentacular crown, and a sac-shaped *cystid*. The cystid is composed of an *endocyst* and its secretory product, the *ectocyst*. This group is divided into the Gymnolaemata and the Phylactolaemata.

I **Gymnolaemata**

The lophophore is circular and the tentacles grow in a single ring. There is no epistome, and no fusion takes place among individual zooids. Most of these animals are marine forms, with the exception of the two fresh-water genera *Paludicella* and *Victorella*.
Ctenostomata: *Alcyonidium, Flustrella, Farrella, Paludicella, Victorella*
Cheilostomata: *Membranipora, Flustra, Bugula, Schizoporella, Retepora*
Cyclostomata: *Crisia, Diastopora, Frondipora, Tubulipora, Lichenopora*

II **Phylactolaemata**

The lophophore is horseshoe-shaped, with the tentacles borne peripherally. There is always an epistome covering the mouth. The individual zooids are all of the same shape; these fuse together and have a common body cavity. All are fresh-water forms.

Fredericella, Plumatella, Stephanella, Hyalinella, Gelatinella, Lophopus, Lophopodella, Rectinatella, Cristatella

The methods of reproduction found among these forms are various: larvae are produced by sexual reproduction on the one hand, and colonies are formed by asexual budding on the other. In some cases these species also proliferate asexually by means of dormant buds of a peculiar type.

Embryological studies of this group were begun in the latter half of the nineteenth century by Nitsche, Metschnikoff and others, and during the twenty years around 1900 major articles were contributed by Barrois, Vigelius, Prouho, Braem, Oka, Kraepelin, Calvet, Robertson, Pace, Seeliger and others. Among later investigations may be mentioned the work of Marcus. The general outline of the developmental processes has been clarified by these investigators, but many problems concerning the details remain unsolved.

[1] This term formerly included the Entoprocta as well as the Ectoprocta, but is now limited to refer only to the Ectoprocta.

1. SEXUAL REPRODUCTION

(1) FERTILIZATION AND THE STORAGE OF EGGS

The individual zooid is generally hermaphroditic. The testes usually arise as botryoidal structures from the epidermal layer of the *funiculus*[2], while the ovaries are formed from the oral side of the inner epithelium[3] of the endocyst. Both spermatozoa and eggs are liberated into the body cavity, where fertilization takes place.

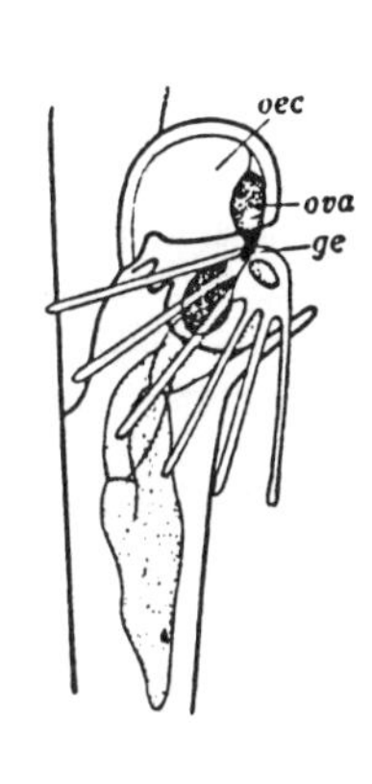

Fig. 8.7 Migration of ova in *Bugula avicularia* (Gerwerzhagen)

oec: ooecium, *ova:* ovum, *ge:* genital pore.

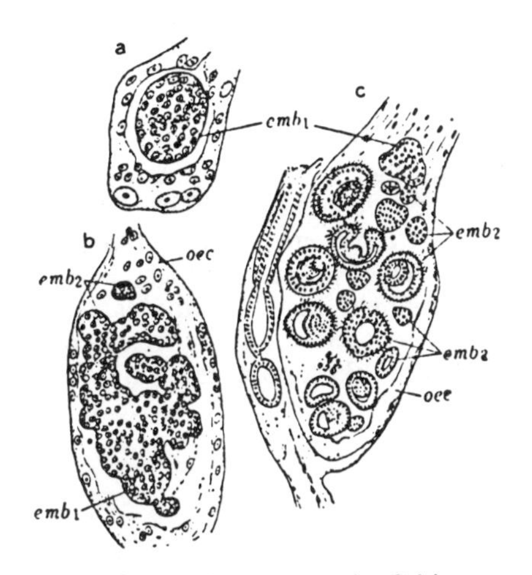

Fig. 8.8 Polyembryony in *Crisia* (Robertson)

a. ooecium containing morula-like primary embryo; b. ooecium with constricted primary embryo; c. ooecium with many secondary embryos. *oec:* ooecium, *emb*$_1$*:* primary embryo, *emb*$_2$*:* secondary embryo.

This is usually by self-fertilization, but in species in which the gonads mature at different times, sperm of other zooids or of another colony must be brought in. In such cases the spermatozoa are said to enter through the *inter-tentacular organ*, which opens to the outside of the body, or through small tubules at the tips of the tentacles. There still remain a number of obscure points in connection with the mechanism of fertilization in these species.

The fertilized eggs are liberated to the outside, or retained within the parent body, where they begin to develop. In species with a well-developed *ooecium*, the fertilized eggs are transferred through the genital pore to the ooecium, where they are stored for long periods (Fig. 8.7). In the Cyclostomata, large numbers of embryos are sometimes found in the ooecium. This is due to *polyembryogeny*; i. e., after an embryo has reached the morula stage, it divides into several, each of which develops independently into a *secondary embryo*. A repetition of this process results in the formation of *tertiary embryos*. For this reason more than one hundred embryos in various developmental stages can sometimes be found

[2] A cord of tissue connecting the blind end of the stomach with the body wall. With respect to the origin of this structure, see the sections dealing with budding and statoblast germination.

[3] See sections on budding and statoblast germination for description of its formation.

within one ooecium (Fig. 8.8). In the Phylactolaemata, the fertilized egg is taken into a cystic ooecium located among the zooids; here it continues its development, the embryo receiving nutrient substances through a *placenta* formed between the wall of the ooecium and the embryo (Fig. 8.12).

(2) DEVELOPMENT OF THE EMBRYO

a. Gymnolaemata

In eggs with small amounts of yolk, such as those of *Membranipora, Alcyonidium, Bugula* and *Paludicella,* the cleavage furrow divides the whole egg almost equally. The 16-cel stage consists of four regular rows of four cells each. This plate of cells then divides horizontally so that the 32-cell stage is composed of two cell-plate layers (Fig. 8.10 b). Soon these cell layers bulge outward, a cavity is formed between them and the embryo grows into a convex lens-shaped blastula. At the next stage, granules appear in the four central blastomeres of the vegetal side, making them easily distinguishable from the other cells; they then move inward and completely fill the blastocoel. This constitutes the origin of the endoderm. The blastopore resulting from this process is soon occluded by the surrounding cells; the surface where it was formed is called the *oral face.*

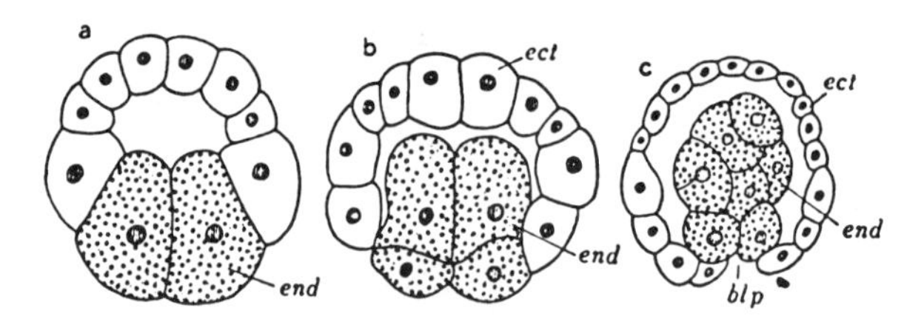

Fig. 8.9 Gastrulation in *Flustrella hispida* (Pace)
a. blastula with blastomeres of unequal size; b. early gastrula; c. late gastrula.
blp: blastopore, *ect:* ectoderm, *end:* endoderm.

The yolky *Flustrella* egg cleaves unequally, so that there are four large and four small blastomeres at the 8-cell stage. At the 32-cell stage a large blastocoel is formed, and the gastrula results from the ensuing migration of the large cells into the blastocoel (Fig. 8.9). Gastrulation is thus accomplished by a combination of invagination and epiboly.

In *Alcyonidium,* as the development proceeds further, two characteristic cells appear at the two sides of the median plane; these are the *mesoblasts,* which will later form the larval musculature. No such cells have been observed in *Membranipora,* but it is assumed that they appear in a similar way, since muscle tissue is formed in the larva. Prouho (1892) did not definitely determine the origin of these cells, but he suggested that they are probably derived from the micromeres at the oral face of the embryo — in other words, from the ectoderm — since they lie in front of the mass of endoderm cells.

During this period the embryo as a whole is laterally compressed so that it takes a conical form, with the oral face as the base. The ectoderm at the tip of

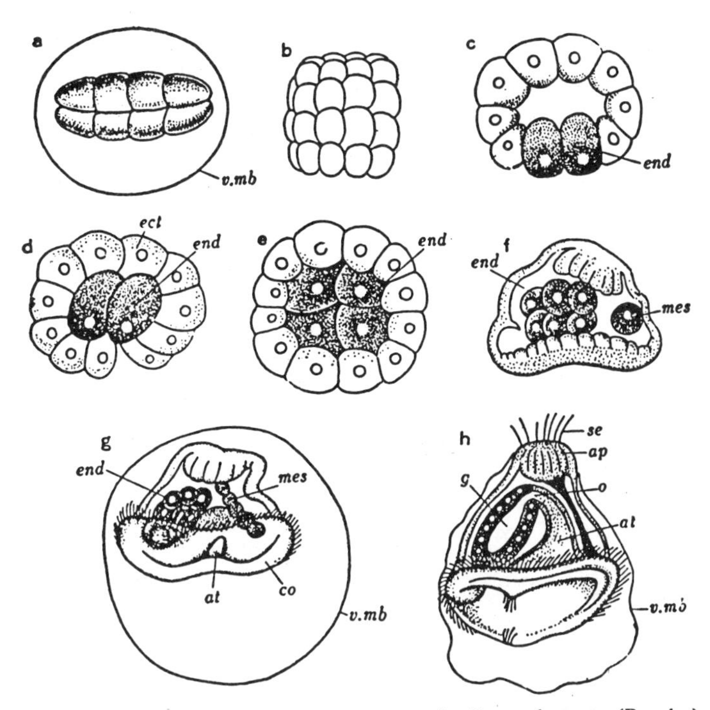

Fig. 8.10 Early embryonic development in Gymnolaemata (Prouho)

a. 8-cell stage; b. 32-cell stage; c. blastula; d. gastrula; e. gastrula (from vegetal pole); f~h. formation of cyphonautes larva. *ap:* apical organ, *at:* atrium, *co:* corona, *ect:* ectoderm, *end:* endoderm, *g:* gut, *mes:* mesoderm, *o:* mouth, *se:* sensory hair, *v.mb:* vitelline membrane.

the cone thickens markedly, to form an *apical organ,* and a large cavity develops in the oral face; this deepens and pushes the mass of endoderm cells posteriorward, as the first step in the formation of the *stomodaeum.* At this stage the embryo still adheres to the vitelline membrane by the margin of the oral face and the apical organ, but a considerable degree of morphological change is about to take place. Sensory hairs appear on the apical organ, the ectoderm around the oral face thickens into a *mantle,* the cells of which produce cilia to form a locomotory organ. When the mantle is folded, this portion is called the *mantle fold.* The whole of the oral face within the encircling mantle becomes depressed to form the *atrium,* the deepest part of which becomes the stomodaeum. The ciliated cell-ring is called the *corona* (Fig. 8.10).

An embryo which has acquired such a structure breaks out of the vitelline membrane and enters the free-swimming stage. It is only after this time that the formation of the *larval stomach* begins: the endodermal cell-mass which lies in the blastocoel becomes hollow, and then this cavity opens into the atrium when the tip of the stomodael invagination makes contact with its wall. Part of the endoderm which has become cylindrical constitutes the *digestive tract,* and its opening develops into a *mouth.* Another ectodermal invagination, the *proctodaeum,* which occurs in the posterior part of the oral face, elongates until it reaches the stomach; the opening formed at the point of contact is the *anus.* As soon as the digestive tract has thus been completed, the larva begins to take in food. The mesoderm cells anterior to the stomodaeum proliferate and form a muscular *string* connecting the apical organ with the oral face. This is the rudiment of the larval *dorsal muscles*. Two thin, triangular *shells* secreted by the body wall envelop the larva; between them the apical organ can be seen. Among the Ctenostomata and the Cheilostomata, a larva which has reached this stage of development is generally called a *cyphonautes* (Fig. 8.11).

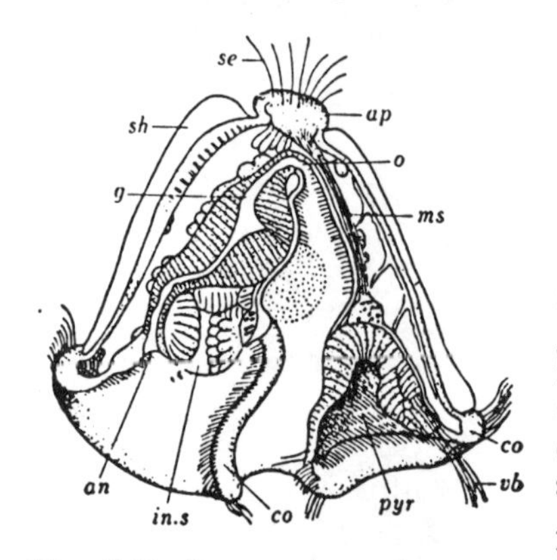

Fig. 8.11 Cross-section of cyphonautes (Prouho)

an: anus, *ap:* apical organ, *co:* corona, *g:* gut, *in.s:* internal sac, *ms:* muscular system, *o:* mouth, *pyr:* pyriform organ, *se:* sensory hair, *sh:* shell, *vb:* vibratile plume.

Two kinds of organs are subsequently formed in the cyphonautes larva. One is the *pyriform organ*, which arises as an ectodermal invagination in front of the mouth. It consists of columnar cells, each of which has a glandular structure and carries on secretion. The surface of the organ is covered with fine cilia. Although these cells are contiguous to the cell layers of the ciliary ring, their origin is different and their union with them is a secondary phenomenon. Anterior to this organ there is a group of long cilia bent like a hook, called the *vibratile plume*; the muscular string described above attaches at the base of this structure. Kupelwieser (1905) has suggested that the vibratile plume may be a kind of sensory organ with which the larva selects a place to attach when it enters the benthic stage.

The other kind of organ is the *internal sac* or *sucker*, which is formed anterior to the anus. According to Kupelwieser, this results from an ectodermal thickening. Prouho held a different opinion, suggesting that it originates from an ectodermal invagination anterior to the anus. This structure is small at first, but develops gradually during the larval period into a large, sac-like organ with a thin upper wall and a glandular lower portion that secretes a mucous substance. These larval organs play important roles at the time of metamorphosis.

b. **Phylactolaemata**

The fertilized egg is taken into the ooecium, which is located between the individual zooids, and development proceeds there. The investigations of Kraepelin (1892) show that in *Plumatella*, the number of blastomeres first increases by equal cleavage, forming a blastula. Later the cells in the upper part of this embryo proliferate into its interior and fill the blastocoel, so that a coeloblastula results. These cells then proceed to shift to the periphery and line the inner surface of the ectoderm, giving rise to an embryo with two cell layers and a central cavity. The inner of these layers is mesodermal; the endoderm is believed to degenerate. The cavity which has appeared in the embryo therefore corresponds to the *body cavity*.

Surrounded by the wall of the ooecium, the embryo has been elongating. Its outer layer develops a band of enlarged cells, and the cells of the ooecium lying opposite to them also protrude to meet them and form the characteristic placenta, through which the embryo takes in nutrition (Fig. 8.12 d, e). Somewhat later a fold encircling the body wall of the embryo at the upper edge of the placenta forms a *ring fold*; this corresponds to the mantle fold which appeared in the larva of the Gymnolaemata. Above the ring fold the outer layer of the embryo invaginates and gives rise to a *polypide rudiment*. Before this primary polypide is completely formed, the outer layer invaginates again and forms a *secondary polypide*. At this stage periodic contractile movements begin near the polypide rudiments and propagate to the two ends of the embryo.

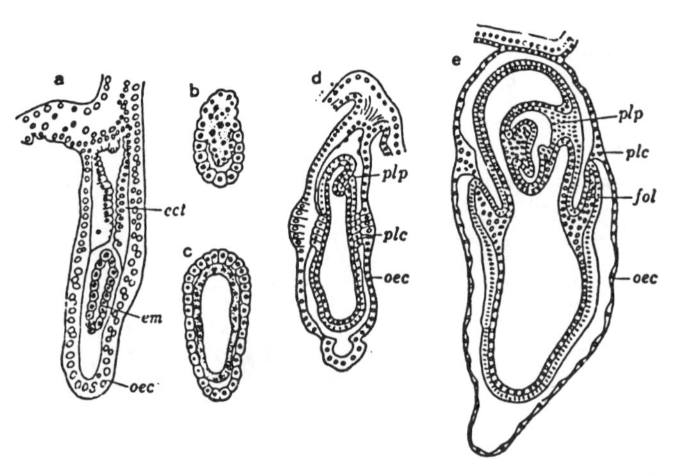

Fig. 8.12 Early embryonic development in *Plumatella* (Kraepelin)
a. blastula in ooecium; b. proliferation of cells into blastocoel; c. embryo with two layers of cells; d. formation of polyp rudiment and placenta; e. fully developed embryo. *ect:* ectodermal cell, *em:* blastula, *fol:* ring fold, *oec:* ooecium *plc:* placenta, *plp:* polyp rudiment.

The polypide rudiments differentiate the various polypide organs by a process similar to that which will be described below in connection with budding. The fully developed embryo is nearly oval and already bears two polypides. Entering the larval stage, it swims out through openings left where zooids have died or in the attachment region of the ooecium. In *Lophopodella carteri*, the wall of the ooecium inverts as the larva escapes, protrudes outside of the body, and swells up like a balloon. It later shrinks gradually into a rod-like shape, and finally disintegrates (Oka and Oda 1948).

(3) LARVA

Swimming larvae of various species can be caught with a plankton net. It is also a simple matter to obtain large numbers of *Bugula* larvae: if colonies with well-developed ooecia are collected, kept in the dark for a while, and then suddenly exposed to light, the larvae all begin to swim simultaneously (Lynch 1947).

As was mentioned above, the cyphonautes larva of the Gymnolaemata differs markedly from the larva of the Phylactolaemata; however, comparison of various larvae shows that the gap between these two forms is filled by a graded series of developmental types and their associated morphological patterns.

The larva of *Membranipora* is a typical cyphonautes (Fig. 8.11), and conforms in many respects to the fundamental structural formula peculiar to the bryozoan larva. The three germ layers are established early in its development, and the larval organs derived from the respective layers are formed as the larva begins its free-swimming life. The larva as a whole has a conical shape, with two shells. At its top there is an apical organ, and its lower margin consists of the ciliary ring. Having a complete digestive tract, it can take in the food it requires. Muscles and nerves also develop, and the larva has a pyriform organ and an internal sac. These structures are analogous in many respects to those of the trochophore and veliger. Viviparous larvae, on the other hand, show many morphological variations which are correlated with the degree of viviparity.

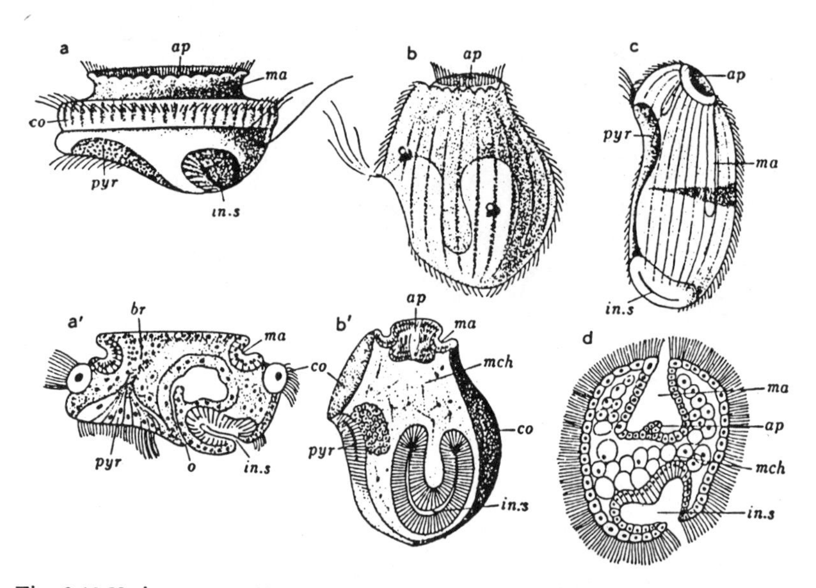

Fig. 8.13 Various types of larvae among Gymnolaemata (Barrois, Harmer, Vigelius)
a. larva of *Alycyonidium*; a'. cross-section of larva shown in a; b. larva of *Bugula*; b'. cross-section of larva shown in b; c. larva of *Serialaria*; d. larva of *Frondipora*. *ap:* apical organ *br:* brain *co:* corona *in.s:* internal sac, *ma:* mantle cavity, *mch:* mesenchyme, *o:* mouth *pyr:* pyriform organ.

The larva of *Alcyonidium* (Fig. 8.13 a, a') loses its shells, its apical organ expands and becomes disc-shaped, and the mantle turns inward, forming a deep depression called the *mantle hole*. The oral face curves outward, and the pyriform organ and sucker protrude. The digestive tract is extremely degenerate, consisting of only a *stomodaeum* and a *stomach*.

A further degree of degeneration is observed in the barrel-shaped larva of *Bugula* (Fig. 8.13b, b'). This larva has a ciliated ring composed of a thick cellular layer covered with cilia, which encircles the body parallel to the main axis. The digestive tube has completely disappeared, and the larval interior is filled with mesenchyme. The pyriform organ and internal sac, however, persist. In the *Seria-*

laria larva (Fig. 8.13c), the portion bearing the ciliated ring becomes greatly extended longitudinally, and the larva as a whole acquires a very long and slender form.

In the *Frondipora* larva (Fig. 8.13d), the tendency toward degeneration is still more conspicuous: the body is ovoid, there is no trace of a digestive tract, and the apical organ has lost its sensory cells and become a deep depression. The part corresponding to the ciliary ring has spread to cover the whole body, and furthermore has been replaced by small cilia-bearing cells. Although the larva lacks a pyriform organ, the internal sac still persists.

The larvae of the Phylactolaemata are spherical or ovoid; they already have well-developed buds, and differ substantially from the larvae of the Gymnolaemata. The observation that they lack any digestive tract is borne out by consideration of the process by which they develop. The larva is completely covered with a broadly extended ring fold, and moves sideways as well as up and down by means of the fine cilia on its surface.

The differences among these larval types may be said to depend on the degree of degeneration in the endodermal region, which in turn is related to the environment in which the larva develops. That is, increasing degeneration of the digestive tract is generally observed in the following order: Ctenostomata→Cheilostomata→Cyclostomata, and in this same order the other larval organs also become simpler in structure. The Phylactolaemata, which constitute the extreme projection of this tendency, begin asexual reproduction earlier than do the other groups. When their larvae are being compared with the larvae of these groups, therefore, the discussion should be based on the cyphonautes, which has the most fundamental structure.

There are some Phylactolaemata in which no larva has yet been discovered.[4] In such species it seems probable that the sexual mode of reproduction has in effect degenerated and been replaced by an asexual process.

(4) METAMORPHOSIS

The length of the larval free-living period is correlated with the degree of development of the digestive tract. In such a form as the cyphonautes, which has a well-developed digestive tract, there is a long free-living period of from a week to a month, during which the development of the incomplete organs proceeds. In larvae without a digestive tract, however, this period is generally short: in the larvae of the Phylactolaemata it is finished within from 2—3 hours to half a day.

The cyphonautes which has reached the end of its free-swimming life begins to sink with its oral face downward, and when it arrives at the bottom, it creeps about and selects a place to attach by moving the vibratile plume of its pyriform organ. Soon a strong contraction of the muscles causes the larval body to shrink; this brings about the extrusion of the internal sac, and the larva becomes fixed by it to the substrate. The layer of cells constituting this internal sac, or sucker, expands and its outer margin turns upward and makes contact with the edge of the mantle, while the larva itself becomes flattened and acquires two covering shells (Fig. 8.14). After this, disintegration of the larval organs begins; the ciliary ring, digestive tract, pyriform organ, muscle fibers and so on break up and their

[4] Both the ovary and testis of *Pectinatella gelatinosa* develop, but its larva has not yet been observed (Oka 1890; Oka and Oda 1948).

fragments accumulate in the lower part of the larva, where they are eventually consumed by phagocytes.

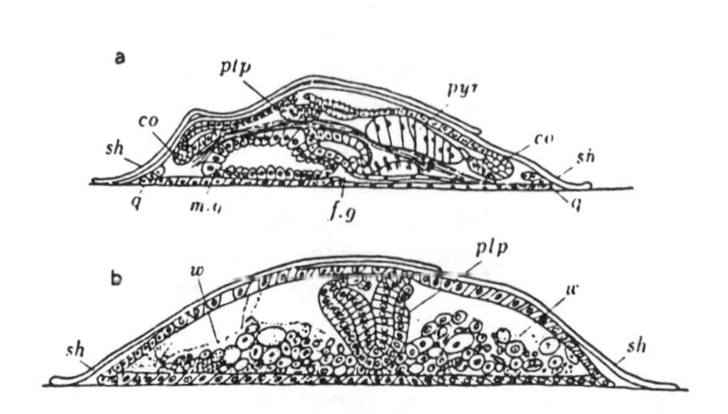

Fig. 8.14 Metamorphosis in *Membranipora pilosa* (Kupelwieser)

a. flattened larva after attachment; b. larva in metamorphosis stage. *co:* corona *f.g:* fore-gut, *m.g:* mid-gut, *plp:* polyp rudiment, *pyr:* pyriform organ, *q:* basal disc, *sh:* shell, *w:* degenerating tissue mass.

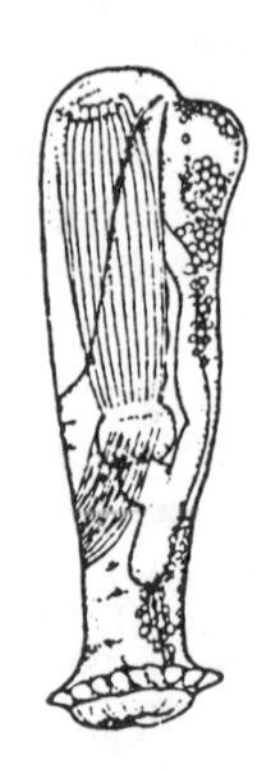

Fig. 8.15 Ancestrula of B. *veritina* (Mawatari)

The larva as a whole thus acquires the form of a thinwalled sac, into the interior of which the layer of cells constituting the apical organ extends. The upper part of this structure is closed off, and it becomes the *polypide rudiment.* This is at first composed of a single layer of ectodermal cells, but it soon acquires a second layer. Although many investigators have dealt with the question of whether this second layer of cells separates off from the first (ectodermal) layer or arises by the adhesion to it of mesodermal cells which had been suspended in the body cavity, the problem has not yet been solved.

The polypide rudiment develops into the characteristic polypide form by a process similar to that found in the budding which occurs at a later stage, and the completed polypide, gradually everting its tentacular sheath, eventually emerges to the outside as a young zooid. The first zooid to make its appearance is called the *ancestrula;* in the Gymnolaemata usually only one ancestrula arises from the metamorphosed larva (Fig. 8.15).

In experiments performed by Yamada (1940), larvae of *Bugula neritina* were sectioned horizontally at various stages of metamorphosis. His results show that any cut part will metamorphose; regeneration, however, takes place only in the basal portion. Since 1947, Lynch has been conducting a series of investigations into the factors controlling the metamorphosis of *Bugula* larvae. These indicate (1949, 1952) that heating, hypertonic sea water, an appropriate amount of light, lack of magnesium and excess of copper and of calcium accelerate metamorphosis, whereas cold, hypotonicity, lack of light and excess of magnesium and of potassium suppress it. In other words, he believes that the accelerating and inhibiting agents are mutually antagonistic. Lynch's further research has been concerned with the histochemical aspects of the mechanism of metamorphosis.

Since marine bryozoans are among the animals responsible for fouling ship bottoms, investigations into their larval metamorphosis have been made from the standpoint of fouling-prevention. Mawatari (1951) has reported that mercuric compounds are the most effective agents for poisoning *Bugula neritina*.

In the Phylactolaemata, as described above, the larva already contains two polypides, the rudiments of which are present even in its embryonic stage, indicating that the shift toward an asexual mode of reproduction is already recognizable during ontogeny. After a short free-swimming life, the larva attaches to the bottom with its aboral pole, and its ring fold slowly reverses, exposing first the first polypide and then the second (Fig. 8.16). The ring fold itself gradually shrinks and becomes a lump of yellowish tissue which is taken into the body cavity and gradually absorbed as it circulates with the blood stream.

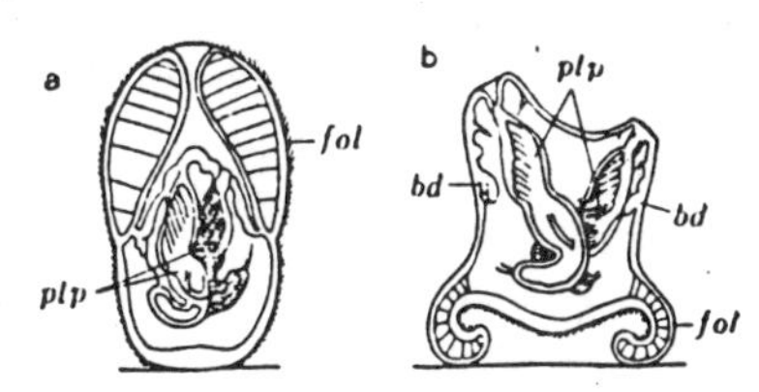

Fig. 8.16 Metamorphosis in *Plumatella* (Braem)

a. attaching larva; b. larva in metamorphosis. *bd:* bud, *fol:* ring fold, *plp:* polyp body.

2. ASEXUAL REPRODUCTION

(1) BUDDING

As the zooids grow, asexual reproduction brings about an increase in their number. The process of budding in these forms has been the subject of many papers (Nitsche, Barrois, Prouho, Oka, Kraepelin, Braem); in the following section the description by Oka (1890) of budding in *Pectinatella gelatinosa* will be cited to give an outline of the process.

The body wall (endocyst) of the zooid is composed of an outer layer and an inner lining epithelium, or *peritoneum.*[5] A bud first appears as a knob-like process, directed inward, which forms on the oral side of the endocyst near its upper margin (Fig. 8.17a *bd*), and consists of a protuberance of some cells from the outer layer, surrounded by peritoneum. The outer layer cells proliferate, and soon a cavity is formed at their center with the cells arranged around it. The bud thus grows into a cyst composed of an outer and an inner layer. Later this closed cyst develops an external opening (Fig. 8.17c, d), and extends downward to become elongate and slender. Such a bud is still attached to the outer layer at one side, through the peritoneum. The portion of this membrane which lies between the outer layer and the bud gradually becomes thinner and at last breaks. The bud is then attached to the endocyst by its basal part and the portion adjacent to its external opening. A cord of tissue connecting the base of the bud with the endocyst is the rudiment of the funiculus (Fig. 8.17d, e *fun*).

This long, slender cyst then proceeds to constrict in the middle and become separated into an *inner chamber* near the base and an *outer chamber* adjacent to the opening. Later, the inner chamber forms a digestive tube and the constriction between it and the outer chamber becomes a mouth. The outer chamber differentiates into a conical *vestibulum,* the wall of which will become the *tentacular sheath.* The cellular layer constituting the basal part of the outer chamber eventually thickens and bulges upward, and from each side of it a process arises toward the anal side. These extend in a semicircle and become the rudiments of the *lopho-*

---

[5] As mentioned above, there is some question as to the origin of the peritoneum; theoretically, it should be thought of as mesodermal.

phore arms (Fig. 8.17f, g *loph*); soon small fingerlike processes, the rudiments of the *tentacles,* appear on their upper circumferences. In the meantime, the inner chamber becomes still more deeply extended and differentiates into an *oesophagus* and a *stomach.* An invagination from the stomach is directed upward, forming a *rectum* (Fig. 8.17f, g *re*). This makes contact with the wall of the tentacular sheath and then breaks through it to form the *anus.* The rudiment of the *brain* appears as a small invagination on the wall of the oesophagus, just below the mouth at the anal side. After this invagination is closed off, it separates from the oesophagus and forms the brain. From the two sides of this organ, horn-like processes arise and extend into the lophophore arms, to become the *nerve trunks.*

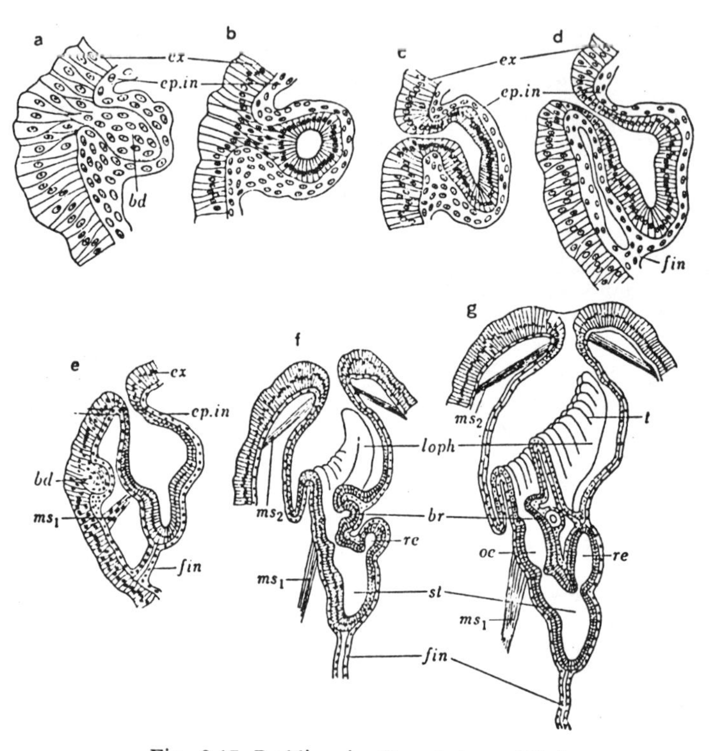

Fig. 8.17 Budding in *P. gelatinosa* (Oka)

bd: bud, *br:* brain *ep.in:* inner lining epithelium, *ex:* outer layer, *fun:* funiculus, *loph:* lophophore arm, ms_1*:* retractor muscle, ms_2*:* sheath muscle, *oe:* oesophagus, *re:* rectum, *st:* stomach, *t:* tentacle.

Immediately after the formation of the funiculus rudiment, the *lophophore retractor muscles* appear (Fig. 8.17f, g ms_1). A few cells of the peritoneum between the cystic bud and the outer layer become spindle-shaped and separate from the mother layer, forming muscle fibers which bind the middle part of the bud to the body wall. These later develop into the retractor muscles. Similarly, muscles located between the tentacular sheath and the body wall differentiate from the peritoneum.

In the meantime, the rudiment of the funiculus gradually elongates, a space arises in its interior and it develops into a tube. The growth of muscle fibers around this tube completes the formation of the funiculus. In this case, it may therefore be said that the muscles and funiculus arise from the peritoneum of the endocyst.

As the bud takes on the polypide form, it emerges to the outside by turning its tentacular sheath inside out. In a young zooid which has just emerged, the tentacles are still immature, appearing as mere bumpy outgrowths at the tips of the arms. As the zooid feeds and continues to grow, the tentacles also develop extensively and the arms become beautiful horseshoe-shaped lophophores.

Each zooid bears pairs of buds in various developmental stages.[6] Since, however, the two buds which constitute a pair differ somewhat from each other in their rate of development, they do not bud at the same time. Thus each fully grown zooid has two young zooids one large and one small, and consequently, as budding becomes active, the number of zooids, increases gradually, giving rise to a fan-shaped colony. Further increase in size leads to division of the colony, and these divided colonies clump together into colonies of colonies.[7] The colony of *P. gelatinosa* forms a gelatinous mass which may be as large as a human head.[8] Among fresh water forms such as *Plumatella,* there are many species which have a chitinous ectocyst and form branching, vine-like colonies. In many marine species, on the other hand, each tubular or box-shaped cystid secretes a calcareous substance around itself, forming a container called a *zooecium;* the colony as a whole may be lichenous or cristate. There are also some species, such as those of *Bugula,* which stand upright with an attachment at one end, resembling hydroids or seaweeds. Since the shape of the colony thus differs from species to species, the type of budding can be used as a taxonomic character.

(2) DORMANT BUDS

Fresh water bryozoans not only form colonies by asexual budding but also produce over-wintering buds, called *hibernacula* and *statoblasts,* which are protected by a characteristic coat that enables them to endure severe environmental changes.

a) Hibernaculum

The hibernaculum is a specially modified "outer bud"[9], which is protected by a thick sclerotic coat. It is found only in the fresh-water gymnolaemates *Paludicella* and *Victorella* (Fig. 8.18). Its function is similar to that of the statoblast.

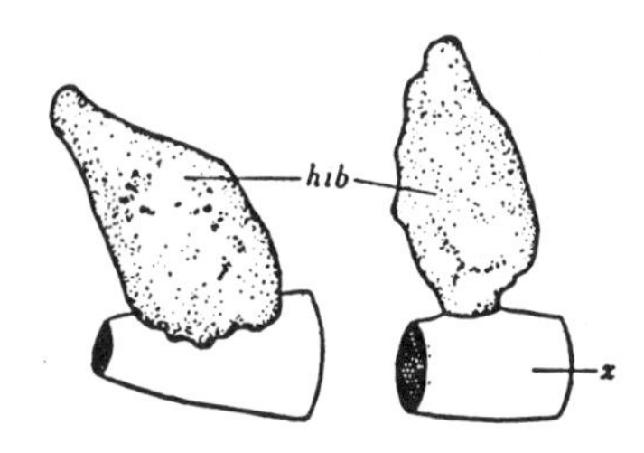

Fig. 8.18 Hibernacula of *Paludicella articulata* (Rogick)
hib: hibernacula, *x:* part of old body.

---

[6] In *L. carteri,* typically two buds are formed, but a third may appear after a long delay.

[7] Braem (1911) gave the name of "Stock" to such colonies.

[8] Colonies attached to reeds and other water plants grow in late autumn to be 10 cm in diameter and 30 cm in length.

[9] Buds which form on the outside of the cystid, as described in the section on Budding, are called "outer buds".

b) Statoblast

The statoblast, a bud produced in the funiculus of the zooid,[10] is protected by two chitinous shells. Statoblast formation is characteristic of the Phylactolaemata. Important investigations into the mode of formation and development of these structures have been made by Verworn (1888), Braem (1889, 1890, 1912), Oka (1890) and Kraepelin (1892), using *Plumatella, Pectinatella* and *Cristatella* as material. In order to compare this process with the previously described course of budding, the account presented here will deal mainly with *P. gelatinosa.*

(i) *Formation of Statoblast*

To understand the nature of the statoblast, it is highly important to establish clearly its mode of origin. Verworn's (1888) account maintains that the wall of the funiculus thickens and liberates one egg cell, which develops parthenogenically into a statoblast. Oka (1890), however, was quite unable to recognize any such process; according to his observations, at least eight cells first appear in the funiculus. Although he did not ascertain the place of origin of these statoblast rudiments, he suggested that since they arise at successively lower positions in the funiculus, they could not well originate from the inner wall of the stomach, and must therefore stem from the (ectodermal) body wall connected with the lower end of the funiculus, or from the (mesodermal) funicular wall, or from both.

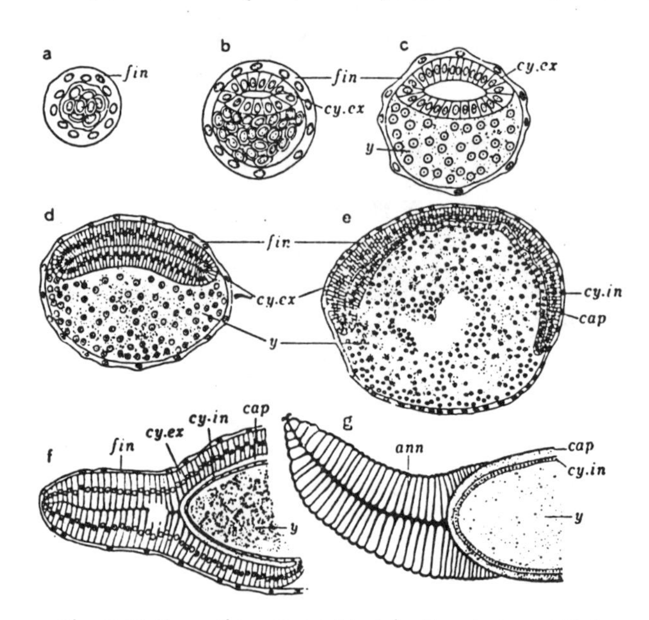

Fig. 8.19 Formation of statoblast in *P. gelatinosa* (Oka)
ann: annulus, *cap:* capsule, *cy.ex:* outer wall of cystogeneous cup, *cy.in:* inner wall of cystogeneous cup, *fun:* wall of funiculus, *y:* yolk mass.

[10] Such buds, which form within the body, are called "inner buds". The gemmule of fresh water sponges is also a kind of inner bud.

These cells at first lie scattered free in the lumen of the funiculus; they gradually increase in number and come together into a morula-like formation (Fig. 8.19a), causing a local distension of the funiculus. Soon certain special cells, which are clearly distinguishable from the others, accumulate at one side of this mass. This group of cells is at first spherical, and then as its number increases, a space appears at its center. Gradually it flattens and curves, finally taking the shape of a double-walled bowl, the *cystogenous cup* (Fig. 8.19b-d). Meanwhile, granules appear in the rest of the cells, so that they acquire the nature of yolk cells (Fig. 8.19c *y*). The cystogenous cup further increases in size and begins to include the yolk cells within its rim. During this time its outer wall has been gradually thickening, while the inner wall becomes extremely thin; between these two walls a chitinous substance is secreted. Finally the margins of the cup come together and the thin cell layer constituting its inner wall completely encloses the mass of yolk cells. In the completed statoblast, this thin layer surrounding the yolk mass is called the *outer layer* (Fig. 8.19f).

Shortly before the complete closure of the cystogenous cup, a crest begins to form on its outer wall; this gradually extends around the equator of the ellipsoidal statoblast, forming an *annulus* (Fig. 8.17g *ann*), which consists of two layers of columnar cells arranged end-to-end. The statoblast rudiment is white at first, but as the annulus develops and a chitinous substance begins to be deposited in the cell walls it takes on a yellowish tint which increases in intensity until the mature statoblast finally becomes dark brown.

Several statoblasts are formed in an orderly downward succession in membranes derived from the funiculus; eventually it leaves the funiculus and stays for some time in the body cavity.[11] In general, statoblasts which have been stored in the colony are released when it dies and disintegrates. However, it has also been observed that in *Lophopodella carteri* they break out through the body wall of the colony (Oka and Oda 1948), and in *Storella* and *Plumatella* they are released through a *vestibular pore* located on the cystid of the zooid (Marcus 1941; Wiebach 1953). When the released statoblast is dried, each cell of the annulus becomes an air chamber, making the statoblast buoyant.

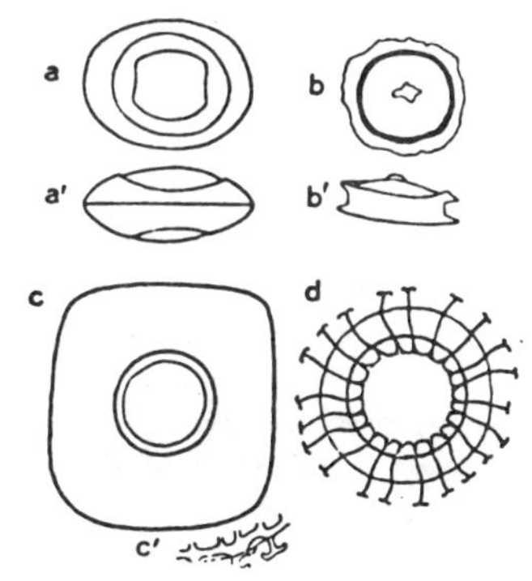

Fig. 8.20 Various types of statoblasts (Toriumi)

a. *Pl. repens*, dorsal view; a'. *Pl. repens*, lateral view; b. *G. toanesis*, dorsal view; b'. *G. toanesis*, lateral view; c. *P. gelatinosa*, dorsal view; c'. hooks at periphery of statoblasts of *P. gelatinosa*; d. *C. mucedo*, dorsal view.

The statoblast always has a capsule surrounding its contents and an annulus; its shape, however, varies markedly according to the species, and this provides an important key for classifying these forms. Even within the same species there may occur considerable variation (Braem 1911, 1912; Oda 1955). The statoblast of *Fredericella* completely lacks an annulus, and also in *Plumatella, Stepha-*

---

[11] Five or six statoblasts are usually formed in *P. gelatinosa*, although there may be as many as eight. In some cases a testis and statoblasts occur in the same funiculus, the testis located in the upper portion and the statoblasts below it.

nella and *Gelatinella* statoblasts may be formed without an annulus. On the other hand, the annulus in *Lophopodella, Pectinella* and *Cristatella* is conspicuously developed and bears spines around its periphery; these spines sometimes have several small hooks[12] (Fig. 8.20).

The statoblast is so constituted that it can withstand desiccation and low temperature. The power of resistance to desiccation differs according to the species (Rogick 1938, 1940, 1941). In *L. carteri* almost all the statoblasts which have been desiccated and kept at room temperature for one or two years are able to germinate; after periods longer that this, however, the germination rate falls off abruptly. On the other hand, statoblasts can survive a considerable amount of cold. In nature, the members of these genera over-winter as statoblasts, the colonies dying off in this season. Since, however, statoblasts are extensively formed also in summer, it is thought that their formation is not necessarily only for hibernation. In spite of being under conditions favorable for germination, the statoblasts formed in summer do not germinate until the following spring, suggesting a phenomenon of dormancy (Oda 1959). These various considerations indicate that the fresh-water bryozoans have adapted to their environment by making statoblast formation their principal method of reproduction. Since the statoblasts provided with an annulus always float, once they have been dried, they can be carried great distances by water and wind currents, or by attaching to water-fowl; this must be effective in widening the distribution of the species.

(ii) *Statoblast Germination*

In nature, the statoblasts of any given species germinate simultaneously in spring or early summer.[13] Dried statoblasts readily germinate in the laboratory at any time if they are placed under favorable conditions.

In *P. gelatinosa* the first sign of germination is the separation of the two shell valves. Within the statoblast capsule a milky, granular mass, the yolk mass,[14] is enclosed in a thin outer layer. Before long the cells of this outer layer lying along the equator of the ellipsoid begin to thicken, and this thickening gradually propagates toward the polar sides. At two opposite places in the equatorial plane the thickening of the cell layer is particularly conspicuous. In one of these regions the cells become markedly columnar, form vacuoles, and take on the character of the outer layer in the zooid endocyst. The cells at the center of the other region proliferate and extend into the yolk mass, producing a rod-like process in which they become regularly arranged around a space that opens in its center, to form a cyst. This is the polypide (Fig. 8.21a, b, c).

According to Braem's (1890, 1912) observations on *Cristatella* and *Pectinatella*, the outer layer on the dorsal side of the statoblast thickens to form a disc, on the surface of which a crescent-shaped hollow appears. When this hollow has developed into a deep invagination, the surface closes over it to form a cyst, which

[12] Rogick (1943), classifying the statoblasts, called those which have a reduced annulus and tend to sink, *sessoblasts*; those with a well-developed annulus which float, *floatoblasts*: and those with well-developed annulus having spines on its periphery, *spinoblasts*.

[13] In Japan, *L. carteri* germinates at the beginning of April, *P. gelatinosa* in mid-June and *Cristatella mucedo* at the end of May.

[14] In the dried statoblast this is in a jelly-like state.

secondarily opens to the exterior. The early developmental processes thus appear to differ more or less in different species. In the meantime, a certain number of granule-containing cells have become arranged around the outside of the cystic rudiment, forming a peritoneum. The primordium of the polypide has thus acquired its two cellular layers.

The cystic primordium continues to extend more deeply into the mass of yolk, and then a median constriction divides in into inner and outer chambers. Each chamber differentiates as it does in the case of budding, to form a polypide; that is, the outer chamber becomes the vestibule, from the base of which the lophophore differentiates, while the inner chamber elongates to form the oesophagus and stomach. An invagination extending upward from the stomach develops into the rectum, completing the formation of the crooked digestive tract (Fig. 8.21 d, e). Only the mode of formation of the muscles and funiculus differs from that found in budding.

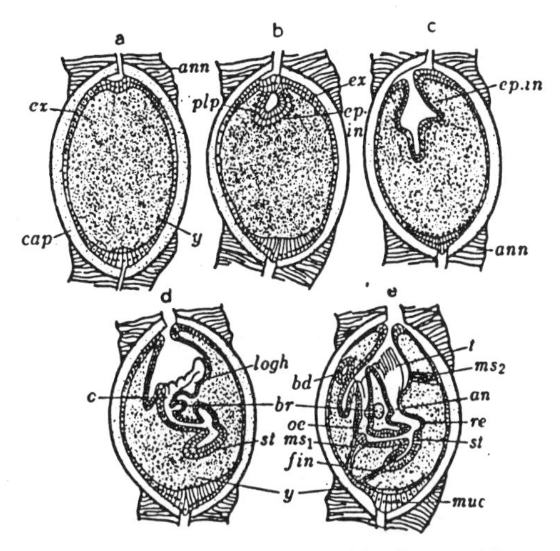

Fig. 8.21 Development of polyp in statoblast of *P. gelatinosa* (Oka)

an: anus, *ann:* annulus, *bd:* bud, *br:* brain, *cap:* capsule, *ep.in:* inner lining epithelium (mesoderm), *ex:* external layer (ectoderm), *fun:* funiculus, *loph:* lophophore, ms_1: retractor muscle, ms_2: sheath muscle *muc:* mucous pad *o:* mouth, *oe:* oesophagus, *plp:* polyp rudiment, *re:* rectum, *st:* stomach, *t:* tentacle, *y:* yolk mass.

At the time when the anus is being differentiated, the body cavity is still filled with a mass of granules; after that time, however, some granular cells between the tentacular sheath and the body wall, and others between the lower part of the oesophagus and the body wall, lose their granules and become spindle-shaped. It is these cells that develop into muscles. The former differentiate into the *parieto-vaginal muscles,* the latter into the *retractor muscles.* The muscles of the digestive tract and body wall differentiate from the peritoneum in a similar way. Simultaneously with the formation of the muscles, other granular cells become arranged in a cylindrical formation between the end of the stomach and the body wall. These cells are clearly distinguishable from the others by their marked stainability. These later lose their granules and a space appears among them, forming the primordium of the tubular funiculus. These granular cells are thus responsible for differentiating the muscles and the funiculus rudiment.

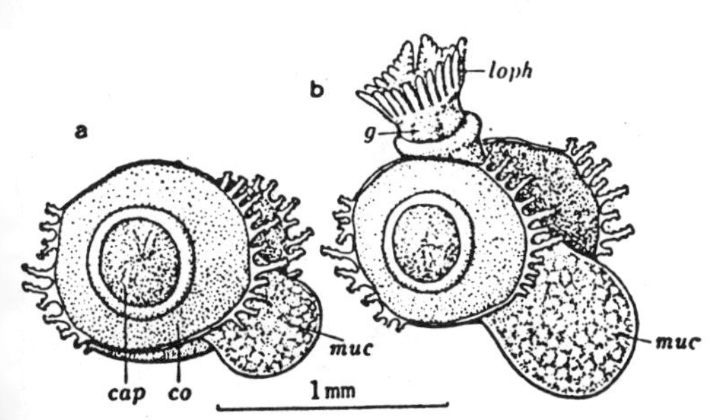

Fig. 8.22 Germination of statoblast of *L. carteri* (Oda)

a. appearance of mucous pad between shells; b. ancestrula having two shells, anterior and posterior. *cap:* capsule, *co:* annulus, *g:* gut, *loph:* lophophore, *muc:* mucous pad.

By the time polypide formation is completed, two cystic buds, a large and a small one, have already developed on the oral side of the body wall. The granular cells which have been filling the body cavity gradually disappear, although some of them acquire the nature of *blood corpuscles*. The body wall continues to grow, increasing the size of the body cavity, and a *mucous pad* protrudes between the two shell valves and attaches the germinal body to some object (Fig. 8.22a). This process in which the valves separate and the mucous pad protrudes through the space between them is called *germination*.[15] On the following day the young polypide (ancestrula) appears from the side of the statoblast opposite the mucous pad (Fig. 8.22b). At this stage, the zooid resembles a brachiopod, with anterior and posterior valves. The arms and body wall are still dotted with milky-white masses of yolk; these, however, gradually disappear as growth continues. This ancestrula which has arisen asexually from the statoblast also forms a colony by a regular process of budding.

3) POLYMORPHISM AND OTHER PHENOMENA

The individual zooids constituting a colony typically proliferate at the expense of nutrients which they procure by the activity of their own tentacles. Among the marine bryozoans, however, some of the zooids differentiate along special lines. The atypical zooids thus formed are divided into *kenozooids* and *heterozooids*.

The kenozooid is a degenerated independent zooid which has lost its polypide. Inside such a zooid there still remain some epidermal cells and muscles. The chitinous or calcareous ectocysts of these zooids accumulate to form complicated structures which are useful for attaching the colony to various substrates.

The heterozooids include *avicularia, vibracularia* and *gonozooids*. The avicularium is a cystid, the bud of which develops in a special manner to resemble a forceps; this is seen among the Cheilostomata. The basic part of the cystid forms a stalk, and held upright by this, the distal part bifurcates into upper and lower mandibles. Inside the cystid there are well developed muscles (Fig. 8.23). The vibracularium is similar to the avicularium, with a characteristic whip-like lower mandible. Both these heterozooids seem to have the function of protecting the other zooids against external enemies. The gonozooids are certain zooids of the colony which are modified for brooding the embryos. These have been described above under the name of "ooecia"; they enclose the fertilized eggs, which proceed to develop within them.

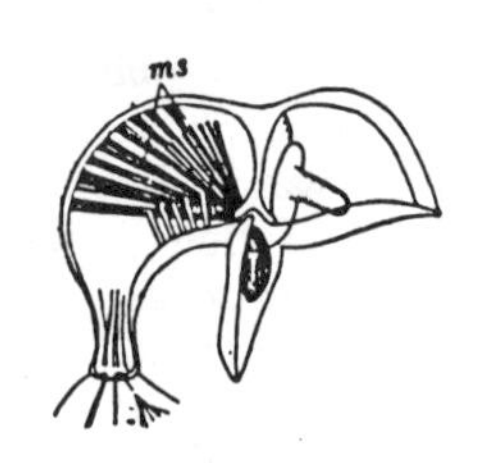

Fig. 8.23 Avicularia of *Bugula* (Calvet) *ms:* muscle

As this brief description indicates, polymorphism similar to that found in hydrozoan colonies is also seen among the Bryozoa.

Among the marine bryozoans, a moribund polypide often contracts into a brownish mass of disintegrating tissue within the cystid. This has been given

[15] In *L. carteri*, germination occurs within four to five days after the dried statoblasts have been put into water at 25°C.

the name of *brown body*.[16] From the cystid wall a new polypide regenerates, and in some cases this gradually ingests the brown body into its stomach and uses it as nutriment. In this way the dead polypide is replaced by a new one, constituting a sort of "total regeneration".

REFERENCES

II. Bryozoa

Barrois, J. 1877: Recherches sur l'embryologie des Bryozoaires. Trav. Inst. Zool. Lille: **1**—305.

Braem, F. 1889: Die Entwicklung der Bryozoenkolonie im keimenden Statoblasten. Zool. Anz. **12** : 675.

———— 1890: Untersuchungen über die Bryozoen des süssen Wasser. Bibliotheca Zoölogica, **6** : 1—134.

———— 1892: Die Keimblatter der Bryozoenknospe. Zoöl. Anz. **15** : 113—115.

———— 1897: Die geschlechtliche Entwicklung von *Plumatella fungosa*. Bibliotheca Zoölogica, **23** : 1—96.

———— 1911: Die Variation über den Statoblasten von *Pectinatella magnifica*. Arch. Entw.-mech., **32** : 314—348.

———— 1912: Nachträgliches über die Variation der Statoblasten von *Pectinatella magnifica*. Arch. Entw-mech., **35** : 46—55.

———— 1912: Die Keimung der Statoblasten von *Pectinatella* und *Cristatella*. Bibliotheca Zoölogica, **67** : 35—63.

Buchner, P. 1924: Polymorphism der Bryozoa. Zoöl. Jahrb. Syst. **48.**

Buddenbrock, W. v. 1910: Beiträge zur Entwicklung der Statoblasten der Bryozoen. Zeitschr. wiss. Zoöl. **96** : 1—50.

Calvet, L. 1900: Contribution a l'histore naturelle des Bryozoaires ectoproctes marins. Théses Fac. Sci. Paris, Montpellier, 1—488.

Cornea, D. 1948: A embryologia de *Bugula flubellata*. Zoölogica, Sao Paulo, Brazil. **13** : 7—17.

Dawydoff, C. 1928: Traité d'embryologie comparés des invertébrés. Paris.

Harmer, S. F. 1893: Embryonic fission in cyclostous Polyzoa. Quart. Jour. Micr. Sci., **34.**

Korschelt, E. 1936: Vergleichende Entwicklungsgeschichte der Tiere. Jena.

Kraepelin, K. 1892: Die deutschen Süsswasserbryozoen, II. Entwicklungs-geschichtlicher Teil. Abh. Nat. Ver. Hamburg, **12** : 1—68.

Kupelwieser, H. 1905: Über den Bau und die Metamorphose des Cyphonautes. Bibliotheca Zoölogica, **46.**

Lynch, W. F. 1947: The behavior and metamorphosis of the larvae of *Bugula neritina* (Linnaeus). Biol. Bull., **92** : 115—150.

———— 1949: Acceleration and retardation of the onset of metamorphosis in two species of *Bugula* from the Woods Hole region. J. Exptl. Zool., **111** : 27—54.

———— 1949: Modification of responses of two species of *Bugula* larvae from Woods Hole to light and gravity. Biol. Bull., **97** : 302—310.

———— 1952: Factors influencing metamorphosis of *Bugula* larvae. Biol. Bull., **103** : 369—382.

MacBride, E. 1914: Text-book of Embryology. London.

Marcus, E. 1925: Bryozoa, Moostiere. P. Schulze's Biologie der Tiere Deutschland, Lief. **14** : Berlin.

———— 1926: Beobachtungen und versuche an lebenden Meeresbryozone. Zoöl. Jahrb. Abt. Syst., **62** : 1—102.

———— 1926: Beobachtungen und Versuche an lebenden Süsswasserbryozoen. Zoöl. Jahrb. Abt. Syst., **52** : 279—350.

Marcus, E. 1934: Über *Lophopus crystallinus* (Pall.). Zoöl. Jahb. Anat., **68** : 501—606.

———— 1941: Sobre Bryozoa do Brazil, I. Bol. Fac. Filos. Cienc. Sao Paulo (Zool.) **5** : 3—208

[16] Rogick (1936) has used the term "brown body" in connection with disintegrating tissue masses, found in fresh water bryozoans, which originate from degenerating polypides and circulate with the blood stream within the common body cavity. Use of this term should be limited, however, to the phenomenon of total regeneration observed among the marine forms.

Mawatari. S. 1951: The natural history of a common fouling bryozoan, *Bugula neritina* (Linnaeus). Misc. Rep. Res. Inst. Nat. Res., **19-20** : 27—54.
——— 1951: On *Tricellaria occidentalis* (Trask), one of the fouling bryozoans in Japan. Misc. Rep. Res. Inst. Nat. Res., **22** : 9—16.
——— 1952: On *Watersipora cucullata* (Busk) II. Misc. Rep. Res. Inst. Nat. Res., **28** : 17—27.
Nitsche, H. 1868: Beiträge zur Anatomid und Entwicklung der Süsswasserbryozoen. Reichert und Du Bois-Reymands Arch. Anat.
——— 1870: Beiträge zur Entwicklungsgeschichte der Chilostomen Bryozoen. Zeitschr. wiss. Zoöl. **20** : 1—27.
Oda S. 1955: Variability of the statoblast in *Lophopodella carteni.* Sci. Rep. Tokyo Kyoiku Daigaku, Sec. B, **8** : 1—22.
——— 1959: Germination of the statoblast in freshwater Bryozoa. Relation of statoblast formation stage to dormancy. Sci. Rep. Tokyo Kyoiku Daigaku, Sec. B, **9** : 90—132.
Oka, A. 1890: Observation on freshwater Polyzoa. Jour. Coll. Sci. Imper. Univ. Tokyo, **4** : 89—150.
Oka, H. & S. Oda 1948: Observations on freshwater Bryozoa. Collecting and Breeding, **10** : 39—48. (in Japanese)
Pace, R. M. 1906: Early stages in the development of *Flustrella.* Quart. Jour. Micr. Sci., **50.**
Prouho, H. 1890—92: Contribution a l'histoire des Bryozoires. Arch. Zool. exp. gén., **8—10.**
Robertson, A. 1904: Embryology and embryonic fission in genus *Crisia.* Publ. Univ. California, **1.**
Rogick, M. D. 1935: Studies on fresh-water Bryozoa, III. The development of *Lophopodella carteri* var. *typica.* Ohio Jour. Sci., **35** : 467—467.
——— 1936: Studies on fresh-water Bryozoa, IV. On the viability of statoblasts of *Lophopodella carteri* var. *typica.* Trans. Amer. Micr. Soc., **55** : 327—333.
———1938: Studies on fresh-water Bryozoa, VII. On the viability of dried statoblasts of *Lophopodella carteri* var. *typica.* Trans. Amer. Micr. Soc., **57** : 178—199.
——— 1939: Studies on fresh-water Bryozoa. XI. The viability of dried statoblasts of several species. Growth, **4** : 351—322.
——— 1940: Studies on fresh-water Bryozoa. VIII. Larvae of *Hyalinella punctata* (Hancock) 1850. Trans. Amer. Micr. Soc., **58** : 199—209.
——— 1941: The resistance of fresh-water Bryozoa to desiccation. Biodynamica, **3** : 369—378.
——— 1945: Studies on fresh-water Bryozoa, XV. *Hyalinella punctata* growth data. Ohio Jour. Sci., **45** : 55—79.
——— 1953: Bryozoa, in: Pennak, R. W. Fresh-water Invertebrates of the United States. New York; 256—277.
Seeliger, O. 1906: Über die Larven und Verwandschaftsbeziehungen der Bryozoen. Zeitschr. wiss. Zoöl., **84** : 1—78.
Toriumi, M. 1941: Studies on fresh-water Bryozoa of Japan, I. Sci. Rep. Tohoku Imp. Univ. 4th ser., **16** : 193—215.
——— 1956: Taxonomical study on fresh-water Bryozoa, XVII. Sci. Rep. Tohoku Univ. 4th ser., **22** : 57—88.
Vigelius, W. J. 1886—88: Zur Ontogenie der marien Bryozoen, Mitt. Zoöl. Stat. Neapel, **6—8.**
Yamada, M. 1940: Regeneration of *Bugula neritina* Linné during metamorphosis. Zool. Mag., **52** : 372—379. (in Japanese)
Verworn, M. 1888: Beiträge zur Kenntniss der Süsswasser Bryozoen. Zeitschr. wiss. Zoöl., **45** : 99—130.
Wiebach, F. 1953: Über den Ausstoss von Flottoblasten bei Plumatellen. Zoöl. Anz., **151** : 266——272.

III. BRACHIOPODA

INTRODUCTION

The Brachiopoda are divided into two orders: the Testicardines, in which the left and right valves of the shell are articulated, and the Ecardines, which lack such a connection. The lamp shell, *Terebratulina,* belongs to the first order, and *Lingula,* to the second. The embryology of the Testicardines has been studied by various workers in Europe and America, but the investigations of Yatsu (1902) into the developmental processes of *Lingula* at Misaki provide the only connected account relating to the embryology of the Ecardines.

1. DEVELOPMENT OF ECARDINES *(LINGULA)*

(1) SPAWNING

The breeding season of *Lingula* at Misaki extends from the middle of July to the middle of August, with a peak of spawning early in August. Spawning occurs daily, after sunrise and sunset; since change in light intensity is an important factor in stimulating spawning, it is possible to induce the animals to shed even in the daytime by placing them in the dark. The females are induced to spawn by previous shedding of sperm by the males, but independent spawning by females is also observed on occasion (Kumé 1956).

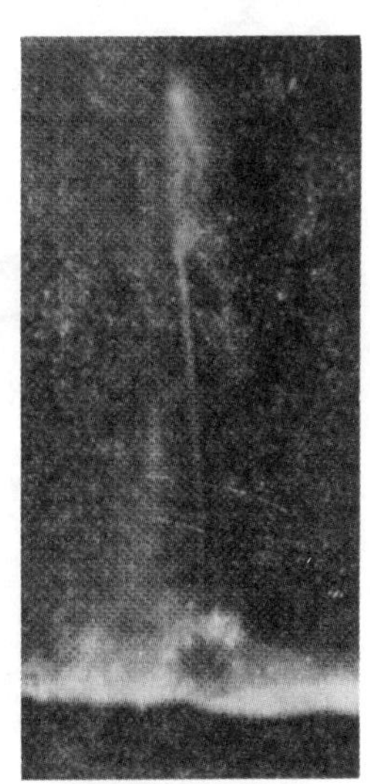

Fig. 8.24 Spawning of *Lingula* (Kumé)

The shedding of sperm and eggs provides a spectacular sight. If 10—15 animals are kept together in a glass container, one male will usually begin to spawn, and keep it up for about an hour. The ejected spermatozoa rise in a white line for eight centimeters or more (Fig. 8.24) and if the animal is in a slanting position, the white line may extend in a parabola to a distance of about ten centimeters. The gametes are expelled to the outside through the median one of three ciliated ducts which open along the anterior margin of the valves, after reaching this duct by way of the left or right nephridium; as a result the white line of spawned sperm can be seen to be formed by the union of two fine streams.

After this shedding of spermatozoa has gone on for 20—30 minutes, the rest of the animals begin to spawn, until practically the whole population of the container is shedding eggs or sperm. The eggs are shed in clumps, surrounded by a mucous substance; these clumps appear as a succession of dots when they are ejected from the ciliated duct. Each clump contains several pale yellow eggs.

(2) FERTILIZATION AND CLEAVAGE

The eggs, as spawned, are spherical, 0.09—0.10 mm in diameter, surrounded by a transparent vitelline membrane, and filled with pale yellow yolk so that they are opaque. There are radial streaks on the vitelline membrane, and the cortex of the egg contains from two to four layers of vacuoles. According to Yatsu, the sperm enters the vegetal side of the egg, and fertilization takes place at the telophase of the second meiotic division. The diploid number of chromosomes is 16.

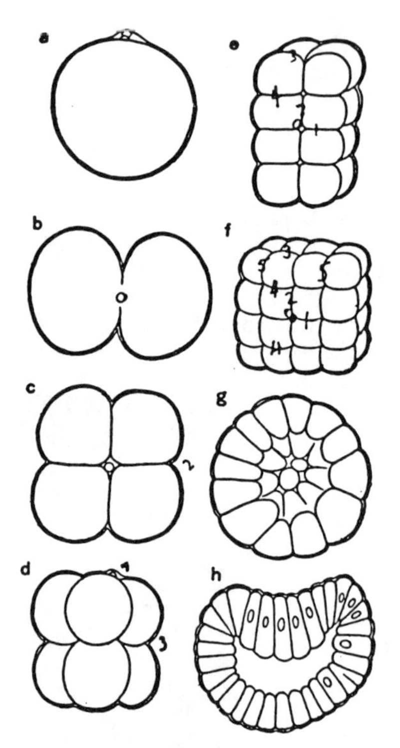

Fig. 8.25 Cleavage and gastrulation in *Lingula* (Yatsu)

a. fertilized egg; b. 2-cell stage; c. 4-cell stage; d. 8-cell stage; e. 16-cell stage; f. 32-cell stage; g. blastula; h. gastrula (Upper side of sketch is vegetal side of embryo).

Since the polar bodies can be traced until at least the 32-cell stage, it is possible, by using them as landmarks, to determine the egg axis and the order and direction of the cleavages. The mode of cleavage is extremely unusual; in general it resembles the cleavage of the bryozoan *Bugula*, and in its characteristic of biradial symmetry is similar to that of the ctenophores. After fertilization, and before the first division begins, a marked constriction appears in the egg, but this is not a true cleavage. The first cleavage plane includes the egg axis, and the second is perpendicular to this, also passing through animal and vegetal poles. The third cleavage plane is horizontal, and therefore perpendicular to both first and second planes; it divides the egg into eight blastomeres of about equal size (Fig. 8.25a-d). The peculiar cleavages are the succeeding fourth and fifth: at the fourth, two cleavage planes appear, parallel to the first, producing two regular rows of four blastomeres each in both animal and vegetal hemispheres (Fig. 8.25e). The two planes of the fifth division are parallel to the second cleavage plane, forming in each hemisphere a regularly aranged layer consisting of four rows of four cells each, which gives the effect of a square floor laid with tatami (Fig. 8.25f). The first cleavage begins about an hour after fertilization at 27°C, and later cleavages take place at intervals of about 15 minutes.

The cleavage process after this stage has not been observed, but eventually the embryo expands into a spherical shape, and becomes a *coeloblastula* with a spacious blastocoel. The blastular wall is rather thick on what is believed to be the vegetal side; an invagination presently appears in the center of this region, and the embryo becomes a gastrula. Invagination takes place in the typical way, and the surface where it occurs corresponds to the anterior part of the future embryo. The cells making up the lateral walls of the invaginated archenteron proceed to divide and increase in size, to give rise to the mesoderm. They do not, however, separate from the archenteron, but remain for some time as a common cell mass with the endoderm. Yatsu (1902) called this the *meso-entoblastic cell-mass;* it is possible to differentiate between the two types of cells by differences in the staining properties of their nuclei. This meso-entoblastic cell-mass occludes the blastocoel and presses against the archenteron so that its lumen becomes indistinct, and the blastopore is closed for a time.

(3) EARLY LARVA

It is after the gastrula stage has been completed that the embryo displays the special form peculiar to these species. First a circular fold of ectoderm rises around the embryo; this is equatorial and directed outward, and causes a shortening of the antero-posterior axis so that the embryo is flattened into the shape of a disc (Fig. 8.26a, b). This ectodermal fold is the *rudiment of the mantle,* while the mound-shaped bulge which it surrounds on the anterior side of the embryo is the *arm-ridge,* the rudiment of the arm region. Eventually a groove is formed on the anterior surface of this rudiment, which tends to divide it dorso-ventrally.

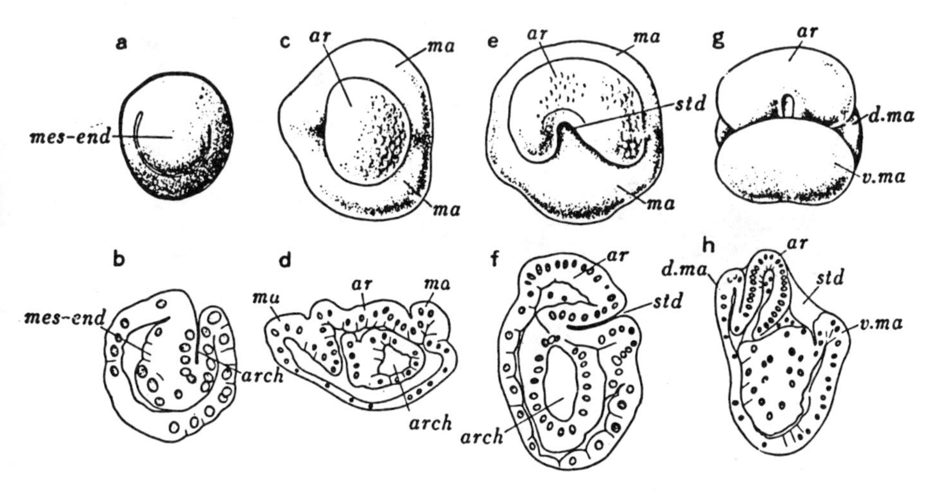

Fig. 8.26 Embryology of *Lingula* (1) (Yatsu)

a. late gastrula; b. longitudinal section of embryo in a; c. frontal view, d. longitudinal section, of embryo at stage of formation of mantle rudiment; e. frontal view, f. longitudinal section, of embryo at stage of formation of stomodaeum; g. frontal view, h. longitudinal section, of embryo undergoing marked growth of arm ridge and mantle. *ar:* arm ridge, *arch:* archenteron, *d.ma:* dorsal mantle, *ma:* mantle rudiment, *mes-end:* meso-endodermal cell-mas, *std:* stomodaeum, *v.ma:* ventral mantle.

As development proceeds, this groove extends to the posterior surface of the rudiment, completing the division of the mantle rudiment into *dorsal mantle rudiment* and *ventral mantle rudiment* (Fig. 8.26c, d). This change makes evident the difference between the dorsal and ventral surfaces of the embryo. The lower part of this divided mantle rudiment later changes to a flattened crescent shape as the result of extensive growth that takes place at the anterior edge of the embryo.

About the time when the mantle rudiment begins to become separated into dorsal and ventral parts, a new invagination appears at the place where the original blastopore closed. This elongates in a dorso-posterior direction and eventually its tip makes contact with the anterior end of the archenteron. The newly invaginated part is the *stomodaeum,* which is of ectodermal origin, and will later form the larval oesophagus. The stomodaeal invagination causes the ventral part of the arm-ridge to assume a spread U-shape when viewed from the anterior side, but the whole arm-ridge grows along with the growth of the mantle, causing marked changes in its form. As the arm-ridge increases in size, its front end expands into a mushroom-shape, and becomes flattened dorso-ventrally. At the same time it shifts to a somewhat more dorsal position, coming to look as though it belonged to the dorsal mantle (Fig. 8.26g, h). At this stage, the anterior edge of the ridge becomes densely covered with *cilia;* as these begin to move vigorously, the membrane which has been surrounding the embryo is broken, the rupture appearing first in the neighborhood of the cilia.

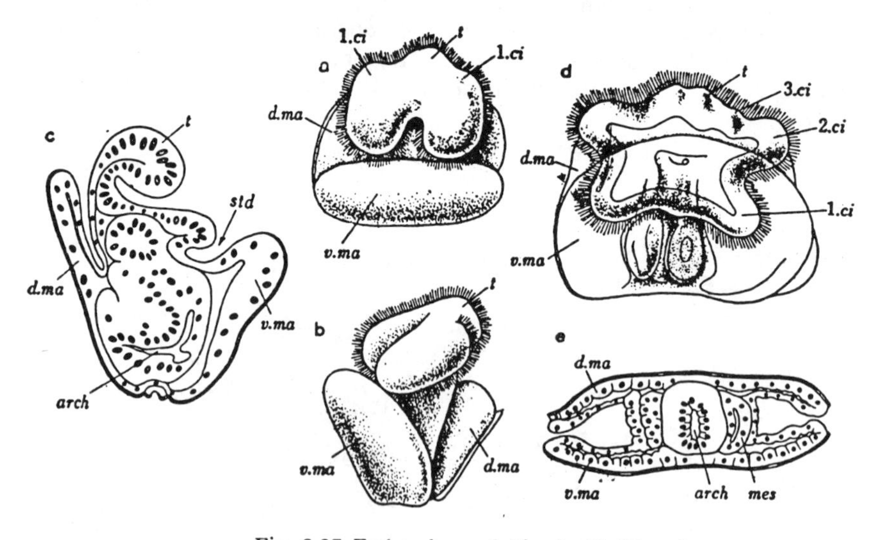

Fig. 8.27 Embryology of *Lingula* (2) (Yatsu)

a. appearance of tentacle and first pair of cirri; b. lateral view of same embryo; c. longitudinal section of same embryo; d. appearance of third pair of cirri; e. cross-section of the embryo shown in d. *1.ci:* first pair of cirri, *2.ci:* second pair of cirri, *3.ci:* third pair of ciri, *arch:* archenteron, *d.ma:* dorsal mantle, *mes:* mesoderm, *std:* stomodaeum, *t:* tentacle, *v.ma:* ventral mantle.

About this time bumps appear on the anterior edge of the ridge; the protrusion which appears in the center is the *tentacle rudiment,* and those at the left and right corners are the rudiments of the first pair of *cirri* (Fig. 8.27a-c). In the interior the differentiation of muscle tissue is already taking place, so that the arm can be expanded and contracted; the secretion of *chitin* is beginning on the outside of the mantle. Internally, the lumen of the archenteron, which has established connection with the stomodaeum, begins to re-expand, and the differentiation of the archenteron wall produces a definite demarcation between *stomach* and *oesophagus*. At the two sides, the masses of mesoderm cells which separated from the wall of the archenteron are still lying without change, in contact with the digestive tract. In this stage, the embryo sheds its membrane and takes up a free-swimming life.

During this stage, the number of cirri gradually increases, the new ones always appearing between the tentacle and the cirri which were formed first at the two sides. The number of cirri thus gradually increases from two pairs to three, four and so on, but the central tentacle is a larval organ which begins to retrogress after the larva reaches a stage with 15 pairs of cirri. While the tentacle degenerates and the number of paired cirri increases, the ventral side of the arm-ridge begins to form lateral processes which meet centrally to give rise to a sort of shelf extending horizontally across the anterior side of the mouth opening, the future *epistome*. The larval body at this stage is colorless and transparent, except for a pale yellowish color which is beginning to appear at the tips of the cirri.

Internally, the mesoderm cells have become completely separated from the endoderm, and *coelomic cavities* are forming at the centers of the mesodermal masses (Fig. 8.27e). As the cavities enlarge, their walls form thin cellular layers which surround the digestive tract and become attached to the inner side of the ectoderm. In the region which forms the anteriormost margin of the body cavity, however, some of the mesodermal layer breaks up into mesenchyme. These cells invade the arm rudiment and part of them differentiate into the *muscle* tissue referred to above. When larvae are raised in the laboratory, their development stops at about this level, with not more than 5~6 pairs of cirri.

(4) LATER LARVA

The rudiments of the major organs in general put in their appearance after the stage in which the larva has 5~6 pairs of cirri (hereafter referred to as '5~6 pair stage'). When it acquires ten pairs, a *stalk (peduncle)* projects from the posterior part of the mantle; at this stage the larva puts an end to its free-swimming life and becomes a bottom-dweller. The main changes which take place during this period can be summarized as follows.

The *setae* on the edges of the mantle first appear at the 7-pair stage; these are secreted by the walls of ectodermal sacs formed as invaginations of the mantle (Fig. 8.29a, b). Conspicuous changes in the configuration of the digestive tract are chiefly caused by the formation of the three *liver lobes*, which arise as swellings of the stomach wall; the *anterior* and *posterior dorsal lobes* are formed from the dorsal side, and the *ventral lobe*, from the ventral side. The first of these to appear is the posterior dorsal, while the ventral lobe is formed last. The walls of the liver lobes are made up of polynucleate glandular cells, and judging from the fact that diatoms and similar objects are found embedded in their cytoplasm, *intracellular digestion* is believed to be going on there. An intestinal tube connects

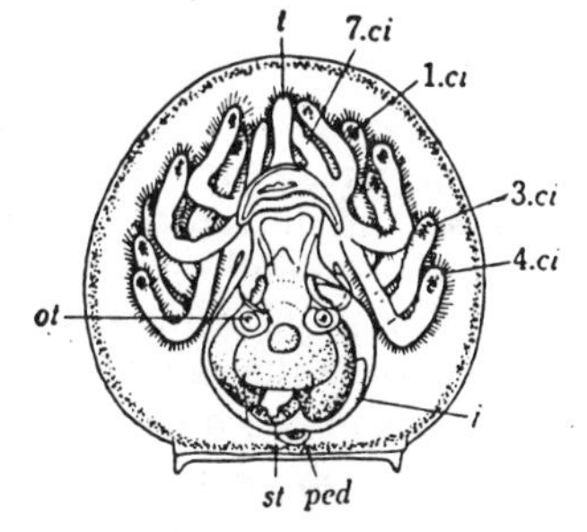

Fig. 8.28 Larva of *Lingula* with seven pairs of cirri (Yatsu)

1.*ci*: first pair of cirri, 3.*ci*: third pair of cirri, 4.*ci*: fourth pair of cirri, 7.*ci*: seventh pair of cirri, *i*: intestine, *ot*: otocyst, *ped*: stalk rudiment, *st*: stomach, *t*: tentacle.

with the stomach; this opens to the outside by way of an *anus* in the 8 ~ 9-pair stage. As stated above, the ectodermal stomodaeum gives rise to part of the oesophagus; not only is the nuclear arrangement different in this part from that in the portion of endodermal origin, however, but the cilia of the two regions beat in opposite directions, the former beating toward the anterior while the latter beat toward the posterior.

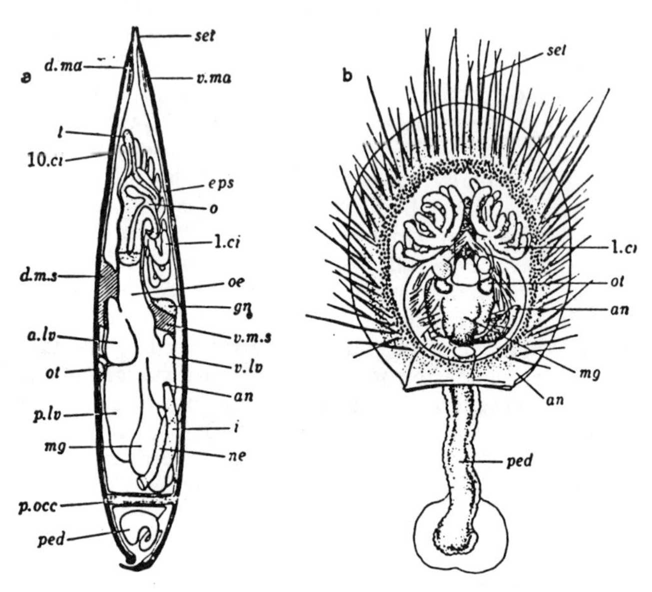

Fig. 8.29 Larvae of *Lingula* with 10 pairs and 12 pairs of cirri (Yatsu)

a. longitudinal section of larva with 10 pairs of cirri; b. larva with 12 pairs of cirri, *1.ci:* first pair of cirri, *10.ci:* 10th pair of cirri, *a.lv:* anterior dorsal lobe of liver. *an:* anus, *d.ma:* dorsal mantle, *d.m.s.:* dorsal mesentery, *eps:* epistome, *gn:* ventral ganglion, *i:* intestine, *mg:* mid-gut, *ne:* nephridial duct, *o:* mouth, *oe:* oesophagus, *ot:* otocyst, *ped:* stalk, *p.lv:* posterior dorsal lobe of liver, *p.occ:* posterior occlusor, *set:* setae, *t:* tentacle, *v.lv:* ventral lobe of liver, *v.ma:* ventral mantle, *v.m.s.:* ventral mesentery.

A ganglion appears as early as the 5~6-pair stage, as a thickening in the anterior part of the body wall; from its center the *infra-oesophageal ganglion (ventral ganglion)* is differentiated. The *nephridia* arise at about the same stage, as ridges on the body wall close to the ventral mantle; these have become tubular by the 7~9-pair stage, with one end of each opening through a funnel into the body cavity, while the ducts run upward along the left and right sides of the body wall and open to the outside in the anterior part of the animal (Fig. 8.29a *ne*).

A study of the *Discinisca* larva led Fritz Müller (1860) to the discovery that the Testicardines possess *otocysts*; somewhat later Morse (1881) and Brooks (1878) found such organs in both the adult and larval forms of *Lingula*. After the 5-pair stage these appear, one at each side of the oesophagus just above the anteriormost end of the stomach (Fig. 8.28), as spherical pouches 45 ~55 μ in diameter, in which about 40 minute otoliths are freely suspended. The cuboidal epithelium which makes up the walls of these vesicles arises by invagination of the ectoderm.

Muscle tissue of two types develops in the larva; the *dorsal* and *ventral muscles* belong to the larval type, while perpendicular muscles connecting the dorsal and ventral valves of the shell *(anterior occlusors, posterior occlusors)*, and oblique muscles between the valves *(lateral, obliquus internus, obliquus medius, obliquus externus)* persist into the adult stage (Fig. 8.30). The ventral muscles are already developing at the two sides of the oesophagus in the 3-pair stage; in the 5~6-pair stage the dorsal muscles, anterior occlusors and obliquus internus muscles are added, and in the 7~9-pair stage the posterior occlusors, lateral muscles, and obliquus medius and externus are formed.

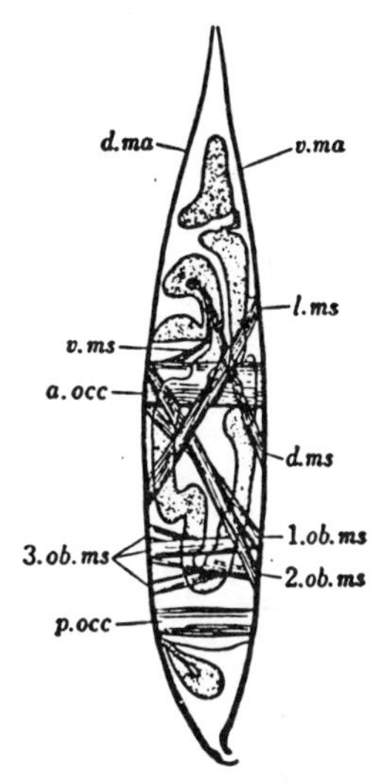

Fig. 8.30 Muscular system of *Lingula* larva (stage with 7~9 pairs of cirri) (Yatsu)

1.*ob.ms*: obliquus internus muscle, 2.*ob.ms*: obliquus externus muscle, 3.*ob.ms*: obliquus medius muscle, *a.occ*: anterior occlusor muscle, *d.ma*: dorsal mantle, *d.ms*: dorsal muscle, *l.ms*: lateral muscle, *p.occ*: posterior occlusor muscle, *v.ma*: ventral mantle *v.ms*: ventral muscle.

The stalk rudiment arises from the inner surface of the ventral mantle at its posterior end as a spherical bud which later grows into a twisted stem-like rod. The coelom penetrates into the interior of this structure to form the *stalk cavity*. Since the stalk is simply an extension of the body cavity, its wall is composed of no more than an ectodermal epidermis and the coelomic epithelium, a small part of which differentiates into longitudinal muscle tissue. As described above, this rudiment protrudes to the outside in the 10-pair stage; the most striking change that occurs after this takes place in the ectodermal cells of the outer layer, particularly at the tip of the stalk, where they take the form of columnar epithelium, and secrete an adhesive substance at the surface (Fig. 8.29 *ped*).

2. DEVELOPMENT OF THE TESTICARDINES

Unfortunately no studies have yet been made of the Japanese members of this group. Among the Testicardines are some genera, such as *Cistella* and *Lacazella,* which have a brood chamber in the body wall, where the larvae develop for a certain period. In the fossil genus *Stringocephala,* young animals which have already formed shells can be seen in the mantle cavity.

(1) EARLY DEVELOPMENT

As in *Lingula* the eggs are spherical *(Cistella,* 0.12 mm diam); although there are few detailed descriptions, it is known that they cleave equally and form a coeloblastula. The blastular wall in *Terebratulina* (Morse 1871~1873; Conklin 1902) is composed of tall cells, and the blastocoel is consequently rather small. In any case, invagination takes place in a typical manner to form a gastrula, the blastopore of which then begins to close, starting at the posterior side and proceeding anteriorly, until only a trace of it remains open at the front end, from which the closed part extends backward as a groove.

One point which should be noted is the mode of formation of the *coelomic pouches* which appear together with the mesoderm. As both Kowalevsky (1883) and Plenk (1915) ascertained in *Cistella,* two swellings of the archenteron wall

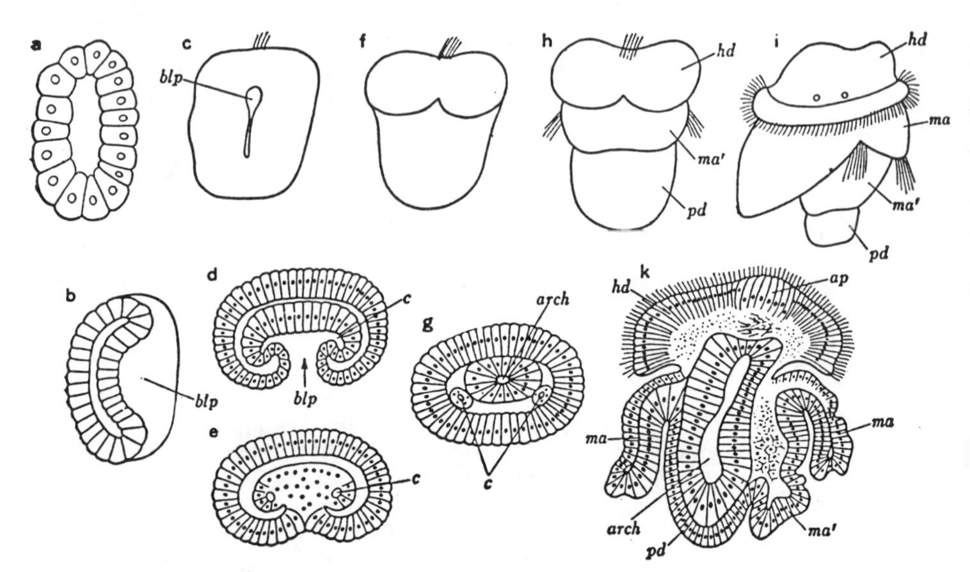

Fig. 8.31 Embryology of *Cistella neapolitana* (Plenk)

a. blastula; b. gastrula; c. ventral view of gastrula; d. cross-section of gastrula through region behind blastopore; f. embryo at stage in which formation of anterior mantle groove separates head from rest of body; g. cross-section of this embryo; h. embryo at stage in which formation of posterior mantle groove separates mantle from trunk; i. differentiation of mantle rudiment; k. longitudinal section of embryo shown in i. *ap*: apical plate, *arch*: archenteron, *blp*: blastopore, *c*:coelomic *sac*, *hd*: head region, *ma*: mantle, *ma'* :mantle region,*pd*: pedal region.

project into the blastocoel about the time the blastopore is closing; these are separated off to become the coelomic pouches (Fig. 8.31d, e, g). This method is in essence not different from that found in the Chaetognatha, Echinodermata, etc. According to Conklin (1902), the mode of coelom formation in *Terebratulina* has undergone a certain amount of modification, but the end result is the same. In these species, a shelf-like process begins to extend from the anterior wall of the archenteron, dividing this structure into upper and lower parts. The smaller, upper part will form the *midgut* (Fig. 8.32 *mg*), while the lower part spreads laterally to surround this from the sides and differentiate into the *coelomic sac* (Fig. 8.32 *c*). When the embryo later undergoes considerable vertical flattening, the left and right sides of the lower structure become separated from

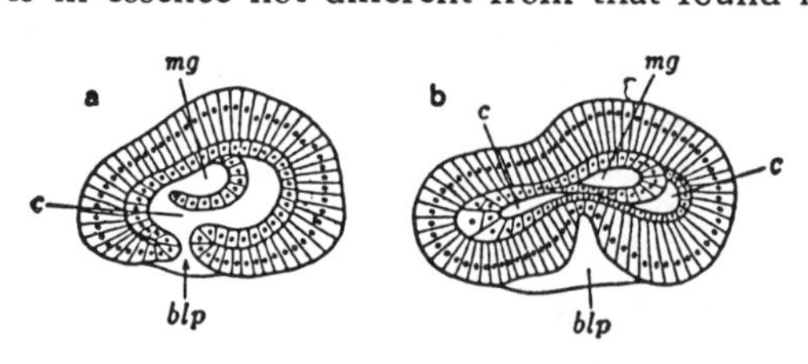

Fig. 8.32 Formation of mesoderm in *Terebratulina sepententrionalis* (Conklin)

a. archenteron about to be separated into upper and lower parts; b. archenteron and right and left coelomic sacs are independent of each other, and blastopore is closed. *blp*: blastopore, *c*: coelomic sac, *mg*: mid-gut.

each other as independent coelomic sacs. At the same time the upper structure is pushed against the ectoderm, and the blastopore is closed.

The subsequent developmental changes will be described on the basis of the process as it occurs in *Cistella*. The anterior part of this embryo (Fig. 8.31c) is rather broader than the posterior; presently a transverse groove makes an appearance on its dorsal side and extends laterally *(anterior mantle groove)*, at a position somewhat posterior to the place where the stomodaeal invagination will later occur (Fig. 8.31f). This is followed by the formation of another groove behind the first one *(posterior mantle groove)*, so that the embryo is divided into three parts, consisting of a *head* portion, a central part enclosed in the mantle, and a posterior *foot* (Fig. 8.31h). The mantle region elongates, forming the mantle rudiment, and overgrowing the foot; sooner or later, however, the dorsal and ventral sides of the mantle grow much more markedly than the sides, so that the mantle rudiment becomes divided into a dorsal and a ventral lobe (Fig. 8.31i). The setae which are formed on the borders of these lobes are larval organs, and will be shed at the time of metamorphosis.

The head portion eventually grows into a parasol-shaped structure with two *eye-spots*. Around the margin of the parasol there is a ring of particularly long cilia, while the ectoderm cells at its apex enlarge and form an *apical plate*. These structures and the way in which the cells at the base of the apical plate give rise to the *superoesophageal ganglion* are extremely reminiscent of the trochophore larva. The *suboesophageal ganglion* differentiates from the ectoderm near the stomodaeal invagination, in the basal part of the ventral mantle.

In both *Cistella* and *Terebratulina*, the re-opening of the stomodaeum occurs at the site where blastopore closure took place; internally an extension from the midgut unites with the stomodaeum, and this part becomes the *oesophagus*. No anus is formed in these species. Together with metamorphic changes of this kind, the coelom becomes greatly expanded, the cavities of the left and right sides meeting in the center and uniting to give rise to a *mesentery*. Since the coelomic epithelium of the pedal region differentiates into muscle tissue, the coelom becomes indistinct here. When the foot later changes into a stalk, this muscle tissue forms the dorsal and ventral stalk muscles. On the ventral side of the embryo just below the ectoderm are formed the *nephridia*; these are differentiated from the coelomic epithelium as blind tubes which open to the outside at the base of the mantle.

(2) METAMORPHOSIS

In general, the free-swimming period is brief. According to Morse, the *Terebratulina* larva begins to attach to the bottom within 24 hours. The development of the arm organs is consequently delayed, so that they fail to appear in as early a stage as they do in *Lingula*, and the only means of locomotion is provided by the ciliated band girdling the head region. As the tip of the foot begins to become attached to the bottom, the foot region transforms into a stalk, and the mantle, discarding its cilia, turns inside-out so that it is directed upward to surround the head region, while the foot region is left exposed. At this time the setae which were present on the mantle are shed; species which have setae as adults form them anew after this stage. Finally, the shell begins to be secreted from the outer lateral surface of the reversed mantle.

Eventually the head region begins to degenerate, although a *preoral process* is formed from part of it. Next a horseshoe-shaped ridge appears on the inner surface of the dorsal mantle, and its two ends extend around the mouth to become the rudiments of the arm organs. On this ridge the cirri will presently be formed,

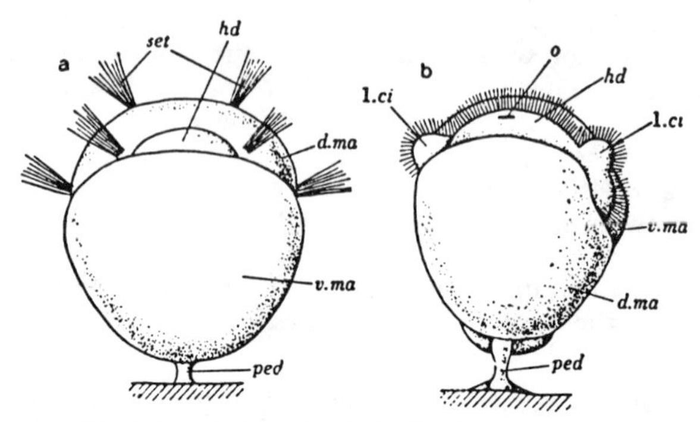

Fig. 8.33 Metamorphosis in *Terebratulina sepententrionalis* (Morse)

a. immediately after metamorphosis; b. appearance of first pair of cirri. 1.*ci:* first pair of cirri, *d.ma:* dorsal mantle, *hd:* head, *o:* mouth, *ped:* stalk, *set:* setae, *v.ma:* ventral mantle.

bringing the shape closer to that of the adult arm. Korschelt u. Heider emphasize the unique mode of formation of the preoral process and arm in these species, but as MacBride points out, considering the arm as having its origin in the head region, in this group also, makes it easier to draw a parallel with *Lingula*.

(3) COMPARISON WITH ECARDINES

As described above, the formation of the preoral lobe and arm in the Testicardines is a rather different process from that of the Ecardines; there are, however, still more fundamental differences to be found in the development of these two orders, to such an extent that some people even doubt that they are particularly closely related. The most important point of difference is concerned with the mode of formation of the mesoderm. *Lingula* is the only available representative of the Ecardines; as we have already seen, its mesoderm arises as thickenings of the two side walls of the meso-entoblastic cell mass, and the coelom appears later as *schizocoels* within these thickenings (Yatsu 1920). In contrast to this, the coelom of the Testicardines is an extremely clear example of an *enterocoel*. Furthermore, the two orders have different modes of cleavage: while that of the former is somewhat special, it nevertheless clearly shows radial symmetry, rather resembling bryozoan cleavage. The cleavage formula of the Testicardines, on the other hand, is difficult to decipher, but has been said to show some traces of spiral cleavage (Conklin 1902).

The amount of research which has been devoted to the developmental problems of the Brachiopoda is still inadequate, making it difficult to formulate definite judgements on the basis of present knowledge. More investigation is necessary to complete the account.

REFERENCES

III. Brachiopoda

Conklin, E. G. 1902: Embryology of a brachiopod (*Terebratulina*). Proc. Amer. Phil. Soc., **41**.

Kumé, M. 1956: The spawning of *Lingula*. Nat. Sci. Rep. Ochanomizu Univ., **6**.

Morse, E. S. 1871—1873: Embryology of *Terebratulina*. Mem. Boston Soc. Nat. Hist., **2** and **3**.

———— 1902: Observation on living Brachiopoda. Nem. Boston Soc. Nat. Hist., **5**.

Percival, E. 1944: A contribution to the life history of the brachiopod, *Terebratella inconspicula* Sowerby. Trans. Roy. Soc. New Zealand, **74**.

Plenk, H. 1915: Entwicklung von *Cistella* (*Argiope*). Arch. Zoöl. Inst. Wien., **20**.

Yatsu, N. 1902: Development of *Lingula anatina*. Jour. Coll. Sci. Tokyo Univ., **17**.

———— 1902: Notes on histology of *Lingula anatina*. Jour. Coll. Sci. Tokyo Univ., **18**.

———— 1920: On the habits of the Japanese *Lingula*. Ann. Zool. Jap., **4**.

ECHINODERMA

Chapter 9

INTRODUCTION

The phylum Echinoderma includes the classes Echinoidea (sea urchins, sand dollars), Ophiuroidea (brittle stars), Asteroidea (starfishes), Holothuroidea (sea cucumbers) and Crinoidea (sea lilies). While this group is considered to be so highly evolved that it is placed close to the Vertebrata, its body structure, at first glance, seems incredibly similar to the radially organized coelenterates and other primitive animals. Once the embryology of these animals is known, however, it is apparent that their radial organization is secondarily acquired as the result of their derivation from sessile ancestors.

In view of the wide experimental use to which sea urchins have been put in recent years, one part of this article will be devoted to a rather detailed description of their development; the other part attempts to review the five echinoderm classes on the basis of their systematic relationship.

1. ECHINOIDEA

(1) EARLY DEVELOPMENT

a. Cleavage stages

Since the sea urchin egg has served so frequently as material for studies of fertilization and cleavage that it might seem to represent the whole animal kingdom in this respect, a description of these processes as they occur in sea urchins has been presented in the Introductory Chapter. In Figure 9.1, the various changes which precede the first cleavage, such as the formation of the fertilization membrane and hyaline layer, the appearance of the sperm aster and syngamy, the streak stage and the development of the diaster, are gathered into a diagram which summarizes the early developmental processes in the sea urchin.

The polar axis is included in the first cleavage plane, which divides the egg into two blastomeres; in the ensuing resting stage these come into close contact with each other so that each is virtually a hemisphere. As the next cleavage approaches, however, the blastomeres round up and tend to separate from each other. Actual measurements (Brown 1934; Mitchison and Swann 1955) show that this results from an increase in the tension of the blastomere surfaces. The same change, in fact, also occurs just before the first cleavage, but it is hard to detect an increase in sphericity of the already round uncleaved egg. Just at this time, however, the hyaline layer undergoes a sudden increase in thickness; the two phenomena taken together provide an adequate indication that cleavage is about to begin.

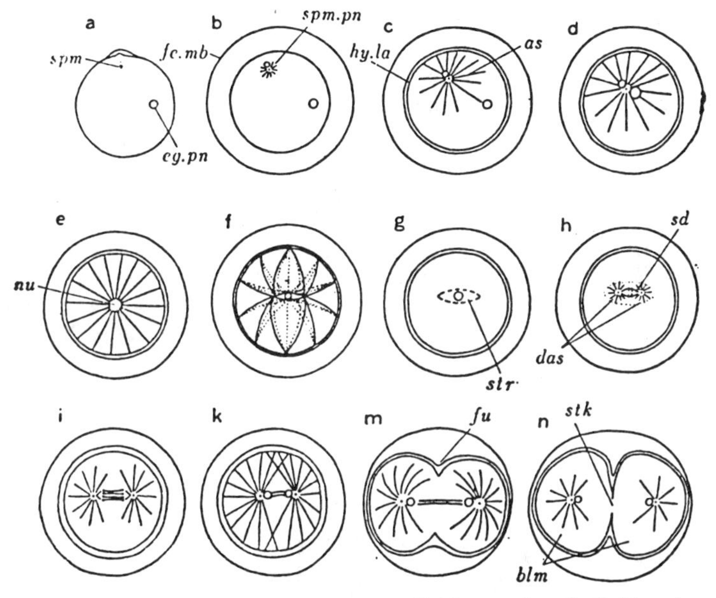

Fig. 9.1 Fertilization, syngamy, nuclear division and cell division in sea urchin egg

a. sperm entrance, formation of fertilization membrane and fertilization cone; b,c. sperm monaster; d,e. syngamy and completion of monaster formation; f,g. streak stage; h. metaphase spindle and early diaster; i. anaphase; k. telophase, completely formed diaster; m,n. successive stages of first cleavage. *as:* astral ray, *blm:* blastomere, *das:* diaster, *eg. pn:* egg pronucleus, *fe.mb:* fertilization membrane, *fu:* cleavage furrow, *hy.la:* hyaline layer, *nu:* fusion nucleus, *sd:* spindle, *spm:* sperm head and middle piece, *spm.pn:* sperm pronucleus, *stk:* stalk, *str:* streak.

The second cleavage is also equal, and also includes the egg axis, so that the resulting four blastomeres are evenly grouped around it (Fig. 9.2c). The central space between the blastomeres is the beginning of the blastocoel. The third cleavage is again equal, but this time the plane of cleavage appears at right angles to the egg axis, dividing the four blastomeres into upper and lower halves and thus giving rise to two layers of four blastomeres each (Fig. 9.2d). Since these eight blastomeres are all of the same size, it is still impossible to distinguish between the animal and vegetal poles.

Next, however, comes the highly characteristic fourth cleavage, which provides the first difference in the behavior of the blastomeres at the two poles of the egg. The four cells of the upper layer divide equally and vertically, forming eight blastomeres of the same size arranged in a ring, while the lower four cells cleave horizontally into two layers. Their cleavage, moreover, is extremely unequal, so that the upper layer consists of large cells, and the lower, of four very small ones (Fig. 9.2e). In the 16-cell stage, then, the blastomeres are arranged in three layers: an upper layer containing eight medium-sized *mesomeres*, a middle layer containing four large *macromeres* and a lower layer composed of four small *micromeres*.

At the next cleavage the ring of eight mesomeres divides into upper and lower layers (Fig. 9.2f). All sixteen of the cells thus formed are destined to become ectoderm, but for convenience in tracing their later distribution, each layer has been given a particular designation: the upper layer is thus known as an_1, and the lower, as an_2 (Hörstadius 1935). The macromeres divide vertically, forming a single ring of eight cells, and the micromeres cleave horizontally into two layers. The embryo as a whole thus consists of a total of 32 cells arranged in five layers.

At the sixth cleavage an_1 and an_2 both divide horizontally to make a total of four layers, but the most important characteristic of this cleavage is that the macromeres are divided horizontally for the first time, into an upper layer of veg_1 and a lower layer *of* veg_2 cells (Fig. 9.2g). Although their labels are similar, the embryological destinies of these two layers are extremely different: the veg_1 layer will become ectoderm, while the veg_2 cells are endodermal in nature. In this cleavage cycle the micromeres fail to divide, so that the number of cells is $32 \times 2-8=56$.

At this stage the prescribed arrangement of the embryo is roughly completed, and can be summarized as:

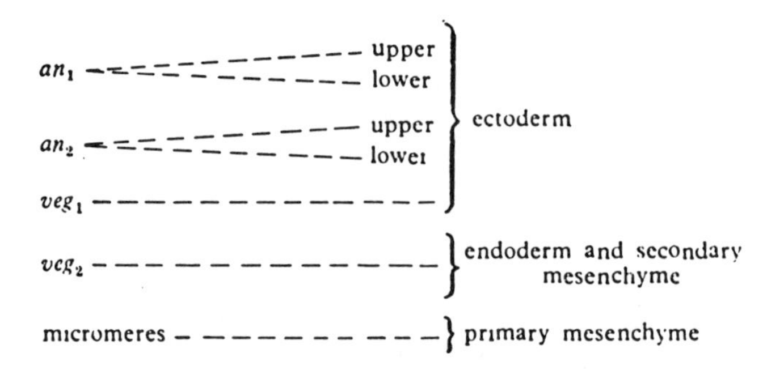

The stage from the $2^5 = 32$-cell stage to about the $2^6 = 64$-cell stage is known as the *morula stage*. The reason for giving it this name (which means 'mulberry') is that the embryo, composed of a rather large number of still discrete blastomeres, presents a bumpy appearance resembling this fruit. But even in the morula stage, the blastomeres continue, as described above, to become rounder immediately before and after cleavage, and to flatten together in the resting intervals, so that the surface of the embryo is especially bumpy during each cleavage, and rather smooth between times.

I think there is little occasion to refer to the cleavage formula of the seventh cleavage. In this phase the cells grow successively smaller with each doubling of their number, and as the surface of the embryo becomes gradually smoother, the term 'morula' loses its applicability.

b. Blastocoel

After it begins to divide, the embryo increases in size at each cleavage. But this apparent enlargement does not represent an increase in the amount of protoplasm; it mainly reflects an expansion of the blastocoel (Fig. 9.2h). In *Clypeaster* and *Mespilia* the blastocoel is especially well developed, the blastomeres surrounding it in a single layer of flattened cells so that the embryo presents the appearance of a golf ball. In *Anthocidaris*, on the other hand, the presence of a blastocoel can hardly be recognized until after the embryo hatches out of the fertilization membrane.

It has been mentioned above that the thickness of the hyaline layer increases just before the first cleavage; species in which the blastocoel is highly developed show this thickening to an extreme degree, while it is hardly perceptible in those with very small segmentation cavities. There is thus a clear correlation between the two phenomena. The author believes that the apparent thickening of the hyaline layer before cleavage is due, not to a thickening of the layer itself, but to the accumulation of fluid between the hyaline layer and the egg surface. Since such accumulation of fluid occurs at each cleavage, the volume of material within the hyaline layer increases each time as though a rubber balloon were being inflated. On the other hand, since the blastomeres are attached to the hyaline layer by fine processes on their surfaces, they are held at the periphery as the layer expands, leaving a space at the center, and thus leading to the formation of a blastula (Dan and Ono 1952; Dan, K. 1952b). In consequence, species in which hyaline layer thickening is exaggerated tend to expand more, and in turn, the process affords an explanation of blastocoel enlargement.

The synchrony of cleavage continues beyond the seventh cleavage, and can be followed until the time of hatching at the tenth cleavage. The size of the blastocoel, however, stops increasing at the seventh cleavage, and remains constant until the embryo hatches.

c. Hatching

No conspicuous changes in the embryonic morphology occur in connection with the eighth, ninth and tenth cleavage cycles, but the physiological process of hatching enzyme secretion takes place during this period. While this secretory activity shows a peak at the time of the ninth cleavage, the embryo undergoes one more cleavage (the tenth) and forms cilia before the fertilization membrane dissolves and lets it swim away. The time required from fertilization to hatching is 5 hours at about 30°, 10—12 hours at 20° and 30 hours at about 10°C.

The newly hatched embryo, seen as a whole, resembles the ciliated protozoa in having its whole body thickly covered with cilia; considered from the cellular viewpoint, it is very different, since each cell bears only one cilium. However, the beating of these organelles is clearly ciliary in nature, consisting of an *effective stroke* followed by a *recovery stroke,* and making it impossible to think of them as flagella. Before the fertilization membrane dissolves in some species, the embryo begins to rotate within the perivitelline space as the result of its ciliary activity, giving the impression that mechanical force is required to tear the membrane. The *Pseudocentrotus* embryo, however, shows no rotational activity whatsoever before hatching, and *Hemicentrotus* embryos will hatch even if cilium formation is completely suppressed by the removal of calcium from the medium (Endo 1953).

It therefore appears that destruction of the fertilization membrane is a purely chemical process.

Judging from the position of the nuclei in the cells during this period, Ishida (1954) suggests that if the general rule governing the relation between location of nucleus and direction of secretion in secretory cells also applies in this case, the hatching enzyme must be liberated first into the blastocoel. But regardless of whether the original direction of secretion is inward or outward, it is obvious that in order to reach the perivitelline space the enzyme must pass through the hyaline layer. Since, however, this is a semipermeable membrane and therefore impervious to colloids, some kind of special provision must be made for an outlet. There is an idea that such an outlet is located at the vegetal pole, but its nature is still completely unknown. At any rate, the enzyme passes by some means into the perivitelline space, where it attacks the inner side of the fertilization membrane.

The method of Ishida (1936) for obtaining the hatching enzyme consists in collecting a large number of *Hemicentrotus pulcherrimus* embryos in a small volume of sea water just before they begin to hatch. When most of them have hatched, the enzyme-containing supernatant is separated by centrifugation. If young embryos or recently fertilized eggs are placed in this supernatant, their membranes are dissolved within about 30 minutes. It is also possible to investigate the chemical properties of the enzyme itself by using this supernatant (Sugawara 1943). With most other species, however, the necessary crowding at the time of hatching exerts an injurious effect on development, so that this method has unfortunately been found practicable only with *Hemicentrotus* in Japan and *Arbacia punctulata* in America (Kopac 1941).

d. **Blastula**

The embryo continues to develop from morula to blastula, without any clear boundary between the two stages. The embryo of *Mespilia,* as stated above, attains the dimensions of the early swimming stage at the seventh cleavage, and can thereafter be called a blastula if size is the only criterion. If, on the other hand, the free-swimming state is held to be an important characteristic of a blastula, hatching obviously constitutes the boundary between the two stages. Unfortunately the two usages are mixed indiscriminately in the literature.

If, for the time being, we accept hatching as the starting point of the blastula, we have an embryo which has undergone ten cleavage cycles and is covered with cilia. But do we have any idea how many cells we are dealing with? According to MacBride (1914) there are 808, and this number agrees exactly with that found by Endo (unpublished). If we simplify the arithmetic by supposing that our egg has cleaved into 1000 cells of equal size, the diameter of these small cells will be 1/10 the original diameter of the egg, since their volume is 1/1000, or $(1/10)^3$, of its original volume. This being so, the surface area of the small cells will be $(1/10)^2=1/100$ of the original area. But since the final number of cells is 1000, their total surface area will be $1000 \times 1/100$, or 10 times the area of the original egg surface. This bit of calculating presents us with the problem of accounting for the origin of enough extra surface to cover nine uncleaved eggs. It may be imagined that the original surface material is spread out uniformly, or alternatively, that it remains unstretched on some parts of the blastomeres while new surface is formed alongside of these areas.

The work of Motomura (1935) has provided the solution of this problem. He found that the orange pigment extracted from the eggs of *Hemicentrotus pulcherrimus* appeared red through a Wratten No. 49 blue filter, and this led him to use such a filter for observing the intact eggs. He found that the orange pigment,

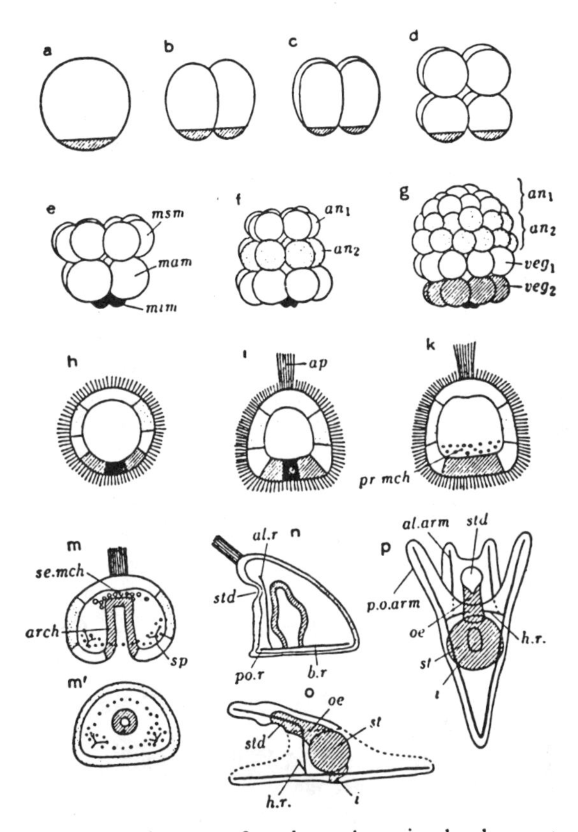

Fig. 9.2 Diagram of early embryonic development in the sea urchin (Fertilization membrane and hyaline layer are omitted) (Hörstadius)

a. 1-cell stage; b,c. 2-cell and 4-cell stages; d. 8-cell stage; e. 16-cell stage; f,g. cleavage stages; h,i. blastula stage. Dotted, cross-lined and black areas show localization in blastula of cells descended from blastomeres of fifth and sixth divisions; k. formation of primary mesenchyme; m,m'. gastrula. Appearance of bilaterality; n. pyramid stage. Primary and secondary mesenchyme cells are omitted from sketch, but skeleton formed by primary mesenchyme is shown in black; o. pluteus stage (lateral); p. frontal view of pluteus in natural swimming posture. *al.arm:* anterior lateral arm, *al.r:* anterior lateral rod, an_1, an_2: see text, *ap:* apical tuft, *arch:* archenteron, *b.r:* body rod, *h.r:* horizontal rod, *i:* intestine, *mam:* macromere, *mim:* micromere, *msm:* mesomere, *oe:* oesophagus, *p.o.arm:* post-oral arm, *po.r:* post-oral rod, *pr.mch:* primary mesenchyme cell, *sc.mch:* secondary mesenchyme cell, *sp:* spicule, *st:* stomach, *std:* stomodaeum, veg_1, veg_2: see text.

which until then was thought to be diffused uniformly through the cytoplasm, was actually localized in special granules lying close to the cortex. If Wratten No. 49 filters are difficult to obtain, an adequate substitute can easily be made by staining a sheet of cellophane with 0.5% trypan blue dye. When Motomura extended the use of this method to a later stage, he found the orange granules which had covered the surface of the unfertilized egg still at the surface in the swimming blastula, and entirely absent from the inner cell walls. That the material making up the original egg surface remains in this position in spite of the increase in cell surface area is a matter of considerable embryological significance.

The eggs of the American *Arbacia punctulata* and the Mediterranean *Arbacia lixula* contain conspicuous red pigment granules which can easily be observed without the use of any special device (see Dan and Dan 1940, Pl. II B, C). Since these *Arbacia* granules behave in the same way as the pigment granules of *Hemicentrotus* in the early cleavage stages, it is very likely that they are also arranged at the surface of the blastula, but so far as the author is aware, no one before Motomura has pointed out this fact.

To return to the matter of following the developmental process — the blastula immediately after hatching is practically spherical (Fig. 9.2h), and it is impossible to distinguish the position of the primary egg axis, but very soon it elongates in the direction of the axis. The cell layer at the anterior end thickens and changes into a sensory element provided with a tuft of long, immotile cilia (Fig. 9.2i). This structure, which obviously enables the embryo to sense the presence of obstacles that it meets, is called an *apical organ*. The immotile cilia alone are referred to as the *apical tuft*, and the cellular thickening underlying them is called the *apical plate*.

In the blastula of *Hemicentrotus*, a wedge-shaped extension of the blastocoel pushes through the cellular layer in the center of the posterior body wall exactly opposite the apical organ, almost reaching the outer surface. There are some who think that this might be the outlet where the hatching enzyme is released, but since no such structure can be detected in other species, it remains no more than a possibility.

e. **Primary Mesenchyme Cells**

About an hour and a half (at 30° C) after hatching — i. e., about six and a half hours after fertilization — a change can be observed in the blastular wall at the vegetal pole side. Some of the blastomeres, which until this time have been arranged in an orderly row, begin to slip out of line into the blastocoel. This makes the hitherto smooth lining of the blastocoel in this region look like a patch of ground through which young mushrooms are pushing their heads. As the heads grow larger, the parts of the cells still in the row become more and more attenuated, and finally they slip out altogether from the wall and are free in the blastocoel (Fig. 9.3). According to Gustafson and Kinnander (1956), the separation of these cells takes place at the center of the blastular wall on the vegetal pole side, and those which are released first are pushed in regular order toward the periphery. The total number of cells thus released amounts to several dozen, all of which are descendants of the micromeres formed in the 16-cell stage, and mesodermal in nature. After their separation from the blastular wall they are known as *primary mesenchyme cells*, and their immediate function is to form the spicules.

In *Clypeaster* and *Mespilia*, which have large blastocoels, the primary mesenchyme cells first spread out in the blastocoel and then reassemble in the posterior region, but in *Hemicentrotus* they remain in place, lined up along the posterior blastular wall (Fig. 9.2k, 9.3).

It should be pointed out here that although a large number of cells have escaped into the blastocoel, this does not mean that a hole has been left in the blastular wall. When, as described above, these cells gradually slip out of the lineup, the surrounding cells draw together and replace them. In order to avoid the formation of gaps, the cell wall in general has to spread out toward the vegetal side. It should be firmly kept in mind that as a result of this shift in position, the posterior wall of the embryo after this time is composed of cells descended from the veg_2 blastomeres (Fig. 9.2k). At this stage the embryo is rather pointed at the anterior side, while the posterior side formed by the veg_2 cells is quite flat. It is for this reason that the posterior side is referred to as the posterior wall.

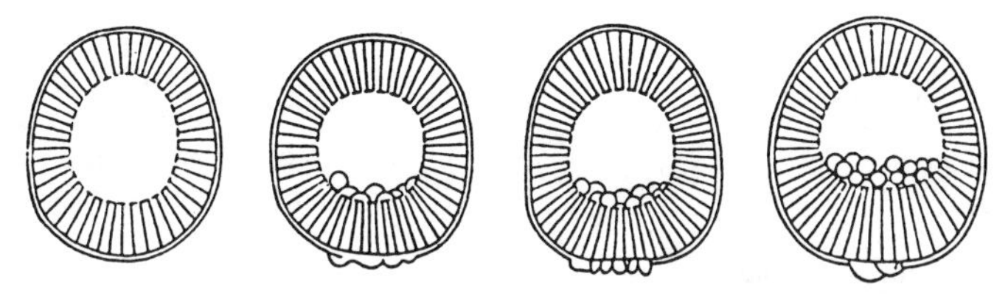

Fig. 9.3 Formation of mesenchyme and change of hyaline layer in the sea-urchin *Anthocidaris crassispina*

One problem which arises at this point concerns the situation with respect to the hyaline layer after the release of the primary mesenchyme cells. It is obvious that since the hyaline layer is closely attached to the outer surface of the blastula, the parts of the layer which were in contact with the lost cells should be lifted up as wrinkles when the neighboring cells draw together to close the gaps. This is precisely what happens: as the primary mesenchyme cells move out of the blastular wall, the posterior part of the hyaline layer separates as blisters, some of which unite into pouches (Runnström 1935). This is most strikingly seen in *Anthocidaris* (Fig. 9.3).

The same thing can also be observed in *Pseudocentrotus, Hemicentrotus* and *Mespilia,* although some of them form only small processes instead of the pouches. In these cases it may be that pouches are formed but rupture immediately. *Clypeaster* generally shows no such effect, or at most process-like roughenings of the surface; these are believed to represent the traces of ruptured blisters, since swimming blastulae of this species can sometimes be found trailing scraps of debris behind them (Dan, K. 1952b).

f. Gastrula

The process of gastrulation begins about ten hours after fertilization in the summer sea urchins, or after about twenty hours at 20°C (Kumé 1929). Simultaneous with this activity, however, is the formation of spicules by the primary mesenchyme cells; these two phenomena will be considered separately.

i) *Spicule Formation*

After becoming disengaged from the blastular wall, the primary mesenchyme cells in *Hemicentrotus* and *Anthocidaris* move sideways and gather in two places, in contact with the inner surface of the vegetal body wall. These two places are never exactly on opposite sides of the blastular periphery — actually, their radii form an angle of about 140° at the center. No one has yet discovered why the mesenchyme cells choose such unlikely positions, but at any rate the side subtended by the 140° angle is the ventral side of the embryo, and the 220° angle subtends its dorsal side (Fig. 9.2m, 9.4b). With the choice of these two places, the embryo at one time acquires not only bilaterality but also dorsoventrality. In the midst of each of these accumulations of mesenchyme, a certain number of the cells become arranged in a triangular formation, and eventually a triradiate *spicule* appears at their midpoint. The spicules have the appearance of being formed in the space between the mesenchyme cells, but it is believed that these cells use their pseudopodia to form a *matrix*, and the spicule is actually laid down within this matrix sac (Okazaki 1956b).

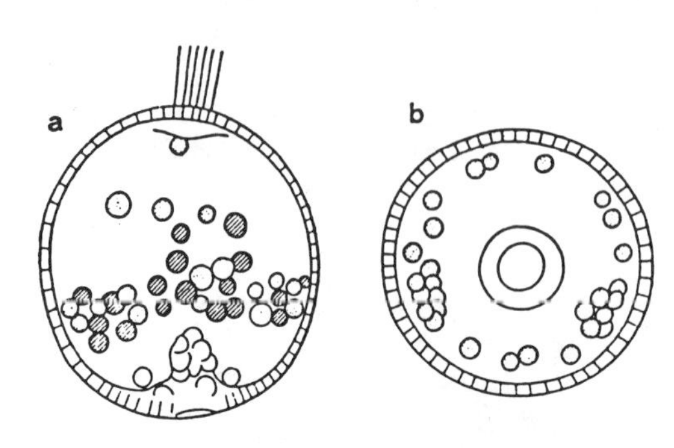

Fig. 9.4 Primary mesenchyme of blastula in the sand dollar *Clypeaster japonicus*

a. view perpendicular to egg-axis. Primary mesenchyme cells are arranged in a circle at about 1/3 of the distance from the vegetal pole. Invagination of archenteron beginning in vegetal pole region; b. view parallel to egg-axis. Circle in central region shows endodermal plate. Primary mesenchyme cells forming clumps at two peripheral points. Upper side of sketch shows dorsal side of embryo.

The above description applies to *Hemicentrotus* and *Anthocidaris,* in which the primary mesenchyme cells do not migrate away from the posterior wall of the blastula. However, in *Clypeaster* and *Mespilia,* which develop a spacious blastocoel, the primary mesenchyme cells first form a perfect ring at a level some distance away from the posterior wall, and only then do they aggregate at two points (Fig. 9.4). Such cases, in which the problem of estimating the distance from the posterior wall is superimposed on the determination of the 140° angle, point out a new factor involved in skeleton formation. Although the primary mesenchyme cells move about by means of their pseudopodia, their movement is restricted to traveling along the inner surface of the ectodermal wall. This fact alone is enough to suggest a close relation between mesenchyme and ectoderm. According to Hörstadius's analysis (1939b), the primary mesenchyme cells form perfect spicules only when they are in contact with ectoderm derived from veg_1. In other words, some adverse situation seems to arise if they are combined with an_1 or veg_2. If this is so, the ability of *Clypeaster* and *Mespilia* mesenchyme cells to recognize a definite position which is seemingly very difficult to define is actually the consequence of tactic response to an attraction exerted by certain ectodermal elements. However, the mechanism of this effect is utterly unknown.

There is a general belief that a mutual influence between the physiological properties of the animal and vegetal poles plays a highly significant role in sea urchin development (Runnström 1928, 1935). Pushing this concept a step further, the developmental fate of any part of the larva must be determined by the ratio of the two polar properties involved in that part. When the primary mesenchyme cells are in contact with an_1, spicule formation will be impaired by an excess of the animal pole characteristics possessed by the an_1 cells, and when they are in contact with veg_2, by the too strongly vegetal character of the veg_2 cells. The successful skeleton formation occurring in collaboration with veg_1 must be due to the appropriate polarity characteristics of the veg_1 cells. The interesting fact has been known for a long time that inorganic salts such as LiCl and NaSCN enhance the characteristics of one pole and suppress those of the other, thus tipping the balance to one side (Herbst 1892). Emphasizing the animal character and diminishing the vegetal character is called *animalization,* while the opposite process, increasing the vegetal character at the expense of the animal, is called *vegetalization* (Lindahl 1936). When the larva is vegetalized by LiCl, the position taken by the primary mesenchyme cells is correspondingly shifted toward the animal pole and in extreme cases, the mesenchyme cells are pushed directly against the animal pole where they manage, with much difficulty, to form a defective skeleton. Such facts can be explained by thinking that the polarity characteristic of the normal veg_1 is shifted as the result of vegetalization by the reagent.

ii) *Gastrulation*

Since it has been stated that the cells constituting the posterior wall of the blastula after the departure of the primary mesenchyme cells are the descendants of veg_2 (see p. 287), it may not be difficult to anticipate that the gastrulation process is performed by these cells. The first morphological change associated with gastrulation is the closer aggregation and growth in height of the veg_2 cells as distinguished from those of the veg_1 group, which remain consistently flat; the second change is a gradual sinking of the veg_2 group as a whole into the blastocoel. Since the aggregate has the shape of a circular plate, it is called the *endodermal plate.* As a consequence of this change, a longitudinal section of the *primitive gut (archenteron)* at the beginning of invagination resembles a low, flat-topped table (Fig. 9.5a). As will be explained below, the secondary mesenchyme cells are included with the endoderm in the endodermal plate (cf. p. 282).

Soon after the endodermal plate begins to sink into the blastocoel, pseudopodia appear on the cells situated in the center of the plate. These are the secondary mesenchyme cells. However, these pseudopodia of the secondary mesenchyme cells, unlike those of the primary mesenchyme cells which extend only along the inner surface of the blastular wall, traverse the blastocoel and reach the opposite ectodermal wall (Fig. 9.5b). With the formation of the pseudopodia, the flatness of the endodermal plate is lost and the archenteron changes to a tall, slender cylinder. Although the pseudopodia at first extend radially as if in search of something, they are eventually directed toward the animal pole, and following the lead of these pseudopodia, the archenteric tip establishes connection with the animal pole (Fig. 9.5c). In forms with a blastocoel too large to be traversed by the pseudopodia of a single cell, some of the cells leave the archenteric tip precociously and form bridges across the space to make connection with the animal pole.

That the pseudopodia contract to bring the archenteric tip into contract with the animal pole is indicated by the following three facts. (1) The part of the ectodermal wall where the pseudopodia are concentrated is often indented as the result of their inward pull. (2) The blastula of *Pseudocentrotus* has a roundish contour (Fig. 9.5a) but as soon as it gastrulates, it becomes flattened in the antero-posterior direction so that the width exceeds the height (Fig. 9.5c) (Okazaki 1956a).

(3) In all the species so far studied, the longitudinal section of the tip of the completed archenteron is squarish, resembling a suspended canopy (Dan and Okazaki 1956) (Fig. 9.5d, e, Fig. 9.2m).

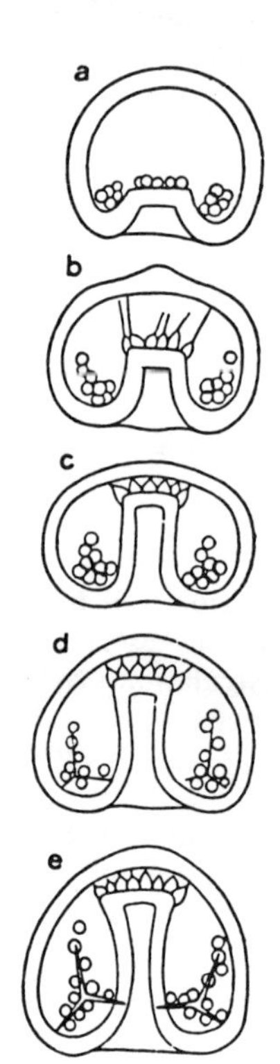

Fig. 9.5 Gastrulation and behavior of secondary mesenchyme cells in the sea-urchin *Pseudocentrotus depressus*

a. secondary mesenchyme cells before formation of pseudopodia, embryo rounded; b. appearance of pseudopodia; c. connection formed between tip of archenteron and animal pole region by pseudopodia of mesenchyme cells. Note that embryo has become flattened; d, e. archenteron is stretched as blastocoel enlarges.

Can it be said, then, that the entire process of gastrulation from the inward sinking of the endodermal plate to the elongation of the primitive gut is brought about by the contractility of the pseudopodia? This is by no means the case. In the first place, when the endodermal plate begins to invaginate, the secondary mesenchyme cells show no trace of pseudopodia. In the second place, even if the pseudopodia are torn loose soon after their appearance, the archenteron can elongate autonomously to a certain extent. To these points, which remain open to question, must be added the further puzzle of how the secondary mesenchyme cells locate the animal pole. The study of Gustafson and Kinnander (1956), which immediately followed our paper (Dan and Okazaki 1956), used micro-cinematographic analysis and arrived at the same conclusion with respect to the importance of the pseudopodia in invagination.

As the subject of artificial severance of the pseudopodia was touched upon in the preceding paragraph, it will be added that under this condition, after the archenteron elongates to about half the blastocoelic diameter, it is eventually everted so that the larva becomes an *exogastrula*. There is a tendency, however, to associate this term with larvae vegetalized by LiCl, which are obviously highly abnormal larvae resulting from suppression of the animal pole characteristics. When the pseudopodia are mechanically severed, the larva develops into a normal-looking pluteus, except for its everted gut, since the gradient systems have not been disturbed. In this sense, it would be more proper to distinguish between the two abnormalities by calling the latter a "hernia larva".

(2) SWIMMING LARVA

a. Pyramid stage

The *pyramid stage* follows the gastrula. As is clear from Figure 9.2n, the body axis of the gastrula tilts to one side to give rise to this stage. Because of this tilting or bending of the body axis, the larva assumes a pyramid shape with the aboral side as the apex (Fig. 9.2n right side) and the oral side as the base (Fig. 9.2n left side). This is an important period during which the establishment of the body organization and the growth of the spicular skeleton, the formation of the stomodaeum, the tissue differentiation of the digestive system and the separation of the coelom are taking place.

i) *Skeleton formation*

If the triradiate spicule mentioned above (p. 288) is superimposed on Figure 9.2n, two of the three branches of the spicule are seen to fall on the plane of the page (the two branches drawn with solid lines in Figure 9.2; *b. r.*, *al. r.*), while the last branch stands perpendicular to the page and is directed toward the median plane of the larval body. In Figure 9.2o and p, this is shown as *h. r.* Of the two branches shown in Figure 9.2n, the *basal rod (b. r.)*, which extends along the base toward the right, represents a pair of skeletal rods which run along the posterior body wall toward the aboral side, including the blastopore between them. (Figure 9.2n is a view of m′ as it appears from the right side; the upper side of the drawing in m′ corresponds to the right side of n.) The mutual positions of the pair of rods are such that they converge aborally and diverge orally. Since the two aggregates of primary mesenchyme cells are situated on the inner surface of the ectodermal wall on an arc which embraces a 140° angle at the center (Fig. 9.2m′), the body rods will naturally converge on the aboral side when they elongate tangentially to the archenteron.

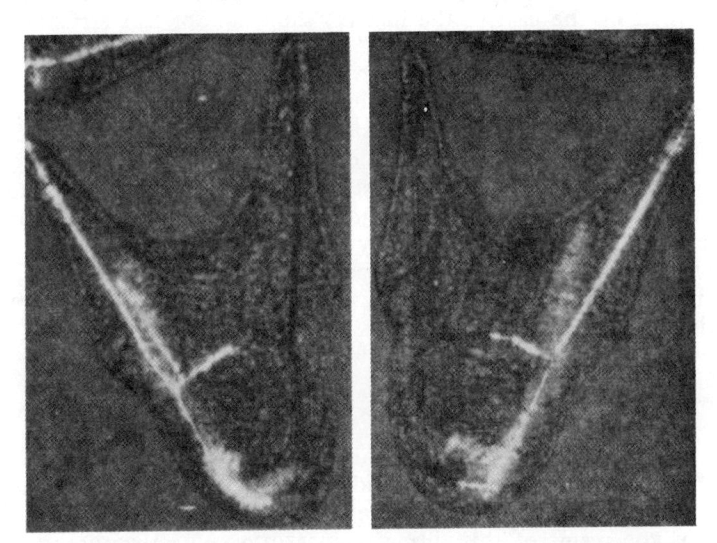

Fig. 9.6 Birefringence of the skeleton in the sea-urchin *Pseudocentrotus depressus* (Photographed by K. Okazaki)

When a larva is rotated between crossed nicol prisms, the right and left sides of the skeletal system brighten and darken independently. The fact that these optical changes occur simultaneously in each side of the system suggests that all the molecules making up the skeleton on one side are arranged in the same direction. The left side of the skeleton has an extra branch, but its optical behaviour is exactly the same as that of the right. This phenomenon is evidence that the arrangement of molecules in the skeletal system is unaffected by abnormality in structure.

Each triradiate spicule soon acquires a new branch which is in line with the body rod, but directed in the opposite sense. These are the *post-oral rods*. If the body rods and the post-oral rods elongate on the same line, the natural consequence will be that their ends will push out the body wall, forming projections. Since the two body rods come together on the aboral side. However they form a single common projection, while the two post-oral rods, diverging on the oral side, give rise to two projections (Fig. 9.2p). These two projections are the *post-oral arms*. Referring once more to Figure 9.2n, the third branch of the original spicule *(al. r.)* is seen to extend up along the page and bend orally. This is the *antero-lateral rod*. After the larva reaches a fairly advanced stage, the antero-lateral rods of the two sides separately extend to form a pair of *antero-lateral arms*.

Although the formation of these arms has so far been described as though it resulted simply from an out-pushing caused by the elongation of the rods, the r al situation seems to be somewhat more complicated. For instance, if the development of the post-oral rod is suppressed, the ectoderm destined to form the arm proliferates and gives rise to wrinkles or vesicles in spite of the absence of the rod. However, these vesicles fall away as long as the rod formation is completely suppressed. In other words, the skeleton and the ectoderm cooperate to produce the arm (Okazaki 1956a, Fig. 4).

Here the mechanism underlying skeleton formation will be discussed briefly. It has already been stated that triradiate spicules are formed within two matrices lying among the primary mesenchyme cells (p. 288). Consequently, as the spicule develops into a skeleton, the direction of its elongation, its shape and its branching are all governed by the cast of the matrix; the shape of this structure, in turn, is determined by the pseudopodia of the primary mesenchyme cells. A very interesting fact is that if the skeleton thus elaborated is examined with a polarizing microscope, the entire skeleton is found to be sharing a common optical axis, irrespective of the intricacies in its shape. In other words, no matter how the skeleton may bend or bifurcate, the molecules of calcium carbonate constituting it are all arrayed uniformly. Although such an arrangement of the molecules is practically the same as that in the crystalline lattice, the consequence is not a simple crystal but a skeleton which is delicately adapted to the mode of life of the sea urchin larva. This may be taken to mean that the process of steletal growth is probably the same as that of crystal growth, and we see the living organism cleverly turning this inanimate force to its own purpose (Fig. 9.6).

ii) *Tissue differentiation*

As the archenteron tilts to one side, a shallow depression appears in the ectodermal wall on that side. This is the stomodaeum. The blind end of the archenteron, as the result of this tilting, comes into contact with the depression, and an opening is made through the two layers to form the mouth of the larva. A band of strong cilia differentiates on the body surface surrounding the larval mouth. The stomodaeum, derived from an_1 blastomeres of an early stage, was for some time believed to be induced by the tip of the archenteron, because partial larvae consisting of an_1 cells fail to form this structure. This idea has been disproved by Hörstadius, who succeeded in showing that an an_1 group isolated after being left in contact with the vegetal elements for several hours is able to form a stomodaeum. He interpreted this result as indicating that direct contact by the archenteric tip is not necessary for the formation of the stomodaeum: if the vegetal influence is exerted on an_1 cells for some time, this suffices to endow the an_1 group with the capacity to form a stomodaeum. Such an example again indicates that polarity, rather than induction, forms the basis of harmonic development in sea urchins.

After the formation of the mouth, the rest of the digestive tract ceases to be a simple cylinder. The front part, continuous with the mouth, becomes narrow and slender as the *oesophagus;* the middle part expands to become the *stomach* and the rear part changes into the slender *intestine,* thus establishing the so-called tripartite digestive tract. The blastopore of the gastrula stage remains unchanged and serves as the *anus.*

To form the *coelom,* the mass of secondary mesenchyme cells which lies between the archenteric tip and the animal pole is divided into two groups when the archenteric tip unites with the stomodaeum; these give rise to a pair of small *coelomic vesicles* on the sides of the oesophagus, the rudiments of the future coelom. Before this appearance of the coelom, the free space within the larval body is provided by the blastocoel.

The last point to be added to this description of the pyramid stage is that when the larvae of this stage swim, they show a queer rolling motion quite different from the regular spiralling of the blastula and gastrula. This is because the pyramid stage represents a transition between the gastrula and pluteus stages, during which the axis of swimming motion is changing.

b. **Pluteus stage**

i) *External appearance*

The *pluteus stage* is essentially only a continuation of the pyramid stage. But because of the enormous elongation of the skeletal rods, the aboral end is stretched to a more acute point, and the postoral and antero-lateral arms extend directly outward from the sides of the stomodaeum. When the larva reaches this shape, the body axis deviates from that of the gastrula by nearly 90°, and the larva begins to swim with the pointed aboral end down and the four arms up. If Figure 9.2n, o, and p are compared, this change in posture will be understood. From the pluteus stage, the side including the mouth and anus is called the ventral side. The apical tuft, which is shown at the top of Figure 9.2n, has become the upper brim of the stomodaeum in p, where it spreads along the front edge and gives rise to a band of strong cilia.

The larva at this stage has the shape of a pyramid with four legs, or perhaps of an inverted stepladder. Johannes Müller, however, named this larva the "pluteus", because of the similarity which he saw to an easel.

The pluteus stage is reached within a few days following fertilization. Since the larvae do not feed until they become plutei, it is relatively easy to rear them to this stage. On the other hand, more than a month is required before metamorphosis takes place; in order to see the various changes occurring during this long period, it is absolutely necessary to feed the pluteus larvae with planktonic diatoms. In ordinary cases, therefore, encounters with late pluteus stage larvae are most likely to occur during examination of plankton hauls. One common denominator of the various changes in external morphology which occur during this period is an increase in the floating capacity of the larva. There are two ways of achieving this purpose: (1) to acquire strong cilia and (2) to increase the body surface which supports these cilia. As to (1), bands of cilia running horizontally on the dorsal side and around the bases of the arms develop during the late pluteus

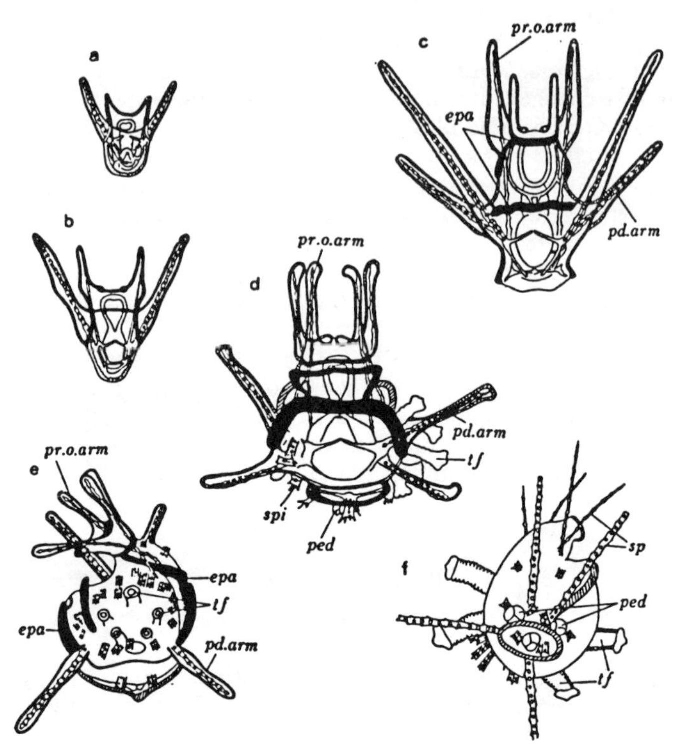

Fig. 9.7 Late larval stages in the sea-urchin *Anthocidaris crassispina* (Onoda)

a. pluteus, 72 hours (ventral view); b. 8 days (ventral view); c. 15 days (ventral view). Pluteus with pre-oral and postero-dorsal arms and epaulette; d,e,f. successive stages observed during metamorphosis on 33rd day; d. ventral view. Tube feet project from echinus rudiment at right side of larva; spines and pedicellaria at lower side; e. left side view; f. young sea-urchin beginning locomotion after discarding larval structures. *epa:* epaulette, *pd.arm:* postero-dorsal arm, *ped:* pedicellaria, *pr.o.arm:* pre-oral arm, *sp:* remnant of pluteus skeleton, *spi:* spine, *tf:* tube foot.

stage. These bands are formed on ridges which project from the general level of the body surface so that the structure as a whole resembles an epaulette of a full-dress French uniform of the last century, and is actually named after it (Figs. 9.7, *epa.*, 9.8, *epa.*). Increase in body surface (2), is accomplished by increasing the number of arms. Although there are some variations among different species, most often two more pairs of arms are added. One pair appears between the two antero-lateral arms in front of the mouth *(pre-oral arms)*, and the other pair is formed laterally *(postero-dorsal arms)* (see also Fig. 9.23).

With respect to the late larval development of Japanese sea urchins, it is advisable to consult the original papers by Mortensen (1921, 1931, 1937, 1937), Kumé (1929), and Onoda (1931, 1936, 1938), although a few representative figures taken from Onoda's study will be reproduced here as a suggestion of the sort of changes that take place.

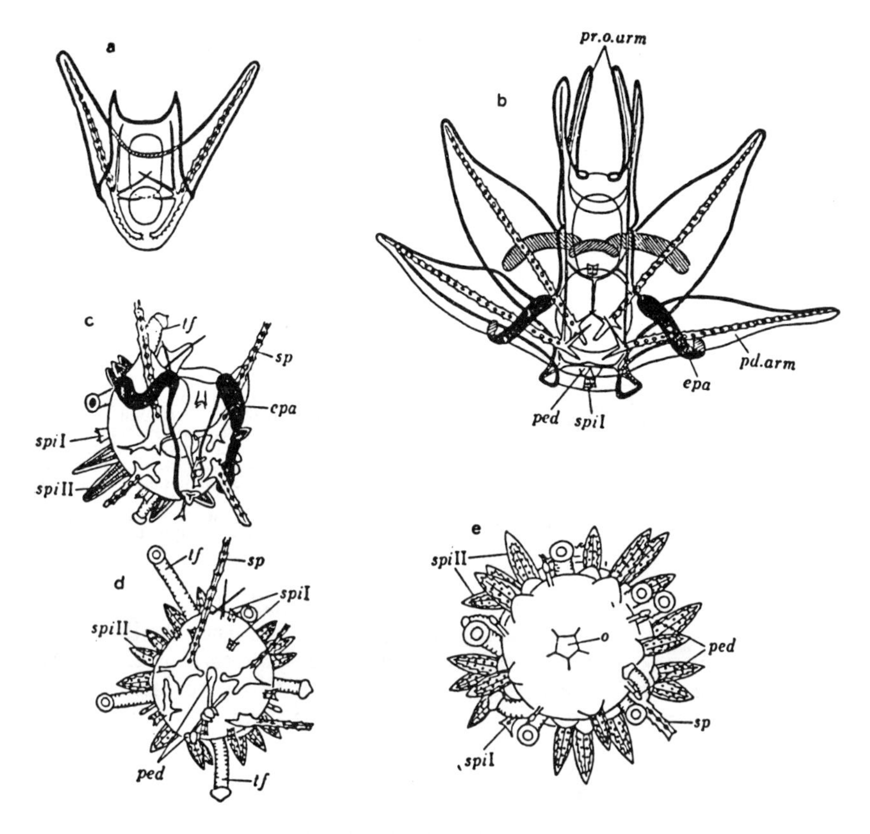

Fig. 9.8 Late larval development in the sea-urchin *Mespilia globulus* (Onoda)
a. pluteus larva, 6 days; b. 30 days; c. 35 days (right side view of metamorphosing larva). Tube foot and spine appearing from amniotic cavity at left side; d. 35 days (dorsal view of young sea-urchin just after metamorphosis); e. ventral view of larva at same stage as d; part where adult mouth will open appears as pentagonal structure. *epa:* epaulette *o:* mouth, *pd.arm:* postero-dorsal arm. *ped:* pedicellaria, *pr.o.arm:* pre-oral arm, *sp:* remnants of larval skeleton, *spi*I: spine, *spi*II: sharp spine formed in amniotic cavity, *tf:* tube foot.

ii) *Coelom and hydrocoel*

It was mentioned above that in the pyramid stage, when archenteron and stomodaeum become connected, the secondary mesenchyme cells shift their position and form two lateral masses on the archenteric tip (p. 293). These two masses develop into two slender tubes running along the oesophagus and stomach; these are the *coelomic vesicles,* the rudiments of the future coelom and hydrocoel. Soon each slender tube acquires a constriction in the middle, separating it into *anterior* and posterior *chambers* (Fig. 9.9 *l.a.c, l.p.c, r.a.c, r.p.c.*). In the meantime, the anterior chamber of the left coelom constricts off a new chamber in an obliquely posterior direction. This is the *hydrocoel* which will give rise to the adult *water vascular system* (Fig. 9.9, *hyd. c.*). However, the hydrocoel and the left anterior chamber remain connected by a communicating tube, the *stone canal* (Fig.

9.9, *st. ca*), and a swelling at the junction between the anterior chamber and the stone canal is the rudiment of the *axial sinus*. Although the anterior chamber of the right coelomic vesicle (Fig. 9.9, *r.a.c*) likewise constricts off a small chamber a little later than the left side, the communication between the two is severed, and the small chamber becomes isolated (Fig. 9.9, *md, vs*) and is eventually pushed to the dorsal side of the adult to become the *madreporic vesicle*.

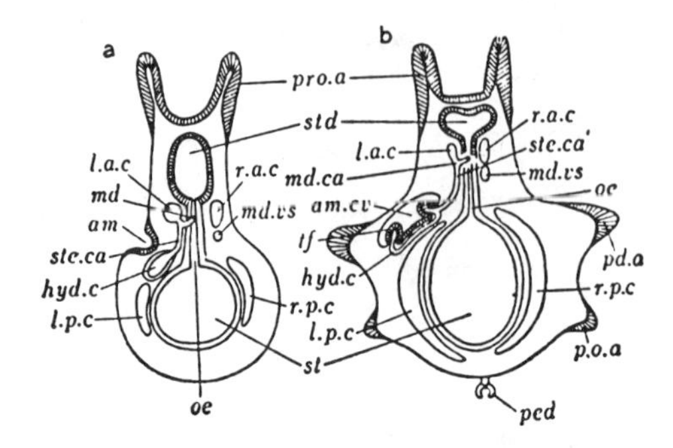

Fig. 9.9 Diagrams showing relation among coelom, hydrocoel and amniotic cavity

am: amniotic invagination, *am.cv:* amniotic cavity, *hyd.c:* hydrocoel, *l.a.c:* left anterior coelom, *l.p.c:* left posterior coelom, *md:* madrepore, *md.ca:* madreporic canal. *md.vs:* vestige of right madreporic canal, *oe:* oesophagus, *pd.a:* postero-dorsal arm, *ped:* pedicellaria, *p.o.a:* postero-oral arm, *pro.a:* preoral arm, *r.a.c:* right anterior coelom, *r.p.c:* right posterior coelom, *st:* stomach, *std:* stomodaeum, *ste.ca:* stone canal, *tf:* tube foot rudiment.

In addition to the stone canal and the hydrocoel, the anterior chamber of the left coelom sends out toward the dorsal side of the larva another tube which opens to the outside. Its opening constitutes the *madrepore* or *dorsal pore*, and the tube leading to it is the *pore canal*. Not only is this pore canal widely found among echinoderms, but its presence is believed to indicate a kinship between the Echinoderma and Hemichordata (Fell 1948).

(3) METAMORPHOSIS

a. Echinus rudiment

Preparation for metamorphosis begins at the body surface of the pluteus. A shallow depression appears in the ectoderm between the left postero-dorsal arm and the post oral-arm of the same side — that is, at the base of the arms on the left side of the body, exactly above the hydrocoel. This depression is the *amniotic invagination*, which later expands toward the inside like a sac and comes into contact with the hydrocoel (Fig. 9.9, a, *am*; b, *am.cv.*).

Although the hydrocoel is at first a simple circular structure as seen from directly above on the left side, it eventually extends at two places, taking a crescent shape. Later the two horns of the crescent bend inward as they elongate, finally fusing by their tips to form a complete ring. The amniotic invagination settles onto this ring, in the meantime closing off its opening and sinking below the surface (Fig. 9.10b, c). It must be remembered that the structure which has thus sunken inward is purely ectodermal, and the space included within it was originally part of the outside environment.

This cavity is the *amniotic cavity* and its thin ectodermal ceiling is called the *amnion*. The bottom of the cavity is made up of columnar ectoderm cells which lie directly over the hydrocoel (Fig. 9.9b). The amniotic cavity and the hydrocoel are collectively called the *echinus rudiment*, because this unit is the center of formation of the adult mouth and its surrounding structures.

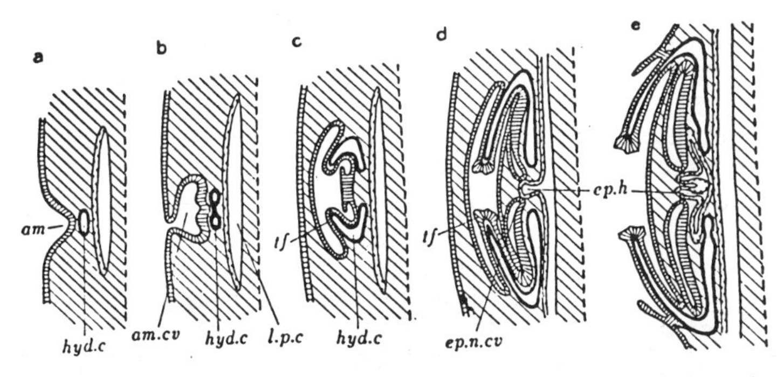

Fig. 9.10 Schematic sketches showing relation among amniotic cavity, hydrocoel and left posterior coelom through course of development of echinus rudiment. Cross-lined area: epithelium. Blacked area: wall of hydrocoel. Dotted area: wall of left and right coeloms. Oblique-lined area: connective tissue. (MacBride)

am: amniotic invagination, *am.cv:* amniotic cavity, *ep.h:* perihaemal system, *ep.n.cv:* epineural space, *hyd.c:* hydrocoel, *l.p.c:* left posterior coelom, *tf:* tube foot rudiment.

Better to follow the course of differentiation of the body organization beyond this stage, the reader should consult Figure 9.11 at this point, in order to have a clear image of the anatomy of the adult sea urchin.

The next change is the formation of five outgrowths from the ring of the hydrocoel, extending obliquely upward and converging toward the center somewhat like five fingers trying to hold a small object. Since such extensions of the hydrocoel push up the bottom of the amniotic cavity, they project into the cavity as five processes (Fig. 9.9b *tf.*, 9.10, *tf*). These are the rudiments of five *tube feet (podia)*, representing the tips of the five *ambulacral zones* of the adult. In the adult, the tube feet found in the ambulacral zones are usually paired, but the first tube feet formed in the larva are unpaired ones which lie at the very tips of the radial canals *(azygous tube feet)*. Their presence is a feature common to all echinoderms.

There next follows the formation, on the lower surface of the cavity, of five radial ridges alternating with the tube feet. The top of each ridge then begins

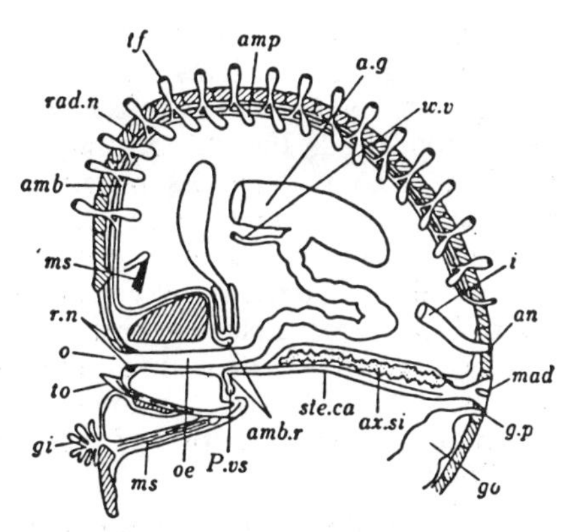

Fig. 9.11 Median cross-section of adult sea-urchin (mouth at left of sketch) (Ijima)

a.g: anterior gut, *amb:* radial ambulacral canal, *amb.r:* ambulacral ring, *amp:* ampulla, *an:* anus, *ax.si:* axial sinus, *gi:* gill, *go:* gonad, *g.p:* genital pore, *i:* intestine, *mad:* madrepore, *ms:* muscle, *o:* mouth, *oe:* oesophagus, *P.vs:* Polian vesicle, *rad.n:* radial nerve, *r.n:* nerve ring, *ste.ca:* stone canal, *tf:* tube foot, *to:* tooth, *w.v:* water vessel.

to extend horizontally toward the sides like a pair of wings or fins, and these fins of adjacent ridges unite so that it is as if the floor of the cavity is covered by another layer, as by a carpet. Between the floor and the carpet is the *epineural space,* which will contain the *nerve ring* when this is formed from the ectodermal floor under the cover. In the meantime, mesodermal cells derived from the posterior coelomic chamber of the left side migrate into the five radial ridges (Fig. 9.10d). These elements constitute the *perihaemal system,* the main role of which is to form the *haemal system* of the body. However, at the center of the floor of the amniotic cavity where ridges converge and the *oral field* is eventually established, they specially form *teeth.*

Enumerating the structures described above from the cavity side toward the inside of the body: there are five tube feet protruding into the cavity; one layer lower, in the area circumscribed by the bases of the tube feet, there is the epineural space; under this is the ring nerve; on the next lower level, there are five teeth alternating in position with the tube feet; and finally, at the lowest level, there lies the ring of the hydrocoel which will form the water vascular system.

Another glance at the adult anatomy, (Fig. 9.11), shows that immediately beneath the outer covering of epidermis, there lies the nerve ring (*r. n.*), next the teeth (*to*) and under these the *ambulacral ring* (*amb. r.*); it is thus evident that practically all the important structures of the adult mouth region are already present within the echinus rudiment.

After a while, *spines* make their appearance on the side wall of the amniotic cavity. To which part of the adult body, then, do the spines and the tube feet protruding into the cavity correspond? It is apparent that these two kinds of structures represent the rest of the adult surface other than the mouth region. Expressed in another way, it can be said that the body skin is rolled inside of the amniotic cavity.

Carrying the story back to the larva: the internal expansion of the left coelom is a striking feature. This structure invades the right half of the larva where it amalgamates with the right coelom, the development of which has consistently been suppressed. For this reason, the axial sinus and the gonads originate from the left coelom alone. As metamorphosis approaches, the stomach temporarily changes into a blind sac when it loses its connections with the stomodaeum and the anus; it then sends out a new extension toward the left which extends through the ambulacral ring and nerve ring and reaches the center of the amniotic cavity floor. Only after metamorphosis, however, does it break through to the outside and become the adult mouth. The echinus rudiment immediately before metamorphosis is fairly large, to the extent that it even presses the stomach flat; since it contains an accumulation of green pigment, it is easy to recognize in the living condition.

The echinus rudiment is not the only part of the body in which preparations for metamorphosis are proceeding. On the body surface of the advanced pluteus, spines and *pedicellariae* and a part of the adult *test* are formed (Figs. 9.7, 9.8, *spi I, ped*). It may be mentioned that the spines which are found in the amniotic cavity have pointed ends, while those appearing on the larval surface are square rods with flat ends (Fig. 9.8, *spi I, spi II*).

b. Metamorphosis

Once the adult organs are prepared to this extent, it only remains to break open the roof of the amniotic cavity and expose the structures within it. Indeed, the process of sea urchin metamorphosis consists simply in bringing the echinus rudiment to light by the rupture of the amnion, and the best way to describe it is by way of illustrations (Figs. 9.10e, 9.7d,e, 9.8c). The entire change takes at most only a few hours. The amount of epidermis making up the inner surface of the amniotic cavity is obviously insufficient to cover the entire larval body after inversion, so that enormous expansion of this layer is necessary. Together with this expansion, the five branches of the ambulacral ring reaching into the tube feet are also greatly extended. As these branches elongate, the azygous tube feet are always pushed ahead at the distal ends of the branches, while behind them paired tube feet are laid down to form the radial ambulacral zones of the adult (Fig. 9.11). When a sea urchin reaches the completely mature adult state, therefore, the azygous tube feet are found to have been pushed all the way around to the dorsal side. It may help in visualizing this state of affairs to think of a sea urchin as corresponding to a starfish with its five arms held together on the dorsal side.

With the eversion of the amniotic cavity, the arms of the pluteus are immediately resorbed; the skeletal rods, which are absorbed only slowly, are left naked on the body surface until they break off and are discarded, enabling the young sea urchin gradually to round up (Fig. 9.7f, 9.8d). At this point, the body axis again rotates through 90°(p. 293) to bring the left side of the animal down to the substratum; it then walks away on its five tube feet. The perforation of the mouth and anus take place at a still later time (Fig. 9.8e). The strangest feature in this whole metamorphosis of the sea urchin is the fact that the left half of the pluteal body is a solo player which monopolizes the stage.

c. Metamorphosis of Clypeastroida

Since it requires a long period of culturing before the metamorphosis of the Regularia occurs as described above, probably because of the technical difficulties there are scarcely any studies of sea urchin metamorphosis besides the paper of Shearer, de Morgan and Fuch (1914) on the metamorphosis of hybrids, and those of Harvey (1949, 1956) on the metamorphosis of centrifuged fragments of *Arbacia* eggs. On the other hand, sand dollars belonging to the Clypeastroida metamorphose in a much shorter time than do the Regularia. Only about two weeks are required in *Astriclypeus manni* (Mortensen 1921, Kumé 1929) and a few days are sufficient for *Peronella japonica* (Onoda 1938, Okazaki and Dan 1954). The developmental course of *P. japonica* includes many aberrant features, because of its exceptionally fast rate of development. The pluteus stage is reached on the day following fertilization, the pluteus has only one pair of post-oral arms and the larval mouth never opens. The amniotic invagination appears on the median plane (instead of on the left side) in the part where a mouth would usually be formed; this determines the median position of the echinus rudiment between the arms, and may have something to do with the failure of the mouth to open. It is significant that even in this case, the left coelom maintains its dominance. The swimming stage lasts for only one day and metamorphosis is complete on the third day.

It has been known since the time of Driesch (1891) that partial larvae derived from isolated blastomeres of the 2- or 4-cell stage develop into perfect but miniature plutei; partial larvae of *P. japonica,* moreover, are able to complete the pluteus stage and metamorphose (Okazaki and Dan 1954). In *Astriclypeus manni,* which has a more orthodox type of development, Okazaki (1954) further showed that the two half-larvae developing from the sister blastomeres of the 2-cell stage both form echinus rudiments on their left sides. This result has interesting implications in connection with the mechanism of determination of bilaterality.

d. **Inversion of symmetry**

As has been suggested several times, one very impressive aspect of sea urchin development is the predominance of the left side. This may be a secondary modification which occurred in the hypothetical bilaterally symmetrical ancester of the Echinoderma, the *dipleurula.* But if so, there ought to be some right-handed exceptions among a great number of left-handed individuals. In fact, Ohshima (1922) reported a larva in which the echinus rudiment was found on the right side, and in another paper (1921) he maintains that when something goes wrong with the left pore canal, the right one may develop compensatorily.

In concluding this description of sea urchin development, a word of explanation seems to be in order to justify placing it at the head of the chapter. The reason is partly that sea urchin development has been more thoroughly studied than that of any other group, and partly that sea urchin eggs are widely used in physiological as well as in biochemical experimentation. However, when surveyed together with the embryology of the Echinoderma as a whole, the mode of development of the echinoids is seen to be a rather aberrant type having many strange features. The significance of these aberrations will be better understood when sea urchin development is compared with that of the other classes.

2. OPHIUROIDEA

As far as the writer is aware, there are practically no reports on the development of Japanese brittle stars except that of Murakami (1937, 1940, 1941) on the formation of the calcareous plates. The main reason may lie in the non-availability of convenient materials. However, this misfortune is not necessarily limited to this country, as it is generally said that artificial insemination of the brittle star is almost impossible. As a result, the contents of this chapter will mainly follow McBride's account (1914) on *Ophiothrix.*

(1) EARLY STAGES

The early cleavages take place in close succession. The 16-cell stage shows no special cleavage pattern like that of the sea urchin, consisting of four tiers, each made up of four equally sized cells. The blastula frees itself from the fertilization membrane by dissolving it. Between the stage in which the primary mesenchyme cells arise at the vegetal pole and the gastrula stage, there is no essential difference from sea urchin development except for the minor point that the cells on the animal pole side become vacuolated.

The first deviation from the sea urchins is encountered in the mode of coelom formation. While the sea urchin secondary mesenchyme cells at the tip of the archenteron first separate into two masses and then form the coelomic vesicles, the wall of the archenteric tip in *Ophiothrix* becomes thin, is pushed out and separates as an independent sac which later divides into two vesicles. In other words, this is a case of typical enterocoel formation.

At the same time, a pair of arms supported by triradiate spicules formed by the primary mesenchyme cells develop as in the sea urchin larva. However, the first pair of arms of the brittle star larva is not homologous with the first post-oral pair of the pluteus. In fact, the situation here is just the opposite: the first pair in the brittle star turn out to be the postero-lateral arms which appear last in sea urchins or, often, are omitted entirely. These larvae eventually acquire four pairs of arms; although they resemble sea urchin plutei in appearance, they fail to develop pre-oral arms (Fig. 9.23). For these reasons, the larva of the brittle star is called an *ophiopluteus* and is distinguished from that of the sea urchin, which is called an *echiopluteus*.

The changes in the internal organization of the larva closely resemble those of sea urchins; these include the connection of the archenteric tip to the stomodaeum, the formation of a tripartite digestive tract, the transformation of the blastopore to an anus, the separation of the coelomic vesicles into anterior and posterior chambers and finally, formation of the hydrocoel by the left anterior chamber only. The madrepores and pore canals of both sides develop equally for a short time, but since the right ones soon degenerate, the final result is the same as that in the sea urchins.

(2) METAMORPHOSIS

The first step in the preparation for metamorphosis is the growth of the left coelom. The posterior chamber of the left side invades the right half in the form of two projections (Fig. 9.12 c, *l. p.c.*), which fuse to occupy most of the posterior half of the larva. With the enlargement of the left anterior chamber,

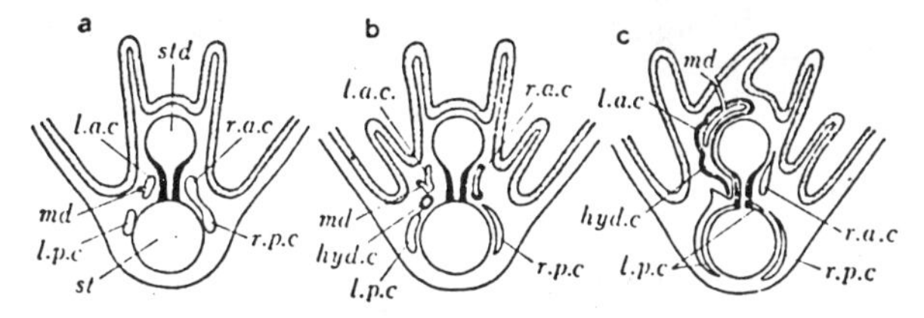

Fig. 9.12 Formation of coelomic system in brittle star (MacBride)
hyd.c: hydrocoel, ***l.a.c:*** left anterior coelom, ***l.p.c:*** left posterior coelom, ***md:*** madreporic canal and its opening, ***r.a.c:*** right anterior coelom, ***r.p.c:*** right posterior coelom, ***st:*** stomach, ***std:*** stomodaeum.

the left madrepore is pushed to the right as far as the median line, while the hydrocoel extends along the left side of the oesophagus, assuming a crescent shape with five finger-like projections (Fig .9.12c). As described above, the sea urchin hydrocoel first passes through a crescent stage to become a ring, and then forms five projections (p. 297); the form change of the brittle star hydrocoel can be derived by superimposing these two processes of sea urchins. The two horns of the crescent in the Ophiuroidea further lengthen into the right half of the body, one on the dorsal side and the other on the ventral so that after the ends of the horns fuse, the completed ring canal encircles the stomodaeum. This expansion of the left hydrocoel causes the median axis of the larva to bend toward the right (Fig. 9.12c).

Sooner or later five chains of mesodermal elements destined to form the haemal system are sent out from the left posterior coelom to alternate with the five lobes of the hydrocoel. Five ectodermal ridges appear along these five chains; their upper parts extend horizontally like fins and later fuse to form the epineural space. These changes are practically the same as those occurring an sea urchins. A marked difference, however, is that in sea urchins, all the organs are formed within the echinus rudiment on the left side of the body, while in the Ophiuroidea, the hydrocoel begins on the left side but moves to a median position.

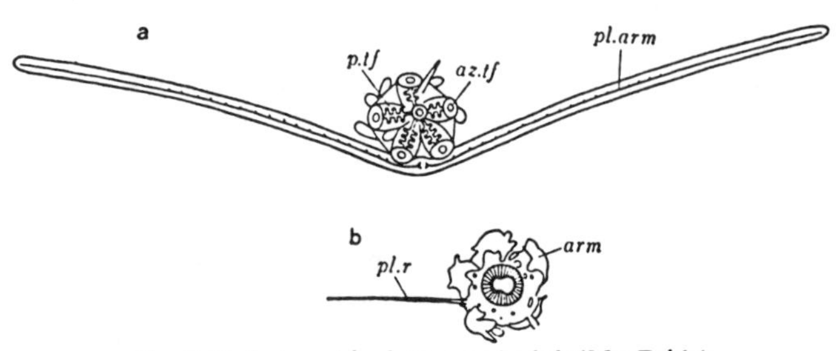

Fig. 9.13 Late ophiopluteus and adult (MacBride)

a. ventral view of late ophiopluteus larva; b. dorsal view of young brittle star. *arm:* adult arm, *az.tf:* azygous tube foot, *pl.arm:* postero-lateral arm, *pl.r:* postero-lateral rod, *p.tf:* zygous tube foot.

At the end of the section on sea urchins (p. 299), the development of *Peronella* was said to be an exception in that a stomodaeum fails to form and an amniotic invagination appears in its place on the median line. A consideration of ophiuroidean development together with that of the Echinoidea suggests that the amniotic invagination of sea urchins is essentially a modification of the stomodaeum.

The later development of the brittle stars follows a different course from the explosive change by rupture of the echinus rudiment characteristic of sea urchins, proceeding gradually toward metamorphosis through the swimming stage. The larva loses its arms, except for the postero-lateral pair, a pentagonal adult test is formed between the arms, and the left coelom is pulled into the test by the contraction of the epidermis. The pentagon, which represents the disc of the future brittle star, already appears in this stage as a little star suspended from the pair of arms. By the time five short tube feet begin to protrude from the disc, the ophiopluteus loses its capacity to float, sinks to the bottom and becomes an adult by casting off the postero-lateral arms (Fig. 9.13).

Fig. 9.14 Larva of *Ophiura brevispina* (Grave)

az.tf: azygous tube foot, *cil:* ciliary band. The larva normally has five such bands but in this sketch only four are shown. *o:* area of future mouth *p.tf:* zygous tube foot.

Like *Peronella* among the echinoids, which develops at an extraordinarily accelerated rate, is the ophiuroid species, *Ophiura brevispina,* whose yolk-laden eggs exhibit an aberrant mode of development. This larva, which never passes through an ophiopluteus stage, is provided with a projecting anterior end and five circles of ciliated bands (Fig. 9.14). A fact of the greatest interest is that this peculiar larva, which looks so strange among the Ophiuroidea, finds resemblances in the larvae of other echinoderm classes. The projecting front end filled with mesenchyme is like that of the starfish larva, and five rings of ciliated bands are common features in the larvae of the Holothuroidea and Crinoidea. That seemingly anomalous cases like this often reveal remnants of now hidden kinship among remotely separated classes is a most exciting phase of biology.

3. ASTEROIDEA

Any Japanese student of embryology who thinks about starfish embryogenesis will remember the classical papers of Goto (1897, 1898), although this investigation was performed on *Asterias pallida (A. vulgaris)* of North America. Concerning Japanese species, there are two papers by Kubo (1948, 1951) on *Leptasterias ochotensis similispinis* (Clark); M. Kojima is also currently studying the development of *Henricia nipponica.* Since some fragmentary observations have been made in the author's laboratory concerning *Asterias amurensis amurensis, Asterina pectinifera* and *Astropecten scoparius,* these will be used as the main basis of the present description.

(1) EARLY STAGES

a. Fertilization

Asterias amurensis spawns from February to April in Tokyo Bay and in May-July in Hokkaido; in Sagami Bay, *Asterina pectinifera* matures in April-May and *Astropecten scoparius* in July. If the ovaries are excised and the eggs examined immediately in sea water, they are all found to be in the pre-reduction division stage, with large germinal vesicles and conspicuous nucleoli. In starfish, there seems to be no relation between the primary egg axis and the point at which the ovum is attached to the ovarian wall (Yatsu 1910). These eggs fail to begin the maturation divisions as long as the medium is contaminated with body fluid, but washing with sea water causes the ger-

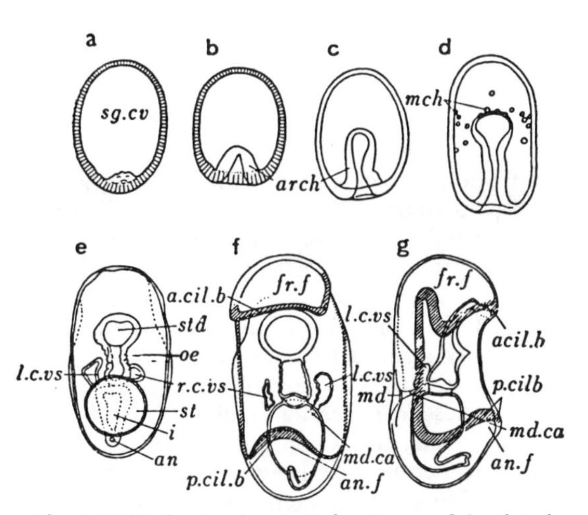

Fig. 9.15 Early development in the starfish *Asterias amurensis amurensis* (Sketches from photographs by Kayo Okazaki)

a. blastula; b,c. successive stages of young gastrula; e. young larva (dorsal); f. ventral view of early bipinnaria larva; g. lateral view of same larva. *a.cil.b:* anterior transverse ciliary band, *an:* anus, *an.f:* anal field, *arch:* archenteron, *fr.f:* frontal field, *i:* intestine, *l.c.vs:* left coelomic vesicle, *mch:* mesenchyme cell, *md:* madrepore, *md.ca:* madreporic canal, *oe:* oesophagus, *p.cil.b:* posterior transverse ciliary band, *r.c.vs:* right coelomic vesicle, *sg.cv:* segmentation cavity, *st:* stomach, *std:* stomodaeum.

minal vesicles to break down spontaneously within an hour, and the first polar bodies are extruded. Although second polar body extrusion follows if the eggs are simply left in sea water, the best results in terms of later development are obtained if the eggs are inseminated during the period between germinal vesicle breakdown and first polar body formation. As a matter of fact, the newly spawned eggs in induced spawning are found to be in this stage.

Since starfish spermatozoa have round heads and the jelly layer around the egg is rather firm, the spermatozoa cannot swim through the jelly to reach the eggs. But, as was described in the first chapter, they are able to establish contact with the eggs by means of very long filaments produced by the acrosomes (Dan, J. C. 1954) (Fig. 1.18).

The fertilization membrane, which appears after fertilization, is formed in a slightly different way from that of sea urchin eggs. In the latter, the presence of the vitelline membrane, which is the precursor of the fertilization membrane, can be established indirectly by experiments but cannot be visually detected; for this reason the impression is created that *de novo* membrane formation takes place on sperm entry. In starfish eggs, not only is the vitelline membrane visible before fertilization, but the rate of its elevation as the fertilization membrane is very slow. It will be noticed that the membrane continues to rise almost until the first cleavage. When the first polar body is already formed and lying under the vitelline membrane when the eggs are inseminated, it adheres to the inner surface of the membrane when this rises away the egg surface, and becomes stretched flat as the fertilization membrane expands. If the second polar body is extruded after the membrane has been elevated, it is not subjected to this strain, and lies free and spherical in the perivitelline space. A hyaline layer, thinner than those of sea urchin eggs, can also be seen surrounding the egg surface.

b. **Morula, blastula, gastrula stages**

The first, second and third cleavages are much like those of sea urchin eggs, but in the fourth cleavage, no size differences occur among the 16 blastomeres similar to those among the macro-, meso- and micromeres of sea urchins. As cleavage proceeds, a blastocoel appears at the center of the embryo and the swimming blastula breaks out of the fertilization membrane, as in sea urchins. It is reported that the Mediterranean species, *Astropecten aranciacus,* forms a morula with many surface foldings and wrinkles which are smoothed off by the time it becomes a blastula (Hörstadius 1939a). Apparently the same thing happens in the Japanese *Leptasterias* (Kubo 1951). A point worth noting is that according to Hörstadius, this smoothing is not due to a simple stretching by the growth of the larva, because the larva is rather losing bulk while becoming smoother on the surface. No such phenomenon can be observed in *Asterias amurensis* or *Asterina pectinifera* (in connection with wrinkling of the blastular wall, see section on Holothuroidea).

The starfish blastula proceeds directly to the gastrula stage with no intervening formation of primary mesenchyme cells such as was seen in the Echinoidea and Ophiuroidea (Fig. 9.15a, b, c). While the gastrula has a spacious blastocoel, the archenteron remains relatively small, extending barely halfway across the blastocoel (Fig. 9.15d). After a time, mesenchymal cells provided with pseudopodia creep out from the tip of the archenteron into the blastocoel. The wall of the archenteric tip then thins and bulges and two vesicles separate from it, one on each side. These are the coelomic vesicles, and with their formation the starfish larva for the first time acquires bilateral symmetry. In *Astropecten aranciacus*, studied by Hörstadius, a large hole is made in the archenteric tip by the simultaneous departure of so many mesenchyme cells; this does not prevent the formation of the two coelomic vesicles, however, and the hole is closed over only after their separation. Hörstadius ascertained that such larvae develop quite normally.

The next process is the tripartition of the digestive tract; the foremost part of the oesophagus is connected with the stomodaeum, the middle part expands as the stomach, the slender posterior part forms the intestine and the blastopore functions as the anus. All these structures are similar to those of sea urchin larvae (Fig. 9.15e, f). At this stage, the coelom exists as two vesicles on the right and left sides of the oesophagus.

(2) BIPINNARIA

a. External morphology

The swimming larval stage of the starfish which corresponds to the pluteus of the Echinoidea and Ophiuroidea has no spicular skeleton whatsoever and is oval in shape (Fig. 9.15f, g). There seems to be a correlation between this lack of a skeleton and the absence of the primary mesenchyme cells which are responsible for spicule formation in the other two classes. A ciliary band makes a complete

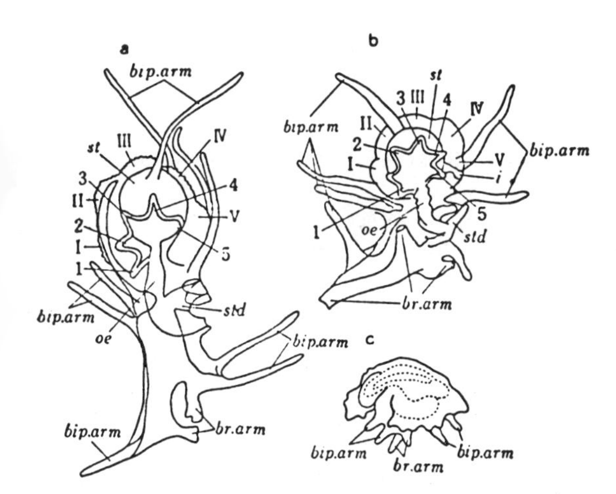

Fig. 9.16 Metamorphosis in brachiolaria larva (Goto)

a. fixed brachiolaria (left side view); b. left side view of late brachiolaria in which stalk is being absorbed; c. young starfish metamorphosis almost complete. I~V: five primordia of adult arms. 1~5: primordia of radial ambulacral zone. *bip.arm*: bipinnaria arm, *br.arm*: brachiolar arm, *i*: intestine, *oe*: oesophagus, *st*: stomach, *std*: stomodaeum.

circuit on the ventral surface of the larva, its anterior side passing in front of the stomodaeum and the posterior side in front of the anus following a squarish course (Fig. 9.15f). As development advances, the body wall in front of the stomodaeum and anus bulges out in two separate protrusions. The cilia are located on the edges of these protrusions, so that the complicated development of their shape results in a sinuously winding ciliary band. In the simple, squarish, early lárva, the front part is called the *anterior transverse band* and the rear part, the *posterior transverse band*, while the remaining two side sections are called the *longitudinal bands* (Fig. 9.15f). The part of the larval body in front of the anterior transverse band is the *frontal field* and that behind the posterior transverse band is the *anal field.* These fields protrude more and more, and since the frontal field bends back and the anal field bends forward, the ciliary band takes the shape of two apposed horse-shoes. The gap between the ends of the horse-shoe surrounding the frontal field becomes narrower and narrower, until it is finally cut off from the rest of the band to make an independent circuit. This double-circle of ciliary band (Fig. 9.17a, 9.23c) is a diagnostic character distinguishing the bipinnaria larva of the starfish from the auricularia of the sea cucumber. The starfish larva proceeds to form several pairs of lateral protrusions, which are named in accordance with the system of nomenclature adopted for the pluteus (Fig. 9.24c). Since these protrusions are also rimmed with the ciliary band, the over-all pattern formed by the band has somewhat the look of a bipinnate fern leaf; hence the starfish larval name of *bipinnaria.*

Coelom formation in the bipinnaria follows the typical course common to the other members of the Echinoderma. In *Asterias vulgaris* the coelomic vesicles are long tubes paralleling the digestive tract and running almost the whole length of the larva, although their anterior communication is a peculiarity of the species. The left vesicle sends out a pore canal toward the dorsal side near the border between the oesophagus and the stomach, and communicates with the outside through the madrepore. At its junction with the pore canal, this left coelomic vesicle constricts slightly, demarcating the separation of the anterior and posterior chambers. Soon the anterior chamber forms the hydrocoel which, as in the brittle star, is crescent-shaped with five projections. From the right coelom, the starfish being more loyal than the other classes to their bilateral ancester, a certain proportion of the larvae form pore canals which open to the outside, although they are soon closed (Gemill 1912).

(3) SESSILE STAGE (BRACHIOLARIA) AND METAMORPHOSIS

In preparation for metamorphosis the larva becomes fixed to the substratum. The advanced bipinnaria acquires a new and very prominent projection in the anterior frontal region between the pre-oral and median-dorsal arms (cf. Fig. 9.23c). The tip of this projection is divided into three branches, each of which is provided with a *fixing disc* at the end. Since this projection includes the anterior coelom, it differs fundamentally from the arms of the bipinnaria, and is called the *brachiolar arm.* The bipinnaria fixes itself to the substratum by this arm, in an inverted position. During the sessile stage which follows fixation, the larva is referred to as a *brachiolaria*; the brachiolar arm is called the *stalk* and the rest of the body, the *disc* (Fig. 9.16a). The stalk includes the stomodaeum and the oesophagus and surrounds the anterior coelom. By this time, the right and left anterior chambers have joined and the posterior chambers of the two sides are mutually united, so that the entire coelom consists of two parts.

Sooner or later, the digestive tract becomes a blind sac, losing its connections with stomodaeum and anus. About this time, five ectodermal bulges appear on the right side of the body surface. These bulges, arranged in a circle, are the rudiments of the future arms of the starfish (Figs. 9.16, I, II, III, IV, V; 9.17).

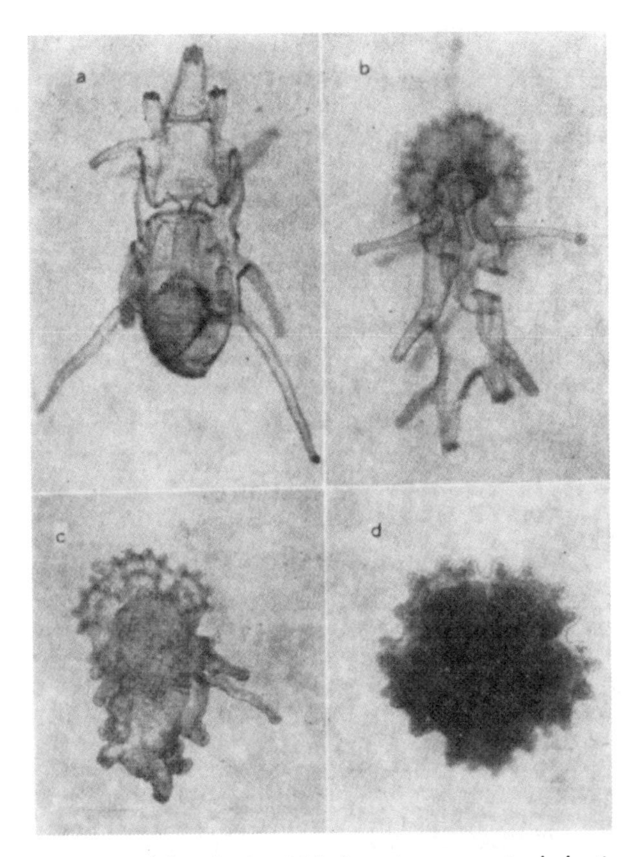

Fig. 9.17 Bipinnaria, brachiolaria and metamorphosis in the starfish *Asterias amurensis* (Photos by Shoko Kikuchi)

a. ventral view of bipinnaria larva; b,c. larvae at stages corresponding to a and b of Fig. 9.16; d. young starfish.

On the left side of the body, the crescent-shaped hydrocoel is almost a complete ring, with its five branches so directed as to fit into the five bulges of the opposite side (Figs. 9.16, 1, 2, 3, 4, 5; 9.17b). As the ectodermal pentagon increases in size, it comes to house the posterior coelom and the stomach beneath it; the tips of the five branches of hydrocoel develop into azygous tube feet. The blind sac of the digestive tract begins to push out two horns, one to form the future oesophagus and the other, the adult intestine.

With these changes, the lateral bipinnaria arms become less and less conspicuous and the stalk is also reduced in length. As the stalk shrinks (Fig. 9.16b; 9.17c), the brachiolaria, while remaining fixed, turns its axis by 90° so that the pentagonal test comes around to the posterior side. By the time the test is fully grown, the whole body is contained within it and the hydrocoelic pentagon fits over it like a lid (Fig. 9.17c). After complete absorption of the stalk, its place is taken by the oral surface of the adult starfish.

(4) ABERRANT MODES OF DEVELOPMENT

The inclusion of a sessile stage sets apart the development of the Asteroidea as clearly different from those of the Echinoidea and Ophiuroidea. But even among the starfish, two types of cases can be found which have secondarily given up the habit of fixation. One type precociously forms a pentagonal test on the posterior side of the swimming bipinnaria. Such a mode of development closely parallels that of the Ophiuroidea (Fig. 9.18a). The other type seems to be the result of extreme adaptation, an unusually generous amount of yolk permitting an abbreviation of the developmental process which omits fixation and even feeding. For example, in *Asterina gibbosa* (Fig. 9.18b, c) the adult body is immediately formed, without any intervening stages, under the preoral lobe which maintains sufficient floating capacity. Although belonging to this genus, the Japanese *Asterina pectinifera* develops through the bipinnaria stage, while *Leptasterias ochotensis* follows he same course as *A. gibbosa* (Kubo 1948).

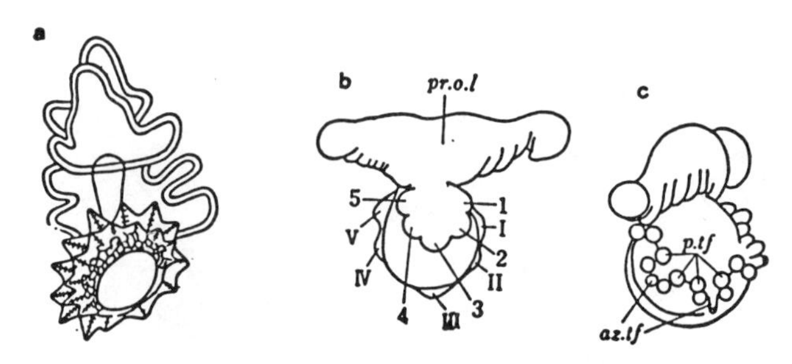

Fig. 9.18 Development of free-swimming starfish larva
a. larva metamorphosing in free-swimming state (J.Müller); b,c. *Asterias gibbosa* (Dawydoff). I ∼ V: rudiments of dorsal arms. 1 ∼ 5: ventral ridges formed by finger-like protrusions of hydrocoel. *az.tf:* azygous tube foot, *pr.o.l:* pre-oral lobe, *p.tf:* zygous tube foot.

Among the starfishes, *Solaster* usually has about ten arms. Interestingly enough, this deviation does not appear in the larval development until after a hydrocoel with the usual five branches has first been formed.

(It seems to be the custom in embryology books to place the Holothuroidea next to the Asteroidea. The Holothuroidea is the only class among the Echinoderma which shows a return to bilateral symmetry from its secondarily acquired radial symmetry. Readers who wish to follow the orthodox sequence of echinoderm embryology should read next the article by Inaba.)

4. CRINOIDEA

One crinoid which has been well investigated is the European species, *Antedon rosacea*. This form can be collected by a bottom trawl, and if mature individuals are selected and kept in a tank, they are said to spawn daily around seven o'clock in the morning (Bury 1888; Seelinger 1892). At spawning, the eggs of *A. rosacea* become attached to the arms of the mother by a mucous secretion. While this characteristic makes it easy to trace the development of this species, there is a high degree of probability that such a brooding habit has introduced considerable specialization into its embryology, such as a delay in the onset of the swimming stage and particularly its shortening to only a few hours. On the other hand, the Japanese crinoid, *Comanthus japonica*, is the second form, after *Tropiometra carinata* of Mortensen (1920), to be reported in which the eggs are liberated into the sea water without being attached to the mother animals. Unfortunately, however, the spawning habit of *Comanthus* is very curious: it is confined to one or two days in the middle of October, and even the spawning hour is limited to 3—4 o'clock in the afternoon (Dan, K. and J. C. Dan 1941). For this reason, our endeavor has mainly been directed toward the prediction of the spawning date, and observations of the development are still incomplete. The following account of crinoid embryology is based on *Comanthus*, supplemented with information obtained from *Antedon*.

(1) EARLY DEVELOPMENT

In crinoids, the gametes are produced within pinnules which are arranged in two rows on each arm. However, since the pinnules have no definite openings, spawning takes place by the rupture of their walls (Dan, J. C. and K. Dan 1941). The eggs begin to undergo maturation a little after noon on the spawning day (Dan 1952a), and are liberated within 3—4 hours. The sperm is usually released shortly before the eggs, which are shed together with a large quantity of mucus. Since *Comanthus* females normally spawn their eggs in large lots, these can readily be collected and artificially inseminated. The fertilization membrane of crinoids in general is a spiny polygon with a large number of facets (Fig. 9.19a).

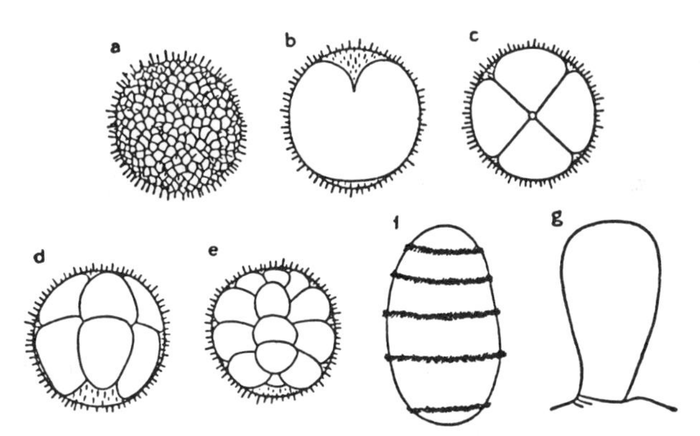

Fig. 9.19 Early development of *Comanthus japonica*

a. spiny fertilization membrane; b. "heart-shaped" first cleavage; c. 4-cell stage; d. 8-cell stage, from side. 4 cells of animal hemisphere (upper side in sketch) are slightly smaller than those of vegetal hemisphere; e. 16-cell stage. Large intercellular space remains at vegetal pole; f. free-swimming larva, anterior end of embryo at upper side of sketch; g. larva attached by former anterior end.

The first cleavage takes place after about an hour; this is heart-shaped as in *Astriclypeus manni*, the furrow starting from the animal pole (Fig. 9.19b). The second and third cleavages follows the usual pattern, resulting in an 8-cell stage like that of sea urchins. In the description of *Antedon*, however, the four cells of the animal-pole side seem to be a little smaller, and at the morula stage, these animal cells adhere together more compactly, causing the vegetal-pole side to gape open for a longer time. This tendency can be detected also in *Comanthus* (Fig. 9.19d,e).

(2) SWIMMING LARVA

Eggs which have been fertilized around three o'clock in the afternoon hatch out on the following morning by dissolving the fertilization membrane. They soon develop into the typical crinoid larva *(doliolaria)* provided with an apical tuft and five rings of ciliary bands (Fig. 9.19f; 9.20a). Strictly speaking, the anteriormost ring is incomplete on the ventral side. The swimming stage of *Comanthus* lasts for 2 — 3 days; and although this is longer than that of *Antedon*, neither larva takes in food during this time. While it is swimming, a fixing disc is formed within the anteriormost ciliary ring, and a stomodaeal invagination appears between the second and third rings (Fig. 9.20a). Gastrulation and coelom formation cannot be observed in the living state in *Comanthus* because of the opacity of the larva. Information about these processes as they occur in *Antedon* can be found in MacBride (1914, p. 545).

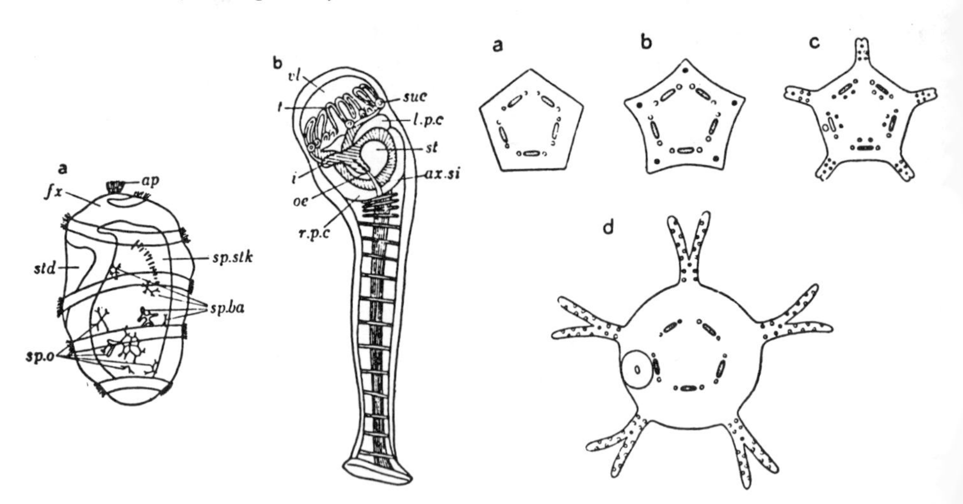

Fig. 9.20 Free-swimming and attached larval stages of *Antedon rosacea* (MacBride)

ap: apical tuft, *ax.s:* axial sinus, *fx:* fixing disc, *i:* intestine, *l.p.c:* left posterior coelom, *oe:* oesophagus, *r.p.c:* right posterior coelom, *sp.ba:* spicule forming rudiment of basal plate, *sp.o:* spicule forming rudiment of oral plate, *sp.stk:* spicule forming rudiment of stalk skeleton, *st:* stomach, *std:* stomodaeum, *suc:* sacculus, *t:* tentacle, *vl:* vestibule.

Fig. 9.21 Order of formation of tentacles in *Comanthus japonica*

o: zygous tentacle. •: azygous tentacle. Five central plates constitute oral valves; five peripheral processes are arm rudiments. Circle at left indicates anus.

(3) SESSILE STAGE

If the *Comanthus* larvae are kept in a glass jar, they do not attach for 4—5 days and when they do, they settle at the meniscus of the best lighted side. This behavior pattern, however, is induced only by the conditions existing in the jar, because once a proper stone brought from the sea is introduced into the jar, they immediately attach to the stone, regardless of the light conditions. Since the fixing disc is situated in the frontal area, the larva at first attaches obliquely upside-down, but within 2 — 3 days, the posterior part of the body stands perpendicular to the substratum. While this change in angle is taking place, the stomodaeum shifts from the ventral to the posterior side, simultaneously sinking from the surface so that when the larva stands erect, the stomodaeal invagination forms an independent blind sac buried in the posterior (upper) end of the body. Although this occluded space is a kind of amniotic cavity, it is called the *vestibule* in the crinoid (Fig. 9.20b).

Within 2 — 3 weeks, the roof of the vestibule bursts open and ten *tentacles* make their appearance. These tentacles are located at the sides of five *oral valves* which are situated in the interradial zones surrounding the mouth (Fig. 9.21a). Within four weeks, five *azygous tentacles* are formed peripherally on the radii, thus making a total of fifteen tentacles (Figs. 9.21b; 9.22a). The larva is able to swing them about and contract them vigorously. In the fifth week, the contour of the disc corresponding to the radial zones begins to be extended into five arms with the azygous tentacles remaining at their tips. As the arms grow longer, paired tentacles are added proximal to the azygous ones. The number of pairs added is probably two. By this time, ten other small tentacles can be seen directly encircling the mouth at the inner sides of the oral valves (Figs. 9.21c; 9.22b). The larva of this stage is called a *pentacrinoid*.

In the seventh week, the tips of the arms bifurcate and continue to elongate. Although this gives the impression that the number of arms has increased to ten, close examination makes it evident that the azygous tentacles have left been behind at the points of bifurcation. This fact suggests that the arm proper probably stops at the azygous tentacle, and what appears to be a fork may actually be the point of division of the arm into two side branches. On this interpretation it follows that each branch should have only a single row of tentacles if the branch represents only one-half of the original arm. Owing to the small size of the branches and the consequent crowding of the tentacles, judgement is rather difficult. At present, the author believes that the tentacles are alternately arranged. Although alternating arrangement could be considered a modification of a single arrangement, no conclusion can be reached at present. It is certain, however, that when pinnules appear on the branches, they have an alternating arrangement.

In contrast to the outward shift of the azygous tentacles, the positions of the *perioral tentacles* do not change. Considering the anatomy of the Echinoderma as a whole, it appears that the water vascular system of the tentacles around the mouth is directly connected to the ring canal, while that of the azygous tentacles connects with the tips of radial canals which lead off from the ring canal.

After the fifth week, an anus develops at the peripheral side of one of the oral valves, the ectoderm rising up around it to form a prominent papilla (Figs. 9.21c, d; 9.22c,d). Since the anus is situated in an interradial zone, it does not affect the symmetry of the arms.

Two significant changes occur in the eighth week. One is that the arms, which until this time could only roll inward toward the mouth, suddenly become able to bend outward (Fig. 9.22d). This alternate extension and flexion constitutes the most common movement of the adult animal. The second change is the appearance of the cirri. When they first appear, the cirri are directed upward, and like the cirri of *Metacrinus*, it is quite impossible for them to grasp the bottom.

In Figure 9.22d, one cirrus is indicated at the center of the drawing. Since the animals have acquired the capacity to bend the arms outward by this time, they cling to the bottom by their arms (not by the cirri) and begin to creep about within the radius of the stalk-length.

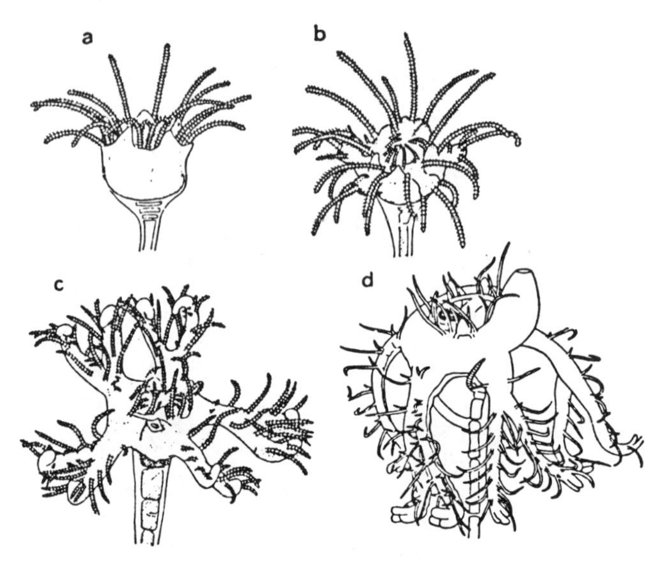

Fig. 9.22 Late development in *Comanthus japonica*
a. 4 weeks; b. 5 weeks; c. 7 weeks; d. 8 weeks.

(4) FREE STAGE

By the tenth week, the two branches of each arm have grown longer and produced pinnules which are arranged alternately from the base to the tip. The young animal assumes the dark purplish color of the full grown adult, and finally cuts itself loose from the stalk. Whether the stalk is cast off or absorbed has not been determined, although the author is inclined to think that the former is more likely.

Within the limit of four months' rearing of young *Comanthus*, the general body size increased, the anus moved to the center of the disc, and the cirri increased in number. However, the arms did not divide more then once, the number of branches falling far short of the forty found in the adult.

5. CONCLUDING REMARKS

In the foregoing pages, the processes comprising the development of various classes of Echinoderma have been summarized. On reflection, it will be noticed that in spite of many superficial differences, there is a similarity running through them which can be thought of as either a common basis or a general trend. The phylum Echinoderma is often divided into two groups; the Pelmatozoa, which includes the essentially sessile animals comprising the Crinoidea, and the Eleutherozoa, which are further removed from the sessile habit, including the Echinoidea, Asteroidea, Ophiuroidea and Holothuroidea. Let us then reconsider the whole situation in the light of this classification.

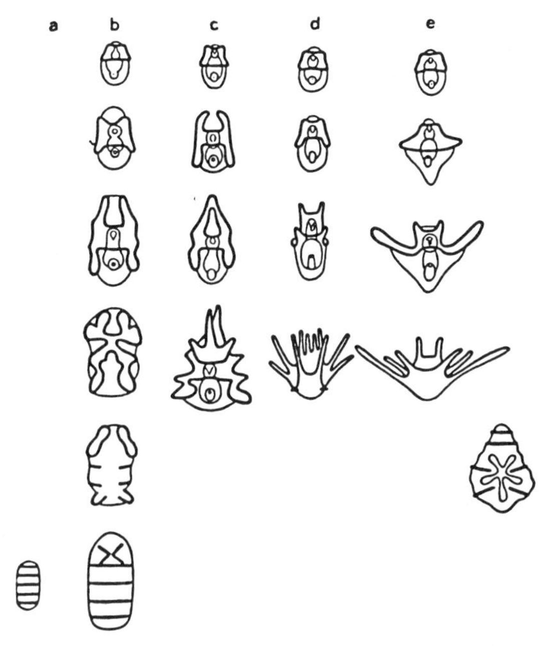

Fig.9.23 Comparison of ciliary bands in echinoderm larvae

a. Crinoidea; b. Holothuroidea; c. Asteroidea; d. Echinoidea; e. Ophiuroidea.

In the first place, the diagram in Figure 9.23 summarizes the patterns of the ciliary bands in the larvae of the Echinoderma.

As is evident in this figure, the ciliary band patterns can be roughly divided into two types. One type is characterized by five rings of cilia (lowest row) and the other has but one sinuous circuit of cilia (first to fourth horizontal rows). Although the bipinnaria alone has two circles, which show exceptionally complicated windings, it can easily enough be included in the latter type; while *Ophiura brevispina*, with its five rings of cilia, is an exception among the brittle stars. Holothurian development presents a case of the utmost interest, for it starts with a single ring in the auricularia larva and ends with five rings in the doliolaria. After all, the two types of ciliary banding must be only two segments of a continuous series.

Limiting our attention to the Eleutherozoa for a moment: if the various types of ciliary bands are traced back to the beginning of their formation, the differences among the four classes become slighter and slighter and finally it is possible to include them in a single common pattern (Fig. 9.23, *top row*), consisting of a squarish band lying between the frontal and anal fields. If the more advanced stages are grouped, we find the pluteus type on the one hand and the bipinnaria-auricularia type on the other. However, the difference between them is due to the rather superficial reason that while the body of the bipinnaria or auricularia remains relatively flat, in the pluteus it has been bent and doubled upon itself toward the ventral side. In other words, protrusion of the frontal and anal fields occurs equally in all these larvae. In the bipinnaria and auricularia, however, this protrusion involves only the body wall, the body axis being left unbent; while in the pluteus, the stomodaeum and anus are extended together with the surface areas, bending the body axis into a V. In order to accomodate its structure to this bending, the pluteus has to turn its axis by 90° from the direction of the gastrula stage.

Reversing the sense of the above analysis, if a pluteus could be held by the frontal and anal fields and stretched out, the result would approximate a bipinnaria or auricularia larva. Since this is so, it is not impossible to think that homology exists among the larval arms of the Eleutherozoa. Although it may be extremely difficult to prove such an homology, it would be a convenient practice to name them by a unified system of nomenclature.

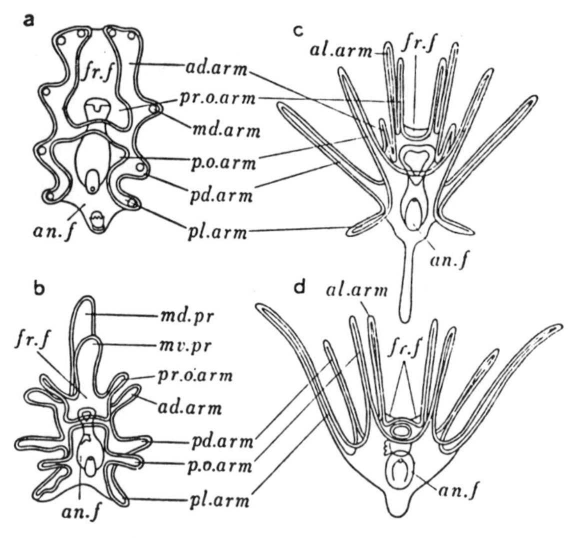

Fig. 9.24 Comparison of larval arms in the Eleutherozoa (Ohshima)

a. auricularia; b. bipinnaria; c. echinopluteus; d. ophiopluteus. *ad.arm:* anterior dorsal arm, *al.arm:* anterior lateral arm, *an.f:* anal field, *fr.f:* frontal field, *md.arm:* mid-dorsal arm, *md.pr:* mid-dorsal protrusion, *mv.pr:* mid-ventral protrusion, *pd.arm:* postero-dorsal arm, *pl.arm:* postero-lateral arm, *p.o.arm:* postero-oral arm, *pr.o.arm:* pre-oral arm.

Figure 9.24 is a scheme taken from Ohshima (1911). Incidentally, this report includes a key for identifying the larvae of the Eleutherozoa.

Our reconsideration must still deal with the larval type having five ciliated bands. Among the five classes treated in this book, the Crinoidea are acknowledged to be the most ancient group; as we have seen, their larvae develop directly into a stage with five bands, without passing through the previous stages seen among the larvae of the Holothuroidea. Assuming that the five-banded type is ancestral, and using the Holothuroidea as a stepping stone, it is possible to postulate that the evolution of the Eleutherozoa may have taken place by a kind of neoteny, involving an expansion of young, inconspicuous stages in crinoid development preceding the establishment of the five ciliary bands. This possibility should not be lightly disposed of, because without resource to some drastic method as accentuation of juvenile stages, the Eleutherozoa could not have achieved their emancipation from the sessile habit.

It is also possible to reason in an exactly opposite way. It is a geologically well established fact that the crinoids flourished during the Palaeozoic era. At that time, the average larval form of the Pelmatozoa may have resembled those of the present day Eleutherozoa. As the result of wide distribution in their heyday, some races may have evolved which began to store much yolk in the egg to enable the larva to avoid the necessity of early feeding, as can be found at present in aberrant cases. Further, it is possible to imagine that when most of the crinoids became extinct with later geological and meteorological changes of the earth, those races which had an aberrant mode of development happened to find a better chance for survival, to give rise to the present members of the Pelmatozoa. In other words, it is possible to think that the groups with the formerly typical larval form were lost and only an extremely specialized type survived to constitute the Pelmatozoa of the present. This interpretation can cover without much difficulty the occurrence of the doliolaria form in the late stage of Holothurian development as well as the exceptional case of *Ophiura* larvae with five ciliary rings. At any rate, the later developmental history has probably not gone beyond these extremes.

Next let us turn our attention to the internal structure. If the phylum Echinoderma as a whole is regarded as moving along the path from a sessile to a free-moving life, among the Eleutherozoa the Asteroidea with their brachiolaria stage can be said to be the most conservative. As has been pointed out, the Asteroidea become free by first moving the structures of the left side toward the stalk and stomodaeum to acquire radial symmetry, and then absorbing the stalk. By this shifting of the internal organs, the larva executes an internal revolution of 90°.

This invention of the Asteroidea was adopted by the Holothuroidea and Ophiuroidea. The Echinoidea, however, follow a formula which is one step further removed from the tradition of sessile life. The stalk, which still held some meaning for the previous evolutionary generation, has lost it entirely for a group as thoroughly accustomed to free-moving life as are the Echinoidea. Having come so far, it is not only very inconvenient but even dangerous to turn the viscera by 90° toward the position of the stalk, especially since the same result can be more simply achieved by making a substitute stomodaeum on the left side. The structure which accomplishes this substitution is the echinus rudiment. Moreover, this method is doubly effective, since it also preserves the larval stomodaeum and permits a longer feeding period.

The Holothuroidea as a group are very true to tradition in their early stage of coelom- and hydrocoel-formation. Along the way, however, they turn rebellious and make an attempt to acquire bilateral symmetry. These features are reflected in their development as various modifications of the general rules.

Certainly some common under-current can be seen in the development of the Eleutherozoa, but a great paradox is encountered in that of the Pelmatozoa. Although the developmental changes of the Pelmatozoa can be said to parallel those of the Eleutherozoa, closer examination shows that the direction of the shift in the internal structures is quite opposite to that of the Eleutherozoa; i. e., the stomodaeum, accompanied by the left coelom, moves away from the stalk. Although this offers a good reason for separating the two groups, nothing is known about the cause of the opposite movement.

6. HOLOTHURIA

INTRODUCTION

The study of holothurian embryology began in 1850, when Joh. Müller discovered the auricularia larva; further reports by Danielssen and Koren (1856), Baur (1864), Kowalevsky (1867) and Metschnikoff (1868) followed, so that by the time MacBride's textbook of invertebrate embryology appeared in 1914, a series of complete studies concerning the development of the synaptid species *Labidoplax (Synapta) digitata* and *Synapta vivipara*, as well as of the dendrochirote *Cucumaria planci*, had been achieved by the efforts of many investigators. Since then, the development of a considerable number of species has been described by Newth (1916), Ohshima (1918, 1921), J. and S. Runnström (1918—19), S. Runnström (1927), Mortensen (1921, 1931, 1937, 1938) and Inaba (1930, 1933, 1937, 1943). In general, however, because of difficulties in collecting material, lack of information about breeding seasons and the frequent failure of artificial insemination, it can even now be said that our knowledge of holothurian development is scanty in comparison with the information which has been accumulated concerning other echinoderm classes such as the Echinoidea and Asteroidea. Moreover, many of the investigations have been done on the Synaptidae, which have no tube feet and thus can hardly be said to represent the majority of holothurian species. This report will therefore attempt to fill some of these gaps with data gained from observation of the developmental processes of such Japanese species as *Stichopus japonicus*, *Cucumaria echinata* and others.

(1) EARLY DEVELOPMENT

The mature holothurian egg is spherical or rather ovoid (Fig. 9.25). In most cases the eggs are nearly transparent, but there are also some opaque ones, such as the yellowish-green eggs of *Cucumaria* or the yellow-brown *Paracaudina* eggs. In general they are larger than those of sea urchins or starfish; the egg of *Leptosynapta* (diam. 0.2 mm) is one of the smaller, those of *Cucumaria echinata*, *Paracaudina* and *Cucumaria japonica* are about 0.5 mm, those of *Stichopus* and some species of *Cucumaria* measure form 1.0 mm to almost 1.5 mm, and the ovarian eggs of *Enypniastes eximia* reach a diameter of 3.5 mm. These are all homolecithal eggs. In most holothurians, fertilized eggs are obtained by placing a number of animals in an aquarium and letting them spawn, but it is also possible to secure eggs and sperm for the usual artificial insemination procedure. In this case, it should be noted that four types of eggs are found in the ripe ovary: (1) small, irregularly shaped eggs with a prominent germinal vesicle; (2) regularly shaped eggs with a germinal vesicle; (3) regularly shaped eggs without a germinal vesicle; (4) irregularly shaped eggs without a germinal vesicle, which sometimes occur in clusters. Of these four types, only the eggs

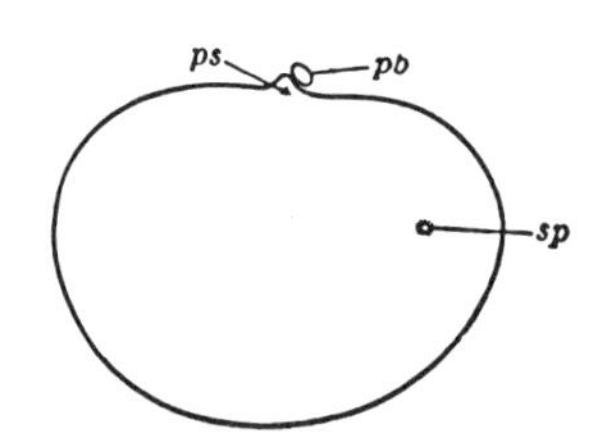

Fig. 9.25 Fertilized egg of *Cucumaria planci* (Ohshima)

pb: first polar body, *ps*: second polar body, *sp*: sperm pronucleus.

of the third type are fertilizable (Inaba 1937). Hörstadius (1925) was able to obtain large numbers of fertilized eggs of *Holothuria Poli* by using the method which is often successful with bivalve molluscs, of raising the pH of the sea water by means of NaOH or KOH; Mortensen (1937), however, reports that this method was ineffective with Mediterranean holothurians.

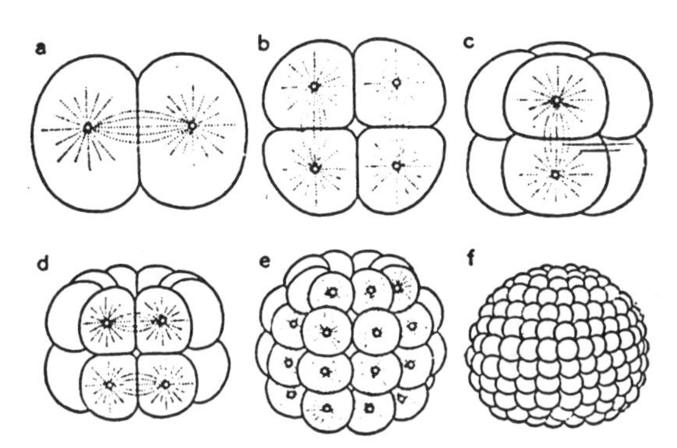

Fig. 9.26 Cleavage in *Labidoplax digitata* (Selenka, from Wilson)
a ~ f. successive stages showing schematic regularity of radial cleavage.

By whatever method ripe eggs are obtained, they will give off the first and second polar bodies from the animal pole on insemination, and begin to cleave (Figs. 9.25, 26). Cleavage is total and equal and the arrangement of the blastomeres is schematic in most cases; in particular the radial cleavage of *Labidoplax digitata*, as observed by Selenka (1883), is famous as being the most schematic

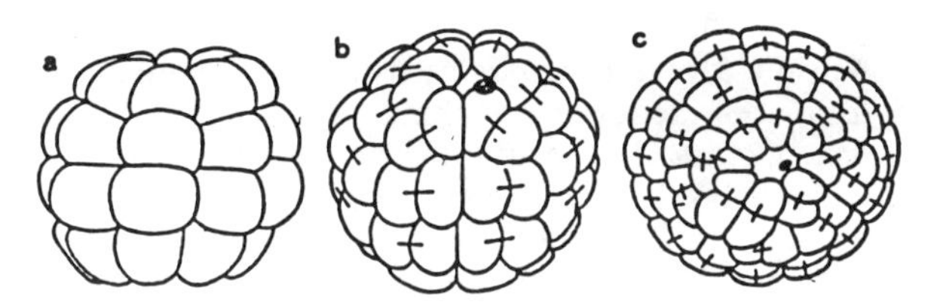

Fig. 9.27 Cleavage in *Leptosynata inhaerens* (S. Runnström)
a. 32-cell stage; b. 64-cell stage; c. 130-cell stage. Lines show positions cleavage spindles.

example of equal, total cleavage (Fig. 9.26). In the holothurians meridional cleavage alternates with horizontal from the third cleavage onward; there is no such reversal of the cleavage direction as is seen in the fifth cleavage of sea urchin eggs when the four vegetal pole cells cleave meridionally. After the 32-cell stage, however, the time of cleavage becomes different for different blastomeres, the direction of cleavage loses its uniformity and the radial arrangement becomes more or less disordered in many cases, leading to slight size differences among the blastomeres. These are not, however, as marked as those characterizing echinoid cleavage (Fig. 9.27). In *Stichopus japonicus*, cleavage takes place about every

30 minutes at 20° C, giving rise to a coeloblastula after about 11 hours. *Cucumaria normanii* and *Cu. saxicora* (Newth 1916), as well as *Paracaudina chilensis marenzeller* (Inaba 1930), form *wrinkled blastulae* like those seen in the starfishes *Solaster*, *Porania* and *Cribrella* (Fig. 9.28). These wrinkles are formed by local invaginations of the blastular surface; their number and depth are not fixed, but six or seven such branching indentations can be seen on one side of the blastula of *Paracaudina*.

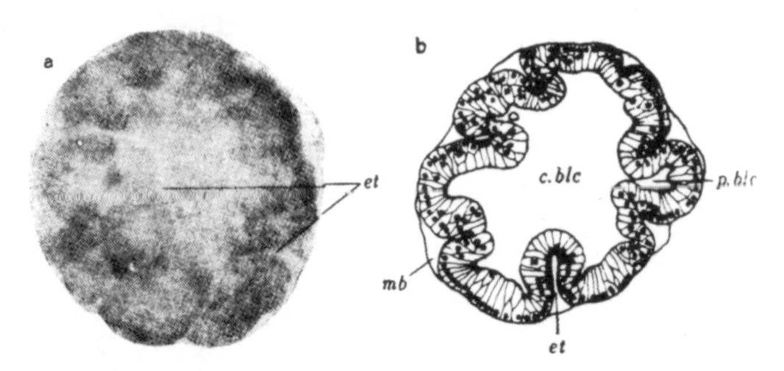

Fig. 9.28 Wrinkled-blastula of *Paracaudina chilensis* (Inaba)

a. external appearance; b. cross-section of the embryo. *c.blc:* central blastocoel, *et:* egression tract, *mb:* egg membrane, *p.blc:* peripheral blastocoel.

The presence of these wrinkles causes the blastular cavity to be divided into a large *central blastocoel* and small indented *peripheral blastocoels* (Fig. 9.28, *c.blc., p.blc.*). The surface streaks made by these wrinkles are called *egression tracts.* The parts of such egression tracts which are mutually in contact eventually fuse so that the blastula regains its smooth surface. It is still not known why the embryo must undergo this change, but as a result of it blastulae which have gone through a wrinkled stage are double- or triple-layered when other blastulae have only a single cell layer.

In other species, cleavage proceeds as usual, and about the time the blastula is fully formed each cell acquires a cilium. According to Selenka (1883), this occurs in *Labidoplax digitata* at the ninth cleavage, when the number of blastomeres reaches 512. At about this time the vegetal side of the embryo becomes somewhat flattened, and in *Cucumaria* and *Paracaudina* some of the cells from this region are freed into the blastocoel where they give rise to mesenchyme. The time of mesenchyme formation, however, is variable: in *Labidoplax digitata,* it occurs first during the gastrula stage, while in *Stichopus* it takes place simultaneously with gastrulation. After the cilia develop, the blastula begins to rotate within its membrane, and two or three hours later it breaks the membrane and hatches. The direction of rotation is mainly dextral as seen from the animal pole, but this may be suddenly reversed. Some species of *Cucumaria* do not hatch until the gastrula stage.

The recently hatched larva of *Stichopus japonicus* is nearly oval, about 190μ in height and 170μ in width; as its size gradually increases it grows more in height than in width. In *Synapta vivipara* the whole developmental process takes place inside the body cavity of the mother animal; no cilia are formed, and consequently there is no swimming movement (Clark 1898). Eventually the vegetal side flattens, an invagination at its center leads to the formation of an *archenteron,* and a typical gastrula is formed. *St. japonicus* reaches this stage in about 24 hours; the long axis has increased to about 220μ and the width to about 170μ, and the larvae mostly swim about near the surface of the water. Subsequently the archenteron elongates together with the body, and the amount of mesenchyme in the blastocoel increases, but there is no formation of a larval spicular skeleton like those seen in the sea urchins and brittle stars. In this respect holothurian development resembles that of the asteroids.

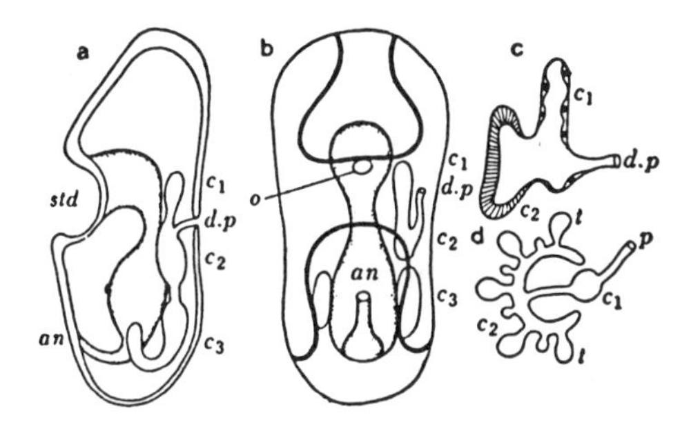

Fig. 9.29 Auricularia larva (Korschelt u. Heider)

a. lateral view; b. ventral view; c. differentiation of hydrocoel from entero-hydrocoel; d. sketch showing mode of formation of tentacles, radial-canals and pore canal from hydrocoel. *an:* anus, c_1—c_3: fore-, mid- and hind-enterocoels, *d.p:* dorsal pore, *o:* mouth, *p:* pore canal, *std:* stomodaeum, *t:* primary tentacle.

The developmental changes in the archenteron after this stage are different depending on whether or not the larva passes through an *auricularia* stage; there are also marked differences in the ciliation and mode of formation of the hydrocoel. In species having an auricularia stage, the archenteron grows until it reaches about half the body length. Its anterior tip then turns toward a depression which has appeared in the center of the body wall slightly anterior to the midpoint. The bent part of the archenteron elongates, unites with this stomodaeal invagination and becomes the *oesophagus*. The straight part of the archenteron, which still lies parallel to the body axis, enlarges and differentiates into a *stomach* and a long, slender *intestine*. The blastopore persists while the intestine is being developed and changes into an *anus*, and its opening gradually shifts round toward the side where the mouth is forming. These changes practically complete the development of the digestive tract.

The internal changes are more complicated: cells proliferate within the body cavity near the turning of the archenteron, some of them extending as a tube toward the mid-dorsal line, while others form a pouch near the base of the esophagus. The tubular part gives rise to an opening in the dorsal surface, the *dorsal pore*, while the intervening tube becomes the *pore canal*. The rest of the structure increases in size and differentiates into three regions: one running along the anterior dorsal side of the stomach, another spreading laterally near the anterior part of the stomach, and a third extending longitudinally parallel to its lateral surface. These are called the *anterior*, *central* and *posterior enterocoels* (Fig. 9.29c_1, c_2, c_3). Since these will later develop into the body cavity and hydrocoel, they are given the collective name of *enterohydrocoel* before such differentiation takes place.

Externally the body, as it continues to elongate, becomes somewhat angular at the vegetal pole side and rounded anteriorly; most of the cilia which until this time have been covering the whole surface disappear, leaving only those on marginal ridges which outline the characteristic shape of the larva. The cilia are borne on a single, continuous band which is slightly thicker than the rest of the body wall, and curves to form ring-shaped borders around the anterior and posterior ends of the larva. The side-view of such a larva rather suggests the shape of an ear, leading Johannes Müller to give it the name of 'Auricularia'.

In the forms which do not pass through an auricularia stage, such as *Leptosynapta inhaerens* (Runnström 1927), *Cucumaria echinata* (Ohshima 1918, 1921), *Paracaudina chilensis* (Inaba 1930) and various other species of the Cucumariidae (Kowalevsky 1867; Selenka 1876, 1883), the gastrula stage is externally almost indentical with that described above, but the internal development includes a conspicuously different stage. This stage, which has been suggested by Ohshima (1921) to correspond to the hypothetical dipleurula larva of Bather (1900), involves a striking change in the archenteron and a rapid proliferation of mesenchyme cells. At its beginning, the archenteron elongates and its anterior end turns slightly toward the left ventral side; at the same time a separate pouch-like structure is formed. This is the entero-hydrocoel, which continues to grow for a while in connection with the archenteron, but eventually separates from it (Fig. 9.30d, e).

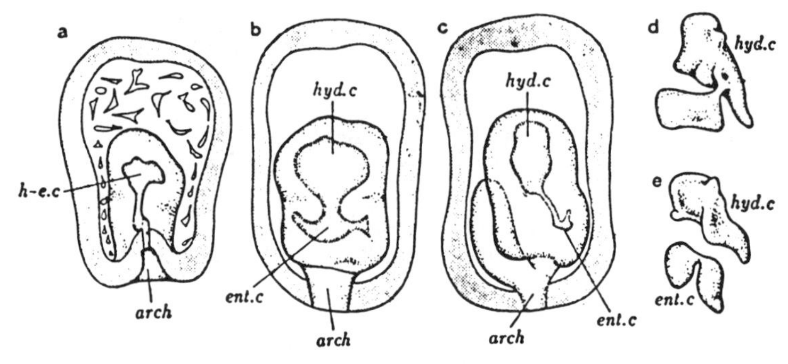

Fig. 9.30 Modes of formation of coelom from entero-hydrocoel in Holothuroidea lacking auricularia stage

a ~ c. *Leptosynapta inhaerens* (Runnström); d ~ e. *Cucumaria planci* (Ohshima) *arch:* archenteron, *ent.c:* enterocoel, *h-e.c:* enterohydrocoel, *hyd.c:* hydrocoel.

When the archenteron has reached a length equal to about half that of the body, it turns ventrally and elongates to make contact with the stomodaeal invagination which has developed in the center of the ventral surface. Soon after these two structures come together, a *mouth* opening is formed. The blastopore becomes the anus, which unlike that of the auricularia remains in its original position at the posterior end of the larva. The part of the archenteron continuous with the oesophagus widens to some extent and becomes the stomach. The shape of the entero-hydrocoel varies somewhat from species to species, but is usually shorter than the archenteron and extended laterally (Fig. 9.30a). This becomes differentiated into a wide *hydrocoel* and an *enterocoel (somatocoel)* which elongates posteriorly (Fig. 9.30b). The connection between these structures is finally lost, and the hydrocoel gives rise to a slender rudiment of the pore canal which extends to the dorsal mid-line, as well as to three bulges which are the rudiments of tentacles. The enterocoel divides into left and right parts which grow downward around the archenteron. This larva, which externally resembles a gastrula, has thus advanced in its internal differentiation to the later part of the auricularia stage.

(2) LARVAL STAGE

a. Auricularia

This larva, widely known as the most characteristic representative of the holothurian development stages, has been described in more than 20 species from all the genera, including *Lapidoplax, Stichopus, Holothuria, Opheodesoma, Synaptura,* etc. Furthermore, a great many auricularia larvae have been collected in plankton hauls but not yet identified with their parent species; these all show characteristic variations in their morphology. Of these, *Auricularia nudibranchiata*[1] (Chun 1895), which is very large and has a complexly sinuous ciliated band (Fig. 9.31), and *Au. paradoxa*[2] of Mortensen (1898), in which the posterior processes are conspicuously elongated (Fig. 9.32), are uniquely holothurian in structure. The others resemble the bipinnaria larvae of the Asteroidea, but can be distinguished from them by two characteristics; (1) however complicated it may be, the ciliated band is single and continuous; (2) the processes extending from the peripheral ridges are always short (Fig. 9.31—33). In larvae which have just reached this stage, the ciliated band forms a small, circular *preoral loop* above the

Fig. 9.31 *Auricularia nudibranchiata* (Chun, Ohshima)
sp: spicules

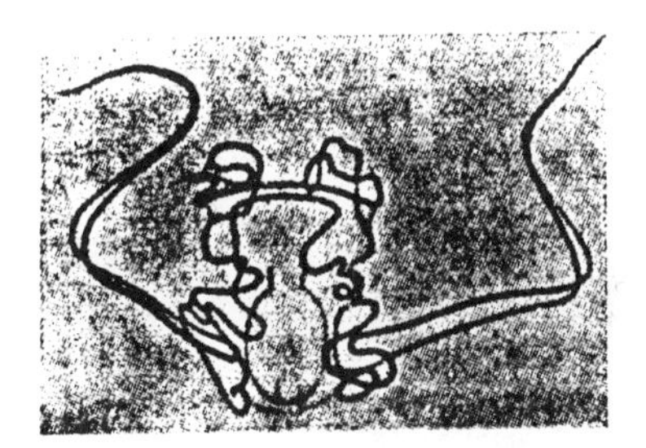

Fig. 9.32 *A. paradoxa* (Mortensen)

mouth and another small *anal loop* anterior to the anal opening; the processes are developed hardly at all (Fig. 9.29). For this reason the depressed region surrounded by the ciliated band takes the shape of a letter H. In *St. japonicus* this stage is reached in 45 — 46 hours. From about this time the larva begins to feed, and its rate of growth depends in large measure upon the suitability of its food. In *St. californicus,* for example, some larvae may be reared for more than three weeks without metamorphosing. The colorless flagellate *Monas* is an ideal food

[1] Also collected at Misaki. Length 15 mm. Mortensen (1938) suggested that it may be the larva of *Synapta maculata.*

[2] Collected in the South Atlantic. Lenght 0.28 mm.

for St. *japonicus* larvae; if they are fed with this organism, they develop normally under artificial conditions (Imai et al. 1950).

As time passes the auricularia larva increases in size, and its ciliated band becomes more convoluted. The course of its winding is bilaterally symmetrical; the projections are much shorter than those of sea urchin or starfish larvae, but they are given the same names. Altogether there are six pairs of processes: the *pre-oral arms* near the mouth, the *post-oral arms* near the anus, the *postero-lateral arms* at the rear end of the body; and dorsally, there are *antero-dorsal, mid-dorsal* and *postero-dorsal arms* (see Fig. 9.24). Figure 9.33 shows the most fully grown auricularia larva of *St. japonicus*. At this stage another pair of protuberances develop between the mid-dorsal and postero-dorsal arms. There is no pigment in these arms or in the ciliated band, although in some species the latter may have yellowish-green spots. Table 9.1 shows the growth of artificially cultured *St. japonicus* larvae during the auricularia stage.

Table 9.1 **Growth of artificially cultured larvae of Stichopus japonicus.**

Culture period (days)	Number of examinations	Average size	
		Length	Width
2	3	421.5 μ	244.4 μ
4	5	603.0	424.1
6	5	728.4	463.5
7	3	764.4	487.6
11	5	842.5	661.2
13	5	938.1	701.2

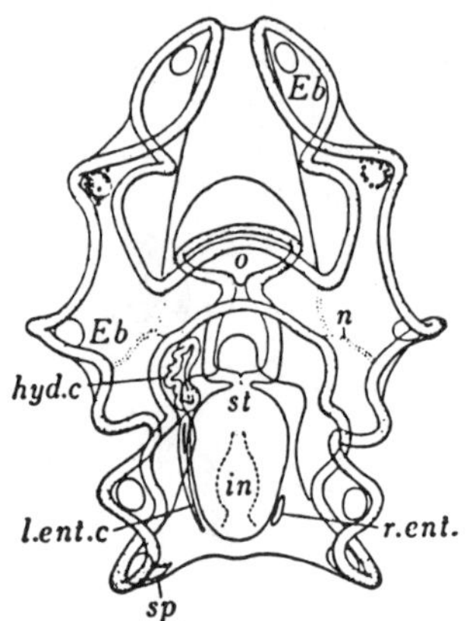

Fig. 9.33 Fully developed auricularia larva of common Japanese sea cucumber (Dorsal view) (Inaba)
Eb: spherical body *hyd.c*: hydrocoel, *in*: intestine *l.ent.c*: left enterocoel, *n*: larval nerve, *o*: mouth, *r.ent.c*: right enterocoel, *sp*: larval spicule, *st*: stomach.

Calcareous larval spicules appear for the first time several days after the beginning of the auricularia stage, in most cases one spicule being formed at the tip of each postero-lateral arm. In *Synapta* there may be two in these locations, or in other forms there may be many in other positions (Fig. 9.31), while in the genus *Holothuria* there are species with spicules at the center of the posterior end of the body. In *St. japonicus* as well as other species of *Stichopus* and *Holothuria*, five pairs of spherical structures are formed, first on the postero-lateral processes and then on the postero-dorsal, mid-dorsal, and antero-dorsal processes and on the antero-dorsal part od the depressed oral region (Fig. 9.33). These spheres, formed by groups of cells which gather on the inner side of each process, correspond to the structure which Müller (1852) named "elastiche Kugel"; they were also described by Mortensen in 1937. Sometimes there are three or four pairs, and they may occasionally occur singly. These hyaline, strongly refractive bodies become ovoid at the time of metamorphosis, and persist in the doliolaria stage. Since they finally disappear together with the doliolarian ciliary rings, they are believed to play some role in the formation of the latter.

b. **Hydrocoel and Body Cavity**

The hydrocoel and enterocoel, which separated from the entero-hydrocoel at the end of the gastrula stage, grow and undergo changes in form, while the digestive tract simply increases in size. The hydrocoel lies at the left side of the oesophagus and bends upward to surround it dorsally; the water vascular system of the tentacles and the rudiment of the ring canal are formed as large and small vesicles, respectively, so that by the end of this larval stage five lobes representing the *primary tentacles* and a *Polian vesicle* are recognizable. As a result of this development, the larva loses its bilateral symmetry internally, while retaining it externally. The pore canal extends to the mid-dorsal line; partway along it cells accumulate and form an anterior chamber. In *St. japonicus* and *Holothuria impaticus* a calcareous precipitate is laid down in this part; before metamorphosis this is organized into a reticulate structure which later develops into the *madreporic body.*

The enterocoel at first forms only on the left side, but one or two days later another vesicle separates off on the right side. The left enterocoel extends from the anterior side of the stomach, and the right enterocoel from near its middle, both growing downward and spreading anteriorly to surround the stomach. The mouth is mushroom-shaped, with a large opening which is bordered by cilia. In the depressed oral field to the left and right of the mouth region are slender, bent, rather opaque raised lines which represent the so-called *larval nerves.* As the larva grows, a branch extends outward from the point of bending, forming a Y-shaped structure.

When larvae of *St. japonicus* are artificially reared at temperatures between 20.5° and 26° C, the most rapidly growing individuals achieve the greatest degree of growth as auricularia larvae on the tenth day, while the slowest ones require 15 days. The largest specimens at this stage are about 1 mm. in length.

(3) **METAMORPHOSIS**

Holothurian metamorphosis from the auricularia stage involves a severing and rearrangement of the ciliated band and a change in form, and gives rise to a barrel-shaped *doliolaria* or *pupa stage* encircled by hoop-like rings of cilia. This striking transformation is completed both externally and internally within a very short time. *St. japonicus* completes the process within 30—36 hours, although this is a most difficult period, requiring the greatest care in culturing.

As the result of metamorphosis, the larval length is abruptly reduced, becoming about 1/3 of the length of the auricularia in *Labidoplax digitata* (Semon 1888), and about 1/2 in *St. japonicus* (Imai et al. 1950) and *Holothuria* (Mortensen 1937). The disappearance and recombination of parts of the ciliated band are due to unequal growth of the epidermis; since accounts of this process differ from species to species and also among the various investigators, the description of Bury (1889) has been selected to give a general idea of the changes involved. (The white parts of the ciliated band in Figure 9.34 represent those which disappear, the black parts those which persist.) In the preoral loop, a central section and two lateral sections persist, and three similar sections represent the anal loop. At the sides, there are three fragments of the ciliated band from the antero-dorsal, mid-dorsal and post-dorsal processes, as well as one from the mid-dorsal depression between the mid-dorsal and postero-dorsal processes, making a total of four. These sections join together to form five ciliated rings, each uniting with other sections which are brought close to it by the sudden shrinkage of the peripheral processes at the time of metamorphosis. Describing the five rings in order from the posterior end: the fifth ring is formed by the union of fragments from the two postero-lateral processes. The fourth ring is made up of two fragments from the postero-dorsal process and two from the sides of the anal loop. The third ring consists of pieces of the band from the mid-dorsal process; the second, of two sections from the antero-dorsal process and one from the right side of the pre-oral loop. The first is formed of sections of the ciliated band from the oral region and the left side of the pre-oral loop. In *St. japonicus*, it appears that the five pairs of laterally located hyaline spheres serve as organizing centers for the process by which the pieces of ciliated band are recombined into ciliary rings, these latter being formed in positions corresponding to the locations of the spheres.

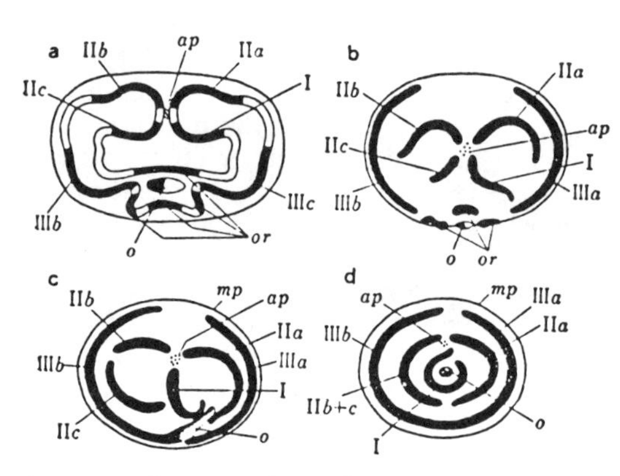

Fig. 9.34 Diagrams of anterior aspect of auricularia showing changes in ciliary bands as larva metamorphoses into doliolaria (Bury)

I ~ V: ciliary bands and ciliary rings. *ap:* thickened apical region, *o:* mouth, *or:* fragments which give rise to oral ring

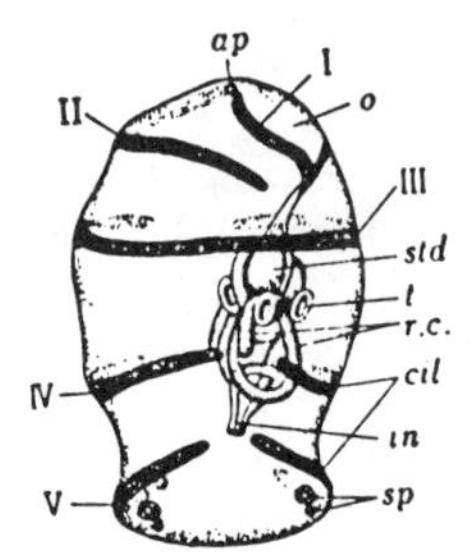

Fig. 9.35 Doliolaria larva immediately after completion of auricularia stage (Bury)

I~V: ciliary rings. *ap:* apical thickening, *cil:* hoop-like ciliated ring, *in:* intestine, *o:* mouth, *r.c:* vestigial radial canal, *sp:* larval spicule, *std:* stomodaeum, *t:* primary tentacle.

In the holothurian species which do not pass through an auricularia stage — species of the genera *Cucumaria, Paracaudina, Leptosynapta* — the cilia, which have been covering the whole surface of the gastrula, begin to disappear from parts of the ventral side, and this process continues toward the dorsal side until several ciliary rings are formed. These are different in different species, but four or five is the usual number (Fig. 9.35, 9.36).

As a result of metamorphosis the mouth, which was in about the center of the ventral surface during the auricularia stage, shifts to the anterior end, and the anus moves to the posterior end. In *Cucumaria, Paracaudina, Leptosynapta,* etc., the anus is located at the posterior extremity from the first, and does not change its position, while the mouth is still on the ventral side at this time and only gradually shifts to the anterior pole (Fig. 9.36).

While these external changes are going on, the internal organs are also developing. The hydrocoel grows around the oesophagus and finally its left and right sides unite ventrally to form the circular *ring canal,* which accompanies the mouth as it moves anteriorly. At the same time the larval tentacles or primary tentacles elongate, and the Polian vesicle increases in size. The radial canals arise from the ring canal between the five tentacles and elongate posteriorly, in the opposite direction from the tentacles. Between these the pore canal extends to the midline in the center of the dorsal side, and its position determines the positions of the tentacles, Polian vesicle and radial canals. These spatial relationships are different in different species, and are particularly easy to observe just before the right and left sides of the hydrocoel unite. According to Clark (1896), the radial canals persist as mere vestiges in *Synapta vivipara* and *Labidoplax digitata,* while *Leptosynapta inhaerens* resembles the adult Synaptidae in having no radial canals. The place at which the two sides of the ring canal unite varies: in some species their union takes place at the mid-ventral line, as in *L. inhaerens,* while in others such as *Lab. digitata, Hol. floridana* and the Cucumariidae, it occurs on the left ventral side, at the position of the first radial canal. In most species the Polian vesicle develops at the left tip of the ring canal, but in *Paracaudina chilensis* it is found on the right, at the inner side of the fifth tentacle. Such radical specific and genetic differences in the spatial relationships within the water vascular system make it necessary to exercise great caution in investigating holothurian embryology.

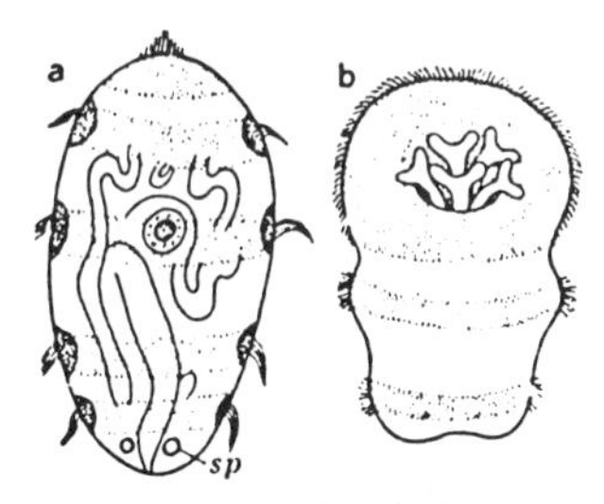

Fig. 9.36 Doliolaria larvae

a. *Leptosynapta inhaerens* (Runnström); b. *Paracaudina chilensis marenzeller* (Inaba). *sp:* larval spicule.

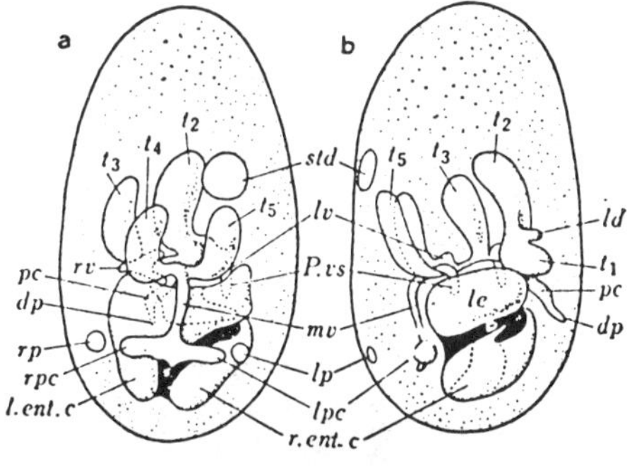

Fig. 9.37 Changes in hydrocoel and enterocoe during doliolaria stage of *Cucumaria planc* (Ohshima)

dp: dorsal pore, *l.ent.c:* left enterocoel, *lp:* pore of left tube foot, *lpc:* left tube-foot canal, *lv:* left ventral radial canal, *P.vs:* Polian vesicle, *r.ent.c:* right enterocoel, *rp:* pore of right tube foot, *rpc:* right tube-foot canal, *rv:* right ventral radial canal, *std:* stomodaeum, $t_1 \sim t_5$: primary tentacles, *mv:* mid-ventral radial canal.

The pore canal becomes reduced, the dorsal *hydropore* is drawn inside the body and by the end of this stage is gradually becoming calcified to form the *stone canal,* while it makes connection with the mid-dorsal point of the ring canal. The radial canals elongate along the body wall; in *Cucumaria, Stichopus* and other of the holothurians having tube feet, the mid-ventral radial canal develops faster and to a greater extent than the others, and bifurcates distally. At the same time two small pores appear in the posterior ventral body wall, and the branches of the radial canal grow toward these, eventually to form the *primary tube feet* (Fig. 9.37 *lpc, rpc, lp*).

The enterocoel extends posteriorly in the ventral region, its left and right sides surrounding the stomach and intestine, finally meeting and uniting both dorsally and ventrally. The dorsal plane of union practically coincides with the longitudinal body axis, but the ventral plane is diagonal, since the left enterocoel has developed faster in the posterior part while the right enterocoel is larger anteriorly. The membrane formed by the union of the two enterocoels persists in the adult as the *mesentery*.

The mouth, surrounded by the tentacles, gradually becomes protuberant; the oesophagus and stomach shorten when the body length is reduced, while the intestine grows longer and begins to bend to some extent. The anus opens at the posterior end of the body. In the sea cucumbers which have the spherical structures described above, these lie at the sides of the body in close connection with the ciliary rings. They elongate into an oval shape and eventually nearly come in contact with each other.

The nerves arise at the bases of the tentacles on the inner face of the ring canal and elongate along the outer side of the radial canals, finally forming a *nerve ring* and five *radial nerves*. X-shaped calcareous *spicules (ossicles)* begin to be formed in the body wall. At about the same time calcareous spicules of the same shape appear in the bases of the tentacles. The former are the *endoskeletal spicules*, while the latter are the rudiments of the calcareous *ring*.

No conspicuous changes in the external morphology take place during this stage, but at its end the tentacles become covered with ectodermal tissue and parts of them can be seen from the outside, through the mouth opening; the *vestibule*, which surrounds the mouth, becomes smaller. Since these striking internal changes occur in the end of this stage, Ohshima (1921) distinguished it from the doliolaria stage, giving it the name of *metadoliolaria*. In *St. japonicus* the doliolaria stage is completed within about three days. Among the Synaptidae, special structures described as *otocysts* are formed as swellings of the epidermal nerves. The larval spicules persist at the posterior end of the body as they were formed in the auricularia stage. When many of these larval spicules are present, as among the synaptids, they are found throughout the body wall, mixed with the newly formed spicules. During the earlier part of the doliolaria stage, the larvae swim vigorously, but as the end of this period approaches, their ciliary movement becomes sluggish and they begin the change toward a more sedentary way of life.

In a short time the rings of cilia disappear and the young animal becomes completely bottom-dwelling. The vestibular invagination is still more reduced so that the mouth is shifted almost to the anterior end of the body, and the five primary tentacles move freely in and out of it. Semon (1888) differentiated between the doliolaria stage and this period, which he named the *pentacula*. In *St. japonicus* this stage is reached on the fourth day after metamorphosis; about ten days more are required for the young holothurian to pass through a series of gradual changes before it finally achieves an appearance like that of the adult. The most striking of these is the formation of the endoskeletal spicules. As the numbers of X-shaped spicules increase, each spicule also extends its branches and joins with the branches of other spicules to form fenestrated plates, with a characteristic structure for every species. By the end of this period enough spicules have been produced to nearly cover the body surface. *St. japonicus* forms spicules of a *table* shape, as well as large, fenestrated *spicular plates* in the vicinity of the anus and between the bases of the primary tentacles. When the tube feet are fully developed and appear outside the body, these plates can be seen within them. The spicular plates near the anus continue to increase in size and later grow into the *anal teeth*. Except for those special species in which the calcareous framework of the madreporite is formed in the auricularia, this structure appears as a reticulate sphere in the late pentactula stage. The spicules constituting the calcareous ring are first found early in this stage as elongated, -shaped units which gradually increase in size. The units which form the *radial pieces* grow faster and their ends divide into two or more projections; the central part of each is perforated and the perforated plate as a whole is slightly curved. The less developed *interradial pieces* alternate with these larger units, the whole forming a ring around the bases of the primary tentacles.

In the holothurian orders which form *tube feet (podia)*, the midventral radial canal grows more rapidly than the others, divides at its tip into two *podial canals* which continue to grow and finally protrude to the outside of the body through two pores formed in the posterior ventral body wall. In the Cucumariidae, the two podia develop almost simultaneously, but in *St. japonicus* the left podium appears two or three days before the right one, and clearly exceeds it in size. Together with the primary tentacles, these podia form the locomotory system of the pentactula and early juvenile stages. In artificially reared *St. japonicus*, the function of locomotion is centered in these structures alone for the next two months, no other tube feet making their appearance during this period.

In the meantime, the left and right sides of the enterocoel have united completely to form the *coelom*, which surrounds the *digestive tract*. This is a tubular structure of approximately uniform diameter throughout its length, which extends from the mouth nearly to the posterior end of the body; there it turns back on the left side and runs anteriorly for some distance before again turning posteriorly to connect with the anus. The *water vascular system* consists of the five primary tentacles, the ring canal and the stone canal; in addition, except among the Synaptidae, the typical holothurian radial canals extend back along the outer surface of the coelom to reach its posterior limit. The *nervous system* is made up of the nerve ring, which surrounds the oesophagus at the bases of the primary tentacles, and five radial nerves which run along the outer side of the radial canals. These five nerves are present even in species like those of the Synaptidae which lack radial canals. Finally, in species which retained external dorsal hydropores from the doliolaria stage, these are withdrawn to an internal position at this time.

(4) JUVENILE HOLOTHURIAN

Early in this period the appearance of the young animal is like that of the pentactula, but as time passes there is a noticeable growth in body size, accompanied by an active formation of the characteristic spicules which will form the basis of classification within the class. There is also an increase in the number of secondary tentacles and tube feet. Artificially cultured juveniles of *St. japonicus* measure 0.3 — 0.4 mm at this stage, but show extreme individual differences in their rate of development. Exceptionally fast-growing specimens may reach such a size on the fifteenth day, while slower ones require 23 days and the slowest may take as long as a month. The color of the body wall is whitish and somewhat opaque, although it is possible to see the intestine through it. The oral region opens widely and a mesentery connects the five primary tentacles so that, extended, they present a parasol-like appearance. Spicules are formed in the tentacles and tube feet as well as in the body wall. These variously shaped spicules, having the form of rods, circular plates and "tables", are distributed at random over the whole body surface; except for increased size and thickness, the adult spicules are practically the same as those of the juvenile. The anal teeth grow and surround the anus, and the five large spicular plates between the primary tentacles become still larger, acquiring the shape of shields. These project to the outside, covering and protecting the oral region when the tentacles are contracted. The spatial relations between the ring canal of the primary tentacles and the radial canals are different from one genus and species to another, and determine the conditions regulating the appearance of the secondary tentacles.

In all holothurian species the primary tentacles at first arise from the ring canal, but a shift takes place later in some species so that two types are formed: one in which the tentacular canals take their exit from the radial canals, and another in which they continue to leave the ring canal at points between the exits of the radial canals. The system of radial canals includes some which give off two tentacular canals, others which have only one and some with none at all. The secondary tentacles usually originate from radial canals lacking primary tentacles. *Cucumaria echinata* (Ohshima 1921) is a species in which the five primary tentacles retain their connection with the ring canal for a long time; when the secondary tentacles develop, one arises from the right ventral radial canal, followed by another (the seventh tentacle) from the left ventral radial canal. The eighth is formed from the right dorsal radial canal, the ninth opposite the sixth on the right ventral radial canal, and the tenth opposite the seventh on the left ventral radial canal. In this stage with ten tentacles there is thus none directly connected with the left dorsal radial canal. In *Holothuria floridana* (Edwards 1909) there are two (left and right) primary tentacles on the mid-ventral radial canal, and one each connected with the right ventral, left ventral and left dorsal canals; the secondarily formed sixth tentacle is formed on the right ventral, the seventh on the left ventral, the eighth on the right dorsal canal, and the ninth and tenth appear simultaneously on the left and right ventral canals. In this case the left and right ventral radial canals connect with three tentacles each, the mid-ventral with two, while the right and left dorsal radial canals have one tentacle apiece.

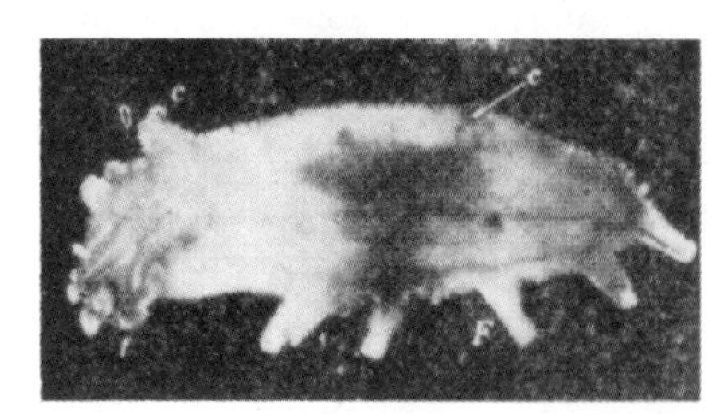

Fig. 9.38 Common Japanese sea-cucumber at 2 months (Inaba)
c: papilae, *F:* tube foot, *t:* tentacle.

In the holothurians having tube feet, these increase in number, the positions of the newly added podia differing in different genera and species. In many cases, they are located one somewhat behind the other, to left and right on the ventral surface. Those appearing later are generally formed in pairs on opposite sides of the mid-ventral line. Species having papillate podia form one on each side of the anterior part of the body early in the juvenile stage; for some time there are only these two, but later one is formed on each side rather near the posterior end of the body (Fig. 9.38).

Each of these various structures has branched off from a corresponding radial canal, and remains attached to it. As the body increases in size they become more numerous and take on the shape characteristic of the adult organ. In species having arborescent *dendritic tentacles,* or finger-like *digitate* ones, the characteristic branching and development of the tentacles takes place in this period.

Cultured *St. japonicus* juveniles reach a length of 4 to 5 mm after two months, and come to prefer attached diatoms as food. Pigmentation of the body surface begins to appear after four to five months, when the body length is about one centimeter, and the intestinal tract takes on a yellowish-red color. This rate of growth under artificial conditions, however, is extremely slow compared with that in nature, as estimated from the size of juvenile holothurians collected at Misaki by Mitsukuri (1903). The feeding habits of these immature sea cucumbers therefore remain as a problem for future investigation.

REFERENCES

Echinoidea, Asteroidea, Ophiuroidea, Crinoidea.

Brown, D. E. S. 1934: The pressure coefficient of 'viscosity' in the eggs of *Arbacia punctulata.* J. Cell. Comp. Physiol., **5** : 335.

Bury, H. 1888: The early stages in the development of *Antedon rosacea.* Phil. Trans. Roy. Soc., **179** : 257—301.

Dan, J. C. 1954: Studies on the acrosome. II. Acrosome reaction in star-fish spermatozoa. Biol. Bull., **107** : 203—218.

Dan, J. C. & K. Dan 1941: Early development of *Comanthus japonica.* Jap. J. Zool., **9** : 565—574.

Dan, K. 1952a: Meiosis in the egg of the crinoid, *Comanthus japonica.* Annot. Zool. Jap., **25** : 258—264.

——— 1952b: Cyto-embryological studies of sea urchins. II. Blastula stage. Biol. Bull., **102** : 74—89.

——— 1954a: Further study on the formation of the "new membrane" in the eggs of the sea urchin *Hemicentrotus* (*Strongylocentrotus*) *pulcherrimus.* Embryologia, **2** : 99—114.

——— 1954b: The cortical movement in *Arbacia punctulata* eggs through cleavage cycles. Embryologia, **2** : 115—122.

Dan, K. & J. C. Dan 1940: Behavior of the cell surface during cleavage. III. On the formation of new surface in the eggs of *Strongylocentrotus pulcherrimus.* Biol. Bull., **78** : 486—501.

——— & ——— 1941: Spawning habit of the crinoid, *Comanthus japonica.* Jap. J. Zool., **9** :555—564.

Dan, K. & K. Okazaki 1956: Cyto-embryological studies of sea urchins. III. Role of the secondary mesenchyme cells in the formation of the primitive gut in sea urchin larvae. Biol. Bull., **110** : 29—42.

Dan, K. & T. Ono 1952: Cyto-embryological studies of sea urchins. I. The means of fixation of the mutual positions among the blastomeres of sea urchin larva. Biol. Bull., **102** : 58—73.

Delage, Y. 1901: Etudes éxpérimentales sur la maturation cytoplasmique chez les Echinodermes. Arch. Zool. éxp. gén., ser. 3, **9** : 285—326.

Driesch, H. 1891: Entwicklungsmechanische Studien. I. Der Wert der beiden ersten Furchungszellen in der Echinodermenentwicklung. Experimentelle Erzeugung von Teil- und Doppelbildungen. Zeitschr. wiss. Zoöl., **53** : 160—178.

Endo, Y. 1953: Secretion of hatching enzyme in Ca-free sea water in sea urchin embryos. Zool Mag., **62** : 106. (in Japanese)

Fell, H. B. 1948: Echinoderm embryology and the origin of Chordates. Biol. Rev., **23** : 81—107,

Gemmill, J. F. 1912: The development of the star-fish, *Solaster endeca* Forbes. Trans. Zool. Soc. **20** :1—58.

Goto, S. 1897: The metamorphosis of *Asterias pallida* with special reference to the fate of the body cavities. Contr. Zool. Lab. Mus. Comp. Zool., no. **98**.

——— 1898: Some points in the metamorphosis of *Asterina gibbosa*. J. Coll. Sci. Tokyo Imp. Univ., **12** : 227—242.

Gustafson, T. & H. Kinnander 1956: Microaquaria for time-lapse cinematographic studies of morphogenesis in swimming larvae and observations on sea urchin gastrulation. Exp. Cell Res., **11** : 36—51.

Harvey, E. B. 1949: The growth and metamorphosis of the *Arbacia punctulata* pluteus, and late development of the white halves of centrifuged eggs. Biol. Bull., **97** : 287—299.

——— 1956: The American *Arbacia* and Other Sea Urchins. Princeton University Press.

Herbst, C. 1892: Experimentelle Untersuchungen über den Einfluss des veränderten chemischen Zusammensetzung des umgebenden Mediums auf die Entwicklung der Tiere. I. Versuche an Seeigeleiern. Zeitschr. wiss. Zoöl., **55** : 446—518.

Hörstadius, S. 1953: Über die Determination im Verlaufe der Eiachse bei Seeigeln. Pubbl. Staz. Zoöl. Napoli., **14** : 251—429.

——— 1939a: Ueber die Entwicklung von *Astropecten aranciacus* L. Pubbl. Staz. Zoöl. Napoli., **17** : 221—321.

——— 1939b: The mechanics of sea urchin development, studied by operative methods. Biol. Rev., **14** : 132—179.

Ishida, J. 1936: An enzyme dissolving the fertilization membrane of sea urchin eggs. Annot Zool. Jap., **15** : 453—457.

——— 1954: Function of egg surface during fertilization in sea urchins. Cyto-chemistry Symposium, **2** : 65—80. (in Japanese)

Kopac, M. J. 1941: Disintegration of the fertilization membrane of *Arbacia* by the action of an "enzyme". J. Cell. Comp. Physiol., **18** : 215—220.

Kubo, K. 1948: Development of *Leptasterias ochotensis similispinis* Clark. Zool. Mag., **58** : 34. (in Japanese)

——— 1951: Some observations on the development of the sea-star *Leptasterias ochotensis similispinis* (Clark). J. Facul. Sci. Hokkaido Univ. ser. 6, **10** : 97—105.

Kumé, M. 1929: On the development of sea urchins from Misaki. Zool. Mag., **41** : 100—105. (in Japanese)

Lindahl, P. E. 1936: Zur Kenntnis der Entwicklungsphysiologie des Seeigeleies. Acta Zool., **17** :179—365.

MacBride, E. W. 1914: Text-book of Embryology. London.

Mitchison, J. M. & M. M. Swann 1955: The mechanical properties of the cell surface. III. The sea urchin egg from fertilization to cleavage. J. Exp. Biol., **32** : 734—750.

Mortensen, T. 1920: Studies in the Development of Crinoids. Carnegie Inst. Washington.

——— 1921: Studies of the Development and Larval Forms of Echinoderms. Copenhagen.

——— 1931: Contributions to the study of the development and larval forms of Echinoderms I, II. Mem. Roy. Acad. Sci. Let. Denmark, **4** (1) : 1—39.

——— 1937: Contributions to the study of the development and larval forms of Echinoderms III. Mem. roy. Acad. Sci. Let. Denmark, **7** (1) : 1—65.

——— 1938: Contributions to the study of the development and larval forms of Echinoderms IV. Mem. roy. Acad. Sci. Let. Denmark, **7** (3) : 1—59.

Motomura, I. 1935: Determination of the embryonic axis in the eggs of Amphibia and Echinoderms. Sci. Rep. Tohoku Imp. Univ. ser. 4, **10** : 211—245.

Murakami, S. 1937: On the development of the calcareous plates in an ophiuran larva, *Ophiopluteus serratus*. Annot. Zool. Jap., **16** : 135—147.

——— 1940: On the development of the calcareous plates of an ophiuran, *Amphipholis japonica* Matsumoto. Jap. J. Zool., **9** : 19—33.

——— 1941: On the development of the hard parts of a viviparous ophiuran, *Stegophiura sculpta* (Duncan). Annot. Zool., Jap., **20** : 67—78.

Ohshima, H. 1911: Larva of Echinoderma. Zool. Mag., **23** : 377—394. (in Japanese)

——— 1921: Reversal of asymmetry in the plutei of *Echinus miliaris*. Proc. roy. Soc ser. B, **92** : 168—178.

——— 1922: The occurrence of situs inversus among artificially-reared echinoid larvae Quart. J. micro. Sci., **66** : 105—150.

Okazaki, K. 1954: Metamorphosis of half-larvae in *Astriclypeus manni* Verrill. Zool. Mag. **63**: 166. (in Japanese)

——— 1956a: Exogastrulation induced by calcium deficiency in the sea urchin, *Pseudocentrotus depressus*. Embryologia, **3** : 23—36.

——— 1956b: Existence of matrix for spine formation in sea urchin larvae. Zool. Mag. **65** : 77—78. (in Japanese)

Okazaki, K. & K. Dan 1954: Metamorphosis of partial larvae of *Peronella japonica* Mortensen, a sand-dollar. Biol. Bull., **106** : 83—99.

Onoda, K. 1931: Notes on the development of *Heliocidaris crassispina* with special reference to the structure of the larval body. Mem. Coll. Sci. Kyoto Imp. Univ. ser. B, **7** : 103—134.

——— 1936: Notes on the development of some Japanese echinoids with special reference to the structure of the larval body. Jap. J. Zool., **6** : 637—654.

——— 1938: Notes on the development of some Japanese echinoids with special reference to the structure of the larval body. Report III. Jap. J. Zool., **8** : 1—13.

Runnström, J. 1928: Zur experimentelle Analyse der Wirkung des Lithiums auf den Seeigelkeim. Acta Zool., **9** : 365—424.

——— 1935: An analysis of the action of lithium on sea urchin development. Biol. Bull., **68** : 378—383.

Seeliger, 1892: Studien zur Entwicklungsgeschichte der Crinoiden (*Antedon rosacea*). Zoöl. Jahrb. (Anat), **6** : 161—444.

Shearer, C., W. de Morgan & H. Fuchs 1914: On the experimental hybridization of Echinoids. Phil Trans. Roy. Soc. London, **203**.

Sugawara, H. 1943: Hatching enzyme of the sea urchin, *Strongylocentrotus pulcherrimus*. J. Fac. Sci. Tokyo Imp. Univ., sec. 4, **6** : 109—127.

Yatsu, N. 1910: A note on the polarity of the primary oocyte of *Asterias forbesii*. Annot. Zool Jap., **7** : 219—221.

Holothuroidea

Baur, A. 1864: Beiträge zur Naturgeschichte der *Synapta digitata*. Nova Acta Acad. Leop. Carol. **31**.

Bury, H. 1889: Studies in the embryology of the echinoderms. Quart. Jour. Micr. Sc., **29**.

Bury, A. 1895: The metamorphoses of echinoderms. Quart. Jour. Micr. Sc., **38**.

Clark, H. L. 1898: *Synapta vivipara*: A contribution to the morphology of echinoderms. Mem. Boston Soc. Nat. Hist., **5** (3).

——— 1910: The development of an apodous holothurian (*Chiridota rotifera*). Jour. Exptl. Zool. **9**.

Danielssen & Koren, 1856: Observations sur le développement des Holothuries. Fauna littoralis Norvegiae.

Edwards, Ch. L. 1909: The development of *Holothuria floridana* (Pourtalès) with special reference to the ambulacral appendages. Jour. Morph. Boston. **20**.

Inaba, D. 1930: Notes on the development of a holothurian, *Caudina chilensis* (J. Müller). Sci. Rep. Tohoku Imp. Univ. 4th Ser. **5**.

——— 1934: On some holothurian larvae and young from New Guinea. Bull. Jap. Soc Sci. Fish., **2**.

——— 1937: On the artificial fertilization of sea-cucumber eggs. Suisan Kenkiu-shi **32**. (in Japanese)

——— 1937, 1943: On the development of sea-cucumbers. I, II. Zool. Mag. **49, 55**. (in Japanese)

Imai, T., D. Inaba et al. 1950: On the artificial breeding of Japanese sea-cucumber, *Stichopus japonicus* Selenka. Bull. Inst. Agricul. Res. Tohoku Univ., **2**.

Korschelt u. Heider 1936: Vergleichende Entwicklungsgeschichte der Tiere, Jena.

Kowalevsky, A. 1867: Beiträge zur Entwicklungsgeschichte der Holothurien. Mém. de L'Acad. impér. Sci. de St. Pétersbourg, Ser. **7, 11**.

Ludwig, H. 1891: Zur Entwicklungsgeschichte der Holothurien. Sitz. der Kgl. Preuss. Akad. Wiss. Berlin, I, II.

——— 1898: Brutpflege und Entwicklung von *Phyltoporus urna* Grube. Zoöl. Anz., **21.**

MacBride, E. W. 1914: Textbook of Embryology. London.

Metschnikoff, E. 1869: Studien über die Entwicklung der Echinodermen und Nemertinen. Mém. de l'Acad. Impér. des Science de St. Pétersbourg, Ser. 7, **11.**

Mitsukuri, K. 1903: Notes on the habits and life-history of *Stichopus japonicus* Selenka. Annot· Zool. Japon., **5** (1).

Mortensen, Th. 1898: Die Echinodermenlarven der Plankton-Expedition. Kiel und Leipzig.

——— 1921: Studies of the development and larval forms of Echinoderms. Copenhagen.

——— 1931, 1937, 1938: Contributions to the study of the development and larval forms of Echinoderms. I-II, III, IV. Mém. de l'Acad. Royale. d. Sci. et Lettres de Danemaık, Copenhagen.

Müller, Joh. 1848—1853: Ueber die Larven und die Metamorphose der Echinodermen. Abh. Berliner Akad. Wiss. Berlin.

Newth, H. G. 1916: The early development of *Cucumaria.* Preliminary Account. Proc. Zool. Soc. London,. Part II.

Ohshima, H. 1918: Notes on the development of *Cucumaria echinata.* Annot. Zool. Japon., **9.**

——— 1921: On the development of *Cucumaria echinata* v. Marenzeller. Quart. Jour. Micro. Sci., **65.** Part II.

——— 1925: Note on the development of the sea-cucumber, *Thyone briareus.* Science **61.**

——— 1930: Echinoderma. Iwanami Biological Course, Iwanami Shoten, Tokyo. (in Japanese)

Runnström, J. & Runnström, S. 1919: Ueber die Entwicklung von *Cucumaria frondosa,* Gunnerus und *Psolus phantapus* Strussenfelt. Bergens Mus. Åarbok, 1918—1919.

Runnström, S. 1927: Ueber die Entwicklung von *Leptosynapta inhaerens* (O. F. Müller). Bergens Mus. Åarbok, 1927.

Selenka, E. 1876: Zur Entwicklung der Holothurien. Zeitschr. wiss. Zoöl., **27.**

——— 1883: Studien über Entwicklungsgeschichte der Tiere. 2 Heft, Die Keimblätter der Echinodermen. Wiesbaden.

Semon, R. 1888: Die Entwicklung der *Synapta digitata* und die Stammesgeschichte der Echinodermen. Jen. Zeitschr., **22.**

Vaney, Clement, 1906: Deux nouvelles Holothuries incubatrices. Compt. rend. Assoc. Française Avancem. Sci., 1906.

Vaney, Clement, 1925: L'incubation chez les Holothuries, Trav. d. la St. Zool. de Wimereux. **9.**

ARTHROPODA

Chapter 10

I. CRUSTACEA

INTRODUCTION

The Crustacea are classified as follows:[1]

Subclass I.	Phyllopoda
Order 1.	Branchiopoda. *Branchinella, Artemia, Apus, Estheria.*
Order 2.	Cladocera. *Daphnia, Bosmina, Moina, Holopedium.*
Subclass II.	Ostracoda. *Cypris, Cypridina.*
Subclass III.	Copepoda
Order 1.	Eucopepoda. *Calanus, Diaptomus, Cyclops.*
Order 2.	Branchiura. *Argulus.*
Subclass IV.	Cirripedia
Order 1.	Eucirripedia. *Balanus, Mitella, Lepas.*
Order 2.	Rhizocephala. *Sacculina.*
Subclass V.	Malacostraca
Series A.	Leptostraca. *Nebalia.*
Series B.	Eumalacostraca
Division a.	Syncarida. *Anaspides.*
Division b.	Peracarida.
Order 1.	Mysidacea. *Mysis, Mesopodopsis, Hemimysis.*
Order 2.	Cumacea. *Cuma.*
Order 3.	Tanaidacea. *Tanais, Apseudes.*
Order 4.	Isopoda. *Cymothoa, Asellus, Jaeropsis, Armadillidium, Porcellio.*
Order 5.	Amphipoda. *Gammarus, Caprella.*
Division c.	Eucarida
Order 1.	Euphausiacea. *Euphausia.*
Order 2.	Decapoda
Suborder .	Natantia. *Penaeus, Lucifer, Leander, Alpheus.*
Suborder .	Reptantia.
Section (1)	Palinura. *Panulirus, Jasus.*
Section (2)	Astacura. *Astacus, Homarus.*
Section (3)	Anomura. *Eupagurus, Petrolisthes, Callianassa, Gebia.*
Section (4)	Brachyura. *Maja, Neptunus, Cancer.*
Division d.	Stomatopoda. *Squilla, Gonodactylus.*

[1] The table of classification inserted here has been abridged to include only the orders requisite to the descriptions on the following pages. The Phyllopoda, Ostracoda, Copepoda and Cirripedia are occasionally included under the general term Entomostraca for the sake of contrast with the Malacostraca.

The Crustacea, a class which shows great variety in the modes of living and morphological structure of its members, is also diversified with respect to their developmental processes. These animals, however, possess some noteworthy characteristics in common. One of these is that during the course of their development, the Crustacea without exception go through a larval stage known as the nauplius (Fig. 10.1). In species having ova with little yolk, this stage, which is furnished with three pairs of appendages and a median eye, represents the first free-living larva, while it is passed through during the embryonic life of species with yolky eggs. Among all the crustacean larvae, the nauplius is the form with the smallest number of body segments which is adequate to support an independent existence.

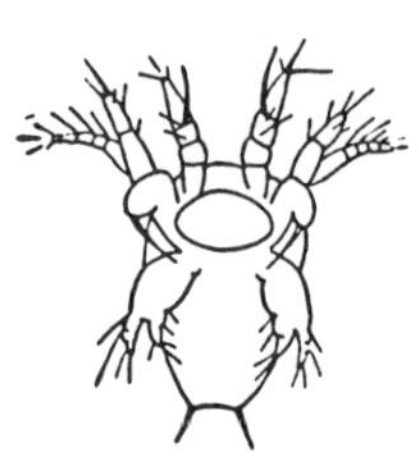

Fig. 10.1 Eucopepodian nauplius larva (Calman)

In the second place, the mode of cleavage of crustacean eggs retains vestiges of the spiral type found in the Plathelminthes, Annelida and Mollusca. Although the Arthropoda in general undergo superficial cleavage, only the Malacostraca among the Crustacea conform to this rule, the majority of the others cleaving totally. Not only do the totally cleaving eggs follow the spiral pattern, but the superficially cleaving ones also show traces of this type to a certain extent in their modes of cleavage furrow formation.

A third characteristic is that the teloblasts (primordial cells which give rise to the posterior part of the trochophore larva in the above phyla) also play an important role in forming the body region posterior to the first maxillary segment, at least in the Malacostraca.

Finally, it should be borne in mind that the various complicated structures of all the adult Crustacea are built up in conformity with the general ontological principle that their bodies are constructed on a fundamental plan consisting of a succession of segments, each of which carries a pair of biramous limbs. Besides these properties common to all the Crustacea, the appearance of such a habit as brooding the young exerts a definite influence on the amount of yolk in the ova, and this, in turn, causes variations in the mode of embryonic development as well as the time of hatching. Furthermore, the evolutionary gradations met with in the adult forms add various degrees of complexity to the processes of metamorphosis.

(1) SPAWNING SEASONS AND METHODS OF COLLECTING EGGS

The habit of caring for their offspring is highly developed among the Crustacea; in the majority of cases, the young are carried on the bodies of the mother animals, and can easily be obtained by collecting such brooding adults. To secure uncleaved eggs, it is necessary to study the spawning behavior of females kept in confinement together with some male animals. Generally speaking, crustacean eggs are laid during the spring and summer. In lower crustaceans such as the Cladocera or Eucopepoda, the eggs are obtainable all year round, so long as the mother animals are detectable.

The adults of the Branchiopoda make their appearance in May or June, swarming in the paddy fields, but disappear within a short time. All the females carry eggs which, however, do not develop immediately. They are set free on the death of the mother shrimp, and lie dormant in the soil of the paddy fields, undergoing desiccation during the winter, and only beginning their development the following spring, with the return of warm weather and adequate water.

Among the Cirripedia, such forms as *Balanus* and *Mitella* contain spawned eggs within their mantle cavity from June or July to September (Hudinaga 1940). The earlier developmental stages of *Panulirus* are obtainable in June or July (Terao 1929; Shiino 1950). The breeding season of *Leander* is much longer, lasting from the middle of May to the middle of September (Kajishima 1950). *Paralithodes* lays its eggs from the middle of May to the middle of June, and is said to carry them for the unexpectedly long period of more than ten months. *Nephrops* lays its eggs in March.

In contrast to the ease with eggs carried by the mother can be obtained is the difficulty of collecting eggs shed free in the water, such as those of *Penaeus*. This species spawns at midnight, from the first week in May to the end of September, and the eggs sink to the sandy bottom of the sea. Since they are rather small and hardly distinguishable from the sand grains, it is almost useless to search for them in the areas where they breed naturally. The only althernative is to prepare aquarium equipment particularly designed for this purpose (Hudinaga 1935). *Sergestes* also sheds its eggs free in the water, from June to August, and there is a record that these were taken in Suruga Bay from a depth of 180 m with a vertical collecting net (Nakazawa 1919). The egg mass of *Squilla* is merely held by the mother under its head with the maxilipeds during the breeding season (June); when an animal is caught in a fishing net, it nearly always drops the egg mass. The best way of procuring developmental stages in this species, therefore, is the more difficult one of finding the animals, together with their eggs, in their burrows on a muddy beach at low tide.

(2) CARE OF THE YOUNG

Among the Crustacea it is rather usual for the mother animal to protect, in some way or other, not only the embryos but also the larvae after hatching. *Estheria* carries the eggs in the spaces between the body and the lateral shell (Fig. 10.2a), while the Cladocera take care of their embryos in a *brood-chamber* formed by the carapace and the dorsal side of the body (Fig. 10.2b). When the eggs are provided with only a small amount of yolk, the skin of the mother's body may even secrete nutriments into such a chamber to feed the larvae. Among the Ostracoda there are two types of spawning behavior: in one the ova are protected in the space between the carapace and the dorsal side of the hind part of the body; in the other they are either deposited on water plants or shed free into the water. Free egg-spawning is also met with at times in the Eucopepoda, but in the greater part of this order the eggs are shed into egg sacs secreted by the oviducts, and carried thus attached to the body of the mother animal until they hatch (Fig. 10.2c). The egg sacs may be either paired or unpaired, according to the species. Various species of *Argulus* lay their eggs on small stones or other objects. The Eucirripedia spawn their eggs into a flattened sac, secreted by the oviducts, which passes over the back and surrounds the mother's body within the mantle cavity; the young are carried in this until they hatch.

Among the Malacostraca, the Peracarida form a brood-chamber under the thorax by the imbrication of foliaceous *oostegites* which develop from the bases of the maternal pereiopods and extend to the midventral surface (Fig. 10.2d, e). Although their number varies, they are usually present in five pairs. The ova of the terrestrial Isopoda are not only protected by these oostegites, but in certain cases are also bathed in a nutritive fluid secreted by the mother animal. The Leptostraca, which lack oostegites, form the brood chamber by the median interlocking of plumose setae borne on the pereiopoda of the two sides. In the Euphausiacea, a transient 'cradle' is similarly formed by an overlapping of these setae, or the egg mass may simply be attached to the posterior pairs of pereiopoda.

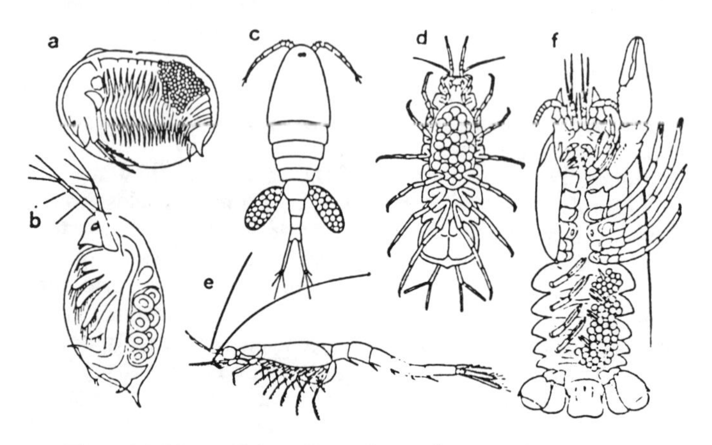

Fig. 10.2 Maternal brooding of eggs in several crustaceans
a. *Estheria*; b. *Daphnia*: c. *Cyclops*; d. *Asellus communis*; e. *Mesopodopsis* (Nair); f. *Astacus*.

Except for Penaeus and allied species, which shed their eggs free into the water, all the members of the Decapoda carry their offspring attached to their pleopoda (Fig. 10.2f). In many small prawns and shrimps, as well as in *Callianassa*, a secretion from the dermal glands on the pleopoda adheres to the egg-membranes and coagulates, forming threads which fasten the eggs to the setae of the pleopoda. Such glands are also found in *Astacus* on the sternal surface of the abdomen, with ducts extending to the interior of the pleopoda and opening to the exterior at the bases of the pleopodal setae. In *Eupagurus*, similar glands are distributed on the soft ventral surface of the abdomen. In many crabs, the secretion from these glands is replaced by a similar secretion from the receptaculum seminis. The mantis-shrimp lumps the eggs into a flattened, disc-shaped mass and guards it in its burrow, or holds it under the head with the maxillipeds until the larvae hatch.

Hand-in-hand with the development of the brooding habit, there generally goes a retardation in the stage at which the larva begins an independent life, although the time of hatching is not always similarly affected. For example, the larvae of the Peracarida, which emerge from the egg membranes in an early stage of development, do not leave the brood-chamber for some time; when they do finally depart, they have already acquired a structure not essentially different from that of the adult.

It appears to be a general rule that the females lay their eggs immediately after molting. This fact has been ascertained in *Caridina* (Nair 1949), *Penaeus* (Hudinaga 1942), *Jasus* (a near ally of *Panulirus* occurring in South Africa) (von Bond 1936), *Gammarus* (Bergh 1849) and *Mesopodopsis* (Nair 1939).

(3) EGGS

Crustacean eggs are rather large and yolky. The size of an egg depends mainly on the amount of yolk it contains, and bears no relation to the size of the mother animal; it is, however, definitely correlated with the stage at which hatching takes place. Within the order Decapoda are both *Astacus*, in which the egg is about 2.5 mm in diameter and the larva closely approaches the adult state before hatching, and *Penaeus*, with an egg only about 0.3 mm in diameter, from which the larva hatches as a nauplius.

The eggs of the Crustacea are centrolecithal, irrespective of their size. The protoplasmic portion of an uncleaved egg consists of a thin superficial layer of *peripheral protoplasm*, and a *protoplasmic island* isolated in the center of the mass of yolk which occupies the greater part of the egg (Figs. 10.12a and 10.13a). The island is a stellate mass of cytoplasm, from which issue a number of processes. In the oocytes of the Isopoda, this mass is at first connected with the peripheral protoplasm by a number of fine protoplasmic threads, but as the accumulation of yolk increases, the connection is broken (McMurrich 1895). In the newly laid eggs of *Astacus*, a protoplasmic reticulum can still be recognized traversing the yolk so as to connect the peripheral layer with the central protoplasmic island (Zehnder 1934).

The accumulation of yolk within the oocyte proceeds uniformly throughout the entire cell in *Squilla wood-masoni*, while in *Palaemon idae* the yolk granules make their first appearance in the peripheral region and gradually move toward the center (Aiyar 1949). In both these cases the nucleus lies in the center of the cell, which is centrolecithal throughout the whole process. In contrast to these examples, the nucleus of the *Mesopodopsis* oocyte lies in the peripheral protoplasm at one pole of the cell, and the accumulation of yolk proceeds as in telolecithal eggs. After fertilization, however, by the time the synkarion enters the first division, it has moved toward the center and the egg becomes centrolecithal (Nair 1939). No similar case is known among the other crustaceans.

In the centrolecithal eggs of crustaceans it is generally rather difficult to distinguish the egg axis except as it is indicated by the location of the polar bodies. Occasionally, however, cases like that of *Calanus* occur, in which the synkarion lies somewhat closer to the animal pole (Grobben 1879). In *Moina*, two kinds of differentially-staining yolk granules surround the nucleus concentrically, but they have a tendency to be more abundant near the vegetal than the animal pole (Grobben 1879). The *Holopedium* egg has a large, eccentric oil drop as well as a concentric arrangement of oil drops of much smaller but varying sizes and a similar arrangement of yolk granules (Baldass 1937). The summer egg of *Polyphemus* is accompanied during its formation by three nutritive cells. When they have been almost completely absorbed by the egg, their remnants remain as a small mass within the egg at one side, and the polar bodies emerge at the opposite side (Fig. 10.6a). Among the Cirripedia, the newly laid egg has one blunt and one more pointed end, like a hen's egg; the blunt end includes the animal pole.

Before the maturation divisions come to an end, the oocyte of *Balanus amphitrite rosa* is centrolecithal, and the granular yolk filling the space between the peripheral protoplasm and the central protoplasmic island contains comparatively uniformly distributed oil spherules. About the time when the polar bodies are given off, the two protoplasmic regions unite and the confluent mass flows toward the animal pole; after the emission of the polar bodies the nucleus returns to the center of the egg, the oil spherules disperse and disappear, and the yolk structure of the egg becomes homogeneous (Kajishima 1951). In *Mitella mitella,* on the other hand, when cleavage begins, the oil spherules concentrate toward the vegetal pole, making the egg appear to be telolecithal (Fig. 10.9a-c). The peripheral protoplasm becomes somewhat thicker in the region slightly dorsal to the animal pole, thus giving the first indication of bilateral symmetry. The polar bodies are always given off from the blunt end, showing that the primary egg axis connecting the animal and vegetal poles in the stage just before the beginning of cleavage coincides with the longer axis. The polar axis in the obovate eggs of the Decapoda does not necessarily agree with their longer axis, however, frequently making an angle with it.

It is well established that the Cladocera lay summer and winter eggs. The former develop parthenogenetically, while the latter must be fertilized, and the two types differ further in their mode of development, even though they belong to a single species. *Artemia salina* has two local strains, one multiplying by parthenogenesis and the other by bisexual reproduction. Parthenogenesis is known also in the Ostracoda. The eggs of this order, and those of the subclass Phyllopoda, which are able to endure desiccation, are transported long distances by the wind, by aquatic birds to whose bodies they are attached, or by other means. In a certain species of *Diaptomus,* which belongs to the Eucopepoda, it is said that the eggs laid during the summer have a thin egg membrane and hatch out within a short time, while those laid in autumn have a thick membrane and rest under the mud in a dormant state until the spring of the following year. In this species, however, reproduction is always bisexual even in the warm seasons.

(4) EGG MEMBRANES

Some confusion seems to exist in the terminology used to define the egg laembranes in crustaceans. Although the term 'chorion' appears frequently, espeiemy with respect to the malacostracan egg, this refers in most cases to the tertiary cgg membrane secreted by the oviduct rather than to the true chorion which is, in the sense of Korschelt and Heider, a secondary membrane derived from the follicle cells.

The Crustacea usually have one or two egg membranes. In *Panulirus* and *Astacus,* the tertiary membrane secreted by the oviduct is enclosed in another, which is a product of the dermal glands of the pleopoda or pleomeres (Shiino 1950; Zehnder 1934). This has been termed an 'exo-chorion' by Zehnder. The secretion from the dermal glands, however, is sometimes no more than a string fastening the so-called chorion to the setae of the pleopods; in *Caridina, Alpheus, Maja, Eupagurus,* etc, the eggs are protected by this membrane alone (Nair 1949; Brooks-Herrick 1880; Mayer 1877). The 'chorion' of *Palaemon* is asserted to be a product of the dermal glands (Aiyer 1949). The coverings enclosing the egg of *Homarus* correspond respectively to secondary and tertiary membranes, the latter being formed by a secretion of the oviducts, but there is some room for doubt about their origin. It is claimed that the eggs of *Paratya* and *Eriphia* have, inside the chorion, a vitelline membrane which has been detached from the egg surface (Lebedinsky 1890). In any case, decapod eggs have, on their outer membranes, a string by means of which they are attached to the hairs of the pleopoda, separately or in bundles. The *Squilla* egg is enclosed in two membranes; the outer of these abelrs on its surface five to ten strings, which connect it with the membranes of the adjoining eggs so that the entire batch forms a discoidal mass (Shiino 1942). Since no formation of a membrane around the ovarian ovum is detectable in this species, the inner membrane must represent a product of the oviduct. In the Decapoda, Stomatopoda and some others, a third membrane separates from the embryonic body during the course of development, and takes a concentric position under the existing membranes. This is a case of molting during embryonic life rather than the formation of a vitelline membrane (Shiino 1942).

The eggs of the Isopoda, Amphipoda and Mysidacea are enclosed in a vitelline membrane and chorion (McMurrich 1895; Bergh 1894; Manton 1928). It is believed that the chorion of the Isopoda is derived from the follicle cells and is a true chorion, although this has not been fully proved. In other forms a membrane secreted by the oviduct is called by this name. Among the Isopoda, *Cymothoa* (Buller 1879) and *Apopenaeon* (Hiraiwa 1936), among the Mysidacea, *Hemimysis* (Manton 1928), the Copepoda (Grobben 1881) and the Cladocera (Kühn), have a vitelline membrane only. In the Cladocera, winter eggs are laid within the cast-off shell of the mother; or one or two eggs, without a special membrane, are deposited within a much thickened portion of the brood chamber. This thickened portion is detached from the rest of the cast-off covering after ecdysis, to become a protective membrane called the *ephippium*, within which the eggs are able to withstand freezing and desiccation. In spring, the ephippium, charged with an air bubble, rises to the surface where it is exposed to various dispersing influences (Calman 1909).

The formation of the egg membranes in *Penaeus japonicus* presents some peculiar features. When the ovarian oocytes are approaching maturity, they completely absorb the contents of the follicle cells which encircle them. These then develop into vacuoles containing a jelly-like substance, and become arranged in the peripheral part of the cell. As soon as the eggs are spawned, they undergo a rapid contraction which forces out the contents of the vacuoles to give rise to an enveloping jelly layer. This layer accordingly corresponds to a chorion in its origin. After the elevation of the vitelline membrane from the egg surface as a fertilization membrane, and before the commencement of the first cleavage, the jelly layer dissolves completely, leaving only the vitelline (fertilization) membrane intact around the egg (Hudinaga 1935).

(5) SPERMATOZOA

Generally speaking, crustacean spermatozoa are atypical; even such forms as are flagellated have unusual structural features. Among the Cladocera, the spermatozoa are not very different in their constitution from ordinary cells. They have abundant cytoplasm, in which the large nucleus is embedded, and exhibit rod-like, sickle-like, or other similar shapes, to which are sometimes added a number of protoplasmic processes (Fig. 10.3a, b, c). In *Polyphemus* they are even able to perform amoeboid movement (Fig. 10.3d). In every case, the whole spermatozoon appears to unite with the ovum.

Spermatozoa which are to be introduced directly into the oviduct are simply spherical in shape and only small numbers of them are produced. When fertilization is to take place within the brood chamber, however, much larger numbers of smaller, but more complex, spermatozoa are formed (Korschelt und Heider 1936). A spermatozoon of the spherical type is also found in *Squilla* (Fig. 10.3e).

This consists of a vesicular cytoplasmic body, against one side of which the nucleus is pressed to form the so-called head. The anterior half of this head, which represents the acrosome, has as its central axis a styliform structure emanating from a centrosome; the posterior, nuclear part includes the granulate centriole (Komai 1919). Although these spermatozoa are introduced into the oviduct, they are formed in large quantities since the females produce a great abundance of ova.

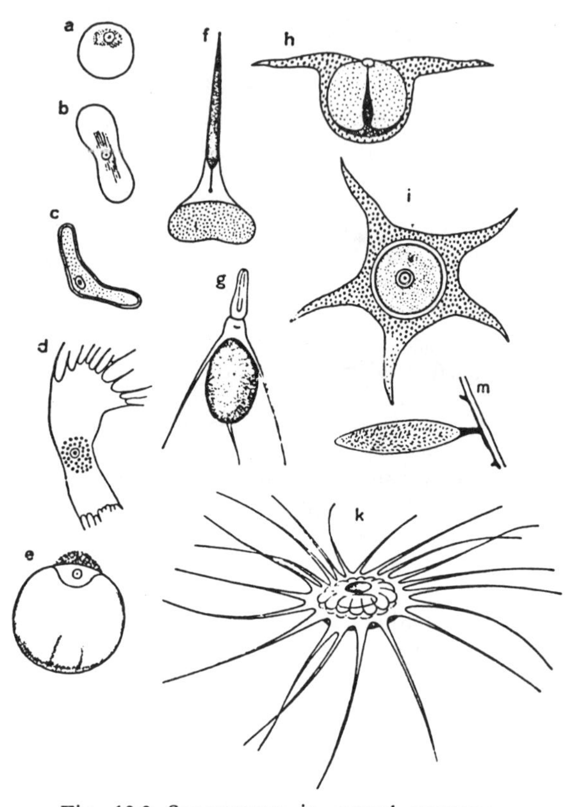

Fig. 10.3 Spermatozoa in several crustaceans

a ~ d. Cladocerans; a. *Latona setifera*; b. *Daphnella brachyura;* c. *Moina paradoxa*; d. *Polyphemus pediculus;* e. *Squilla oratoria;* f. *Leander aspersus;* g. *Galathea squamifera;* h. *Ethusa mascarone* (lateral view); i. *E. mascarone* (upper side); k. *Astacus fluviatilis;* m. spermatophore of *Galathea.* (a, d, f,h,i,k. from Korschelt u. Hieder; e. from Komai; g,m. from Pike)

In the Euphausiacea, also, the sperm cells are spherical or ovoid. In the Ostracoda, Cirripedia, Mysidacea, Amphipoda and Isopoda they are filiform, with a cylindrical head and a long flagellum, but differ considerably in their morphology from ordinary flagellated spermatozoa. The spermatozoa of the Cirripedia have two flagella, while those of the Ostracoda are distinguished by the disproportionate greatness of their length in comparison with the size of the adult animals. For example, the *Pontocypris monstrosa* spermatozoon is said to attain a length of 5 — 7 mm, while that of the adult does not exceed 0.6 mm (Calman 1909)

Most curious of all are the immotile decapod spermatozoa. In the Natantia the spermatozoon is composed of a round, vesicular head and an elongated spine; the nucleus is enclosed within the head, pressed against the side opposite the spine (Fig. 10.3f). The spine contains within it a centrosome of a similar shape and has a granulate centriole at its base. Judging from its behavior at fertilization, the spine appears to correspond to the flagellum of ordinary spermatozoa (see Fig. 10.4).

The spermatozoon of the Reptantia consists of a head formed from the nucleus and a cytoplamic tail which is called a *capsule*. In the Palinura and Anomura, these two parts are connected by a middle piece which contains a centriole and gives rise to three elongate processes (Fig. 10.3g) (Pike 1947). The tail has a chitinous double wall. It is believed that when the head comes in contact with an egg, the sperm stands upright on its surface by means of the three processes, and the osmotic swelling of a substance contained in the tail causes it to explode, forcing the head and middle piece into the egg (Korshelt und Heider 1936) (see Fig. 1.14). In the sperm of the Brachyura, the nucleus is cupshaped and enclosed in a capsule bearing a number of processes (Fig. 10.3h, i). The spermatozoon of the Astacura resembles that of the Brachyura, but is encircled by a larger number of slenderer processes (Fig. 10.3k).

The spermatozoa of the Malacostraca and Copepoda are passed to the female packed into *spermatophores*, which are enclosed in a substance secreted by the terminal part of the vas deferens. The spherical, ovoid or sausage-shaped spermatophore, of a chitin-like texture, is stuck to the gonopore of the female. Besides the spermatozoa, it also contains a secretion from the vas deferens which swells within the female receptaculum seminis and causes the spermatophore covering to burst. Spermatophores are found among the Malacostraca, in *Nebalia*, *Euphausis*, and the Decapoda. Enclosed by the vas deferens secretion, the sperm of the Natantia is made up into a spermatophore which is filiform, or soft and of indefinite shape. In *Penaeus japonicus* and its allies, the spermatophore is composed of a soft main portion and a hard chitinous plug. The main portion is received in the female receptaculum seminis, which is formed on the under surface of the thorax between the bases of the pereiopoda, and the plug is fixed into the entrance of the sac to close it. Several days after copulation, when the female spawns its eggs, the sperm is discharged from another small pore of the sac and distributed over the eggs (Hudinaga 1940-1a).

The spiniform spermatophores of the Anomura are formed in large numbers and lie side by side in the ductus ejaculatorius. In copulation they are fastened to the outside surface of the female body (Fig. 10.3m) (Pike 1947). The spermatophores of the Brachyura are transferred into the oviduct. In Neptunus they are thrust into the receptaculum seminis in autumn, when copulation takes place; while the external covering is completely absorbed, the spermatozoa are preserved in the sac until the following spring, when egg deposition takes place (Ohshima 1938).

(6) FERTILIZATION

Our knowledge concerning fertilization in the Crustacea is rather limited, since observation is made difficult by the opacity of their eggs, owing to the abundance of yolk, and the rare occurrence of free egg-laying, as a result of the highly developed brooding habit. With regard to the penetration of the spermatozoon into the egg, however, an excellent study has been made on *Penaeus japonica* by

Hudinaga (1942). In this species, sperm entrance occurs almost simultaneously with the formation of the first polar body, which takes place immediately after the egg is laid (Fig. 10.4). When a number of spermatozoa reach the egg after penetrating the jelly layer, the contact of each results in the elevation of a fertilization (or entrance) cone on the egg surface. These cones, with the apices of which the sperm heads are in contact, are at first transparent cylindrical elevations, but soon become filled with inflowing ooplasm. The cone in which this cytoplasm first reaches the sperm head begins to contract, drawing the entire sperm, including its tail, into the interior, until the original smooth contour is finally restored. All the other cones then disappear without engulfing their spermatozoa. The second polar body is discharged after the formation of the fertilization membrane, which separates it from the first polar body.

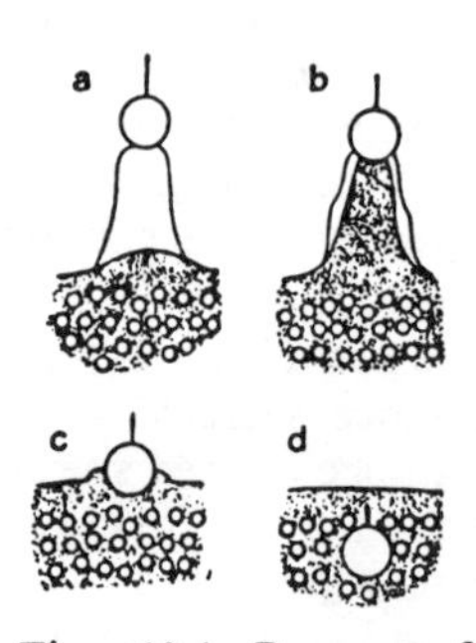

Fig. 10.4 Process of fertilization in *Penaeus japonicus* (Hudinaga)

a. formation of fertilization cone; b. influx of cytoplasm into fertilization cone; c. retraction of fertilization cone; d. entry of spermatozoon into egg.

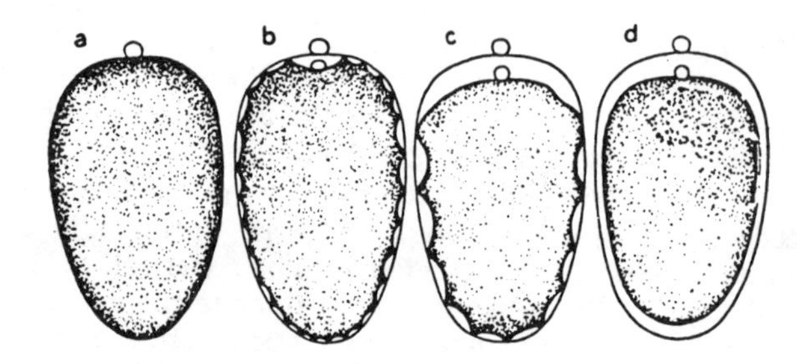

Fig. 10.5 Formation of polar bodies and fertilization membrane in *Balanus amphitrite albicostatus* (Hudinaga)

a. formation of first polar body; b. formation of second polar body and beginning of separation of fertilization membrane; c. appearance of perivitelline space; d. completion of fertilization membrane elevation.

The egg of *Balanus amphitrite albicostatus* has no surrounding membrane in the stage immediately following its deposition. The fertilization membrane makes its appearance all around the egg at once, after the formation of the first polar body (Fig. 10.5). In contrast to the smoothness of the membrane, the peripheral layer of cytoplasm has an irregular surface, retaining points of contact with the membrane. Following the emission of the second polar body, however, this layer is completely detached from the membrane and its surface again becomes smooth. At the same time, the egg undergoes a change in outline, becoming shorter than before; it is surrounded by a broad perivitelline space which is particularly wide at the animal and vegetal poles (Hudinaga 1940-1b). In *Leander pacificus* also, the polar bodies are formed after the egg is laid (Kajishima 1950). According to observations made on the egg of *Mesopodopsis* (Mysidacea), the germinal vesicle breaks down within the ovary after yolk accumulation is complete. The egg is laid in the metaphase stage of the first maturation division and the first polar body is discharged within the brood-pouch. Only a spindle is formed in this division, instead of the usual amphiaster. Before the end of the second division, which follows immediately, fertilization membrane formation is evoked by the entrance of a spermatozoon. It appears, however, that the second polar body is not actually constricted off from the egg, but is rather resorbed (Nair 1939). Observation of the fact that, as in other animal groups, the germinal vesicle disintegrates in advance of meiosis, leaving the nucleoplasm exposed in the cytoplasm, has been made in *Astacus, Cacridina* and *Palaemon*. Union of the sperm and egg pronuclei takes place at the center of the egg (Zehnder 1934; Nair 1949; Aiyar 1949).

It occasionally happens that the time relation of formation of the polar bodies and detachment of the fertilization membrane varies even in a single species. Although the polar bodies of *Jaera* lie in the space between the chorion and the vitelline membrane in most cases, one or both can sometimes be found inside the latter. Fertilization in this species takes place in either the ovary or the oviduct, but both the fertilization membrane and the polar bodies make their appearance much later (McMurrich 1895). In *Armadillidium* and *Porcellio* the membrane is recognizable only after cleavage begins (Goodrich 1939). Although *Apoenaeon* belongs to the Isopoda, like the above forms, its fertilization takes place externally (Hiraiwa 1936). Polyspermy has been observed in *Porcellio*, but the fate of the extra spermatozoa has not been determined (Goodrich 1939).

The opaqueness of the yolky eggs, and the difficulties met with in securing appropriate stages and in sectioning have largely prevented the making of observations on the union of the egg and sperm pronuclei. Syngamy in the *Cyclops* egg is known to take place immediately after insemination, although this may be an unusual case. Not only do the two pronuclei remain distinct during the first cleavage, but the chromosome groups derived from the two parents are completely separated from each other by well-defined nuclear membranes in the blastomeres of the two-cell stage, although they lie close together. Such a condition is said to be maintained until the eight-cell stage, or, in extreme cases, even to the stage in which the primordial germ cells are differentiated (Fig. 10.7a-c) (Amma 1911). Incidentally, in the maturation division of parthenogenetic eggs, e.g., *Artemia*, there are two different series, one forming only a single polar body and the other forming two. In the latter series, the nucleus of the second polar body reunites with the egg nucleus (Wilson 1924).

1. EARLY DEVELOPMENT

With a few exceptions on both sides, the eggs of the Entomostraca, which have relatively less yolk, undergo total cleavage, whereas superficial cleavage is the rule in the yolky eggs of the Malacostraca.

(1) HOLOBLASTIC CLEAVAGE

The cell-lineage has been precisely worked out for the totally cleaving eggs of the Entomostraca, with the result that their modes of cleavage are found to be more or less mutually similar. In the summer egg of *Polyphemus pediculus*, for example, each ovarian ovum is accompanied by three nutritive cells (see p. 337). By the time the ovum is laid, the remnants of these cells have been incorporated into its vegetal part; later they will be received by the blastomere destined to be the primordial germ cell (Fig. 10.6a).[2]

---

[2] Such bodies, as well as the ectosomes to be described later, are called 'germ-cell determinants'.

The first and second cleavages are meridional. The first cleavage spindle lies at an angle with the equatorial plane so that the plane of cleavage, missing the two poles, divides the egg into two somewhat unequal blastomeres. At the second cleavage the spindles appearing in the two blastomeres also form an angle

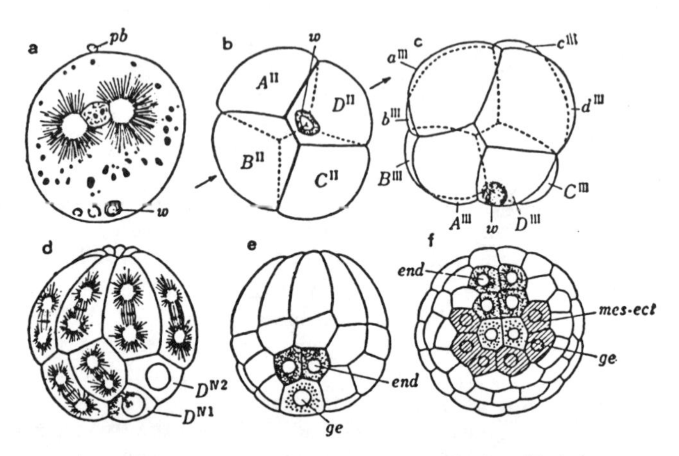

Fig. 10.6 Cleavage in *Polyphemus pediculus* (Kühn)

a. first cleavage (cross-section); b. 4-cell stage (from vegetal pole); c. 8-cell stage (lateral view); d. 16 ~ 32-cell stage (lateral view); e. 31-cell stage (frontal view); f. 62-cell stage (from vegetal pole). $A^{III} \sim D^{III}$: blastomeres in vegetal half; $a^{III} \sim d^{III}$: blastomeres in animal half. Arrows in b indicate anteroposterior axis of embryo. *end:* endoblast, *ge:* primordial germ cell, *mes-ect:* mes-ectoblast, *pb:* polar body, *w:* remnant of nurse cell.

with the equatorial plane; viewed directly from the side they can be seen to intersect each other. Owing to this inclination of the spindles, the second cleavage furrows do not meet the first at one point so as to form a cross at the poles, the right and left furrows intersecting the first at separate points. The blastomeres subsequently change their shape to some extent, so that the first furrow describes a zigzag line across the two polar faces (Fig. 10.6b).

At the close of the four-cell stage, the embryo consists of three blastomeres of equal size (A^{II}, B^{II}, C^{II}), and the somewhat smaller D^{II}, which contains the remnants of the nutritive cells. B^{II}, lying opposite D^{II}, is in contact with it at the vegetal side, but separated at the animal side, whereas with A^{II} and C^{II} this relation is reversed. Since D^{II} is smaller than the other blastomeres, the bilateral symmetry of the embryo is already determined at this stage, the line connecting the centers of B^{II} and D^{II} corresponding to the future median plane, and thus indicating the antero-posterior axis. The inclination of the spindles, which is most marked at this stage, the characteristic short line where the opposing blastomeres make contact at the animal and vegetal poles, and the resulting zigzag conformation of the furrows are all suggestive of spiral cleavage.

The third division occurs on a horizontal plane lying nearer the vegetal pole, and results in the formation of four larger blastomere ($a^{III} — d^{III}$) in the animal hemisphere and four smaller vegetal cells ($A^{III} — D^{III}$). All the descendants of $a^{III} — d^{III}$ give rise to ectoderm (Fig. 10.6c) and are called *primary ectoblasts* to distinguish them from the ectoderm cells which will be derived from $A^{III} — C^{III}$.

These blastomeres, termed *mesectoblasts,* later form ectoderm and mesoderm, whereas D^{III} develops into endoderm and the primordial germ cell. Although cleavage is partial in the earlier stages, the cleavage furrows grow deeper and deeper as development proceeds; at the close of the third cleavage all the blastomeres are completely separated from each other for the first time, and a small segmentation cavity is formed at their center. This time the vestiges of spiral cleavage are hardly perceptible, either in the direction of the spindles or the arrangement of the blastomeres. The fourth cleavage is meridional in all the blasto-

Table 10.1 **Cell-lineage of** *Polyphemus pediculus.*

	2-cell stage	4-cell stage	8-cell stage	16-cell stage	32-cell stage	64-cell stage	118-cell stage
egg	AB'	A''	a''' — prim. ect.				
			A'''	A^{IV1}	A^{V11}	A^{VI111}	$A^{VII1111}$ — mes.
							$A^{VII1112}$ — sec. ect.
						A^{VI112} — sec. ect.	
					A^{V12} — sec. ect.		
				A^{IV2}	A^{V21}	A^{VI211}	$A^{VII2111}$ — mes.
							$A^{VII2112}$ — sec. ect.
						A^{VI212} — sec. ect.	
					A^{V22} — sec. ect.		
		B''	b''' — prim. ect.				
			B'''	B^{IV1}	B^{V11}	B^{VI111}	$B^{VII1111}$ — mes.
							$B^{VII1112}$ — sec. ect.
						B^{VI112} — sec. ect.	
					B^{V12} — sec. ect.		
				B^{IV2}	B^{V21}	B^{VI211}	$B^{VII2111}$ — mes.
							$B^{VII2112}$ — sec. ect.
						B^{VI212} — sec. ect.	
					B^{V22} — sec. ect.		
	CD'	C''	c''' — prim. ect.				
			C'''	C^{IV1}	C^{V11}	C^{VI111}	$C^{VII1111}$ — mes.
							$C^{VII1112}$ — sec. ect.
						C^{VI112} — sec. ect.	
					C^{V12} — sec. ect.		
				C^{IV2}	C^{V21}	C^{VI211}	$C^{VII2111}$ — mes.
							$C^{VII2112}$ — sec. ect.
						C^{VI212} — sec. ect.	
					C^{V22} — sec. ect.		
		D''	d''' — prim. ect.				
			D'''	D^{IV1} — prim. germ cell			
				D^{IV2} — endoblast			

meres except D^{III}. D^{III} divides later than the others and latitudinally, into D^{IV_1} and D^{IV_2}, the former lying nearer the vegetal pole (Fig. 10.6d). D^{IV_1}, which includes all the remnants of the nutritive cells, is now differentiated into the *primordial germ cell,* and all its descendants give rise to generative cells. D^{IV_2} represents the *primordial endoderm cell.* The 16-cell stage is thus composed of eight ectoblasts derived from a^{III} — d^{III}, six mesectoblasts descended from A^{III} — C^{III}, as well as D^{IV_1} and D^{IV_2}. The fifth cleavage is latitudinal; the 32-cell stage consists of 16 ectoblasts in the animal hemisphere, six secondary ectoblasts resulting from the divisons of A^{III} — C^{III}, and nearer the vegetal pole, six mesectoblasts which are the sister cells of the secondary ectoblasts, and four cells produced by the delayed meridional divisions of D^{IV_1} and D^{IV_2} (Fig. 10.6e). The tendency shown by the descendants of D^{III} to lag behind those of A^{III} — C^{III}, and the still more retarded cleavage of the primordial germ cell lead to the appearance of intermediate stages (viz., 30-, 31-, 32-, 62-cell). After the sixth cleavage, the blastomeres of the animal hemisphere divide before the vegetal blastomeres, and in later stages even the direction of division becomes irregular. In the 118-cell stage, six *mesoblasts* are differentiated for the first time from among the descendants of A^{III} – C^{III}. By the eighth cleavage ,the segmentation period is about at an end, and the embryo may now be called a blastula.

This blastula consists of 128 primary ectoblasts, 84 secondary ectoblasts, 12 mesoblasts, 8 endoblasts and 4 generative cells — a total of 236 cells enclosing a relatively spacious blastocoel. The mesoblasts surround the primordial germ cells in a U-shaped formation, and the entoblasts lie between the germ cells and the equator. Eight ectoblasts situated at the animal pole are larger than the others; these give rise to the anlage of the *apical plate,* which will form the brain. Gastrulation, which begins about the time of the ninth cleavage, is effected by a sinking of the mesoblasts, endoblasts and germ cells into the blastocoel, rather than by a clear-cut invagination. Details concerning the fates of the blastomeres in *Polyphemus pediculus* are given in Table 10.1.

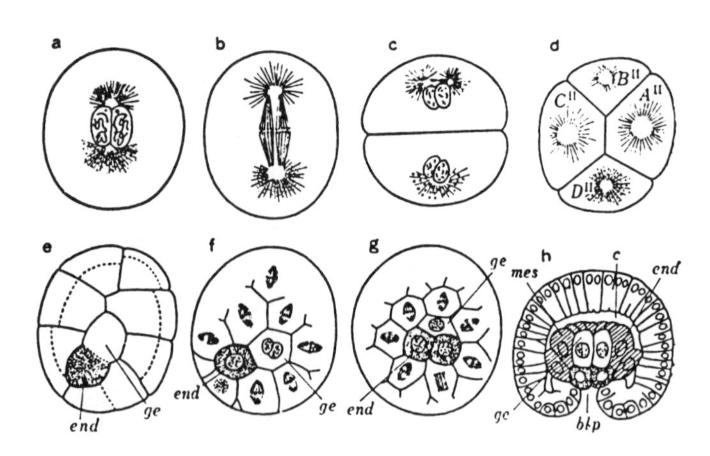

Fig. 10.7 Cleavage in *Cyclops viridis*

a. uncleaved egg; b. first division; c. 2-cell stage; d. 4-cell stage; e. 16~32-cell stage; f. sixth division; g. seventh division; h. cross-section of gastrula. (a~d, h. cross-sections; e~g. vegetal polar view) ***blp:*** blastopore, c: segmentation cavity, ***end:*** endoblast, ***ge:*** primordial germ cell, ***mes:*** mesoblast. (a~d. Amma; e~h, Fuchs)

The egg of *Cyclops virides* (Copepoda) undergoes total and equal cleavage from the outset. In the 4-cell stage, the zigzag furrows described above are formed in nearly the same manner (Fig. 10.7a-d). At the time of the first cleavage, granules known as ectosomes crowd around one of the asters; in succeeding divisions these will always enter only one of the blastomeres. In the fifth cleavage the blastomere containing the ectosomes divides into a primordial germ cell, which receives all the granules, and a primordial endoderm cell (Fig. 10.7e). When the seven blastomeres surrounding these cells undergo the sixth cleavage, they give rise to the mesoblasts (Fig. 10.7f, g). After the eighth cleavage, the four endoderm cells and two germ cells, together with the mesoderm cells, move into the blastocoel where they form a plug-like mass. The blastopore invagination which results from their sinking inward is closed after the end of the ninth cleavage (Fuchs 1914).

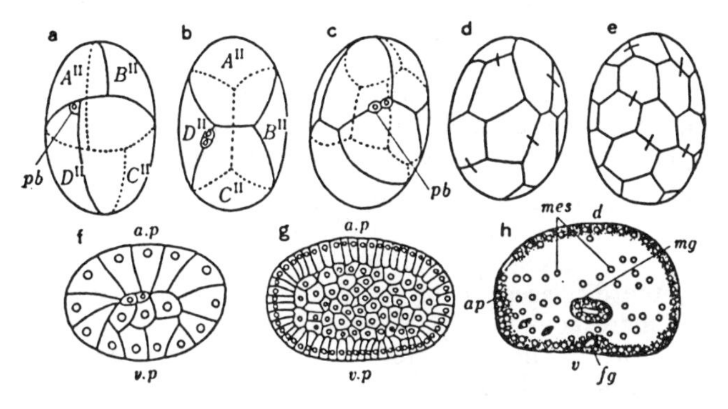

Fig. 10.8 Cleavage in *Cypris incongruens* (Müller)

a. 4-cell stage, early cleavage furrow; b. late 4-cell stage; c. 8-cell stage; d. 16-cell stage; e. 32-cell stage; f. 134-cell stage; g. blastula; h. gastrula. (f~h. cross-sections) *ap:* apical plate, *a.p.:* animal pole, *d:* dorsal side of embryo, *fg:* fore-gut, *mes:* mesoderm, *mg:* mid-gut, *pb:* polar body, *v:* ventral side of embryo, *v.p:* vegetal pole.

The egg of *Cypris incongruens* is oblong ovate, with its primary axis vertical to the long dimension (Fig. 10.8). Its cleavage is total and equal from the beginning. The second cleavage spindles are parallel at first, but then rotate slightly in the same direction with relation to the egg axis, so that their cleavage planes intersect when viewed from the side. The second cleavage furrows therefore meet the first at two separated points on each polar surface (Fig. 10.8a). With the ensuing change in the configurations of the blastomeres, the first furrow acquires a zigzag course; the embryo thus comprises two blastomeres which lie on the long axis and come in contact at the animal pole, and two on the short axis which meet at the vegetal pole (Fig. 10.8b). There are neither size differences among the blastomeres, nor such peculiar granules as are found in the *Cyclops* egg.

The traits of spiral cleavage are more conspicuous in this species than in the two examples cited above. In the succeeding stages the new furrows continue to meet the previously formed ones unilaterally at separate points, giving rise to zigzag lines which divide the whole surface into polygonal areas (Fig. 10.8c-e). The cleavage spindles in all the blastomeres are tilted in a similar manner with respect to the polar axis, and the direction of this inclination alternates from left to right and vice versa at each succeeding cleavage.

With the third cycle, cell division tends to begin first in the blastomeres at the animal pole, and successively later towards the vegetal pole. This time-lag in division according to the location of the blastomeres becomes increasingly marked as development progresses, until finally in the eighth cleavage cycle, division begins first at the vegetal pole and proceeds toward the animal pole. At this time the division spindles of a limited number of cells at the vegetal pole are oriented perpendicular to the egg surface. As a consequence, five to eight endoderm cells are constricted off into the interior of the embryo, which is thus transformed into a morula-like blastula (Fig. 10.8f). The embryo subsequently undergoes an increase in the number of cells making up its outer layer as well as its inner mass (Fig. 10.8g), and during a long resting stage the endoderm cells, which have lost their cellular boundaries, accumulate at the ventral side while the yolk becomes concentrated at the dorsal side (Fig. 10.8h). The endoderm cells then differentiate into a small, sac-like *midgut (mesenteron) anlage,* and the remaining inner cells become mesoderm. In the region where it is in contact with the mesenteron, the ectoderm invaginates to form the primordium of the *stomodaeum*; at the anterior end of the ventral surface an *apical plate* differentiates, and posterior to this along the median line the primordium of the *ventral nerve cord* takes shape. Even in this stage it is not yet possible to recognize any differentiation of the germ cells.

The development of the Cirripedia is noteworthy for its distinctly unequal cleavage and its gastrulation effected by epiboly. As soon as the egg of *Mitella mitella* enters the first cleavage stage, it begins to rotate inside the egg membrane, as evidenced by the displacement of the polar body along the egg surface from the blunt to the more pointed end. Keeping pace with this change, the mitotic spindle in the interior of the egg, which was first oriented perpendicular to the egg axis, likewise undergoes rotation. The surface furrow, which originally appeared in a position oblique to the long axis, becomes vertical to it by the end of this stage. The cleavage, which thus appears to be equatorial, produces the micromere AB^{I} at the blunter end and the macromere CD^{I} at the sharper end (Fig. 10.9a-d). Since the areas around the asters at the two ends of the spindle are semitransparent, the rotation of the spindle is rather easily observed in living material.

The second cleavage is also meridional, dividing AB^{I} into two micromeres of equal size, and CD^{I} into C^{II}, resembling these in size, and the macromere D^{II}, which receives a much greater part of the total amount of yolk (Fig. 10.9e). The furrows cleaving AB^{I} and CD^{I} are not in a single line, but meet the first furrow at separated points. Following the second cleavage, A^{II} and C^{II} move around the polar axis in a counter-clockwise direction as viewed from the animal pole, so that they become arranged bilaterally in relation to the longer egg axis. As a consequence, A^{II} and C^{II} come in contact with each other at the animal pole and lie above B^{II} and D^{II}, which are in contact at the vegetal pole (Fig 10.9f). The line bisecting B^{II} and D^{II} represents the antero-posterior axis of the embryo, the animal pole corresponding to the future dorsal side and the vegetal pole to the future ventral side.

The third cleavage is primarily latitudinal: A^{II}— C^{II} divide almost synchronously, while D^{II} cleaves somewhat later and unequally, cutting off dorsally d^{III}, which is similar in size to the other micromeres (Fig. 10.9g, h). The blastomeres a^{III}— c^{III}, located at the anterior end of the embryo in this 8-cell stage, are ectoblasts, and A^{III}— C^{III} as well as d^{III} are mesectoblasts. In the fourth cleavage, D^{III} again lags behind the others, producing the dorsal micromere D^{IV_2} and the

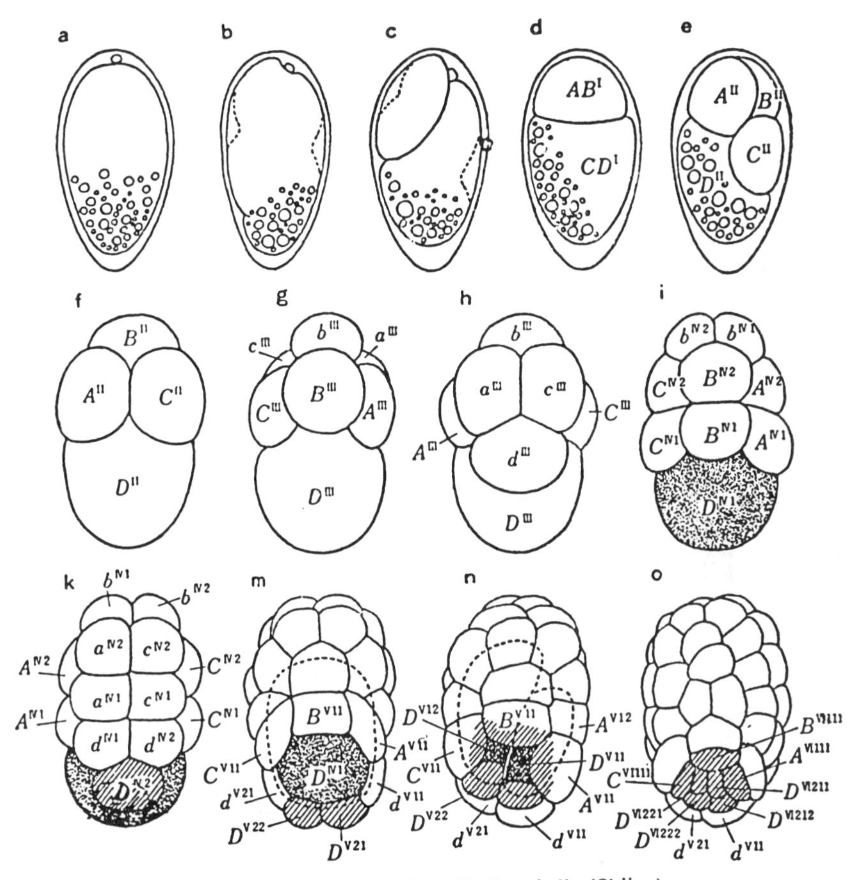

Fig. 10.9 Cleavage in *Mitella mitella* (Shiino)

a. uncleaved egg; b. beginning of first cleavage; c. early 2-cell stage; d. late 2-cell stage. (a~d. lateral views) e. early 4-cell stage (animal pole); f. late 4-cell stage; g. 8-cell stage (vegetal pole); h. 8-cell stage (animal pole); i. 16-cell stage (vegetal pole); k. 16-cell stage (animal pole); m. 31-cell stage, beginning of gastrulation (vegetal pole); n. 32-cell! stage, diminution in size of blastopore; o. 60-cell stage, sinking of mesoderm into interior. Cross-hatched area: endoderm. Oblique-lines: mesoderm.

ventral macromere D^{IV_1} (Fig. 10.9i, k). D^{IV_2}, the descendants of which later become the mesoderm, is called the primary mesoblast to distinguish it from the mesoderm cells which will be derived from the mesectoblasts, while D^{IV_1} is the primordial endoblast. At the close of this division, the 15 micromeres begin to move over the macromere (D^{IV}) towards the posterior, the displacement being more marked on the dorsal than on the ventral side.

In the fifth cleavage cycle, D^{IV_2} divides somewhat after the others, and D^{IV_1} is still more retarded, so that a 31-cell stage occurs (Fig. 10.9m). $D^{V_{21}}$ and $D^{V_{22}}$, the daughter blastomeres of D^{IV_2}, move around the posterior end to the ventral side of the embryo and join with A^{V11}—d^{V11} and d^{V12} in a circular arrangement. These cells are descended from A^{III}—C^{III} and d^{III}, and lie more posteriorly than their sister cells (see Table 10.2). The opening encircled by these blastomeres

Table 10.2 **Cell-lineage of** *Mitella mitella*.

2-cell stage	4-cell stage	8-cell stage	16-cell stage	32-cell stage	62-cell stage
egg	AB'	A''	a''' — prim. ect.		
			A'''	A^{IV1}	A^{V11}	A^{VI111} — sec. mes.
						A^{VI112} — sec. ect.
					A^{V12} — sec. ect.	
				A^{IV2} — sec. ect.		
		B''	b''' — prim. ect.			
			B'''	B^{IV1}	B^{V11}	B^{VI111} — sec. mes.
						B^{VI112} — sec. ect.
					B^{V12} — sec. ect.	
				B^{IV2} — sec. ect.		
	CD'	C''	c''' — prim. ect.			
			C'''	C^{IV1}	C^{V11}	C^{VI111} — sec. mes.
						C^{VI112} — sec. ect.
					C^{V12} — sec. ect.	
				C^{IV2} — sec. ect.		
		D''	d'''	d^{IV1}	d^{V11}	d^{VI111} — sec. nes.
						d^{VI112} — sec. ect.
					d^{V12} — sec. ect.	
				d^{IV2}	d^{V21}	d^{VI211} — sec. mes.
						d^{VI212} — sec. ect.
					d^{V22} — sec. ect.	
			D'''	D^{IV1} — endoblast		
				D^{IV2} — mesoblast		

represents the wide blastopore, within which only a small part of the primordial endoblast D^{VI_1} remains exposed (Fig. 10.9n). After the 32-cell stage is attained by the division of D^{IV_1}, all the blastomeres except D^{V11} and D^{V12} undergo the sixth division, producing the 62-cell stage. Five secondary mesoblasts given off at this time by A^{V11} — C^{V11}, d^{V12} and d^{V12}, which surround D^{V11} and D^{V12}, follow these endoblasts and the primary mesoblasts in migrating into the blastocoel, which is then closed (Fig. 10.9o) (Shiino, unpublished).

Lepas anatifera is very similar to *Mitella* in its mode of cleavage, only differing in that all the descendants of d^{III} develop into ectoderm (Bigelow 1902). As may by seen by comparing Table 10.1 with Table 10.2, the differences between the modes of cleavage of the Cirripedia and of *Polyphemus* consist in (1) that D^{III} gives rise to a primordial endoblast and primary mesoblast in the former, whereas it produces a primordial endoblast and primordial germ cell in the latter, and (2) that the mesoderm cells differentiate in connection with the sixth cleavage in the former, and with the seventh in the latter.

Holoblastic cleavage can be discovered in the Malacostraca as well as in the Entomostraca: among the Pennaeidae and Luciferidae of the Decapoda, and also in the Euphausiacea, although the cell-lineage has been traced only in this latter order. *Euphausia* follows substantially the same pattern as the Entomostraca with respect to its cell-lineage; cleavage is total from the start and traces of spiral cleavage persist in the second cleavage stage in the same way as in the Entomostraca. In the 32-cell stage, two endoderm cells derived from the D^{II} blastomere are prolonged internally to occupy the greater part of the blastocoel. Since their division is temporarily inhibited, the next cleavage cycle produces a 62-cell stage (Fig. 10.20d). These two endoderm cells have different destinies, one giving rise to endoderm only, while the other produces germ cells as well. A ring of eight cells bordering the blastopore differentiates into mesoderm. The invagination of the blastopore attains its greatest extent in the 122-cell stage; the vegetal pole becomes the posterior end of the embryo, and an anus is formed anew after the closure of the blastopore (Taube 1909, 1915).

The egg of *Penaeus japonicus,* belonging to the equal cleavage type, exibits at the second cleavage the same blastomere rotation as that which we have seen in *Mitella mitella* (Fig. 10.10a, b). The blastocoel appearing in this early stage grows more and more spacious as development proceeds, but in the 64-cell stage it is largely occupied by the inward elongation of two cells situated at the vegetal pole. It has been confirmed by means of vital staining that these cells, which have been named the *central endoderm cells,* are derived from the blastomere D^{II}, although it is not certain whether their fate is to become endoderm or germ cells (Fig. 10.10c).

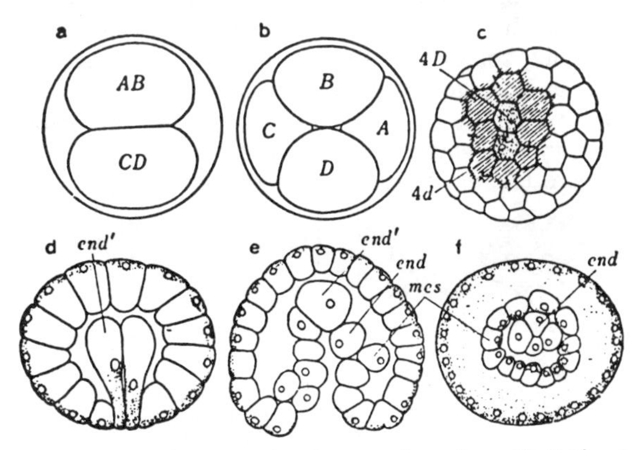

Fig. 10.10 Cleavage in *Penaeus japonicus* (Kajishima)

a. 2-cell stage; b. 4-cell stage; c. 64-cell stage. 4D, 4d: central endoderm. Cross-lined area at periphery of endoderm indicates region which will invaginate; d. longitudinal section of same embryo; e. longitudinal section of gastrula; f. cross-section of late gastrula. *end:* endoderm, *end':* central endoderm cell, *mes:* mesoderm.

After the invasion of the blastocoel by these cells,their immediate neighbors follow them in sinking below the surface, thereby forming the archenteron invagination. Among these cells, the ones which are destined to become mesoderm leave the archenteric complex and take a position in the space between ectoderm and endoderm, surrounding the latter in a ring (Fig. 10.10e, f) (Kajishima 1950).

Lucifer forms a blastula which encloses a spacious blastocoel, and gastrulates by typical invagination with an archenteron composed of large cells. Before the archenteron invaginates, two cells, which are regarded as mesoderm cells containing a large amount of yolk, migrate into the blastocoel. These probably correspond to the central endoderm cells of *Penaeus* (Brooks and Herrick 1880).

(2) SUPERFICIAL CLEAVAGE

Although they may show certain modifications, eggs with much yolk generally undergo superficial cleavage. As mentioned above (p. 8), the synkarion, surrounded by a small mass of cytoplasm of an amoeboid shape, is isolated in the center of the yolk (Figs. 10.12a, 10.14a, etc). With each division the cleavage nuclei, together with the surrounding cytoplasm, move toward the egg surface until they reach the peripheral cytoplasm, with which they unite to form the *blastoderm* (Figs. 10.12—14).

The most typical cleavage of this type is that seen in *Astacus,* which has extremely yolky eggs. Its cleavage consists simply in successive subdivisions of the nucleus and protoplasmic island, which do not involve the yolk mass as a whole. As development proceeds, the nuclei separate from each other to an increasing degree, and at the same time gradually migrate toward the surface, which they reach in the 128-cell stage. Blastomere boundaries, surrounding each nucleus, make their appearance abruptly and synchronously over the whole surface in the 512-cell stage, and under the influence of these cells, the yolk mass is simultaneously divided into a corresponding number of elongate pyramidal masses called *primary yolk pyramids,* the apices of which converge toward the center of the egg (Fig. 10.17a). The yolk, somewhat altered by a digestive process beginning about the 16-cell stage, has become granular; the granules are concentrated near the apices of the yolk pyramids. After the 128-cell stage, the nuclear divisions on one side of the embryo lag behind those on the other side. Gastrulation begins at the 1024-cell stage (Zehnder 1934; Reichenbach 1886).

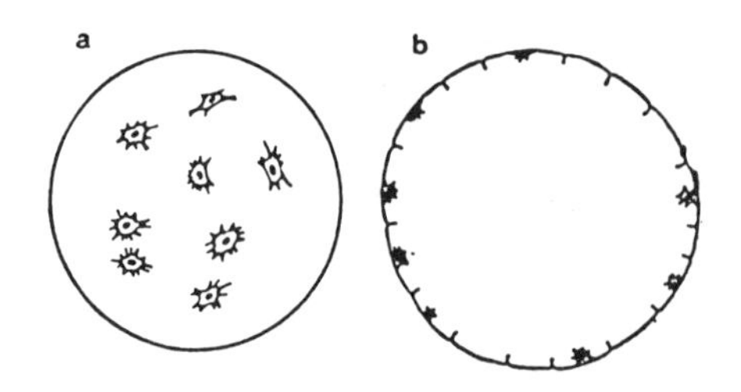

Fig. 10.11 Cleavage in *Squilla oratoria* (Shiino)

a. 8-cell stage, cleared specimen, showing all nuclei and protoplasmic islands; b. ca. 200-cell stage, arrival of protoplasmic islands in periphery (cross-section).

In the egg of *Squilla oratoria,* all the cleavage nuclei reach the peripheral cytoplasm by the 128-cell stage and the cleavage furrows appear toward the end of this stage. The yolk pyramids, however, are very short in this species, the furrows extending only a short distance into the yolk and advancing very little farther even in the succeeding stages, so that the central yolk mass remains homogeneous. At the stage with about 500 cells, regional differences in the distribution of nuclei almost obliterate the yolk pyramids, although the cell boundaries remain visible on the surface (Shiino 1942).

In the egg of *Panulirus japonicus,* the cleavage furrows are abruptly and simultaneously established in the 8-cell stage (Fig. 10.14), describing zigzag lines over the surface exactly as was observed in the Entomostraca. When the 8-celled embryo cleaves, the right and left sides of the new furrows meet the earlier ones at separate points. Subsequent shifting of the blastomeres into a more stable arrangement changes the directions of the furrows so that they form continuous zigzag lines. This behavior can definitely be regarded as a vestige of spiral cleavage (Fig. 10.14b). In the 16-cell stage, the furrows penetrate deeply into the yolk, dividing this into complete blastomeres with distinct inner as well as superficial boundaries (Fig. 10.14c). The major part of the yolk is distributed among these blastomeres, but since the furrows do not reach the very center of the egg, a small amount of yolk which has not been included within the cells remains there. In other words, the space corresponding to the blastocoel is filled with yolk. As the nuclei continue to divide and move out to the surface, the height of the blastomeres gradually decreases, enlarging the blastocoel; by the 128-cell stage all the nuclei have reached the surface. The outer surface of each blastomere is covered by a protoplasmic layer containing the nucleus, and its interior is traversed by a protoplasmic fibrillar reticulum which encloses the yolk spherules in its interstices (Fig. 10.4e,f). At the stage with about 1000 nuclei, these blastoderm cells formed at the embryonic surface have grown extremely thin and flattened; although gastrulation begins at this time, the central yolk mass remains quite homogeneous (Fig. 10.14g, h) (Shiino 1950).

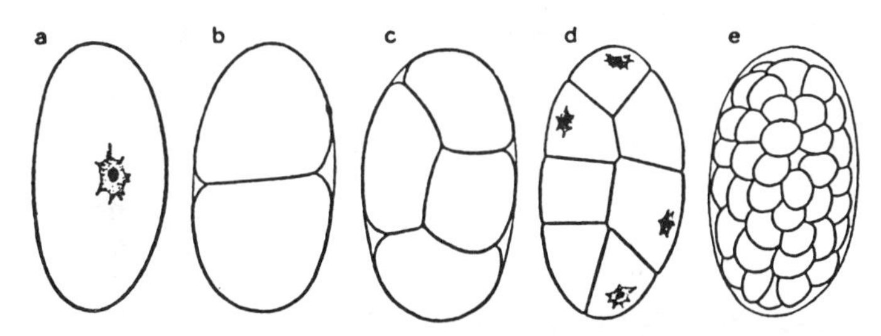

Fig. 10.12 Cleavage in *Caridina laevis* (Nair)
a. longitudinal section of uncleaved egg; b. 2-cell stage; c. 4-cell stage; d. 16-cell stage (longitudinal section); e. 64-cell stage.

The egg of *Leander pacificus* follows a similar cleavage process, but in this species the blastomeres are not so distinctly demarcated from one another at the inner ends as those of *Panulirus* (Kajishima 1950). Early cleavage in *Palaemon idae* is indicated merely by the divisions of the protoplasmic islands, surface furrows failing to appear until the end of the 8-cell stage, when the egg divides totally into eight blastomeres with distinct boundaries. The fourth cleavage is again total, giving rise to a morula without a blastocoel. In the succeeding cleavage stage all the nuclei arrive at the surface, but the new furrows appearing this time are much shallower than before and even the previously formed ones become comparably shallow as their inner portions disappear. The yolk making up the central mass thus formed is freely confluent with that in the yolk pyramids (Aiyer 1949).

The cleavage of *Caridina laevis* represents a modification of that found in *Palaemon,* being total and equal from the beginning, although no blastocoel is formed (Fig. 10.12). At the end of the 16-cell stage all the nuclei reach the surface of the embryo. After the next stage, the cleavage furrows fail to extend all the way inward, leaving the central yolk mass undivided, and from this time onward the pattern of cleavage closely resembles that of *Palaemon.*

In the cleavage of *Hemimysis* and *Mesopodopsis,* no superficial furrows are formed at any time. Blastoderm formation is initiated by the gradual approach of the nuclei to the periphery, and completed by the spreading of their accompanying protoplasmic islands over the surface. As early as the 16-cell stage it can be observed that more of the nuclei are assembled on the side where the blastodisc will later be established. In the 64-cell stage, a little below the surface and parallel to it, they form a disc-like assemblage of cells connected by protoplasmic strands; in the 128-cell stage, they move out of the yolk, forming a *germinal disc* which partly covers the yolk mass like a hood (Manton 1928, Nair 1939).

The cleavage of the Isopoda resembles that of *Leander,* but the nuclei reach the surface relatively earlier, i.e., in the 16-or 32-cell stage, and shallow cleavage furrows appear at the same time. Spindle rotation occurs in the 4-and 8-cell stages (McMurrich 1895). The egg of *Gammarus* undergoes total cleavage during the earlier stages, and the third cleavage is unequal, giving rise to four micromeres and four macromers. Thereafter, as the nuclei rise toward the surface, the yolk is filtered little by little out of the blastomeres and into central blastocoel; the blastomeres thus decrease in height, finally forming a blastoderm as in *Panulirus* (Heidecke 1903).

In the Malacostraca generally, it is rather difficult to follow the polar axis definitely from the uncleaved egg to the stage of blastoderm formation, because the polar bodies are sooner or later absorbed into the embryo, and there are no appreciable polar differences in the composition of the yolk in the uncleaved egg. Furthermore, there is no perceptible variation among the exposed surfaces of the blastomeres to use as an indicator of the egg axis. This situation changes, however, with blastoderm formation. As will be described later, a greater concentration of nuclei in one hemisphere of the embryo induces there the formation of the germinal disc, and the blastopore invagination at one point on the periphery of the disc defines the orientation of the antero-posterior and dorso-ventral axes. The site of the blastopore corresponds to the posterior end, and the line between it and the opposite side of the disc represents the antero-posterior axis. The disc itself is destined to become the ventral side, while the opposite hemisphere, with sparser nuclei, will be the dorsal side (Fig. 10.15a). Nevertheless, nothing is known about the relation between these axes and the original polar axis. Since the first cleavage plane cuts the longer axis perpendicularly in oval eggs such as that of *Caridina,* the polar axis is believed to lie in this plane, although the exact positions of the two poles remain unknown. The antero-posterior axis of the embryo coincides with the original longer axis.

(3) DISCOIDAL CLEAVAGE

Although it has been claimed that certain members of the Mysidacea and Isopoda undergo discoidal cleavage, it is not clear to what extent the claims are based on valid grounds. The only conclusive evidence hitherto put forward is that relating to *Nebalia bipes.* The uncleaved egg of this species is centrolecithal and possesses at its center a protoplasmic island containing a nucleus (Fig. 10.13a). Before the first cleavage begins, the island rises to the surface at one pole of the egg, there undergoing a horizontal cleavage and forming a protoplasmic mass distinctly separated from the yolk (Fig. 10.13b). Horizontal division continues in subsequent stages, the cells formed in this way spreading peripherally over the surface as their number increases (Fig. 10.13c), although a certain region which is regarded as the vegetal pole is left uncovered for a considerable period. The relation between the polar axis thus indicated and the antero-posterior axis of the established germinal disc is uncertain.

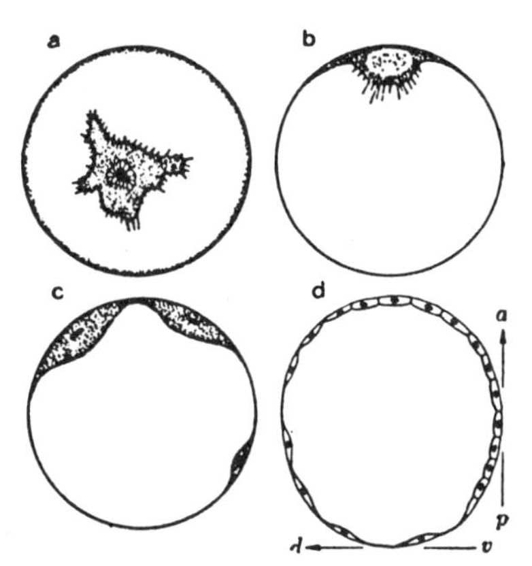

Fig. 10.13 Cleavage in *Nebalia bipes* (cross-section) (Manton)

a. uncleaved egg immediately after spawning; b. beginning of first cleavage; c. early cleavage stage; d. blastula. *a:* anterior, *d:* dorsal side, *p:* posterior, *v:* ventral side.

In addition to this normal process, various abnormalities occur in the cleavage of *Nebalia.* Sometimes the germinal disc may be divided into a number of disc-like groups of cells instead of forming a continuous layer, or the cells may be arranged in a double layer before completion of the disc. These represent only transient phases, however, since the final result in every case is a blastula enclosed in a single-layered blastoderm, as in the other malacostracans (Fig. 10.13d) (Manton 1933).

(4) GERM-LAYER FORMATION

As cleavage among the Crustacea includes various types, so also does their gastrulation show varieties of modes according to differences in the cleavage types as well as in systematic groupings. Since gastrulation and germ layer differentiation in the holoblastic eggs have already been dealt with, the following description will be confined to these processes as observed in the eggs which undergo superficial cleavage, exemplified by the embryo of *Panulirus japonicus* (Shiino 1950).

As soon as the blastoderm is established, its constituent cells begin to accumulate in the region destined to be the future ventral side, so that the surface

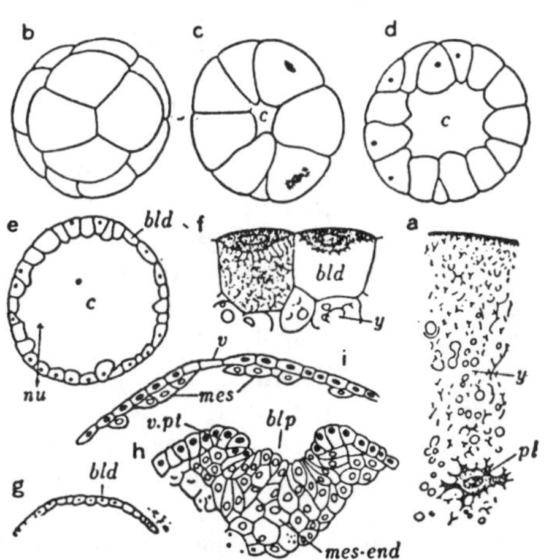

Fig. 10.14 Cleavage and formation of germ layer in *Panulirus japonicus* (Shiino)

a. cross-section of part of egg at 2-cell stage; b. surface view of 16-cell stage; c. cross-section of same; d. cross-section of 64-cell stage; e. cross-section of 350-cell blastula; f. enlarged sketch of blastodermal cells of same blastula; g. cross-section of part of blastula; h. cross-section of part of gastrula; i. cross-section through blastoderm of gastrula. *bld:* blastodermal cell, *blp:* blastopore, *c:* segmentation cavity filled with yolk, *mes:* nauplius mesoderm, *mes-end:* mes-endoderm cell complex, *nu:* regressing nucleus, *pl:* protoplasmic island, *v:* ventral side, *v.pt:* ventral plate rudiment, *y:* yolk.

of the embryo becomes divided into two hemispheres, one densely and the other sparsely nucleated. The densely nucleated hemisphere, which represents the primordium of the germinal disc, forms at one point on its periphery a small area particularly crowded with nuclei. This area then sinks to form a shallow invagination, from which the cells actively migrate into the interior (Fig. 10.14h). This constitutes the blastopore, which lies at the posterior end of the embryo (Fig. 10.15a).

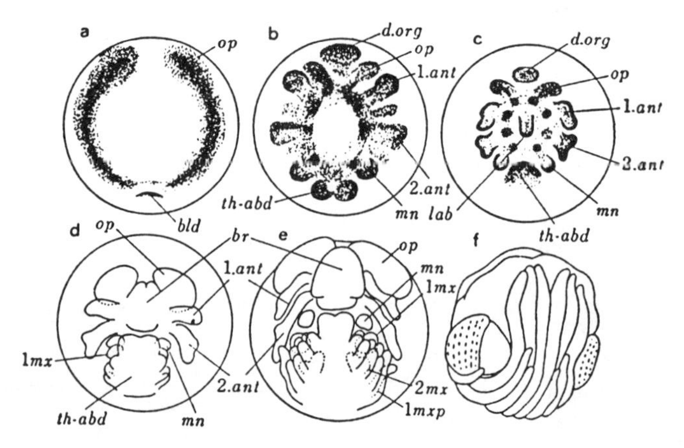

Fig. 10.15 External appearance of *Panulirus japonicus* embryo (Shiino)

a. embryo at formation of blastoderm; b. pronauplius stage; c. nauplius stage; d. stage with 7 pairs of appendages; e. stage with 11 pairs of appendages; f. embryo shortly before hatching (frontal view). *1.ant:* first antenna, *1.mx:* first maxilla, *1.mxp:* first maxilliped, *2.ant:* second antenna, *2.mx:* second maxilla; *br:* brain; *blp:* blastopore; *d.org:* dorsal organ; *lab:* labrum; *mn:* mandible; *op:* optic lobe; *th-abd:* thoracico-abdominal process (ventral plate). Arrow shows anterior side of embryo.

Within a short time, a horseshoe-shaped accumulation of cells appears on the germinal disc, extending anteriorly from the blastopore. Each of the free ends is a mass of especially tall cells; these form two small areas called the *optic lobes*, which will develop later into the compound eyes. The appendages and their respective ganglia will differentiate on the rest of the band. Of the cells migrating inward from the blastopore, some form a pair of cellular bands proceeding forward along the under-side of the horseshoe (Fig. 10.14i, *mes*), while others accumulate in a plug-like mass beneath the blastopore (Fig. 10.14h, *mes- end)*. The former represent the *naupliar mesoderm* and the latter, the so-called *undifferentiated mesendoderm cell-complex.*[3] These cells gradually enlarge as they take in yolk granules from the surrounding yolk mass; during this process the amount of their cytoplasm fails to increase fast enough to keep pace with such rapid enlargement, and is stretched into a thin layer, like a membrane. At times, the cytoplasm is recognizable only in the immediate neighborhood of the nucleus, as a stellate mass, and in extreme cases numbers of such stellate cells seem to be freely wandering through the yolk. The cells containing yolk granules, as well as these stellate cells

[3] As the undifferentiated state of this term suggests, the mesoderm is not clearly separable from the endoderm at this stage.

scattered in the main mass of yolk, are usually included in the general term *yolk cells*. After the stage described above, the blastopore closes and on either side of it the rudiments of the *ventral plate,* which gives rise to the posterior body segments, make their appearance as pair of ectodermal elevations (Figs. 10.14h, *v. pt,* and 10.15b, *th-abd)* which very soon unite (Fig.10. 15c, *th-abd)*. The vestige of the closed blastopore takes a position on the ventral plate and becomes the *anal rudiment*.

About this stage, on the two arms of the horseshoe-shaped ectodermal band in the region between the blastopore and the optic lobes, three pairs of *limb rudiments* are raised from the surface as transverse ridges; these acquire distinct shape in the sequence: *mandibles* (Fig. 10.15b, *mn), antennules* (1. *ant)* and *antennae (2. ant)*. In front of the optic lobes on the median line the *dorsal organ* is formed as an assemblage of tall ectoderm cells (*d. org)*. By inward development of the ectodermal folds, these limb rudiments grow into tubular processes projecting over the germinal disc (Fig. 10.15c), and the naupliar mesoderm extends into their interior to constitute the *limb mesoderm.* Near the bases of these limbs and optic lobes, the ganglion primordia are formed as ectodermal thickenings arranged in pairs on either side of the median line. The *stomodaeum* invaginates in the region between the antennules and the antennae, and its anterior wall grows backward over the oral aperture to cover it as a *labrum.* Since the stage with these three pairs of appendages is equivalent to the free-living nauplius larva, it is called the *egg-nauplius*. The germinal disc of this embryo corresponds to the ventral surface of the nauplius.

By the time the limbs assume their several definitive shapes — viz., antennules, uniramous; antennae, biramous; mandibles, wart-like — the development of the so-called *metanaupliar segments* (the *maxillularly segment* and all those posterior to it) has already begun in the region posterior to the limb rudiments. In order to explain this process, it is necessary to refer to the change which the mesendoderm cell-complex has previously undergone. The constituent elements of this complex, which has pushed its way into the yolk from the blastopore region as a plug-like mass, absorb yolk from their surroundings and gradually change into vesicular cells, which develop as a whole into a sac-like mass termed the *yolk sac* (Fig. 10.16a, *y*). However large it may grow, the sac remains connected with the blastopore; even after the latter has been converted into the *proctodaeum,* the yolk sac retains connection with the inner end of this through a slender neck. Its constituent cells, which represent the endoderm, continue to grow gradually into elongate pyramids, and finally take in the whole yolk mass. Called the *secondary yolk pyramids,* these cells have as boundaries only very thin protoplasmic membranes, and their nuclei are situated on the periphery of the yolk mass as in the primary yolk pyramids (Fig. 10.22 *y*). The yolk sac represents a transient *mesenteron* which will later be replaced by epithelial endoderm.

Before this, in an earlier stage of yolk sac development, two characteristic cells appear on either side of the neck of the sac, within the still-undifferentiated sub-blastoporic mesendoderm cell-complex. No yolk granules are included in their cytoplasm, they have large, strongly-staining vesicular nuclei, and lie close beneath the ectoderm of the ventral plate. These are the *mesoteloblasts,* which produce the mesoderm of the metanaupliar segments; their number will later increase to four on each side (Fig. 10.16a, b *Mes)*. The ectoderm cells which lie directly over the meso-teloblasts subsequently differentiate into the *ecto-teloblasts*; these closely resemble the meso-teloblasts and form a transverse semi-circular line along the anterior border of the ventral plate *(Ect.)*. In the suceeding divisions, groups of both ecto- and meso-teloblasts send out daughter cells in an anterior direction, and these in turn at once begin to proliferate to give rise to the maxillulary and more posterior segments. By such addition of new segments and their growth, the ventral plate develops into the *thoracico-abdominal process,* which arises from the posterior edge of the germinal disc and gradually elongates forward over its sufrace (Fig. 10.15 c-e, *th-abd)*. As a result, the main portion of the embryo after this stage corresponds to the anterior part of the head, and all the more posterior segments are formed on this process, which is folded over the germinal disc, or ventral side of the head (Figs. 10.21 and 10.22).

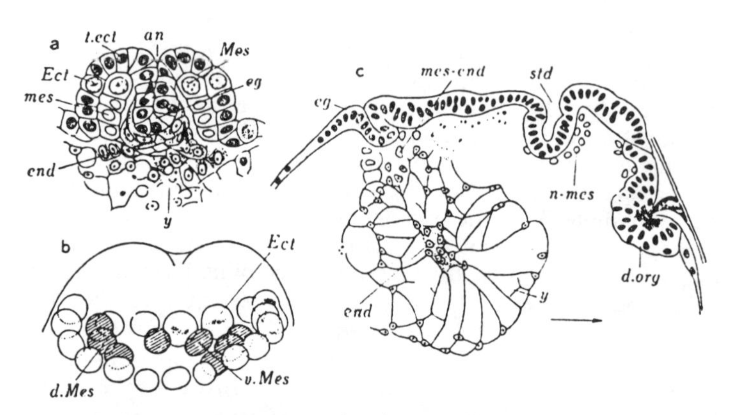

Fig. 10.16 Germ layer formation in *Planulirus japonicus* (Shiino)

a. nauplius larva, cross-section of ventral plate through hindgut; b. metanauplius stage, sketch of part of thoracico-abdominal process showing only teloblasts; c. longitudinal section of larva at nauplius stage. *an:* anus, *d.Mes:* dorsal mesoderm cell, *d.org:* dorsal organ, *Ect:* ecto-teloblast, *eg:* hind-gut, *end:* endoderm, *Mes:* meso-teloblast, *mes:* daughter cell of meso-teloblast, *mes-end:* mes-endoderm cell complex, *n-mes:* nauplius mesoderm, *std:* stomodaeum, *t.ect:* tail ectoderm, *v.Mes:* ventral meso-teloblast, *y:* yolk. Arrow points anteriorly.

The growth of the thoracico-abdominal process begins with cell proliferation at the ends of the ecto-teloblast line; this increase continues until 18 cells encircle the process. Their daughter cells are given off anteriorward (in the anatomical sense: i.e., cephalad along the body axis which makes a U-turn at the base of the thoracico-abdominal process). The ecto-teloblasts themselves continue to retreat toward the tip of the process, always encircling it at its apex (Fig. 10.16a-b). (Figure a shows an early stage in the elevation of the process, in which the daughter cells are given off vertically downward toward the surface of the germinal disc.) The anus is located on the process, always lying distal to the teloblastic ring. On each side of the process, three meso-teloblasts lie on the dorsal, and one on the ventral, side of the *mesenteric tube,* which will be described below. These cells give rise to a dorsal and a ventral pair of cellular bands which extend anteriorly (Fig. 10.16a). Since the ring of ecto-teloblasts is incomplete at first, at least a small medio-dorsal portion of a few anterior segments is formed by an extension of the dorsal head ectoderm.

Both the ecto- and meso-teloblasts lose their characteristic properties after they have finished supplying the material of the sixth abdominal segment. The *telson ectoderm* is derived from ectoderm cells occupying the area posterior to the teloblastic ring, or, to trace their origin back to the stage before the appearance of the teloblasts, from the peri-blastoporic ectoderm surrounding the closed blastopore. The *telson mesoderm* is formed by inward migration of cells from the telson ectoderm.

Another important change occurring in the egg-nauplius is the secretion of a *chitinous membrane* from the entire ectodermal surface, a phenomenon associated with the activity of the dorsal organ. In the cells of this organ, an active nuclear metabolism and the secretion of a substance from the outer surface take place (Fig. 10.16c). The separation of the chitinous membrane begins at this organ and extends posteriorward, involving even the inner face of the stomodaeum and the proctodaeum, although the membrane maintains a single connection in the mid-dorsal region for a considerable period. When this state is reached, the dorsal organ, which has thus fulfilled its temporary, embryonic function, degenerates completely. Later, however, a *median dorsal organ,* which shows marked nuclear activity, makes its appearance and at the same time the chitinous membrane becomes entirely separated from the ectoderm in this area also. As a consequence, the embryo is now enclosed in three coverings, including the original egg membrane. The formation of this third membrane represents a precocious embryonic molt taking place as early as the nauplius stage.

The Leptostraca, Mysidacea and Stomatopoda, as well as the other Decapoda, form the thoracico-abdominal process in the same way as *Panulirus.* In the first two orders, in which the embryo emerges early from the egg membranes in a structurally incomplete stage and lives within the brood-pouch of the mother animal until it becomes a free-living larva, the embryonic body straightens on hatching, and even bends somewhat dorsally, in contrast to its earlier posture. In the Isopoda, on the other hand, the primordia of all the segments are laid down on the blastoderm, and the embryonic body, forming no thoracico-abdominal process, shows a dorsal curvature from the beginning. Since many of the important developmental processes in these orders occur after the straightening of the body, the organogeny of various parts can be made out more easily in them than in the Decapoda or the Stomatopoda, the embryos of which remain folded ventrally until they reach a free-living stage.

Such processes of gastrulation and germ layer formation as we have seen in *Panulirus* also take place in other decapods in a substantially similar way, although they may show more or less variation. In both *Palaemon* and *Caridina,* the blastopore is represented by an elongate longitudinal slit, the inward migration of cells taking place along its whole length. As this migration proceeds, blastopore closure begins at the anterior end and extends backward, although a shallow depression persists for some time before the posterior end closes completely. The invaginated endoderm cells of *Palaemon* are indistinguishable from the mesoderm cells in the early stages, but they presently take in yolk granules and differentiate into yolk cells. The corresponding cells of *Caridina* can easily be recognized from the beginning of invagination by their inclusion of yolk. As their yolk content increases, these yolk (endoderm) cells are converted in both genera into the stellate form by the breakdown of the protoplasmic membrane. Penetrating the yolk mass, they disperse, even reaching the side opposite to the blastopore (dorsal side of embryo), and form rudimentary secondary yolk pyramids which, however, do not develop sufficiently to construct a yolk sac such as that seen in *Panulirus.* The head mesoderm and endoderm differentiate from the antero-lateral and posterior parts, respectively, of the mesendodermal cell-complex (Aiyer 1949, Nair 1949).

In *Maja squamifera* the endodermal cells become stellate and scatter in all directions from the bottom of the small invaginated blastopore; within the yolk mass they arrange themselves on the periphery of a sphere concentric with its surface. With the gradual enlargement of this sphere, they finally reach the yolk surface and form secondary yolk pyramids (Mayer 1887).

In the gastrula of *Astacus fluviatilis*, the blastopore makes its first appearance as a large circular area, the contraction of which causes the epithelial layer

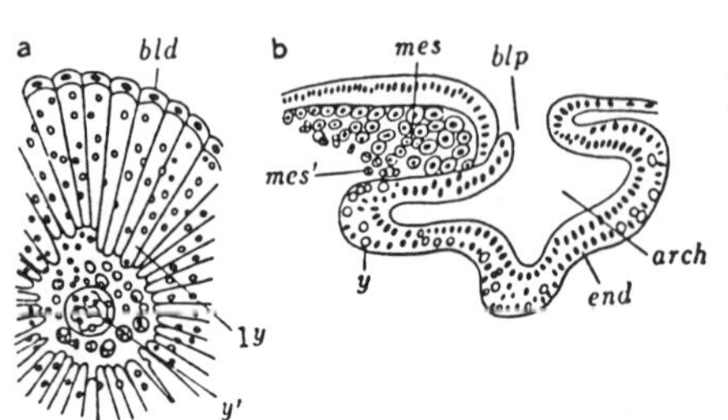

Fig. 10.17 Embryo of *Astacus fluviatilus* (Reichenbach)

a. cross-section through part of blastula; b. longitudinal section through blastopore of gastrula. *1y*: primary yolk pyramid, *arch*: inner cavity of archenteron, *bld*: blastoderm, *blp*: blastopore, *end*: endoderm, *mes*: mesoderm, *mes'*: degenerated nuclei of secondary mesoderm, *y*: yolk granule, *y'*: central yolk body.

to invaginate without losing its cellular continuity and results in the formation of a distinctly invaginated archenteron (Fig. 10.17b). By the gradual incorporation of yolk granules, the cells constituting the archenteron grow taller and cylindrical, and together form a yolk sac enclosing a central cavity. After they have taken in the whole yolk mass, they finally become completely typical secondary yolk pyramids. Some of the epithelial cells at the base of the thoracico-abdominal process, however, retain their epithelial characteristics; repeated division of these cells together with reversion of yolk sac cells to their original epithelial nature effects the expansion of the endodermal epithelium. The naupliar mesoderm originates in ectoderm cells which migrate inward from the anterior lip of the blastopore (Reichenbach 1886). The differentiation of the germ layers in *Squilla* proceeds in exactly the same way as in *Panulirus*, except that the endoderm cells, instead of forming a yolk sac, spread out over the yolk surface as stellate yolk cells (Shiino 1942).

The Malacostraca named above form the thoracico-abdominal process in quite the same way as does *Panulirus*: i.e., by cell-budding on the part of the ecto- and meso-teloblasts. The ecto-teloblasts of *Palaemon*, *Caridina* and *Squilla* number 21 and include a central teloblast which is located on the medio-ventral line (while *Panulirus* has 18 and lacks a central member). In all cases, a few of these cells differentiate first from the ectoderm just in front of the blastopore on either side and increase in number to encircle the thoracico-abdominal process as a complete ring. The meso-teloblasts, always eight as in *Panulirus*, differentiate in pairs from the sub-blastoporic mesendodermal cell-complex. In *Palaemon*, *Caridina* and *Squilla* the inward migration of ectoderm cells takes place in the region between the optic lobes and the antennulary segment. In the first two genera, such cells, which are called *pre-antennulary mesoderm*, enter the labial cavity, where they even form coeloms, but in *Squilla*, their nuclei disintegrate as soon as they leave the surface, and eventually the cells degenerate completely. No such mesoderm ever occurs in *Panulirus*, even in a rudimentary way. In *Caridina*, in the region just anterior to the optic lobes, the *preoptic mesoderm* sinks from the surface into the interior, and soon afterward disappears. The telson ectoderm and mesoderm of *Squilla* have the same origin as those of *Panulirus*. The problem has not been studied in *Palaemon* and *Caridina* (see Table 10.3).

Table 10.3 Summary of observations related to germ-layer formation in several representative crustaceans.

	SPECIES	*Nebalia bipes*	*Hemimysis*	*Jaera*	*Caridina laevis*	*Palaemon idae*	*Panulirus japonicus*	*Squilla oratoria*
	INVESTIGATOR	Manton	Manton	McMurrich	Nair	Aiyer	Shiino	Shiino
ECTODERM	Origin of naupliar ectoderm	blastodisc ectoderm	blastodisc ectoderm	blastodisc ectoderm	blastodisc ectoderm	blastodisc ectoderm	blastodisc ectoderm	blastodisc ectoderm
	Origin of metanaupliar ectoderm (no. of ectoteloblasts)	15	15	23—25	21	21	18	21
	Origin of telson ectoderm	periblastoporal ectoderm	ectoderm posterior to blastoporal ring	ectoderm posterior to blastoporal ring	periblastoporal ectoderm	periblastoporal ectoderm	periblastoporal ectoderm	periblastoporal ectoderm
MESODERM	Preoptic mesoderm	absent	absent	?	present	absent	absent	absent
	Pre-antennulary mesoderm	present	present	?	present	present	absent	degenerate
	Origin of naupliar mesoderm	mesendodermal cell-complex	periblastoporal ring	mesendoderm cell-complex	mesendoderm cell-complex	mesendoderm cell-complex	anterior lip of blastopore	anterior lip of blastopore
	Origin of metanaupliar mesoderm (number of meso-teloblasts)	8	8	8	8	8	8	8
	Origin of telson mesoderm	mesendodermal cell-complex	ectoderm posterior to blastopore	?	?	?	telson ectoderm	telson ectoderm
ENDODERM	Origin of endoderm (yolk cells)	mesendodermal cell-complex	periblastoporal ring	mesendoderm cell-complex	periblastoporal ring	mesendoderm cell-complex	mesendoderm cell-complex	mesendoderm cell-complex
	Characteristics of yolk cells	sac-shaped	sac-shaped	stellate	stellate	stellate	sac-shaped	sac-shaped
	Mode of yolk-sac formation (yolk cell migration)	yolk cells cover yolk mass	yolk cells cover yolk mass	yolk cells penetrate yolk mass	yolk cells penetrate yolk mass	yolk cells penetrate yolk mass	yolk cells expand to include yolk granules	yolk cells cover yolk mass

In the malacostracans cited above, as well as in other decapods, including *Alpheus* (Brooks and Herrick 1880), it occasionally happens, by the time the germinal disc is formed, that some of the superficial cells from regions of the disc other than the blastopore, or from the surface outside the disc, migrate into the yolk in a scattered fashion. Even earlier, in the blastula stage, it may happen that some of the blastoderm cells spontaneously sink below the surface, or send a daughter cell into the yolk by a division oriented vertical to the egg surface; or some blastomere nuclei may fail to emerge from the yolk. Such cells, which are also called yolk cells, show signs of degeneration in their nuclei, and are believed to play a part in converting the yolk to usable substances (Fig. 10.14e, *nu*). In *Panulirus* and *Squilla,* nuclear disintegration is also seen in the lowest cells of the mesendoderm cell-complex, just under horseshoe-shaped mesodermal band, or even within it. Close under the anterior lip of the blastopore of the *Astacus* embryo there are vacuolated cells known as *secondary mesoderm,* which may also represent a kind of disintegrating yolk cells (Fig. 10.17b, *mes'*). In every case, however, the disintegration figures are most frequently seen before or after the establishment of the naupliar limb rudiments, and disappear by the time the endodermal epithelium is formed.

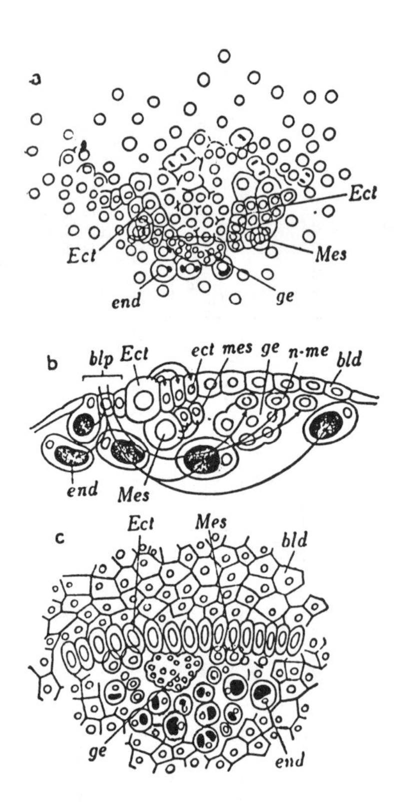

Fig. 10.18 Gastrulation in two mysid species

a. blastopore region of embryo of *Hemimysis lamornae;* b. longitudinal section of embryo at same stage as a, showing sinking inward of meso- and endodermal cells; c. external appearance of *Mesopodopsis orientalis* embryo. *bld:* blastodermal (nauplius) ectoderm, *blp:* blastopore, *Ect:* ecto-teloblast, *ect:* daughter cells of ectoteloblast, *end:* endoderm (yolk cell), *ge:* germ cell, *Mes:* meso-teloblast, *mes:* daughter cells of meso-teloblast, *n.me:* nauplius mesoderm cell. (a, b. Manton; c. Nair)

In contrast to the gastrulae of the Decapoda and Stomatopoda, which form a larger or smaller but distinctly invaginated blastopore, those of the Leptostraca and Peracarida exhibit no invagination. Gastrulation in these groups is effected by inward cell migration from the blastoporic area, which is at the same time covered over as the anterior lip of the blastopore grows backward. The mesoderm and endoderm cells, as well as the primordia of the gonads, are occasionally distinguishable from the other elements before they move inward. For example, several cells with small granular nuclei and strongly staining cytoplasm differentiate at the center of the blastoporic area on the germinal disc of *Hemimysis* (Manton 1928) and *Mesopodopsis* (Nair 1939); these later give rise to the germ cells (Fig. 10.18a-b, *ge*). On either side of them, a number of cells rich in cytoplasm make their appearance, forming two or three backwardly curving, crescentic lines, the anteriormost of which represents the forerunner of the ecto-teloblastic line to be completed soon afterward (Fig. 10.18a, *Ect*). Endoderm cells with a yolk vesicle differentiate just behind the germ cells. In front of the blastopore, a concentration of ectoderm cells initiates the formation of the germinal disc, which will become the ectoderm of the naupliar segments.

Unlike the Decapoda and Stomatopoda, the Peracarida form a V-shaped germinal disc. While the germ cells move inward and then anteriorward under the disc (Fig. 10.18b), the paired rows of ecto-teloblasts on either side of them unite in the middle to form an arch of 15 cells, which are soon rearranged on a transverse line. In *Mesopodopsis,* the number of cells in this line varies from 14 to 16; the full number is completed earlier than in *Hemimysis,* and actually before the inward migration of the germ cells and endoderm (Fig. 10.18c, *Ect*).

In an earlier stage of *Hemimysis,* just behind the paired rudiments of the ectoblastic lines, some of the cells which are larger than the others differentiate into meso-teloblasts (Fig. 10.18a, c, *Mes*). These cells increase in number to eight, and move inward one by one to a position directly below the ecto-teloblastic line (Fig. 10.18b). Even after the differentiation of the teloblasts is accomplished, the inward migration of certain smaller cells continues to take place in the blastoporic area. These proceed forward along the lower surface of the germinal disc to become the naupliar mesoderm (Fig. 10.18b, *n.mes*). In its location, the ectoblastic line therefore corresponds to the anterior lip of the blastopore, and is shifted backwards so that it overlaps the blastoporic area as its constituent teloblasts bud off their daughter cells anteriorwards.

Repeated teloblastic divisions result in a regular arrangement of the daughter cells in longitudinal and transverse rows on the ventral surface of the embryo. In both ectoderm and mesoderm, one division of the teloblastic line corresponds to the formation of one segment; that is, one transverse row of daughter cells furnishes all the material for forming one segment. After these cells bud off the third thoracic segment, the thoracico-abdominal process begins to grow and bend forward; when the budding of the cells to form the second abdominal segment comes to an end, the embryo tears the egg membrane and escapes from it, and the body straightens.

The teloblasts produce segments as far as the seventh abdominal, but this segment is a transient structure and later coalesces with the sixth. Since the ecto-teloblasts are simply arranged along a transverse line in one plane, they produce only the ventral surface of the segments; the dorsal surface as well as the epithelium of the telson are derived from the ectoderm of the extra-germinal region. The telson mesoderm in *Hemimysis* is said to originate in cells which belatedly sink inward from the posterior border of the blastopore. In *Neomysis,* however, it is asserted that this tissue arises from the mesoderm of the seventh abdominal segment which has shifted to the interior of the telson, and that no mesoderm formation occurs posterior to the teloblasts (Needham 1937).

When the optic and antennulary rudiments are laid down, numbers of cells in the region between them on each side migrate inward to form the pre-antennulary mesoderm as it was formed in *Palaemon* and other species. The endoderm cells which left the surface and became vesicular yolk cells continue to incorporate yolk and increase in size. They are scattered along the underside of the germinal disc over the yolk surface (Fig. 10,18b, *Mes),* which they finally enclose completely to form a yolk sac. Although this formation of the sac in the Mysidacea resembles that of *Panulirus,* it differs in that the yolk cells envelop the whole yolk mass by moving along its surface instead of penetrating it, thus leaving its central part unchanged throughout the process. This behavior of the yolk cells is closely similar to that found in *Squilla,* although the cells retain their vesicular nature and never become stellate. As the embryo hatches, the production of new yolk cells ceases, and the already formed cells move into the posterior part of the body, which is now straightened. In explanation of the hatching process, it is suggested that the yolk absorbs water and swells, exerting enough pressure on the egg membrane to rupture it (Manton 1928).

Germ layer formation in *Nebalia bipes* resembles that of the Decapoda in some respects, and of the Mysidacea in others. As in the Decapoda, the mesoderm cell-complex is formed beneath the blastoporic area, although germ cell differentiation cannot be recognized here in the early stages; on the other hand, the blastopore does not exhibit any invagination. The mesendoderm cell-complex diferentiates into naupliar mesoderm which spreads forward under the germinal disc, and into endoderm which occupies the posterior portion of the complex. Some of these cells take in a certain amount of yolk and grow into vesicular cells like those in the Mysidacea; they then scatter over the yolk surface until they form a continuous layer temporarily enclosing the whole yolk mass. Later they join the endodermal epithelium which has differentiated directly from the posterior portion of the mesendoderm cell-complex. The ento-teloblasts make their first appearance in an arched line along the anterior lip of the blastopore, as in the Decapoda, and later encircle the thoracico-abdominal process as a complete ring of 19 cells. Eight meso-teloblasts differentiate from the mesendoderm cell-complex, as in the Decapoda. Several cells remaining in the complex posterior to the meso-teloblasts later develop into the telson mesoderm. The pre-antennulary mesoderm is derived from an inward migration of extra-blastoporic cells which takes place in the head region (Manton 1933).

In the Isopoda, which do not form a thoracico-abdominal process, the behavior of the teloblasts can be observed rather clearly, since the development of the embryo takes place entirely on the surface. As early as the cleavage stages in *Jaera,* it can be demonstrated by differences in staining properties and cytological characteristics that the yolk cells occupy the vegetal pole, the mesendoderm cells surround them and the ectoderm cells lie at the animal pole (Fig. 10.20e). As the ectoderm cells are concentrated toward the ventral side to form the germinal disc, the mesendoderm cells also shift from the dorsal side, severing their cellular connections there and forming a transverse band across the ventral surface. The constituent cells then sink one by one beneath the surface to form a many-layered band. Several cells situated on the mid-ventral line in the posterior part of this band later develop into *hepatic mesoderm.* The ectoderm cells lying along the anterior margin of the band differentiate into ecto-teloblasts, which lie in an arched line. As they give off their daughter cells anteriorward, they gradually move toward the posterior, and together with the daughter cells overlie the mesendoderm and yolk cells. The ecto-teloblastic line in this case, also, thus represents the anterior lip of the blastopore, and gastrulation is effected by a process of epiboly. In contrast with the regular longitudinal and transverse arrangement of the teloblastic descendants, the V-shaped germinal disc is composed of a disorderly array of cells (Fig. 10.19b). The ectoderm in the region posterior to the teloblastic line gives rise to the telson ectoderm and the primordium of the *proctodaeum* When the teloblasts reach their full number of 23—25 cells, they form a straigh transverse row, with the central teloblast protruding slightly. Beneath this line eight meso-teloblasts form a similar transverse line (McMurrich 1895).

The meso-teloblasts of *Asellus* differentiate from cells lying between th yolk cells and the ecto-teloblasts, which are already aligned in a crescent (Fig

10.19a). Inside the embryo they are arranged in bilaterally symmetrical sets, each with one cell close to the median line and three lateral to this a short distance away. It is probably appropriate to regard these cells as corresponding respectively to the ventral and dorsal teloblasts of the Decapoda. Since the foundations

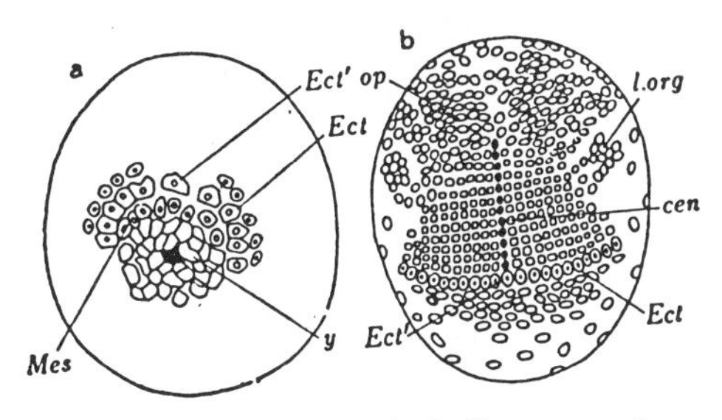

Fig. 10.19 Embryo of *Asellus communis* (Korschelt u. Heider)

a. gastrula, differentiation of ecto- and meso-teloblasts; b.formation of nauplius segment, showing regular alignment of ecto-teloblast. *cen:* central row of cells, *Ect:* ecto-teloblast, *Ect′:* central teloblast, *l.org:* lateral organ, *Mes:* meso-teloblast, *op:* optic lobe, *y:* yolk cell.

of all the segments are laid down on the germinal disc, only a narrow portion of their ventral surfaces and appendages is teloblastic in origin; the rest, representing a far greater part of their total area, is derived from the ectoderm outside the germinal disc. Two divisions of each ecto-teloblast and a single division of the meso-teloblasts furnish all the material for one segment. The segments as far as the sixth abdominal are formed in this way. No seventh segment is formed, and the telson takes its origin from the ectoderm in the region posterior to the terminal disc (McMurrich 1895).

In the embryos of *Porcellio* and *Armadillidium,* the blastoderm of the region external to the germinal disc becomes for a while entirely lacking in nuclei, since its constituent cells migrate into the yolk, where they disintegrate. This state of affairs has led some workers to the erroneous conclusion that the Isopoda undergo discoidal cleavage. However, when the germinal disc has developed to such an extent that it occupies four-fifths of the ventral area, the nuclei in its marginal zone begin to migrate back over the egg surface toward the dorsal side. About this time, the embryonic exuviae become separated from the disc, the separation gradually extending dorsally in the wake of the migrating nuclei. The most dorsal region, in which the molting takes place last, represents the so-called dorsal organ. As nuclear migration proceeds, this organ is gradually reduced in area until it becomes only a short, slender connection between the newly formed dorsal ectoderm and the molted skin. In the mesendodermal cell-complex, on the other hand, a vigorous proliferation has taken place and numerous stellate yolk cells scatter through the yolk mass to form a temporary yolk sac connecting the stomodaeum with the proctodaeum in the dorsal part of the embryo. The rest of the complex forms a pair of endodermal epithelial cells which lie side by side in the posterior part of the head; the mass of germ cells, distinguishable from the other elements by their characteristic cytoplasm and nuclei, also differentiate early in this complex Goodrich 1939).

Since the available information concerning the development of the Amphipoda is based on rather old-fashioned studies, a re-examination of this material should be made. In view of the facts, however, that gastrulation does not involve invagination, and that a regular longitudinal and transverse arrangement of cells is continuous with the germinal disc, it is highly probable that the teloblasts play an active part in forming the segments, as in the Isopoda. In *Gammarus*, the embryo begins to develop at right angles to the longer axis of the egg, but later rotates so that the two long axes coincide. Some workers claim that the yolk cells arise from the posterior end of the germinal disc and move into the interior of the yolk mass (Bergh 1894), while others believe that inward cell migration occurs over the whole ectodermal surface (Heidecke 1903). In any case, the Amphipoda bear a close resemblance to the Isopoda in forming a pair of endodermal epithelial cells.

The record is again ambiguous with respect to the origin of the mesoderm. It is said to be derived from the ectoderm at the bases of the limb rudiments, either by delamination or by inward cell movements, but this explanation may well have resulted from incomplete observation. No distinction has been made between naupliar mesoderm and teloblastic, metanaupliar mesoderm, nor has the existence of teloblasts even been noted. The dorsal organ which is formed in front of the germinal disc resembles that of the Decapoda rather than of the Isopoda.

(5) PRESUMPTIVE GERM LAYER AREAS ON EMBRYONIC SURFACE

So far as it is possible to follow the cell-lineage, as in the examples described above, the endoderm cells are found to arise from the D^{II} blastomere of the four-cell stage. This blastomere is one of the two which make contact at the vegetat pole, and is distinguished from the others by some characteristic such as an abundance of yolk or the possession of certain peculiar granules. If the latter is the case, the descendant blastomere which contains these granules will give rise to the primordial germ cell. In *Penaeus*, a malacostracan showing total cleavage, D^{II} can hardly be distinguished from B^{II}, and the origin of the germ cells is not known. In *Euphausia*, however, both germ cells and endoderm originate from a single blastomere of the four-cell stage. In the malacostracans with superficia, cleavage, on the other hand, the germ cells take their origin from the mesoderm.

Manton (1928) has stated that in the holoblastic eggs of the Entomostraca, the endoderm always differentiates in front of the mesoderm, while in the Malacostraca it originates behind it (Fig. 10.20). Among the Malacostraca, *Astacus*, the Mysidacea, the Euphausiacea and the Isopoda constitute a group in which the differentiation of the germ layers has been ascertained in relation to the surface of the embryo (Fig. 10.20d-g); in eggs having a small blastoporic area and a large amount of yolk, however, it is rather difficult to establish this connection. The existence of such a definite spatial relationship has been asserted in *Palaemon* (Aiyer 1949), *Caridina* (Nair 1949), and *Leander* (Kajishima 1950). As is seen in *Nebalia* (Manton 1933), *Panulirus* (Shiino 1950) and *Squilla* (Shiino 1942), it often happens that when the mesoderm is derived from the mesendodermal cell-complex, the meso-teloblasts originate in the antero-lateral part of the complex and the endoderm arises from its posterior part. Moreover, if consideration is confined to the naupliar mesoderm — this corresponds to a far greater proportion of the mesoderm in entomostracan embryos, which hatch in the form of a nauplius — it has been established in *Panulirus* and *Squilla* that this tissue migrates into the interior from the anterior lip of the blastopore. One of the fundamental differences between the Entomostraca and the Malacostraca thus appears to be that the spatial relation of the presumptive areas of mesoderm and endoderm is reversed.

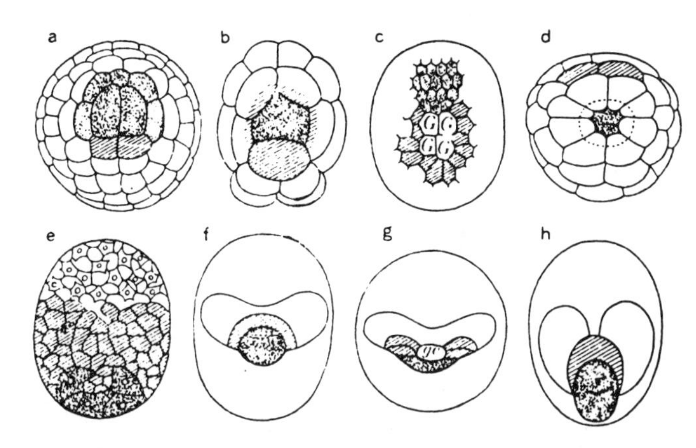

Fig. 10.20 Presumptive area of germ layer formation in several crustaceans (a-g. Manton; h. Kajishima)

a. *Calanus;* b. *Lepas;* c. *Moina;* d. *Euphausia;* e. *Jarea;* f. *Asellus;* g. *Hemimysis;* h. *Leander.* Dotted area; endoderm. Oblique-lined area; mesoderm. *ge:* germ cell.

Finally, the pre-antennulary mesoderm, which has been found in the Mysidacea, Nebaliacea and Decapoda, has not attracted attention in other orders except the Stomatopoda, where a rudimentary formation of this segment has been noticed. It remains for future investigation to solve the problem of how to bring the existence of this supernumerary mesodermal segment into harmony with the constitution of the nauplius, which is generally regarded as consisting of the three segments from the antennulary to the mandibular. The different modes of germ layer formation which are observable in the malacostracans cited above are summarized in Table 10.3.

(6) **ORGANOGENY**

As may be obvious from what has been reported in connection with germ layer formation, the establishment of the segment primordia in the Malacostraca proceeds posteriorward in a regular way, but the time at which these segments complete their development and form their appendages does not necessarily follow this regular sequence. In the thoracico-abdominal process of *Panulirus,* the *nerve ganglia* are first formed in the pereiomeres, and *pari passu* with this activity, the limb-bearing segments swell laterally to produce the *limb rudiments* (Figs. 10.15d-f, 10.21b). The segments from the maxillulary to the second maxilliped shift from the base of the thoracico-abdominal process to the germinal disc during the later embryonic period, before hatching takes place. Among the appendages, the antennules and the antennae develop an elongate form, the latter becoming temporarily biramous and later uniramous again (Fig. 10.15c-e). The mandibles are short and finger-shaped. The maxillulae are even shorter, and triangular, with the two members of the pair close together at the median line. The maxilla, although small, exhibits a distinctly biramous structure. The first maxilliped, which shows a temporary, abortive bifurcation, develops into an unbranched limb, as does the second maxilliped; the epipodite of the former is derived from a projection of the body wall. The third to sixth pereiopoda are represented by slender, elongate, biramous limbs, although the exopodite of the sixth is no more than a vestige. The remaining segments are all limbless, and the caudal furca is vestigial (Shiino 1950). In *Paratya compressa* the second antenna is typically biramous. By the time of hatching, the appendages down to the third maxilliped are well developed, while the fourth to the seventh pereiopoda remain in the state of distinctly biramous limb-buds (Ishikawa 1885).

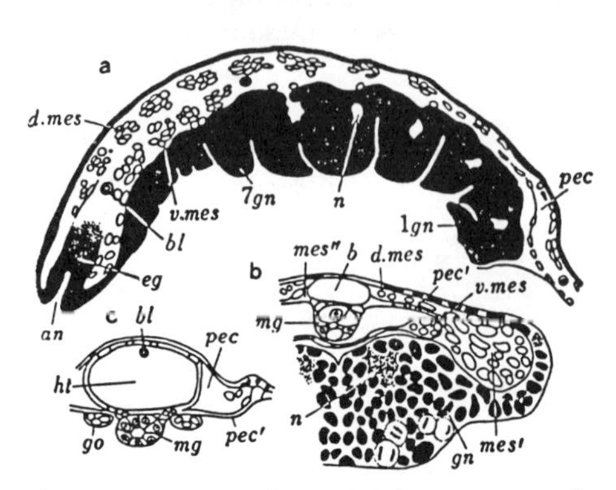

Fig. 10.21 Thoracico-abdominal process in *Panulirus japonicus* embryo (Shiino)

a. longitudinal section of thoracico-abdominal process in embryo forming compound eye; b. cross-section of thoracico-abdominal process in embryo at same stage; c. formation of heart in embryo prior to hatching. *1gn:* third thoracic ganglion, *7gn:* seventh thoracic ganglion, *an:* anus, *b:* large posterior artery, *bl:* blood corpuscle, *d.mes:* dorsal mesoderm, *eg:* hind-gut, *gn:* ganglion, *go:* gonad, *ht:* heart, *mes′:* leg-cavity mesoderm, *mes″:* mesoderm surrounding mid-gut, *mg:* mid-gut, *n:* nerve fiber, *pec:* pericardium, *pec′:* floor of pericardium, *v.mes:* ventral mesoderm.

In the Decapoda, the first and second antennae extend postero-laterally; the pereiopoda, which lie in a similar direction in the structural sense, are directed forward in relation to the embryonic body as the result of the bend in the thoracico-abdominal process (Fig. 10.15d-f). In stomatopod embryos, which resemble those of the decapods in the disposition of their appendages, the antennule provisionally shows a rudimentary bifurcation in the primordium stage, but later becomes uniramous like the antenna. By the time of hatching, the appendages as far as the second maxilliped, and the first to fourth biramous pleopoda have made their appearance; the segments interposed between the two series of limb-bearing ones remain limbless. The caudal furca, which is usually well developed in the Decapoda, is represented by a transient embryonic structure in the Stomatopoda (Shiino 1942).

The embryos of the Isopoda differ from those of the above orders in being curved dorsally and having the limb rudiments directed from the sides toward the mid-ventral line; above all, they are characterized by the absence of appendages on the seventh pereiomere before the time of hatching (McMurrich 1895). The first and second antennae of the Mysidacea show a markedly higher degree of development than the other appendages. At hatching, which takes place in an early developmental stage, the embryo straightens its ventrally folded body, and even assumes a dorsal curvature like that of the Isopoda (Nair 1939). The Nebaliacea resemble the Macrura in the development of their external appearance, and follow the Mysidacea in hatching at an early stage. The embryo is not yet supplied with the full number of appendages in this stage, although it has all the segments in the state of primordia. The Nebaliacea are distinguished from the other malacostracans bv retaining the seventh pleomere throughout the adult stage.

To elucidate the developmental processes undergone by the internal organs let us take *Panulirus* as an example. About the time when the limb rudiments make their appearance on the germinal disc, a vigorous multiplication of cells causes the ectoderm to become multilayered in each segment on either side of the midventral line. At about the same sime, neuroblasts differentiate in the superficial layer of these regions; their activity leads to the formation of cells which give rise to the paired primordia of the nerve ganglia (Figs. 10.21a-b; 10.22). The ganglia of the consecutive segments are separated from each other by relatively distinct boundaries; those, however, which differentiate at the bases of the optic lobes and of the first and second antennae (designated respectively as *protocerebrum*, *deuterocerebrum* and *tritocerebrum*) soon unite to form the *brain* (Figs. 10.15d, e; 10.22*br*). The proto- and deuterocerebra of the two sides are in contact at the median line, but the stomodaeum is interposed between the two tritocerebra, so that the brain as a whole forms an inverted V. The ganglia from the mandibula to the second maxillary segment also coalesce into the *suboesophageal ganglion* (Fig. 10.22 *oe. gn.*). On either side of each segment, the ganglion forms within its interior a bundle of neurofibrils. Although the cellular parts of the ganglia are bordered segmentally by very thin cellular membranes, the fibrillar part of each ganglion is continuous anteriorly and posteriorly as well as to right and left with the corresponding parts of the adjoining ganglia, so that together the fibrillar bundles form a continuous, ladder-like nervous system. The hind end of the Λ-shaped brain sends off paired fibrillar bundles which extend backward, passing around the stomodaeum to connect with the bundles of the mandibular ganglionic pair. These bundles, together with the *transverse fibrillar commissure* of the tritocerebra, which lies just behind the stomodaeum, constitute the *circum-oesophageal nerve ring*. In the Decapoda, the ganglia are formed in each segment as far as the sixth abdominal. *Squilla*, however, forms the seventh abdominal ganglia, in this respect resembling *Nebalia*, in which a seventh mesodermal segment develops. Neuroblasts have also been found in *Squilla* and *Mysis*; especially in the latter it has been recorded that the ganglionic cells show a regular arrangement in lines extending from the outside inward, exactly as observed in the daughter-cell rows of the teloblasts (Bergh 1894).

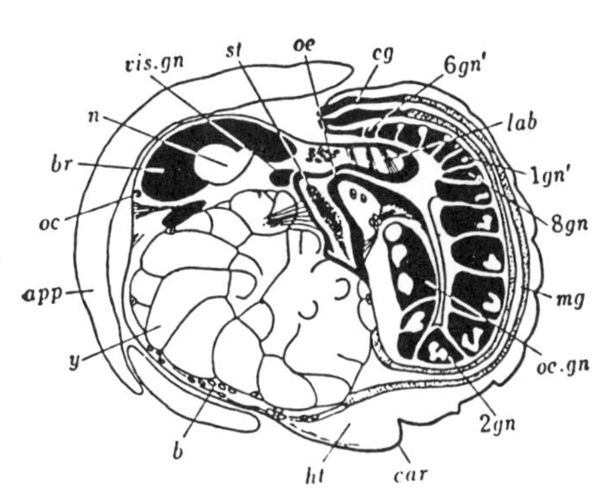

Fig. 10.22 Longitudinal section of embryo prior to hatching (Shiino)

1gn': first abdominal ganglion, *2gn*: second thoracic ganglion, *6gn'*: sixth abdominal ganglion, *8gn*: eighth thoracic ganglion, *app*: appendage, *b*: large anterior artery, *br*: brain, *car*: carapace *rudiment*, *eg*: hing-gut, *ht*: endo-cardial cavity, *lab*: labrum, *mg*: mid-gut, *n*: nerve fiber, *oc*: median ocellus, *oe*: oesophagus, *oe.gn*: suboesophageal ganglion, *st*: stomach, *vis.gn*: visceral ganglion, *y*: yolk sac.

The optic lobes give rise to the *compound eyes* and their associated ganglia, in addition to the protocerebrum. After the latter differentiates, each lobe divides into an apical portion, representing the visual area, and a basal portion, which lies directly beside the brain and represents the primordia of the *optic ganglia*. The apical portion again divides into two layers, the outer producing the *visual epidermis*, and the inner, the first and second ganglionic segments associated with the compound eyes. The basal part develops into the third and fourth ganglionic segments. The nerve fibrils appearing within these segments establish connection with the fibrils enclosed in the protocerebrum.

The visual epidermis subsequently becomes multi-layered and forms all the constituent elements of the *ommatidia* by the morphological differentiation of its constituent cells. As the differentiation of these elements proceeds, each optic lobe, which has already thickened markedly, develops around its entire circumference a double-walled fold of ectoderm which advances toward the base of the lobe. By the confluence of the inner borders of these folds along the longitudinal axes of the lobes, they become separated from the body surface and extend as *eye stalks*, enclosing a compound eye and the first to fourth optic ganglia of each side. The *median eye* is formed by some *retina cells* derived from the ectoderm just anterior to the brain, as well as certain nerve cells which separate from the protocerebra to participate in the process. Neurofibrils formed by these nerve cells provide connection with the protocerebra. The *carapace* develops on either side of the germinal disc from ectodermal folds, into which mesoderm cells migrate to form fibrillar bundles connecting the inner walls.

The stomodaeum, which originates as an invagination of the ectoderm in the region between the first and second antennae, is partitioned into an *oesophagus* and *stomach*. A *visceral ganglion* enclosing a mass of neurofibrils differentiates from the anterior wall of the stomodaeum (Fig. 10.22 *vis gn*). Most of the intestine and all the liver are endodermal in origin, only a very short terminal part of the former being derived from the proctodaeum. As stated above (p. 357), the yolk sac, which was formed by yolk cells, represents merely a temporary structure. In the neck region of the sac, in contact with the base of the proctodaeum, there are several endodermal cells which from the beginning fail to assume the form of yolk cells (Fig. 10.16a *end*). These arrange themselves around the base of the proctodaeum, forming an epithelium known as the *posterior endodermal plate*. This plate gradually enlarges by the proliferation of the cells composing it, as well as by the addition at its periphery of yolk cells which have resumed their original epithelial nature. In *Squilla*, an anterior endodermal plate is also formed at the lower side of the stomodaeum by the yolk cells assembled there. In *Panulirus*, however, this plate remains rudimentary, and the development of the mid-gut results chiefly from the activity of the posterior plate. As the thoracico-abdominal process elongates, this plate grows into a tube which, maintaining a connection with the proctodaeum, enters the process to form the major part of the intestine; at the same time it extends antero-laterally in a U-shape over the ventral surface of the yolk sac to give rise to the primordium of the *liver epithelium* (Fig. 10.22). Within the head cavity the mesodermal fibers become tightly stretched so that they divide the yolk sac, in its transitory role of mesenteron, into three pairs of *coeca*: the *antero-median*, the *antero-lateral* and the *postero-lateral*. Only the last pair of these coeca become surrounded by endoderm and, after a further division of each, develop into two pairs of *liver lobes*. The remaining, anterior pairs of yolk sac coeca undergo a gradual shrinkage and degenerate entirely by the time of hatching. The endodermal epithelium unites with the stomodaeum near the bases of the liver lobes, and the dorsal wall of the gut at this point of junction forms folds which develop into a pair of *anterior midgut coeca* (Shiino 1950).

The anterior and posterior endodermal plates of *Squilla* become connected with each other before the embryo hatches, but they develop only to cover the ventral surface of the yolk sac, and the greater part of the intestine derives from the posterior plate, as in *Panulirus*. Into the interior of the thoracico-abdominal process, the plate also sends a pair of tubular posterior liver lobes, which lie on either side of the intestine and exceed it in diameter. The yolk sac is divided temporarily into a pair of anterior liver lobes and a pair of postero-lateral mid-gut coeca, but these organs are only partially covered by endoderm and are finally entirely absorbed by the mid-intestine after hatching (Shiino 1942).

The development of the mid-intestine and its accessory glands in *Nebalia* takes place in a manner similar to that seen in *Squilla*. The posterior liver lobes arising from the yolk sac as a pair of coeca in the maxillulary segment extend posteriorly as far as the abdomen; anterior liver lobes are also formed (Manton 1933).

In *Hemimysis* and *Mesopodopsis* among the Mysidacea, it has been reported that clumps of cells separate from the mandibular mesodermal segments and extend posteriorly to form a pair of mesothelia, each of which then rolls into the form of a tube, makes connection with the anterior endoderm at the lower side of the stomodaeum and finally becomes a liver lobe (Fig. 10.23). The anterior endodermal plate forms a pair of short mid-gut coeca at its fore end. Absorbing the whole yolk sac, the posterior plate formed at the base of the short proctodaeum grows into a tubular mid-intestine, which is much longer than the corresponding part derived from the anterior plate (Manton 1928, Nair 1939). Needham (1937), however, working with *Neomysis*, insists that the primordia of the liver lobes are entirely different in their histological characteristics from the mesoderm of the mandibular segment, and can find no reason for assigning to them an origin different from that of the other endodermal elements.

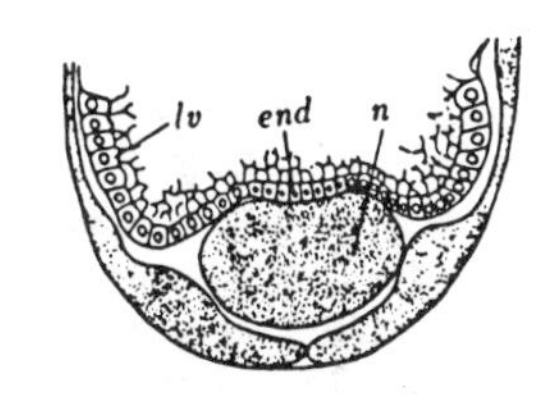

Fig. 10.23 Cross-section of part of *Mesopodopsis orientalis* embryo, showing liver rudiment and formation of endodermal plate (Nair)

end: anterior endodermal plate, *lv:* liver rudiment, *n:* ganglion

In *Porcellio* and *Armadillidium* among the Isopoda (see p. 365), yolk cells liberated from the mesendoderm cell-complex constitute a temporary yolk sac connecting the fore-gut with the hind-gut. The remaining cells of the complex form, as in the Mysidacea, a pair of endodermal epithelia which later unite on the mid-ventral line. The united epithelia come in contact with the ventral side of the yolk sac along their free margins, which are now turned dorsally, while a longitudinal fold appears on the ventral side of the epithelium and divides it again into two lobes. By the fusion of the fold with the lateral margins of the paired lobes which are now turned inward, a pair of tubes are formed. Communicating anteriorly with the short duct which has developed from the stomodaeum, each of the tubes further subdivides into two, forming the dorsal and ventral liver lobes. With the complete absorption of the yolk sac, the stomodaeum makes a direct connection with the proctodaeum; the larval endoderm thus remains only in the liver (Goodrich 1939).

As will be gathered from this description, the development of the alimentary canal and its accessory glands appears, at first sight, to very considerably from one order to another. If, however, it is assumed that the condensation of endodermal epithelium primarily occurs in four separate places — at the lower sides of the stomodaeum and proctodaeum and at either side of the yolk sac — it may be possible to reconcile the different reports concerning the modes of development as follows. In the Mysidacea all the condensation sites participate in the development of the alimentary system; in *Panulirus, Squilla* and *Nebalia* the lateral sites are replaced by an extension of the subproctodaeal epithelium; and in the Isopoda, where only lateral condensations occur, the complete elimination of the anterior and posterior sites results in an ectodermal derivation of the whole intestine.

The greater part of the naupliar mesoderm of *Panulirus* enters the limb cavities to become limb mesoderm. The mesoderm in the labral cavity and on the circumference of the stomodaeum, as well as the oral muscles, are all derived from the mesoderm of the antennulary segments. Some slender longitudinal bands left behind in the region extending anteriorward from the stomodaeum are finally transformed into connective tissue on the frontal surface of the head, after have played a part in constricting the yolk sac into right and left halves to form the anterior pair of yolk sac coeca. In *Squilla* these bands pass around the frontal side of the head and extend to its dorsal side, where they form paired longitudinal walls connecting the yolk sac with the ectoderm, and enclosing between them a median tubular cavity. This unites with the tubular elongation advancing forward from the posteriorly situated heart (p.[373]), and gives rise to the primordium of the *anterior dorsal aorta.* In *Squilla* the pre-antennulary mesoderm is formed as described above, but its constituent cells fail to undergo further development, and disintegrate entirely before the appearance of the aorta.

In *Hemimysis, Nebalia, Palaemon* and *Caridina,* the pre-antennulary mesoderm forms a pair of narrow, transitory coeloms; this tissue is said to extend posteriorly and change later into labral mesoderm (Manton 1928, 1933; Aiyer 1949; Nair 1949). In these species, paired coeloms are also formed in the antennulary, antennary and mandibular segments. In *Panulirus,* however, the inner cavity of the antennary gland is the only coelom which appears. A spherical mass of cells, which has differentiated within the limb mesoderm of the antenna, develops at its center a small cavity, a vestige of the coelom; this is the primordium of the *end sac* of the *antennary gland* (Fig. 10.24a). The mesoderm cells surrounding this sac also delaminate to form a cavity which unites with the end sac and becomes the duct of the gland (Fig. 10.24b). In other crustaceans, also, this gland is always mesodermal in origin. *Squilla,* in which the adult is provided with maxillary glands, does not give any indication of their differentiation before hatching, although it shows rudimentary formation of antennary glands (Shiino 1942). Both kinds of glands persist in *Nebalia* also, but most of their development takes place in the free-living larva (Manton 1933). In *Squilla,* a *labral gland is* formed in the mesoderm of the labral cavity.

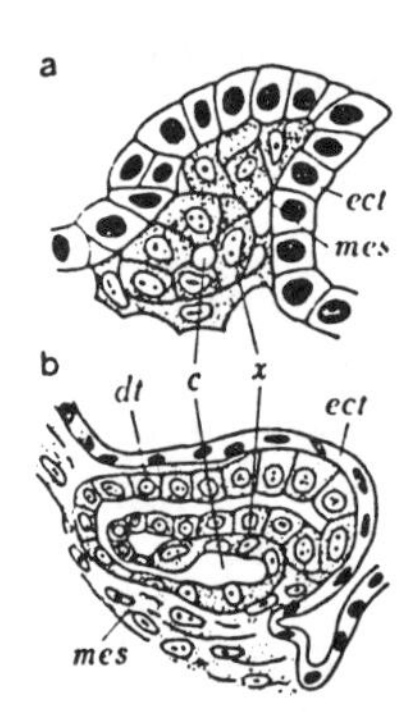

Fig. 10.24 Development of antennal gland in embryo of *Panulirus japonicus* (Shiino)

a. appearance of end-sac cavity; b. development of duct of antennal gland. *c:* vestige of coelom, *dt:* duct, *ect:* ectoderm, *mes:* mesoderm *x:* end-sac.

The metanaupliar segments following the mandibular always enclose a pair of *dorsal mesodermal bands,* each derived from three dorsal meso-teloblasts, and another pair of *ventral bands* each originating from one ventral meso-teloblast. Exhibiting signs of metamerism, the dorsal bands aggregate into a pair of cell masses for each segment (Fig. 10.21a, d *mes*). In *Hemimysis, Nebalia, Caridina* and *Palaemon,* each of these mesodermal somites forms a coelomic cavity, whereas no coeloms make their appearance in the dorsal bands of *Panulirus* and *Squilla* although it has been claimed that series of paired coeloms are found in some species of the latter such as *S. raphidea* and *S. wood-masoni* (Nair 1949). There is some doubt with respect to this point in the Decapoda and Stomatopoda. The dorsal mesodermal bands of *Panulirus* spread to the interior of the first and, second maxillae and into the limb-buds from the first to the sixth pereiopoda, furnishing these with *limb mesoderm* as well as with the material of the *extensor* and *flexor muscles* to be inserted at the bases of the limbs (Fig. 10.21b). The major portion of each band begins to take part in forming a trunk extensor muscle; this, however, soon degenerates almost completely and changes into connective tissue, leaving a feeble muscle extending from the first maxilliped segment to the third thoracic. The ventral mesodermal bands also largely degenerate into connective tissue after their provisional formation of a trunk flexor muscle, since the posterior thorax and the abdomen are too feebly developed to require strong musculature. In *Squilla* both extensor and flexor are well developed, even forming myofibrils.

In *Panulirus*, again, the dorsal bands also send out a connective tissue membrane to surround the intestine and give rise to the primordium of the *posterior dorsal aorta*, which is enveloped dorsally with ectoderm, laterally with the dorsal bands and ventrally with this membrane (Fig. 10.21b, *b*). The bands then spread along the midline of the dorsal body wall as another thin membrane, and complete the wall of the blood vessel by the union of its right and left sides. The *heart* is formed in the dorso-posterior part of the head in a similar way. Since the two bands are rather far apart in this region, the connective tissue membrane extends to a considerable distance from them in a V-shaped configuration on each side. One arm of the V, uniting the intestine with the dorsal body wall, constitutes the lateral wall of the heart, while the other arm, which is connected with the lateral body wall, contributes to the formation of the pericardial floor (Fig. 10.21c).

In the crustaceans which develop coeloms, the *heart cavity* is produced by the union of the inner walls of the paired coeloms at the dorsal and ventral angles. The telson mesoderm gives rise to the *rectal sphincter*, the mesoderm surrounding the rectum and the connective tissue within the telson cavity. In *Squilla*, this type of mesoderm forms, on either side of the rectum, a pair of spherical cell masses which develop into the primordia of the *anal glands* with the formation of a secretory cavity in the center of each. These glands do not communicate with the rectum until the time of hatching (Shiino 1942).

Manton (1928) states that in *Hemimysis* the extensor and flexor muscles inserted at the bases of the appendages, the carapace muscles from the second maxillary to the third thoracic segment, and the labial muscles are derived from the ectoderm, but in *Neomysis* all of them are regarded as mesodermal in origin (Needham 1937). Manton (1933) also considers the labial muscles and the intersegmental musculature of *Nebalia* to be ectodermal derivatives.

Development of the gonads takes place in a variety of ways among the Malacostraca. In *Euphausia*, as mentioned above (p. 351), the primordial germ cell differentiates during the cleavage stage as the sister cell of the endoblast. This divides into four cells, which form the paired *gonad rudiments* situated on the upper sides of the liver lobes (Taube 1909, 1915). The germ cells of *Hemimysis*, which differentiate on the surface of the germinal disc, move after gastrulation to the first pereiomere, where they divide into right and left groups. Enclosed in dorsal mesoderm, these cell masses are suspended from the pericardial floor. The *gonoducts* arise from hollow thickenings developed on the pericardial floor (Manton 1928).

The germ cells in *Porcellio* originate in the mesendoderm cell-complex and differentiate somewhat later than those of *Hemimysis*. They form a pair of cell masses on the ventral side of the mesoderm in the first pereiomere and eventually move back to the posterior pereiomeres, to take their final position on the dorsolateral side of the intestine (Goodrich 1939). The gonad rudiments make their appearance still later in *Nebalia*, forming metameric pairs on the pericardial floor. As their cells proliferate, they project downward from the floor of the pericardium, and finally become a single pair of gonads when the consecutive rudiments of each side fuse together (Manton 1933). The gonads of *Panulirus* also begin to develop at a late stage, and differentiate from the mesoderm in the second maxillary segment as paired rudiments. These later move to the first pereiomere where they hang from the lower surface of the pericardium (Fig. 10.21c *go*). In *Cuma* the gonads arise from the mesoderm as a pair of rudiments lying above the intestine (Butschinsky 1893); in *Gebia* nothing is known except that they are a pair of mesodermal rudiments located beneath the heart (Butshinsky 1894); in *Gammarus* they are said to originate in a concentration of migrating mesoderm cells near the dorsal side of the mid-intestine (Heidecke 1903). No signs of developing gonads are detectable in *Squilla* before hatching.

On the extra-germinal ectoderm of malacostracan embryos there occurs the development of the so-called dorsal organs (Fig. 10.15 b, c *d. org*), or lateral organs (Fig. 10.19b *l.org*); these are embryonic structures which disappear before the time of hatching (see p. [38]). The terms include various organs of different histological constitution, and many points with respect to their function await future confirmation. The median dorsal organ present in front of the optic lobes of *Panulirus* and *Gammarus*, the corresponding organ in a similar location in the Mysidacea, and the lateral organs formed dorsally on the sides of the head in the latter, are all represented by an assemblage of tall cylindrical cells; it is certain only that they have some sort of secretory function. It is especially suggestive that they show the most vigorous activity previous to the embryonic ecdysis, followed by rapid degeneration and complete disappearance. As a result, they are considered by some authors to correspond to the dermal glands of adult crustaceans (Manton 1928). The correlation between the appearance of the dorsal organ and molting in *Panulirus* and *Porcellio* has been discussed above (pp. 359, 366). In *Squilla*, an ectodermal thickening is produced by an accumulation of cells at the center of the dorsal surface; as extensive nuclear degeneration begins in its constituent cells, the exuviae become detached from the body. The anterior dorsal organ of *Hemimysis* forms an invagination before hatching occurs. The strong pull exerted on the dorsal epithelium by the development of this invagination results in a mechanical straightening of the bent embryonic body; this in turn causes the egg shell to burst. The molting begins together with the appearance of secretory substance in the invagination, and ends simultaneously with the formation of the lateral organs. It is possible that the anterior dorsal organ may secrete the material of the exuviae, while the lateral organs provide a new cuticle for the hatching larva (Manton 1928).

In concluding this description of organogenesis, which has been chiefly concerned with the Malacostraca, let us consider briefly the development of the mesoderm in the nauplius larva of *Estheria*, which may serve to represent the Entomostraca (Cannon 1924). In the larva just after hatching, the body region posterior to the mandibles does not yet show any segmentation, and the space between ectoderm and endoderm in this region is occupied simply by a cluster of mesoderm cells still containing many yolk granules. The anterior end of this cluster is distinctly separated from the mandibular muscles and possesses on its ventral side a narrow lumen, posterior to which the paired gonad rudiments lie inserted between the ectoderm and endoderm. The endoderm has already become continuous with the rectum.

The mesodermal cluster then divides into right and left halves, which separate from each other both dorsally and ventrally and form the blood-vessel space along the median line above the intestine, and the peri-intestinal space below it. The mesodermal masses of the two sides again divide, this time into dorsal and ventral halves; the ventral halves show provisional metameric divisions which soon disappear, and the dorsal halves subsequently form seven pairs of *coelomic sacs,* one pair for each segment. The *dorsal longitudinal muscles* are formed from the walls of these coeloms, as well as the heart wall from the maxillulary segment to the second thoracic. The lower wall of the heart then differentiates into the pericardial floor and the cardiac muscles, and eventually the coeloms disappear entirely. All the trunk muscles other than the extensors are derived from the ventral mesoderm.

The *maxillary gland* is wholly of mesodermal origin, the mesoderm in the second maxillary segment becoming thickened on each side and differentiating into a coiled tube. About the time when the *central nerve cord* is developing in the ventral ectoderm, one end of the tube extends downward through the ectoderm lateral to the cord to serve as the duct of the gland. The other end enlarges into a terminal sac, and establishes connection with the coelom of the first maxillary segment. The ventral mesoderm of this segment forms on each side a V-shaped mass, with the two arms of the V directed laterally; the posterior arm gives rise to the adductor of the carapace, and the anterior arm produces the accessory muscles of the maxillulae. The *ovaries* lie ventral to the coeloms in the second maxillary segment, surrounded by the coelomic walls. They later elongate posteriorly and each develops a lumen; these become continuous with the oviducts which originate from the coelom and take shape on the wall of the pericardium. The *antennary gland,* which develops much later, is composed of a mesodermal terminal sac and a duct formed by a tubular invagination of the ectoderm.

As may be seen from this account, no special cells which can be called teloblasts have been discovered in entomostracan embryos, and it has not been ascertained whether the naupliar and metanaupliar mesoderms originate from different blastomeres. There is a record of a pair of primitive mesoderm cells which are found near the caudal end of the nauplius larva, but no details have been reported with respect to their origin or behavior (Grobben 1779, 1881). The denominations given to the primary and secondary mesoderm in describing the cleavage stages (p. 350) are based on differences in the stage at which they differentiate and the blastomeres from which they arise, and do not imply a difference in their fates. They have, accordingly, no connection with the distinction between naupliar and metanauplear mesoderm.

2. LARVAE AND MODES OF METAMORPHOSIS

The larval stages of the Crustacea usually display a remarkable series of form changes. In most cases the Entomostraca hatch in the form of a nauplius, whereas the young of the Malacostraca assume the free-living habit at a more advanced stage, or undergo direct development. This difference depends mainly upon whether or not the egg is supplied with enough yolk to build up the necessary complexity of body structure. For example, although *Penaeus* belongs to the Malacostraca, a secondarily imposed reduction in the yolk content of its egg causes it to become free-living as a nauplius larva. The stage in which the larva becomes free-living is also intimately related to the brooding habit of the mother animal. The young of the Peracarida, which are reared in a brood-pouch, hatch from the egg membrane at an early stage, but their structure is close to the adult form by the time they leave the pouch. The embryos of the Decapoda, on the other hand, which are simply fastened to the maternal pleopoda, are set free as zoea larvae, or slightly later.

As has been mentioned repeatedly, when a larva hatches in a fairly advanced stage, it has always passed through a stage corresponding to the nauplius at some time in its embryonic life. There is, however, no form in which the larva becomes free-living before it attains the nauplius stage. This larval stage may be regarded as representing one of the primitive features of the Crustacea.

Among the nauplii of the various orders, those of the Copepoda are the most schematic. They are provided with three pairs of appendages on an oval body, with a median eye *(naupliar eye)* at the anterior end and a labrum covering the oral aperture in the region between the first two pairs of appendages (Figs. 10.1; 10.26a). The first appendage, representing the antennule, is composed of three or four segments and is uniramous. The second and third appendages, which correspond to the antenna and the mandible, respectively, are both biramous, with *exopodite* and *endopodite* attached to a *protopodite*. All the appendages are covered with setae, and move synchronously to propel the body through the water; ahe two posterior pairs, each member of which carries an inwardly-directed process trmed with bristles, are used also for feeding. The carapace covering the body lacks any external segmentation, and has a definite and characteristic dorsal structure.

In this larva, the addition of new segments and their accompanying appendages occurs in the region between the mandible and the caudal end. This is a gradual change, however, taking place under the integument, and in most cases a single ecdysis causes a sudden change in the external structure of the larva. Such a change consists not only in the appearance of several new segments and appendages, but also in alterations in the constitution of those previously formed. There are, however, some cases in which ecdysis results only in some inconspicuous change, such as a small increase in the number of bristles or body length. In the primitive orders, the formation of the appendage rudiments proceeds regularly toward the posterior in parallel with the appearance of the segments, but in the Malacostraca the posterior appendages are sometimes formed before the anterior ones. Occasionally the anterior segments are much retarded in their development, or remain in an undifferentiated condition while better developed segments posterior to them may even bear appendages. Even in such cases, the anterior segments exist at least as primordia, only their elaboration being suppressed. On rare occasions, already formed appendages degenerate, or disappear temporarily, only to reappear in a later stage.

Although the constitution of the crustacean body has already been referred to in various connections, it may be well to give here the nomenclature of the segments and their appendages before beginning to describe metamorphosis. The head is composed of five united segments, and joined to the trunk. The cephalic appendages comprise two pairs of *antennae*, one pair of *mandibles* and two pairs of *maxillae*. Except for the first antennae, these are all biramous from an ontological viewpoint. Each trunk segment is also theoretically provided with a pair of biramous appendages. The number of segments constituting the trunk is indeterminate in the Entomostraca, varying according to the order; when the abdomen is distinguishable from the thorax, it is without appendages. In the Malacostraca the thorax is eight-segmented, and the abdomen consists of six segments (seven in *Nebalia)* and a telson. Some or all of the first three pairs of thoracic appendages *(pereipoda)* join the oral parts and are termed *maxillipeds*. The six abdominal segments are furnished with paired *pleopoda*; the last of these are called *uropoda*.

(1) ENTOMOSTRACA

Among the Phyllopoda, the Branchiopoda, which are characterized by the large number of segments in the adult, begin their larval life in either the nauplius or the *metanauplius* stage. The latter have somewhat more segments and several additional pairs of appendages or their rudiments posterior to the mandibles.

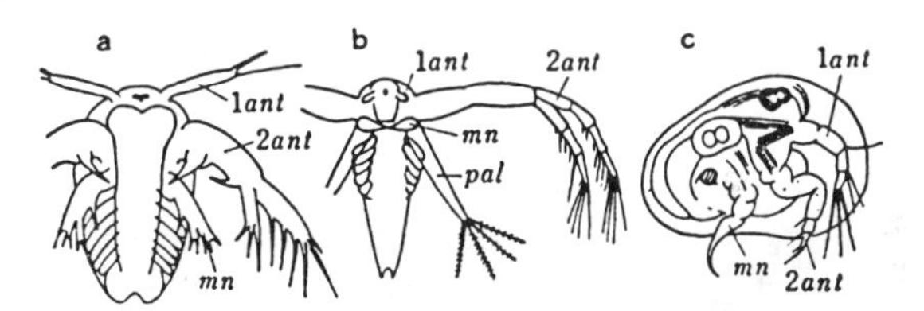

Fig. 10.25 Nauplius larvae of three phyllopodian species (Calman)

a. metanauplius of *Apus;* b. metanauplius derived from winter egg of *Leptodora;* c. nauplius of *Cypris.* 1 *ant:* first antenna, 2 *ant:* second antenna, *mn:* mandible, *pal:* palpus.

Branchinella and *Apus*, for example, hatch as metanauplii. In contrast to their particularly well formed second antennae, the development of the two pairs of maxillae is suppressed until the succeeding appendages grow enough to become foliaceous (Fig. 10.25a). The development of the trunk and its appendages proceeds caudally in regular sequence. The nauplius of *Estheria* has much reduced first antennae and a very large labrum, the more advanced larva resembling the adult of a cladoceran. Adult specimens of the Cladocera bear only four to six pairs of trunk appendages. Their summer embryos, which are reared in a brood-chamber, hatch in an early stage, but the winter embryos remain within the egg membrane through the winter, secreting in the egg-nauplius stage an external cuticle which is cast off at the time of hatching. Two pairs of maxillae put in their appearance after the trunk appendages have been formed; the second pair later degenerate and are not found in the adult. Even when the adult has only five pairs of trunk appendages, the embryo forms six pairs of primordia, of which the last pair subsequently undergo degeneration. The summer egg of *Leptodora* develops in a brood-pouch, while the winter form hatches as a metanauplius (Fig. 10.25b). In contrast to the rudimentary antennules, the antennae are conspicuously prolonged to serve as the principal locomotive organs, the mandibles carry a long styliform, unjointed palp, and six pairs of appendage rudiments are formed on the trunk. Although the compound eyes are not yet developed, the larva is provided with a median eye, which persists throughout life in individuals arising from winter eggs, but is entirely absent in the summer form (Calman 1909).

The brooding habit is also found in some of the Ostracoda, the mothers carrying their hatched larvae between their shell-folds. In the forms which deposit their eggs, the hatching larva is provided with bivalve-like shell folds exactly as in the adult, but since it has only three pairs of appendages, it is still a nauplius (Fig. 10.25c). The second antennae and mandibles are both uniramous and each pair shows a characteristic form. The remaining appendages appear in regular sequence with successive molts. This serial development of the appendages is temporarily interrupted in the stage between the formation of the first maxillae and the formation of the succeeding appendages. For this reason it has been claimed that the larva lacks the pair of appendages corresponding to the second maxillae (Calman 1909).

The Eucopepoda begin their free-living lives as nauplii and pass through a comparatively gradual metamorphosis. By two ecdyses, the nauplius of the fresh-water genus *Diaptomus* metamorphoses into a metanauplius which bears several additional pairs of appendage rudiments and the distinct primordium of a carapace (Fig. 10.26b). With each succeeding molt the metanauplius increases in size and acquires more paired appendage rudiments in the newly developing posterior region, which is bent somewhat ventrally. After molting four times, the larva shows an abrupt change in structure as it enters the *copepodid*, or *cyclops*, stage (Fig. 10.26c): it acquires elongate, many-jointed antennules, and its bent body straightens and becomes divided, as in the adult, into a thicker anterior, and a slenderer posterior region. In spite of its close resemblance to the adult, this larva at first still has fewer segments and swimmerets, and only attains the adult form after six molts.

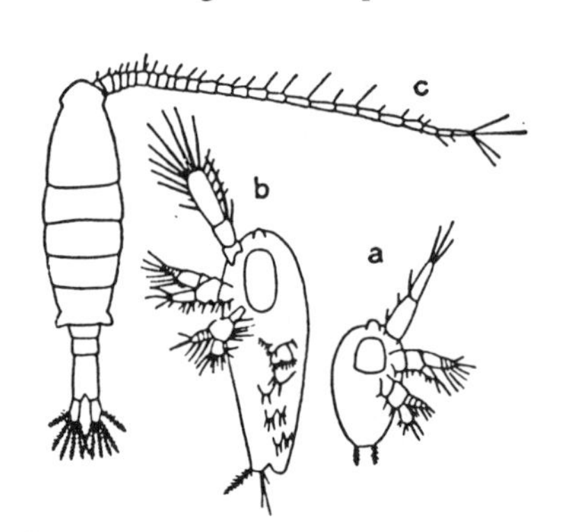

Fig. 10.26 Larval stages of *Diaptomus chaffanjoi* (Horasawa)
a. nauplius; b. metanauplius; c. copepodid.

The first copepodid larva[4] has an anterior region composed of four, and a posterior region, of two segments, and carries two pairs of swimmerets in addition

[4] Although the same general term is used to describe them, the consecutive stages of crustacean larvae always manifest some morphological difference with every molt. The term 'copepodid' applies to several different stages which are called 'first copepodid', 'second copepodid', etc., according to the number of molts they have undergone.

to the cephalic appendages and two pairs of maxillipeds. Each molt adds one segment to the body as well as a pair of swimmerets, and displaces the boundary between the anterior and posterior regions caudally by one segment. The larva achieves the full number of segments and swimmerets by the fourth copepodid stage, although the constitution of the appendages still differs somewhat from that of the adult (Horazawa 1929). Coalescence or suppression of some of the segments often occurs in the adults of the Eucopepoda, leading to an alteration in the process of metamorphosis. The larva of *Argulus* (Branchiura) is supplied with a full number of appendages which resemble those of the adult in composition, except for the first maxillipeds; these are represented by three-jointed, hook-like structures which are later transformed into a pair of adhesive discs. In certain genera, however, larvae very different in structure from the adult can be discovered.

It is one of the characteristics of the Eucirripedia that their nauplii have a triangular body and carry a horn-like process furnished with secretory glands at either antero-lateral angle of the carapace. The just-hatched nauplius has a median eye and a large labrum; its rounded caudal end soon develops a posteriorly elongated spine (Fig. 10.27a). Another marked peculiarity of this order is that the appendages bear setae which are continuous with their external cuticle. At the first molt a pair of sensory hairs make their appearance at the anterior end, and during several succeeding molts the trunk develops as a linguiform process which arises from the ventral side near the caudal end and bears appendage rudiments (Fig. 10.27b). The carapace of this metanauplius then gradually bends ventrad in the lateral regions and under the external cuticle are formed the rudiments of six pairs of pereiopoda on the trunk, of paired eyes on the head and adhesive discs on the first antennae. When it passes the sixth or seventh molt after hatching, the metanauplius suddenly changes to enter the *cypris* stage, which is so named oJr its outward resemblance to an ostracod (Fig. 10.27c). The carapace is represented by a bivalved shell, with its valves united on the dorsal margin and separate ventrally. From the anterior end protrude the antennules, which bear an adhesive disc containing cement glands on the third segment. The antennae and labrum have degenerated, and the oral parts are also reduced to inconspicuous papilliform processes, whereas the six pairs of pereiopoda show vigorous activity. The four-segmented abdomen is extremely small and without appendages except for a caudal furca. After actively swimming about, the cypris larva attaches itself to some object by means of the adhesive disc on the antennules, and secretes cementing substance around the place of attachment. As the antennules shorten, the carapace opens widely and the larva finally molts (Fig. 10.27d). During this process a fold of integument develops under the carapace, extending anteriorly from the posterior end, so that eventually the carapace maintains connection with the body only in the anterior region, and at right angles to its main axis. Attachment occurs with the larva standing on its head, directing its pereiopoda upward, and the carapace region is now transformed into a mantle which begins to secrete *shell-plates* on its surface (Hudinaga and Kasahara 1942). In *Lepas* and *Mitella* the pre-oral region of the larva elongates to form a *peduncle*; five shell-plates are

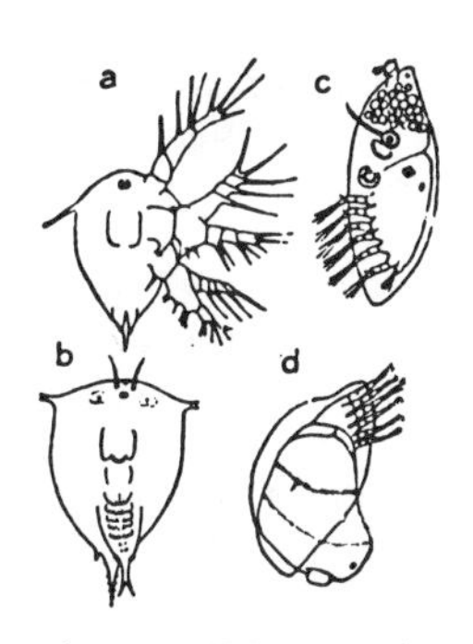

Fig. 10.27 Metamorphosis in *Balanus amphitrite hawaiiensis* (Hudinaga and Kasahara)
a. nauplius; b. metanauplius; c. cypris larva; d. cypris at attachment stage.

secreted on the mantle in the former, and a larger number in the latter. The Balanidae and allied families do not form a peduncle, but attach themselves directly to the substratum by shell-plates which also coalesce with each other. In addition to these, they form two pairs of movable apical shell-plates protecting a mantle aperture which permits the extension and retraction of the appendages. The Rhizocephala, which parasitize various species of prawns, shrimps, crabs, hermit-crabs, etc., retain the nauplius and cypris larval stages in their ontogeny, although their body structure shows an extreme degree of degeneration.

(2) MALACOSTRACA

As has been mentioned several times, the larvae of the Leptostraca and Mysidacea leave the brood-pouch of the mother in practically the adult form. The larvae of the Cumacea, Tanaidacea and Isopoda closely resemble the adult animals when they enter the free-living stage, although the eighth pereiopoda are undeveloped. In *Limnoria* (Isopoda), the larva which has just left the brood-pouch must molt twice before it acquires eighth pereiopods with the complete number of joints and properly disposed accessory setae. Although these appendages emerge with the first molting, they have only three segments in this stage, lack any armature except a small terminal ungulus, and are useless for locomotion since they are close to the ventral surface of the thorax and directed inward. In the first free larva of the Tanaidacea, even the pleopoda are still undeveloped. The adult carapace of *Apseudes,* also an isopod, is firmly attached to the sternum by the downward bending of its lateral regions, so that closed branchial chambers like those of the crabs are formed between carapace and body wall. In the carapace of the young, however, these regions are free from the sternum and expanded laterally like a pair of wings (Lang 1953).

Among the Amphipoda the larvae usually remain in the brood-pouch until just before the maternal molt which precedes the next spawning. They leave the mother resembling her closely in structure, and their subsequent changes are too trivial to be described as metamorphosis, consisting merely in differences of secondary importance such as changes in the shape of the legs and the number of spines, and further development of the sexual characters.

The larva of the Euphausiacea hatches as a nauplius, with an incompletely developed intestine, lacking basal hairs for capturing prey on the second antennae, and without a median eye. The development is gradual, bringing about the change in body shape little by little with each molt, but it is divisible into a number of stages based on certain distinctive features. Each of these stages is again divided into several substages according to the number of molts through which the larva has passed. The first substage (metanauplius) is furnished with three pairs of appendage rudiments, from the maxillulae to the first pereiopoda (Fig. 10. 28a). The mandibles of the later metanauplius, which have lost the swimming function, consist of a masticatory part and a short palp; the carapace projects forward and backward, covering the cephalic region like a hood. The metanauplius is followed by the *calyptopis* larva (Fig. 10.28b). In this stage (corresponding to the protozoëa of the Decapoda to be described below), the cephalic region becomes distinguishable from the trunk, which elongates posteriorly as its segments differentiate in regular sequence. In contrast to the short thorax, which is crowded with segments, the abdomen is longer but devoid of segmentation in the earlier substages. During the period covering several molts, the three pairs of appendages develop further, the first antennae becoming biramous; later the rudiments of the sixth pleopoda make a precocious appearance and a pair of compound eyes begin to develop under the carapace (Fig. 10.28c). A late calyptopis practically corresponds to the decapod zoëa stage, differing in lacking the second pereiopoda.

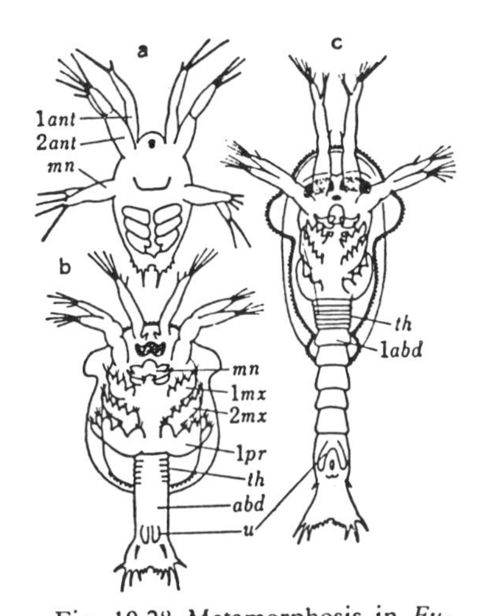

Fig. 10.28 Metamorphosis in *Euphausia* (Korschelt u. Heider)
a. metanauplius; b. early calyptopis stage; c. late calyptopis stage. 1 *abd:* first abdominal segment, 1 *ant:* first antenna, 1 *mx:* first maxilla, 1 *pr:* first thoracic segment, 2 *ant:* second antenna, 2 *mx:* second maxilla, *abd:* abdominal segment *mn:* mandible, *th:* thoracic segment, *u:* uropod.

In the following *furcilia* stage, the compound eyes become movable, emerging from under the carapace, and all the pleopoda are formed (Fig. 10.29a). Only after this do the pereiopoda begin to appear as buds on the thoracic segments, commencing from the more anterior ones. After molting a few more times, the larva attains the stage called *cyrtopia;* this differs from the previous stage in that the first antenna is greatly prolonged and the second, abandoning its swimming function, undergoes a differentiation of the *ramus* into *scale* and *flagellum* (Fig. 10.29b). Through repeated moltings, by which the appendages achieve their final form, the cyrtopia approaches step by step to the adult structure (Fig. 10.31).

Although the larval development of the Decapoda varies according to the suborder, its course is divisible into several stages, the names of which are applicable to all suborders. The body of the *protozoëa,* which follows the metanauplius, is already divided into cephalothorax and abdomen and carries a pair of compound eyes but lacks the pereiopoda posterior to the first two pairs, which are biramous, and does not yet show abdominal segmentation (Fig. 10.31c). The *zoëa* represents an advance over the protozoëa in having distinct abdominal segments (Fig. 10.31d), and the *metazoëa* carries the three anterior pairs of pereiopoda and may also have the rudiments of succeeding pairs (Fig. 10.31e). The *mysis* stage is provided with eight pairs of biramous pereiopoda and six pairs of similarly biramous pleopoda (Fig. 10.31f). In the *macrura* stage, the fourth to eighth pereiopoda have returned to the uniramous form, but the larva is still structurally distant from the adult, and may be given the special name of *post-larva* (Fig. 10.31g). These larval stages show more or less variation according to their suborder and may be omitted in some cases.

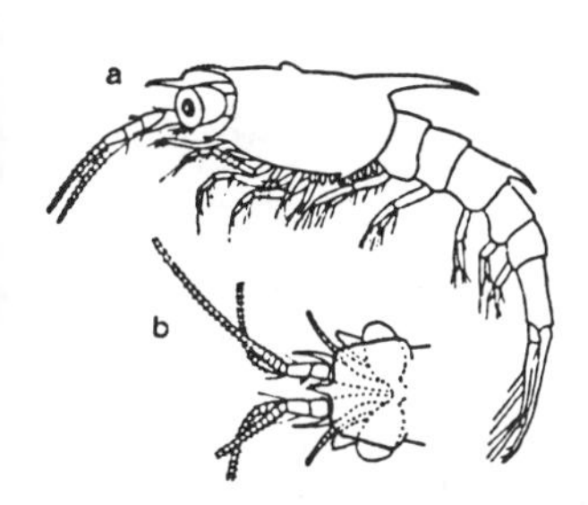

Fig. 10.29 Larva of *Euphausia longirostris* (Tattersall)
a. furcilia larva; b. head part of cyrtopia larva.

Among the Decapoda, *Penaeus* hatches in the nauplius stage, as described above (p. 376). This larva is ovoid, with a median eye, the typical appendages and two caudal setae, but the carapace is not yet developed (Fig. 10.30a). The metanauplius has four additional pairs of appendage rudiments and is furnished with a molar process on the mandibles, the swimming rami of which begin to be reduced. Two molts of the nauplius and five of the metanauplius are followed by the protozoëa stage (Fig. 10.30b). In the anterior part of the body this larva has a distinct cephalothorax, which carries two pairs of pereiopoda as well as the cephalic appendages, and the primordia of the compound eyes can be seen beneath

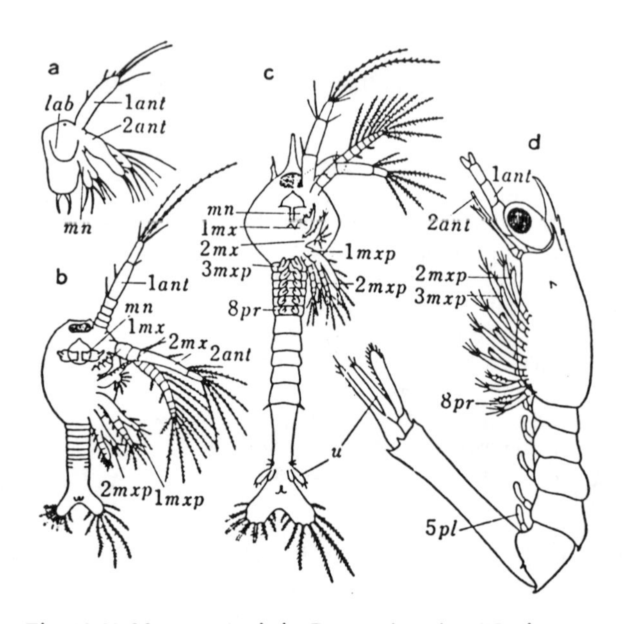

Fig. 10.30 Metamorphosis in *Penaeus japonicus* (Hudinaga)

a. nauplius; b. protozoëa; c. zoëa; d. mysis stage. 1 *ant:* first antenna, 1 *mx:* first maxilla, 1 *mxp:* first maxilliped (first thoracic leg), 2 *ant:* second antenna, 2 *mx:* second maxilla, 2*mxp:* second maxilliped (second thoracic leg), 3 *mxp:* third maxilliped (third thoracic leg), 5 *pl:* fifth pleopod, 8 *pr:* eighth thoracic leg, *lab:* labrum, *mn:* mandible, *u:* uropod.

the integument. The slender trunk bears six distinct pereiomeres, but no metamerism is apparent on the abdomen, although this ends in a caudal furca. With the succeeding molt, five pleomeres and the rudiments of the third pereiopoda make their appearance. One more molt takes the larva to the zoëa stage (Fig. 10.30c), in which the compound eyes are elevated so as to be movable, all the pereiopoda are formed as biramous rudiments and the sixth pleopoda develop precociously, before the more anterior ones. The antennulary peduncle, which was four-jointed in the protozoëa, is now reduced to a single segment; the mandible loses its palp, and a rostrum is formed on the cephalo-thorax. In the following mysis stage, the carapace reaches its full development, covering the pereiomeres down to the eighth; the pereiopoda grow into biramous swimming legs; the rudiments of the first five pleopoda are formed and a sixth unites with the telson to form the *tail fan* (Fig. 10.30d). The antennule consists of a three-jointed peduncle and two flagella; the flagellum and scale of the antenna arise from the inner and outer ramus, respectively; and the mandible regains its palp. After three molts the mysis larva reaches the macrura stage; in this stage the exopodites of the pereiopoda are reduced, and the well developed pleopoda serve as the chief locomotive organs. This is the last larval stage before the attainment of a fully adult structure (Hudinaga 1935).

The allied genus *Sergestes* follows a similar course of larval development, although it hatches as a metanauplius. The carapace in the protozoëa and zoëa

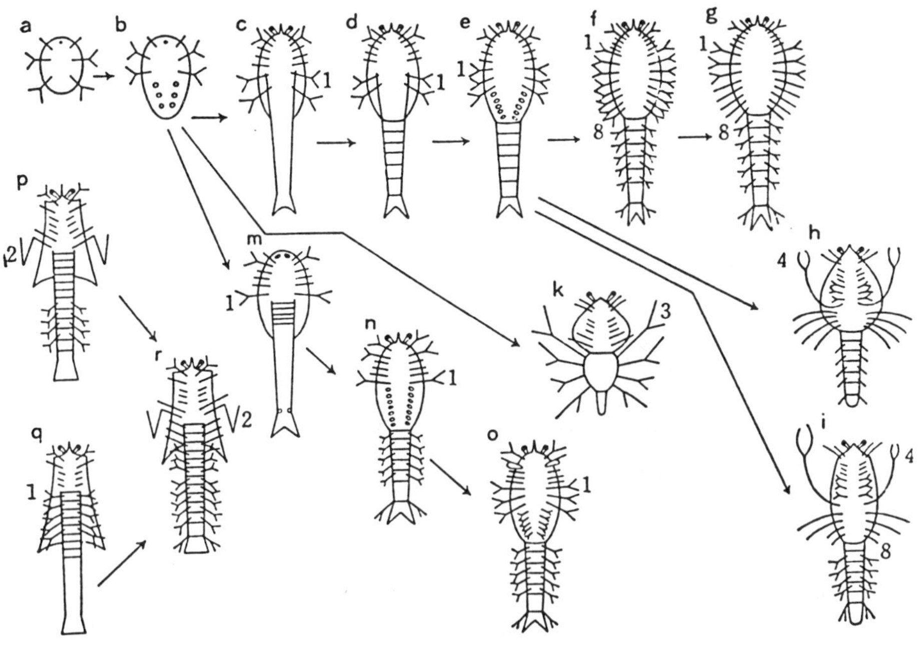

Fig. 10.31 Metamorphosis in several malacostracan larvae (Shiino)

a~g. process of metamorphosis in Natantia; a~e, h. metamorphosis in Brachyura; a~e, i. metamorphosis in Anomura-a, b, k. metamorphosis in Palinura; a, b, m~o. metamorphosis in Euphausiacea; p~r, q~r. metamorphosis in Stomatopoda; a. nauplius; b. metanauplius; c. protozoëa; d. zoëa; e. metazoëa; f. mysis stage; g. macrura stage; h. megalopa; i. glaucothoë; k. phyllosoma; m. calyptopis; n. furcilia; o. cyrtopia; p. alima; q. erichthus; r. synzoëa.

stages is armed with *rostral, supra-ocular, lateral* and *dorso-posterior spines*; all of these as well as the terminal spines on the caudal furca are greatly elongated and thickly covered with fine spicules (Nakazawa 1919). In certain species of the genus, these stages are elaborately provided with double rows of long secondary spinules along the entire length of the rostral and lateral spines. (Their possession of these structures indicates that the larvae are well adapted for pelagic life. This genus pasess from the mysis stage to the adult form by way of the *mastigopus* stage, in which the last two pairs of periopoda disappear provisionally. *Lucifer* never develops beyond this stage, even as an adult.

In the other decapods there is no larva younger than the zoëa. As stated above, this stage begins with only two pairs of biramous pereiopoda and immovable compound eyes. In the Natantia, excluding *Peneaus* and *Sergestes,* the hatched larva has the third pereiopod already developed; it is thus actually a metazoëa, although it is called a zoëa. The carapace is furnished with a rostrum and supra-ocular and antennal spines. In some cases, the hatched larva may be further advanced, with rudiments of more posterior appendages and movable eyes (Fig. 10.32). Except for the Uropoda, which develop well from the early stages, the fourth to the eighth pereiomeres are retarded, while the other appendages develop in regular sequence. In general, the mysis stage lacks the eighth pereiopoda. Deep-sea and fresh-water forms pass through most of the larval stages during their embryonic life. In the Astacura, which similarly undergo direct development, the hatched young lacks the first and sixth pleopoda. The young of *Homarus* emerges from the egg as a mysis larva without any pleopoda.

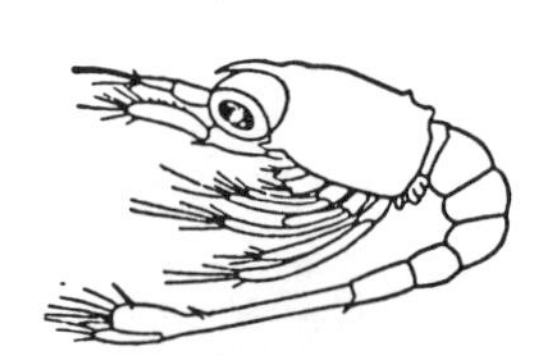

Fig. 10.32 Metazoëa larva of *Spirontocaris spinus* (Lebour)

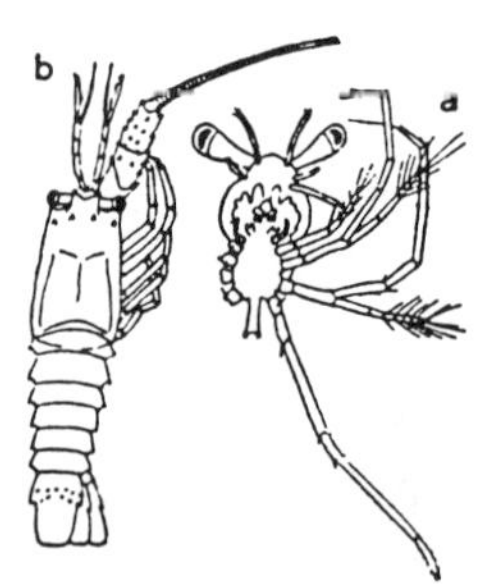

Fig. 10.33 Larva of *Panulirus japonicus* (Nakazawa)
a. phyllosoma stage;
b. puerulus stage.

The Palinura hatch as a peculiar larval form called *phyllosoma* (Fig. 10.31k; 10.33a), which is regarded as a modified early mysis stage. In *Panulirus* this larva has a flattened, disc-like carapace composed of cephalon and first two pereiomeres; the succeeding pereiomeres make up a smaller disc carrying four pairs of biramous pereiopoda, the last of which have rudimentary exopodites. The first and second pereiopoda are short and uniramous. The last two pereiomeres and the abdomen lack appeandages and constititute a small process at the posterior end (Nakazawa 1917). In the allied South African genus, *Jasus*, the phyllosoma stage is preceded by a pre-naupliosoma and a naupliosoma stage, both resembling the phyllosoma externally. The pre-naupliosoma has relatively long antennules and well developed antennae which are biramous and fringed with plumose setae; it lacks setae on the exopodites of the swimming legs. The pre-naupliosoma molts several hours after hatching, and metamorphoses into the naupliosoma, in which the antennae are much smaller and without setae. After about a week, the naupliosoma molts and becomes a phyllosoma larva. Although none of the efforts to rear *Palinurus* larvae have yet been rewarded with success, it is known that before they reach the adult form they must pass through a *puerulus* stage, which is a transparent miniature of the adult, resembling it in body shape as well as in disposition of appendages (Fig. 10.33b).

In the metamorphosis of the Anomura and the Brachyura there is no stage representing the mysis larva. In addition to two pairs of biramous pereiopoda, the zoëa of the Anomura bears the rudiments of a third pair. Its carapace is produced posteriorly into two postero-lateral spines and anteriorly to form an elongate rostrum. The metazoëa has biramous third pereiopoda, and uniramous rudiments of the succeeding pereiopoda and of the pleopoda (Fig. 10.34). To reach the adult form, the metazoëa of *Eupagurus* bypasses the mysis stage and forms a peculiar *glaucothoë* larva (Fig. 10.31i). In this larva, which bears a striking resemblance to the adult, the first three pairs of pereiopoda are transformed into maxillipeds and the succeeding pairs are reduced to uniramous appendages, of which the fourth pair forms a chela on either side, the next two remain elongated, and the last two are reduced in size. The larva differs from the adult only in the possession of a symmetrical abdomen on which the pleopoda are also symmetrically arranged. The zoëa of *Petrolisthes* is remarkable for its possession of a rostrum and postero-lateral spines which attain several times the length of the body.

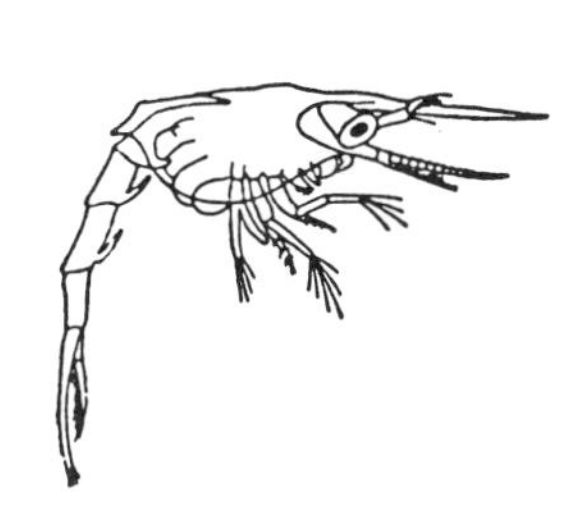

Fig. 10.34 Metazoëa larva of *Eupagurus* (Korschelt u. Heider)

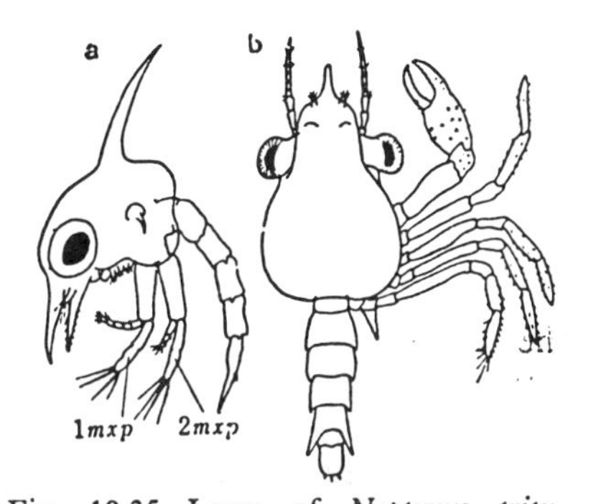

Fig. 10.35 Larva of *Neptunus trituberculatus* (Ohshima)

a. zoëa stage; b. megalopa stage. 1 *mxp:* first maxilliped, 2 *mxp:* second maxilliped.

The zoëa of the Brachyura carries on its carapace a rostrum, a dorso-median spine and a pair of lateral spines, all of which are very sharply pointed (Fig. 10.35a). In the newly hatched larva, these spines are often concealed beneath the cuticle, from which they emerge at the first molt. The larva also frequently hatches as a metazoëa already furnished with the rudiments of the more posterior pairs of pereiopoda. In any case, this group differs from the Anomura in that the third pereiopoda do not develop into swimming legs. The larva next reaches the *megalopa* stage, which has a complete set of appendages morphologically similar to those of the adult. The constitution of the body also resembles that of the adult, but the abdomen extends straight backward, and the five pairs of uniramous walking legs are used for swimming (Figs. 10.35b; 10.31h). The larva of *Neptunus* passes through two molts in the zoëa stage and two more as a metazoëa before reaching the megalopa stage.

The larvae of the Stomatopoda pass through morphologically peculiar stages which are not comparable with those of the other malacostracans; the processes of metamorphosis also differ according to the genus. In *Squilla, Gonodactylus* and some others, the larva hatches in the *alima* or *pseudozoëa* stage (Figs. 10.31p; 10.36). This has stalked compound eyes, biramous antennules, uniramous appendages on the first two pereiomeres, and first to fourth or fifth biramous pleopoda on the abdomen, which is composed of six distinct segments. The first larva of

Lysiosquilla and others, named *erichthus* or *antizoëa* (Fig. 10.31q), has nonpedunculate compound eyes, uniramous antennules, biramous pereiopoda on the first five pereiomeres and an indistinct metamerism of the abdomen, which carries either a single pair of pleopoda or none. Through about five molts, both types of larva reach a stage *(synzoëa)* which has five pairs of maxillipeds, three pairs of pereiopoda, five pairs of pleopoda and one pair of uropoda (Fig. 10.31r), or substantially all the structural elements found in the adult. Molting about ten times during their pelagic life, they develop into the type of young animal characteristic of each species.

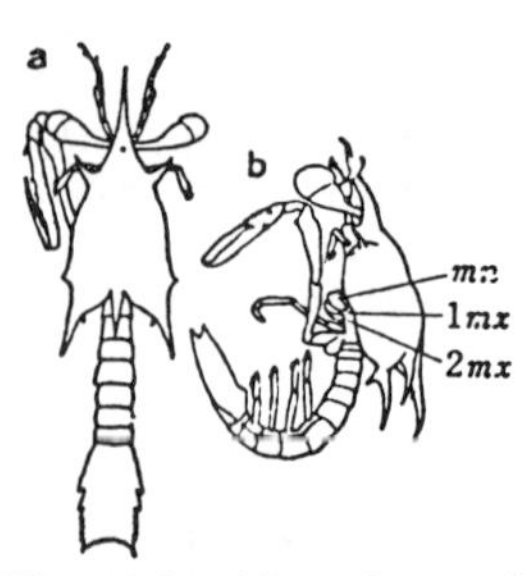

Fig. 10.36 Alima larva of *Squilla oratoria* (Komai and Tung)
1 *mx:* first maxilla, 2 *mn:* second maxilla, *mx:* mandible

The youngest alima larva of *Squilla oratoria* (Fig. 10.36a, b) has, on its more or less quadrate carapace, a spine-like rostrum, a pair of aciform antero-lateral spines, a similar pair of postero-lateral ones and a medio-dorsal one. Four pairs of pleopoda are present and the maxillipeds are well developed, prehensile appendages. In the second larval stage, the rudiments of the fifth pleopoda make their appearance; in the third stage the third and fourth maxillipeds, together with the primordia of the uropoda, are formed. In the fourth stage the pereiopoda and uropoda are nearly completed by the appearance of the last three pereiopod rudiments. With the fifth stage, in which a boundary appears between the sixth pleomere and the telson, the larva becomes a synzoëa. The stages from the sixth to the ninth are distinguishable from one another according to the subsequent changes in body shape and the degree of growth of the appendages. The outline of the adult body is complete under the cuticle of this last larva (Komai and Tung 1929).

Figure 10.31 illustrates diagrammatically the larval forms and the courses of metamorphosis in the Euphausiacea, the suborders of the Decapoda, and the Stomatopoda.

REFERENCES

Aiyer, R. P. 1949: On the embryology of *Palaemon idae* Heller. Proc. Zool. Soc. Bengal, **2** : 101—147.

von Baldass, F. 1937: Entwicklung von *Holopedium gibberum*. Zoöl. Jahrb., Abt. Anat. Ontog., **63** : 399—453.

Bergh, R. S. 1894: Beiträge zur Embryologie der Crustaceen. II. Die Drehung des Keimsreifens und die Stellung des Dorsalorgans bei *Gammarus pulex*. Zoöl. Jahrb., Abt. Anat. Ontog., **6** : 491—528.

Bigelow, M. A. 1902: The early development of *Lepas*. A study of cell-lineage and germ-layers. Bull. Mus. Harvard, **40** : 60—144.

von Bond, C. 1936: The reproduction, embryology and metamorphosis of the Cape Crawfish, *Jasus lalandii* (Milne-Edwards) Ortmann. Inves. Rep. Dept. Fish. Mar. Biol. Surv. Div., Union of South Africa, **6** : 1—25.

Brooks, W. K. & F. H. Herrick 1880: The embryology and metamorphosis of the Macrura. Proc. Nat. Acad. Sci., 4th Mem., **5** : 325—577.

Buller, J. E. 1879: On the development of the parasitic Isopoda. Phil. Trans. Roy. Soc. London, **169** : 505—521.

Butschinsky, P. 1893: Zur Embryologie der Cumaceen. Zoöl. Anz., **16** : 386—387.

———— 1894: Zur Entwicklungsgeschichte von *Gebia littoralis*. Zoöl. Anz., **17** : 253—256.

Calman, W. T. 1909: A treatise on zoology. E. R. Lankester (ed), Pt. 7, Appendiculata, Fasc. **3**, Crustacea. London.

Cannon, H. G. 1924: On the development of an Estherid crustacean. Phil. Trans. Roy. Soc. London, **212** : 395—430.

Eaxon, W. 1883: Selections from embryological monographs. I. Crustacea. Mem. Mus. Comp. Zool., Harvard Coll., **9**.

Fuchs, K. 1913: Die Zellfolge der Copepoden. Zoöl. Anz., **62**.

Goodrich, A. L. 1939: The origin and fate of the entoderm elements in the embryology of *Porcellio laevis* Latr. and *Armadillidium nasatum*. J. Morph., **63** : 401—423.

Grobben, C. 1879: Die Entwicklungsgeschichte der *Moina rectirostris*. Arb. Zoöl. Inst. Wien, **2** : 1—66.

——— 1881: Entwicklungsgeschichte von *Cetochilus septentrionalis*. Arb. Zoöl. Inst. Wien, **3** : 1—40.

Hanaoka, T. 1952: On the nauplii of free-living copepods. Rep. Bur. Fish. (Naikai Ku), **1** : 1—36. (in Japanese)

Heidecke, P. 1903: Untersuchungen über die ersten Embryonalstadien von *Gammarus locusta*. Inaug.-dissert., 1—48.

Hiraiwa, Y. K. 1936: Studies on a bopyrid *Epipenaeon japonica* Thielemann. III. Development and life-cycle, with special reference to the sex differentiation in the bopyrid. J. Sci. Hiroshima Univ., B Div. 1, **4** : 101—141.

Horasawa, Y. 1929: Metamorphosis of *Diaptomus chaffanjoi* Richard. 1,2. J. Nat. Hist. Soc., **38, 39**. (in Japanese)

Hudinaga, M. 1935: Studies on *Penaeus*. I. Development of *Penaeus japonicus* Bate. Rep. Hayatomo Fish. Inst. , **1** : 1—51. (in Japanese)

——— 1940-1a: On the nauplius stage of *Penaeopsis monoceros* and *P. affinis*. Proc. Sci. Fish. Assoc., **8** : 282—289. (in Japanese)

——— 1940-1b: Early development of *Balanus amphitrite albicostatus*. Proc. Sci. Fish. Assoc., **8** : 298—299. (in Japanese)

——— 1942: Reproduction, development and rearing of *Penaeus japonicus* Bate. Jap. J. Zool., **10** : 305—392.

——— & K. Kasahara 1942: Culture and metamorphosis of *Balanus amphitrite*. Zool. Mag., **54** : 108—118. (in Japanese)

Ishikawa, C. 1885: On the development of a freshwater macrurous crustacean, *Atyephira compressa* de Haan. Quart . J. Micr. Sci., **25** : 391—428.

——— 1902: Über das rhythmische Auftreten der Furchungslinie bei *Atyephira compressa* de Haan. Arch. Entw.-mech., **15** : 535—542.

Kajishima, T. 1950: The development of *Leander pacificus* Stimpson. II. Early development. Zool. Mag., **58** : 82—86. (in Japanese)

——— 1950: The development of *Leander pacificus* Stimpson. III. Late embryonic period. Zool. Mag., **58** : 108—111. (in Japanese)

——— 1950: Mechanism of germ-layer differentiation in *Leander pacificus* Stimpson, Exp. Morph., **6** : 19—38. (in Japanese)

——— 1951: Development of isolated blastomeres of *Penaeus japonicus* Zool. Mag. **60** : 258—262. (in Japanese).

——— 1951: Development of isolated 1/2 blastomeres of *Balanus*. Zool. Mag., **61** : 18—21. (in Japanese)

——— 1952: Experimental studies on the embryonic development of the isopod crustacean *Megaligia exotica* Roux. Annot. Zool. Japon **25** : 172—181.

Kamoashi, S. 1914: Freshwater shrimps and their larvae. Zool. Mag., **26** : 183—187. (in Japanese)

Kinoshita, T. 1934: Puerulus larva of *Leander pacificus* and the following metamorphosis. Zool. Mag., **46** : 391—399. (in Japanese)

Komai, T. 1919: Spermatozoon and development of *Squilla oratoria*. Zool. Mag., **32** : 399—400. (in Japanese)

——— 1924: Development of *Squilla oratoria* de Haan. I. Change of external form. Mem. Coll. Sci. Kyoto Imp. Univ., Ser. B, **1** : 273—283.

——————— & Y. M. Tung 1929: Notes on the larval stages of *Squilla oratoria* with remarks on some other stomatopod larvae found in the Japanese seas. Annot. Zool. Japon., **12** : 187—219.

Korschelt, E. & K. Heider 1902: Lehrbuch der vergleichenden Entwicklungsgeschichte der wirbellosen Tiere. Jena.

——— & ——— 1936: Vergleichende Entwicklungsgeschichte der Tiere. Jena.

Lang, K. 1953: The postmarsupial development of the Tanaidacea. Ark. f. Zoöl., ser 2,**4**: 409—422.

Lebour, M. V 1937: The newly hatched larva of *Spirontocaris spinus* (Sowerby) var. *lilljeborgi* Denielssen. J. Mar. Biol. Assoc., **22** : 101—104.

Lebedinsky, J. 1890: Einige Untersuchengen über die Entwicklungsgeschichte der Seekrabben. Biol. Zentralbl., **10** : 178—185.

Manton, S. M. 1928: On the embryology of a mysid crustacean, *Hemimysis lamornae*. Phil. Trans. Roy. Soc. London, ser B, **216** : 363—462.

——————— 1933. On the embryology of the crustacean *Nebalia bipes*. Phil. Trans. Roy. Soc. London, ser. B, 223: 163—238.

Mayer, P. 1877: Zur Entwicklungsgeschichte der Dekapoden. Jena. Zeitschr., **11** : 188—269.

McMurrich, J. P. 1895: Embryology of the isopod Crustacea. J. Morph., **11** : 63—154.

Müller, K. 1913: Über die Entwicklung von *Cypris incongruens*. Zoöl. Jahrb., **36**.

Nair, K. B. 1939: The reproduction, oogenesis and development of *Mesopodopsis orientalis* Tatt. Proc. Ind. Acad. Sci., **9** : 175—223.

——————— 1949: The embryology of *Caridiana laevis* Heller. Proc. Ind. Acad. Sci., **29** : 211—288.

Nakazawa, K. 1917: Metamorphosis in the spiny lobster: On the ecology of the attached larvae. Zool. Mag., **29** : 259—267. (in Japanese)

——————— 1919: The embryology of *Sergestes lucens*. Zool. Mag., **31** : 141—212. (in Japanese)

Needham, A. E. 1937: Some points in the development of *Neomysis vulgaris*. Quart. J. Micr. Sci., **79** : 559—588.

Ohshima, N. 1938: A survey of the Inland Sea *Neptunus*. J. Imp. Fish. Expt. Sta., Tokyo, **9** : 141—212. (in Japanese)

Pike, R. B. 1947: *Galathea*. L. M. B. C. Memoirs. London-Liverpool.

Reichenbach, H. 1886: Studien zur Entwicklungsgeschichte des Flusskrebses. Abh. Senckenberg. Naturf. Ges., **4** : 1—137.

Shiino, S. M. 1942: Studies on the embryology of *Squilla oratoria* de Haan. Mem. Coll. Sci. Kyoto Imp. Univ. Ser. B, **17** : 77—174.

——————— 1950: The embryonic development of *Panulirus japonicus*. Rep. Coll. Fish., Mie Univ., **1** : 1—168. (in Japanese)

Taube, E. 1909: Beiträge zur Entwicklungsgeschichte der Euphausiden. I. Die Furchung des Eies bis zur Gastrulation. Zeitschr. wiss. Zoöl., **92** : 427—262.

——————— 1915: Beiträge zur Entwicklungsgeschichte der Euphausiden. II. Von der Gastrulation bis zum Furciliastadium. Zeitschr. wiss. Zoöl., **114** : 577—656.

Terao, A. 1914: On the development of *Panulirus japonicus*. Zool. Mag., **26** : 473—476. (in Japanese)

——————— 1919: Embryology of *Panulirus japonicus*. J. Imp. Fish. Expt. Sta. Tokyo, **14** : 1—79. (in Japanese)

——————— 1929: The embryonic development of the spiny lobster *Panulirus japonicus* (von (Siebold). Jap. J. Zool., **2** : 287—449.

Wilson, E. B. 1924: The Cell in Development and Heredity. New York.

Yasugi, R. 1937: The free-swimming larva of *Mitella mitella* (L.) Bot. and Zool., **5** : 792—796 (in Japanese)

Yokoya, Y. 1931: On the metamorphosis of two Japanese freshwater shrimps, *Paratya compressa* and *Leander paucidens*, with reference to the development of appendages. J. Coll. Agric. Imp. Univ. Tokyo, **11** : 75—150.

Zehnder, H. 1934: Über die Embryonalentwicklung des Flusskrebses. Acta Zool. Stockholm **15** : 261—408.

II. ARACHNIDA

INTRODUCTION

The Araneinae are classified into three sub-orders: Liphistiomorphae, Mygalomorphae and Arachnomorphae. For a long time the materials used for embryological studies of these groups were limited to the species belonging to the Arachnomorphae, as represented by *Agelena*. The only embryological studies of Mygalomorphae were those of Montgomery (1909), and of L. W. Shimkewitsch (1911). Even among the Arachnomorphae, only a very few of the many species included in the group have served as objects of study. As a result, this field contains many conspicuous blank spots.

Recently, however, Holm (1954) has been proceeding to fill in some of these blanks by investigating the development of an orthognath spider, *Ischnothele karschi*, and Yoshikura (1954, 1955) has studied for the first time the development of *Heptathela kimurai*, which belongs to the Liphistiidae. The works of these two investigators are considered to have made significant contributions to the general knowledge of spider embryology.

Some methods, the knowledge of which is essential in studying the embryology of spiders, are described as follows.

The chorion of the spider egg is an opaque membrane (see below) which obscures the interior of the egg under ordinary conditions. However, when the eggs are immersed in liquid paraffin or glycerine, the processes of development going on inside the chorion can easily be followed with a binocular microscope. This method was early used by Herold (1824), and by a few other investigators at the end of the century, but their failure at that time to obtain satisfactory results, as well as the introduction of sectioning, caused the immersion method to fall into disuse until recently, when it was taken up independently by Holm, Monterosso, Yoshikura and the author. The method consists in immersing several eggs in a few drops of liquid paraffin in a hollow slide (ca. 2 mm deep) and observing tehm with a binocular microscope. In this quantity of liquid paraffin, embryonic development usually proceeds to the prehatching stage.

As fixatives for spider eggs, Carnoy's fluid, alcohol-Bouin's and Petrunkevisch's fluids are listed. Carnoy's fluid in particular has been used by many investigators, while Yoshikura recommends the use of a mixture of 15 parts absolute alcohol, 5 parts formalin and 1 part glacial acetic acid. Since it is very difficult to obtain good sections through the eggs of the early stages because of the large amount of yolk which they contain, it is necessary to elaborate a plan to suit each case. For total observation, the fixed embryo is readily and satisfactorily stained with Ehrlich's hematoxylin. To make preparations for microscopical studies, the embryos are overstained with borax carmine for 5—6 days, differentiated slowly with 1% acid alcohol, and then sectioned. Good results have been obtained by using toluol in the embedding process; the paraffin block is cut until the embedded material is exposed at the cut surface and then immersed in water for one day before sectioning.

1. STRUCTURE OF EGGS[1]

The largest diameter found among spider eggs is about 1.8 mm *(Heteropoda, Agelena, Dolomedes)*, while there are many species which have small eggs, about 0.5 mm in diameter. The eggs are usually spherical or oval, but sometimes they show an irregular shape as the result of being crowded together.

There are two egg membranes: an outer *chorion* secreted by the oviduct, and an inner *vitelline membrane* secreted by the egg itself (Fig. 10.37). The irregular surface of the chorion makes the membrane opaque like frosted glass, so that it is difficult to see inside the egg in air; the vitelline membrane, however, is a thin, elastic, transparent structure.

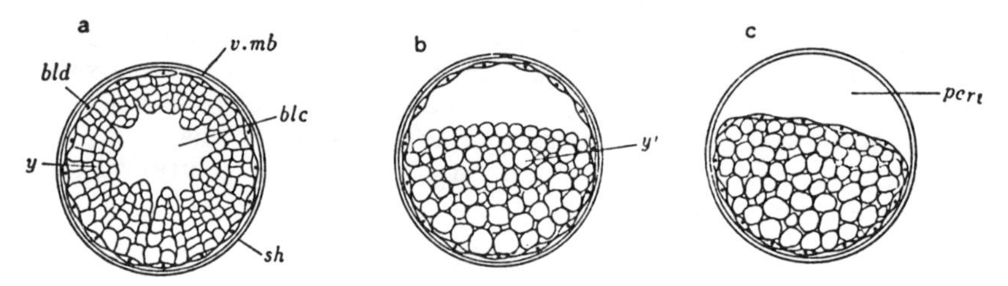

Fig. 10.37 Schematic sketches of longitudinal sections of spider eggs, showing egg structure and contraction of blastula (Sekiguchi)

blc: blastocoel, *bld:* blastoderm, *perv:* perivitelline space, *sh:* shell, *v.mb:* vitelline membrane, *y:* yolk granule, *y':* yolk mass segregating from blastoderm.

Within the vitelline membrane is a large yolk mass consisting of small, globular yolk granules and large irregular *yolk bodies* formed of accumulations of yolk granules. The eggs immediately after deposition have only yolk granules, but within a few hours the yolk bodies appear, especially at the underside of the egg.

According to Holm (1952), at the time of oviposition the nucleus is situated between the center and the circumference of the egg; after completion of the reduction divisions the pronucleus migrates to the center of the yolk mass. The cytoplasm consists of the *centroplasm,* which contains the egg pronucleus, the *periplasm,* which envelops the yolk mass, and the *protoplasmic reticulum,* which connects the centroplasm with the periplasm through the yolk mass, forming a complicated delicate network between the yolk granules.

[1] Although the breeding seasons of spiders vary somewhat with the species, they generally extend from June to October in Japan. The eggs are usually deposited together with a drop of fluid on a specially prepared mat of silk, and covered with more threads to form a cocoon, or egg sac. The fluid evaporates rapidly, leaving whitish crystalline particles on the surfaces of the eggs.

2. FERTILIZATION, CLEAVAGE AND BLASTODERM FORMATION

Previous investigators have agreed that the sperm and egg pronuclei unite 3 or 4 hours after oviposition. The tempo of the subsequent developmental processes differs according to species and temperature. The following description will be based chiefly on Holm's investigation (1952) of the egg of *Agelena labyrinthica*, developing at a temperature of 22°C.

In this spider the first nuclear cleavage takes place 6 hours after oviposition; as the subsequent cleavages occur, the nuclei, surrounded by cytoplasm derived from the protoplasmic reticulum, migrate slowly between the yolk bodies to the surface of the egg. Sixteen hours after oviposition, the nuclei, their number increased to 8, have reached a position midway between the center and the periphery of the egg. The segmentation of the yolk accompanies these nuclear cleavages; the yolk bodies, arranged radially from center to periphery, compose *yolk columns*, and several of these join to make a *yolk pyramid*. At the center of the egg, a small cavity, enclosed within the central ends of the yolk pyramids, forms a blastocoel which enlarges as the yolk is consumed (Fig. 10.38 b,c). At the 64-nucleus stage (after the 6th cleavage), the nuclei approach the surface of the egg, and the cyto-

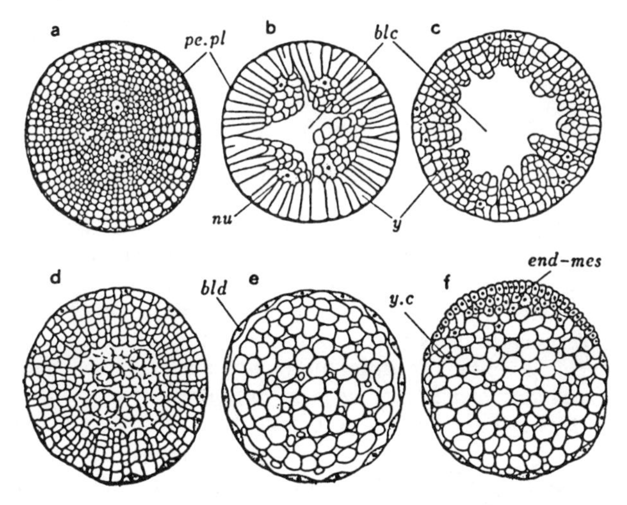

Fig. 10.38 Cleavage and blastulation in the spider *Theridium maculatum* (a~e. Morin; f. Dawydoff)

blc: blastocoel, *bld:* blastoderm, *end-mes:* endo-mesoderm cell, *nu:* cleavage nucleus, *pe.pl:* peripheral cytoplasm, *y:* yolk pyramid, *y.c:* yolk cell.

plasm enveloping each nucleus extends pseudopodia and establishes connection with the periplasm[2]. Eventually the nuclei reach the surface of the egg (Fig. 10.38c,d; 10.39b); previous investigations have generally agreed that the nuclei arrive at the egg surface when their number reaches 60—80.

At the surface, the nuclei continue to cleave once or several times more, while they form a layer and envelop the surface of the egg. This layer is the *blasto-*

[2] If the egg is observed under a binocular microscope with strong direct illumination, the cleavage nuclei can be recognized before they reach the surface of the egg.

derm, and with its completion, 35 hours after oviposition, the embryo enters the blastula stage (Figs. 10.37a; 10.38e; 10.39c).

There is a question as to whether, at the time of the general migration of cleavage nuclei to the surface, some of them remain in the yolk mass as the *primary yolk cells*. According to Holm, a few primary yolk cells can be recognized in the yolk mass before the formation of the blastoderm. However, as the number of yolk cells increases rapidly after the formation of the blastoderm, he deduces the presence of *secondary yolk cells* which penetrate into the yolk mass from the blastodermal layer.

Soon after the formation of the blastoderm, the whole yolk mass contracts, and the yolk granules of the upper side of the egg separate from the blastoderm and sink downward, obliterating the blastocoel (Fig. 10.37b). Accompanying this contraction, the fluid of the blastocoel permeates between the blastoderm and the yolk surface. It is considered that the contraction is due to shrinkage of the protoplasmic reticulum between the yolk pyramids.

The blastoderm then begins to contract, separates from the vitelline membrane and sinks slowly to the surface of the yolk mass (Fig. 10.37c). As the result of the sinking of the blastoderm, the fluid which accumulated between it and the yolk mass passes through the blastoderm and comes to lie under the vitelline membrane as the *vitelline fluid*. The embryo in this stage thus appears as a hemisphere with the upper surface flattened, and the cavity formed between it and the vitelline membrane contains a large amount of vitelline fluid.

3. GERM LAYER FORMATION

Soon after this, small cells gather into a white ring[3] on the upper surface of the egg, which will later become the ventral side of the embryo (Fig. 10.39d). The ring is then filled in to form a round plate, the *germ disc*, in which many small cells are densely distributed (75 hrs after oviposition). Ten hours later, a white spot consisting of several layers of cells appears in the center of the thin germ disc. This spot has been given many different names such as *procephalic lobe, Blastodermverdickung, cumulus primitivus, cumulus anterior, Primitivplatte*, etc. Here we shall call it the *primary thickening*, according to Kishinouye (1891).

A shallow groove of irregular shape which appears in the center of this thickening is the *blastopore*. According to the results of local vital staining studies by Holm, most of the cells forming the early germ disc invaginate inward through the blastopore, and spread along the inner side of the germ disc. He also observed that the cells which invaginate early become endoderm, whereas those invaginating later become mesoderm[4] (cf. pp. [10—11]). The process is therefore a sort of gastrulation, which results in the formation of the primary thickening. As development proceeds, the germ disc spreads to cover the whole upper surface of the yolk mass.

[3] As Holm has pointed out, the white ring is not often recognizable in species other than *Agelena*. In most cases, a white spot appears on the egg surface and develops directly into the germ disc.

[4] Opinion is divided on this question. According to Montgomery (1909) and others, the cells invaginated through the blastopore first become undifferentiated mesentoderm; then some of them enter the yolk to become yolk cells, later transforming into endoderm. According to Yoshikura, the mesentoderm is formed "by vertical mitosis, or by the inrolling of the blastoderm cells", not only in the blastoporic region, but in every part of the germ disc.

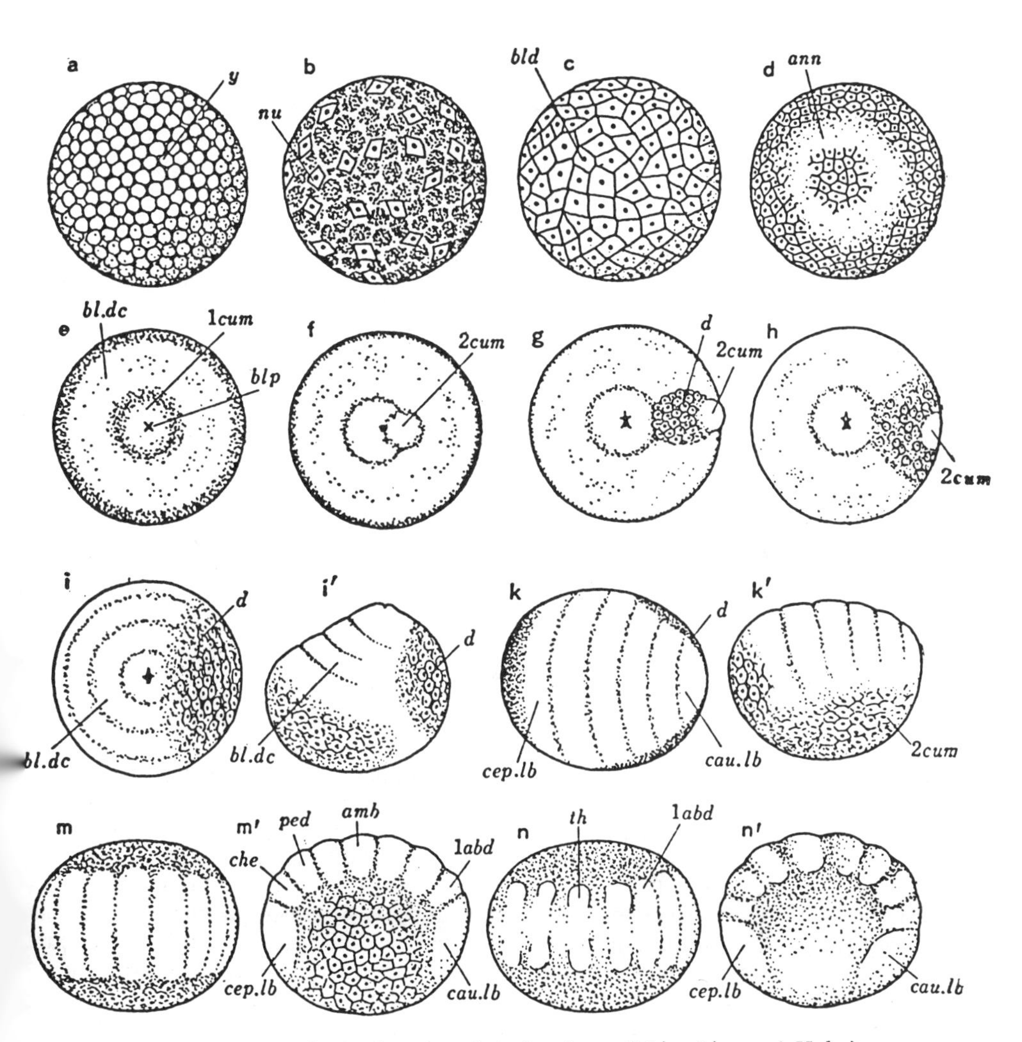

Fig. 10.39 Stages in development of *Agelena* (a~c. Sekiguchi; c~n'. Holm)

d~n. ventral views of embryo; i'~n . lateral views. Outermost edge of sketches a~c represents edge of vitelline membrane. Vitelline membrane omitted in sketches d~n'. 1 *abd:* first abdominal segment, 1 *cum:* primary thickening-2 *cum:* secondary thickening, *amb:* ambulatory leg, ann: annular zone, *bld:* blastoderm, *bl.dc:* blastodisc, *blp:* blastopore, *cau.lb:* caudal lobe. *cep.lb:* cephalic lobe. *che:* cheliceral segment, *d:* dorsal area, *nu:* cleavage nucleus at surface of v *ped:* pedipalpal segment. *th:* thoracic extremity appendage. *y:* yolk granule.

The primary thickening at first does not extend above the egg surface, but within a short time one part of it rises to form a hemispherical protuberance (100 hrs after oviposition) which separates from the primary thickening and migrates toward the margin of the germ disc (Fig. 10.39 f).

This protuberance was first called the *cumulus primitivus* by Claparède (1862); since then, however, many workers have given it a variety of names, such as *cumulus, cumulus posterior, Primitivhügel, caudal thickening,* etc., and it has often been confused with the primary thickening. In order to avoid confusion, we shall follow Kishinouye and call it the *secondary thickening.*

As this secondary thickening migrates away from its original position, its height increases steadily, attaining a maximum about 120 hours after oviposition, when the center of the projection reaches the margin of the germ disc (Fig. 10.39 g).

Several investigators in the past have discussed the significance of the secondary thickening. Recently Holm, by culturing vitally stained embryos in liquid paraffin, found that the secondary thickening migrates along the body axis, from the tail end to the dorsal region[5]. From the fact that the chief role of the cells making up the secondary thickening is the formation of the intestinal wall, he also proved that they are endodermal cells. These and other experimental results convinced him that the migration of the secondary thickening plays a very important part in forming the dorsal area.

GERM-BAND FORMATION

The area left in the wake of the migrating secondary thickening consists of a single layer of large, flat, ectodermal cells. As the migration proceeds, this part extends in the shape of a fan, and when the secondary thickening reaches its maximum development, the fan is widened to include an angle of about 60°, with its apex at the blastopore (Fig. 10.39 h). This is the true dorsal area of the future embryo. The fan continues to open, and about the time when its angle reaches 90°, the secondary thickening gradually flattens into a white patch and is finally dispersed and disappears (Fig. 10.39 h, i).

The dorsal area extends its angle to 180° and continues to widen still further. The germ disc, on the other hand, follows a course of development just opposite to that of the dorsal area: its at first almost circular shape is gradually reduced to a semicircle, and finally it becomes the nearly triangular *ventral plate* (144 hrs after oviposition). Holm calls this phenomenon *inversion* (Fig. 10.39 g,i,k). At first glance the ventral plate resembles a bivalve shell laid inside downward on the yolk surface. The ventral margin of this "shell" consists of a thin cellular layer, while the part corresponding to the umbo is occupied by the cells of the primary thickening, and is elevated to some extent. The ventral margin represents the cephalic end of the embryo, and the elevated part will form its caudal end[6]. Already in this stage several shallow grooves, like growth lines of the shell, somewhat vaguely indicate 3—4 incipient segments. As these segments become definite, the posterior parts of the embryo also widen, and the embryo takes the form of a *germ band* which covers the upper surface of the egg (Fig. 10.39k). Before long the foremost segment of this germ band becomes a semicircular *procephalic lobe*, and the primary thickening of the caudal end forms a *caudal lobe*, which is pushed around to the under surface of the egg as the number of segments increases. Five segments are recognizable between these two lobes at this stage (155 hrs after oviposition). The segment nearest the procephalic lobe is the *pedipalpal segment*; posterior to this are the *first, second, third* and *fourth ambulatory leg segments.*

---

[5] Holm's experiments suggest that the separation and migration of the secondary thickening may be interpreted as the migration beneath the blastodermal cell layer of a group of cells derived from the disc, which takes place independently of the outer layer.

[6] The body axis becomes evident after the formation of the ventral plate; since, however, the body axis coincides with the migration path of the secondary thickening, its direction is apparent as soon as the secondary thickening is formed.

The procephalic lobe next divides into a *cephalic lobe* and a *cheliceral segment*, so that the *cephalothorax* consists of a cephalic lobe and six subsequent segments. Simultaneously with the formation of the cheliceral segment, the *first abdominal segment* is separated from the caudal lobe (Fig. 10.39 m, m'). As these segments appear, the mesoderm differentiates under the ectoderm in coordination with it. In other words, the appearance of the segments indicates that the mesodermal cells have gradually proliferated toward the anterior and ventral parts of the embryo, accumulating into rectangular masses associated with each segment.

Before long, a pair of *appendages* is formed on each of the cephalothoracic segments. The caudal lobe separates off 11 *abdominal segments* and a *telson*, and on each of the second to fifth abdominal segments appears a pair of button-like projections, the rudiments of the *abdominal appendages* (Fig. 10.39n; 10.40a-c).

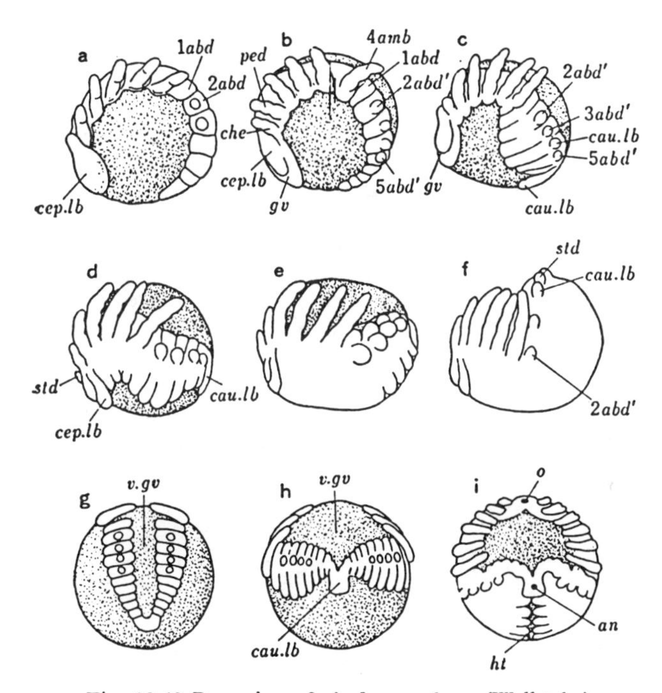

Fig. 10.40 Reversion of *Agelena* embryo (Wallstabe)

a~f. lateral views of several embryonic stages from most extended to early pyriform embryo. Sketches g, h and i are respectively back views of embryos at same stages as a, c and e. 1 *abd:* first abdominal segment, 2 *abd:* second abdominal segment, 2 *abd':* second abdominal extremity appendage, 3 *abd':* third abdominal extremity appendage, 4 *amb:* fourth ambulatory leg, 5 *abd':* fifth abdominal extremity appendage, *an:* anus, *cau.lb:* caudal lobe, *cep.lb:* cephalic lobe, *che:* cheliceral segment, *gv:* cerebral groove, *ht:* heart, *o:* mouth, *ped:* pedipalpal segment, *std:* upper wall of stomodeaum, *v.gv:* ventral groove.

The germ band, from the pedipalpal segments to the posterior end of the abdominal segments, is separated into left and right strips by a shallow median furrow, the *ventral sulcus*, which also divides the mesoderm lying beneath each segment.

These two strips of segments next shift laterally, causing a marked widening of the ventral sulcus; at the same time they elongate toward the dorsal side. This causes the germ band, which has been bent backward around the surface of the yolk mass with its cephalic and caudal ends almost in contact, to rise slowly from the yolk surface, while the two extremities draw away from each other. The abdomen migrates gradually in an antero-ventral direction; the left and right bands of segments extend around the sides to reach the dorsal median line, and then they also extend ventrally and almost completely surround the yolk mass. The embryo thus finally becomes bent forward into the shape of a Dharma (Fig. 10.40a-i). This phenomenon, which is regarded as a specific feature of the developmental process in spider eggs, is called *reversion* (See Supplement).

5. PRESUMPTIVE REGIONS OF SPIDER EGG

Holm (1952), studying in detail the cellular migration on the surface of the *Agelena* egg by the method of local vital staining, has mapped the presumptive organ-forming regions. As he says, the map is approximate and can serve only as a starting point for further study. However, his opinions, supported by the substantial achievement of his experiments, raise some important questions, in the light of which the views of past investigators must be re-examined.

According to Holm, it is clear that a striking cellular migration toward the blastopore occurs after blastoderm formation in *Agelena*, and the germ disc surrounding the blastopore does not represent the rudiment of the embryo as in most animals, but is the part which will invaginate through the blastopore and later become the mesentoderm, while the future ectoderm takes its origin from the blastular cell layer at the periphery of the germ disc. The so-called germ disc in this stage, therefore, does not correspond to the usual germ disc, and the region surrounding it does not represent an extra-embryonic area as in other animals. The true extra-embryonic area of the spider embryo appears only after the embryonic dorsal region has been formed by the separation of the secondary thickening.

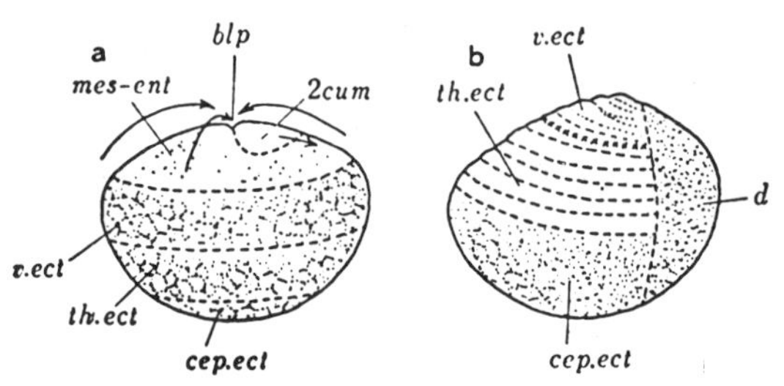

Fig. 10.41 Map of presumptive regions in the *Agelena* egg (Holm)

a. embryo at gastrulation. 2 *cum:* presumptive area of secondary thickening, which invaginates through blastopore (*blp*), changes into mass of entoderm cells and migrates backward (in direction of short arrow), *cep.ect:* presumptive area of cephalic ectoderm, *mes-ent:* presumptive area of mesentoderm, invagination occurs inside periphery of this area, *th.ect:* presumptive area of thoracic ectoderm, *v.ect:* presumptive area of ventral (abdominal) ectoderm (long arrows show direction of migration of mesentoderm). b. embryo just after completion of invagination. At this stage, secondary thickening has already disappeared, mesentoderm has invaginated and ectoderm remains at surface. Presumptive ventral ectoderm lies near blastopore with thoracic ectoderm ventral to it and head ectoderm farthest from blastopore. At this stage of development, presumptive head ectoderm consists of single layer of blastula cells. *d:* dorsal region.

Holm was not able to determine the boundary between endoderm and mesoderm. But as the stained cells which make up the secondary thickening later appear in the intestinal wall and constitute the bulk of the abdominal endoderm, he concluded that the cells in the anterior half of the primary thickening which invaginate first similarly become the endoderm of the cephalo-thorax, while the cells invaginating later form the mesoderm.

These results are incorporated into the presumptive map of the early gastrula stage shown in Figure 10.41a. The true germ disc at this stage is seen to consist of that part of the presumptive secondary thickening which spreads out over the prospective dorsal area (*2 cum*), together with the remaining area of presumptive mesentoderm *(mes-ent)* surrounding it. The germ disc is enclosed within an area of presumptive ectoderm. In this stage the boundary between the presumptive mesentoderm and presumptive ectoderm corresponds roughly to the edge of the germ disc. The blastodermal layer outside the germ disc is divided into three areas; the *cephalic ectoderm (cep. ect)*, which lies on the opposite side of the egg, farthest from the blastopore; an annulate area of *thoracic ectoderm (th. ect)* surrounding the former; and the *abdominal ectoderm (v. ect)*, which lies between the germ disc and the thoracic ectodermal area.

A map of the presumptive regions at the stage immediately after invagination, in which the dorsal area covers about 180° of the surface posterior to the germ disc, is shown in Figure 10.41b. In this map, the margin of the germ disc corresponds roughly to the boundary between the cephalic lobe *(cep. ect)* and the anterior margin of the thoracic cheliceral segment *(th. ect)*. Next to the cheliceral segment, the pedipalpal segment and four ambulatory leg segments are arranged in concentric arcs, and posterior to these there is an abdominal region also consisting of concentric arcs which have as their center a point slightly posterior to the blastopore. The extent of this abdominal area corresponds roughly with that of the primary thickening.

After this stage, the conspicuous axial elongation of the germ band and simultaneous contraction, especially of its anterior part, in a direction perpendicular to the body axis, take place as has been described above. According to Holm's experiments, the original position of the blastopore corresponds to the sixth abdominal segment in the germ-band stage.

6. ORGAN FORMATION

Accompanying these external developments, some important organ-forming processes have been going on inside the egg. The mesoderm underlying the ectoderm of each segment increases independently to form a *coelomic sac*. As the reversion of the embryo proceeds, these sacs grow laterally to enwrap the yolk mass from the sides. The cavities of the coelomic sacs later disappear and their walls become muscles, while their ends which extend from the two sides toward the back of the embryo grow to reach the dorsal median line. There they meet to form the *heart* in the abdominal region and the *dorsal aorta*, continuous with the heart, in the cephalothoracic region[7] (Fig. 10.40 i; 10.43).

[7] Opinion with respect to the origin of the blood cells is divided. Kishinouye, Kautsch and others claim that they are derived from the yolk cells; Morin, Schimkewitsch and others advocate a mesodermal origin, while Montgomery derives them from the ectoderm. Recently Yoshikura has found that the blood cells originate in the ectoderm which constitutes the carapace at the stage of brain formation in *Heptathela*.

Each coelomic sac which has developed in a segment bearing appendages forms a projection into the base of the appendage; these cavities also disappear in time and the tissues transform into muscles. However, in the four segments bearing the ambulatory legs, small saccular vestiges which remain in the base of each segment later unite with each other in a row on each side. These open into a pair of ectodermal ducts which are formed by invagination at the bases of the first ambulatory legs, to form the *coxal glands*.

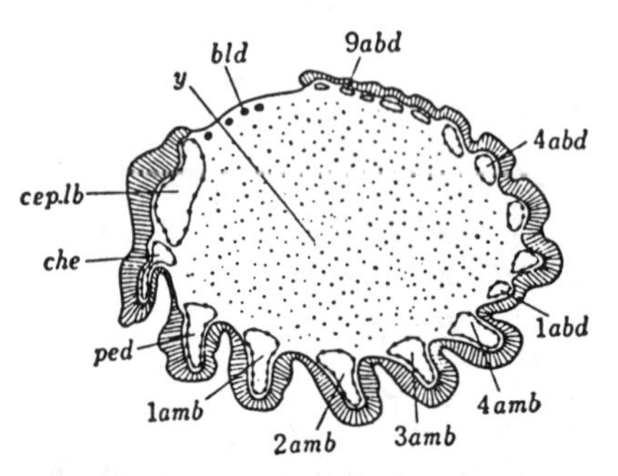

Fig. 10.42 Longitudinal section of *Agelena* embryo at extended stage (Ventral side of embryo at lower side of sketch) (Wallstabe)

1 *abd:* first abdominal segment, 1 *amb:* first ambulatory leg, 2 *amb:* second ambulatory leg, 3 *amb:* third ambulatory leg, 4 *amb:* fourth ambulatory leg, 4 *abd:* fourth abdominal segment, 9 *abd:* ninth abdominal segment, *bld:* blastoderm, *cep.lb:* cephalic lobe, *che:* chelicera, *ped:* pedipalp (structures indicated by reference lines in each segment are coelomic sacs), *y:* yolk.

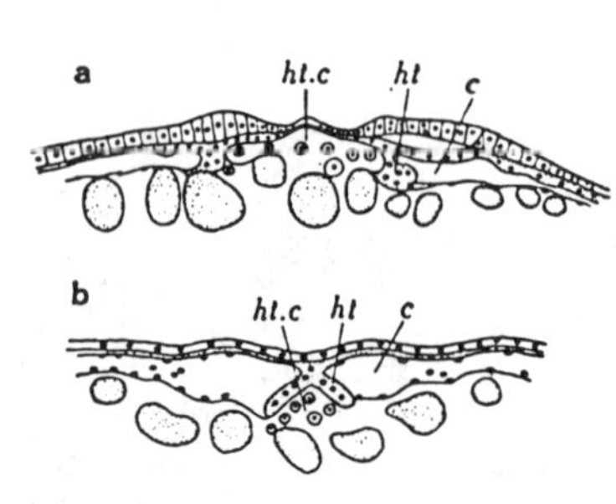

Fig. 10.43 Development of heart in *Agelena* embryo (Transverse sections) (Kautsch)

Wall of heart formed by thickenings of wall of coelom which grows from sides toward median dorsal line. *c:* coelomic sac, *ht:* heart wall, *ht.c:* heart cavity.

About the time when the embryo achieves its maximum length, the cephalic lobe has the appearance of two semicircular plates laid side by side; invagination of the ectoderm now takes place along their margins, forming the *cerebral grooves* (Fig. 10.40b, c *gv*). The grooves become deeper, and their walls thicken and acquire a complex structure to form the main parts of the *brain*, or *supra-oesophageal ganglia*. On the other hand, paired ectodermal thickenings appear on the two sides of the median ventral line, in each of the segments lying posterior to the cephalic lobe. These thickenings later separate from the ectoderm, become con-

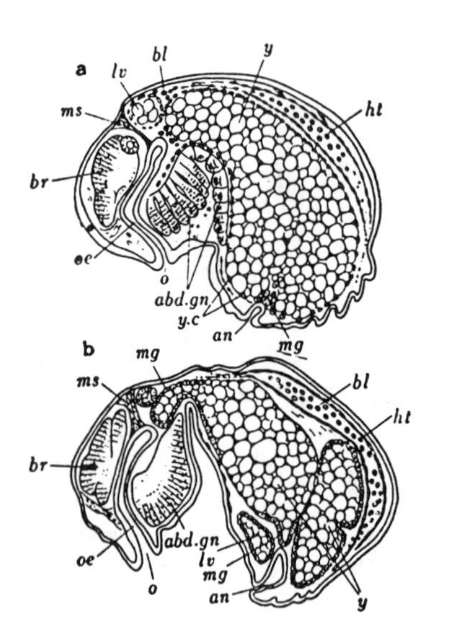

Fig. 10.44 Sagittal sections through embryos of the spider *Theridion maculatum* in succeeding stages of organ formation, showing migration of ganglia and formation of digestive tract (Morin)

abd.gn: abdominal ganglion, *an:* anus, *bl:* blood corpuscle, *br:* brain, *ht:* heart, *lv:* liver, *mg:* mid-gut, *ms:* muscle, *o:* mouth, *oe:* oesophagus, *y:* yolk granule, *y.c:* yolk cell.

nected with each other by a nerve cord, and form the *ventral nerve chain*. In time the ganglion of the cheliceral segment unites with the supra-oesophagial ganglion, while the ganglia of the other segments migrate forward as development proceeds and come together into a mass in the cephalothorax before hatching, as the sub-oesophagial ganglion (Fig. 10.44).

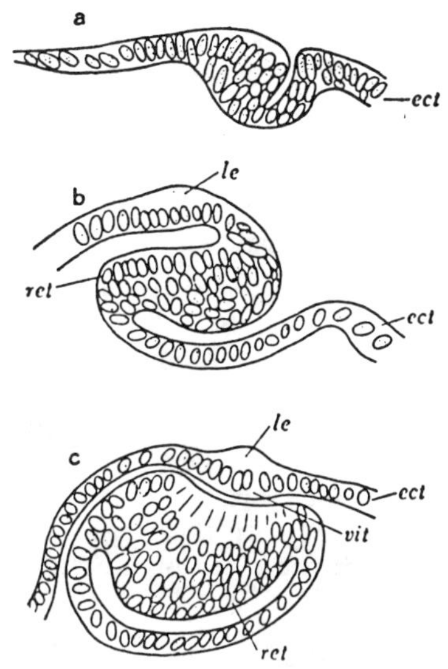

Fig. 10.45 Development of central eye in *Agelena* embryo (Locy)
ect: ectoderm, *le:* lens, *ret:* retina, *vit:* vitreous body.

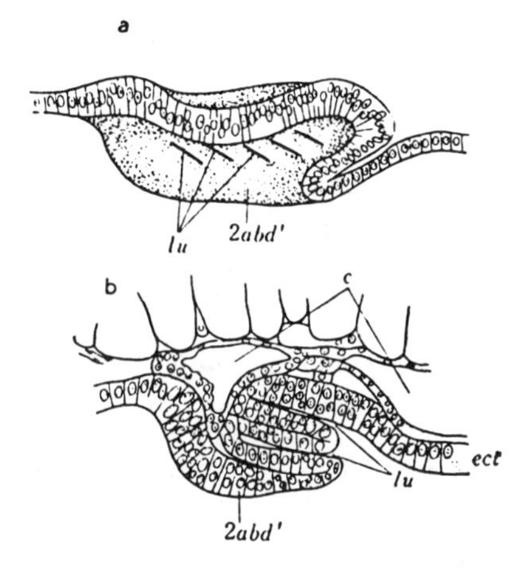

Fig. 10.46 Lung sac (Purcell)
a. inner side, viewed from posterior; b. cross-section. 2*abd'*: second abdominal extremity appendage, *c:* coelomic sac, *ect:* ectoderm, *lu:* infolding of lung sac wall.

The eyes of spiders consist of a pair of *central eyes* and three pairs of *lateral eyes*. The central eyes are formed first, behind the cephalic lobes near the median ventral line, as invaginations of the ectoderm closely connected with the brain. As the result of a complicated process of infolding, the ectoderm becomes double-layered and differentiates into a *lens* and a *retina* (Fig. 10.45). The lateral eyes appear in the dorso-lateral regions of the cephalic lobes as simple, ring-like hollows of the ectoderm; they differ from the central eyes in that the bottom of each hollow is transformed directly into a retina.

At about the time when the reversion of the embryo occurs, marked changes also take place in the abdominal appendages. Those of the second and third abdominal segments become flattened, and ectodermal invaginations appear in the posterior parts of their bases. These are the rudiments of the *lung sacs*; in the inner wall of each sac arise many folds which develop into the *book lungs* (Fig. 10.46). Two pairs of such book lungs are formed in the tetrapneumon spiders, represented by the Liphistiidae. In the Arachnomorphae *(Agelena)*, however, only a single pair of these structures is formed in the second abdominal segment, while the invaginations of the appendages on the third abdominal segment become *tracheae*.

Each appendage of the fourth and fifth abdominal segments divides into an *endopodite* and an *exopodite*. In the Liphistiidae, these all persist and become four pairs of *spinnerets*. In the Arachnomorphae, however, the paired endopodites of the fourth segment fuse and change into the *colulus*, which does not perform the function of spinning. The exopodites of this segment become the *anterior spinnerets*, the endopodites of the fifth segment form the *intermediate spinnerets*, and the exopodites, the *posterior spinnerets*. The ectoderm at the tips of these spinnerets invaginates and differentiates into *spinning glands*.

As the cerebral grooves are deepening, a *stomodaeum* is formed by the invagination of the ectoderm on the median line behind the cephalic lobes and just anterior to the bases of the palpi. This extends inward and later differentiates into a *pharynx*, an *oesophagus* and a *sucking stomach*. By means of this large blind sac, which sends branches into each pair of legs, the spider is able to generate the negative pressure necessary to suck the body fluids from its prey.

Shortly after the appearance of the stomodaeum, a *proctodaeum* is formed by the invagination of the ectoderm in front of the caudal lobe; this becomes the *rectum*. These ectodermal parts of the digestive tract at the anterior and posterior ends of the embryo are connected by the *mid-gut*. The developmental history of this structure is complex, and includes a good many points which are still matters for debate, but it may in general be regarded as derived from the endoderm. Posteriorly, a sheet like accumulation of yolk cells appears near the caudal lobe; this soon becomes the *sterocoral pocket* and is connected with the rectum. A pair of anteriorly directed *Malpighian tubes* are formed at the sides of the sterocoral pocket.

Another accumulation of endodermal cells also occurs posterior to the stomach; a little later these spread over and enwrap the yolk, which has by this time been divided by mesodermal septa into four pairs of blocks. The mid-gut is eventually formed in this region (Fig. 10.44).

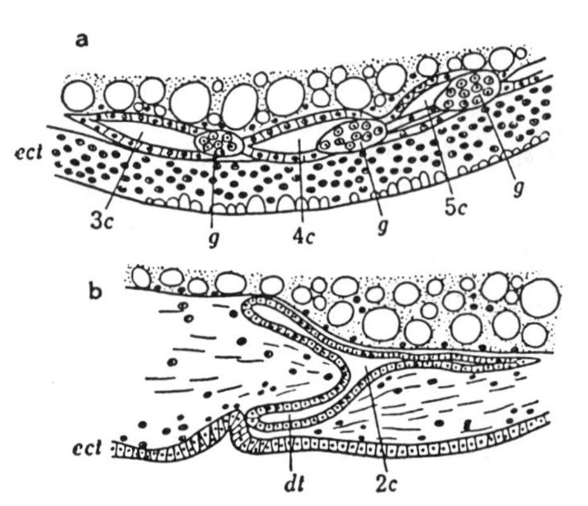

Fig. 10.47 Longitudinal sections of part of *Agelena* embryo at reversion stage

2*c*, 3*c*, 4*c*, 5*c*: second, third, fourth and fifth abdominal coelomic sacs, *dt*: gonadal duct growing toward ectoderm, *ect*: ectoderm, *g*: masses of germ cells appearing in coelomic wall of each abdominal segment.

In the spiders, the germ cells can be recognized only in a relatively late stage, when germ-band formation is complete and the embryo is undergoing reversion, as large cells in the wall of the coelomic sacs of the third, fourth, fifth and sixth abdominal segments. When the cavities of the sacs disappear, these cells in each segment come together into lateral clumps which give rise to the *genital glands*. The coelomic sacs of the second abdominal segment elongate and connect with ectodermal invaginations to form the *genital ducts*. These processes by which the reproductive organs develop are quite similar to those which give rise to the coxal glands (Fig. 10.47).

7. HATCHING AND LARVAE

The Dharma embryo breaks open its chorion, along a line connecting a pair of *egg-teeth* which are formed on the outer sides of the bases of the palpi, and escapes from it by expanding and contracting movements of its body. The egg-teeth are shed during this process and remain stuck to the discarded vitelline membrane. In certain species, immediately after hatching some of the larvae bite at the undeveloped eggs in the same egg sac. This phenomenon, which can be considered as a sort of adelphophagy, affords evidence that the mouth parts have opened into the digestive tract. Not a few of the other organs, however, are completed after hatching and begin to function only after the first moulting.

For example, the spinning organs of *Nephila* are incompletely developed before the first moult, and do not function at all. After this, however, they rapidly acquire a complicated structure which closely resembles that of the adult.

The hatched larvae generally leave the nest after moulting once within the egg sac, although in many cases they pass the winter in the egg sac and do not emerge from the nest until the following spring. Spiders usually attain their full growth after moulting six to eight times; some species, however, such as those of *Heteropoda,* have to moult more than ten times before they reach maturity.

(SUPPLEMENT) DEVELOPMENT OF THE PRIMITIVE SPIDERS

As a rule, the early development of the Liphistiomorphae and Mygalomorphae is similar to that of *Agelena,* except for the mode of formation of the abdomen, which shows the striking peculiarity described below.

In *Heptathela,* according to Yoshikura (1954, 1955), a small, button-like prominence appears on the edge of the germ disc. Although this seems to correspond to the secondary thickening of *Agelena,* it neither separates from the germ disc nor disappears, but remains in its original position and proceeds to develop and give rise to the abdominal segments. While the germ disc is differentiating into the cephalothoracic segments, (which are divided into lateral bands by the longitudinal groove of the ventral sulcus and then form the appendage rudiments),

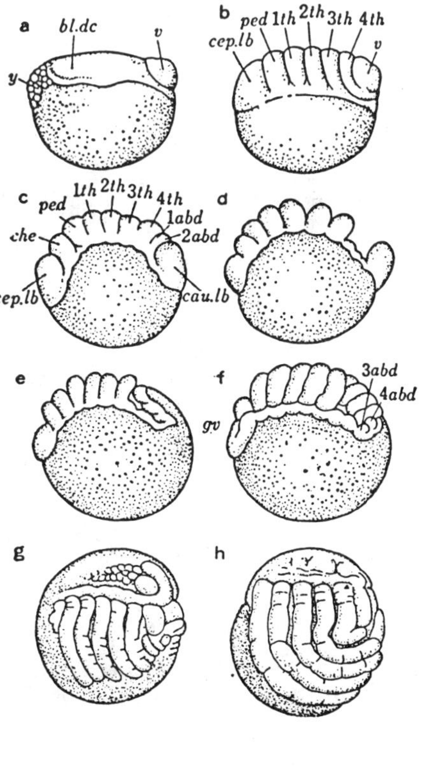

Fig. 10.48 Embryonic development in *Heptathela kimurai* (Yoshikura)

1 *abd:* first abdominal segment, 1 *th:* first thoracic segment, 2 *abd:* second abdominal segment, 2 *th:* second thoracic segment, 3 *abd:* third abdominal segment, 3 *th:* third thoracic segment, 4 *abd:* fourth abdominal segment, 4 *th:* fourth thoracic segment, *bl.dc:* blastodisc, *cau.lb:* caudal lobe, *cep.lb:* cephalic lobe, *che:* cheliceral segment, *gr:* cerebral groove, *ped:* pedipalpal segment, *v:* ventral process, *y:* yolk.

this prominence also separates off segments anteriorly, forming the abdomen in the region between itself and the posterior edge of the cephalothorax. The process is thus shifted slowly backward; about the time when it is separating off the third abdominal segment, its tip gradually rises away from the yolk and begins to bend forward toward the ventral side of the cephalothorax. A total of twelve abdominal segments are eventually formed in this way (Fig. 10.48).

Comparing the relation between yolk mass and germ band in the *Agelena* embryo with that of *Heptathela* in these stages: the *Agelena* embryo bends backward around the yolk mass with its caudal end also directed posteriorly, while the embryo of *Heptathela* bends forward, carrying the yolk mass on the upper half of its back, and its abdominal end, which is free from the yolk surface, bends forward over the ventral side (Fig. 10.48; 10.49).

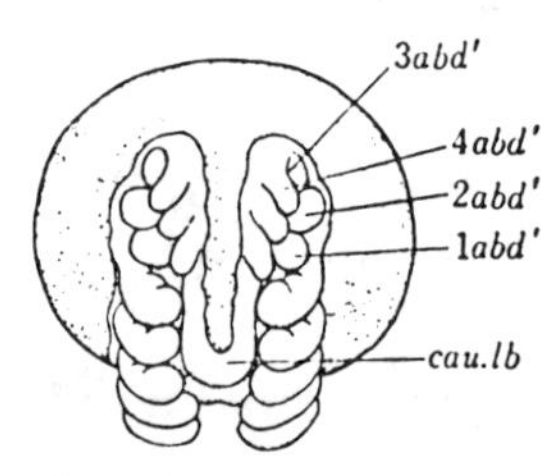

Fig. 10.49 Postero-ventral view of abdominal region with extremity appendages in embryo of *Heptathela kimurai* (Stage of embryo corresponds to that in Fig. 10.48 f) (Yoshikura)

1 *abd* : first abdominal extremity appendage 2 *abd'*: second abdominal extremity appendage, 3 *abd'*: third abdominal extremity appendage, 4 *abd'*: fourth abdominal extremity appendage, *cau.lb*: caudal lobe.

These differences naturally lead to an equally striking contrast in the subsequent process of reversion. As described above, in *Agelena* the right and left bands of the ventral plate, as they elongate, carry out extensive movements on the surface of the yolk mass, and then surround and enwrap it, to form the characteristic forward-bending Dharma embryo. In *Heptathela,* on the other hand, the migration of the yolk from the dorsal side of the cephalothorax to the abdomen, as the abdominal segments take shape, causes a marked forward elongation of this abdominal region. In the cephalothorax, the migration and extension of the right and left ventral plate bands toward the dorsal side bring about the envelopment of the yolk mass, and the embryo finally acquires the Dharma shape, as in *Agelena.*

In other words, both types of development eventually give rise to a Dharma-shaped embryo, but the courses followed are different: in *Agelena* the embryo proper carries out the autonomous movements which result in the phenomenon of reversion, the yolk playing only a passive role, whereas in *Heptathela* the activity of the yolk is rather more pronounced than that of the true embryonic areas.

The type of development observed in *Heptathela* was first discovered by Claparède (1862) in *Pholcus*; later it was found to be characteristic of many primitive spiders such as the Liphistiomorphae, the Mygalomorphae and the Haplogynae[8] belonging to the Arachnomorphae. On the other hand, the *Agelena* type of development characterizes the Entelegynae, with the single exception of Pholcus.

As described above, the *Heptathela* type of development appears at first glance to differ from the *Agelena* type. However, the study of some spiders belonging to the Mygalomorphae and Haplogynae has made it clear that these two

[8] Taxonomically the *Heptathela* type of development covers a wide range of spider groups. The Entelegynae, however, which develop according to the *Agelena* type, include a much greater number of species than any of the other true spiders (about 85% of all the Araneinae). The spider embryos that we usually encounter, therefore, generally belong to the latter developmental type.

types are part of a continuous series. For example, the developmental processes of *Ischnothele* (Holm 1954), a mygalomorph, are similar to those of *Agelena* up to an early stage in germ-band formation, but the sixth, seventh and eighth abdominal segments are formed successively as ring-like ridges surrounding the caudal lobe, and then separate to the sides as right and left bands. The segments lying posterior to the sixth abdominal, therefore, do not extend dorsally, as in *Agelena,* but are directed ventrally, with the caudal lobe in the lead (Fig. 10.50). These developmental processes of the abdomen obviously belong to the *Heptathela* type. If the developmental type of *Ischnothele* is arranged between those of *Heptathela* and *Agelena,* it is quickly apparent that the *Ischnothele* type represents a transitional stage between them.

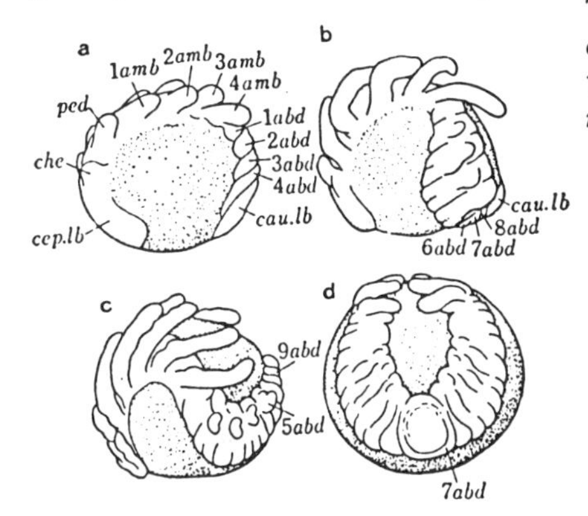

Fig. 10.50 Formation of abdominal region in *Ischnothele* embryo (Holm)

a. fourth abdominal segment stage; b. eighth abdominal segment stage; c. ninth abdominal segment stage (mid-reversion stage); d. seventh abdominal segment stage, a~c. viewed from side; d. posterior view. 1 *abd*~9 *abd*: first~ninth abdominal segments, 1 *amb*~4 *amb*: first~fourth ambulatory legs, 5 *abd'*: fifth abdominal extremity appendage, *cau.lb:* caudal lobe, *cep.lb:* cephalic lobe, *che:* chelicera, *ped:* pedipalp.

That, however, these developmental types do not shift according to any phylogenetic steps which have so far been elucidated can be deduced from the example of *Segestria,* which belongs to the Haplogynae, but develops according to a process closely similar to that of *Heptathela.*

Generally speaking, it is clear that the Mygalomorphae and Haplogynae among the Arachnomorphae are transitional groups between the *Heptathela* and *Agelena* types from an embryological point of view. Studies in these still unexplored fields should contribute to the solution of important questions concerning both the relations between the types of development and the phylogeny of these spiders, and the mechanism of the peculiar phenomenon of reversion, which is a unique characteristic of spider development.

REFERENCES

Claparède, E. 1862: Recherches sur l'évolution des Araignées. Natur. Verh. Prov. Utrecht Genoot. Kunst. Wiss. Deel. 1, Utrecht.

Dawydoff, C. 1949: Développement embryonaire des Arachnides. Pierre-P. Grassé: Traité de Zoologie, **6** : 320—285. Paris.

Holm, A. 1940: Studien über die Entwicklung und Entwicklungsbiologie der Spinnen. Zool. Bidrag, Uppsala, **19** : 1—214.

——————— 1952: Experimentelle Untersuchungen über die Entwicklungsphysiologie des Spinnenembryos. Zool. Bidrag Uppsala, **29** : 293—424.

——————— 1954: Notes on the development of an orthognath spider, *Ischnothele karschi* Bös. & Lenz. Zool. Bidrag Uppsala, **30** : 199—221.

Kautzsch, G. 1909: Über die Entwicklung von *Agelena labyrinthica* Clerck. Zoöl. Jahrb. Anat., **28** : 477—538.
Kishinouye, K. 1891: On the development of Araneina. J. Coll. Sci. Imp. Univ. Tokyo, **4** : 55—88.
——— 1891: On the lateral eyes of spiders. J. Coll. Sci. Imp. Univ. Tokyo, **5** : 101—103.
——— 1894: Note on the coelomic cavity of the spider. J. Coll. Sci. Imp. Univ. Tokyo, **6** : 287—294.
Korschelt, E. & K. Heider 1936: Vergleichende Entwicklungsgeschichte der Tiere. Jena.
Locy, W. A. 1886: Observation on the development of *Agelena naevia*. Bull. Mus. Comp. Zool. Harvard, **12** : 63—103.
MacBride, E. W. 1914: Textbook of Embryology. London.
Monterosso, B. 1948—1949: Note Araneologiche. XV. I primi stadi embrionali di *Teutana triangulosa* (Walck) studiata in "vivo". Rend. Soc. Ital. delle Sci., Ser. III, **27** : 135—145.
Montgomery, T. H. 1909: On the spinnerets, cribellum, colulus, tracheae and lung books of Araneads. Proc. Acad. Nat. Sci. Phila., **61** : 229—320.
Morin, J. 1886—1887: Zur Entwicklungsgeschichte der Spinnen. Biol. Zbl., **6** : 658—663.
Purcell, W. F. 1910: Development and origin of the respiratory organs in Araneae. Quart. J. Micr. Sci., N.S., **54** : 1—110.
Schimkewitsch, L. & W. 1911: Ein Beitrag zur Entwicklungsgeschichte der Tetrapneumones I. Bull. Acad. Sci. St. Pétersbourg, **5** : 637—654.
Ein Beitrag, usw. II. ibid., **5** : 685—706.
Sekiguchi, K. 1953: On the egg-teeth of spiders. Atypus , **2** : 3—6. (in Japanese)
Wallstabe, P. 1908: Beiträge zur Kenntnis der Entwicklungsgeschichte der Araneinen. Die Entwicklung der äusseren Form und Segmentierung. Zoöl. Jahrb. Anat., **26** : 683—712.
Yoshikura, M. 1952: Preliminary note on the development of the liphistiid spider, *Heptathela kimurai* Kishida. J. Sci. Hiroshima Univ., Ser. B, Viv. 1, **13** : 71—74.
——— 1953: Formation of the egg-sac and oviposition in *Heptathela kimurai*. Atypus, **3** : 71—74 (in Japanese)
——— 1954: Embryological stydies on the liphistiid spider, *Heptathela kimurai*. Part I. Kumamoto J. Sci,. Ser . B, **1** : 41—48.
——— 1955: Embryological studies on the liphistiid spider, *Heptathela kimurai*. Part II. Kumamoto J. Sci., Ser· B, **2** : 1—86.

III. INSECTA

1. FERTILIZATION TO GERM BAND FORMATION

(1) MORPHOLOGY OF THE EGG

The Class Insecta, the largest group in the animal kingdom, includes a vast number of members. The morphology of their eggs consequently shows an extreme diversity. The insect egg of the most fundamental type is, as shown in Figure 10.51, covered with a *chorion* and a *vitelline membrane*, and contains a large quantity of yolk.

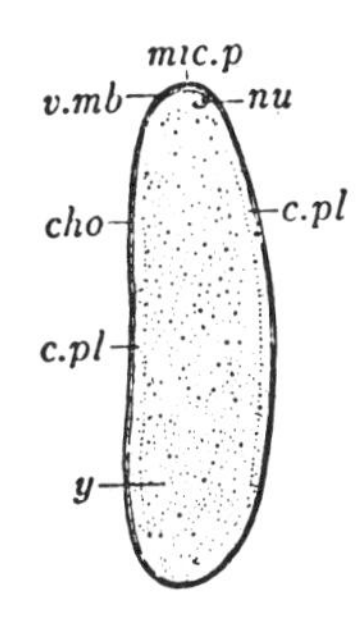

Fig. 10.51 Longitudinal section of insect egg (Korschelt u. Heider)

c.pl: cortical oöplasm, *cho*: chorion *mic.p*: micropyle, *nu*: egg pronucleus, *v.mb*: vitelline membrane, *y*: yolk.

The chorion is secreted by the *follicle cells* of the adult female (Machida 1940). Unlike the egg shell in fowls, it is a relatively early product of oogenesis, already present in a fully constituted

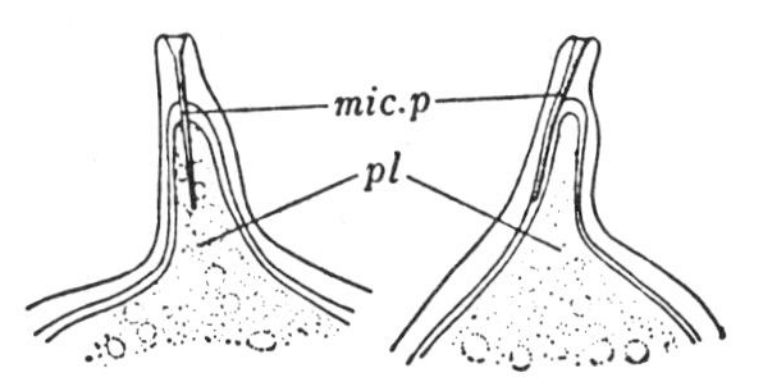

Fig. 10.52 Micropylar cone in *Drosophila melanogaster* (Nonidez 1920)

left, ventral view; right, lateral view. *mic.p*: micropyle, *pl*: protoplasm.

state on the surface of ovarian eggs. The chorion is provided with a *micropyle* (Fig. 10.52), which permits the entrance of spermatozoa, and *respiratory canals* (Fig. 10.53), which facilitate gas exchange. The micropyle is never a simple pore penetrating the chorion. It generally branches inside the chorion into several subtybes, the number of which varies with

Fig. 10.53 Chorion of *Bombyx mori* egg (Wigglesworth and Beament 1950)

ex: external layer, *in*: inner layer, *in'*: innermost layer, *m*: middle layer, *x*: respiratory canal.

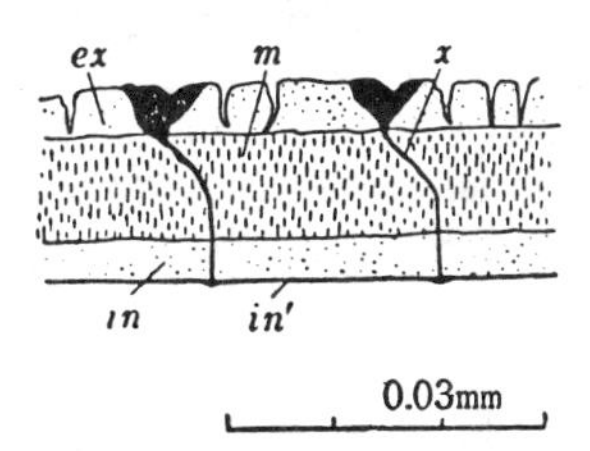

the species; there are three on an average in the silkworm *(Bombyx mori)*, with a range of from two to five (Hayashi 1937). Some maximum numbers are: eight in *Antheraea yamamai*, nine in *Dictyoploca japonica*, eleven in *Antheraea pernyi* and thirteen in *Attacus atlas* (Kawaguchi 1926a; Oba 1934). Observations made on microscopical sections reveal that the chorion has a stratified structure (Fig. 10.53). In many species it is ornamented with notches or processes of various shapes.

In grasshoppers, a chorion-like secondary structure appears in an advanced stage of development. This is a kind of cuticle secreted by the serosa and lines the chorion from within, forming a solid shell. This cuticle possesses a *hydropyle*, a structure for obtaining water from the external milieu, at the posterior end of the egg.

The vitelline membrane, formed by the ectoplasm of the oocyte (Machida 1940), is a structureless thin membrane lying just beneath the chorion and in direct contact with the egg contents. It is clearly separated from the chorion except at the micropyle, to which it is connected by trumpet-shaped tubules (Hayashi 1937). The chorion and vitelline membrane can be regarded as protective structures surrounding other components of the egg, but themselves making little contribution to the various metabolic processes accompanying embryonic development.

As shown in Figure 10.51, the insect egg contains much yolk, and just beneath the vitelline membrane, there is frequently a thin layer of protoplasm free from yolk. This layer is variously called the *Keimhautblastem* (Weismann 1863), *periplasm, cortical layer*, etc., but in some insects no such morphologically clear Keimhautblastem can be recognized. The egg of *Platycnemis pennipes*, for example, is of this sort, the cortical protoplasm containing yolk granules and showing a reticular structure, although the granules are much smaller than those found in other portions of the egg (Seidel 1929b). But despite such morphological differences, all the cortical protoplasm is believed to be of the greatest importance in the process of development. After fertilization the cleavage nuclei migrate into this Keimhautblastem, and form the *blastoderm*, which envelops the yolk, giving the egg its *centrolecithal* character. In all the parts other than the Keimhautblastem, the protoplasm surrounds each yolk granule, forming a protoplasmic network.

At the anterior end of the egg, just beneath the micropyle, the Keimhautblastem thickens in the region where the egg nucleus lies. This thickened protoplasm is called the *anterior pole plasm*. In Diptera and Coleoptera, the posterior end of the egg is also differentiated as the *posterior pole plasm*, which contains deeply staining granules called *posterior polar granules* or simply, *polar granules* (Huettner 1923). In the Colorado potato beetle *(Leptinotarsa decemlineata)*, these granules aggregate to form a granular disc, the *pole-disc*. As the cleavage nuclei which enter this pole-disc later give rise to the reproductive cells, these granules are known also by the name of *germ-cell determinants* (Hegner 1911) (cf. p. 446). The yolk is distributed in a granular state; according to its chemical composition, this can be roughly divided into albuminous and fatty types of yolk (Machida 1940).

(2) SPERM ENTRY AND FERTILIZATION

The entry of the spermatozoon into the egg takes place when the egg, after having descended along the oviduct, passes the vagina. The vaginal structure and the position of the receptacular opening are appropriately arranged to facilitate their approach to the egg as it makes its way downward, with the micropyle pointing backward (Fig. 10.54). In many insects, there have been found mechanisms in which contracting muscles squeeze out the sperm toward the micropyle.

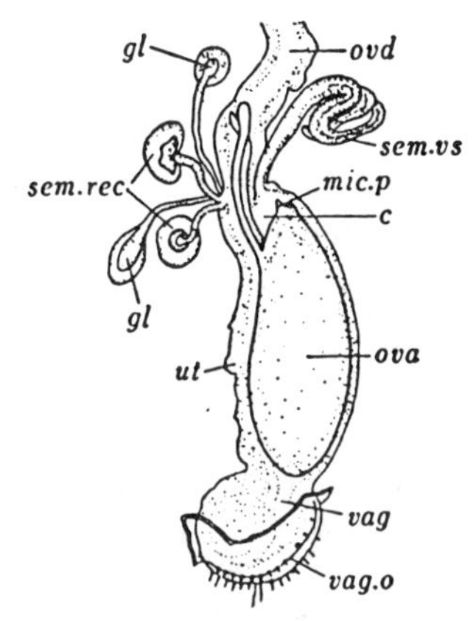

Fig. 10.54 Egg of *Drosophila melanogaster* within uterus (Nonidez 1920)

c: fertilization cavity, *gl:* accessary gland, *mic.p:* micropyle, *ova:* egg, *ovd:* oviduct, *sem. rec:* spermatheca, *sem.vs:* sperm receptacle, *ut:* uterus, *vag:* vagina, *vag.o:* vaginal opening.

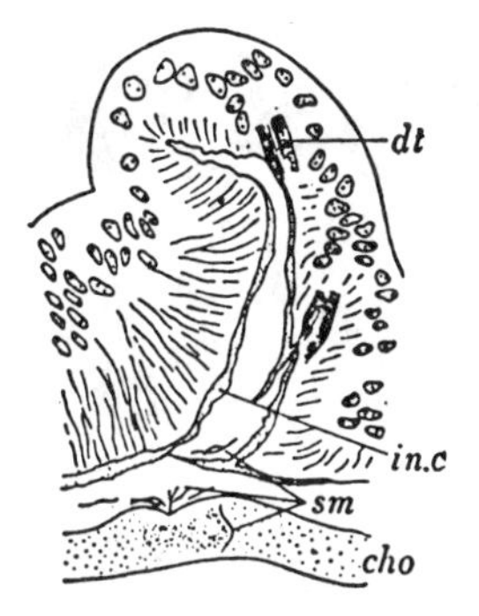

Fig. 10.55 Longitudinal section through vestibular ridge of silkworm oviduct shows spermatozoa entering egg during oviposition (Omura 1938)

cho: chorion, *dt:* fertilization tubule, *in.cu:* inner chitinous membrane, *sm:* sperm.

The spermatozoa themselves are believed to show a kind of chemotaxis, which enables them to reach and penetrate the micropyle (Wigglesworth 1950). However, the possibility that silkworm spermatozoa are chemotactically oriented toward the egg seems to be very limited; in some silkworm races with an hereditary abnormality of the oviduct, in which the eggs assume a peculiar position in passing through the receptacular opening, the percentage of unfertilized eggs is found to be significantly high. In normal races, moreover, it has been proved that many of the unfertilized eggs have been laid in reverse (Miyazaki 1927). Omura (1938) has made a detailed study of the process of sperm entry in the egg of the silkworm (Fig. 10.55).

The moth of the silkworm does not lay all the eggs which are formed; usually considerable numbers of eggs are left unlaid. According to Matsunaga (1933), some of these eggs seem to have been fertilized; these are able to hatch, grow and spin a cocoon, and these cocoons frequently show some paternal characters, which precludes the possibility of parthenogenetic activation. It thus appears that the spermatozoa may also ascend the ovarioles and fertilize ovarian eggs. When the vaginal opening of a moth is cauterized after copulation, preventing oviposition, the eggs obtained from it by dissection nevertheless include some fertilized ones. This interpretation of Matsunaga would be confirmed by a finding that some of the eggs which have already descended to the vagina are fertilized. More work must be done before it can be concluded that the spermatozoa actually ascend the ovarioles and fertilize eggs in the ovary.

Among insects, *polyspermy* — the entry of many spermatozoa into one egg — is of common occurrence. A case in which 11 sperm entered a single egg is known in the silkworm, and in *Drosophila melanogaster*, the number of spermatozoa may exceed 30, and may at times reach even some hundreds (Kawaguchi 1926; Huettner 1927; Wigglesworth 1950).

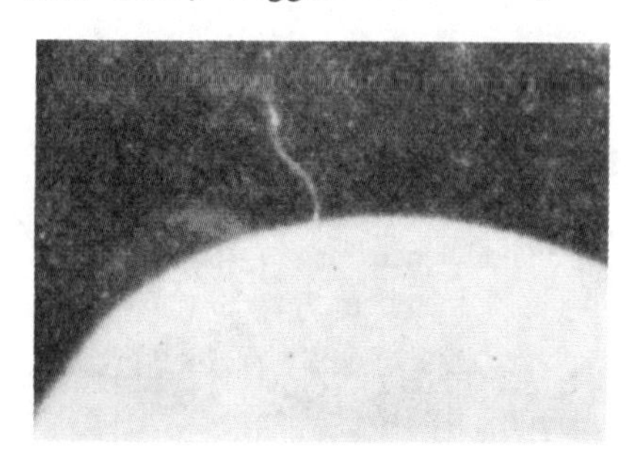

The spermatozoa pass through the micropyle (Fig. 10.56), then through the subtybes (Fig. 10.55) and reach the interior of the egg. The egg nucleus rests in the metaphase of the first maturation division, and only proceeds to the later stages of meiosis after sperm entry. It is therefore obvious that this latter event must act as the stimulus for finishing the maturation process. The egg nucleus thus changes into the *egg pronucleus*, with the reduced number of chromosomes, while the sperm nucleus also undergoes a marked transformation: its tail is shed and its head swells to become the *sperm pronucleus*. This gradually approaches the egg pronucleus, and eventually unites with it. The time of this union of the two pronuclei — *syngamy* — does not follow immediately upon sperm entry through the micropyle, nor does it coincide with the time of oviposition. If, for any reason, an inseminated egg spends a long time in the vagina, syngamy may easily take place within the egg while it thus still remains in the body of the mother moth, and it is not unusual for further development to proceed within the maternal body. Consequently, in insects with a short egg stage — in *Drosophila* for example (about 24 hours at 25°C) — the deposited eggs are all at different developmental stages, some being still unfertilized, while others are ready to hatch (Huettner 1923). In normally laid silkworm eggs the union of egg and sperm pronuclei occurs 1.5 hours after oviposition at 25°C.

In the viviparous insects, fertilization takes place in the ovary. In the bedbug *Cimex*, for example, the spermatozoa ascend the oviduct, reach the ovariole and there fertilize young eggs which have not yet acquired a chorion (Wigglesworth 1950). This type of fertilization, however, is rather exceptional among insects in general, and will not be discussed in further detail. Although, as mentioned above, many spermatozoa enter a single egg, only one participates in normal fertilization, the others degenerating. Occasionally, however, some of the remaining sperm nuclei begin to divide, either singly, or after having fused with another sperm nucleus or one of the polar bodies, and take part in the further embryonic development together with the cleavage nuclei derived from normal fertilization. In ordinary cases, the presence of such an abnormal process cannot be detected, even if it has taken place, but experiments making use of appropriate crossing between two different mutants for body, eye or serosa coloration give evidence of the participation of additional spermatozoa in development. Some mosaic individuals and sexual mosaics have been shown to develop in this way (cf .p. [458]). For this reason, the possibility of so-called 'polyfertilization' (Glushtchenko 1950) and other similar phenomena cannot be denied. In *Drosophila* and the silkworm, where many mutants are known, the formation of mosaic individuals has frequently been induced by the action of chemical agents or of high temperatures before or after fertilization. The entrance of excess spermatozoa may lead to abnormal cleavage (Huettner 1927).

Since the insect egg is covered with a thick chorion and contains much yolk, and sperm entry takes place within the maternal body, the vital observation of the delicate changes which must naturally occur in the cortical layer following sperm entry, and of the fertilization process, is rendered almost impossible. On the other hand, the insect egg offers instead a special material for experiments which cannot be done with the eggs of other groups of animals: e.g., experiments such as that mentioned above on mosaic formation in *Drosophila, Habrobracon, Ephestia, Bombyx,* etc., in which genetical research has made great progress.

(3) CLEAVAGE STAGE

a. Cleavage nuclei

The nucleus derived from union of the male and female pronuclei is called the *first cleavage nucleus*. It divides, and its daughter nuclei continue to divide within the yolk-containing protoplasmic network, gradually increasing in number. In most insect species, this series of divisions is limited to the nuclei and a small amount of dense protoplasm surrounding each nucleus, and does not extend to effect division of the egg as a whole. Opinions fail to agree as to whether these divisions are mitotic or amitotic, but it is generally admitted that in silkworm eggs, the nuclei participating in the formation of the future blastoderm divide mitotically, whereas the yolk nuclei which differentiate later divide by amitosis (Toyama 1902). This is also the case in *Antheraea pernyi* (Saito 1937).

In the early stages the nuclei usually all divide simultaneously (synchronous division). Later, however, division becomes asynchronous. The duration of the synchronous division period varies with the species: in *Ephestia*, synchronized division continues until the 512-nucleus stage (Sehl 1931), but in *Calandra* (Coleoptera, Curculionidae) division is asynchronous from the beginning (Tiegs and Murray 1938). In *Platycnemis pennipes*, the first eight divisions are synchronous, the ninth is still wholly or nearly synchronous, and further divisions become asynchronous; the egg enters the blastoderm stage at the end of the ninth division (Seidel 1928b). In *Drosophila melanogaster*, the divisions are synchronous throughout the stage of nuclear cleavage, and the divisions of the somatic cells continue to be synchronous even after blastoderm formation has begun at the 256-nucleus stage. Synchronized division of as many as 1100 somatic cells has been observed, while the cells which have entered the pole plasm region, i.e., the presumptive germ cells, begin to divide at a different pace (Huettner 1923). The transition from synchronized to unsynchronized division is evidence that some differentiation has appeared among the cleavage nuclei, which have hitherto been, at least apparently, isopotent.

The cleavage nuclei, each surrounded by a small mass of dense cytoplasm, are arranged to form a sphere. As their number increases, the diameter of the sphere also increases, and the nuclei approach the egg surface. Each nucleus drags behind it a tail-like mass of cytoplasm, as if the nucleus is moving actively toward the egg surface. It is not yet clear, however, whether they are actually performing an active movement or are merely being carried by protoplasmic streaming. On the other hand, it is observed in *Pieris* that the cytoplasm is less dense within the sphere of nuclei than outside it (Fig. 10.57).

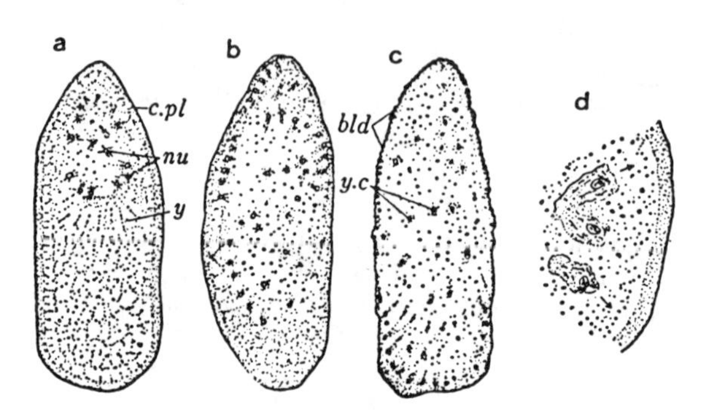

Fig. 10.57 Migration of cleavage nuclei into cortical layer in eggs of *Pieris* (a ~ c) and *Calliphora* (d) (a ~ c. Eastham; d. Strasburger; from Wigglesworth 1953)
a. 6 hours after oviposition; b. 8 hours; c. 10 hours; d. part of *Calliphora* egg showing cleavage nuclei approaching cortical layer.

As is shown in Figure 10.57, the arrival of the cleavage nuclei at the Keimhautblastem (cortical layer) is not simultaneous in all parts of the egg surface. In *Ephestia* and the silkworm, the nuclei clearly arrive earlier at the anterior pole than at the posterior, whereas in *Platycnemis*, the cleavage nuclei not only fail to assume the spherical disposition, but some of them reach the egg surface at the posterior pole as early as the 4th, or at latest 5th, division (i.e. during the 32-nucleus stage). By the 128-nucleus stage, the nuclei have arrived at the surface all over the egg (Seidel 1932). In the silkworm, the arrival of the nuclei at all parts of the egg surface is observed about 10 hours after oviposition at 25°C.

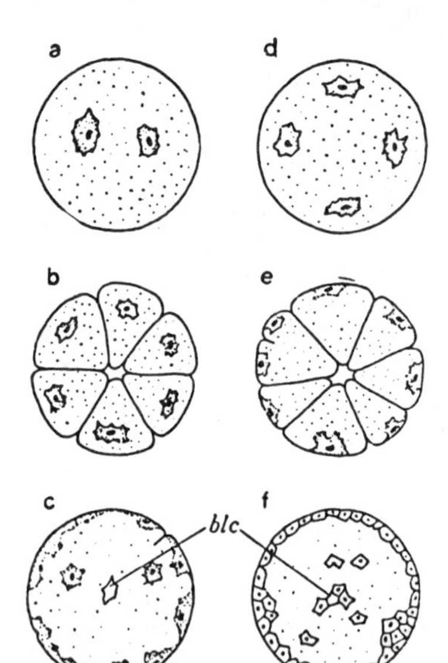

Fig. 10.58 Temporary holoblastic cleavage in the egg of *Macrotoma vulgaris* (Korschelt u. Heider 1909)
blc: blastocoel.

As to which region of the surface first receives the nuclei, Marshall and Dernehl (1905) have proposed the classification shown in Table 10.4.

In insect eggs, as already stated, there is in general no true cleavage accompanying nuclear division. Instead, the migration of the nuclei to the egg surface brings about the so-called '*superficial cleavage*'. However, exceptions are known among the apterygote insects like *Isotoma*, *Tomocerus* (=*Macrotoma*),

Table 10.4 **Modes of arrival of cleavage nuclei at the egg surface**

	author	*species*
Arrive simultaneously at all parts of surface	Blochmann	Diptera *(Musca)*
	Bobretzky	Lepidoptera *(Porthesia)*
	Heymons	Dermaptera *(Forficula)*
	Noack	Diptera *(Calliphora)*
	Voeltzkow	Coleoptera *(Melolontha)*
	Weismann	Diptera *(Chironomus)*
Arrive first at posterior end	Ganin	Hymenoptera *(Formica)*
	Graber	Diptera *(Lucilia)*
	Grimm	Diptera *(Chironomus)*
	Heider	Coleoptera *(Hydrophilus)*
	Heymons	Orthoptera *(Gryllus)*
	Kowalevsky	Diptera *(Musca)*
	Weismann	Hymenoptera
Arrive first at anterior end	Ayers	Orthoptera *(Oecanthus)*
	Bobretzky	Lepidoptera *(Pieris)*
	Bütschli	Hymenoptera *(Apis)*
	Carrière and Bürger	Hymenoptera *(Chalicodoma)*
	Dickel	Hymenoptera *(Apis)*
	Grassi	Hymenoptera *(Apis)*
	Kowalevsky	Hymenoptera *(Apis)*
	Weismann	Diptera *(Musca)*
Arrive first at equatorial part of surface	Kulagin	Hymenoptera *(Platygaster)*
	Schwartze	Lepidoptera *(Lasiocampa)*
Arrive first at ventral surface	Heymons	Orthoptera *(Gryllotalpa)*
	Melnikow	Coleoptera *(Donacia)*
Nuclei appear at surface in groups	Brandt	Odonata *(Callopteryx)*

etc., in which cleavage temporarily proceeds holoblastically (Fig. 10.58). In the earlier stages, each division of the nucleus is followed by ordinary cytoplasmic division, and blastomeres of the same size are produced. As the nucleus migrates to the surface of each blastomere, the inner cell partitions between the blastomeres gradually disappear, and further cleavage is superficial. Some of the parasitic Hymenoptera are reported to be holoblastic, but such cases are always related to a special phenomenon called *polyembryony*. It is not certain whether true holoblastic cleavage exists among insects except in connection with polyembryony (Hirschler 1928).

b. **Yolk nuclei**

In *Ephestia* and the silkworm, not all the nuclei migrate to the egg surface: some of them remain in the central part of the egg and later become *yolk nuclei (vitellophages)*. This is the first morphological differentiation observable among the cleavage nuclei. It would be interesting to know whether the vitellophages originate from nuclei which are accidentally slow in migrating and are consequently left behind, or from certain ones which are actually differentiated as presumptive vitellophages. Unfortunately, this problem remains unsolved. In *Ephestia*, the cleavage nuclei destined for the future blastoderm and those which are to be vitellophages are morphologically distinguishable as early as the 128-nucleus stage

Table 10.5 **Classification of insect orders according to mode of vitellophage differentiation.** (Sehl 1931)

	Intra-ovular differentiation	Secondary inward migration of cleavage nuclei
Diptera	*Musca* (Kowalewsky)	*Musca* (Graber)
	Lucilia (Noack)	*Calliphora* (Noack)
		Miastor (Kahle)
Coleoptera	*Melolontha* (Voeltzkow)	
	Lina (Graber)	
	Hydrophilus (Heider)	
	Meloe (Nussbaum)	
	Tenebrio (Saling)	
	Donacia (Hirschler)	
	Doryphora (Wheeler)	
Lepidoptera	*Lasiocampa* (Schwartze)	
	Pieris (Bobretzky)	
	Pieris (Eastham)	
	Endromis (Schwangart)	
	Catocala (Hirschler)	
	Eudemis (Huie)	
	Ephestia (Sehl)	
Hymenoptera	*Polistes* (Marshall & Dernehl)	
	Chalicodoma (Cariere & Bürger)	
	Apis (Dickel)	
	Vespa (Strindberg)	
Dermaptera	*Forficula* (Heymons)	
Orthoptera	*Gryllus* (Heymons)	*Gryllotalpa* (Heymons)
	Oecanthus (Ayers)	*Periplaneta* (Heymons)
		Periplaneta (Wheeler)
		Periplaneta (Weismann)
		Dixippus (Hammerschmidt)
Hemiptera	*Pyrrhocoris* (Seidel)	
Odonata	*Platycnemis* (Seidel)	
Thysanura	*Lepisma* (Heymons)	*Campodea* (Uzell)
	Machilis Heymons)	

(Sehl 1931). This observation suggests that a morphological differentiation is already present during the period of synchronized nuclear cleavage (cf. p. 409). In *Platycnemis* also, ten cleavage nuclei remain inside as presumptive vitellophages at the 128-nucleus stage while synchronized cleavage is still going on (Siedel 1929b). The vitellophages can be discriminated from the presumptive blastoderm nuclei by an absence of nucleoli, larger size and other features (Sehl 1931; Saito 1934).

In most insect species, the vitellophages are thus derived from nuclei left inside the yolk mass. In other words, they are products of intra-vitelline separation. In some insects, however, all the cleavage nuclei first migrate to the egg surface, so that none are left in the interior at an early stage of blastoderm formation. Some of them later leave the blastoderm and migrate toward the interior to give rise to vitellophages. The various modes of vitellophage differentiation in insect eggs have been summarized by Sehl (1931), as shown in Table 10.5. Some reexamination of this table may, however, be necessary. It is curious, for example, that the genus *Musca* is classed in both groups. It may be added here that *Ephemera strigata* likely belongs to the migrating type (Ando and Kawana 1946).

(4) BLASTODERM AND GERM BAND

a. Formation of blastoderm and germ band

The cleavage nuclei just after their arrival at the cortical layer are large and sparsely distributed, although protoplasmic threads connect neighboring nuclei. As their number increases as the result of actively repeated division, septa appear between adjacent nuclei. Similar septa are also formed between the cortical nuclei and the inner yolk-containing mass, and a layer of cells thus appears, which covers the egg surface and constitutes the blastoderm (Fig. 10.59). These inter-nuclear septa are believed to originate as invaginations of a structureless membrane formed over the cortical surface (Iwasaki 1931, 1932). The silkworm egg reaches the stage of blastoderm formation 14 to 15 hours after oviposition (at 25°C). In *Platycnemis* the nuclei number 528 at this stage (Seidel 1929b).

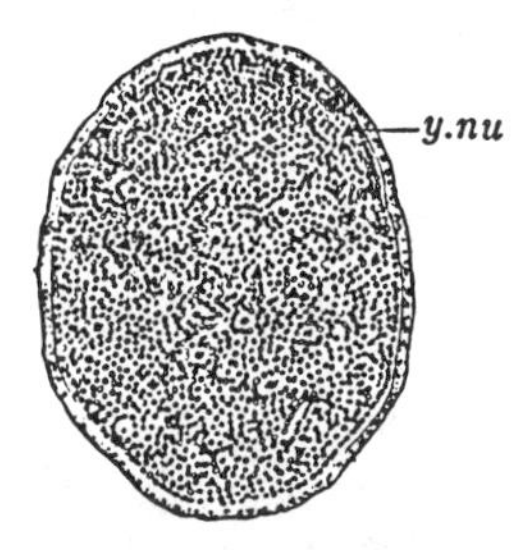

Fig. 10.59 Blastoderm of egg of *Bombyx mori* (longitudinal section) (Ikeda 1913)

y.nu: yolk nucleus.

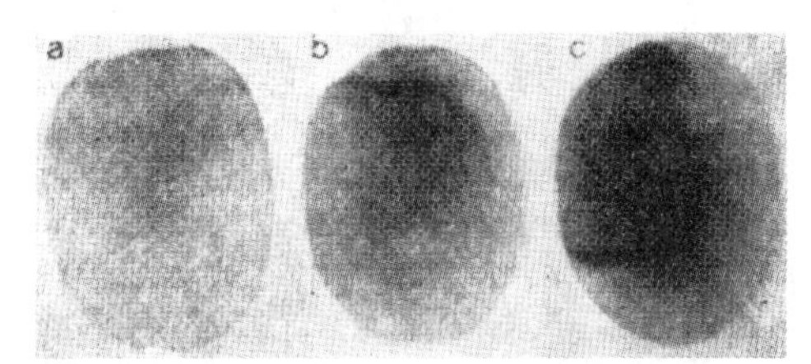

Fig. 10.60 Formation of germ band in *Philosamia cinthia ricini* (Takami)

a. 15 hours after oviposition, blastoderm stage; b. 20 hours; c. 24 hours. (Eggs were dechorionated, stained and cleared) The germ band of the *Philosamia* egg migrates toward the anterior pole as development proceeds.

The place where blastoderm formation begins differs from one species to another. In some insects, it starts in the anterior part of the egg, in others it is median, and there are even some species in which it is initiated at the posterior end of the egg. Such variety seems to depend upon several factors: the shape of the egg, the position of the cleavage center (cf. p. 417), the place at which the cleavage nuclei first arrive at the egg surface, etc. In the silkworm, for instance, nuclear cleavage begins near the anterior pole and the nuclei disperse outward from here. Blastoderm formation, consequently, also starts at the anterior pole of the egg and proceeds toward the posterior.

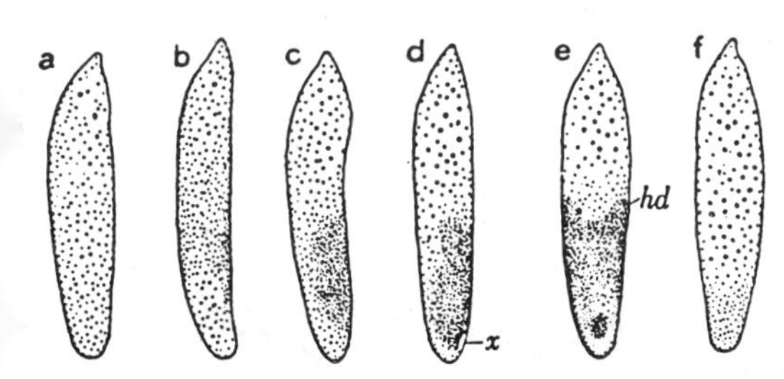

Fig. 10.61 Formation of germ band in *Platycnemis pennipes* (Seidel 1926b)

a. egg at about 650-nucleus stage (lateral view); b. cleavage nuclei are beginning to aggregate to form germ band (lateral view); c. two germ band primordia separately formed at sides of egg are beginning to fuse (lateral view); d. formation of germ band completed (lateral view); e. ventral view of same stage; f. dorsal view of same stage. *hd:* head rudiment, *x:* region where germ band will turn inward at blastokinesis.

In an early stage, the cells which constitute the blastoderm are similar in shape over the whole surface of the egg. Soon, however, those situated in a definite region of the blastoderm become taller and densely packed. This indicates the beginning of *germ band* (or *ventral plate)* formation, and occurs about 18 hours (25°C) after oviposition in the silkworm.

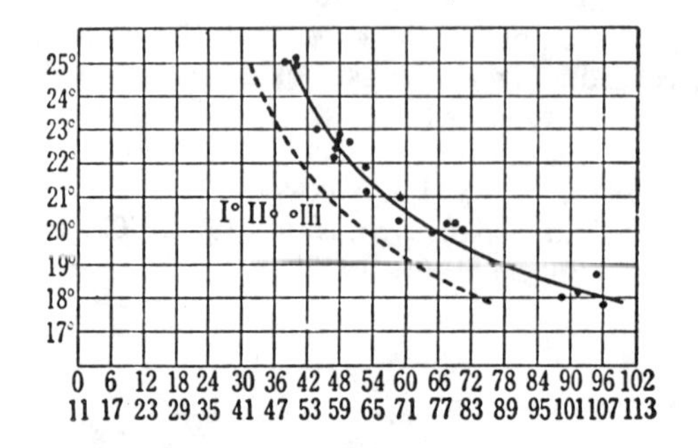

Fig. 10.62 Relation of time required for aggregation of cleavage nuclei and germ band formation to temperature in *Platycnemis* egg (Siedel 1929b)

Ordinates: temperature (°C). Abscissae: time in hours required for development (Numbers in upper row: time after oviposition). Unbroken line: time required to form two germ band primordia (cf. Fig. 10.61c). Dotted line: time when cleavage nuclei begin to aggregate. I: 128~256- nuclei stage. II: 528-nuclei stage. III: about 570-cell stage.

The formation of the germ band indicates that the blastoderm has differentiated into an *embryonic area* which directly participates in embryogenesis and an *extra-embryonic area* which gives rise to the serosa.

The transparent chorion of the *Platycnemis* egg permits vital observation of the behavior of the cleavage nuclei. Siedel (1929b), using these eggs, followed and traced in detail the processes of cleavage, migration and the convergent and divergent movements of the nuclei leading to germ band formation (Fig. 10.61), and showed the temperature-dependent modification of this process (Fig. 10.62).

b. **Formation of serosa and amnion**

Once the germ band is established, the extra-embryonic blastoderm grows into an outward-directed fold from its margin, forming the *amniotic fold* (Fig. 10.63), which extends to envelop the germ band from the outside. After its two sides meet and unite along the median ventral line, it splits into inner and outer layers, an outer *serosa*, and an inner *amnion* (Toyama 1896). The serosa, lying just beneath the vitelline membrane, envelops the whole egg; the amnion lies inside the serosa and covers the embryo. Both are thin membranes consisting of flat cells. In the grasshoppers, the serosa secretes a kind of cuticle (cf. p. [406]),

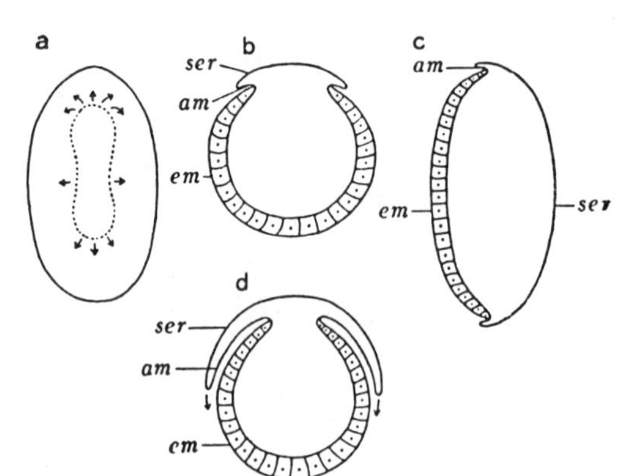

Fig. 10.63 Schematic sketches showing development of amniotic fold (Sehl 1931)

a. dorsal view of egg showing direction of growth (→) of amniotic fold; b. transverse section of embryo at early stage of this process; longitudinal section of embryo at same stage; d. transverse section of embryo at more advanced stage, *am:* amnion, *em:* germ band, *ser:* serosa.

but the function of this membrane in the silkworm is not clear. On the other hand, it is generally admitted that the serosa of nonhibernating eggs is not pigmented, while that of hibernating eggs has a characteristic pigmentation, associated with a special tryptophane metabolism (cf. p. [470]). It is therefore beyond doubt that the serosa plays an important role in the physiology of the egg.

The way in which the amnion and serosa are formed is related to the morphogenetic movements of the germ-band, although its extent differs from species to species, depending especially upon the embryonic movements to be described later (p. [451]). In the silkworm, for example, the movements of the germ-band are less marked, and it is the amniotic folds which show a positive extension, whereas in *Platycnemis* the germ-band executes intense embryonic movements by which it invaginates into the interior of the egg from the posterior end *(anatrepsis*; see p. [451]), stretching the extra-embryonic blastoderm to form the amnion. In this case, the serosa undergoes only an incidental extension (Fig. 10.79A). The mode of formation of the serosa and amnion in the aquatic beetle *Hydrophilus* represents a type intermediate between these two extremes (Fig. 10.64). With respect to the mechanism by which the amniotic folds expand when the germ-band performs less intense movements, Wheeler postulated a mechanical force generated by the partial thickening of the blastoderm at the time of germ-band formation and acting so as to push the yolk toward the egg interior; the reaction to this force would then cause the formation of the amniotic folds (Saito 1937).

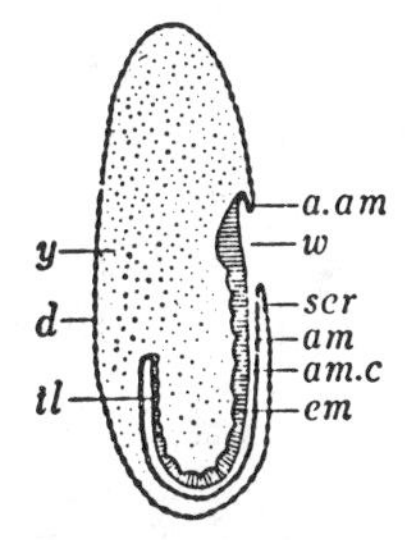

Fig. 10.64 Formation of amnion and serosa in *Hydrophilus* (Korschelt, from Ishikawa 1921)

a.am: anterior edge of amniotic fold, *am*: amnion, *am.c*: amniotic cavity, *d*: dorsal side of egg, *em*: embryo, *ser*: serosa, *tl*: caudal part of embryo embedded in yolk, *w*: part of embryo not covered by serosa, *y*: yolk.

Some objections have been made to these proposed explanations of amnion and serosa formation. Although *Pieris* (Pieridae) and *Porthesia* (Orgyidae) belong to the Lepidoptera like the silkworm, the amnion and serosa appear separately in these forms, and the serosa is said to undergo its expansion earlier (Bobretzky 1874). A similar mode of formation has been observed in the Arctiid moth *Diacrisia virginica*, by Johannsen (1929). Even in the silkworm, there are reports that the serosa and amnion have a common origin, as observed by Toyama (see above), in the amniotic fold, but the serosa subsequently separates from the amnion and develops earlier (Ikeda 1912, 1913; Takahashi 1912). These contradictory observations, however, may well result from technical difficulties caused by the thinness and close contact of the developing membranes. If an egg is partially cauterized and the growth of the amniotic fold observed in this region, it is found that at least in the silkworm, there appears to be no separation of the two membranes throughout the process of their expansion (Takami 1944b). The amniotic fold of *Antherea pernyi* also develops without separating into two membranes (Saito 1937).

Some insects lack an amnion. The Insecta may therefore be classified into two groups: Insecta anamnia (amnionless insects) and Insecta amniota (amniote insects). Among species which have no serosa may be cited the strepsipteran *Xenos Bohlsi*, which parasitizes the wasp *Polistes canadensis*. There are also many species of apterygote insects such as *Campodea, Sminthurus, Arnurophorus, Anurida, Isotoma*, and some ants — *Leptothorax, Tetramorium, Tapinoma* — in which the membrane anlage appears, but fails to develop in the usual way, so that both serosa and amnion are lacking (Hirschler 1928).

(5) YOLK SEGMENTATION

In the germ-band stage of silkworm eggs, the yolk, which has hitherto been homogeneously packed in the egg, begins to divide *(yolk segmentation)*. Iwasaki (1931, 1932) has suggested that the membrane, which earlier invaginated to form the blastoderm (p. [413]), again invaginates further into the interior. The segmentation begins at the egg periphery and gradually extends inward. The large initial blocks formed in this way, containing many nuclei, are gradually cut up, into smaller masses (Fig. 10.93) and result finally in mononucleate *yolk cells*. Later the number of nuclei again increases, so that mononucleate cells become rather rare (Toyama 1902). The number of nuclei also varies with the species.

Since the yolk cell, which may also be called a *yolk spherule, yolk ball* or *yolk segment*, is an independent structural unit consisting of nuclei, cytoplasm and yolk, enclosed in a thin membrane, and even endowed with motility (see p. [471]) (Takami 1953b, 1954a, b), it seems convenient to give it the name of yolk cell, to distinguish it from the vitellophage. The latter, however, is not simply a nucleus, but rather a unit including a surrounding mass of dense cytoplasm; for this reason the term 'yolk cell' is often used interchangeably with 'vitellophage'.

It should be noted here that the yolk, after the completion of yolk cell formation by the segmentation process, is no longer a mere deposit of nutriment containing vitellophages, but now constitutes a *yolk tissue* composed of cells with inclusions of yolk, analogous to the inclusions of fat in fatty tissue. Whereas fat, however, is a direct metabolic product stored in the cells, the yolk was produced by maternal metabolism without relation to the development of the embryo, and is enveloped in the yolk cells. When the question of the embryo-yolk relationship is under consideration in connection, for example, with the phenomenon of diapause (see p. [473]), it should be kept well in mind that the relationship between nuclei and yolk has been made much closer in consequence of the yolk segmentation.

2. FACTORS DETERMINING EARLY STAGES OF EMBRYONIC DEVELOPMENT

The formation of the germ-band indicates the completion of the framework of embryogenesis. The processes of further development, although they may attract attention by their striking and complicated morphological changes, can in one sense be described as merely additions applied to a finished frame. The greatest significance of development is therefore hidden within the early processes leading to germ band formation; this section will deal with these points.

Since every cleavage nucleus has received a set of genetic factors from the parents, and since these genetic factors play an incontestably important role in the realization of biological characters, each stage of development seems to be an expression of the actions of the genes contained in the cleavage nuclei. However, in addition to such genic activity, the condition of the egg cytoplasm is of undeniable importance, at least during early development.

(1) POLARITY

Owing to the presence of the tough chorion, the place of sperm entry into the insect egg is strictly limited to the micropyle, the position of which is determined in the course of egg formation within the ovary. Furthermore, the head of the embryo is always directed toward the end of the egg where the micropyle is located. There is thus no doubt that the *cranio-caudal axis* of the embryo is determined while the egg is still within the maternal ovary. The end of the egg to which the head will be directed is called the *anterior pole*, and the opposite end, the *posterior pole*. In the more advanced stages of embryonic development, however, the initial head-tail orientation is often lost as the result of the embryonic movements (p. [451]) (Fig. 10.79).

How the cranio-caudal axis is determined is not clear, but dissection shows that all the ovarian eggs are arranged so that their anterior poles point upward. In ovaries of *Drosophila hydei* isolated in Ringer's solution, Child (1943) observed an intensity difference in the formation of indophenol blue, which occurs in the presence of an oxidase, when a mixture of dimethyl-*p*-phenyldiamine and α-naphthol was introduced into the living tissue. By this method he recognized the presence of a physiological gradient of indophenol-blue formation, which is highest in the anterior part of the ovary, and gradually diminishes posteriorly. Child attributed the origin of a similar gradient in the egg to this ovarian gradient.

The eggs of some insects, such as the silkworm (Fig. 10.70) and the dragonfly *Platycnemis* (Fig. 10.61), are dorso-ventrally asymmetrical and they therefore have a *dorso-ventral axis* — i.e., a morphological differentiation between the ventral, germ-band-forming side and the dorsal extra-embryonic area-forming side. This dorso-ventral axis also is determined while the egg is still in the ovary. The dorsal side is represented by the convex side of the egg in the silkworm (Takahashi and Yagi 1926; Takami 1943), and by the concave side in *Platycnemis*. According to Child (1943), the dorso-ventral physiological gradient also reflects the gradient found in the ovary.

(2) CENTERS OF DEVELOPMENT

Paralleling in interest the problem of polarity is the question of the centers of development. Seidel (1926, 1928, 1929a, b, 1932) performed a series of delicate experiments on developing eggs of *Platycnemis* by means of cauterization, 'Strahlenstichmethode (a technique for selectively destroying a particular part by ultraviolet irradiation), ligaturing, etc., which enabled him to extirpate or isolate various portions of the egg from the overall process of development, and demonstrate the presence of two development centers, each of which plays a remarkably important role in early development. These are known, respectively, as the Bildungszentrum and the differentiation center (Differenzierungszentrum). The Bildungszentrum has also been called 'activation center' by Wigglesworth (1950) and 'Determinationszentrum' by Reith (1931).

In addition to these two centers, there is also a *cleavage center*.

a. **Cleavage center**

In the fertilized egg of *Tachycines asynamorus* (Orthoptera), the first cleavage nucleus is situated at a distance equal to 43% of the longer diameter from the posterior pole, and cleavage begins here. It is also in the area overlying this position that the cleavage nuclei first arrive at the egg surface, and here blastoderm formation is initiated. This, then, is where development takes its start. The name of cleavage center was given to this area by Krause (1939).

The position of the cleavage center varies with the species. In the cricket *Gryllus*, like *Tachycines*, a member of the Orthoptera, it lies at 40% of the distance from the posterior pole; in *Platycnemis* (Odonata), 60% (Fig. 10.79A); in the backswimmer *Notonecta* (Hemiptera), 75—80%; in *Tenebrio* (Coleoptera), 33%, in the aphid-lion *Chrysopa* (Neuroptera), 80%; in *Ephestia* (Lepidoptera) 85% (Fig. 10.79B); and in *Drosophila* (Diptera), about 70% (Krause 1939).

b. Activation center

When an egg of *Platycnemis pennipes* is completely ligatured in the central zone of the prospective embryonic area (Fig. 10.65b, around *DZ*) in the early blastoderm stage, the formation of the embryo is limited to the part posterior to the ligature. In the anterior part, yolk segmentation takes place, and migration of cells occurs, but embryogenesis is completely suppressed; only a structure showing some resemblance to the extra-embryonic blastoderm is formed. If the ligature is not sufficiently tight, a head may be formed anterior to it, and a thorax and abdomen, posteriorly.

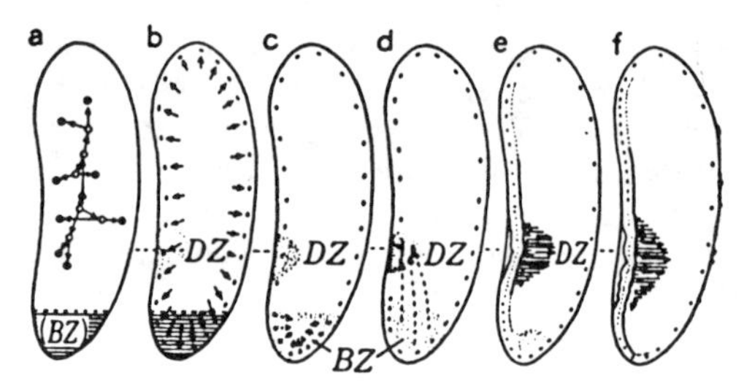

Fig. 10.65 Foci of development in *Platycnemis pennipes* (Weber 1938)

a. arrows show course of nuclear division to 8-nucleus stage. (BZ): Activation Center before activation; b. cleavage nuclei form blastoderm and enter presumptive area of Activation Center; c. activation of Activation Center by cleavage nuclei; d. substance diffusing from activated Activation Center affects yolk in area of presumptive Differentiation Center (DZ); e. yolk of this area contracts and separates from chorion, blastodermal cells aggregate toward this area and differentiation begins here; f. these morphogenetic movements extend from this region toward anterior and posterior ends.

The *Platycnemis* egg has an average length of 37 units (1 unit= 24μ). When the posterior end of the egg is partially cauterized in the early blastoderm stage, the undamaged part of the egg forms a complete embryo as long as the lesion does not exceed 3—4 units. If it is more extensive, the remaining part alone carries out yolk segmentation, and embryogenesis fails to occur, the developmental process progressing no further than the formation of an extra-embryonic blastoderm-like structure. On the other hand, a lesion applied at the anterior end or in a lateral part of the egg does not influence germ-band formation, although it may occasionally cause some abnormalities in the resulting embryo. These experimental results point to the presence at the posterior end of the insect egg of an important center of development — i. e., the activation center, in the absence of which the blastoderm is unable to form a germ-band.

The next experiment is performed at an earlier stage — i. e., before the cleavage nuclei arrive at the posterior end of the egg. If a ligature is then applied between the nuclei and the posterior pole, the formation of the embryo is completely inhibited except when the ligature is extremely close to the posterior end. But if the ligature is loosened and the cleavage nuclei are allowed to migrate to the posterior pole, or when the ligature is made, after nuclei have been observed to reach the posterior end, in such a way that it permits a certain degree of cytoplasmic communication, embryo formation takes place. It is therefore evident that the posterior portion of the egg has achieved a particular state, before the arrival of the cleavage nuclei, which is not replaceable by other cytoplasmic components of the egg. This activation center is not, however, activated before the arrival of the cleavage nuclei. It follows that the capacity for determination is not due either to the posterior periplasm or to the cleavage nuclei migrating into it, but to an interaction of the two. It is believed that such interaction produces some substance which diffuses anteriorly and determines the developmental fate of each part of the blastoderm.

When one of the nuclei is destroyed by ultraviolet irradiation at the 2-nucleus stage, the surviving nucleus divides and proliferates, and its descendants migrate to the egg surface. Although the activation center thus receives some nuclei which would normally have been destined for other parts of the egg, and the number of nuclei entering it is also less than usual, embryogenesis proceeds normally. It may therefore be concluded that the cleavage nuclei are still equipotent at this stage of development, and that the action of the activation center does not depend directly upon the number of nuclei entering it.

As the result of various experiments, it is now evident that the anterior border of the activation center corresponds to the *Einrollungsstelle*, the place where involution of the embryo later occurs (Fig. 10.61). As development proceeds, the center expands, and germ-band formation is no longer disturbed even when a larger lesion is made in this region. By the latter part of the blastoderm-forming stage, each area of the blastoderm has become capable of self-differentiation in accordance with its presumptive fate (Fig. 10.66).

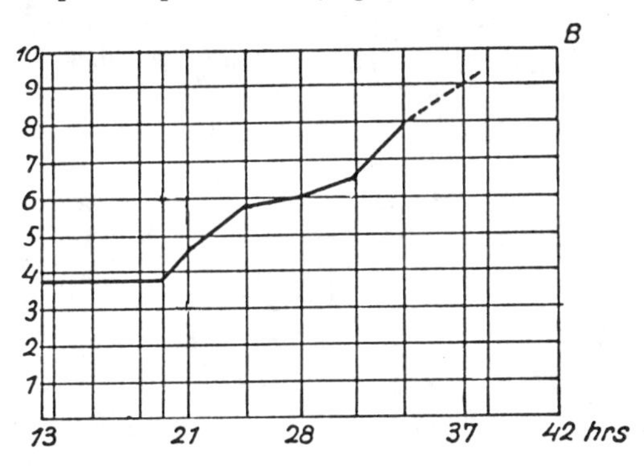

Fig. 10.66 Change in mode of action of Activation Center accompanying development in *Platycnemis pennipes* (modified from Siedel 1929b)

Ordinates: distance from posterior pole of the egg (1 unit = 24μ). Abscissae: time in hours after 4-nucleus stage at 21.5° C. The curve shows the change with development in the position of the posterior-most level of the egg where cauterization will not inhibit subsequent germ band formation. B: stage at which nuclei begin to aggregate to form germ band.

Unlike the amphibian organizer, the activation center is not a center of differentiation, nor does it show a particularly high rate of nuclear division. The existence of an activation center has been demonstrated in *Camponotus* (Hymenoptera; Reith 1931), *Sitona* (Coleoptera; Reith 1935), *Bruchus* (Coleoptera; Bauer and Taylor 1936), in addition to *Platycnemis*.

c. **Differentiation center**

The differentiation center is situated in the thoracic part of the presumptive germ-band. In the *Platycnemis* egg it lies 10—15 Seidel units from the posterior end, and begins to operate only after it has been acted upon by the activation center (Fig. 10.65d, e, f). If a ligature is applied posterior to the differentiation center just after each part of the blastoderm has been 'determined' by the activity of the activation center, embryo formation is limited to the part anterior to the ligation. A ligature anterior to the differentiation center restricts the formation of the embryo to the region posterior to the ligature. Even when the tying is done loosely enough to permit the passage of diffusible substances from the differentiation center, the embryo-forming capacity is not restored; for embryogenesis to take place it is necessary that at least part of the differentiation center be left untied. If a ligature is applied at the 10—15 unit level (i. e., at the middle of the differentiation center), both anterior and posterior portions will include some part of the differentiation center, and embryo formation is observed in both parts.

The mode of action of the differentiation center differs, however, from that of the activation center; even if its blastodermal cells are all destroyed by cauterization, its activity is not affected so long as its yolk cells remain undamaged. In other words, the effect of the differentiation center does not depend upon cellular activity; its essential role consists rather in its influence on the yolk in its vicinity. Reacting to the active substance from the differentiation center, the yolk contracts and withdraws from the chorion, leaving a space into which blastodermal cells migrate. If contraction of the yolk is artificially provoked by heat treatment or ultra-violet irradiation, such concentrations of blastodermal cells can be induced in any part of the egg, without reference to the differentiation center (Wigglesworth 1950). The wave of these contractions and concentrations spreads forward and backward, causing the blastodermal cells to migrate in sequence toward the differentiation center and take part in the formation of the germ-band.

The existence of a differentiation center has been established in many species of insects other than *Platycnemis:* these include *Pyrrhocoris* (Hemiptera) (Wigglesworth 1950; Krause 1939), *Chrysopa* (Neuroptera) (Wigglesworth 1950; Krause 1939), *Apis* (Hymenoptera) (Schnetter 1934; Wigglesworth 1950; Krause 1939), *Sitona* (Coleoptera) (Reith 1935), *Tenebrio* (Coleoptera) (Ewest 1937; Wigglesworth 1950; Krause 1939), *Ephestia* (Lepidoptera) (Maschlanka 1937; Krause 1939).

As has been outlined above, the differentiation center exerts its effect only after it has been acted upon by the activation center, and the latter is not activated until it has reacted with the cleavage nuclei. The importance of the nuclear substance in connection with the early stages of development it thus indisputably clear; at the same time, the fact that the topographical location of the activation center is restricted to the posterior end of the egg suggests that there may be an embryologically important state of the egg protoplasm already determined before fertilization, and that this state depends on the determination of the egg axis.

(3) PRE-DETERMINATION OF DEVELOPMENT

In the *Platycnemis* egg, the prospective fate of each region of the blastoderm is thus seen to be determined only after it has been affected by the activation center. From this time onward, the egg is capable of self-differentiation: even the removal of a part of the embryo has no influence on the development of the remainder. The time and degree of determination differ, however, in various species of insects (Krause 1939; Oka 1935). The egg of *Tachycines asynamorus*, for example, possesses such a high degree of regulative power that if a germ-band is cut in half along its median line, each half retains the capacity to develop into a complete embryo (Krause 1930). The egg of the cricket *Gryllus mitratus* is also highly regulative. Cauterization of the anterior 1/5 before the formation of the blastoderm causes some delay in development but does not prevent the formation of a complete embryo. Destruction of the same part shortly after the formation of the blastoderm is followed by the appearance of a small but complete embryo. Destruction of the posterior 1/10 at this stage does not influence embryogenesis — the resulting embryo is almost normal (Oka 1934). Seidel (1929a) ligatured the egg of *Platycnemis pennipes* in the 4-nucleus stage at about the middle, and observed that the posterior part, consisting of 19 units including the determination and differentiation centers, formed a small but complete embryo. This result indicates that the egg of this stage has the necessary regulative power to construct a complete embryo from insufficient materials. In an early stage of blastoderm formation, the egg is capable of producing an entire embryo even if the posterior 3—4 units (ca 1/9 of the total length) are cauterized. In the late blastoderm stage, the maximum damage which can be overcome is 5.5 units (ca. 1/7 of the length); this greater regulative power in the later stage is due to the increased range of the activation center. The formation of embryos with multiple heads or intestines (Seidel 1928) and the double embryo showing emboitement (Fig. 10.67) (Seidel 1929a) attest to the greater regulative potency of the early stages.

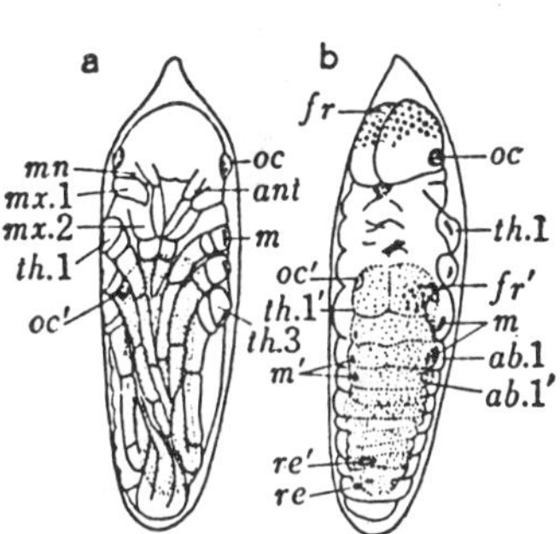

Fig. 10.67 Dwarf internal embryo in *Platycnemis* (Seidel 1929a). Such an embryo results from transverse section made 4 or 5 units from the anterior pole (cf. Fig. 10.66) in the early nuclear cleavage stage.

a. ventral view; b. dorsal view. *ab.*1: first abdominal segment, *ant:* antenna, *fr:* frontal plate, *m:* chitinous structures to which muscles are attached, *mn:* mandible, *mx.*1: first maxilla, *mx.* 2: second maxilla, *oc:* ocellus, *re:* end of rectum, *th.* 1: prothorax, *th.* 3: metathorax. Symbols with dash show organs in small inner embryo. Section of such double monsters show that organs in inner embryo are everted (cf. p. 82).

An egg possessing regulative power is called a *regulative egg*, in contrast to the *non-regulative egg* which lacks this capacity. A non-regulative egg is generally unable to form a whole embryo from a part; each region develops independently of the others. Eggs of this type are therefore also called *mosaic eggs;* the fate of each part of such eggs is already determined before oviposition. It might seem an easy matter to distinguish eggs of this type from regulative eggs, but in fact the difference is not distinct. Even a highly regulative egg, such as that of *Platycnemis*, is no longer regulative in an advanced stage, each part of the embryo forming its structures independently of the others (p. [419]), and the same thing can be said of all insect eggs. At 31—35 hours after oviposition, the egg of *Tenebrio moliter* possesses the regulative capacity to construct a complete embryo even if the posterior 31% of it is removed, but a little later it can form only a partial, defective embryo (Ewest 1937).

Twelve hours after oviposition (at 33—34°C and 80—90% RH), the egg of the honey bee *Apis mellifica* is able to form a small but complete embryo in the part posterior to a ligature applied at a level 21% of the length from the anterior end, but by 24 hrs after oviposition, the degree of determination has advanced so that ligaturing leads to the production of partial embryo before and behind the site of ligature. At the 12-hr stage, a ligature applied more posteriorly (than 21%) results in the appearance of a headless embryo; if the site of tying passes the 27% level, the resulting embryo lacks the gnathous region, and if it is moved more toward the posterior than 33%, the thorax is lost as well. These defects arise discontinuously; there are no records of intermediate defects such as absence of the anterior half of the head or of the head together with part of the gnathous region. A survey made to determine the distribution on the egg surface of the anlages of these unit parts — i. e., the frons, gnathous region, thorax and abdomen — showed that 12 hours after oviposition they are concentrated around a center situated at the boundary between the gnathous and thoracic regions (Fig. 10.68), about 76% of the length from the posterior end. This location coincides with the position of the differentiation center. Twenty-four hours after oviposition, there is a diminution in the degree of concentration of these anlages (Schnetter 1934).

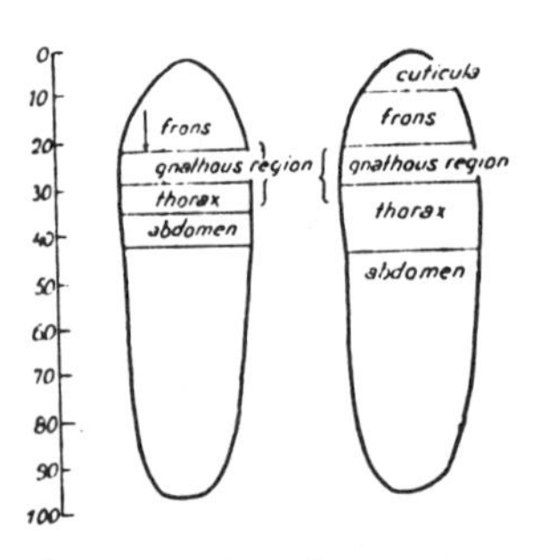

Fig. 10.68 Map of presumptive organ anlages in the honeybee embryo (Schnetter 1934, from Motomura 1941)

left: 12 hours after oviposition; right: 24 hours after oviposition. The transverse lines show the posterior boundary of each presumptive area; if a section of an embryo includes all the material of a given area, the presumptive organs will be formed *in toto*. These lines therefore represent the boundaries of regulative capacity.

The discontinuous manner in which the defects appear suggests the existence of an exclusive regulative power which is fully effective within one *unit region* but cannot extend its influence to adjacent regions. A similar situation has been demonstrated in *Ephestia kühniella* (Maschlanka 1938). The egg of this moth can form a complete embryo even if its posteriormost 20% is cauterized in the cleavage or blastodermal stage. More extensive damage results in the formation of defective embryos in which the defects are more extreme as the area of lesion increases. On the other hand, cauterizing the anterior 20% at the cleavage stage induces partial or total defects in the frontal region, involving the cranium, antennae, labrum, stomodaeum, etc. When the cauterized area even slightly exceeds 20%, however, the defects abruptly take the form of the absence of the whole head, frequently accompanied by some abnormality of the thorax; no intermediate grade of deficiency is found. It is therefore believed that the gnathous region behaves as a unit region (Organkreis).

Although the silkworm *Bombyx mori*, like *Ephestia*, belongs to the Lepidoptera, no such unit regions can be detected in its development. Its egg also possesses less regulative power; burning of the posterior 20% in the early (2 hrs after oviposition) or late (7 hrs) cleavage stage invariably causes the formation of a partial embryo (Takami 1942). As development proceeds, cauterization of the posterior 20% becomes less deleterious: some perfectly formed embryos appear after such an operation made in the blastoderm stage (14 hrs after oviposition), and a still greater number when the operation is performed 10 hours later, in the germ-band stage. This does not mean, however, that there is an increase in regulative power or in the range of effectiveness of the activation center in the course of development. It is more likely due simply to the fact that in the egg of these stages, the embryonic area has undergone a considerable contraction, and since it lies at a greater distance from the posterior end, is less damaged by the cauterizing. When the anterior end is cauterized, the destruction of 20% in both blastoderm and germ-band stages causes the production of partial embryos. The maximum amount of damage that would permit complete embryo formation was found to be about 15%.

In the silkworm, then, cauterizing as early as 2 hours after oviposition — i. e., very early in the cleavage stage, since fertilization usually takes place 1.5 hours after oviposition — also causes the formation of partial embryos. It is thus evident that determination has already been accomplished before the arrival of the cleavage nuclei at the egg surface. Precise examination of these partial embryos, however, brings to light some notable exceptions: for example, the formation of integument along the border zone between the burnt and uninjured parts; frequent completion of mid-intestine formation after cauterization of the stomodaeum, which cannot be expected from the principle that the mid-intestine is formed as the result of contact between two endodermal projections taking their respective origins from the lower sides of the stomodaeum and proctodaeum; the occasional formation of embryos without the longitudinal ventral fracture representing the cauterized area, which usually occurs in some degree after ventral cauterization in the blastodermal stage (p.[428]). These facts sugest that the silkworm egg is endowed with a certain amount of regulative power, and cannot be regarded as a strictly mosaic egg.

Various abnormalities were found to result when silkworm eggs were centrifuged at 11,200 g for 30 to 50 seconds, one to ten hours after oviposition (Sakaguchi 1952a). The type of the abnormality was dependent upon the direction in which the centrifugal force acted on these eggs. When centrifugation was parallel to the long axis of the egg and the posterior pole of the egg was directed centrifugally, the resulting abnormalities were always limited to the part anterior to the 8th abdominal segment. With the direction of the axis reversed, the abnormalities were found only in the abdomen. Centrifugal force acting parallel to the short axis from either the ventral or the dorsal side, however, produced abnormalities of both thorax and abdomen. In all cases, the abnormalities occurred most frequently near the 2nd—3rd and the 6th abdominal segments, extending anteriorly and posteriorly from these loci. Similar results characterize the mode of appearance of certain genetic malformations and of abnormalities induced by acid treatment (p. [472]), suggesting that abnormal formation depends in some way upon the mechanism of segmentation. These effects have been explained as disturbances in the differentiation pattern following centrifugation which dislocates the predetermined parts of the pattern. It is interesting to note that double monsters often occurred, with laterally reduplicated sets of caudal horns, thoracic legs, dorsal vessels, testes, etc.; this may indicate that the silkworm egg possesses a certain amount of regulative capacity. Similar monsters are also formed following centrifugation of the egg of *Attacus ricini*. In this species, however, double formation is much more prevalent than in the silkworm, where each part of an organ tends to show a high degree of independent development. Centrifugation of

Attacus eggs five hours after oviposition resulted in longitudinally cleft embryos which, by stretching the concept a bit, could be regarded as double embryos (Sakaguchi 1952a, b). Since the determination of lepidopteran eggs has been believed to occur very early, it is of great interest to know whether these embryos are really to some extent double, or merely cleft. These results, as well as those obtained in *Chironomus* (p.[426]) by Yajima (1956), suggest many interesting problems in connection with the embryonic development of insects. Even in *Attacus*, however, some parts of the embryo may complete their normal differentiation independent of other parts of the body, so that it is no less reasonable to believe that embryonic differentiation takes place with a considerable degree of independance in each segment.

The ant, *Camponotus ligniperda*, is another insect in which determination takes place in an early stage. Cautery experiments carried out on *Camponotus* eggs have shown that the developmental fate of each region of the egg is already determined when the cleavage nuclei arrive at the egg surface, 6—18 hours after oviposition (Reith 1931). Centrifuging before the egg surface was determined suppressed germ-band formation, but had no suppressive effect if performed after determination was complete. Centrifugation experiments on the eggs of a related species, *Lasius niger*, gave similar results (Reith, cited by Oka 1935). In the egg of *Sitona lineata*, also, determination takes place at the cleavage stage (Reith 1935).

Well-known among mosaic eggs are those of the Colorado potato beetle *(Leptinotarsa decemlineata)* and the house fly *(Musca domestica)*. When any region of the *Leptinotarsa* egg was killed by cautery, the resulting embryo was always partial, regardless of whether the treatment was applied before or after blastoderm formation (Hegner 1911). Since, in this case, cauterizing the egg surface prior to the arrival of the cleavage nuclei was sufficient to induce the formation of partial embryos, the determination must already have taken place within the cytoplasm without relation to the nuclei, as in the case of the silkworm. When the eggs of some chrysomelid beetles such as *Leptinotarsa, Calligrapha multipunctata, C. bigsbyana, C. lunata, Lema trilineata*, etc., were subjected to centrifugation, their contents were divided into four layers: grey cap, yolk, cytoplasm and fatty material. The embryos formed, nevertheless, were practically normal if the centrifugation was performed after blastoderm formation, but some dwarf embryos resulted from centrifugation before blastoderm formation. These dwarfs proved to have been formed within the cytoplasmic layer, indicating that predetermination had already taken place in the cytoplasm (Hegner 1909). Partial cautery or ligature experiments in the flies *Musica domestica* and *Calliphora erythrocephala* always resulted in forming partial embryos, showing that these species really belong among the mosaic eggs, a view which was supported by the results of centrifugation experiments (Reith 1925; Pauli 1927). However, the egg of *Drosophila melanogaster*, also believed to be one of the typical mosaic eggs, was capable of normal development even after being punctured by a fine glass needle which caused a partial outflow of the cytoplasm, provided this was done before fertilization, within 20 minutes after oviposition (Howland and Sonnenblick 1936).

According to Geigy (1932), mortality of the eggs of *Drosophila melanogaster* following ultraviolet irradiation is low before blastoderm formation, and increases as the latter proceeds (Stage I). Three to six hours after oviposition the mortality rises to almost 100%, decreasing again seven hours after oviposition (Stage II).

Following ultraviolet irradiation at this stage, no deficiencies were observed in the resulting larvae, although various abnormalities appeared in the imagines after metamorphosis. When irradiation was administered early in Stage II, such imaginal abnormalities were mainly restricted to the wings and thoracic region, never affecting the legs. When embryos were irradiated in the middle of Stage II, the deficiencies appeared only in the legs: most frequently in the third thoracic, less in the second, and least in the first. Irradiation in the latter part of Stage II caused markedly fewer deficiencies; those which did appear were concentrated in the third thoracic legs, and the degree of abnormality was lower than in earlier periods.

That these imaginal abnormalities following irradiation during Stage II involved not only malformations but also partial double monsters suggests an incomplete determination of the imaginal anlages at this stage. The imaginal determination appears to proceed from anterior to posterior, like larval determination in *Platycnemis,* which originates at the determination center and extends anteriorly and posteriorly.

From these results, Geigy concluded that in holometabolous insects such as *Drosophila,* the differentiation of the larval organs is already determined at fertilization or earlier, while the imaginal organs are determined much later, with a distinct interval separating the two processes. In hemimetabolous insects like *Platycnemis,* however, he supposed that these two kinds of determination take place in a continuous succession. According to Lüscher (Bodenstein 1953), who treated eggs of the moth *Tineola* with ultraviolet rays, determination of the larval organs is still incomplete at the blastoderm stage, so that organ duplication may be induced by the irradiation. At the stage of amnion formation (about 16 hours after oviposition), however, the determination has been completed, and irradiation after this time results not in duplication, but only in organ deficiencies. Double formation of the imaginal organs, on the other hand, may be evoked by irradiation even after the germ band has been fully formed (about 24 hrs after oviposition), indicating that their determination is not yet complete at this time. It is not until about 72 hours after oviposition, according to Lüscher, that determination is finally completed. The puncture experiments of Howland and Child (1935) on *Drosophila* with respect to the time of imaginal determination gave results similar to those of Geigy.

As this short review shows, the terms "regulative" and "nonregulative eggs" refer only to the two extremities of one continuous series arranged according to whether determination occurs earlier or later, so that it is difficult to set up precise criteria. Krause (1939), and Seidel, Bock and Krause (1940) insist that eggs with a generous amount of well-differentiated periplasm (Keimhautblastem), which form a large embryo, are highly mosaic, determination in them taking place much earlier than it does in eggs with less periplasm and a smaller embryo. Among the Diptera, the eggs of which have the most highly developed periplasm of any insect order, development of the periplasm originates at the two poles of the egg as the result of yolk contraction following fertilization, and gradually spreads to cover the whole egg surface (Weismann 1863). Considerations of this kind make it necessary to include a time factor in any attempt to explain differentiation.

The results of partial cautery or ligature experiments, designed to observe the development of an embryo from which some part has been eliminated, have ed to the concept of "predetermination of development". They do not, however, provide evidence for considering the periplasm of a particular region to have some definite and unchangeably fixed developmental fate, such that if transplanted to another region it would self-differentiate independently of the over-all development of the host. Although the periplasm is distinguishable from the inner protoplasm, it is nevertheless a part of this fluid substance, and there is not sufficient reason to regard it as a distinct structure, clearly separate from the ordinary endoplasm.

Although the periplasm of some particular region may seem to have a definite prospective fate, this may well be only an apparent specificity consequent upon the isolation of the region from its normal situation in a physiological gradient involving the egg as a whole. It follows, therefore, that the behavior of a part of the periplasm which has been cut out and transplanted into a different gradient can never be the same as it would have been *in situ*. In practice, the impossibility of transplanting the periplasm in its natural form makes it difficult to verify this point, but a reconsideration of centrifugation experiments from this point of view might contribute to a solution of the problem. Some of the earlier work of this kind has been referred to above (p. [423]); more recently Yajima (1956) has observed the formation of embryos with double cephalon or double abdomen when *Chironomus* eggs (usually classified as mosaic) are centrifuged before the nuclei arrive at the egg surface. Double cephalic embryos were obtained when the centrifugal force was directed toward the posterior end of the egg, while centrifuging in the opposite sense gave bicaudal embryos.

(4) PRESUMPTIVE REGIONS IN DEVELOPMENT

From many experimental results such as those described above, Seidel (1935) constructed the presumptive map of the organ anlages in the egg of *Platycnemis* shown in Figure 10.69. Since the movement of the nuclei can be traced from the outside in the living *Platycnemis* egg, it provides an excellent material for research, and the arrangement of the presumptive anlages has been more fully studied in this insect than in any other. A similar mapping of the presumptive regions in the honey-bee embryo, as described above (p.[422]), has disclosed the existence of unit areas of organogenesis (Fig. 10.68).

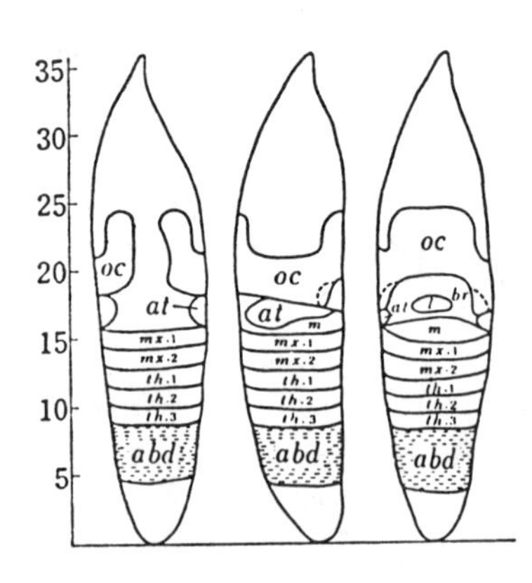

Fig. 10.69 Map of presumptive organ anlages in late blastoderm stage of *Platycnemis pennipes* (Seidel 1953)

left: ventral, middle: lateral, right: dorsal. *abd:* abdominal part, *at:* antenna, *br:* brain, *l:* labrum, *m:* mandible, *mx.*1*:* first maxilla, *mx*2*:* second maxilla, *oc:* ocellus, *th:*1~*th.*3: pre-, meso- metathorax.

Some presumptive areas can be recognized morphologically, as, for example, the posterior pole plasm in the eggs of certain Diptera, Coleoptera and other groups: only the cleavage nuclei entering here finally participate in the differentiation of the germ cells (p. [406]).

No precise knowledge with respect to presumptive regions in the silkworm egg has yet been accumulated, although Takami (unpublished) has proposed a tentative map of the silkworm embryo at the blastoderm stage, on the basis of his experimental results (Fig. 10.70a). The anterior and posterior limits of the embryonic area proposed in this map practically coincide with those experimentally

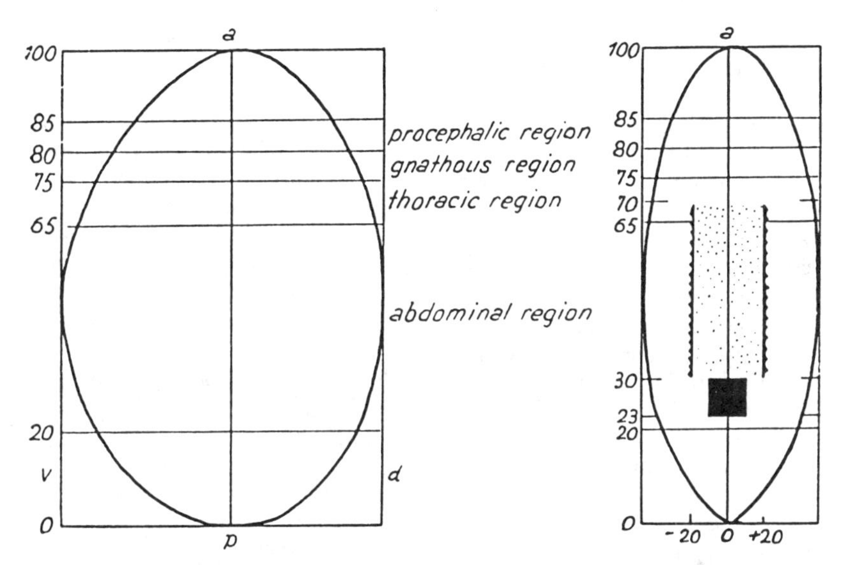

Fig. 10.70 Presumptive areas of *Bombyx mori* egg (Takami)
:::::: mesodermal area, —: nerve area, ■: germ cell area. a. lateral view; b. ventral view. *a:* anterior pole of egg; *d:* dorsal side, *p:* posterior pole, *v:* ventral side.

determined by Maschlanka (1938) in another lepidopteran, *Ephestia* (Fig. 10.79B). Partial cautery applied to the ventral side of the silkworm embryo before the germ-band stage induces the formation of a longitudinally cleft embryo (Takami 1943). Since such an embryo either has or lacks mesoderm and ganglia according to the size of the burnt area, it is possible to estimate the extent of the presumptive mesoderm. From the results of such experiments on embryos in the early cleavage (2—3.5 hrs after oviposition at 25° C) and late blastoderm (17.5—18 hrs at 22—28° C) stages, it can be assumed that so far as the central region lying between 30% and 70% of the long axis is concerned, the presumptive mesoderm extends to a distance of 20% on both sides of the short axis (Fig. 10.70b). Lateral to this band of presumptive mesoderm lie narrow zones of presumptive neural tissue (Takami 1946).

Similar experiments performed in the period between cleavage and blastoderm formation (4—10 hrs after oviposition) have demonstrated that the presumptive germ-cell region occupies a definite area of the presumptive mesoderm (Miya 1949). In the blastoderm stage (12—13 hrs after oviposition at 27—28°), this extends laterally on each side of the median ventral line to a distance of 4.5% of the half-circumference (Miya 1951a, 1953). On the long axis (in the 10-hr embryo), it lies between 23% and 30% of the distance from the posterior pole (Miya 1950).

Kawaguchi (1938) succeeded in producing laterally mosaic silkworms in which the patterns of the right and left halves of the body were under the control of different genes. Using this material, he studied the relationship between the type of mosaicity and mode of gene segregation in filial generations. From the results of this study, he estimated the number of cells which are first recognizable as germ-cells to be about twenty. He further deduced that if any germ-plasm is present in silkworm eggs, its distribution may be one-sided with respect to the median line. It is not impossible, however, for the genic constitution of the ventral integument of the embryo (ectoderm) to differ from that of the germ-cells just beneath it in such a mosaic embryo, since the two kinds of tissue arise from different germ layers at different sites, the germ-cells having their origin in the mesodermal region which invaginates along the mid-ventral line of the blastoderm (Takami 1946). This problem of the non-median location of the germ plasm remains unsolved, although many cases have been observed in which part of the germ cells are actually located on one side or the other of the median line (Miya 1947).

The silkworm egg is believed to belong among the so-called non-regulative eggs. Sometimes, however, a few embryos appear without any longitudinal fissure even after cautery on the ventral side. It therefore seems likely that these embryos are endowed with some degree of regulative capacity, at least within a single segment. It is interesting to note that such non-cleft embryos seem to possess a greater capacity for regulation of ganglion formation than do the cleft embryos. As can be understood from Figure 10.70, neither mesoderm nor ganglia will be produced in the cleft embryo if the burnt area exceeds 20% of the short axis on both sides of the median line. In the absence of a fissure, however, paired ganglia may be produced even if the injury extends beyond the mesodermal limit of one side, provided that a considerable part of the mesoderm is left intact on the other side. The reason for this curious phenomenon is still obscure, although it appears likely that the mesoderm of the less damaged side is able to develop across the median line and exert a nerve-inducing effect on the ectoderm of the more severely damaged side (Takami 1946). From this point of view, much interest attaches to the recent work of Suzuki and Ichimaru (1955). These workers find that a lethal mutant of the silkworm, *reniform egg*, fails to form mesoderm. Only abnormal ectodermal organs develop, and no ganglia except that of the brain are ever formed. These may be examples of the situation in which mesoderm induces the differentiation of the ectoderm just above it.

Seidel, Bock and Krause (1940) have reported a case in which ectoderm was found to exert an inductive effect on the mesoderm. Using eggs of the neuropteran *Chrysopa perla*, Bock cauterized the presumptive mesoderm prior to invagination, and observed an increasing degree of deficiency in the mesodermal organs with increase in the treated area, but no loss of ectoderm. (There is unfortunately no description concerning the nervous system.) His results seem to indicate that the ectoderm is capable of self-differentiation without receiving any influence from the mesoderm. However, the form of the ectodermal organs in these embryos proved to be abnormal. When the presumptive mesoderm on one side of the median line was cauterized, the mesoderm of the intact side neither extended beyond the median line nor compensated for the lost part, although some degree of compensation occurred on the same side. If too much of the presumptive mesodermal area was destroyed, the rest was no longer able to perform such compensatory development, so that the mesodermal lining of the surface ectoderm remained conspicuously incomplete. In intact embryos, this mesodermal layer spreads under the whole ectoderm, whereas in the treated ones, it was often restricted to a narrow zone along the mid-ventral line. Such meager mesoderm was able to differentiate into only the mesodermal organs corresponding to the ectodermal part which it was lining.

In another experiment in which the ectoderm of one side was completely killed by cautery, the mesoderm which had invaginated normally and spread beneath it eventually degenerated without forming coelomic epithelium. Destruction of the presumptive ectoderm of one side of the embryo, again, caused different kinds of mesodermal organs to be produced according to whether the site of cautery was median or peripheral. No mesodermal organs were formed beneath the burnt areas, while under the intact ectoderm there were produced organs corresponding to this part.

Recent work of Miya (1956) on the silkworm has led to the hypothesis that an inductive effect originating in the ectoderm must be required to bring about the differentiation of the genital ridge from the mesoderm. These three sets of results, obtained by independent experimentation, seem to suggest an induction phenomenon in insect development which has heretofore not been given full consideration.

(5) NUCLEO-CYTOPLASMIC RELATION

The important role of the cortical cytoplasm in the determination of development, which has been discussed in the preceding sections, does not detract from the significance of the leading part played by the cleavage nuclei in the very early stages.

The normal sex-chromosome constitution of *Drosophila melanogaster* is XX in the female and XY in the male. There is a strain, however, known to have XX fused together to form $\widehat{XX}$, the female having $\widehat{XX}Y$ while the male has XY. A fertilized egg may therefore contain any of four combinations: $\widehat{XX}X$, $\widehat{XX}Y$, XY or YY. Among these, $\widehat{XX}X$ is a superfemale, $\widehat{XX}Y$, an apparently normal female, XY, a normal male, and the combination YY is lethal, causing the early death of the embryo. Poulson (1940, 1945) studied the development of this YY egg (which he called "nullo-X"), and found a quite abnormal distribution of the cleavage nuclei, almost all of them remaining in the anterior half of the egg, and failing to form a blastoderm. The oxygen consumption of nullo-X eggs, as measured by the Cartesian diver technique applied to individual eggs, was found to decrease to one-fifth of normal at the stage in which such an abnormal distribution of the nuclei becomes apparent. Obviously the YY egg is produced by fertilization of a Y egg by a Y sperm. Since the same Y egg will give rise to a normal male if it is fertilized with an X sperm, its cytoplasm cannot be thought of as containing some agent which will disturb normal development. The nuclear origin of the disturbance is further suggested by the fact that the abnormality appears in the behavior of the nuclei themselves, before they reach the cortical cytoplasm.

Similarly, observations made on a translocation of *D. melanogaster*, $T(1:4)$ $A1$, revealed that a disordering effect leading to abnormal blastoderm formation resulted from the loss of the right arm of the *X* chromosome, from the locus of *lozenge* to the kinetochore. When the left arm was lost, however, a normal blastoderm was formed and no disturbance was manifested before germ layer formation. These results emphasize the importance of the cleavage nuclei in the early developmental processes.

On the other hand, Nakamura and Imaizumi (1950) and Imaizumi and Nakamura (1951) measured the voltage required to cause a dechorionated *Drosophila* egg, placed on a plate electrode at the bottom of a small chamber filled with liquid paraffin, to float toward an upper electrode. The voltage necessary to move the egg varied with the developmental stage, first increasing with time during early development (from oviposition to the appearance of the pole cells), reaching a maximum at the blastoderm stage, and then decreasing. Developing nullo-*X* eggs showed no difference in this respect from normal eggs, when their responses were compared at a stage corresponding to the blastoderm stage, whereas unfertilized eggs showed a lower responsiveness than these.

The silkworm mutant *No crescent* is known to be lethal when homozygous, causing early death of the embryo. In such homozygous individuals, the behavior of the cleavage nuclei was found to be abnormal even before their arrival at the egg surface (Ichimaru 1956).

Cytoplasmic division without any participation of the nuclei has been reported in the moth *Phragmatobia*. A blastoderm-like structure was formed, but it contained no nuclei, and failed to develop further (Seiler 1924).

In considering the nucleo-cytoplasmic relation in insect embryology, the determination of early development must be clearly distinguished from the determination of the final expression. In the *Drosophila* egg, for example, the posterior pole plasm (p. [446]) obviously represents a case of predetermination. Only the nuclei which enter here subsequently differentiate into germ-cells. This posterior pole plasm is determined, however, only as the locus of germ-cell differentiation, and the final morphological expression of maleness or femalennes depends entirely upon the genetic constitution of the cleavage nuclei to be introduced into it. Furthermore, not all the nuclei arriving here differentiate into germ-cells; a certain number of them later participate in the formation of the mesenteric (mid-intestinal) epithelium (Poulson 1947). In the silkworm as well, some of the potential germ-cells appear to degenerate before the gonad is formed (p. [445]).

Among silkworm mutants, there are several with various segment abnormalities, and some discussion has taken place with respect to the time of determination of such abnormalities (Tanaka *et al.* 1952). In the author's opinion, however, this problem cannot be solved by simply considering the determination in early development. Since development is a continuum of changes and is never fixed, each particular character must pass through innumerable steps involving genic actions and/or inductive effects from other tissues before its final expression is reached. From this standpoint it is not strange that Ichikawa (1947, 1952) obtained a result which seems contradictory to the ordinary conception of predetermination in the silkworm egg as it has been presented above.

Ichikawa studied androgenetic development, using the sperm of a silkworm with segment abnormalities *(additional crescent* and *new additional crescent).* He found that the embryos showed these abnormalities even when the sperm were introduced into normal egg cytoplasm.

On the other hand, some external factors may cause malformation even when they exert their action at relatively late stages in development, affording further evidence that the determination of early development is different from the determination of final expression.

In this connection, it is interesting to cite here the case of sex determination in *Phylloxera caryaecaulis,* which was pointed out by Morgan (1926). The eggs laid by an androparous female of this aphid are already smaller than those laid by a gynoparous female, even before they acquire a chromosome constitution determining maleness (cf. p. [454]). This suggests that the condition of maleness is determined earlier in the cytoplasm than in the nucleus. If for some reason, however, reduction of the sex chromatin should fail to take place in an egg laid by an androparous female, it would probably develop into a female (conclusive evidence to establish this point has not yet been obtained).

In the silkworm mutant called *kidney egg,* which shows recessive maternal inheritance, the eggs laid by homozygous females are all kidney-shaped, and all the offspring die, regardless of their own genic constitution, because they are unable to form mesoderm. In this case, therefore, the genotype of the mother determines not only the shape of the eggs but also the developmental fate of the resulting embryos.

3. GERM-LAYER FORMATION TO DIFFERENTIATION OF LARVAL ORGANS

(1) FORMATION OF GERM LAYERS

After the germ-band has differentiated from the blastoderm, its tissues begin to invaginate in a region, corresponding to the differentiation center, which lies close to the anterior ventral end of the germ-band. A strip of cells lying along the ventral median line, which are rounder and more irregularly arranged than their neighbors, migrate inward, forming the *primitive groove* (Figs. 10.71, 10.72).

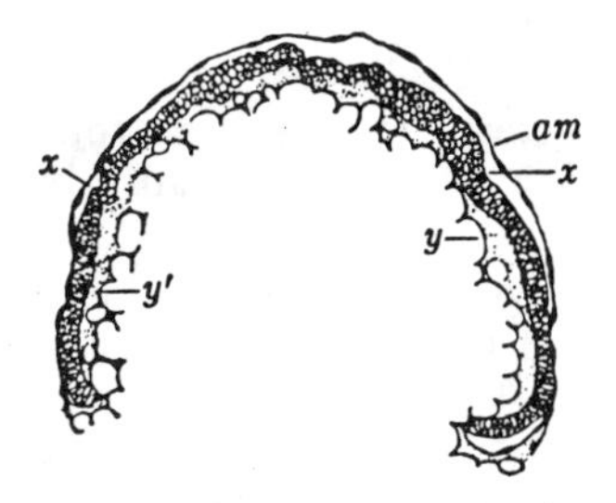

Fig. 10.71 Early stage of primitive groove invagination in *Antheraea pernyi*. Transverse section of germ band (Saito 1937)

am: amnion, *x:* boundary between invaginating part and ectoderm, *y:* yolk, *y':* liquefied yolk. Cells in invaginating part are round and irregularly arranged.

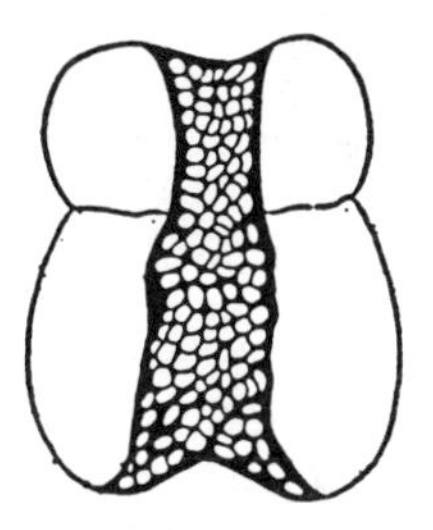

Fig. 10.72 Schematic sketch of *Antheraea pernyi* germ band, showing invaginating primitive groove (Saito 1937)

ventral view; black area represents invaginating part.

Although this invagination has no connection with archenteron formation, it is often called "gastrulation", and some workers have called the resulting pore a "blastopore" (Toyama 1902). Obviously these terms are not appropriate, although Henson (1932) has reported that the anterior and posterior ends of this invagination give rise to the fore- and hind-intestine, respectively (cf. p. [450]). In the non-diapausing egg of the silkworm, the primitive groove appears about 24 hours after oviposition (25°C). The invagination gradually extends antero-posteriorly and deepens; the cells forming the ridges come together, eventually fusing and forming the two-layered germ-band (Fig. 10.73). The outer layer is the *ectoderm* and the inner is the *mesoderm.*

The mode of this primitive groove invagination varies with the species. In the silkworm, invagination in the cephalic region is so deep as to be almost vesicular in cross-section, but it is not more than a shallow furrow in the thoracic region; only the bottom of this furrow separates and forms the mesoderm. In the abdominal region, the invagination is relatively wide, and again becomes shallow in the caudal segment. The time at which the two-layered structure becomes completely established, as the result of closure of the primitive groove by a roof ectoderm, is different for each region of the germ-band, occurring earliest in the central zone of the abdomen (Toyama 1909).

In the honey bee, no invagination actually takes place in mesoderm formation. Instead, the lateral portions grow from both sides of the median longitudinal part of the blastoderm; the two lateral parts become the *upper layer* (ectoderm) and the median part, now covered with this ectoderm, becomes the *lower layer* (mesoderm) (Nelson, cited by Saito 1937).

In some of the Lepidoptera, such as *Endromis versicolora,* the invaginated lower layer (unteres Blatt), instead of directly forming mesoderm, differentiates into endoderm and mesoderm. It is believed that the central part of the lower layer chiefly turns to mesoderm, while the endoderm appears from its anterior and posterior regions (Schwangart 1904). Some earlier workers have reported a similar mode of mesoderm formation in the silkworm (Strindberg 1904), as well as in *Musca* (Kovalevsky 1886) and *Calliphora* (Noack 1901).

In an apterygote insect, *Thermobia domestica,* mesoderm formation takes place as the result of an inward migration of cells from every part of the germ-band (Wellhouse 1954).

These various and often contradictory reports concerning the modes of mesoderm formation have been classified by Saito (1937) into three major types:

1) no distinct invagination,

2) invagination of the primitive groove different in different parts,

3) invagination sufficiently complete to form an almost tubular primitive groove.

Correlating the relationship of these types to insect systematics, Saito finds that Type 1 is characteristic of the Orthoptera, Type 2, of the Lepidoptera and Hymenoptera and Type 3, of the Coleoptera, Trichoptera, Diptera, etc. It is interesting that in the lepidopteran *Antheraea pernyi,* whereas the anterior part of the germ-band invaginates completely, the degree of invagination gradually diminishes toward the posterior end, where only inward proliferation of the germ-disc cells takes place (Figs. 10.73, 10.74). All the transitional phases among the three types described above are thus represented in the different parts of this one embryo. This is suggestive in connection with a general consideration of mesoderm formation in insects, especially since many of the views hitherto expressed concerning this problem seem to have dealt only with quantitative differences in mesodermal invagination.

In the systematically lower Orthoptera, the mesoderm develops immediately after the completion of germ-band formation, while in higher insects such as the Lepidoptera, Coleoptera, Hymenoptera, Diptera, etc., some time intervenes between the completion of germ-band formation and the appearance of mesoderm. All apterygote insects are believed to belong to Saito's Type 1.

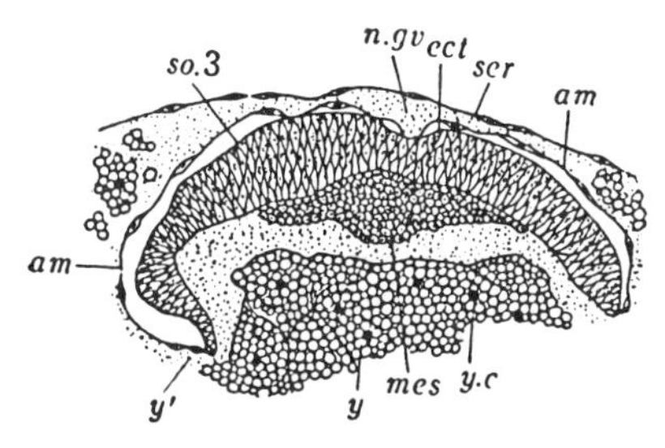

Fig. 10.73 Transverse section of third segment of germ band in *Antheraea pernyi* (Saito 1937)

Embryo at stage when primitive groove has closed and ectoderm and mesoderm are present as two tissue layers. *am:* amnion, *ect:* ectoderm, *mes:* mesoderm, *n.gv:* neural groove, *ser:* serosa, *so.* 3: third body segment, *y:* yolk, *y':* liquefied yolk, *y.c:* yolk cell.

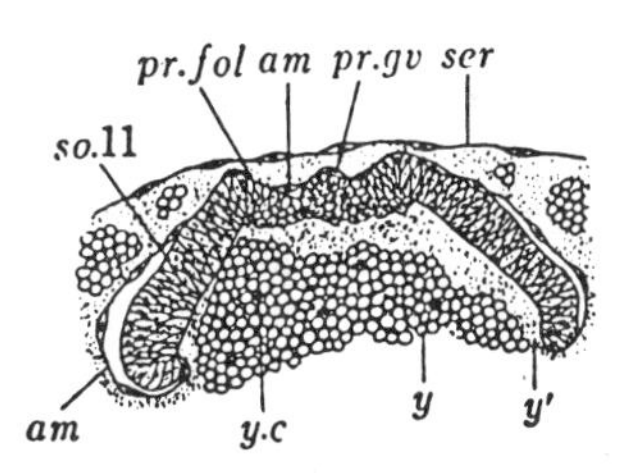

Fig. 10.74 Cross-section of eleventh segment of germ band in *Antheraea pernyi* (Saito 1937)

Same embryo as that shown in Fig. 10.73. Note that mesodermal invagination has not yet occurred at this level. *am:* amnion, *pr.fol:* primitive groove fold, *pr.gv:* primitive groove, *ser:* serosa, *so..*11: eleventh body segment, *y:* yolk, *y':* liquefied yolk, *y.c:* yolk cell.

Some of the contradictory opinions which are revealed by this survey must be due to actual differences in the mode of mesoderm formation in various insect groups; others, which may even deal with a single species, arise from differences in accuracy of observation and the choice of somewhat different developmental stages for study. This latter source of conflict is not confined to the problem of the origin of the mesoderm — not infrequently it underlies lack of agreemeni in other aspects of insect embryology.

While mesoderm formation is going on, the germ-band contracts from the sides toward the ventral surface. Simultaneously, several constrictions appear in the ectoderm indicating that the *body segments* are beginning to differentiate. The first of these distinguishes the *cephalic lobe* from the rest of the body (Fig. 10.72). As it thus contacts laterally, the germ-band elongates and becomes more slender.

The mesoderm lies at first as a narrow band under the ectoderm along the ventral median line; gradually this extends to the sides, to line the ectoderm of increasingly peripheral regions, and eventually acquires segmentally arranged constrictions. In the silkworm, the segmentation makes its first appearance as these mesodermal constrictions, which are later followed by segmentation of the ectoderm (Toyama 1902), whereas in *Antheraea pernyi* the ectoderm segments first, and the corresponding mesodermal constrictions appear later (Saito 1937).

After the segments have differentiated, the ectoderm begins to invaginate at the center of the anteriormost segment, i.e., the cephalic lobe. This invagination, the *stomodaeum*, elongates in a postero-dorsal direction. Then another ectodermal invagination, the *proctodaeum*, occurs in the posteriormost segment, and grows anteriorly along the dorsal side of the embryo. These two invaginations will later constitute the anterior and posterior parts of the digestive tract; at their bottoms there soon appear masses of cells to give rise to the *endoderm* which later forms the wall of the mid-intestine.

Many contradictory views have been held with respect to the origin of the endoderm. In the silkworm, it has been believed to arise either from a part of the mesoderm (Tichomiroff 1879), or from the yolk cells (Tichomiroff, cited by Saito 1937). It is now agreed that its origin is ectodermal, but this agreement still involves two hypotheses; one suggests that it originates from the cells at the tips of the stomodaeal and proctodaeal invaginations (Toyama 1909), and the other ascribes the origin of at least the posterior component of the endoderm to some of the ectodermal cells near the posterior end of the caudal segment, which migrate inward together with the proctodaeal invagination (Nakata 1936). This latter seems to be the case in *Antheraea pernyi*. Saito (1937) has reviewed the various descriptions of endoderm formation, and proposes to group them in the following categories:

1) The mid-intestine is formed exclusively by the yolk cells; there are no other cells which can be described as endodermal (Tichomiroff).

2) In addition to ectoderm and mesoderm, there also exists endoderm in the true sense of the word. This category includes the following four cases:

a) Cells migrate from the blastoderm into the yolk, giving rise to the endoderm (Deegener).

b) Blastodermal cells on the ventral median line invaginate, or the lateral parts grow from both sides in the median part to form the lower layer, of which the anterior and posterior ends become endoderm, and the middle forms mesoderm (Wheeler).

c) Endoderm exists not only at both ends of the lower layer, but also along its whole length (Hirschler).

d) The endoderm anlage appears from both anterior and posterior ends of the primitive groove, without relation to the mesoderm (Eastham).

Nusbaum and Fuliński (1909) have classified the Pterygota into seven types according to when endoderm differentiation begins:

Type 1. Invagination of the stomodaeum and proctodaeum occurs after the complete separation of the anterior and posterior endodermal anlages from the blastoderm (Noack, *Musca*; Hirschler, *Donacia).*

Type 2. Invagination of the stomodaeum and proctodaeum occurs before the anterior and posterior endodermal anlages have completely separated from the blastoderm (Karawaiew: *Pyrrhocoris).*

Type 3. The stomodaeum appears almost simultaneously with the formation of the anterior endodermal anlage, while the proctodaeum is formed much later than the posterior endodermal anlage (Nusbaum and Fuliński, *Gryllotalpa).*

Type 4. Both anterior and posterior anlages appear almost simultaneously with the stomodaeum and proctodaeum. Not only the endoderm, anterior as well as posterior, but also the lower layer between them participates in forming the mid-intestine (Nusbaum and Fuliński, *Phyllodromia).*

Type 5. The anterior and posterior endodermal anlages arise first; then, before they completely separate from the blastoderm, the stomodaeum and proctodaeum are formed. An endodermal strand extending between the anterior and posterior anlages also takes part in forming the mid-intestine (Hirschler, *Gastroides* (Coleoptera)).

Type 6. The anterior and posterior endodermal anlages arise first, and before they completely separate from the blastoderm, the stomodaeum and proctodaeum appear. The innermost portions of these, however, include undifferentiated cells, some of which may differentiate into endoderm (Carrière und Bürger, *Chalicodoma muraria).*

Type 7. The presumptive endodermal regions at the anterior and posterior ends of the blastoderm remain latent for a long period. The tips of the stomodaeal and proctodaeal invaginations formed here are at first undifferentiated; eventually, however, they give rise to endodermal anlages (Heymons, *Forficula*).

In the apterygote insects, the yolk cells generally participate in the formation of the mid-intestine (Heymons 1897).

Additional information in connection with this problem may be found in Section (3) c of this chapter (Organs of Ectodermal Origin).

(2) DEVELOPMENTAL STAGES OF THE EMBRYO

The development of the non-diapausing silkworm embryo (22—23°C; relative humidity, 85—90%) has been tabulated by Nakata (1932) as follows:

Stage 1 (1—4 hours after oviposition) The cleavage nuclei disperse.

Stage 2 (12 hours) The blastoderm is formed.

Stage 3 (20 hours) The germ-band is formed.

Stage 4 (25 hours) A constriction forms across the germ-band, separating the head lobe from the rest of the body. The embryo of this stage is called "pyriform", since the part posterior to the constriction is wider and larger than the anterior part (Fig. 10.91 a). The primitive groove appears along the median line in the posterior part of the region which will become the head lobe.

Stage 5 (35 hours) The embryo is long and slender, with a median depression at the anterior end of the head lobe. The head lobe is larger than the triangular caudal segment; in this segment are paired lateral depressions which will give rise to the anlages of the Malpighian tubules. The mesoderm is in the form of seventeen segmentally arranged masses; the largest of these is the *oral cell-mass* in the head lobe. In the posterior part of the embryo, the mesoderm is still undeveloped and does not line all of the ectoderm. Ectodermal segmentation is still obscure.

Stage 6 (40 hours) Labral and antennal anlages appear on the head lobe. Although the number of primordial segments in the silkworm is generally considered to be eighteen, Nakata recognizes the presence of the so-called *intercalary*

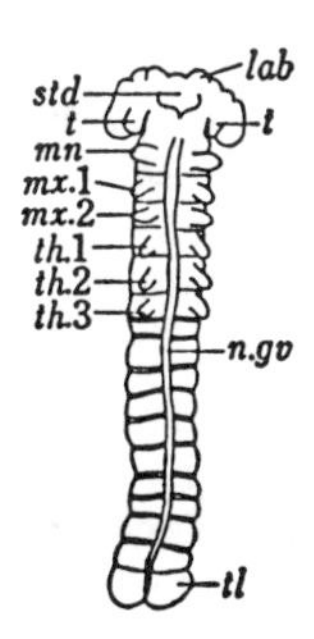

Fig. 10.75 Embryo of *Bombyx mori* in which thoracic appendages are appearing (Toyama 1909)

lab: **labral appendage (process),** ***mn:*** **mandibular appendage,** ***mx.1, mx.2:*** **first and second maxillary appendages,** ***n.gv:*** **neural groove,** ***std:*** **stomodaeum,** ***t:*** **antennal appendage (process),** ***th.*** **1 ~ 3: promeso- and metathoracic appendages,** ***tl:*** **telson.**

segment (reported by Heymons et al. in *Gryllotalpa* and *Lepisma)* as a degenerative intersegmental segment between the head lobe and mandibular segment. He therefore describes the embryo as having nineteen segments. On the six segments following the intercalary are formed the paired anlages of the mandibles, the first and second maxillae and the first to third thoracic legs (Fig. 10.75). The labral and antennal anlages eventually form the *labrum* and *antennae,* and the other six pairs of anlages develop into the *mandibles,* etc. The ectoderm of the head lobe invaginates at the locus of the oral cell mass, giving rise to the stomodaeum, which later forms the *fore-intestine.* A longitudinal furrow, the *neural groove,* appears on the ventral side of the ectoderm along the median line (Figs. 10.75, 10.77a). Anteriorly, the mesoderm is divided by the neural groove into lateral strips, but it retains its median position in the posterior segments. The *Malpighian tubules* invaginate. Since it is at this stage that the embryo attains its maximum pre-revolution length, sericultural terminology refers to it as the "longest stage".

Stage 7 (2 days) Abdominal appendages appear on several segments behind the ninth. The embryo becomes slightly shorter. The intercalary segment is most obvious at this stage. The thoracic legs acquire two segments. An ectodermal invagination appears on the caudal segment and forms the proctodaeum, which later develops into the *hind-intestine.*

Stage 8 (2 days and 12 hrs) The first and second maxillae now have two segments. All the segments of the prospective abdomen except the caudal have acquired appendages; those on the eleventh to fourteenth segments are larger than the others, and will later form the *larval abdominal legs.* The strips of lateral mesoderm differentiate into two layers: an outer *somatopleure* and an inner *splanchnopleure.* The space which forms between them gives rise to the *coelomic sac* (Fig. 10.76).

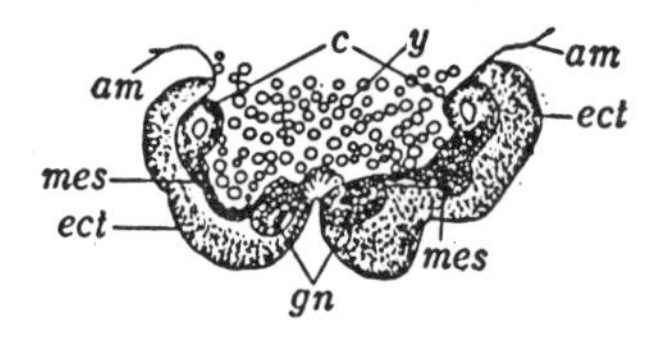

Fig. 10.76 Coelom formation in the silkworm (Toyama 1909)
am: amnion, *c:* coelom, *ect:* ectoderm, *gn:* ganglion, *mes:* part of mesoderm, *y:* yolk.

Stage 9 (3 days) The anterior part of the embryo to, and including, the second maxillae begins movements which will lead to the formation of the head. In the ectoderm, the *silk gland* anlages invaginate at the base of the second maxillary appendage, and *spiracular anlages* form as invaginations on the sixth and the succeeding twelve segments (although Nakata was unable to confirm the occurrence of such invaginations on the last two of these segments — cf. p. [443]). Slight *ganglionic swellings* are seen on the ventral sides of the third, fourth and eighteenth segments. The eighteenth segment, leaving its ventral surface behind, begins to migrate dorsally toward the nineteenth; later this will form the *anal plate.*

Stage 10 (3 days and 12 hrs) The embryo becomes increasingly shorter. The prospective head segments draw together, making it possible to distinguish between the head and thoracic regions. The abdominal appendages other than the prospective abdominal and caudal legs begin to degenerate. Two anlages of the mid-intestine, one originating from the end of the stomodaeum and the other from the end of the proctodaeum, now become completely connected.

Stage 11 a (4 days) Early Revolution Stage 1. The embryo, so far dorsally concave as it lies along the ventrally convex egg surface, now begins to move inward as its body contracts and its posterior half (behind the eighth segment) separates from the egg surface. The segments ranging from the head lobe to the aecond maxillary fuse to form the *head,* and the appendages of each of these segments sssume their definitive positions as head organs. The next three segments form 1he *throrax,* and the *abdomen* is formed from the remainder. The ganglia of the eighteenth segment move forward and fuse with those of the seventeenth. Simultaneously, on the ventral side of this eighteenth segment, the anlage of the *Herold gland* appears as an ectodermal invagination; then the segment itself contracts and migrates to fuse with the seventeenth, finally forming a single segment, the ninth abdominal.

The formation of the dorsal surface which originates in the posterior end of the embryo now reaches the first abdominal segment, although this surface

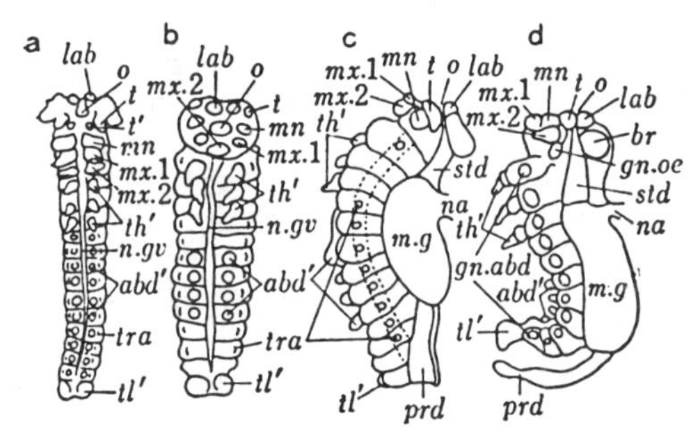

Fig. 10.77 Silkworm embryos before and after blastokinesis (Yanagita 1929)

a. embryo at stage of appendage formation (ventral view); b. contracted embryo before blastokinesis (ventral); c. embryo in transitional stage between dorsal and ventral concavity (lateral); d. embryo shortly after becoming ventrally concave (lateral). Sketches drawn at different magnifications. *abd'*: abdominal leg, *br:* brain, *gn.abd:* abdominal ganglion, *gn.oe:* suboesophageal ganglion, *lab:* labrum, *m.g:* mid-gut, *mn:* mandible, *mx.*1*:* first maxilla, *mx.*2: second maxilla, *n.gv:* neural groove, *na:* navel, *o:* mouth, *prd:* proctodaeum (hind-gut), *std:* stomodaeum (fore-gut), *t:* antennal appendage, *t'*: sub-antennal appendage (probably corresponds to Nakada's "intercalar segment"), *th':* thoracic leg, *tl':* telson, *tra:* tracheal pit.

still consists of only a single layer of cells, except for the anal plate area. The dorsal formation starting in the anterior part extends to the boundary between head and thorax. The change in the curvature of the embryonic body is called *revolution*; the morphogenetic changes associated with this process are indicated in Figure 10.77.

Stage 11 b (4 days and 5 hours) Early Revolution Stage 2. With the change in body curvature, the posterior part of the embryo, behind the sixth abdominal segment, becomes almost straight (Fig. 10.77 c). The dorsal surface, with the exception of the thoracic region, is covered with a layer of cells.

Stage 12a (4 days and 12 hours) Late Revolution Stage 1 (Fig. 10.78 a). The part of the embryo posterior to the seventh abdominal segment is now bent so as to be dorsally convex. The dorsal closure is complete except at the second thoracic segment.

Stage 12b (4 days and 18 hours) Late Revolution Stage 2 (Fig. 10.78 b). The abdomen is now completely convex dorsally, and occupies the dorsal side of the egg, while the head and thoracic regions lie in a nearly straight line. The ventral and lateral walls of the mid-intestine are formed; both *pyloric* and *cardiac anlages* are recognizable. The dorsal surface has now attained almost complete closure, leaving a small opening called the *navel* (Fig. 10.77 c, d) at the posterior part of the second thoracic segment.

Stage 13 (5 days) The revolution of the embryo is now complete, with the whole body dorsally convex (Fig. 10.78 c). Formation of the mid-intestine is complete, and the Herold gland is developing. The ganglia of the seventeenth segment, which arose from the fusion at Stage 11 a of the ganglia of segments 17 and 18, are now situated within the sixteenth segment (eighth abdominal), close to the ganglia of the latter, with which they fuse. The *gonads* appear in the dorsal part of the fifth abdominal segment. On the body surface, *tubercles* become faintly perceptible, appearing first posteriorly. The oral appendages draw together around the stomodaeal opening, taking the arrangement found in the newly hatched silkworm.

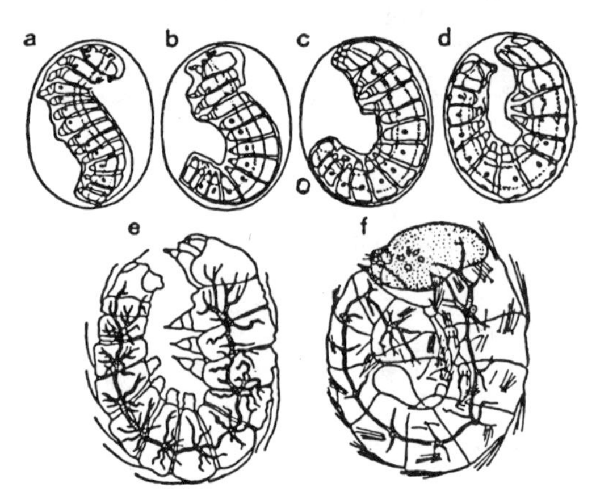

Fig. 10.78 Developmental stages of *Bombyx mori* embryo from late blastokinesis to blueing of egg (Nakada 1932)

a. stage 12a; b. stage 12b; c. stage 13; d. stage 15; e. stage 17; f. stage 18.

Stage 14 (5 days and 12 hours) The boundaries between neighboring segments become somewhat more conspicuous.

Stage 15 (6 days) The tubercles develop further (Fig. 10.78 d). The length of the embryo increases to reach 90% of the egg circumference. Lines of *subdorsal, supraspiracular* and *subspiracular tubercles* appear. The paired subdorsal tubercles of the eigth abdominal segment fuse to give rise to the *caudal horn*. The caudal

legs now have two segments. Ventrally, the number of abdominal segments, theoretically eleven, was reduced to ten by the fusion which took place in Stage 11 a. At the time of hatching, however, the border between the ninth and tenth abdominal segments becomes less conspicuous, so that the embryo seen from the ventral side has apparently only nine segments. Dorsally, on the other hand, the tenth abdominal segment migrates posteriorly to cover the eleventh, as the anal plate, so that there are actually as well as theoretically ten abdominal segments at the dorsal side in this stage. The posteriormost ganglia, which arrived in the eighth abdominal segment during Stage 13, have continued to shift forward and are now near the anterior border of the segment.

Stage 16 (6 days and 12 hours) The tubercles along the subdorsal lines now have *setae*. *Maxillary* and *labial palpi* appear on the maxillae and labrum. The mandibles are serrate; the thoracic legs now bear *claws* at their tips. The navel is almost closed. The posteriormost ganglia enter the seventh abdominal segment.

Stage 17 (7 days) *Taenidia* appear on the *tracheal intima* (Fig. 10.78 e).

Stage 18 (7 days and 12 hrs) Dark-spotted Stage 1 (Fig. 10.78 f). The integument of the head becomes chitinized and dark brown in color. The *eyes (lateral ocelli)* are already distinct. The navel is completely closed.

Stage 19 (8 days) Dark-spotted Stage 2. The embryo takes in through its mouth the serosa which has until now been covering it. As a result, the head of the embryo appears dark and bluish through the chorion; this is the origin of the term "dark-spotted" which is frequently used to refer to this stage in Japanese sericulture.

Stage 20 (9 days) Bluish stage 1. The larval body is now complete. The trunk is colored as well as the head, giving a bluish tinge to the whole egg ("bluish egg").

Stage 21 (9 days and 12 hrs) Bluish stage 2. The embryo is fully pigmented and ready to hatch if given appropriate conditions of temperature, humidity, light, etc.

The foregoing description presents the development of the non-diapausing egg. In the case of the diapausing egg, the embryo enters diapause when it is about to attain the condition described in Stage 5.

The pulsation of the embryonic *dorsal vessel* becomes apparent two to three days before Stage 19, or one day before the tracheae become recognizable (Yokoyama 1929); this corresponds approximately to Stage 16 of Nakata. The dark appearance of the tracheae is due to the air which fills them about one day after the pigmentation of the ocelli, or one day before the head becomes pigmented (Yokoyama 1929, Nagata 1951).

The great importance for sericultural techniques of knowing the morphological changes undergone by the embryo during hibernation and the early spring period has resulted in the development of several systems for classifying the embryological stages of the silkworm (cf. Mizuno 1936, Umeya 1938 a).

The processes of embryonic development in *Platycnemis* and *Ephestia*, from the dispersal of the nuclei to the completion of the embryo, are diagrammatically summarized in Figure 10.79 A and B, with special stress on the dynamic changes in the embryonic area (Krause 1939).

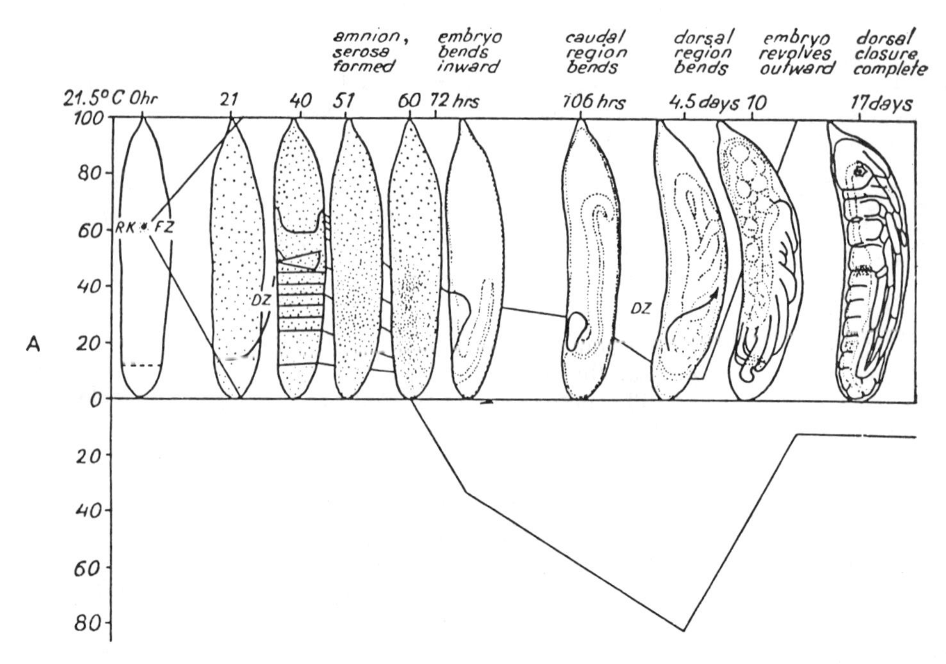

Fig. 10.79 A Embryonic developmental stages in *Platycnemis pennipes* (Krause 1939)

Embryos at successive stages aligned in lateral view; lines indicating anterior and posterior ends of germ band connected to show shifts in position during development. Distance of posterior end from original location shown by curve below 0 line. Development during 18 days at 21.5° C.

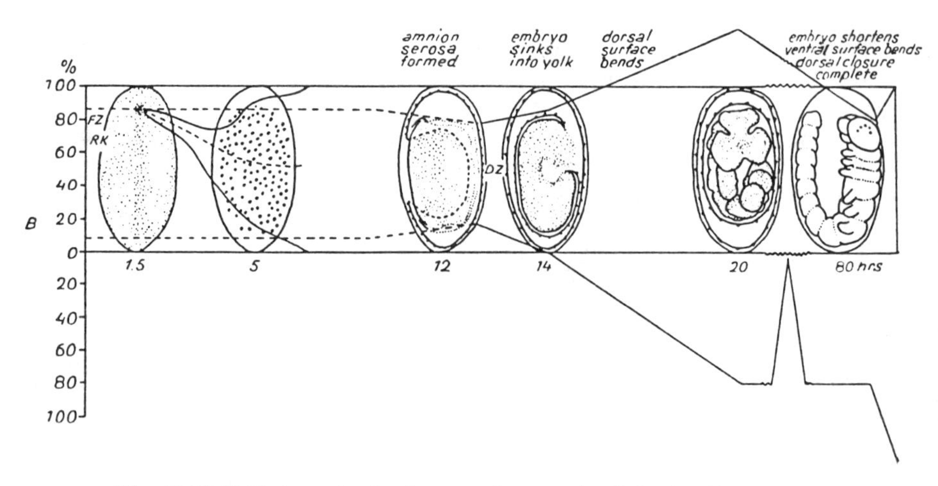

Fig. 10.79 B Embryonic developmental stages in *Ephestia* (Krause 1939)

Treatment of data same as in A. Development during 4 1/2 days at 23.5° C.

(3) DIFFERENTIATION OF LARVAL ORGANS

The mode of organogenesis differs with the species, and the observations of embryologists have not always agreed with each other. We shall therefore deal here mainly with the silkworm, and confine our description to its early differentiation as established by Toyama (1909). Some supplementary notes will be added as necessary.

a. Organs of Ectodermal Origin

The organs of the insect body which have an ectodermal origin outnumber those derived from the other germ layers: they include the integument, appendages, ocelli, salivary glands, prothoracic gland, corpora allata, molting glands, oenocytes, silk glands, fore-intestine, hind-intestine, Malpighian tubules, respiratory system, nervous system, genitalia, etc.

i) *Integument*

The integument is the direct product of the ectoderm. In the earlier developmental stages, it covers only the ventral surface; eventually, however, it extends laterally and dorsally over the whole surface of the embryo.

ii) *Appendages*

As described above, each segment gives rise to its particular appendage processes, which later develop into the larval organs. The appendages make their first appearance as rounded outgrowths of ectoderm (Fig. 10.81), into which mesoderm extends and forms the *muscles*. As already mentioned, not all of the abdominal appendages develop further; excepting those which later become the abdominal and caudal legs, they later degenerate. The *pleuropodia*, which are widely seen in embryos of the Orthoptera, Hemiptera, Coleoptera, Odonata, etc.

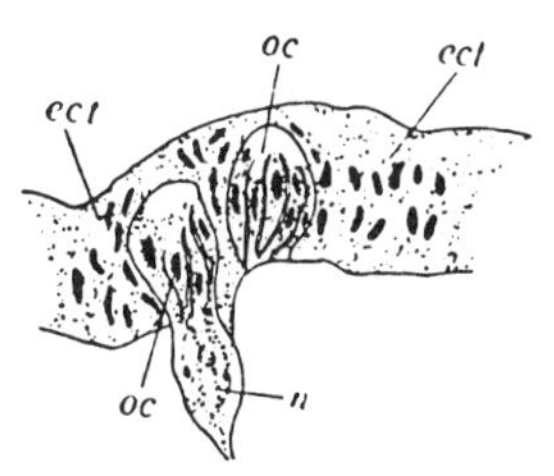

Fig. 10.80 Development of ocellus in embryo of *Bombyx mori* (Toyama 1909)

ect: ectoderm, *n:* nerve, *oc:* tissue of ocellus rudiment

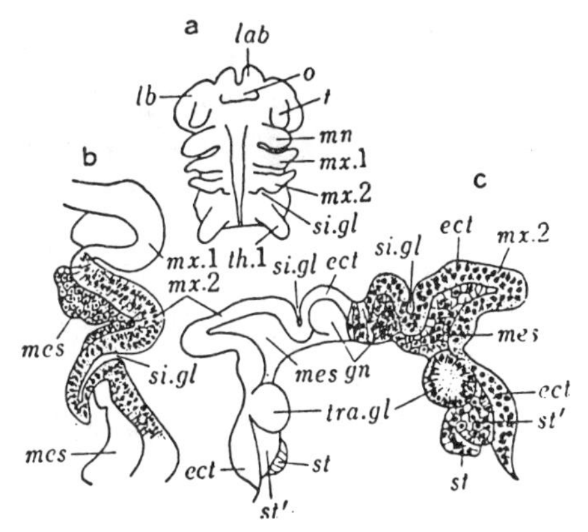

Fig. 10.81 Development of silk-gland in *Bombyx mori* (Toyama 1909)

a. ventral view of embryonic head segments; b. longitudinal section of silk-gland invagination on second maxilla; c. transverse section of same segment. *ect:* ectoderm, *gn:* ganglion, *lab:* labrum, *lb:* head lobe, *mes:* mesoderm, *mn:* mandibular appendage, *mx.1*: first maxillar appendage, *mx.2*: second maxillar appendage, *o:* mouth, *si.gl:* silk-gland invagination, *st:* stomach wall rudiment, *st':* perigastric tissue, *t:* antennal appendage, *th.1*: prothorax, *tra.gl:* subtracheal (prothoracic) gland.

(although not in the silkworm embryo), function as a gland secreting the *hatching enzyme,* and then disappear within the embryonic period. These organs are ectodermal in origin, and are now regarded as being homologous with the appendages of the first abdominal segment (Wigglesworth 1950; Ando 1952, 1953).

In some insects, the labrum appears first as two separate lateral outgrowths which fuse to form a single structure. In others, however, this organ appears to be single from the beginning. In the 55-hour egg of *Leptinotarsa,* the labrum is a single outgrowth with a branched tip. Partial cautery experiments reveal, however, that at about 30 hours (early primitive groove stage), there are two presumptive labral areas, 0.2 mm apart, to the right and left of the median line (Haget 1955).

iii) *Ocelli*

About one day after revolution, circular arrangements of especially large cells can be observed within the ectoderm at the bases of the antennae. These are the anlages of the ocelli (Ikeda 1913), which appear latest among all the organ anlages (Fig. 10.80).

iv) *Glands*

At the base of each mandible is formed an invagination which later divides into two branches. The branch directed backward develops into the *salivary gland,* while the forward growing one forms the *corpus allatum* (Ito 1918). The *prothoracic glands* originate from invagination on the outside of the base of each second maxilla (Toyama 1902, 1909. Toyama called them the "hypostigmatic glands"). A similar origin of these glands has been observed in the hemipteran *Dysdercus cingulatus* (Wells 1954). Its innervation, however, suggests that the prothoracic gland of the silkworm has its origin in two invaginations, one on the second maxillary and another on the first thoracic segment, (or at least on the primordial segment represented by the prothoracic ganglion), which fuse to form a single gland (Yokoyama 1956).

The *silk gland* develops from invaginations at the bases of the second maxillae (Fig. 10.81), which appear a short time before the beginning of the revolution stage. At first they are paired; then they draw together and their anterior parts unite about 24 hours after their appearance (Nunome 1937).

v) *Fore- and Hind-intestine*

The fore-intestine derives from the stomodaeum, and the hind-intestine, from the proctodaeum (Fig. 10.77, 10.82, 10.87).

vi) *Malpighian Tubules*

The origin of the Malpighian tubules has been the subject of much controversial discussion. They have been said to appear on one end of the posterior part of the digestive tract, as a pair of tubular outgrowths, each of which later divides into three tubes (Tichomiroff, cited by Nakata 1935); or as three processes on each side at the end of the fairly elongated proctodaeum (Toyama 1909); or again, as tube-like lateral outgrowths of the inner tissues on the ventral side of the proctodaeal ending, each of which eventually branches into three tubes (Ikeda 1909). Nakata (1935) observed that the Malpighian tubules arise on the ectoderm of the caudal segment, on the ventral or lateral wall of the proctodaeum, as three pairs of independent, pouch-like ectodermal invaginations, which unite at the stage just prior to revolution. This report places the appearance of the Malpighian tubule anlages much earlier than has hitherto been believed; i. e., contemporaneous with the formation of the proctodaeal anlage and the posterior anlage of the mid-intestine.

In *Antheraea pernyi*, also, the Malpighian tubules originate as three pairs of invaginations, a dorsal pair appearing earliest, to be followed by ventral and lateral pairs (Saito 1937).

The Malpighian tubule is reported to have an endodermal origin in *Pieris brassicae* (Henson 1932), and to develop from outgrowths on the wall of the hind-intestine in *Thermobia domestica* (Wellhouse 1954).

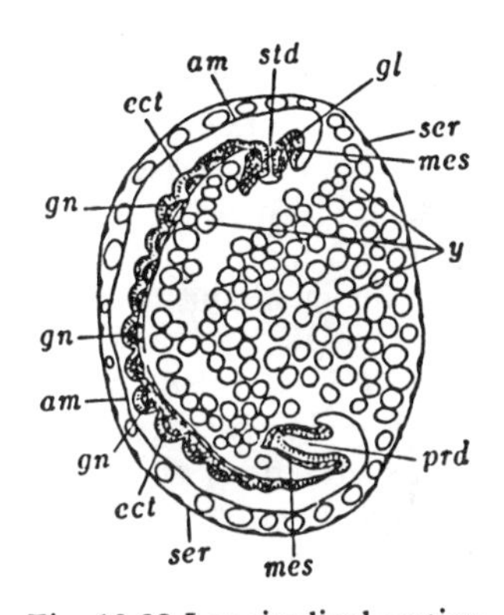

Fig. 10.82 Longitudinal section of silk-worm embryo showing stomodaeal and proctodaeal invaginations (Toyama 1909)

am: amnion, *ect:* ectoderm, *gl.* suboesophageal gland, *gn:* ganglion, *mes:* mesoderm, *prd:* proctodaeum, *ser:* serosa, *std:* stomodaeum *y:* yolk mass.

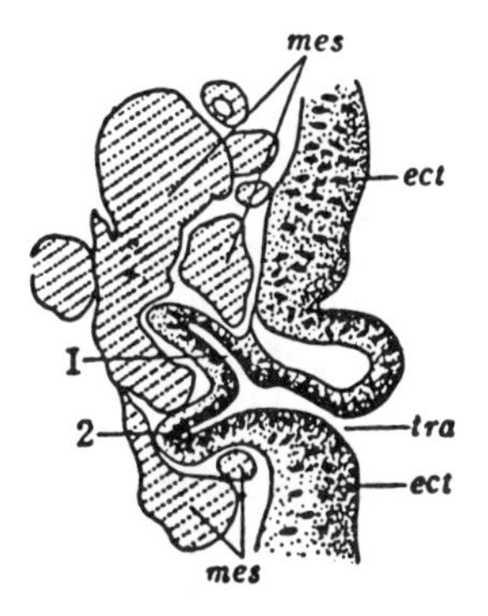

Fig. 10.83 Tracheal invagination in silkworm embryo (longitudinal section of thoracic region) (Toyama 1909)

ect: ectoderm, *mes:* mesoderm, *tra:* trachea, 1,2: branching tracheal tips.

vii) *Respiratory Organs*

Paired spiracular invaginations arise on all segments except the head and the caudal segment. Those on the first thoracic and first to eighth abdominal segments develop into normal spiracles, while the others degenerate except for the second thoracic pair, which persist as vestigial spiracles throughout larval life. In the developing respiratory system, the aperture of the invagination becomes the spiracle, while the part which extends inside branches and forms the tracheae (Fig. 10.83). In *Antheraea pernyi*, spiracular invaginations occur on the first and second thoracic and first to tenth abdominal segments; the second thoracic pair degenerates, and the ninth and tenth abdominal are closed before the embryo hatches (Saito 1937). No invagination at all occurs on the third thoracic segment.

viii) *Nerves*

The anlages of the nerves first appear alongside the neural groove as *neuroblasts* derived from the ectoderm. The neuroblasts divide and proliferate, building up a pair of *ganglia* in each segment except the caudal (Fig. 10.77), and a *nerve cord* connecting these ganglia. As the embryo develops, the ganglia of the three consecutive segments behind the oral cavity fuse to form a single *suboesophageal ganglion*, and those of the last three segments similarly unite (p. [438]). The larva therefore has thirteen pairs of ganglia.

In *Antheraea pernyi* (Saito 1937), the neuroblasts arise from large cells which appear in the ectoderm of each thoracic and abdominal segment at the "longest stage", when embryonic segmentation is becoming distinct and the appendage processes are just appearing. The head neuroblasts giving rise to the brain appear in the dorsal ectodermal wall of the head-lobe at the time when the small antennal processes are growing out from the sides of the stomodaeum. In *Pieris*, the brain consists of three pairs of ganglia: the *protocerebrum* occupying the large area between the anterior end of the embryo and its antennae, *deutocerebrum* lying anterior to the stomodaeum and the *tritocerebrum* behind the antennae. It has therefore been suggested that the part of the head anterior to the mandibular segment is originally composed of three segments (Eastham, cited by Saito 1937). In *Antheraea*, the scattering of the neuroblasts tends to obscure their interrelations, but the cellular arrangement and distribution of the nerv fibers make it possible to recognize a similar tripartite origin of the brain.

In addition to the organs described above, the genitalia also differentiate from the ectoderm.

In a mutant race of the silkworm called *Mottled Translucent*, the integument is a mosaic of two kinds of cells: opaque normal cells and transparent cells described as translucent or oily. When a caterpillar of this mutant is vitally stained with neutral red, a spotty pattern of heavily and lightly stained parts appears on all the ectodermal tissues making up the integument, head organs, silk glands, Malpighian tubules, oenocytes, hypostigmatic glands, salivary glands, tracheal intima, corpora allata, fore-intestine, hind-intestine, nerves, molting glands, Herold's gland, Ishiwata's gland, etc., but never on the mesodermal or endodermal organs (Aruga 1942, 1943). This germ-layer-correlated difference in stainability can be observed in the pupal and imaginal organs as well as in those of the larva.

b. **Organs of Mesodermal Origin**

Among the segmentally arranged masses of mesodermal cells, four major morphological types can be distinguished. These are: the *oral cell mass* in the head-lobe, the *caudal cell mass* in the caudal segment, and the bilateral series of mesodermal tissue blocks which are segmentally arranged along the neural groove, and later divide into *somatopleura* and *splanchnopleura* as they form the coelomic sac (p. [55]). All the mesodermal tissues and organs, including the dorsal vessel, internal reproductive organs, muscles, adipose tissues, suboesophageal glands, hemocytes, etc., are formed from these four types of mesodermal masses.

i) *Dorsal Vessel*

Certain cells become separated from the dorsal margins of the splanchnopleura and form clusters of *cardioblasts* (Fig. 10.84), which grow dorsally as the back of the embryo gradually develops, the two sides finally fusing when the dorsal integument and mid-intestine reach completion. The tube thus formed is the *dorsal vessel* (Fig. 10.87).

Fig. 10.84 Cardioblast in silkworm embryo (transverse section) (Toyama 1909)

am: amnion, *bl:* blood corpuscle, cardioblast, *epi:* epithelial tissue, *end:* mid-gut epithelium, *m:* feather-shaped muscle rudiment, *mes:* muscle-forming mesoderm, *mes':* splanchnic mesoderm of mid-gut, *y:* yolk granule, *y.nu:* yolk nucleus.

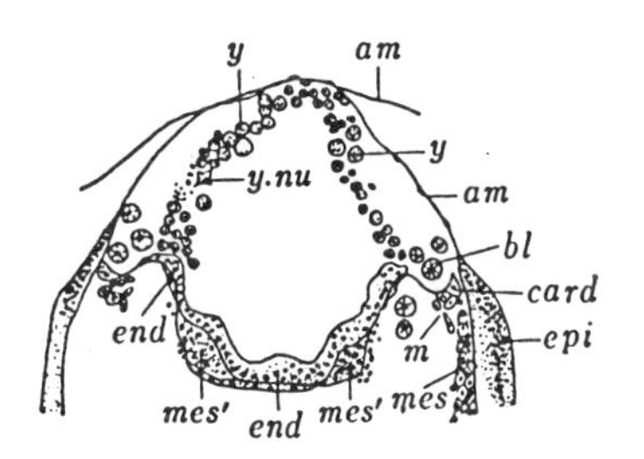

ii) *Internal Reproductive Organs*

In the silkworm, according to Toyama (1909), it is not until the stage of formation of the thoracic leg processes that the germ-cells become morphologically distinct from the other cells. At this time they can be seen in the inner of the two mesodermal layers as weakly staining cells with rather few granular inclusions. They are found in all segments except the caudal. More precise observation has revealed, however, that these cells appear much earlier, and not scattered but in a definite group at a definite location in the egg, later moving to the other regions (Kawaguchi and Miya 1943). The site of their presumptive differentiation is also under investigation (p. [428]).

According to Miya (1947, 1951b, c, 1952b), the morphological differentiation of the germ-cells takes place just after the germ band differentiates from the blastoderm. The other cells of the germ-band divide actively, while the presumptive germ-cells proliferate very slowly or not at all, so that the latter are gradually pushed inward. With the proliferation of the germ-band cells, the presumptive germ-cells are again enclosed within it, and at the stage of mesoderm invagination, they are carried inward together with the surrounding cells. The ratio of the area occupied by the germ-cells to the total length of the embryo remains almost constant from the short germ-band stage to the much elongated stage of primitive segment determination. It is therefore likely that the germ-cells, which appeared as one group at first, are only loosely connected to each other, so that they are mechanically separated and carried into other segments as the embryo elongates. The range of their distribution varies with individuals and with races in the silkworm, usually extending from the mandibular to the sixth abdominal segment, with the highest frequency exhibited in the third to the fifth abdominal segments. Those which lie outside the third-to-sixth abdominal segment range seem to degenerate without participating in gonad formation. The germ-cells do not divide during the period which lasts from the time when they become morphologically distinct to the commencement of gonad formation. There is also an earlier report that the germ-cells appear in every abdominal segment but only those in the fifth develop, the others gradually disappearing (Ikeda 1913).

The germ-cells of *Antheraea pernyi* appear in the abdominal mesoderm at the time of appendage formation. These differentiate in all the abdominal segments except the caudal (Saito 1937). In *Endromis versicolora,* another lepidopteran, the germ-cells arise from the innermost layer of the posterior fourth of the blastoderm, where stratification occurs 2 to 4 hours after blastoderm formation. These cells divide into several groups before the mesoderm forms, and migrate anteriorly. Each group again divides into left and right masses which settle in the coelom of the fourth to sixth abdominal segments (Schwangart 1905).

In eggs with posterior pole plasm, such as those of the Diptera and Coleoptera (p. 406), only the cleavage nuclei entering this special cytoplasm are able to differentiate into germ-cell nuclei. In *Leptinotarsa decemlineata*, strongly staining granules in a disc-like aggregation are observed in the cytoplasm at the posterior pole; these are called the *pole-disc*, or *germ-cell determinants*. Among the cleavage nuclei arriving at the posterior end of the egg, eight enter this pole-disc and become surrounded by the granules. They are thus temporarily separated from the egg itself, and take no part in blastoderm formation. After remaining quiescent for some time, these so-called *primordial germ-cells* again begin to migrate to their definitive position within the now fairly well developed embryo (Hegner 1911).

In *Miastor* and *Chironomus*, also dipterans, only one nucleus enters the posterior pole plasm; this is constricted off as a single *pole-cell*, which is believed to proliferate by later divisions. In *Drosophila*, however, 5 to 11 nuclei appear in the posterior pole plasm. These nuclei cannot be distinguished at the 128-nucleus stage. Even after the 256-nucleus stage, when nuclear migration toward the egg surface begins, the nuclei which will form somatic cells continue to undergo synchronized division (p. 409), but those which have entered the posterior pole plasm behave differently. The entrance of cleavage nuclei into the pole plasm is not simultaneous, and the pole-cells are pushed out one after another. The later-arriving nuclei have the same general appearance as the ordinary blastoderm nuclei, but can be distinguished from them by the presence of the polar granules (Huettner 1923). Huettner, however, has cast some doubt on Hegner's view with respect to the pole granules as germ-cell determinants; he prefers to attribute to the posterior pole plasm itself the capacity to convert the cleavage nuclei into germ-cells.

In the advanced embryonic period, the primordial germ-cells migrate into the coelom, where some of them participate in forming the gonads; others are said to disappear (Huettner 1940), and recent work has indicated that still others take part in forming the epithelium of the mid-intestine (Poulson 1947). The number of primordial germ-cells contained in the gonad anlage of *Drosophila* varies from 5 to 7, although sometimes as many as 13 may be present. The sex differentiation of the gonad is established at about 10 to 11 hours after oviposition at 25° C (Huettner 1940).

We shall now return to the silkworm and follow the further behavior of the germ-cells which lie scattered in a number of the embryonic segments. Those located in the third to sixth abdominal segment congregate into groups in each segment; they are then enclosed in the *genital ridges* arising from the ventral wall of the coelomic sac, to form a pair of long, tubular, *genital cords* which gradually contract and finally settle within the fifth segment. Partial cautery experiments, by disturbing this process, may result in the appearance of several pairs of gonads, formed independently in each segment, but the capacity for gonad formation is limited to the third to sixth abdominal segments — no gonads can be formed outside of this area.

The mesoderm of these four abdominal segments is endowed with the capacity for genital ridge formation even in the absence of any germ-cells; obviously the genital ridge tissue in itself is not able to form germ-cells (Miya 1952a). The E^N mutant of the silkworm is interesting in this respect. The homozygous embryo of E^N dies before hatching. On its first to sixth abdominal segments it has paired appendages which clearly show the characteristics of thoracic legs, and the spiracles on its abdominal segments are identical with those on the thorax.

Furthermore, no gonad formation takes place in any abdominal segment (Itikawa 1944, 1952). An embryological study revealed that the germ-cells differentiate normally, and after remaining quiescent for some time, begin to proliferate as do those in the normal larvae. The mesoderm, however, fails to form genital ridges: it appears that this is the cause of the failure of gonad formation. These facts indicate that the differentiation of the germ-cells and the formation of the genital ridges together constitute a dual condition which is the prerequisite for gonad development (Miya 1955).

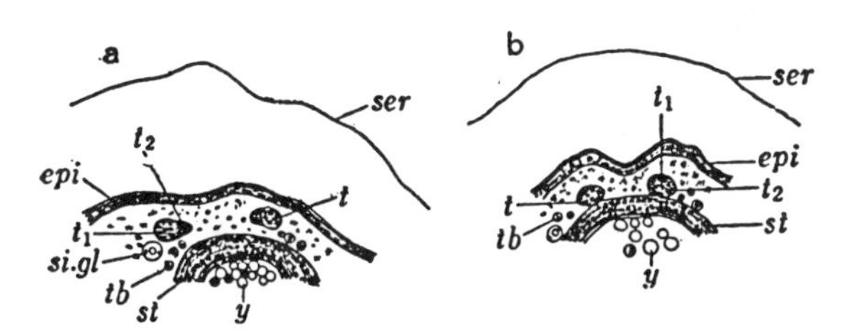

Fig. 10.85 Gonad dimorphism in the silkworm embryo (Ishiwata 1913b)

a. transverse section of male embryo; b. transverse section of female embryo. *epi:* epidermis, *ser:* serosa, *si.gl:* silk gland, *st:* stomach wall, *t:* gonad, t_1*:* macrocellular part of gonad, t_2*:* microcellular part of gonad, *tb:* malpighian tubule, *y:* yolk

Even after a gonad rudiment has settled down in the dorsal part of the fifth abdominal segment, it is still not clear whether this will be a testis or an ovary. However, those with no flexions along the margin when observed in longitudinal section often develop into testes, and those with many flexions become ovaries. In a more advanced stage of development, the sex is clearly distinguished by the arrangement of the large germ-cells and small membrane cells in the central part of the gonad (Ishiwata 1913a, b). In a cross section of a female embryo, the germ-cells lie on the median sides of the paired gonads (Fig. 10.85b), whereas the membrane cells occupy this position in the male embryo (Fig. 10.85a). This difference reflects the different configuration in testis and ovary of the genital duct attachment, which arises from the point of membrane cell aggregation.

iii) *Muscles and Adipose Tissues*

The somatopleura builds up the musculature of the body wall, and the *adipose tissues*. It also extends into the ectodermal outgrowths of the appendage processes and differentiates into their *muscles*. The splanchnopleura gives rise to the muscles of the mid-intestinal outer wall.

iv) *Suboesophageal Gland*

Hatschek, the discoverer of this gland, mistakenly thought it had an endodermal origin; since his time, a number of contradictory views have been held. In the silkworm, it was ascribed to the endoderm by Ikeda (1913), but a mesodermal origin is generally accepted (Toyama 1902; Sakurai 1915; Wada 1955a).

In a longitudinal section of the silkworm embryo at the stage of stomodaeal invagination, part of the mesoderm of the mandibular segment is found lying between the embryonic dorsal surface and the invaginating stomodaeum; as invagination proceeds, this mesoderm grows inward. In a more advanced stage, it forms a mass of cells which moves away from the surface toward the stomodaeum, and as the endoderm is being formed from the bottom of the stomodaeum, it takes on the characteristic shape of the *suboesophageal gland* (Wada 1955a). In the Orthoptera, the anlage of this gland is reported to arise as a pair of cell-masses; in the silk-worm, however, it is single (Sakurai 1915).

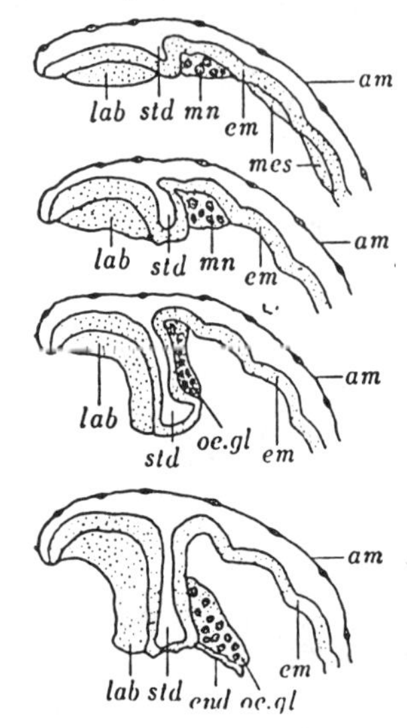

Fig. 10.86 Development of suboesophageal gland in silkworm (Wada 1955a)

Successive stages shown in longitudinal sections of embryonic head region. *am:* amnion, *em:* embryo, *end:* endoderm, *oe. gl:* suboesophageal gland, *lab:* labral mesoderm, *mes:* mesoderm, *mn:* mandible, *std:* stomodaeum.

v) *Peritracheal and Pericardial Glands*

The intercellular connections in the suboesophageal gland begin to loosen early in the revolution stage, and solitary spherical or reniform cells make their appearance. Near the end of the embryonic period, such cells begin to dissociate and migrate forward as well as toward the posterior. Parts of the *peritracheal* and *pericardial glands* are derived from these isolated cells (Sakurai 1915). According to Gamo (cited by Tanaka 1943), however, the pericardial glands are formed postembryonically, at the end of the first instar and the beginning of the second.

vi) *Hemocytes*

The origin of insect hemocytes is still an open question. Toyama (1902, 1909) reported that in the silkworm the *hemocytes* are chiefly derived from cells which separate from the anterior end of the suboesophageal gland, although some of them originate from other mesodermal tissues. According to Iwasaki (1932), the oral cell mass becomes isolated and its cells form the hemocytes, the process of isolation beginning at the front end of the mass and gradually extending to its base. When all the cells at the anterior end of the head-lobe have dispersed, the stomodaeal invagination occurs a short distance behind it. He supposes that Toyama's observation concerning the formation of hemocytes from the post-stomodaeal mesoderm (suboesophageal gland) refers to the stage in which the dispersion of cells from the oral cell mass has proceeded to a point posterior to the stomodaeum. According to him, both the oral cell mass and the post-stomodaeal mesoderm are parts of the single, continuous mesodermal ridge belonging to the first primitive segment. He finds that isolation of the oral cell mass is completed just before the proctodaeum begins to invaginate.

Wada (1955a, b), on the other hand, insists that the suboesophageal gland and oral cell mass originate from the mesoderm of different segments, and that Iwasaki's misunderstanding in this matter was caused by his having confined his observations to longitudinal sections of the embryo. Wada thus supports Toyama's view, that the hemocytes arise from the suboesophageal gland. Hemocytes are also reported to develop from cells which separate from both the oral and caudal cell masses (Ikeda 1913).

There is a report that in *Pieris*, part of the mesoderm, instead of shifting laterally, remains in a median position along the inside of the neural groove and gives rise to hemocytes (Eastham, cited by Saito 1937). In *Antheraea pernyi*, cells separate from the whole inner surface of the mesoderm at about the stage when it splits into two lateral strips. These cells migrate into the yolk and differentiate into hemocytes (Saito 1937). The hemocytes of the cockroach *Phyllodromia* arise from the lower layer (i. e., primary endoderm) (Nusbaum and Fuliński 1906). In the honeybee, the hemocytes are reported to originate from the yolk cells (Will 1888). In another hymenopteran, *Pteronidia ribesii*, the hemocytes are derived from the mesoderm of the median part of the embryo (Shafiq 1954).

In addition to this confusion respecting the origin of the hemocytes, the relationship between embryonic and larval hemocytes also remains obscure.

There are, moreover, several views that the cells isolated from the oral cell mass participate in the formation of structures other than hemocytes. For example, some authors suggest that they enter the yolk mass and assist in its dissociation (Toyama 1909 in the silkworm; Schwangart 1904 in *Endromis*); another reports that part of them are incorporated into the yolk spherules, and part are decomposed (Wada 1954 in the silkworm).

The egg also contains various wandering cells and degenerated cells of unknown origin (Toyama 1909; Iwasaki 1932). On the basis of their distribution and strong positive reaction to tests for nucleic acids, Nittono (1951) suggests that certain of these cells may have some kind of inductive function in histogenesis and organogenesis.

c. Organs of Endodermal Origin

The only organ of endodermal origin is the *mid-intestine*. In the silkworm, formation of the mid-intestine follows the fusion of the two endodermal tissues which develop from the bottoms of the stomodaeum and proctodaeum and proliferate backward and forward, respectively (Fig. 10.87). It has not yet, however,

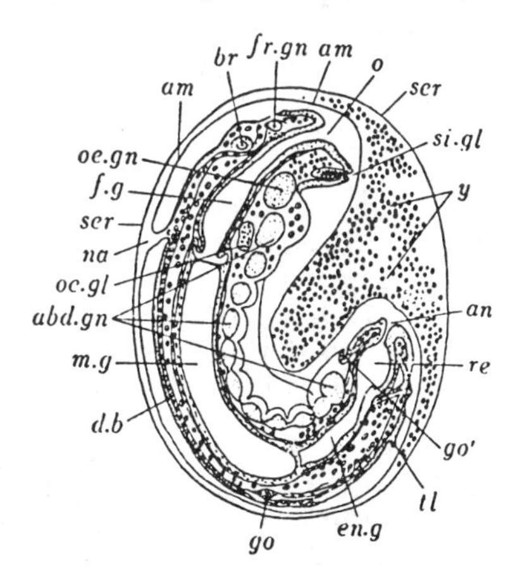

Fig. 10.87 Longitudinal section of fully-developed silkworm embryo (Toyama 1909)

abd.gn: abdominal ganglion, *am*: amnion, *an*: anus, *br*: brain, *d.b*: dorsal vessel, *en.g*: hind-gut, *f.g*: fore-gut, *fr.gn*: frontal ganglion, *go*: gonad, *go'*: attachment of gonadal duct, *m.g*: mid-gut, *na*: navel, *o*: mouth, *oe.gl*: suboesophageal gland, *oe.gn*: suboesophageal ganglion, *re*: rectum, *ser*: serosa, *si.gl*: silk gland invagination, *tl*: tail horn, *y*: yolk.

been conclusively determined that no cells of any other origin take part in forming it. As has been reported above (p. 434), the origin of the endoderm itself is still the subject of unresolved dispute.

The derivation of the mid-intestine in insects other than the silkworm has been the object of a number of studies. In *Pieris brassicae* (Henson 1932), the middle region of the primitive groove invagination (Henson called it "blastopore") is closed early in development, leaving two separate apertures. The stomodaeal invagination is formed at the anterior aperture, and the posterior one deepens to form the proctodaeum. At the sites of these invaginations, there are cell masses of ectodermal origin, which topographically correspond to the oral and caudal cell masses (both mesodermal) of the silkworm. Those of *Pieris*, however, are as yet undifferentiated; later they will differentiate into mesoderm and endoderm. They are therefore called the *anterior* and *posterior mesendoderm rudiments*. The blind ends of the stomodaeum and proctodaeum consist of this undifferentiated tissue, homologous to the blastopore lip of the onichophoran *Peripatus*; later they differentiate into the endoderm which forms the mid-intestine. A similar mode of mid-intestine formation is reported to take place in *Rhodnius prolixus* (Mellanby 1936), as well as in the lepidopteran *Endromis versicolora* (Schwangart 1904). In this species the endoderm differentiates from large masses of cells lying at the anterior and posterior ends of the embryo, migrates inward together with the blind ends of the stomodaeum and proctodaeum, and forms the anterior and posterior anlages of the mid-intestine. Cells which separate from the hypoderm in the central part of the embryo are also likely to participate, together with the yolk cells, in forming the mid-intestine.

In *Gryllotalpa*, the endoderm rudiment similarly invaginates with the stomodaeum and proctodaeum, and forms the mid-intestine (Nusbaum and Fuliński 1909). In *Phyllodromia*, cell proliferation and migration from the inner side of the ectoderm give rise to primary endoderm, from which differentiate mesoderm, hemocytes, etc. The primary endoderm differentiates into the mid-intestine (Nusbaum and Fuliński 1906). The mid-intestine of *Isotome cinera* (Collembola) is formed in a similar way (Philiptschenko 1912).

Tschuproff (cited by Saito 1937), who studied two species of Odonata, *Ephitheca bimaculata* and *Libelluba quadrimaculata*, reported that the terminal portions of the mid-intestine develop from cells derived from the bottoms of the stomodaeal and proctodaeal invaginations, while its median part is formed by the yolk cells. He concluded that this mode of mid-intestine formation is intermediate between those of the Apterygota and Pterygota. Saito (1937) criticized this conclusion as an overly dogmatic attempt to establish the systematic position of the Odonata between the apterygotes and pterygotes, but more recent work seems to support Tschuproff. In *Calopteryx atrata, Epiophlebia superstes* and *Anax parthenope*, all belonging to the Odonata, the terminal portions of the mid-intestine are ectodermal, derived from the stomodaeum and proctodaeum, while the middle part is formed by the yolk cells. No endoderm except the yolk cells seems to exist in these species (Ando 1964). In apterygote insects such as *Lepisma saccharina*, neither stomodaeal nor proctodaeal participation in the formation of the mid-intestine can be observed; the major part of this organ is believed to originate from the yolk cells (Heymons 1897). The same applies to mid-intestine formation in another apterygote, *Thermobia domestica*, in which tissues of yolk-cell origin play a much more important role than do any originating in the stomodaeum or proctodaeum (Wellhouse 1954).

(4) BLASTOKINESIS

As reported above, the process of development of the silkworm embryo includes a reversal in the direction of curvature of its body known as *revolution.* The movements of the developing embryo within the egg, apart from the activity connected with hatching, by which this change is brought about, are generally referred to as *blastokinesis*. Different types of blastokinesis are found in different groups of insects. In the embryos of dragonflies and grasshoppers, there are two conspicuous phases in blastokinesis: a rotation of the completed embryo which carries it into the yolk, and a second rotation by which it rolls back to the surface (Fig. 10.79). Wheeler (1893) called the former *anatrepsis,* the latter *katatrepsis,* and the quiescent stage between them, *diapause*. (The term diapause is at present applied to the period during which growth and development are arrested.)

In contrast to the active and intense movements performed by dragonfly and grasshopper embryos, blastokinesis in such Lepidoptera as the silkworm and *Ephestia* (Fig. 10.79B′) is simply a change in body curvature, with the locomotory movement only incidental.[1] Even among the Lepidoptera, however, are species such as *Diacrisia virginica,* in which the embryo performs a rotatory movement around its long axis in completing the change in curvature (Johannsen 1929).

Several suggestions have been made with respect to the meaning of blastokinesis in the development of the embryo. Wheeler (1893) attributed its significance to the escape of the embryo from its earlier position, where nutritious material had been exhausted and toxic wastes had accumulated. Since the yolk is displaceable, however, it is rather unlikely that blastokinesis takes place for this purpose only (Slifer 1932).

As possible mechanisms for bringing about blastokinesis, the contraction of the serosa (Ayers, cited by Slifer; Wellhouse 1954), and a difference in the developmental velocity of different parts of the embryo have been suggested but observations on the eggs of *Melanoplus femur-rubrum,* in which the movement of the embryo are easily traced from the outside through the transparent chitinous cuticle, have failed to support either of these possibilities (Slifer 1932). In *Melanoplus differentialis,* embryos were found able to develop almost normally without blastokinesis, but were incapable of hatching (Slifer 1932). Tirelli (cited by Slifer) observed that silkworm embryos which did not carry out revolution were not able to hatch. He ascribed the cause of the failure to complete dorsal formation and develop further, which occurred in such cases, to insufficient space and the abnormal curvature of the dorsal region. The significance of revolution, he believed, is that it provides the spatial conditions necessary for successful development. In grasshopper eggs, however, blastokinesis by no means results in an increase in space or in better conditions for dorsal formation (Slifer 1932).

Grandori (cited by Slifer) supposed that the revolution of the *Pieris brassicae* embryo is caused by a force of some sort acting on the ventral surface of the seventh abdominal segment. According to Umeya (1937), physical disturbances due to their concave backs prevents most silkworm embryos which fail to undergo revolution from developing properly. One such abnormality, frequently

[1] The silkworm embryo also sinks temporarily into the yolk after it passes the blastoderm stage. This is probably due to the separation of serosa and amnion, earlier in close contact with one another, and the consequent penetration between them of yolk cells.

encountered when silkworm eggs are exposed to a drastic change in temperature so that the serosa is damaged, is attributed by Umeya to delayed formation of the dorsal region because of unfavorable conditions within the egg, or to failure of the ventral ganglia to fuse (p. 437). Either or both of these causes would tend to disturb the processes of dorsal elongation and ventral contraction required to initiate blastokinesis.

However, cautery experiments show that even partial embryos consisting of only head and thorax are sometimes able to accomplish the normal change in curvature and the movement to the dorsal side of the egg, indicating that the posterior part of the abdomen which appears most likely to initiate this movement in the intact embryo, is not indispensable for its performance (Takami 1944). In this cases, the partial embryo usually accomplishes the change in curvature, without leaving the ventral side of the egg, by means of lateral turning movements. Such lateral turning occurs frequently when the cautery is applied to embryos younger than the "longest" stage (Stage 6). It is therefore likely that during the elongation and contraction of the partial embryo, its position in relation to the egg axis becomes oblique, making it possible for the embryo to reverse its body curvature without dorsal displacement when the two axes subtend a sufficient angle.

This interpretation is supported by the observation of the lateral turning which is induced in normal embryos by centrifugation; i.e., lateral turning occurs rather easily in those embryos which are in the contracted stage after passing the longest stage, probably because they are more easily moved. Since such embryos develop and hatch normally, it can be concluded that the displacement accompanying blastokinesis is not an essential factor in embryonic development. It is certain that an embryo with a ventral lesion induced by partial cautery tends to be ventrally concave, while a dorsal lesion usually leads to dorsal concavity. This strongly suggests that a balance between dorso-ventral growth or tension at a certain stage in embryonic development is necessary to initiate revolution. Fusion of the ventral ganglia can hardly be held responsible for exercising a crucial control over the process of blastokinesis.

Because of incomplete dorsal development, the embryos of the silkworm mutant *Burnt* die early, without carrying out blastokinesis (Aruga 1938). In another mutant, *New Additional Crescents,* revolution is likewise impossible; the obstacle in this case is believed to be "a dorsoventral unbalance due to abnormally stretched dorsal skin" (Itikawa 1943). The difficulty in performing revolution shown by embryos subjected to in vitro culture can be similarly explained (Takami unpublished). These assumptions, however, do not exclude the possibility that other factors control revolution in normal embryos, which obviously develop under very different conditions from those present in the mutants or in vitro cultures.

Blastokinesis in *Antheraea pernyi,* as in the silkworm, is supposed to be mechanically induced, first by a reduction in the dorsal curvature as the embryonic body shortens, and then by the succeeding growth together with the rapid consumption of the yolk in the dorsal zone (Saito 1934).

Some experiments on *Calopteryx strata* and *Cercion hieroglyphicus* provide interesting information with respect to the relationship between blastokinesis and morphogenesis (Ando 1955). When an embryo is prevented from moving by ligaturing it prior to katatrepsis (the rotation which carries it out of the yolk; Fig. 10.79 A), it continues to develop within the yolk and tends to become an everted embryo (Fig. 10.88), with its appendages inside its coelom, surrounded by successive layers of epidermis, mesoderm and yolk. The ventral nerve cord,

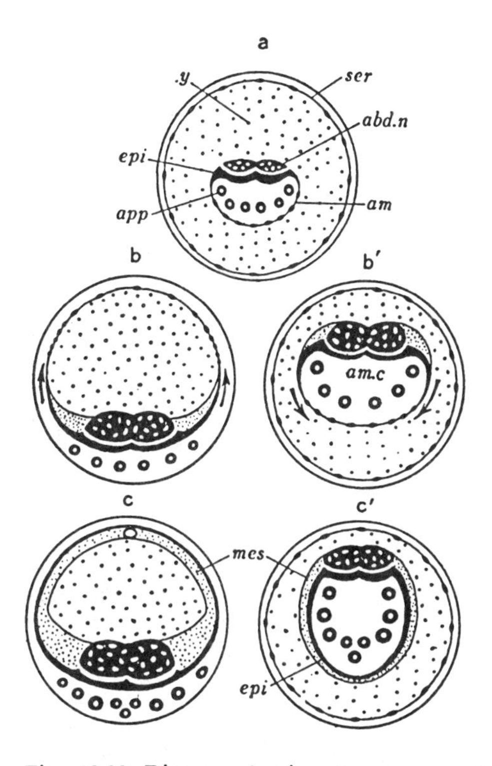

Fig. 10.88 Diagram showing the process of formation of normal and everted embryos (Ando 1955)

a. cross-section through abdomen of normal embryo at pre-revolution stage; b. same sectioning, at post-revolution stage; c. same sectioning, of fully-developed normal embryo; b′. cross-section through abdominal region of everted embryo; c′. same sectioning, of fully-developed everted embryo. *abd.n:* abdominal nerve cord, *am:* amnion, *am.c:* amniotic cavity, *app:* appendage, *epi:* epidermis, *mes:* mesoderm, *ser:* serosa, *y:* yolk.

finally, lies outside the embryo. That such an effect is produced by suppressing blastokinesis is of interest in discussing the "embryo" reported by Seidel (1929a; p. [26]).

4. PARTHENOGENESIS

(1) SPONTANEOUS PARTHENOGENESIS

Spontaneous parthenogenesis is well known among bees and aphids as a natural reproductive process. Doncaster (1924) discriminated five different types of mechanism leading to parthenogenesis; four of these occur naturally in insects.

a. **Maturation with one polar division** (no chromosome reduction)

The parthenogenetic egg produces only one polar body and develops into a female without chromosome reduction (e.,g., *Phyllaphis*).

b. **Conjugation of egg nucleus with polar nucleus**

The egg pronucleus has the reduced number of chromosomes following normal meiosis, but the second polar body sinks back from the surface of the egg and conjugates with the pronucleus, giving rise to a diploid nucleus as though the pronucleus had been "fertilized" by the polar body (e.g., the strepsipteran *Xenos)*.

c. **Maturation with two polar divisions**

i) *Formation of male from unfertilized egg*

The normal process of meiosis produces an egg pronucleus with the reduced number of chromosomes. When this egg is fertilized, it produces a normal female, but if it is not fertilized it develops parthenogenetically into a male (e.g., the Hymenoptera).

ii) *Meiosis without chromosome reduction*

The two meiotic divisions occur, but both are equational, so that no chromosome reduction takes place. The egg formed in this way develops parthenogenetically into a female (e.g., *Poecilosoma luteolum, Rhodites rosae*).

d. **Production of females from diploid, and males from haploid, eggs**

No meiosis takes place during oogenesis in gynoparous females. The eggs are therefore diploid and develop parthenogenetically into females. In the andropara, meiosis occurs, the nuclei have the haploid number of chromosomes, and the eggs develop parthenogenetically to give males (e.g., *Neuroterus lenticularis*).

The eggs of the gynoparous female of *Phylloxera caryaecaulis* are similarly formed without chromosome reduction, and develop parthenogenetically into females. The eggs of the androparous females, however, unlike those of *Neutoterus,* are produced by a special process in which the sex chromosomes are reduced, whereas the autosomes do not undergo reduction (Morgan 1912, 1915). Although this ultimate state of sex determination has not yet been reached at the moment of oviposition, the eggs laid by androparous females are much smaller than those which will develop into females. It follows therefore that the characteristic of maleness is already fixed even before the reduction of the sex chromosomes takes place, since the size of the eggs is determined before the amount of their sex chromatin is definitely established.

These data, then, show that the sex of the embryo to be produced by parthenogenetic development is strictly fixed when parthenogenesis constitutes the normal mode of reproduction. This is not true, however, of incidental parthenogenesis, which produces males or females indiscriminately (see p. [86]).

The dipteran *Miastor* is famous for its *paedogenesis* — the production of larvae within the body of a larva. This is perphaps one of the most extreme cases of parthenogenesis. In *Miastor americanus,* there are two ovaries in the larva, each containing at most 32 oocytes. The egg nucleus undergoes a single, nonreducing polar division, and begins to develop parthenogenetically. After the second cleavage, when there are four cleavage nuclei, one of them enters the pole plasm and the other three eliminate part of their chromatin during subsequent cleavages. The nucleus in the pole plasm divides normally without chromatin-diminution and one of the daughter nuclei moves back into the egg cytoplasm and at the next division eliminates part of its chromatin. The other, maintaining its original quantity of chromatin, is cut off together with the pole plasm from the rest of the egg as a separate cell at the 8-nucleus stage, to form the primordial germ-cell. The primordial germ-cell divides into eight cells, which arrange themselves in groups of four on each side of the embryo. There they undergo three more cleavages and develop into the paired ovaries, each ovary containing 32 oocytes. The nuclei which do not enter the pole plasm migrate to the egg surface and form the blastoderm (Doncaster 1924). Thus, although there are some special features, such as chromatin-diminution in the blastoderm-forming nuclei and the unusual behavior of the nucleus which has entered the pole-plasm, the general outline of development in this insect is not very different from that of ordinary Lepidoptera (Fig. 10.89).

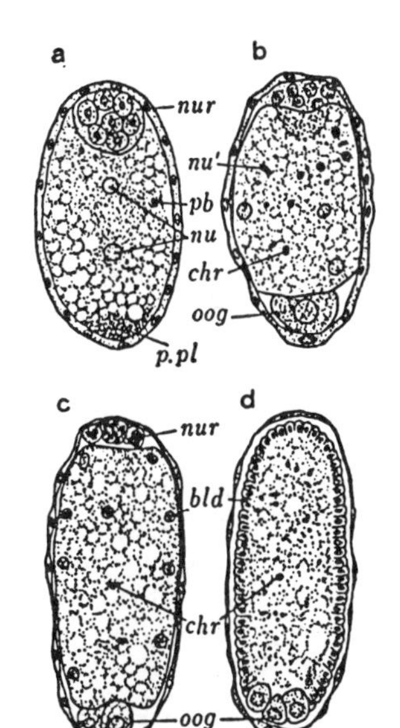

Fig. 10.89 Early embryonic development in *Miaster americanus*

a. 2-nucleus stage; b. more advanced stage; c. early blastoderm stage; d. late blastoderm stage, containing four primordial germ cells. *bld.* blastoderm, *chr:* eliminated chromatin mass, *nu:* cleavage nucleus, *nu':* nucleus eliminating chromatin, *nur:* nurse cell, *oog:* primordial germ cell, *p.pl:* polar plasm containing germ cell determinants, *pb:* polar body in process of formation.

The diapausing eggs of the silkworm can be discriminated macroscopically from unfertilized or non-developing eggs since, with some special exceptions, they become colored when the serosa is formed, owing to the pigmentation of the serosal cells. Among the eggs laid by unmated females, only a few develop to hatching. The others remain either without pigmentation, or with slight pigmentation but no definite structure of the serosa, or again, with pigmentation and serosa formation but abnormal arrangement of the cells. Sections of such eggs reveal various degrees of abnormal development: no initiation of development, formation of yolk-cell-like structures, formation of an abnormal blastoderm, death after complete formation of an embryo, etc. (Toyama 1909, Sato 1924). Many eggs develop sufficiently to show pigmentation; the formation of pigment granules is not limited to the serosal cells as in normal eggs, however, but frequently occurs also in the yolk-cell-like structures. In some cases a serosa-like structure appears inside the egg. Aside from their bearing on the problems of parthenogenesis, these phenomena are also of great interest in connection with the concept of "field" in embryonic development (Sato, Takami and Kitazawa 1956).

Kawaguchi (1933a), studying spontaneous parthenogenesis in the silkworm, observed the hatching of 132 individuals from about 49,000 eggs in one series, and of 32 individuals in about 120,000 eggs in a second. These larvae included both males and females.

(2) ARTIFICIAL PARTHENOGENESIS

When insect eggs are stimulated by some means, the rate of parthenogenesis is much increased, according to the studies of Tichomiroff, Kellogg and others. More recently Koltzoff (1932) investigated the effectiveness of various methods for achieving this result. Sato (1934) has reported that treating silkworm eggs with hydrochloric acid of specific gravity 1.04—1.06 at a temperature of 70—73°F. for 4—6 minutes gave 192 parthenogenetic larvae; Kawaguchi (1933a, 1934) obtained 82 parthenogenetic worms from about 6,800 eggs treated with 15% HCl for 5 seconds.

The meiosis of the silkworm egg begins while the egg is still within the ovary. The eggs in the terminal portion of the oviduct are generally at the metaphase of the first meiotic division. Eggs from non mated females are also at the metaphase of the first division when they are examined just after oviposition. If such unfertilized eggs are treated as described above, they accomplish the first and second meiotic divisions, although both are irregular and variable as compared with normally fertilized eggs. Chromosome reduction usually occurs at the first division; the chromosome number of well developed parthenogenetic silkworms, however, proves to be diploid (as examined in their spermatogonia or oogonia), with 56 chromosomes. This indicates that the number of chromosomes must have doubled at some time during development.

The results of cytological observations suggest that this chromosome doubling is brought about by fusion of the two daughter nuclei formed by cleavage of the reduced egg nucleus, or perhaps by two such fusions at the 4-nucleus stage. Similar fusions between some of the nuclei in a later, multinucleate stage give rise to mixoploidy. The cleavage nuclei in a parthenogenetic embryo usually divide by mitosis as in the normally fertilized egg, although various abnormalities can also be observed: e.g., nuclear cleavage may not be followed by the division of the surrounding cytoplasm; a nucleus may divide amitotically; two or more spindles may be associated with one cleavage nucleus, etc. All these abnormalities are believed to result in increasing the number of chromosomes beyond that found in the haploid nucleus. In fact, counts of the chromosomes in the blastoderm cells of parthenogenetic eggs give various numbers such as 58, 55, 52, 34, 33, 31, 28 (Sato 1934).

The mechanism by which the sex of such parthenogenically developing silkworms is established is still unknown. The sex chromosome constitution of the silkworm is *ZZAA* (*A*= autosomes) in the male and *ZWAA* in the female. If the *ZA* complex of an egg is doubled, it will obviously develop as a male, and the appearance of females cannot be explained. Kawaguchi (1933b, 1934) induced parthenogenetic development in the eggs of females which were heterozygous with respect to an autosomal gene; the female offspring were heterozygous for this gene, while the males were homozygous. Eggs obtained from these heterozygous females mated with homozygous males were centrifuged immediately after oviposition. Among various abnormal caterpillars resulting from this treatment he found four individuals completely devoid of connection with the genic constitution of the sperm nucleus. These were all females, and heterozygous. From these results Kawaguchi concluded that these females were produced by fusion of the egg nucleus with one of the polar bodies. In any case, simple chromosome doubling could not be expected to give rise to heterozygous females, whether or not chromosome elimination was involved.

Tazima (1947) reported that he could obtain individuals of which the genetical constitution was derived from the egg nucleus only, if he submitted the eggs of even normally mated females to high temperature (40°C) for one hour, immediately after oviposition. According to the experiments of Hasimoto (1956a), however, it is difficult to induce parthenogenesis with hot water treatment in normally fertilized eggs even if the treatment is applied before syngamy takes place.

So far we have been dealing with parthenogenesis in eggs which have completed meiosis. In the silkworm, however, it is possible to induce parthenogenesis by stimulating eggs which are still within the ovary: Astaurov, for example, induced parthenogenetic development by treating ovarian eggs with hot water. Hasimoto (1953a, b) likewise treated ovarian eggs of the silkworm moth with water of 40°C for 18 minutes, and obtained 288 parthenogenetic individuals from 43,200 eggs (1.33%) and, in a second series, 115 individuals (0.44%) from 143,000 eggs. Genetical considerations suggest that these individuals resulted from the development of egg nuclei which failed to undergo reduction division; they were all females, including diplonts and tetraplonts, as well as a number of di- and tetraploid mixoplonts. Similar parthenogenesis may be induced by hot water treatment of the ovarian eggs of late pupae or the eggs laid by non-mated moths (Hasimoto 1956a). Various cytological abnormalities can be observed in the nuclei of eggs treated in this way, some of which may contain a polyploid number of chromosomes. This is due to the non-segregation of the chromosomes dividing at the first meiotic division, and it is believed that this non-segregation also underlies the exclusive production of females (Sato 1953).

The behavior of a certain mosaic-inducing gene in the silkworm is very suggestive here, although it has no direct concern with parthenogenesis as such. Cytological evidence shows that, in an egg laid by a female homozygous for this gene, while normal syngamy is taking place a second sperm nucleus fuses with a polar body that has remained within the egg. The participation of such a supernumerary nucleus in the subsequent development is believed to be responsible for the formation of mosaic individuals (Goldschmidt and Katsuki 1928).

(3) ANDROGENESIS

The above discussion has been concerned with true parthenogenesis, in which the nucleus of an unfertilized egg initiates development without the participation of a sperm nucleus. Cases have also been observed in which the sperm nucleus plays a similarly independent role in the development of the egg.

A silkworm gene known as *Kp*, located on Chromosome VI, in the homozygous state induces a pair of supernumerary abdominal legs on the second abdominal segment, as well as a pair of supernumerary crescents on the third abdominal segment. In the heterozygous condition it induces the extra legs but never the extra crescents: the distinction between the homo- and heterozygous states of this gene is therefore easily made. Another gene called *od*, which is sex-linked and recessive, induces a transparent skin. Since the sex-chromosomal constitution of the silkworm is *ZW* in the female and *ZZ* in the male, and *od* is located on the *Z*-chromosome, it manifests itself phenotypically in all the females but only in homozygous males. Hasimoto (1934) mated males homozygous for *od* and heterozygous for *Kp* with normal females, collected the eggs within an hour after oviposition, submitted them to a drastically high temperature (36.5°—41.5°C) for an hour, and then incubated them at ordinary temperatures until they hatched.

The expected phenotype in this case was: all females *od*, all males $+^{od}$ (wild type in respect to *od*), and half of both sexes with supernumerary abdominal legs, which were wanting in the other half. Actually, however, besides these expected types there appeared a few exceptional cases such as *od* males, $+^{od}$ females, *od* mosaic (mosaic of normal and transparent skin), *Kp* homozygous (with both supernumerary legs and crescents). These *od* males comprised a mixture in the ratio of 1 : 2 : 1 of *Kp* homozygous, *Kp* heterozygous and normal. The appearance of such *od* males cannot be explained without assuming that two sperm nuclei fused and induced androgenetic development. Furthermore, 51 of the 54 *Kp* homozygous individuals were males (of 11,374 studied), which supports the view that two sperm nuclei must have fused.

Hasimoto further studied the structure of the *od*-zone skin of the *od* mosaic caterpillars, and found examples of all three phenotypes (*Kp* homozygous, *Kp* heterozygous and normal) representing the composition of the exceptional *od* males. From these results, it can be deduced that the normal skin zone of the mosaic caterpillar is derived from the zygote nucleus produced by normal fertilization, while the *od*-zone (transparent zone) arises from a nucleus produced by the fusion of two sperm nuclei.

Observations made on serial sections of eggs laid by normally mated moths which were exposed to 104°F for one hour, beginning 20—80 min after oviposition, revealed marked disorders in the behavior of both egg and sperm nuclei, such as formation of polyploid nuclei due to fusion of a polar body with the egg and sperm nuclei, fusion of two daughter nuclei after meiotic division of the egg nucleus, fusion of two sperm nuclei, mitotic division of the sperm nucleus (Sato 1942). It is suggested that these abnormalities may underlie the parthenogenesis and androgenesis described above, as well as the formation of mosaics, which will be described below.

Sato and Mesaki (1942) also found that the egg nucleus is much less resistant to high temperature treatment than the sperm nucleus. On the other hand, Hasimoto (1956b) treated eggs of normally mated females with high temperature (40—41°C) for 60 minutes after they had been kept at 2.5°C for 10 days following oviposition. Two percent (134) of the larvae hatched from these eggs; all of them were males and showed recessive characters inherited from the male parent. This result suggests that the egg nucleus was destroyed by the cold treatment, while the more resistant sperm nucleus survived and brought about the androgenetic development of the eggs.

Later studies also elicited various interesting facts. For example, combination of X-ray irradiation and hot water treatment was found to be highly effective in inducing androgenesis. The fusion of two sperm nuclei seems to start about 3 hours after oviposition, but is still observable at 10 hours in delayed cases. The range of motility of the sperm nucleus is limited to an ellipsoidal zone, located at the anterior end of the egg, with a short axis equal to 1/5 or less of the egg's long axis. When fusion of the sperm nuclei occurred early, no departure from normal development could be detected, either in the state of the cleavage nuclei 10 hours after oviposition (9 hrs after treatment), or in the embryo 45 hours after oviposition (Tanaka, Oi and Sato 1955; Tanaka 1956).

The development which follows fusion of two sperm nuclei is termed "merogony" by silkworm geneticists. Since this is a very different phenomenon from true merogony as the term is generally used in embryology, Tazima (1939) proposed that it be called "two-sperm merogony".

Two-sperm merogony can be utilized to compare two groups of individuals in which the genetic constitution is the same, but in one of which the nuclei have been made to develop in cytoplasm of a different genetic constitution. Hasimoto (1951), for example, compared the body weight of two such genetically similar groups, which at the time of hatching showed a marked difference in this respect because of the dissimilar cytoplasms. He found, however, that the weight difference lessened as development proceeded, so that the two sets of caterpillars weighed the same by the end of the larval period.

With respect to the mechanisms by which mosaic individuals arise, at least six possibilities have been suggested (Tazima 1947):

a) Two-egg-nuclei Theory

After the release of the second polar body, the egg pronucleus divides into two. These proceed to develop, one after normal syngamy and the other in the haploid condition. Some parts of the embryo are derived from the former nucleus, and others from the latter (e.g., honey bee). A second egg nucleus may alternatively arise from a polar body which returns into the interior of the egg (e.g., *Habrobracon*, silkworm).

b) Two-sperm Theory

An egg is entered by two spermatozoa, one of which fuses with the egg nucleus, while the other develops by itself. Both participate in development (e.g., honey bee). Three spermatozoa may sometimes enter an egg, one fusing with the egg nucleus, while the other two fuse with each other (silkworm).

c) Two-egg-nuclei-and-two-sperm Theory

Two spermatozoa enter an egg having two pronuclei, and double fertilization takes place. The presence of two egg pronuclei may be accounted for by the fusion of two young oogonia, but in the silkworm the generally accepted explanation is that the second polar body remains within the egg.

d) Chromosome-elimination Theory

Elimination of one chromosome from a daughter nucleus may occur during the first or a succeeding division of the cleavage nuclei. Elimination of the X-chromosome is believed to be the cause of sexual mosaic formation in *Drosophila*. In the silkworm, this process is sometimes observed in trisomic individuals, but the loss of chromosome fragments is much more usual.

e) Autosomal-mutation Theory

This will not be discussed here, since it is not a problem of embryology.

f) Mutable-gene Theory

This also is a problem of genetics rather than embryology.

A curious fact reported in connection with silkworm mosaicism is that larvae and moths which hatched from eggs mosaic with respect to the color of the serosa were found to be laterally mosaic, although the original mosaicism was antero-posterior (Tazima 1942a). Since the serosa arises from the extra-embryonic blastoderm, however, there is no direct embryological continuity between it and the embryo, even though they may be in extremely intimate contact.

Again, the mosaic pattern of the compound eyes in silkworm moths always manifests itself as stripes parallel to the body axis, unlike *Drosophila,* in which the mosaic pattern consists of small, irregularly scattered patches (Tazima 1942b). This is attributed to an orderly arrangement of cell divisions parallel to the body axis along a longitudinal groove which is formed at the center of the compound eye anlage in the course of eye differentiation (Koyama 1956).

5. GROWTH OF THE EMBRYO

(1) GROWTH CURVE

Only a few studies have recorded the increase in weight of the embryo isolated from yolk and other components of the egg. In the silkworm, some workers have fixed eggs with boiling water, isolated the embryos and measured their dry weight (Yamaguchi 1940; Yamaguchi and Kobayashi 1940; Kinoshita 1951). This is found to increase almost logarithmically during 80—90% of the incubation period (from the longest stage), and very rapidly thereafter. The over-all shape of the weight curve agrees well with the sigmoid curve of Robertson's formula (Yamaguchi and Kobayashi 1940). When the growth of the embryo is correlated with the developmental stages, three maxima are found, at 1) about the "longest stage", 2) the revolution stage and 3) the stage of trachea formation (or of mouth-part pigmentation). These three maxima can also be recognized in the curve of daily weight increase.

In the curve of the growth rate:

$$\frac{\text{weight increase in 24 hours}}{\text{weight on the previous day}} \times 100,$$

there are two maxima: one at about the revolution stage (45% development) and the other around the stage of trachea formation (or of mouth-part pigmentation; 70% development), and two minima, one before and the other after revolution (30% and 50—60% development) in non-diapausing eggs. In eggs artificially prevented from diapausing, and in diapause eggs which have passed the winter, one more maximum is added around the stage of appendage-process formation (10—20% development). Kinoshita (1951) has obtained similar results.

The catalase activity of eggs homogenized during the course of development shows a curve which agrees with the curve of embryonic weight increase. This activity increases slowly in the first half of the embryonic period, and then shows a rapid increase beginning at the time of trachea formation (Yamaguchi and Shimizu 1951). The various workers in this field are not, however, in general agreement about the changes in catalase activity during development, except with respect to this abrupt increase in the final stage. Some have reported a rapid rise at about 30—40% development, followed by a gradual increase which ends in an abrupt elevation just before hatching (Bito 1931). Others have observed that it diminishes until about the revolution stage, then increases slightly and finally shows a rapid increase at the "dark-spotted" stage (Ito 1950).

In contradiction to the results of Yamaguchi and others, who find two or three peaks in the growth curve of the silkworm embryo, Tokunaga (1939) has proposed a diphasic course of growth. Calculating the daily output of CO_2 in relation to the total CO_2 output during the embryonic period, from the data of Suzuki (1917), he found that CO_2 output falls off for a time at a point located at 44% of the total embryonic period, in a race called *Godaishu*, and at the 40% point in another race called *Aojuku*. A similar situation is also observed in connection with loss in weight during the development of the embryo. According to Tokunaga, who calculated the daily egg weight loss in relation to the total average loss, from the data of Mizuno and others, a dip appears in the curve at some definite stage. This stage varies somewhat from one race to another: 41.7% total embryonic period in spring eggs of Japanese bivoltine races (average of 3 races), 44.4% in the autumn eggs of Japanese bivoltine races (average of 3 races), and in F_1 eggs of Japanese × Chinese bivoltine hybrids (average of 2 hybrids). If egg volumes are plotted against time, during the development of a beetle *(Phyllotreta vittata)* and a dragonfly *(Epiophlebia suprestes)*, a dip in the former curve appears at 40—50% development, and in the latter, a less pronounced dip at 42.4—47.7%. Tokunaga interpreted these phenomena as suggesting that growth follows a diphasic course, as the result of some diapause-like event at the stage of 40—50% total growth, even in the non-diapausing eggs of insects in which growth seems to be continuous. The observed period of lower growth rate may correspond to a phase of cellular and tissue differentiation which intervenes in the course of cell proliferation, or to a critical point between two types of growth depending on different mechanisms.

(2) GROWTH AND ENVIRONMENTAL CONDITIONS

The environmental conditions considered here are those within the rang normally encountered by the embryo: artificially induced special conditions wil be excluded from the discussion.

a. Temperature

The environmental factor which exerts the most important influence on embryonic development is temperature. The response of the embryo to temperature is different at different stages, as is clearly shown in Table 10.6 (Mizuno 1936).

Table 10.6 **Influence of temperature on length of time required by silkworm embryos to develop at different stages.** (Japanese Strain 8, post-diapause eggs)
(Adapted from Mizuno 1936)

Developmental period	10°C		15°C		20°C		25°C		30°C	
	days	hours	days	hours	days	hours	days	hours	days	hours
diapause to longest stage	11	03	4	17	3	20	3	12	3	12
longest stage to revolution	33	17	9	19	5	02	3	14	3	14
revolution to bluish stage	55	00	16	00	7	16	4	18	4	06

Different lots of test embryos were used in each developmental period. Mizuno's "longest stage" is about one day before the appearance of the labial process.

At the more advanced stage of development, for instance, there is a greater difference between the amounts of time required for growth at low and high temperatures. Another experiment shows that the course of development from the end of diapause to the longest stage becomes highly variable with individuals when the temperature is raised above 17.5°C, whereas a corresponding variability is found in the succeeding period, to the revolution stage, if the temperature falls below

17.5°C. Development during revolution and the bluish stage (embryonal development completed) is always uniform.

The average growth rate (reciprocal of the incubation period) of silkworm embryos developing after overwintering was calculated for ten races of Japanese, Chinese and European stock by Matsumura and Ishizaka (1929). At 72% relative humidity, this was 0.064 at 20°C, 0.090 at 24°C and 0.106 at 28°C.

According to Iwasaki (1930), there is a daily cycle in the mitotic activity of silkworm ectoderm cells. The highest frequency of mitosis, observed at 8 p.m., and the lowest at 4 p.m., do not coincide with the daily temperature fluctuations. There is a lag of 4 to 8 hours between the temperature maxima and mitotic maxima, as well as between the corresponding minima.

The permissable temperature range for hatching in *Saturnia pyretorum* is 10° to 30°C; the larvae do not hatch if the temperature is lower than 9° or higher then 31°C. Temperatures between 15° and 27°C are favorable to hatching. Up to 29°C, increasing the temperature shortens the incubation period, but at higher temperatures it again becomes longer. The reciprocal of the number of days to hatching or to revolution (V) has an approximately linear correlation with temperature (T) : $V = aT + Vo$. A perfectly linear functional relation holds only in the range of 15°—27°C. Calculating the value of T at the points where $aT + Vo = 0$, i.e., where the growth rate becomes 0, gives 9.00°C for development to hatching and 7.81°C for development to the revolution stage, which agree well with the observed experimental results. If the average daily temperature from the day of oviposition is known, the date of hatching can be calculated by the following formula proposed for *S. pyretorum* (Koidzumi and Shibata 1938):

$D = 100/(aT + Vo)$, where D is the number of days required for development, and a and V_o are constants.

As the above example shows, the temperature at which the growth of an embryo is arrested by cold is different for different embryonic stages. The embryo of *S. pyretorum* hatches when the temperature is above 10°C, but fails to do so at 9° or less, although it can develop to the revolution stage even when the temperature is 8° (requiring about 110 days). At 6°, development still proceeds until the stage at which the primitive segments become distinct (in about 100 days), but at 5° or less no development occurs, the egg remaining as it was at the moment of oviposition. If *Saturnia* eggs developing at 20° are transferred to the cold for, say, 30 days and then returned to 20°, they hatch in 50 days (the sum of the 20 days required for hatching at 20° plus the 30 days of cold storage), when the cold storage temperature is 5°. When this is 8°, the eggs hatch in 43—48 days. This result indicates that embryonic development progresses slowly even during storage at 8°. It follows therefore that the developmental 0-point (or threshold) is 9° with respect to hatching, but 5° so far as the earlier development is concerned: in other words, an "effective temperature" calculated on the basis of the requirement for hatching does not coincide with that for embryonic development proper (Koidzumi 1939).

Table 10.7 **Q_{10} of embryonic development in** *Antheraea pernyi* (Kurasawa et al. 1937)

Temperature	Relative humidity	Q_{10}
17.5° to 25.0°C	100%	2.12
"	95	2.0
25.0° to 27.5°C	100	0.90
"	95	1.77

Table 10.7 shows the Q_{10} of the rate of development of *Antheraea pernyi* (Kurasawa et al. 1937).

As in *S. pyretorum,* the minimum temperature for normal development of the silkworm egg lies at about 10°C. Embryos of the early post-diapause stage are able to develop to a very slight degree even at 0°. At 5°, embryos of any stage can develop, although a very long period is required and the embryos die without accomplishing normal development (Mizuno 1936; Totani 1955). Embryos of the early post-diapause stage are not able to hatch at 10° — for this they require at least 12.5°; whereas a low percentage of late-stage embryos (dark-spotted stage) may reach hatching at 10°. Embryos at the bluish stage (embryonal development completed) are able to hatch even at 7.5°. These results may seem to contradict the data of Table 10.6: this table, however, shows the observations of development confined to a limited period. These experiments suggest that the young embryo placed under conditions of low temperature requires such a long period to accomplish normal development that it loses its vitality and is unable to hatch.

From a practical standpoint, the highest temperature at which the silkworm embryo develops normally is 30°C; exposure for three days to 35° is almost invariably fatal to non-diapause eggs (Mizuno 1936).

The output of CO_2 during cold storage is of interest in connection with the developmental 0-point. Eggs of the overwintering silkworm race *Aojuku* laid September 3 were transferred to three different low temperatures: 5°, —2.5° and —5°C on Februrary 12, and their CO_2 output was measured during the next 151 days. The daily excretion per kilogram eggs was calculated as follows (Suzuki 1917):

temp.	(CO_2 /kg eggs/ day)	comments
5.0°C	0.28660 g	during 151 days from Feb. 12
—2.5°	0.00379	,, ,, ,, ,, ,, ,,
—5.0°	0.00322	,, ,, ,, ,, ,, ,,
4.5°	0.20177	(rm temp) 10 days from Feb. 2
25°	4.6058	*transferred from rm temp to 20° for 1 day, then to 25°. Measured on first day at 25°.
30°	12.6894	*via 1 day at 20°; measured on first day at 30°.

* These eggs had already been freed from diapause, and hatching of about 50% took place on February 22 in the 25° group and on February 19 in the 30° group.

In Japanese sericulture, silkworm eggs are often stored at 5° when the average temperature rises in early spring. The main purpose of this cold storage is to synchronize the early development, rather than to preserve the eggs until the culturing period. As stated above, the critical temperature for development varies with the developmental stage. Keeping the embryos at 5°C therefore suppresses further development in the more advanced embryos, whereas it only retards the younger ones, thus tending to synchronize the development of the group as a whole. Since embryos of different stages vary in their resistance to cold treatment, studies of suitable periods, temperatures and methods for storing eggs as long as possible without adverse effects constitute an important aspect of sericulture. Umeya (1955) found that eggs of *Rhodinia fugax* kept at 5°C developed to the stage just prior to hatching in 370 days, but failed to hatch.

Eggs with indefinite diapause, like those of *Melanoplus,* stop developing when the temperature is low, but if the temperature is raised, they resume development even in winter. If such eggs are kept from the beginning at 25°C, some enter diapause and some do not. In this case, the diapausing eggs survive for more than one year at 25° (Bodine, cited by Wigglesworth 1950).

If diapausing eggs of the silkworm, on the other hand, are kept at 25°—27°. from the time of oviposition, they can be freed from diapause by winter cold and hatch normally the next spring only if they are removed from the high temperature conditions within about 90 days. When they are kept at the high temperature for more than 190 days, they almost completely lose the ability to hatch, even if they are returned to room temperature and then exposed to the cold of winter, and eventually they die. Eggs kept after oviposition at 15° begin to hatch in about 120 days, whereas they die within about 140 days at 10°C (Watanabe 1931). In these experiments, however, because of the different lengths of the periods during which the embryos were kept at the respective temperatures, the natural temperatures to which they were returned necessarily differed; in spite of this the observation of hatchability was made the following April. In other words, although the condition of hibernation was controlled, the various lots of eggs differed in degree of diapause, so that it is not possible to ascribe all the different effects directly to the duration of the controlled temperature periods.

Kutsukake (1954) reported that some silkworm larvae hatched after 250 days from eggs kept continuously at 25°C; in the best case, 36.1% hatched between 250 and 390 days after oviposition, at 25°. At 30°, however, no hatching could be observed, although some of the embryos seemed to have survived for 120 days. Kutsukake concluded that diapause of the silkworm egg can be terminated even at these high temperatures, but the embryos mostly die because of physiological damage incurred during the long experimental period.

Temperature not only influences embryonic development but also modifies voltinism. In the silkworm, eggs incubated at 25° or above develop into moths which lay diapausing eggs, while those which are reared at 15° or lower produce non-diapausing eggs, provided that all other conditions of incubation and rearing are the same (Watanabe 1924). Since the effect of temperature on voltinism itself is very different in different silkworm races, however, it cannot be subjected to a simple summarization.

b) **Humidity**

The effect of humidity is inseparable from that of temperature. For example, the maximum temperature for embryonic development of the beetle *Bruchus chinensis* is lowered by a decrease in humidity, being around 32°C at 50—60% humidity, and about 30° when the humidity is lower. The temperature-humidity range for 90%, 70%, 50%, 30% and 0% hatching is shown in Figure 10.90 (Ishikura 1940). The theoretical growth zero point of the egg of *Grapholitha (Cydia) molesta*

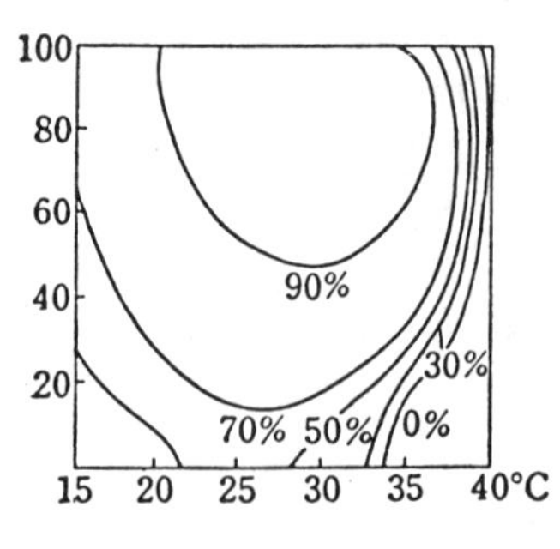

Fig. 10.90 Diagram showing relation between hatchability and the range of temperature and relative humidity in *Bruchus chinensis* (Ishikura 1940)

lies between 40 and 44°F, shifting upward within this range with decreasing humidity (Chaudhry 1956).

As in the case of temperature, the optimum humidity conditions for embryonic growth are different from those required for hatching. A study of *Antheraea pernyi* shows that the temperature required to complete embryogenesis in 90% of the embryos is 16.5° to 29.7° at 53—100% humidity, whereas 90% hatching is only possible between 18.5 and 28.2° when the humidity is between 68 and 100%. The permissable range of temperature and humidity for completing embryogenesis is thus clearly wider than that for hatching (Yamazaki 1939). Table 10.8 shows a comparison of this relation in some different insects. The hatching capacity at 23.3°C of *Philosamia cynthia ricini*, plotted against humidity, gives a sigmoidal curve, with the maximum hatchability at 95—100% humidity (Hodai 1949).

Table 10.8 **Temperature and humidity optima for hatching** (Yamazaki 1939)

Species	Investigator	Expected % of hatching	Temperature	Humidity
Dendrolimus spectabilis	Kojima (1935)	More than 95%	16.0° to 32.2°C	70 to 100%
Bombyx mori	Doke (1935)	"	15.0° to 28.0°	50 to 100%
Antheraea pernyi	Kurasawa et al. (1937)	More than 90%	15.0° to 27.5°	50 to 100%
" "	Yamazaki (1939)	"	18.5° to 28.2°	68 to 100%

As studied in seven silkworm races, the average incubation time at 25°C is: 11.33 days at 60% humidity, 10.98 days at 75% humidity, 11.00 days at 90%. A rise in humidity, from 60% to 75%, was thus found to shorten the incubation period, but no appreciable effect was exerted by increasing it further to 90% (Matsumura, Higuchi and Kitajima 1928).

The percentage of non-diapausing eggs is higher in the eggs laid by moths which develop from eggs incubated under conditions of relatively low humidity (Mizuno 1936).

c. Light

The development of silkworm eggs before the darkspotted stage goes on more rapidly in the light (25 W tungsten lamp at a distance of 1 m) than in the dark. This tendency is clear-cut at 20°, but not so noticeable above 25°. Development after this stage (dark-spotted stage to hatching), on the contrary, is accelerated by darkness, and more markedly at higher temperatures. At a still more advanced stage, darkness has a temporarily inhibiting effect: hatching is delayed by about 9 hours (at 25°) if embryos just about to hatch are transferred to the dark, but after 10 hours some of them, and within 14 hours almost all of them are able to complete hatching in the dark (Watanabe 1934). It is thus possible to synchronize hatching to some extent by adjusting the light conditions. In connection with the slower development observed in the light after the dark-spotted stage, it is still not clear whether light actually exerts a retarding effect, or whether growth is accelerated in the dark.

The light conditions during incubation not only influence the development of the embryo, but also affect the molting character of the larvae from these embryos and the diapause character of the eggs to be laid by these individuals (Kogure 1930, 1933).

d. **Slight changes in atmospheric pressure**

Silkworm embryos at the dark-spotted stage were subjected to atmospheric pressure changes of +100 mmHg, —50 mm, —100 mm, —200 mm and —300 mm for 6, 24 or 48 hours. Exposure to pressures with a comparatively smaller difference from ordinary (+100, —50 and —100 mm) was found to give slightly lower mortality of embryos than the controls (Nunome 1934). The effect of still smaller pressure changes was studied more recently by Yokoyama and Keister (1953), who found that embryos treated in the early dark-spotted stage with —30 mm (730 mm) for 20 or 40 hours showed an eventual hatchability of 99%, and more of the treated than of the control embryos hatched during the first two days. The oxygen consumption was also significantly higher in the low-pressure groups than in the controls. Since the partial pressure of oxygen, water vapor, etc. must be taken into consideration as well as the pressure itself in interpreting these results, the difference in growth rate between the treated and non-treated groups cannot be attributed solely to the pressure effect, but it is nonetheless certainly due to very slight changes in the environmental conditions.

When eggs were exposed to 740 mmHg for 2, 3, 5 and 8 days at stages ranging from very early development to the dark-spotted stage, the treated embryos were always less able to hatch, and required a longer incubation period than the controls (Yokoyama and Takashima 1952).

6. DIAPAUSE

Embryonic development in insects stops under unfavorable conditions of temperature, humidity, etc., but this stoppage is simply a state of *quiescence*, and if this quiescence is within certain permissable limits, development is resumed immediately after the silkworm, however, embryonic development stops at a definite stage even when all the environmental conditions are favorable for its physiological processes. The embryo, in this case, cannot hatch until the following spring, and furthermore, will die if it is not exposed to low temperature (p. 464). Such a cessation of development due to some intrinsic cause is called *diapause*. As originally used by Wheeler (p. 451), the word diapause referred to the resting stage in the middle of blastokinesis; this second meaning was later introduced by Henneguy (1904). The terms "overwintering", or "hibernating", do not necessarily imply such spontaneous cessation of development: eggs during hibernation may either be actually in diapause or in post-diapause (diapasue broken but development still inhibited by low temperature).

1) TYPES OF HIBERNATION

The morphology and diapause state of overwintering embryos vary from one species to another. In a comparative study of such embryos, Umeya (1946, 1952) has made the following grouping, which clarifies the relation between hibernation and diapause.

Group I

The embryo entering hibernation is morphologically in the germ-band stage, without differentiation between the head and posterior end. In the Lepidoptera, the embryo is *pyriform*, with a narrow anterior part and widened posterior

(Fig. 10.91a), while the orthopteran embryo is *reversed pyriform*, wide anteriorly and tapering posteriorly (Fig. 10.91b). To this group belong *Notolophus thyellina* (lymantrid moth), *Homoeogryllus japonicus* (cricket), *Hierodula patellifera, Paratenodera aridifolia* (mantis), etc., These embryos overwinter in a state of complete diapause.

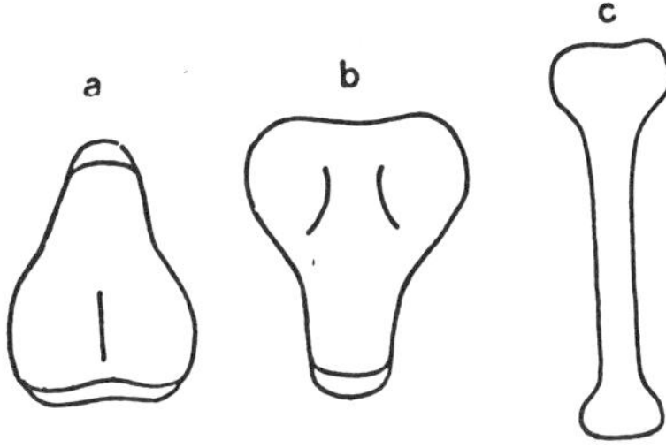

Fig. 10.91 Three types of diapausing embryos (Umeya 1946)

a. pyriform type; b. reversed pyriform type; c. dumb-bell type.

Group II

The overwintering embryo has passed the germ-band stage and become dumb-bell-shaped, with two germ layers (ectoderm and mesoderm) and both head and caudal lobes (Fig. 10.91c). To this group belong *Bombyx mori, Bombyx (Theophila) mandarina, Rondotia menciana, Dictyoploca japonica* (cultured or wild silkworms), *Gryllus mitratus* (cricket). All these insects overwinter in a state of diapause.

Group III

The embryo overwinters in a more advanced state: the stomodaeum is just appearing, the mesoderm is constricted and the primitive segments are easily countable. *Rhopobota naevana* (microlepidopteran moth), *Pterochlorus tropicalis* (aphid) and some others belong to this group. The overwintering insect is in diapause.

Group IV

The overwintering embryo has no definite resting stage of development, continuing to develop slowly at a rate much depressed by the low temperature. Some of the insects in this group, such as *Dendrolimus undans excellens* and *Archips xylosteanus* among the Lepidoptera, and *Oxya japonica* and *Mecopoda elongata* among the Orthoptera, enter hibernation at the germ-band stage, the diapause is broken in November, and the embryo passes the winter developing slowly. The first part of the hibernating period is therefore diapause, and the second, non-diapause. *Paratenodera sinensis* and *Locusta migratoria migratoria* among the Orthoptera belong to the simple, non-diapausing type with no definite diapause period. Some of the eggs of *L. migratoria migratoria,* however, seem to enter diapause when they are exposed to a high temperature (25° C) before the stage of appendage process formation. *Rhodinia fugax* also belongs to this simple non-diapausing type (Umeya 1950).

The embryo of *Melanoplus differentialis* obviously has a resting period which, however, is easily affected by environmental conditions (p. 463-4). The embryo is non-diapausing under natural conditions.

Group V

Differentiation of the larval organs is practically complete within about ten days after oviposition, but the embryo becomes quiescent instead of hatching. *Antheraea yamamai, Liparis (Lymantria) dispar, Malacosoma neustria testacea, Liparis (Lymantria) aurora* (all lepidopterans) represent this group. Although the larval differentiation is almost complete, the embryo is in a typical state of diapause, and unlike the silkworm, cannot be artificially induced to hatch. No orthopteran insect is known to belong to this category.

Campsocleis buergeri (Orthoptera, Saltatoria) lays two kinds of eggs: a one-year egg which hatches the next spring (within 10 months after oviposition) and a two-year egg which hatches a year later (in 22 months). The one-year embryo reaches almost full development in December of the year of oviposition, and the two-year egg, in November of the following year. Since both types overwinter in this state, they can be regarded as similar to Group V, but the embryo is never in a state of diapause, for it hatches within a short and definite time if exposed to high temperature.

Comparing various insects brings to light so much variation that it seems almost impossible to formulate a general principle which will explain the initiation of embryonic diapause. Measurements of the water content, however, reveal a consistent variation correlated with the stage of development, and show a certain connection with diapause. In silkworm eggs, the water content varies with the stage in the way shown in Figure 10.92, with minima at the pyriform, dumb-bell-

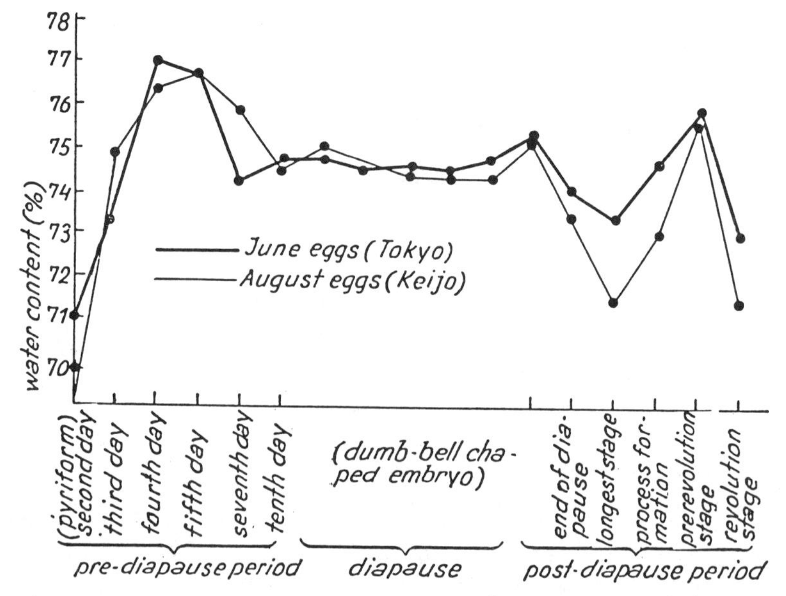

Fig. 10.92 Change in percentage water content with development of diapause egg (Umeya 1946)

shaped, process formation and full development stages. Decrease in the water content suggests reduced physiological and chemical responses, and consequently a decrease in metabolic activity. Embryos in the stages corresponding to these minima should therefore be resistant to changes in their environment, and in an appropriate internal condition for entering diapause. When the water-content curve is considered from this view-point, it is seen that its four minima correspond to one or another of the stages at which embryonic diapause (or hibernation) may begin.

A similar variation in the water content of the egg can also be observed in the mantids, *Hierodula patellifera* and *Paratenodera sinensis*. On the other hand, the measurements of oxygen consumption, CO_2 output, change in water content, etc. made by Boell (1935), Bodine (1923) and others in the hibernating eggs of *Melanoplus differentialis* showed no parallelism with diapause; this may be attributed to differences in the methods of study (Umeya 1946).

In the silkworm, which belongs to Group II, the CO_2 output begins to increase about the middle of January, when no morphological change is yet observable (Suzuki 1917). It is therefore difficult to conclude that the silkworm egg is fundamentally different from the eggs of *Dendrolimus undans excellens* and *Archips xylosteanus*, which shift from diapause to a non-diapause state in the latter part of November (see also p. 474).

On the other hand, the water-content curve of the silkworm embryo shows four low points, whereas diapause always occurs at the second of these, and cannot be shifted to any other. The time of diapause thus seems to be rigidly connected with one definite stage (of low water content), which is characteristic of the species, and the relation which has been said to exist between the water content of the embryo and its resistance to low temperature remains as an unsolved problem.

In the cricket *Gryllulus commodus*, most of the eggs hatch, with only a few entering diapause, when they are incubated after oviposition at about 30° C. When they are incubated at 25°, however, most of the eggs enter diapause, and if they are incubated at 20°, they all enter diapause. When eggs which have completed diapause at a low temperature (13°) and absorbed enough water are desiccated at 29.9° and 90% RH for 48—56 hours and again given water and incubated, those which lost least water during the drying period hatch; those which lost most water die, and those which showed an intermediate water loss enter a diapause-like state (Browning 1953).

2) DIAPAUSE EGGS AND NON-DIAPAUSE EGGS

In an attempt to find a general explanation for this phenomenon — that the embryo stops developing and enters diapause even under conditions favorable for its physiological activities — it has been suggested that some inhibiting substance, transmitted to the eggs from the mother, acts to suppress their development (Watanabe 1924). It is still impossible, however, to discuss the nature of this hypothetical inhibitor, or even to begin searching for such a specific substance in diapause eggs, since nothing is known about the characteristics or mode of action of the hormone which is secreted from the suboesophageal ganglion of the mother and determines the diapausing nature of the eggs (Hasegawa 1951, 1952; Fukuda 1951, 1955). In the silkworm, a difference already exists between diapause-egg-producing and non-diapause-egg-producing pupae with respect to the permeability of the ovary to various substances, or to the absorption by the ovary of the substances being introduced. The difference in serosal pigmentation between diapause and non-diapause eggs is supposed to be due to a differential permeability of the ovary to 3-hydroxykynurenine, kynurenine or their precursors, or to a difference in retention of these substances once they have entered the ovary (Kikkawa 1943; Kato 1952; Yoshitake 1954, 1955, 1956). In the silkworm, except for some peculiar mutants, it is therefore possible to recognize ovarian eggs as diapausing or non-diapausing, according to the presence or absence of 3-hydroxykynurenine (Kikkawa 1943).

The histochemical study of diapause has mainly been focused on glycogen content. Although some contradictions have inevitably arise, agreement has generally been reached that glycogen content is low in diapausing eggs and increases when the eggs are activated (Miura 1932a; Nakata 1952). This fact has also been confirmed biochemically (Nakata 1952). It was then suggested that entry of the silkworm egg into diapause is equivalent to a decrease in the amount of glycogen, while the breaking of diapause represents "a completion of some state" which permits an increase in glycogen content, regardless of whether the stimulus is low temperature or treatment with hydrochloric acid (Akita and Chino 1955). This reduction in the glycogen level is a phenomenon of great interest, in view of the depressed metabolism of the diapausing embryo.

Chino (1956, 1957), in a series of detailed investigations into the fluctuations in the glycogen content of diapausing eggs of the silkworm, has made the important discovery that the glycogen is converted into glycerol and sorbitol when the embryos enter diapause.

The changes in ATPase activity are also interesting as they represent changes in the energy-supplying system accompanying diapause and its termination. A gradual increase in the activity of Mg^{++} -dependent soluble ATPase begins on the third day after oviposition in non-diapause eggs, and reaches 1.5 times the initial activity just before hatching. In diapause eggs, on the other hand, this activity drops rapidly in the 3—6 days after oviposition, then continues to decrease more slowly, finally falling to one-third of the initial on the sixth day (Akita and Kobayashi 1955). A considerable number of studies have also been performed on various other enzymes, but their embryological implications have not yet been fully worked out.

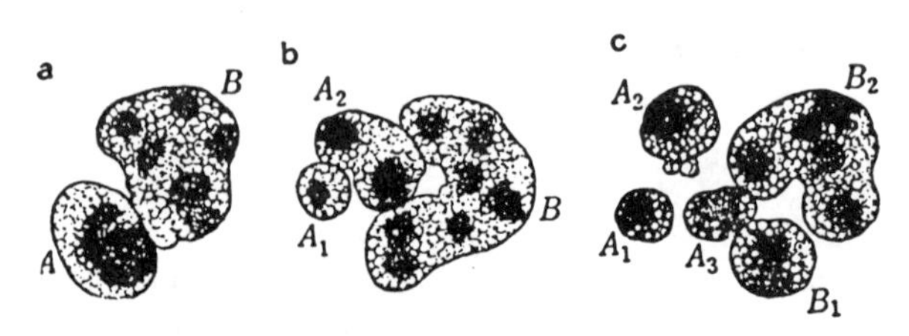

Fig. 10.93 Division of yolk cells in Ringer's solution (Takami 1954b)

A and *B* are large cells, which divide into A_1, A_2, A_3 and B_1, B_2.

In the silkworm, it is possible to distinguish between diapause and non-diapause eggs about 24—25 hours after oviposition, because the serosa of diapause eggs becomes pigmented at this stage. Yolk segmentation begins at about 20 hours after oviposition; in the yolk cells, at first many fatty granules are seen congregating around the nucleus, as shown in Figure 10.93. In non-diapause eggs, these granules are scattered throughout the cell at about 40 hours, while in diapause eggs they remain concentrated for from several days to about two weeks (Takami 1952). This concentration and dispersion of the fatty granules is apparent in serial sections (Miura 1932a; Takahashi 1954). A non-diapause egg may also be distinguished at the moment of oviposition by its milky-white color, which is believed to be due to some state of dispersion, of both fatty and protein granules, different from that in the diapause egg. This supposition has been verified by the observation of sectioned material (Miura 1932a).

Yolk cells removed from eggs about 40 hours after oviposition generally show amoeboid movement when they are placed in an appropriate physiological saline medium. This active movement continues for about one day in cells from non-diapause eggs, while in diapause egg cells it persists for ten days or longer (Takami 1953a, 1954a). In the diapause egg, all the yolk is distributed in the peripheral zone, leaving a yolk-free liquid region at the center (Tsuchiya 1926). This distribution may be the result of locomotory activity on the part of the yolk cells in the intact egg. No similar localization is observed in the non-diapause egg. The yolk cells cease to move, after becoming completely localized in the egg periphery, at the end of the second week after oviposition at 25°. This corresponds in time to the stage at which the so-called tyrosinase activity falls off (Nittono and Takeshita 1953).

The diapausing embryo is tightly enclosed in the mass of peripherally located yolk cells, and this localization persists throughout the period of diapause. When the embryo is awakened from diapause and resumes development in the spring, the yolk cells regain their motility, leave the periphery and become dispersed throughout the interior of the egg. This behavior of the yolk cells makes it possible to estimate the developmental stage of the egg by observing its yolk cells in physiological saline (Yamaguchi 1946).

When an egg is exposed to low temperature, a change seems to occur in the hydration of the albuminous globules contained in the yolk cells (Takami 1955); a similar change is observed in vitellophages (Takami 1953a). According to observations made on fixed preparations, the distribution of stainable particles in the embryonic cell nucleus also undergoes changes on exposure to low temperature (Nakata 1936). These changes may all be regarded as colloidal modifications of the protoplasm or of intracellular structures, induced by a common agent. Although they show an apparent parallelism with embryonic diapause under natural temperature conditions, they can hardly be supposed to have any direct concern with diapause. The pigment granules of the serosa cells congregate as the temperature falls, and disperse again when it rises the following spring (Watanabe 1919).

Changes in the electrical activity of the brain have recently been studied in relation to the changes in enzyme activity associated with diapause in the giant silk moth *(Hyalophora cecropia)*, which is characterized by pupal diapause (van der Kloot 1955). This method could usefully be applied to the study of embryos which enter diapause after reaching full development (p. 468). Jones (1953, 1956a, b), working with embryos of *Locusta migratoria* and *Locustana pardalina*, found that a neurosecretion of the brain induces the secretion of a prothoracic gland hormone which, in turn, participates not only in embryonic molting but also in the further processes of embryonic development, such as the formation of melanin, and even, he believes, activates the pleuropodia to initiate the production of a hatching enzyme.

(3) ACTIVATION OF DIAPAUSE EGGS

The diapause egg of the silkworm can be artificially activated into a non-diapause state by means of various agents such as heat, electricity, friction, treatment with oxygen, etc. In sericulture, hydrochloric acid treatment is now generally practiced: the most routinely used method employs warm (46° C) hydrochloric acid (s.g. — 1.075). Eggs about 20 hours after oviposition are dipped into this solution for 3—5 minutes (depending on the race) and then washed in water. This simple treatment is adequate to induce artificial hatching. As the age of the embryo increases, however, this procedure becomes less and less effective, until at 50 hours after oviposition its effect is practically lost. It is also generally impossible to induce hatching in fully diapausing eggs about one month old by the acid treatment alone.

While this method of treating with acid is useful in a practical sense, the mechanism of its action remains obscure, although Miura (1920) insisted that it could be attributed to the action of Cl ions. Since, according to some authors, the pH of the egg decreases after treatment, it is possible to estimate the effectiveness of the treatment by measuring the subsequent pH (Nagai 1934; Hama 1936; Moritsugu 1937, 1938). In the treated eggs, the fatty granules of the yolk cells disperse as in non-diapause eggs, providing another means of judging effectiveness (Takami 1952). Nunome et al. (1955) have also observed that the water transpiration of acid-treated eggs reaches 2.5 times that of the controls within a few hours. The period after oviposition within which the treatment is successful varies with temperature, being shorter at higher temperatures. If eggs in the stage suitable for this acid treatment are kept cool (5° C), they remain in such an appropriate condition for up to two weeks.

Exposure to low temperature is required to activate eggs which have once entered full diapause. This applies to many other insects in addition to the silkworm. In sericultural techniques, acid-treatment after chilling is utilized to obtain hatching one or two months after oviposition. Eggs about 50 hours after oviposition at 25° C are transferred to a cooling room at 5°, removed from this cold storage at the programmed date and dipped in hydrochloric acid of s.g. = 1.10, at 47.5°, for about five minutes. Chilling alone, not followed by acid treatment, does not induce a high rate of hatchability, and is useless in practice. If the chilling lasts more than 90 days, however, the hatchability reaches more than 90%, and acid treatment is no longer necessary (Kutsukake 1954).

The eggs are fully activated by such exposure to cold, but this type of activation is not simply an abbreviation of the process which occurs during natural diapause, in which the egg is activated by long exposure to low temperature. The diapausing egg 50 hours after oviposition is at the stage in which the yolk cells are beginning to exhibit motility, but they fail to congregate as long as the temperature is kept low, and the egg finally resumes development without passing through the stages in which the yolk cells accumulate peripherally and then lose their motility. This result suggests that chilling before the egg has fully entered diapause has the two-fold effect of retarding the changes which would ordinarily lead to diapause, and accelerating such processes as lead to activation.

This conclusion is supported by the biochemical findings of Chino (1956): the glycogen content of diapause eggs transferred to 5°C on the second day after oviposition decreases much less than that of eggs kept continuously at 25°, and begins to increase again as the egg becomes activated, without dropping to the minimum level normally found in diapause eggs.

Activation is retarded, however, if the temperature is too low. The optimum temperature for activation of the silkworm egg is generally considered to be about 5° C (Kutsukake 1954), although Muroga (1951) has reported an optimum of 7.5°. Temperature lower than 5° has less effect in activating diapause eggs. Activation is still possible at 0°, but it almost never occurs at —2.5° C (Kutsukake 1954). The optimum for eggs of *Lymantria dispar* is 5—12°, with upper and lower limits of 20° and 0°, respectively (Masaki 1956).

Two theories have been proposed to account for the fundamental nature of the diapause egg. One suggests that diapause is due to some condition of the yolk which makes it unsuitable to support the development of the embryo. This implies that the embryo is always ready to carry on the processes of development as soon as the condition of the yolk becomes favorable (Umeya 1937b, 1938a, b, 1939, 1950). In other words, the embryo is passively reduced to the quiescent state characteristic of diapause. The other theory attributes the condition of diapause to the embryo itself, and invokes an autonomous resumption of its developmental processes at the time activation (Miura 1932b, 1938).

The author has carried out in vitro culturing of diapausing embryos in an attempt to elucidate the embryo-yolk relationship. When silkworm embryos in full diapause were dissected out and cultured in vitro in a medium of yolk extracted from naturally non-diapause eggs or eggs artificially prevented from diapausing, very little development was observed. Post-diapause and non-diapause embryos, however, developed well in such extracts, forming appendage processes and often setae, and even when the extract was made from completely diapausing eggs some larvae developed (Takami 1955, 1957). These results seem to indicate that it is the embryo itself which is in diapause, and that its autonomous activity plays an important role in re-initiating development. An embryo cultured in this way, however, has a less intimate connection with the surrounding yolk than it would in its normal location, since it receives its nutrition only through the surface of its body, instead of by direct cellular connection with the yolk cells. It is conceivable that the abnormal experimental conditions might obscure the real influence of the yolk.

In *Melanoplus differentialis*, a small yolkless section of a diapausing embryo is stimulated to terminate diapause and begin to develop simply by being transferred into physiological saline solution. Treatment of the embryo alone is thus sufficient to break diapause in this case; moreover, there is no intervention of the endocrine system (Bucklin 1953). In view of the ease with which diapause in this grasshopper is influenced by environmental factors (p. 463), however, the conclusion of Bucklin requires further confirmation before it can be applied to other insects.

Crossing experiments indicate that the condition of the female parent is the sole factor which determines whether an egg will be diapausing or non-diapausing: a different character of the male parent exerts no appreciable effect (Watanabe 1924). The paternal influence becomes apparent, however, on acid treatment of embryos which have developed beyond the appropriate stage, and in the number of days required for activation by chilling (Kutsukake and Kuroiwa 1951; Fujiwara and Oyanagi 1956). But since it is minor in comparison with the effect of the quantitatively predominant substance derived from the mother, it may be insufficient to change the diapausing character, giving an overall image of maternal inheritance of diapause. Moreover, since the yolk of the insect egg exists not as a mere accumulation of nutrient substance, but rather within and under the control of the vitellophages, which have the same genetic constitution as the cells of the embryo proper, it is erroneous to consider embryo and yolk as separate entities.

When a non-diapause embryo is cultured in vitro in an extract of diapause eggs kept continuously after oviposition at 25°, it develops as well as in an extract of non-diapause eggs, provided that the extracted diapause eggs were less than about two days old. If the extract is made from diapause eggs about thirty days old, embryos cultured in it develop very poorly. On the other hand, good development is again obtained if embryos are cultured in extracts of diapause eggs kept at 25° C for about 150 days (Takami 1957). According to Chino (1956, 1957), the glycogen content of diapause eggs kept at 25° for more than 100 days has begun to increase from the minimum level, and the eggs are advancing toward the activated state even though no morphological changes are yet observable. These results suggest that the widespread belief, that low temperature is indispensable for the activation of diapausing silkworm eggs, is caused by a confusion of activation with hatching. This distinction is highly important, both from the theoretical viewpoint, and as it concerns the handling of eggs in sericulture (p. 464). The longer eggs are kept at a high temperature, up to 100 days, the longer will be the period required for activation by chilling, whereas keeping them at high temperature for more than 100 days shortens the chilling period required for hatching (Ohno 1951; Kutsukake 1954).

The diapause eggs of *Melanoplus* are activated by treatment with lipid solvents such as xylene, because removal of the waxy substance covering the hydropyle region permits the entrance of water. In eggs of this species, suppression of water imbibition by secretion of this waxy substance is believed to be one of the requisite conditions for entering diapause (Slifer, cited by Wigglesworth 1950). The experiments on *Gryllulus* eggs described on page 469 are interesting in this connection.

7. HATCHING

The embryo within the "bluish egg" (p. 439) has broken two (amnion and serosa) of the three membranes which once enveloped it, leaving only the vitelline membrane beneath the chorion. The fate of this vitelline membrane is, however, obscure, because its delicacy and lack of structural characteristics make it difficult to observe. Silkworm embryos older than the dark-spotted stage become less resistant to desiccation, possibly as the result of having lost these protective membranes. The appearance within the chorion of air bubbles has been noticed during observation of the movements made by the embryo as it swallows the serosa (Yanagida 1928). These bubbles are probably derived from the gas filling the tracheae and liberated mainly from the 9th spiracles (Nagata 1951).

The embryo chews an opening in the chorion at some point near the micropyle by means of its mandibles as the last step in the process of hatching. Gastric juice or a saliva-like fluid, released from the mouth at this time, helps to break the chorion by partially dissolving it (Muroga 1934; Nakajima 1938; Takeuchi 1944a). When an embryo just prior to hatching is removed from the chorion and stimulated electrically, it emits a black fluid containing the pigment of the swallowed serosa. If this fluid is applied to a part of the chorion, the affected part becomes swollen. It has not yet been determined, however, whether the normally secreted fluid is gastric juice or saliva, or some specially secreted hatching substance (Takeuchi 1955a).

Observation of the broken chorion after hatching shows that the break is situated somewhat ventral to the mid-line, but always includes the micropyle. The opening is thus located because the embryo after revolution lies at the dorsal side of the egg with its mouth directed ventrally. The black margin seen surrounding the hole in the chorion is attributed by Muroga to the oxidation of the hatching fluid (or gastric juice); Takeuchi suggests it is due to the serosal pigment contaminating the fluid.

Many fragments of the chorion are found in the feces first discharged by the newly hatched larva. It is reported that the larva uses its mandibles to remove this first fecal mass; this behavior may be necessitated by the presence of the stiff chorion fragments which would tend to prevent it from being readily evacuated (Takeuchi 1955a).

In addition to this normal silkworm hatching pattern, there has been described an abnormal mode, in which a fissure-like aperture is made in the chorion as the result of certain peculiar movements of the larva (Moriyama 1935).

Observations of the hatching behavior have also been made in the eggs of *Antheraea pernyi,* which have an opaque chorion, by removing this covering on the micropyle side and replacing it with a sheet of cellophane. The secretion of a large quantity of fluid from the salivary glands and co-ordinated movements of the oral and posterior parts of the body were demonstrated by this method (Takeuchi 1955b).

Various cases are known in which insects hatch without chewing an opening in the chorion. In some species, a cuticle-dissolving fluid is secreted from the pleuropodia (Slifer 1937); in others, a peculiar embryonic cuticle or structures on the skin of the embryo cut an opening by concentrating pressure at a definite region of the shell during its hatching movements (Wigglesworth 1950).

The hatching larva of *Acrydium japonicum* sheds its vitelline membrane after leaving the shell. In *Diestrammena japonica* and *D. apicalis,* the vitelline membrane remains inside the shell, sometimes extending partway outside through the hatching aperture (Asano 1938).

REFERENCES

Akita, K. and Chino, H. 1955. Biochemical studies on the diapause of the silkworm egg. I. Relation between carbohydrate and diapause. Sci. Rep. Tokyo Metropol. Univ. *(***5***)* 35—6. (in Japanese)

Akita, K. and Kobayashi, H. 1955. Biochemical studies on the diapause of the silkworm egg. II. Relation between ATPase and diapause. Sci Rep. Tokyo Metropol. Univ. *(***5***)* 36. (in Japanese)

Ando, H. 1952: On the pleuropodia of the embryo of *Epiophlebia superstes* Selys. Zool. Mag. **61**: 8. (in Japanese)

Ando, H. 1953. Studies on the pleuropodia of Odonata. Sci. Rep. Tokyo Bunrika Daigaku, Sec. B, **7** (108) : 167—181.

Ando, H. 1954: On mid-gut formation in the Odonata. Zool. Mag. **63** : 79. (in Japanese)

Ando, H. 1955: Everted embryos of dragonflies produced by ligation. Sci. Rep. Tokyo Bunrika Daigaku, Sec B, **8** (119) : 65—74.

Ando, H. and Kawana, T. 1956: Embryology of the May-fly, as studied by external observation. Kontyu **24** : 224—232. (in Japanese)

Aruga, H. 1938. Lethality and revolution of the embryo of the dominant silkworm mutant, *Burnt.* J. Sericult. Sci. Japan, **9** : 275—276. (in Japanese)

Aruga, H. 1942. Manifestation of gene action and vital staining in the silkworm. J. Sericult. Sci. Japan, **13** : 225—239. (in Japanese)

Aruga, H. 1943: Pleiotropism of gene action and vital staining in the ectoderm of the silkworm. J. Sericult. Sci. Japan, **14** : 201—203. (in Japanese)

Asano, I. 1938: On the hatching of the eggs of *Diestrammena japonica* and *D. apicalis.* Jap. J. Applied Zool. **10:** 83—85. (in Japanese)

Bito, S. 1931: Changes in catalase activity during the development of insect embryos. Jap. J. Applied Zool. **3** : 314—318. (in Japanese)

Bobretzky, N. 1878: Ueber die Bildung des Blastoderms und der Keimblätter bei den Insecten. Z. Wiss. Zool. **31** : 195—215.

Bodenstein, D. 1953. Embryonic Development. Roeder, K. D., Insect Physiology. New York. Chap. **29** : 780—821

Brauer, A. and Taylor, A. C. 1936. Experiments to determine the time and method of organization in bruchid (Coleoptera) eggs. J. Exp. Zool. **73** : 127—152.

Browning, T. O. 1953: The influence of temperature and moisture on the uptake and loss of water in the eggs of *Gryllulus commodus* Walker (Orthoptera-Gryllidae). J. Exp. Biol. **30** : 104—115.

Bucklin, D. H. 1953: Termination of diapause in grasshopper embryos cultured in vitro. Anat. Rec. **117** : 539.

Chaudhry, G.-U. 1956: The development and fecundity of the oriental fruit-moth, *Grapholitha (Cydia) molesta* (Busck) under controlled temperatures and humidities. Bull. Entom. Res. **46** : 869—898.

Child, C. M. 1943: Indophenol reaction and reduction in living ovaries of *Drosophila hydrei.* Physol. Zool. **16** : 141—161.

Chino, H. 1956: Diapause and carbohydrate metabolism in the silkworm. Comm. to Insect Physol. Disc. Group, 2nd Symposium (in Japanese) Chino, H. 1957. Diapause and carbohydrate metabolism in the silkworm egg — Conversion of glycogen during diapause. Comm. at Feb. Meeting, Kanto Sect., Japan Zool. Soc. (in Japanese)

Doké, N. 1935: Effects of temperature and humidity during the incubation of silkworm eggs on their hatching. Jap. J. Applied Zool. **7** : 63—69 (in Japanese)

Doncaster, L. 1924: An Introduction to the Study of Cytology. London.

Eastham, L. 1927: A contribution to the embryology of *Pieris rapae.* Quart. J. Micr. Sci. **71** : 353—394.

Ewest, A. 1937: Structur und erste Differenzierung im Ei des Mehlkäfers *Tenebrio moritor.* Roux' Arch. **135** : 689—752.

Fujiwara, T. and Oyanagi, M. 1956: Differences in activation, by chilling of dormant eggs, between stock races and their hybrids. Acta Sericologica **15**: 39—41. (in Japanese)

Fukuda, S. 1951: The production of the diapause eggs by transplanting the suboesophageal ganglion in the silkworm. Proc. Jap. Acad. Sci. **27**: 672—677.

Fukuda, S. 1955: On the production of diapause and non-diapause eggs in the silkworm. Jap. J. Exp. Morph. **9**: 22—27. (in Japanese)

Geigy, R. 1931: Erzeugung rein imaginaler Defekte durch ultraviolette Eibestrahlung bei *Drosophila melanogaster.* Roux' Arch. **125:** 406—477.

Glushtchenko, I. E. 1956: The phenomenon of polyfertilization in plants. Acad. Sci. USSR

Goldschmidt, R. and Katsuki, K. 1928: Cytologie des erblichten Gynandromorphismus von *Bombyx mori* L. Biol. Zentralbl. **48**: 685—699.

Haget, A. 1955: Expériences mettant en évidence l'origine paire du labre chez l'embryon du Coléoptere *Leptinotarsa.* Comt. Rend. Soc. Biol. **149**: 690—692.

Hama, T. 1936: Hydrogen ion concentration in the silkworm egg. Jap. J. Applied Zool. **8**: 318—327. (in Japanese)

Hasegawa, K. 1951: Studies on the voltinism in the silkworm, *Bombyx mori* L., with special reference to the organs concerning determination of voltinism. (A preliminary note). Proc. Jap. Acad. Sci. **27**: 667—671.

Hasegawa, K. 1952: Studies on the voltinism in the silkworm, *Bombyx mori* L., with special reference to the organs concerning determination of voltinism. J. Fac. Agric. Tottori Univ., **1**: 83—124.

Hasimoto, H. 1934: Formation of an individual by the union of two sperm nuclei in the silkworm. Bull. Imp. Sericult. Exp. Sta. **8**: 455—464. (in Japanese)

Hasimoto, H. 1951: Replacement of the egg nucleus in the silkworm. J. Sericult. Sci. Japan, **20**: 164—165. (in Japanese)
Hasimoto, H. 1953a: Forigo de vir-sekso, kazo de eksterordinara seksratio ce la morusa silkraupo, *Bombyx mori*. J. Sericult. Sci. Japan, **22**: 175—180.
Hasimoto, H. 1953b: Mixoploids produced by artificial parthenogenesis in *Bombyx mori* L. J. Sericult. Sci. Japan, **22**: 205—210. (in Japanese)
Hasimoto, H. 1956a: Parthenogenesis of the silkworm egg by treatment with hot water soon after deposition. J. Sericult. Sci. Japan, **25**: 215—216. (in Japanese)
Hasimoto, H. 1956b: Androgenesis of the silkworm egg by low temperature treatment soon after deposition. Jap. J. Genet. **31**: 294. (in Japanese)
Hayashi, T. 1937: Morphological studies on abnormally shaped eggs in *Bombyx mori*. Bul. Sci. Fac. Agric. Kyushu Imp. Univ., **7**: 359—372. (in Japanese)
Hegner, R. W. 1909: The effects of centrifugal force upon the eggs of some chrysomelid beetles. J. Exp. Zool. **6**: 507—552.
Hegner, R. W. 1911. Experiments with chrysomelid beetles. III. The effects of killing parts of the egg of *Leptinotarsa decemlineata*. Biol. Bull. **20**: 237—251.
Henson, H. 1932: The development of the alimentary canal in *Pieris brassicae* and the endodermal oiigin of the Malpighian tubules of insects. Quart. J. Micr. Sci. **75**: 283—305.
Heymons, R. 1897: Entwicklungsgeschichtliche Untersuchungen an *Lepisma saccharina* L. Z. Wiss. Zool. **62**: 583—631.
Hirschler, J. 1928: Embryogenese der Insekten, Schröder: Handbuch der Entomologie. I. Jena.
Hodai, T. 1949: Effects of environmental humidity on embryonic development and hatching in the eri-silkworm, with considerations on the mechanism. J. Sericult. Sci. Japan **18**: 171—180. (in Japanese)
Howland, R. B. and Child, G. P. 1935: Experimental studies on development in *Drosophila melanogaster*. I. Removal of protoplasmic materials during late cleavage and early embryonic stages. J. Exp. Zool. **70**: 415—424.
Howland, R. B. and Sonnenblick, B. P. 1936: Experimental studies on development in *Drosophile melanogaster*. II. Regulation in the early egg. J. Exp. Zool. **73**: 109—123.
Huettner, A. F. 1923: The origin of the germ cells in *Drosophila melanogaster*. J. Morph. **37**: 385—423.
Huettner, A. F. 1927: Irregularities in the early development of the *Drosophila melanogaster*. egg. Z. Zellfor. Mikr. Anat. **4**: 599—610.
Huettner, A. F. 1940: Differentiation of the gonads in the embryo of *Drosophila melanogaster* Genetics, **25**: 121.
Ichimaru, M. 1956: Embryological studies on silkworm eggs hereditarily incapable of hatching. IV. On the development of *no-lunule* eggs. Memo. Fac. Educ. Kumamoto Univ. **4**: 205—210. (in Japanese).
Ikeda, E. 1912: Development of the silkworm egg during the first two days after oviposition. Sangyo Shimpo **20** (226): 8—11, (227): 43—47, (228): 19—23. (in Japanese)
Ikeda, E. 1913: Anatomy and Physiology of the Silkworm. Tokyo. (in Japanese)
Imaizumi, T. and Nakamura, K. 1951: Electric charge in the cortical layer of *Drosophila* eggs during early development. Zool. Mag. **60**: 53—54. (in Japanese)
Ishikawa, C. 1921: Lectures on Zoology. III. Tokyo. (in Japanese)
Ishikura, H. 1940: Effects of temperature and humidity on the embryonic development in *Bruchus chinensis*. Jap. J. Applied Zool. **11**: 218—229. (in Japanese)
Ishiwata, S. 1913a. On the sex of the silkworm egg. (cont.) Sangyo Shimpo **21** (243): 5—6, (244): 21—23, (245): 34—38. (in Japanese)
Ishiwata, S. 1913b: Sex of the silkworm egg. Rep. Sericult. Assoc. Japan, No. 256—259. (in Japanese)
Ishiwata, S. 1914: On the reproductive organs in insect embryos. Sangyo Shimpo **22** (250): 153—156, (252): 22—27. (in Japanese)
Itikawa, N. 1943: Genetical and embryological studies on a dominant mutant, "new additional crescent", of the silkworm *Bombyx mori* L. I. Jap. J. Genet. **19**: 182—188. (in Japanese)
Itikawa, N. 1944: Genetical and embryological studies on a dominant mutant, "new additional crescent", of the silkworm *Bombyx mori* L. II. Jap. J. Genet. **20**: 8—14. (in Japanese)
Itikawa, N. 1947: Genetical and embryological studies on a dominant mutant, "new additional crescent", of the silkworm *Bombyx mori* L. III. Embryo formation by development of the sperm nuclei. Jap. J. Genet., Suppl. **1**: 59—66. (in Japanese)
Itikawa, N. 1952: Genetical and embryological studies on the E-multiple alleles in the silkworm, *Bombyx mori* L. Bull. Sericult. Exp. Sta. **14**: 23—91. (in Japanese)
Ito, H. 1918: On the glandular nature of the corpora allata of the Lepidoptera. Bull. Imp. Tokyo Sericult. Coll. **1**: 63—103.

Ito, T. 1950: Relation between incubation temperature and catalase activity in the silkworm egg. J. Sericult. Sci. Japan. **19**: 523—525. (in Japanese)
Iwasaki, Y. 1930: On the periodicity of mitotic divisions in the ectodermal cells of the embryo of *Bombyx mori*. Bull. Kagoshima Imp. Coll. Agr. Forestry, **8**: 285—295. (in Japanese)
Iwasaki, Y. 1931: Concerning the mode of generation and properties of the blastoderm and yolk cells in the egg of *Bombyx mori*. Jap. J. Applied Zool. **3**: 308—313. (in Japanese)
Iwasaki, Y. 1932: On the generation, properties and distribution of blood corpuscles in the egg of *Bombyx mori*. Jap. J. Applied Zool. **4**: 1—7. (in Japanese)
Johannsen, O. A. 1929: Some phases in the embryonic development of *Diacrisia virginica* Fabr. (Lepidoptera). J. Morph. Physiol. **48**: 493—541.
Jones, B. M 1953: Activity of the incretory centers of *Locustana pardalina* during embryogenesis: function of the prothoracic glands. Nature **172**: 551.
Jones, B. M. 1956a: Endocrine activity during insect embryogenesis. Function of the ventral head glands in locust embryos *(Locustana pardalina* and *Locusta migratoria*, Orthoptera). J. Exp. Biol. **33**: 174—185.
Jones, B. M. 1956b. Endocrine activity during insect embryogenesis. Control of events in development following the embryonic moult *(Locusta migratoria* and *Locustana pardalina*, Orthoptera). J. Exp. Biol. **33**: 685—696.
Kato, M. 1952: The permeability of + chromogen in the non-wintering eggs of the silkworm. J. Sericult. Sci. Japan **21**: 252. (in Japanese)
Kawaguchi, E. 1926a: On the micropyle in the silkworm and several other moths. Zool. Mag. **38**: 11—13. (in Japanese)
Kawaguchi, E. 1926b. Polyspermy in the silkworm. Sangyo Shimpo **34**: 38—41. (in Japanese)
Kawaguchi, E. 1933a: Cytological analysis of silkworms produced by parthenogenesis. J. Sericult. Sci. Japan, **4**: 172—173. (in Japanese)
Kawaguchi, E. 1933b. Genetical analysis of silkworms produced by parthenogenesis. J. Sericult. Sci. Japan, **4**: 173—175. (in Japanese)
Kawaguchi, E. 1934: Genetical and cytological analysis of silkworms produced by parthenogenesis. J. Sericult. Sci. Japan, **5**: 1—20. (in Japanese)
Kawaguchi, E. 1938: Genetische Konstruktion der Mosaik-Raupen beim Seidenspinner und ihre Beziehung zur Embryologie der Insekten. Jap. J. Genet. **14**: 262—263.
Kawaguchi, E. and Miya, K. 1943: Embryonale Entwicklung der Geschlechtszellen beim Seidenspinner, *Bombyx mori* L. Jap. J. Genet. **19**: 133—134.
Kikkawa, H. 1934. Origin and appearance of egg-color pigments in the silkworm. Bull. Imp. Sericult. Exp. Sta. **11**: 311—345. (in Japanese)
Kinoshita, M. 1951: Increase in body weight of embryos during incubation and its bearing on the resistance of embryos to high temperature. Sanshikaiho **60** (698): 13—18. (in Japanese)
Kobari, K., Muroga, H. and Maruyama, T. 1929: Studies on the incubation of the silkworm egg. I. Sangyo Shimpo **37**: 591—598. (in Japanese)
Kogure, M. 1930: Studies on voltinism in the silkworm. Bull. Nagano Sericult. Exp. Sta. **(11)**: 1—152. (in Japanese)
Kogure, M. 1933: The influence of light and temperature on certain characters of the silkworm, *Bombyx mori*. J. Dept. Agric. Kyushu Imp. Univ. **4**: 1—93.
Koidsumi, K. 1939: Considerations on the overall effective temperature for the development of insects. Jap. J. Applied Zool. **11**: 1—9. (in Japanese)
Koidsumi, K. and Shibata, K. 1938: Studien über Seidendarm. III. Einfluss der Lufttemperatur auf die Eierentwicklung. J. Soc. Tropic. Agric. **10**: 187—198.
Kojima, T. 1935: Effects of temperature and humidity on the hatching of the egg of *Dendrolimus spectabilis* Butl. Jap. J. Applied Zool. **7**: 211—224. (in Japanese)
Koltzoff, N. K. 1932: Ueber die künstliche Parthenogenese des Seidenspinner. Biol. Zentrbl. **52**: 626—642.
Korschelt, E. and Heider, K. 1902, 1910: Lehrbuch der Vergleichenden Entwicklungsgeschichte der Wirbellosen Thiere. Jena.
Kowalevsky, A. 1886: Zur embryonalen Entwicklung der Musciden. Biol. Zentrbl. **6**: 49—54.
Koyama, N. 1956: Structure of the compound eye and the mechanism of its formation in the silkworm moth. J. Sericult. Sci. Japan, **25**: 244—245. (in Japanese)
Krause, G. 1939: Die Eitypen der Insekten. Biol. Zentrbl. **59**: 495—536.
Kurasawa, Y., Kanazawa, I. and Ikeuchi, S. 1937: Experimental studies on the influence of temperature and humidity upon the development of eggs of *Antheraea pernyi* Guer. Bull. Sericult. Silk-Indust. **10**: 95—114. (in Japanese)
Kutsukake, H. 1954: Hibernation of the silkworm egg. Proc. Tokai Sericult. Soc. Japan, **(2)**: 95—114. (in Japanese)

Kutsukake, H. and Kuroiwa, K. 1951: On acid treatment of hybrid silkworm eggs, with special reference to the differences in results referable to strains of males used for crossing. Gijutsushiryo, Raw Silk Bureau, Japan, (**29**): 19—20. (in Japanese)

Machida, J. 1940: On the formation of yolk and egg membranes in the silkworm, *Bombyx mori* L. Bull. Imp. Sericult. Exp. Sta. **10**: 26—67. (in Japanese)

Marshall, W. S. and Dernehl, P. H. 1905. Contributions toward the embryology and anatomy of *Polistes pallipes* (Hymenopteron). I. The formation of the blastoderm and the first arrangement of its cells. Z. Wiss. Zool. **80**: 122—154.

Masaki, S. 1956: The effect of temperature on the termination of diapause in the egg of *Lymantria dispar* Linné (Lepidoptera: Lymantriidae). Jap. J. Applied Zool. **21**: 148—157.

Maschlanka, H. 1938. Physiologische Untersuchungen am Mehlmotte *Ephestia kuehniella*. Roux, Arch. **137**: 714—772.

Matsumura, S., Higuchi, T. and Kitajima, M. 1928: On the effects of humidity on the silkworm. I. Bull. Nagano Sericult. Exp. Sta. (**4**): 1—104. (in Japanese)

Matsumura, S. and Ishizaka, Y. 1929: On the effects of temperature on the silkworm. I. Bull. Nagano Sericult. Exp. Sta. (**9**): 1—136. (in Japanese)

Matsunaga, N. 1933: On the development of eggs remaining in the bodies of silkworm moths. Bull. Dept. Agr., Chosen Sericult. Exp. Sta., **3**: 85—143. (in Japanese)

Mellanby, H. 1936: The later embryology of *Rhodnius prolixus*. Quart. J. Micr. Sci. **79**: 1—42.

Miura, E. 1929: Studies on the artificial hatching of silkworm-eggs by hydrochloric acid. Bull. Imp. Kyoto Sericult. Coll. **1** (2): 1—44.

Miura, E. 1932a: Studies on the yolk nucleus in hibernating, non-hibernating and artificial non-hibernating eggs of the silkworm. Bull. Imp. Kyoto Sericult. Coll. **1** (3): 1—36. (in Japanese)

Miura, E. 1932b: Principles of the artificial hatching of the silkworm egg. (Prelim. rep.). Bull. Imp. Kyoto Sericult. Coll. **1** (3): 37—42. (in Japanese)

Miura, E. 1938. Questions to Dr. Umeya concerning his criticisms on my opinion of the active and passive development of the silkworm embryo. Sanshikaiho **47**: 35—39. (in Japanese)

Miya, K. 1947: On the location of the embryo and germ cells in the silkworm egg. (Prelim. rep.). Proc. Tohoku Sericult. Soc. Japan, (1): 42—44. (in Japanese)

Miya, K. 1949: Considerations on the region of differentiation of germ cells in the silkworm embryo. Proc. Tohoku Sericult. Soc. Japan, (2): 80—81. (in Japanese)

Miya, K. 1950: Studies on the gonad development of the silkworm, *Bombyx mori* L. I. On the region and period of differentiation of germ cells. Trans. Sapporo Nat. Hist. Soc. **19**: 1—4.

Miya, K. 1951a: On the presumptive region of the differentiation of germ cells at the silkworm blastoderm stage. J. Sericult. Sci. Japan **20**: 49. (in Japanese)

Miya, K. 1951b: On the numerical variation of germ cells in the embryos of four silkworm strains. (Studies on the gonad development of the silkworm. II). J. Sericult. Sci. Japan, **20**: 221—225. (in Japanese)

Miya, K. 1951c: A consideration on the formation of gonads in the silkworm embryo. Gijutsushiryo, Raw Silk Bureau, Japan (29): 49—50. (in Japanese)

Miya, K. 1952a: Studies on the development of the gonad in the silkworm. IV. Gonad formation in cauterized eggs of the silkworm, *Bombyx mori* L. Zool. Mag. **61**: 13—17. (in Japanese)

Miya, K. 1952b: Studies on the development of the gonad in the silkworm. III. Distribution of the germ cells in the silkworm embryo. Jap. J. Genet. **27**: 1—8. (in Japanese)

Miya, K. 1953. The presumptive genital region at the blastoderm stage of the silkworm egg. J. Fac. Agr. Iwate Univ. **1**: 223—227.

Miya, K. 1955: Studies on the development of the gonad in the silkworm, *Bombyx mori* L. V. On the differentiation of the germ cells of the "New additional crescent". J. Fac. Agr. Iwate Univ. **2**: 239—244.

Miya, K. 1956: On the differentiation of tissue in the presumptive abdominal region, with special reference to the formation of genital ridges. J. Sericult. Sci. Japan **25**: 214. (in Japanese)

Miyazaki, S. 1927: The entrance of sperm into the silkworm egg. Chuosanshiho (132): 74—76. (in Japanese)

Mizuno, T. 1936: The Silkworm Egg. Tokyo. (in Japanese)

Morgan, T. H. 1912: The elimination of the sex-chromosome from the male-producing eggs of phylloxerans. J. Exp. Zool. **12**: 479—498.

Morgan, T. H. 1915: The predetermination of sex in phylloxerans and aphids. J. Exp. Zool. **19**: 285—321.

Morgan, T. H. 1926: The Theory of the Gene. New Haven.

Moritsugu, K. 1937: Studies on a method for early discrimination of successful results in artificial hatching of the silkworm egg, with special reference to the hydrogen ion concentration. (Prelim. rep.). Bull. Soies Kinugasa (371): 118—123. (in Japanese)

Moritsugu, K. 1938: Studies on a method for early discrimination or successful results in artificial hatching by the change of pH in the silkworm egg. I. Bull. Soies Kinugasa (378): 42—52. (in Japanese)
Moriyama, T. 1935: An observation on the hatching of silkworm. Bull. Soies Kinugasa (342): 80—84. (in Japanese)
Motomura, I. 1941: Experimental Embryology. Tokyo. (in Japanese)
Muroga, H. 1934: Observations on the hatching of the silkworm. Sanshikaiho **43**: 79. (in Japanese)
Muroga, H. 1951: On the consumption coefficient of inhibitory substance in the silkworm egg. J. Sericult. Sci. Japan **20**: 92—94. (in Japanese)
Nagai, S. 1934: On the hydrogen ion concentration and buffer action in the silkworm egg. Bull. Gunze Sanjisho (1): 57—73. (in Japanese)
Nagata, T. 1951: On the first appearance of gas in the tracheal system of the silkworm embryo. J. Sericult. Sci. Japan **20**: 335—337. (in Japanese)
Nakajima, S. 1938: On the hatching mechanism in silkworm larvae. Bull. Sericult. Silk-Indust. **11**: 56—59. (in Japanese)
Nakamura, K. and Imaizumi, T. 1950: On the changes in the cortical layer of *Drosophila* eggs in early development. Zool. Mag. **59**: 44. (in Japanese)
Nakata, T. 1932. External morphology and resistance to desiccation at unusually high temperatures in the silkworm embryo. Bull. Fukuoka Sericult. Exp. Sta. **1**: 1—26. (in Japanese)
Nakata, T. 1935: Early development of the Malpighian tubules and the posterior part of the mid-gut in the silkworm. Bull. Fukuoka Sericult. Exp. Sta. (3): 1—27. (in Japanese)
Nakata, T. 1936: Nuclear changes in the cells of hibernating silkworm embryos. J. Sericult. Sci. Japan, **7**: 263—265. (in Japanese)
Nakata, T. 1952: Microchemical and biochemical studies of the silkworm egg. Bull. Gunze Kenkyusho. (9): 1—88. (in Japanese)
Nittono, Y. 1951: Histochemical observations on nucleic acid in the silkworm egg. J. Sericult. Sci. Japan, **20**: 182—185. (in Japanese)
Nittono, Y. and Takeshita, H. 1953: Tyrosinase in the eggs of the silkworm, *Bombyx mori* L. I. In the hibernating eggs. J. Sericult. Sci. Japan, **22**: 149—154. (in Japanese)
Noack, W. 1901: Beiträge zur Entwicklungsgeschichte der Musciden. Z. Wiss. Zool. **70**: 1—57.
Nonidez, J. F. 1920: The internal phenomena of reproduction in *Drosophila*. Biol. Bull. **39**: 207—230.
Nunome, J. 1937: Embryonic development of the silk gland in the silkworm. Jap. J. Applied Zool. **9**: 68—92. (in Japanese)
Nunome, J. 1943: On the effect of atmospheric pressure on the egg and larva of the silkworm. (Prelim. rep.). J. Sericult. Sci. Japan, **14**: 228—236. (in Japanese)
Nunome, J., Yawata, Y. and Akagi, Y. 1955: Water-evaporation in silkworms *(Bombyx mori* L.). III. Water evaporated, temperature and heat of vaporization in silkworm eggs treated by artificial method. Bull. Fac. Text. Fibers, Kyoto Univ. Indust. Art. Text. Fibers, **1**: 177—185 (in Japanese)
Nusbaum, J. and Fulinski, B. 1906: Über die Bildung der Mitteldarmanlage bei *Phyllodromia (Blatta) germanica* L. Zool. Anz. **30** (11/12): 362—381.
Nusbaum, J. and Fulinski, B. 1909: Zur Entwicklungsgeschichte des Darmdrüsenblattes bei *Gryllotalpa vulgaris* Latr. Z. Wiss. Zool. **93**: 306—348.
Oba, H. 1934: Investigations on the micropyle and the chrysanthemum-like pattern in the egg of *Dictyoploca japonica* Butler. Bull. Soies Kinugasa (335): 75—85. (in Japanese)
Oba, H. 1936: On the micropyle in *Antheraea yamamai*, *Antheraea pernyi* and *Philosamia cynthia.* Bull. Soies Kinugasa (362): 77—85. (in Japanese)
Oka, H. 1934: Experimental studies on the embryonic development of the cricket. Annot. Zool. Japon. **14**: 373—376.
Oka, H. 1935: Experimental studies on the morphology of insects. VII. Problems in embryonic development. Botany and Zoology, **3**: 2149—2161. (in Japanese)
Okamoto, D. 1939: On the rate of development in *Galerucella distincta* Baly. Jap. J. Applied Zool. **11**: 85—94. (in Japanese)
Omura, S. 1938: Structure and function of the female genital system of *Bombyx mori* with special reference to the mechanism of fertilization. J. Fac. Agr. Hokkaido Imp. Univ. **40**: 111—128.
Omura, S. 1954: The Silkworm. Tokyo. (in Japanese)
Ono, R. 1951: Relation between the duration of exposure to high temperature and the termination of diapause in the silkworm egg. J. Sericult. Sci. Japan, **20**: 297. (in Japanese)
Pauli, M. E. 1927: Die Entwicklung geschnürter und centrifugierter Eier von *Calliphora erythrocephala* und *Musca domestica*. Z. Wiss. Zool. **129**: 483—540.
Philiptschenko, J. 1912: Zur Kenntnis der Apterygotenembryologie. Zool. Anz. **39**: 43—49.

Poulson, D. F. 1940: The effects of certain X-chromosome deficiencies on the embryonic development of *Drosophila melanogaster*, J. Exp. Zool. **83**: 271—325.
Poulson, D. F. 1945: Chromosomal control of embryogenesis in *Drosophila*. Amer. Nat. **79**: 340—363.
Reith, F. 1925: Die Entwicklung des *Musca*-Eies nach Ausschaltung verschiedener Eibereiche. Z. Wiss. Zool. **126**: 181—238.
Reith, F. 1931: Versuche über die Determination der Keimesanlage bei *Camponotus ligniperda*. Z. Wiss. Zool. **139**: 664—734.
Reith, F. 1935: Über die Determination der Keimanlage bei Insekten. Z. Wiss. Zool. **147**: 77—100.
Saito, S. 1934: A study on the development of the tusser worm, *Antheraea pernyi* Guér. J. Fac. Agr. Hokkaido Imp. Univ. **33**: 249—266.
Saito, S. 1937: On the development of the tusser, *Antheraea pernyi* Guerin-Meneville, with special reference to the comparative embryology of insects. J. Fac. Agr. Hokkaido Imp. Univ. **40**: 35—109.
Sakaguchi, B. 1952a: Experimental and embryological studies on silkworm larvae hatched from eggs subjected to high-speed centrifugation. (Prelim. rep.). J. Sericult. Sci. Japan, **21**: 74—83. (in Japanese)
Sakaguchi, B. 1952b: Effect of centrifugation on the embryonic development of the eri-silkworm. J. Sericult. Sci. Japan, **21**: 145. (in Japanese)
Sakaguchi, B. 1952c: The effect of centrifugal force upon the development of silkworm and eri-silkworm eggs. Ann. Rep. Nat. Inst. Genet. Japan, (2): 21—22.
Sakurai, M. 1915: Development of the suboesophageal body. Rep. Sericult. Assoc. Japan, **24**: 5—9. (in Japanese)
Sato, H. 1924: On the silkworm produced by artificial parthenogenesis. Rep. Sericult. Assoc. Japan, **33** (383): 27—32, (384): 110—114, (386): 242—247, (387): 335—337. (in Japanese)
Sato, H. 1929. On the silkworm produced by artificial parthenogenesis. Jap. J. Applied Zool. **1**: 141—158. (in Japanese)
Sato, H. 1934: Genetical and cytological studies on silkworm produced by artificial parthenogenesis. Jap. J. Applied Zool. **6**: 179—187, 225—253. (in Japanese)
Sato, H. 1942: Cytological studies on the silkworm egg exposed to high temperatures. I. Formation of polyploid nuclei and development of the egg- and sperm-nuclei. Bull. Sericult. Silk-Indust. **14**: 19—27. (in Japanese)
Sato, H. 1953: Cytological and embryological studies on silkworm eggs made infertile by exposure to high temperature, with special reference to the formation of diploid egg nuclei by non-disjunction of the chromosomes. J. Sericult. Sci. Japan. **22**: 139—140. (in Japanese)
Sato, H. and Mezaki, M. 1942: Cytological studies on silkworm eggs exposed to high temperature. II. On the causes of failure of syngamy. Bull. Sericult. Silk-Indust. **14**: 28—35. (in Japanese)
Sato, H., Takami, T. and Kitazawa, T. 1956: Pigmentation of unfertilized eggs and its relation to the brown depressed egg in the silkworm. J. Sericult. Sci. Japan, **25**: 273—278. (in Japanese)
Schnetter, M. 1934: Physiologische Untersuchungen über das Differenzierungszentrum in der Embryonalentwicklung der Honigbiene. Roux' Arch. **131**: 285—323.
Schwangart, F. 1904: Studien zur Entodermfrage bei den Lepidopteren. Z. Wiss. Zool. **76**: 167—212.
Schwangart, F. 1905: Zur Entwicklungsgeschichte der Lepidopteren. Biol. Zentrbl. **25**: 721—729, 777—789.
Sehl, A. 1931: Furchung und Bildung der Keimanlage bei der Mehlmotte *Ephestia kuehniella* Zell. nebst einer allgemainen Übersicht über den Verlauf der Embryonalentwicklung. Z. Morph. Okol. Tiere, **20**: 533—598.
Seidel, F. 1926: Die Determinierung der Keimanlage bei Insekten. I. Biol. Zentrbl. **46**: 321—343.
Seidel, F. 1928: Die Determinierung der Keimanlage bei Insekten. II. Biol. Zentrbl. **48**: 230—251.
Seidel, F. 1929a: Die Determinierung der Keimanlage bei Insekten. III. Biol. Zentrbl. **49**: 577—607.
Seidel, F. 1929b: Untersuchungen über das Bildungsprinzip der Keimanlage im Ei der Libelle *Platycnemis pennipes*. I—V. Roux' Arch. **119**: 323—440.
Seidel, F. 1932: Die Potenzen der Furchungskerne in Libellenei und ihre Rolle bei der Aktivierung des Bildungszentrums. Roux' Arch. **126**: 213—276.
Seidel, F. 1935: Der Anlagenplan im Libellenei, zugleich eine Untersuchung über die allgemeinen Bedingungen für defekte Entwicklung und Regulation bei Dotterreichen Eiern. Roux' Arch. **132**: 671—751.
Seidel, E., Bock, E. and Krause, G. 1940: Die Organisation des Insekteneies. Naturwiss. **28**: 433—446.

Seiler, J. 1924: Furchung des Schmetterlingeies ohne Beteilung des Kernes. Biol. Zentrbl. **44**: 68—76.

Shafiq, S. A. 1954: A study of the embryonic development of the gooseberry sawfly, *Pteronidea ribesii*. Quart. J. Micr. Sci. **95**: 93—114.

Slifer, E. H. 1932: Insect development. III. Blastokinesis in the living grasshopper egg. Biol. Zentrbl. **52**: 223—229.

Slifer, E. H. 1937: The origin and fate of the membranes surrounding the grasshopper egg; together with some experiments on the source of the hatching enzyme. Quart. J. Micr. Sci. **79**: 493—506.

Strindberg, H. 1915: Über die Bildung und Verwendung der Keimblätter bei *Bombyx mori*. Zool. Anz. **45**: 577—597.

Suzuki, H. 1917: On the respiration of the silkworm, with special reference to the CO_2 output. Bull. Imp. Sericult. Exp. Sta. **2**: 287—312. (in Japanese)

Suzuki, K. and Ichimaru, M. 1955: Embryological studies on silkworm eggs hereditarily incapable of hatching. III. On the development of kidney-shaped eggs. Memo. Fac. Educ. Kumamoto Univ. (3): 177—197. (in Japanese)

Takahashi, I. 1912: On the serosa and amnion of the silkworm egg. Rep. Sericult. Assoc. Japan, (246): 8—12. (in Japanese)

Takahashi, I. and Sugita, Y. 1912: Experiments on the temperature for embryonic development of silkworm eggs. Tokyo Sangyokoshujo Seiseki (44): 63—89. (in Japanese)

Takahashi, S. and Yagi, N. 1926. Predetermination of the embryonic area in the silkworm egg. Sangyo Shimpo (393): 366—268. (in Japanese)

Takahashi, Y. 1954: Morphological and histological observations on the embryo and yolk cells in the early development of the silkworm egg. J. Sericult. Sci. Japan, **23**: 183—184. (in Japanese)

Takami, T. 1942: Experimental studies on embryo formation in *Bombyx mori*. I. Zool. Mag. **54**: 337—343. (in Japanese)

Takami, T. 1943: Experimental studies on embryo formation in *Bombyx mori*. II. Shift in position of the embryonic area. Zool. Mag. **55**: 220—223. (in Japanese)

Takami, T. 1944a: Experimental studies on embryo formation in *Bombyx mori*. III. Movement and revolution of partial embryos. Zool. Mag. **56**: 62—65. (in Japanese)

Takami, T. 1944b: Experimental studies on embryo formation in *Bombyx mori*. IV. Amnion and serosa. Zool. Mag. **56**: 66—69. (in Japanese)

Takami, T. 1946: Experimental studies on embryo formation in *Bombyx mori*. V. Presumptive mesodermal and neural regions of the egg. Seibutsu **1**: 208—211. (in Japanese)

Takami, T. 1952: Hibernating, non-hibernating and acid-treated eggs of the silkworm, with special reference to the dispersion of fatty globules in the yolk cells. J. Sericult. Sci. Japan, **21**: 245—251. (in Japanese)

Takami, T. 153a: A method to observe the termination of diapause in the silkworm egg. J. Sericult. Sci. Japan, **22**: 120—121. (in Japanese)

Takami, T. 1953b: Studies on the yolk cell in *Bombyx mori*. I. Movement of yolk cells in vitro. J. Sericult. Sci. Japan, **22**: 181—184. (in Japanese)

Takami, T. 1954a: Movement of yolk cells in the silkworm (*Bombyx mori* L.). Science, **119**: 161—162.

Takami, T. 1954b: Studies on the yolk cell in *Bombyx mori*. II. Early division of yolk cells and some related problems. Cytologia. **19**: 299—305.

Takami, T. 1955: Studies on the yolk cell in *Bombyx mori*. IV. Granular staining in the albuminous yolk globules. Exp. Cell Res. **9**: 568—571.

Takami, T. 1955: Studies of diapause in the silkworm egg by in vitro culture of embryos. Comm. at 7th Meeting, Kanto Sericult. Soc. Japan.

Takami, T. 1957: In vitro culture of embryos of the silkworm, *Bombyx mori* L. I. Culture in silkworm egg extract, with special reference to some characteristics of the diapause egg. Bull. Sericult. Exp. Sta. (in press) (in Japanese)

Takeuchi, K. 1955a: Studies on the hatching of the silkworm egg. J. Sericult. Sci. Japan, **24**: 49—56. (in Japanese)

Takeuchi, K. 1955b: Observations on the hatching of the egg of the Chinese tussah silkworm. Acta Serilogica, **14**: 15—20. (in Japanese)

Tanaka, S. 1956: Cytological observations on the union of two sperm nuclei. J. Sericult. Sci. Japan, **25**: 244. (in Japanese)

Tanaka, S., Oi, S. and Sato, H. 1955: Studies on merogony in *Bombyx mori* L. I. Percentages of merogony following different treatments, and the activity of the male nucleus in the egg. Res. Rep. Fac. Text. Sericult. Shinshu Univ. (5): 63—67. (in Japanese)

Tanaka, Y. 1943: Sericology. Tokyo. (in Japanese)

Tanaka, Y. et al. 1952: Genetics in the Silkworm. Tokyo. (in Japanese)
Tazima, Y. 1939: Induction of mosaic embryos in the silkworm *(Bombyx mori)* by means of heat shock. Jap. J. Genet. **15**: 111—117. (in Japanese)
Tazima, Y. 1942a: On spontaneous mosaicism observed in the silkworm egg. Jap. J. Genet. **18**: 44—46. (in Japanese)
Tazima, Y. 1942b: A case of mosaicism in the silkworm egg, probably due to a recessive mutation. Jap. J. Genet. **18**: 305—308. (in Japanese)
Tazima, Y. 1947: Mosaics in the Silkworm. Tokyo. (in Japanese)
Tichomiroff, A. 1879: Ueber die Entwicklungsgeschichte des Seidenwurms Zool. Anz. **2**: 64—67.
Tokunaga, M. 1939: Growth in insect embryos, particularly that of the silkworm. Jap. J. Applied Zool. **11**: 145—159. (in Japanese)
Totani, K. 1955: Abnormal embryos produced by prolonged refrigeration of silkworm eggs. J. Sericult. Sci. Japan, **24**: 184—185. (in Japanese)
Toyama, K. 1896: Structure of the egg and development of the embryo in the silkworm. J. Sci. Agr. Soc. (28): 28—40, (in Japanese)
Toyama, K. 1902: Contributions to the study of silkworm. I. On the embryology of the silkworm. Bull. Coll. Agr. Tokyo Imp. Univ. **5**: 73—118.
Toyama, K. 1909: The Silkworm Egg. Tokyo. (in Japanese)
Tsuchiya, T. 1926: Movement of yolk cells in hibernating eggs of the silkworm, and a simple method for observing it. Sangyo Shimpo **34**: 155—157 (in Japanese)
Umeya, Y. 1937: The revolution of the silkworm embryo, considered from the viewpoint of embryology. Zool. Mag. **49**: 300—302. (in Japanese)
Umeya, Y. 1937: Preliminary note on experiments of ooplasm transfusion of silkworm-eggs with special reference to the development of embryo. Proc. Imp. Acad. **13**: 378—380.
Umeya, Y. 1938a: Studies on the preservation of the hibernating silkworm egg. Bull. Dept. Agr., Chosen Sericult. Exp. Sta. **4**: 1—36. (in Japanese)
Umeya, Y. 1938b: Answers to the questions of Prof. Miura on the development of embryos in the silkworm. Sanshikaiho **47**: 33—38. (in Japanese)
Umeya, Y. 1939: Experiments on the artificial induction of embryonic development in hibernating eggs of the silkworm. Nihon Gakujutsu Kyokai Hokoku **14**: 504—508. (in Japanese)
Umeyo, Y. 1946: Embryonic hibernation and diapause in insects, from the viewpoint of hibernation phenomena in silkworms. Bull. Sericult. Exp. Sta. **12**: 393—480. (in Japanese)
Umeya, Y. 1950: On *Rhodinia fugax* Butler. Zool. Mag. **59**: 278—283. (in Japanese)
Umeya, Y. 1952: Intrinsic Characters and Environment. Tokyo. (in Japanese)
Umeya, Y. 1955. On the developmental zero of the wintering eggs of *Rhodinia fugax* Butler. Zool. Mag. **64**: 201—205. (in Japanese)
Van der Kloot, W. 1955: The control of neurosecretion and diapause by physiological changes in the brain of the Cecropia silkworm. Biol. Bull. **109**: 276—294.
Wada, S. 1954: Development and function of the oral cell mass in the silkworm embryo. J. Sericult. Sci. Japan, **23**: 291. (in Japanese)
Wada, S. 1955a: Zur Kenntnis der Keimblätterherkunft der Subösophagealkörpers am Embryo der Seidenraupe, *Bombyx mori* L. J. Sericult. Sci. Japan, **24**: 114—117.
Wada, S. 1955b: Zur Frage des Subösophagealkörpers als hämopoetisches Gewebes am Embryo der Seidenraupe, *Bombyx mori* L. (Vorläufige Mitteilung). J. Sericult. Sci Japan, **24**: 311—313.
Watanabe, K. 1919: Studies on the relation between the color of the silkworm and environmental factors. I. Relation between egg colors and distribution of pigment in the serosal cells. Bull. Imp. Sericult. Exp. Sta. **4**: 107—116. (in Japanese)
Watanabe, K. 1924: Studies on voltinism in the silkworm. Bull. Imp. Sericult. Exp. Sta. **6**: 411—455. (in Japanese)
Watanabe, K. 1931: On the relation between the temperature after oviposition and the diapause of the silkworm egg. Tech. Bull. Sericult. Exp. Sta. (41): 16—30. (in Japanese)
Watanabe, K. 1934: Embryonic development and hatching in relation to the light during incubation of the silkworm egg. Tech. Bull. Sericult. Exp. Sta. (46): 1—27. (in Japanese)
Weber, H. 1948: Grundriss der Insektenkunde. Jena.
Weismann, A. 1863: Die Entwicklung der Dipteren in Ei, nach Beobachtungen an *Chironomus* spec., *Musca vomitoria* und *Pulex canis*. Z. Wiss. Zool. **13**: 107—158., 159—220.
Wellhouse, W. T. 1954: The embryology of *Thermobia domestica* Packard. Iowa State Coll. J. Sci. **28**: 416—417.
Wells, M. J. 1954: The thoracic glands of Hemiptera, Heteroptera. Quart. J. Micr. Sci. **95**: 231—244.
Wigglesworth, V. B. 1953: The Principles of Insect Physiology. London.

Wigglesworth, V. B. and Beament, J. W. L. 1950: The respiratory mechanisms of some insect eggs. Quart. J. Micr. Sci. **91**: 429—452.

Will, L. 1888: Zur Entwicklungsgeschichte der vivaparen Aphiden. Biol. Zentrbl. **8**: 148—155.

Yamaguchi, T. 1940: On the weight increase of the silkworm embryo. Kyoikunogei, **9**: 28—34. (in Japanese)

Yamaguchi, T. 1946: On the behavior of the yolk cells during the development of the silkworm egg. Sanshikagaku, **1**: 5—6. (in Japanese)

Yamaguchi, T. and Kobayashi, T. 1940: Studies on the growth of silkworm embryos. I. J. Sericult. Sci. Japan, **11**: 204—205. (in Japanese)

Yamaguchi, T. and Shimizu, T. 1951: On embryonic growth and enzyme activity, with special reference to catalase activity in the silkworm egg. J. Sericult. Sci. Japan **20**: 299. (in Japanese)

Yamazaki, T. 1939: Effects of temperature and humidity on the hatching of the Chinese tussah silkworm. Jap. J. Applied Zool. **11**: 167—176. (in Japanese)

Yanagida, T. Embryonic Development of the Silkworm Illustrated. Tokyo. (in Japanese)

Yajima, H. 1956: Effect of centrifugation on embryonic determination of the harlequin fly. Comm. at 27th Ann. Meet. Zool. Soc. Japan. (in Japanese)

Yatomi, Y. and Yamashita, S. 1938: Effects of temperature and humidity on hatching in *Ephestia cautella* Walker. (Prelim. rep.). Jap. J. Applied Zool. **10**: 133—136. (in Japanese)

Yokoyama, T. 1929: On the heart-beat in the silkworm embryo. Jap. J. Applied Zool. **1**: (in Japanese)

Yokoyama, T. 1956: The morphology and innervation of the prothoracic gland of the silkworm, *Bombyx mori*. J. Sericult. Sci. Japan, **25**: 87—94. (in Japanese)

Yokoyama, T. and Keister, M. 1953: Effects of small environmental changes on developing silkworm eggs. Ann. Ent. Soc. Amer. **46**: 218—220.

Yokoyama, T. and Takashima, M. 1952: On silkworm incubated under slightly lower than usual atmospheric pressure. Acta Serilogica, (2): 159—164. (in Japanese)

Yoshitake, N. 1954: Biochemical studies on voltinism in Lepidoptera. I. On the function of the hormone determining voltinism, with special reference to the pigment formation in the serosa of the silkworm egg. J. Sericult. Sci. Japan, **23**: 67—75. (in Japanese)

Yoshitake, N. 1955: Biochemical studies on voltinism in Lepidoptera. IV. On the responses o. staining in the non-hibernating and hibernating eggs of the silkworm, *Bombyx mori* L. J. Sercult. Sci. Japan, **24**: 108—113. (in Japanese)

Yoshitake, N. 1956: Biochemical studies on voltinism in Lepidoptera. VI. On differences in ovarian function between pupae producing non-hibernating, and those producing hibernating eggs in experiments on ovarian implantation in the silkworm, *Bombyx mori*. Jf Sericult. Sci. Japan, **25**: 434—438. (in Japanese)

MOLLUSCA

Chapter 11

I. AMPHINEURA, GASTROPODA, SCAPHOPODA, PELECYPODA

INTRODUCTION

The Mollusca are divided into the five classes, Amphineura, Gastropoda, Scaphopoda, Pelecypoda and Cephalopoda; omitting the Cephalopoda, the development of the first four will be considered here.

Among these the forms belonging to the Amphineura, Scaphopoda and the sub-class Prosobranchia among the Gastropoda are dioecious, while the members of the gastropod sub-classes Opisthobranchia and Pulmonata are monoecious. The great majority of pelecypod animals are dioecious, but it is not unusual to find forms having reproductive organs in which both eggs and spermatozoa are present side by side. Coe (1943, 1944) has presented a detailed account of sex in molluscs.

(1) SPAWNING AND REARING OF YOUNG

Among the Prosobranchia, the primitive Archeogastropoda (abalone, keyhole limpet, *Turbo,* etc.) lay their eggs separately or in egg masses surrounded by jelly; except in a few ovoviviparous species, the eggs of the other orders are laid separately or in various numbers, generally less than a hundred, fixed in *egg cases* of characteristic shapes. Observations have been made on the conditions of spawning and the formation of such egg cases by MacGinitie (1931) in *Alectrion (=Nassarius) fossatus;* by Fretter (1941) in several species of Prosobranchia; by Ino (1950) in *Babylonia japonica;* and by Miyawaki (1953) in *Neptuna arthritica.* In *Babylonia,* 40 eggs, surrounded by a gelatinous substance to form a single mass, are intermittantly extruded from the genital pore; this mass passes along a special groove, temporarily present on the surface of the foot, to the *pedal pore* (also called the *ventral pedal gland*) on the sole of the hypopod, where it is molded into shape and laid.

The shape of the egg case is peculiar to each species; many observations have been made on the morphology of these structures. As representative reports, those of Labour (1936, 1937, 1945), Seno (1907), Thorson (1935, 1940, 1946), Ostergaard (1950) and Nambu (1944, 1953) may be cited. In many species, the eggs develop inside the egg case to the veliger larva stage before they swim out from a fixed region of the case, although forms are not rare in which the larvae develop within the case until they complete metamorphosis and attain the same morphology as the adult before they come creeping out.

It is thought that the embryos utilize the material within the case as a source of nutrition; there are also many forms in which some of the embryos grow by digesting others within the same case. For example, Lebour reports that in *Lucella lapillus,* although several hundred eggs are laid, only 15—25 hatch. This phenomenon is often seen among the Muricidae, Buccinidae and others, in which unfertilized and polyspermic eggs, and embryos which stop developing during the early stages, are used as food by the veliger larvae (Staiger 1950). Furthermore, in these species the larvae hatch after passing through the veliger stage, and their size varies with the number of nurse eggs (Lebour 1937).

There are some workers who claim that the eggs destined to be nurse cells are fertilized by the so-called apyrene spermatozoa (Hyman 1923, 1925; Portman 1931), but Ankel (1930) opposes this idea on the basis of his observation that nurse eggs are found in *Natica catena,* in which sperm dimorphism does not occur.

The most striking egg cases among the Gastropoda are those found in some species of the genus *Cymbium,* in which they form a huge mass 35 cm long, shaped like a pineapple with the core removed; each 1.5 × 2.5 cm egg case encloses one egg. Within a single such mass the developmental stages, from the earliest through the veliger to the young adult measuring more than 1.5 cm, in which the velum has already disappeared, can be found lined up in regular sequence (Wada 1943). This indicates that one animal must have deposited all the egg cases continuously over a long period. The divers on the ships that collect pearl oysters along the Australian coast say that these molluscs assume an inverted position during spawning; it is not known whether the animal maintains such a position during the whole of a spawning process which seems to require such a long time. However that may be, it is interesting that a single egg mass of the animals in this genus will provide a complete series of specimens of the developmental process arranged in the proper order.

Among the Pelecypoda, most of the fresh water species and a very few of the marine species shelter the young in a particular region of the ctenidia (gills). Miyazaki (1936, 1938) has published a general discussion of this breeding habit. Great differences are observable in the duration of the period during which the young are thus carried, from cases like *Ostrea crenulifera* in which it ends in the fairly early veliger stage, to that of *Lasaea nipponica* in which it lasts to the adult stage.

(2) EGGS AND FERTILIZATION, ARTIFICIAL AND NATURAL

Molluscan eggs are all fertilized before extrusion of the first polar body — that is, during the primary oocyte stage. Among the Gastropoda, copulation and internal fertilization are found in all the Opisthobranchia and Pulmonata (Mesogastropoda and Neogastropoda), while the lower Prosobranchia (Archeogastropoda),[1] the Pelecypoda and the Scaphopoda,[2] with a few exceptions are characterized by external fertilization.

---

[1] The Archeogastropoda have no copulatory organs, but in one species of this group, *Lunella coronata coreensis,* at the time of high tide during the summer spring tides the males attach themselves to the shells of the females and the pairs appear near the surface. Holding onto the

Even among the animals showing external fertilization, there are only a few species in which it is possible to observe the phenomenon of fertilization and the developmental processes by using eggs and spermatozoa from extirpated gonads; in most species it is necessary to wait for natural spawning, or use a variety of methods to induce shedding. This point has been discussed to some extent in the Introduction, to which the reader is referred. The most spectacular examples of spawning are found among the Tridacnidae. If egg water is used as the stimulus, the sperm are expelled together with a columm of water two meters high (Wada 1954). During observation of the spawning habits of these animals at the Tropical Marine Station at Palau, the laboratory was occasionally found flooded in the morning as the result of spawning durning the night. The list of forms in which it is easy to bring about artificial fertilization includes rather many species among the Pelecypoda. For example, the oviparous oyster genus, *Crassostrea (=Gryphaea)*, including *Crassostrea gigas, C. nippona;* the clams belonging to the families Mactridae *(Mactra veneriformis, M. sulcataria, Spisula sachalinensis)*, Amphidesmatidae *(Amphidesma striatum, Caecella chinensis)* and Chamidae *(Chama reflexa, C. retroversa)* and the members of the Pholadidae *(Barnea manilensis inornata, Pholadidea penita)* may be cited. In these species, since the eggs taken from extirpated ovaries generally have the germinal vesicles intact, it is believed that fertilization takes place in this stage, although the germinal vesicles have already broken down in eggs which are shed normally. The *Mactra* species provide a few exceptions in that their eggs will not fertilize unless the germinal vesicles are intact (Iwata 1957). In the past, *Crassostrea gigas* and *C. echinata* have been cited as forms in which spawning and fertilization occur before germinal vesicle breakdown, but in these species also, fertilization takes place irregardless of the condition of the germinal vesicle, the breakdown of which normally takes place in the ovary immediately before spawning (Wada, unpublished).

The species among the Gastropoda in which fertilization can be carried out artificially are extremely few, consisting only of the Fissurellidae, which include *Megathura crenulata* and *Scutus scapha.* In some cases artificial fertilization has been successful in the scaphopod *Dentalium* spp.

The cause underlying the failure of attempts at artificial fertilization is thought to be an immaturity of the eggs and spermatozoa. If the eggs and spermatozoa of pearl oysters such as *Pinctada maxima* or *P. martensii* are placed in sea water containing a little ammonia, the eggs mature, the spermatozoa are activated, and fertilization is easily performed; if the fertilized eggs are immediately transferred to normal sea water it is possible to obtain normal larvae (Wada 1941, 1943; Wada and Wada 1953). In the clams *Venerupis semidecussata* and *Meretrix meretrix*, the scallop *Pecten yesoensis,* as well as in *Spondylus cruentis, Pteria penguin* and *Electroma,* observation of developmental stages is possible if this method is followed (Hatanaka, Sato and Imai 1943; Yamamoto and Nishioka 1943; Wada 1953).

In the prosobranchs *Lottia, Acmaea* and *Patella,* the same result can be obtained by treatment with alkaline sea water or benzene (Loeb, Wolfson and MacBride 1914).

---

mouth of the female's shell with the tip of its foot, the male sheds sperm; this is immediately followed by the spawning of the eggs, so that fertilization takes place in an effective manner (Wada, unpublished).

[2] According to Purchon (1941), the boring pelecypod *Xylophaga dorsalis* is a protandrous hermaphrodite, having a sperm receptacle for storing its own sperm, and the eggs appear to be self-fertilized at the time of spawning.

In most cases the mature eggs are spherical or nearly so, but the eggs removed from pelecypod ovaries are often drawn to a point at one side. This pointed region is the part which was attached to the gonadal epithelium, and marks the vegetal pole of the egg. The egg of *Eulota similaris stimpsoni* (Pulmata) has a very peculiar structure (Fig. 11.1). As the egg of this species approaches maturity, the cytoplasm forms many long protuberances, at the tip of which the spermatozoa are said enter (Ikeda 1930).

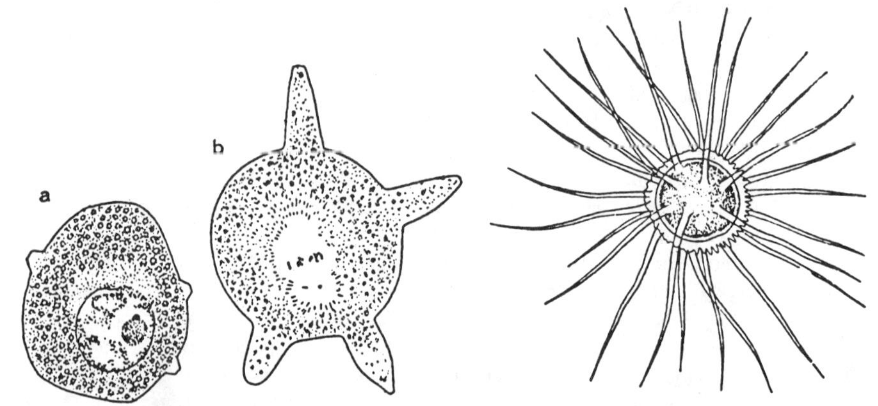

Fig. 11.1 Two stages in the maturation of the *Eulota similaris stimpsoni* egg (Ikeda)

Protrusions through which the sperm enter the egg are formed as it matures.

Fig. 11.2 Egg of *Liolophura japonica* Lischke (Taki)

When the eggs are surrounded by a secondary membrane (chorion), this has an umbilicus-like process (for example, among the Pelecypoda, *Tridacna* spp., *Pecten maximus*; among the Gastropoda, very many of the Archeogastropoda). There are some who consider this structure to be a micropyle (viz., von Medem), but this is incorrect, since spermatozoa are observed to enter through any part of the chorion.

In the Amphineura are to be found species in which the chorion assumes a highly characteristic structure (Fig. 11.2).

Among the Prosobranchia, in some species of the order Archeogastropoda having eggs with chorions, it is known (Dan 1956) that the spermatozoa undergo an acrosome reaction (see Introduction) which is believed to enable the spermatozoa to dissolve the chorion and enter the perivitelline space. It has been found that a striking change takes place in the acrosome region of a large number of pelecypod spermatozoa at the time of fertilization (Dan and Wada 1955); in particular, it is known that a substance which dissolves the egg membrane is contained in the acrosome region of *Mytilus edulis* spermatozoa (Wada, Collier and Dan 1956). Since the eggs are small, it is relatively easy to observe the fertilization process in pelecypod eggs. After sperm attachment several minutes are usually required for the sperm head to disappear inside the egg (3.5 min. in *Mytilus edulis* at 15°C; Wada 1955) and the sperm tail is generally taken in also. Unlike echinoderm fertilization, no sudden change in the nature of the vitelline membrane has been determined to take place immediately after molluscan fertilization, although in pelecypod eggs the space between the viteline membrane and the plasma membrane becomes a little wider after fertilization (Allen 1953; Wada 1955). There is a record that molluscan spermatozoa enter the eggs from the vegetal pole (Korschelt u. Heider 1936), but no fact of this nature can be confirmed in *Crassostrea gigas, C. echinata, Mactra veneriformis* or *Mytilus edulis.*

At a given time after fertilization, the first polar body and then the second are successively extruded at the animal pole. Since the time from fertilization to first polar body extrusion is greatly affected by the condition of the egg nucleus at the time of fertilization, the reports vary according to the investigator, from 10 to 50 minutes at 25°C in such species as *Crassostrea gigas* and *echinata*, in which fertilization occurs without much dependence on the state of the egg nucleus. The first polar body usually does not cleave again, although there are reports of its division (Casteel 1904). In *Crassostrea echinata* it often cleaves when the vitelline membrane is weakened. The second polar body is not as conspicuous as the first in pelecypod eggs; in *Amphidesma* and *Caecella* it cannot even be recognized as a protuberance from the egg surface. At the time of polar body formation, not only the animal pole of the egg but also the vegetal pole undergoes a form change in many cases.

The sperm lies in the egg cytoplasm near the surface without much change in position until the expulsion of the second polar body; when the maturation divisions of the egg are completed, the egg and sperm pronuclei come together near the center of the egg and unite in syngamy. With this the first cleavage begins.

1. EARLY DEVELOPMENT

(1) TERMINOLOGY OF CLEAVAGE

Molluscan eggs belong to the category of so-called "mosaic eggs", and divide according to the spiral mode of cleavage; in order to describe the process of development it is convenient to designate each blastomere by an appropriate term. The formula in general use at the present time is that used by Conklin (1897) to describe the development of *Crepidula*. Taking the 4-cell stage as the standard, the blastomeres as seen from the animal pole side are designated, *A, B, C* and *D* in clockwise order. Since in most cases these blastomeres are larger than those formed later, they are also called *macromeres*. (The designations of the 2-cell stage

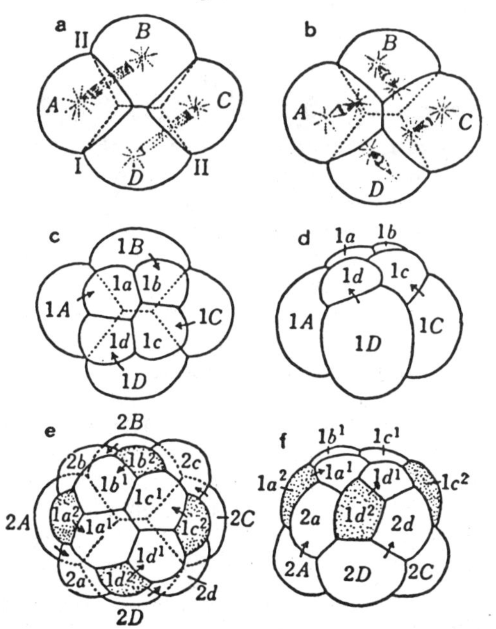

Fig. 11.3 Cleavage in *Trochus* (1) (Robert)
a, b. 4-cell stage (from aminal pole); c. 8-cell stage (from animal pole); d. 8-cell stage (from side); e. 16-cell stage (from animal pole); f. 16-cell stage (from side). Stippled blastomeres are trochoblasts.

blastomeres, $\widehat{AB}$ and $\widehat{CD}$, are formed retroactively from those of the 4-cell stage). In general, the two opposite blastomeres A and C are in contact at the animal pole surface, while B and D are in contact at the vegetal pole (Fig. 11.3b). The four *micromeres* which separate off toward the animal pole side from these large blastomeres are called a *quartet* and labelled with the small letters a, b, c and d; to these are added a numerical coefficient identifying their generation as the first ($1a$, $1b$, etc). and differentiating it from other quartets. Sometimes a quartet is designated collectively by the letter q. Since the quartets are formed from the macromeres in the course of several cleavages, each quartet receives a coefficient specifying its generation. When the blastomeres making up a quartet divide again, the member of the two resulting blastomeres lying on the animal pole side receives the index number 1, while its sister blastomere toward the vegetal pole receives the index number 2. Thus, when the first quartet cleaves it produces $1q^1$ and $1q^2$. At the next cleavage $1q^1$ produces $1q^{11}$ and $1q^{12}$. Conklin retained the large letters A, B, C and D for the four macromeres which lie together at the vegetal pole, but Child (1900) and Treadwell (1901) also added to these coefficients to indicate the generation, and this formula is in wide use at present. According to this system, taking the A macromere as an example, A divides into $1A$ and $1a$; $1A$ next forms $2A$ and $2a$; from $1a$ are formed $1a^1$ and $1a^2$. At the succeeding cleavage $1a^1$ produces $1a^{11}$ and $1a^{12}$ (see Introduction).

Blastomeres which are destined to play a special role in the future are also sometimes given special notation. For example, the $4d$ blastomere is called M or ME, and in pelecypod development, the $2d$ blastomere is designated as X.

Among the Amphineura, the development of the Loricata (chitons) has been investigated by Metcalf (1893), Kowalevsky (1883) and especially Heath (1899), working with *Ischnochition magdalensis*; the mode of development of these animals is found to agree extremely well with that of the Gastropoda. A great deal of research has been done on the Gastropoda; in the first place, beginning with the Prosobranchia, there is the work of Robert (1902) on *Trochus*,[3] of Conklin (1897) on *Crepidula*, Delsman (1914) on *Littorina*, Doutert (1929) on *Paludina*, Clement (1952) on *Ilyanassa*, among many others. Concerning the Opisthobranchia there are the researches of Casteel (1904) on *Fiona marina*, Pelseneer (1911) and many other reports. Concerning the Pulmonata there is the old work of Rabl (1879) on *Planorbis*; that of Meisenheimer (1896) on *Limax*, of Fujita (1904) on *Siphonaria*, Wierzejski (1905) on *Physa*, and Raven (1945, 1946, 1952) and Kubota (1954) on *Limnaea* (for detailed references see Korschelt u. Hieder 1936). The Gastropoda constitute a very large group of animals, but their mode of cleavage is highly consistent.

The Pelecypoda make up the next widest group after the Gastropoda; the number of species is also large, and a good many people have published observations of a general nature concerning the development of these animals. However, studies of cell-lineage and similar detailed investigations are extremely few, so that even at the present time a great deal of our knowledge is due to Meisenheimer's (1901) research on *Dreissensia*. In addition to this are the studies of Lillie (1895) on *Unio*, of Fujita (1929) on *Crassostrea gigas*, and of Woods (1931) on *Sphaerium*. Although the Pelecypoda are thought of as occupying a rather isolated geneological position among the Mollusca, the bases of their development agree well with those of the Gastropoda, and also of the Annelida.

[3] The species which Robert investigated in detail is today called *Cantharidus striatus* (L.), but in the following description it will be referred to as *Trochus*.

Observations on the Scaphopoda have been made by Wilson (1904), Kowalevsky (1883), Schleip (1925) and others; in general the early development of these species agrees well with those of the Gastropoda and other groups.

In accordance with the above viewpoint, this chapter will consider the early development of the Gastropoda as representative of the phylum as a whole, with supplementary descriptions of important differences occurring in the Pelecypoda and other groups.

(2) MODES OF CLEAVAGE

Cleavage in the Amphineura, Gastropoda, Scaphopoda and Pelecypoda is total, and follows the *spiral cleavage* pattern. Among the Amphineura and Gastropoda, except for a few species in the latter group which form a *polar lobe* (or *yolk lobe*), the first two cleavages result in four practically equal-sized blastomeres. Indications of spiral cleavage are already present at the second cleavage, in that the *A* and *C* blastomeres of the four-cell stage are in contact at the animal pole, and *B* and *D* at the vegetal pole (Fig. 11.3b).

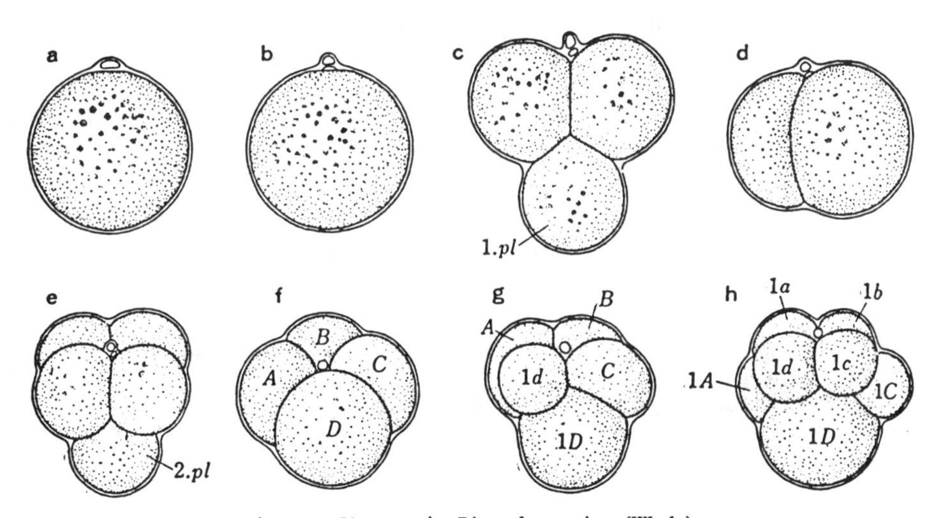

Fig. 11.4 Cleavage in *Pinctada maxima* (Wada)

a. first polar body formation; b. second polar body formation; c. formation of first polar lobe (trefoil stage); d. 2-cell stage; e. formation of second polar lobe; f. 4-cell stage; g. cleavage of *D*- cell; h. 8-cell stage.

The species among the Pelecypoda which form polar lobes are by no means few, but even in those which do not, the first cleavage is unequal, dividing the egg into a smaller *AB* and a larger *CD* blastomere. Moreover, unlike the case of the Gastropoda described above, the first cleavage plane, while passing through the animal pole, does not include the egg axis, but makes a certain angle with it. In other words, the material of the vegetal pole is all contained in the *CD* blastomere. The next cleavage takes place in a spiral manner similar to that in the gastropods, forming equally sized *A*, *B* and *C* blastomeres, and a large *D* blastomere. In the *Mactra* egg, $\widehat{CD}$ often cleaves considerably ahead of $\widehat{AB}$.

In the gastropods *Nassa, Nassarius, Ilyanassa, Modiolaria* and *Aplysia,* the scaphopod *Dentalium* and the pelecypods *Crassostrea, Pinctada, Pteria, Mytilus, Spondylus* and others, a polar lobe is formed during the early cleavages. That is, as the mitotic apparatus for the first cleavage develops in these eggs, the vegetal side bulges out; as the cleavage furrow is formed, the egg is apparently divided into three nearly equal-sized parts (see Fig. 11.4c). This vegetal bulge is the *first polar lobe* — this particular stage is called the *trefoil stage.* There is of course no nuclear substance included in the polar lobe. The first polar lobe is attached to one of the two blastomeres of the animal pole side, and eventually its contents flow into this blastomere; the two-cell stage is thus made up of a large *CD* blastomere containing the material of the polar lobe and a small *AB* blastomere which includes none of this vegetal substance. At the beginning of the next cleavage a polar lobe *(second polar lobe)* is again formed at the vegetal side of the *CD* blastomere; this unites with the *D* cell at the conclusion of the cleavage process. The four-cell stage thus consists of the three practically equal-sized blastomeres, *A, B* and *C,* and a *D* blastomere rather larger than these.

A less conspicuous bulging of the vegetal side is often found accompanying the formation of the polar bodies. In the Pelecypoda the polar lobe appears until the second cleavage, and is not formed at later cleavages, but in *Ilyanassa* and *Dentalium,* it appears also at the third cleavage (Wilson 1904; Clement 1952). In *Dentalium,* part of the polar lobe substance is transferred to the *2d* (=*X*) blastomere at the next (fourth) cleavage.

According to the observations of Meisenheimer (1901) and Stauffacher (1894), a lens-shaped space appears and disappears during the early cleavage stage in the eggs of the pelecypods *Dreissensia polymorpha* and *Sphaerium corneum.*

(3) DIFFERENTIATION OF GERM LAYERS

a. Segregation of ectoderm

In the course of the three cleavages following the four-cell stage, all the material to form the cells of the ectoderm is separated off from the four macromeres. At the first of these cleavages the micromeres are produced in a dextrorotational direction, as is characteristic of spiral cleavage. As explained in (1) above,

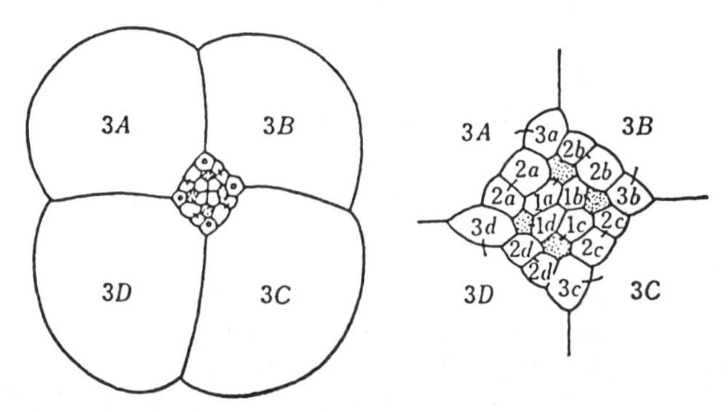

Fig. 11.5 Egg of *Fulgur carica* at 24-cell stage (left: from animal pole. right: enlarged sketch of animal pole region) (Conklin)
Stippled blastomeres are trochoblasts.

these four micromeres, that is, blastomeres 1*a*, 1*b*, 1*c* and 1*d*, form the *first quartet*; they may be referred to collectively by the abbreviation 1*q*. These cleavages do not necessarily occur simultaneously; for example, in *Dreissensia* and *Pinctada maxima* (Fig. 11.4g), the 1*d* micromere is formed ahead of the others (Meisenheimer 1901; Wada 1942).

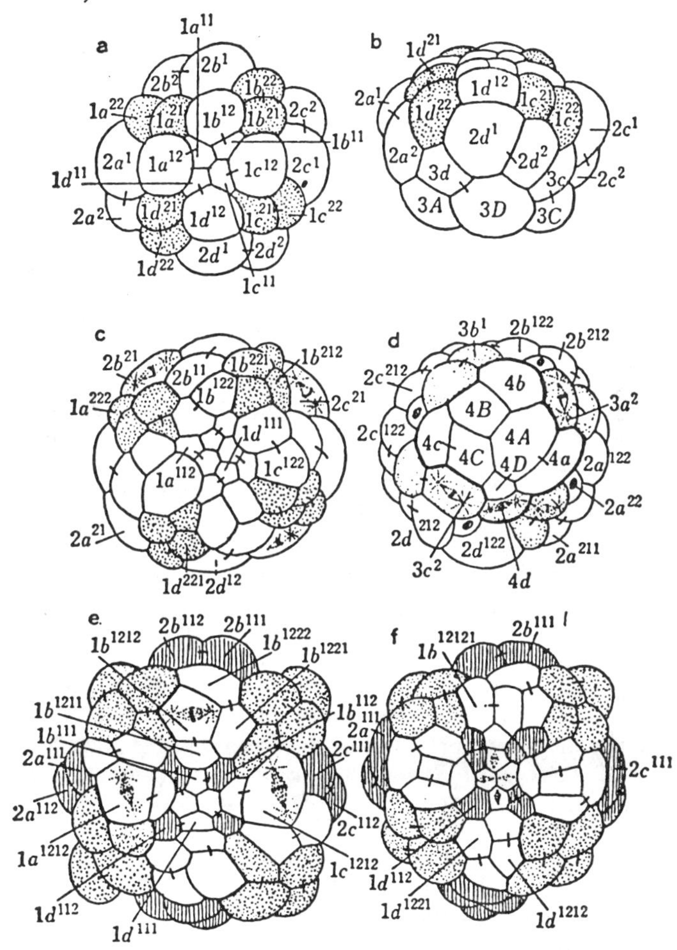

Fig. 11.6 Cleavage in *Trochus* (2)

a. 32-cell stage (from animal pole); b. 32-cell stage (from side); c. 63-cell stage (from animal pole); d. 81-cell stage (from vegetal pole); e, f. more advanced stages, formation of molluscan cross. Trochoblasts: stippled. Tip cells: hatched cells at equator. Peripheral rosette cells: hatched cells near animal pole. Cross cells plus rosette cells: white.

The next cleavage of the macromeres (1*A*—1*D*) is sinistral, giving forth the second micromere quartet (2*q*, 2*a*—2*d*). In the Gastropoda these micromeres are smaller than the macromeres, and in especially yolky eggs size differences may be extreme (Fig. 11.5). In the Pelecypoda, however, the 2*d* micromere is larger than the 2*D* macromere, and is, in fact, the largest of the blastomeres which make up the embryo at this stage. The *2d* blastomere is homologous with the first somatoblast which is seen in annelid development (see Chap. 7, Table 7.1 and Fig. 7.10f); in the Pelecypoda it later produces the *ventral plate, shell gland,* and other ectodermal organs, and Meisenheimer gave it the name of *X blastomere* (Fig. 11.7).

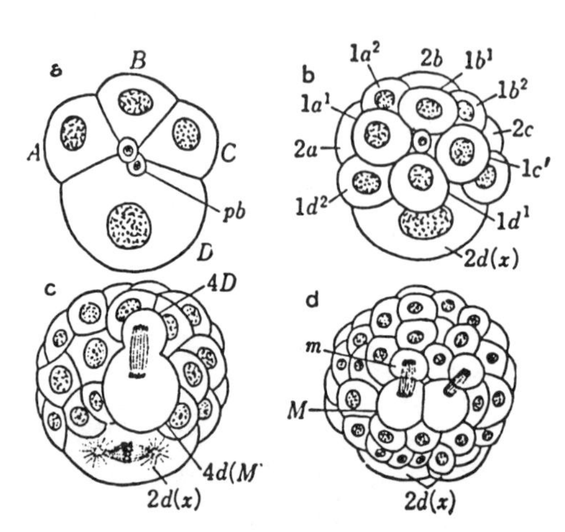

Fig. 11.7 Cleavage in *Dreissensia polymorpha* (Meisenheimer)

a. 4-cell stage; b. 16-cell stage; c. 46-cell stage; d. mesoblasts (*M*) cleaving to form mesomeres (*m*). *2d(x)*: primary somatoblasts

Following the separation of the second quartet (*2q*), the cells of *1q* divide, forming the 16-cell stage. This cleavage is also sinistral, producing $1a^1$—$1d^1$, which occupy the animal pole region, and the four blastomeres, $1a^2$—$1d^2$, which lie directly above the macromeres, held between the blastomeres of the *2q* quartet. These four cells ($1a^2$—$1d^2$) are called the *primary trochoblasts* or *turret cells,* and later produce the *velum.* In *Trochus* these cells continue to cleave, but in *Fiona* their cleavage is suspended until the 60-cell stage.

As the embryo as a whole proceeds from the 16-cell to the 32-cell stage, the cleavage of each blastomere takes place in a dextral direction, although these cleavages are not simultaneous. *Trochus,* for example, passes through a 20-cell stage (cleavage of the four macromeres); a 24-cell stage (cleavage of *2q*); a 32-cell stage (cleavage of *1q*). In the embryo of this stage the blastomeres are regularly arranged so that four of the daughter cells ($1a^{11}$—$1d^{11}$) derived from the cleavage of $1a^1$—$1d^1$ occupy the animal pole, while the positions of the other four ($1a^{12}$—$1d^{12}$) lie alternately with these, forming the "molluscan cross" (Fig. 11.6a). When $1a^{11}$——$1d^{11}$ divide again, they form $1a^{111}$—$1d^{111}$, the *apical cells* or *rosette cells,* and $1a^{112}$—$1d^{112}$, the *peripheral rosette cells* (Fig. 11.6c). The peripheral rosette cells correspond to the "annelidan cross" in annelid embryos. The *tip cells* at the ends of the crossing arms in molluscan cleavage are formed from the $2a^{11}$—$2d^{11}$ cells rather than from the first quartet.

Gastropod embryos maintain a radial symmetry as far as this stage. They are rather flattened, and a small blastocoel can be recognized inside the cell mass. On the surface at the vegetal pole the *polar furrow* is clearly seen, forming, together with the polar bodies, a good indicator of the embryonic axis. In the course of the three macromere cleavages leading to the formation of the 32-cell stage, the ectodermal cells which will form the *ectoblast* have all been separated off from the macromeres; the macromeres (*3A*—*3D*) now contain all the material of the *entoblast* and most of the *mesoblast* material. (A small part of the mesoblast is contained in the *2q* or sometimes the *3q* cells).

Because of the appearance in the Pelecypoda of the large *2d* blastomere and its subsequent cleavage, the transition from radial to bilateral symmetry becomes evident sooner than in the Gastropoda. Since such a size difference exists among pelecypodan blastomeres within the quartets, and their cleavage does not occur simultaneously, the cells making up the embryo are not arranged in regular order, and the rosette and molluscan crosses do not appear. The organs derived from each quartet, however, are the same as in the Gastropoda, indicating that the difference observed in blastomere arrangement is not a fundamental one.

The distribution of the blastomeres among the chitons is extremely similar to that of the gastropods, and the rosette and molluscan crosses are seen.

In the Dentaliidae, as in the Pelecypoda, the *2d* cell is a large blastomere which in both groups is called the *first somatoblast* and given the designation of *X*.

b. Differentiation of ento-mesoblast

All the organ systems of an ectodermal nature are differentiated from the first, second and third quartets of micromeres; from the macromeres (*3A—3D*) arise all the organs of an endodermal nature and practically all the mesodermal organs. Furthermore, from the *2q* (*Crepidula*) or *3q* cells (*Trochus, Fiona, Physa*), besides the ectodermal organs, the *secondary mesoderm*, or *ecto-mesoderm*, is formed.

The division of the *3D* macromere has marked distinguishing characteristics, taking place in a sinistral direction, and producing the *4d* daughter cell which is larger than its sister, the *4D* macromere, and eventually sinks into the blastocoel. The *4d* blastomere is called the *primordial mesoderm cell* or *mesoblast*; since part of the entoderm as well sometimes differentiates from it, it may also be called the *mesentoblast*; or it may be described as the *second somatoblast*. It may be designated by the letters *M* or *ME*. In *Crepidula* and *Fiona*, the cleavage of the *3D* blastomere takes place immediately after the embryo reaches the 24-cell stage, preceding the cleavage of *3A*, *3B* and *3C*. In the gastropods which form equal-sized blastomeres at the 4-cell stage, it is at first impossible to distinguish between *A*, *B*, *C* and *D*, but at the 24-cell stage, for the first time, the correct designations can be assigned to each blastomere. In *Trochus*, the formation of *4d* occurs much later, in the 64-cell stage, lagging behind the cleavage of *3A*, *3B* and *3C*.

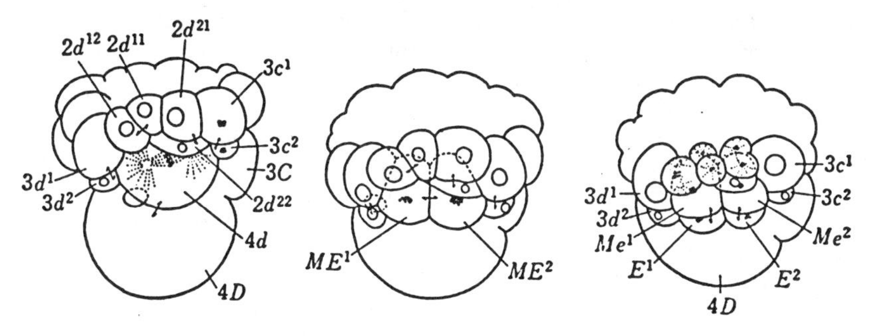

Fig. 11.8 Cleavage in *Ilyanassa* (Clement)
Formation of *4d*-cell and its successive cleavages.

These three blastomeres also cleave very unequally, giving rise to the "micromeres" *4a*, *4b* and *4c* which are larger than the "macromeres" *4A*, *4B* and *4C*. These micromeres are sometimes called *entomeres* or *secondary macromeres*, and together with the micromeres of 5q they form the entodermal organs (digestive system). The *4d*=*ME* cell, which has moved into the blastocoel, very soon divides equally and the paired daughter cells take a position in contact with the embryonic body wall to the left and right of the median plane. These cells ME^1 (right) and ME^2 (left) function as the teloblasts of the *mesodermal bands* which constitute the rudiments of the mesoderm.

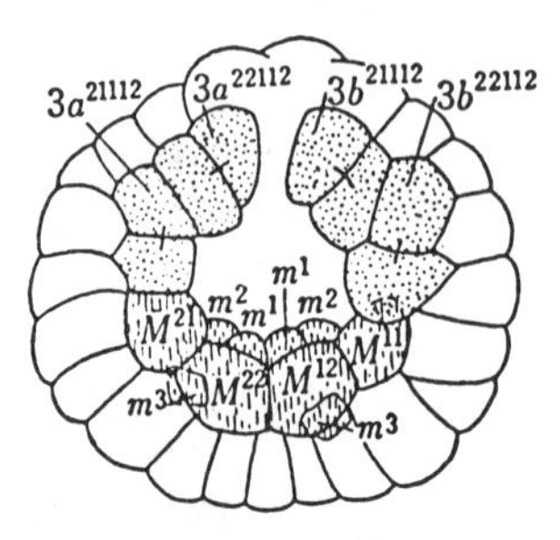

Fig. 11.9 Embryo of *Physa fontinalis*, showing formation of mesoderm (Wierzejski)

Blastomeres of ecto-mesoderm ($3a^{211}$, $3a^{221}$, $3b^{221}$, $3b^{211}$) are stippled. Descendants of mesoblast (M^{12}, M^{22}, m_1, m_2, etc.) are hatched.

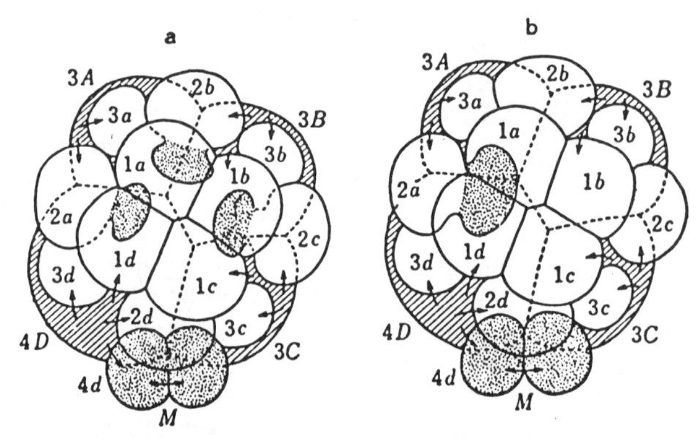

Fig. 11.10 Sketches of embryos of (a) *Crepidula* and (b) *Unio*, showing segregation of ecto-mesoderm (Wilson)

The stage at which the mesentoblast cells separate into mesoblast and entoblast, as well as the details of the process, differ rather widely from one species to another. If we take the opisthobranch *Fiona* as an example (Casteel 1904), $4d = ME$ arises from $3D$ at the end of the 24-cell stage; the next cleavage of $4d$ takes place in the latter part of the 44-cell stage, the resulting ME^1 and ME^2 cleaving together after the 68-cell stage and giving rise to the Me^1 and E^1 and Me^2 and E^2 blastomeres, respectively. E^1 and E^2 are called *primary enteroblasts*. These blastomeres do not cleave for some time, but the Me^1 and Me^2 cells divide very soon (at the 88-cell stage), forming m^1z^1, M^1e^1 and m^2z^2, M^2e^2, respectively. M^1e^1 and M^2e^2 divide at the 128-cell stage, producing M^1, e^1, M^2 and e^2 . e^1 and e^2 are designated the *secondary enteroblasts,* and they unite with the earlier-formed pair, E^1 and E^2, to form a group of four cells. M^1 and M^2 are called the *mesoblastic teloblasts*; eventually they divide to give rise to two irregular rows of mesoderm cells. m^1z^1 and m^2z^2 cleave immediately after M^1e^1 and M^2e^2, forming m^1, z^1, m^2 and z^2. m^1 and m^2 follow the same course as M^1 and M^2, joining the latter in forming the rudiment of the mesoderm. It is said that in *Fiona* the enteroblast cells which arise from the mesentoblast form most of the gut, while the z^1 and z^2 cells take care of forming its anteriormost part.

In *Crepidula* (Conklin 1897), the segregation of the entoblast takes place earlier than in *Fiona*. The above-mentioned Me^1 and Me^2 cells divide, producing the cells m^1 and m^2 which contain only the mesodermal constituents of the parent cells. In any case, facts of this sort serve to indicate the close connection which exists between mesoderm and entoderm.

There are some animals in which the segregation of the mesoblast takes place even earlier. Lillie (1895) concluded that in *Unio* the cells derived from the $4d$ blastomere are exclusively mesodermal in nature. Meisenheimer (1901) makes the same statement with respect to *Dreissensia*. Also in the pulmonate *Planorbis*, the cells derived from the $4d$ blastomere are said to be all mesodermal (Holmes 1900).

c. Ecto-mesodermal Cells

It was Lillie, studying the pelecypod *Unio* in 1895, who first discovered that among the Mollusca, in addition to the mesoderm which originates from entodermal cells, there also exists mesoderm of ectodermal origin. Lillie found that one cell derived from the cleavage of the 2a micromere of the second quartet falls into the blastocoel, and the cells that it later produces form the muscular system of the glochidium larva. Since then, observations of the same sort have been made in various molluscan species. This ectodermal mesoderm is called *ecto-mesoderm* or, since it takes part only in the formation of larval organs, *larval mesoderm*.

Conklin (1897) showed that in the prosobranch *Crepidula*, the ecto-mesoderm has its origin in three of the *2q* blastomeres (*2a, 2b, 2c*). In the opisthobranch *Fiona*, this rudiment can only be seen after gastrulation has already begun. The cells $3a^{211}$ and $3a^{221}$, and also $3b^{211}$ and $3b^{221}$, divide as they sink into the blastocoel; $3a^{2112}$, $3a^{2212}$ and $3b^{2112}$, $3b^{2212}$ are pushed out toward the blastopore, while the large daughter cells are gradually covered over by the growing ectoderm. These cells, $3a^{2111}$, $3a^{2211}$ and $3b^{2111}$, $3b^{2211}$ are the rudiment of the ecto-mesoderm, later dividing and each forming a band composed of several cells which lie at first in the anterior part of the gastrula, but are later found in the head region of the larva (Casteel 1904). In the pulmonate *Physa* also, completely identical cells form the ecto-mesodermal rudiment (Wierzejski 1897). In *Planorbis*, again, eight cells (2111, 2112, 2211 and 2212 of *3d* and *3c*) are said to be the ecto-mesomeres (Holmes 1897).

The fact that in the Mollusca the ecto-mesoderm is seen only during the larval period, disappearing at the time of metamorphosis, is highly interesting from the phylogenetical point of view.

(4) BLASTULA TO TROCHOPHORE LARVA

a. Blastula

Among the various molluscan eggs, blastocoel formation takes place early in those with relatively little yolk, but in general, on account of the large size of the *entomeres* (entoderm-forming blastomeres, entoderm cells) which lie at the vegetal side, the blastocoel is very small in comparison with those of echinoderm and other blastulae. This is particularly true of eggs with a large amount of yolk, which form typical stereoblastulae.

As it begins the blastula stage, the embryo develops cilia at certain regions of its body surface. The *apical cilia* are formed on the *apical cells* at the animal pole, and clumps of cilia resembling those of ctenophore comb-plates are formed on the primary trochoblasts at four points on the sides of the blastula (see Fig. 11.6a, c, e, f). As development proceeds, certain cells in the vicinity of the primary trochoblasts (these are called *secondary trochoblasts)* also produce cilia, forming a complete ring of cilia, which is called the *prototroch*. In many species, especially among the Pelecypoda, the apical cilia develop to become very prominent, to the extent that they are often described as a 'tuft of flagella'. These generally persist through the veliger stage, but in *Crassostrea* they are absent from the beginning, and in the pearl oysters they disappear early in the veliger stage.

After the development of these cilia the embryos begin to swim gracefully through the water or move slowly within the egg case. In species in which the eggs

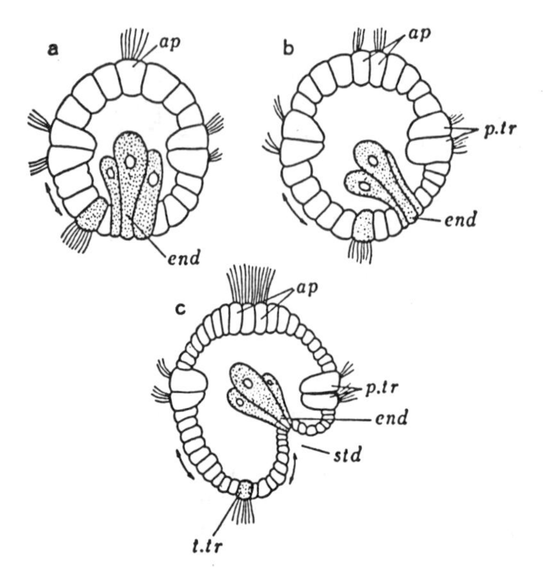

Fig. 11.11 Longitudinal sections of early larvae of *Patella coerulea*, showing shift in position of blastopore and formation of stomodaeum (MacBride)

ap: apical cell, *end:* endoderm, *p.tr:* prototrochal cell, *std:* stomodaeum, *t.tr:* telotrochal tuft.

are surrounded by a chorion, the embryo can be seen to escape through the so-called micropyle. The larvae of oysters, common mussels and many other pelecypods begin their swimming life with the vitelline membrane intact, and gradually shed it later, bit by bit.

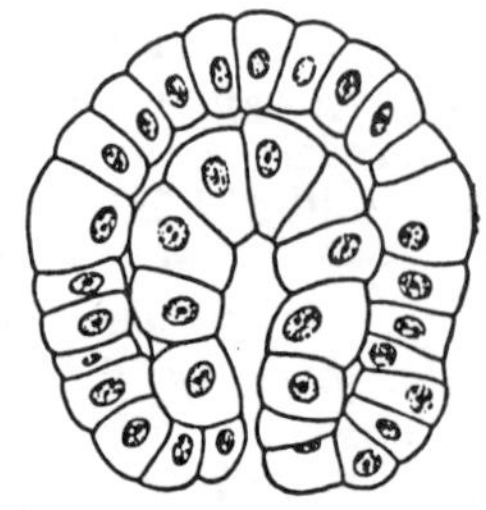

Fig. 11.12 Gastrula of *Paludina vivipara* (Dautert)

b. Gastrulation and Gastrula Stage

Gastrulation takes place in molluscan embryos by emboly as well as by the more common method of epiboly. In typical stereoblastulae like those of *Crepidula,* typical epiboly can be seen (Fig. 11.13), *Paludina* shows typical emboly

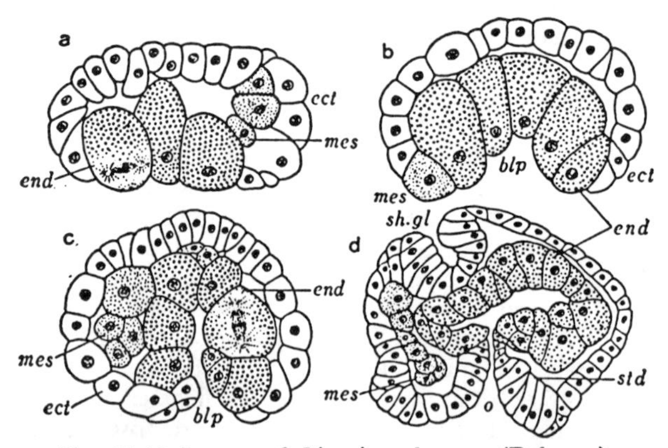

Fig. 11.13 Larvae of *Littorina obtusana* (Delsman)

a. blastula; b,c. gastrulae; d. larva with shell gland. *blp:* blastopore, *ect:* ectoderm, *end:* endoderm, *mes:* mesoderm, *o:* mouth, *sh.gl:* shell gland, *std:* stomodaeum.

(Fig. 11.12), and there are many species which follow a process that might be called intermediate between these two. As early as the 30-cell stage in *Patella,* some of the cells of the vegetal pole region migrate into the blastocoel; this is thought to consitute the beginning of gastrulation (MacBride 1914). In general, however, it begins later than this.

From about the beginning of gastrulation, the cells arising from the 2*q*, especially those derived from 2*d*, proliferate actively in the dorsal part of the embryo; as a result the entodermal cells (entomeres), which until this time have been located near the vegetal pole, are shifted toward the ventral side. The embryo thus becomes rather flattened dorso-ventrally, and with the invagination of the ectoderm cells to form the *shell gland,* assumes a well defined bilateral symmetry.

Invagination in the pelecypod *Dreissensia* occurs by a process of emboly, while in the oysters and many other marine bivalves the blastula lacks a blastocoel, and invagination begins as an overgrowth (epiboly), but very shortly, proliferation of the ectoderm cells causes a blastocoel to form, and the mode of invagination changes over to one of emboly.

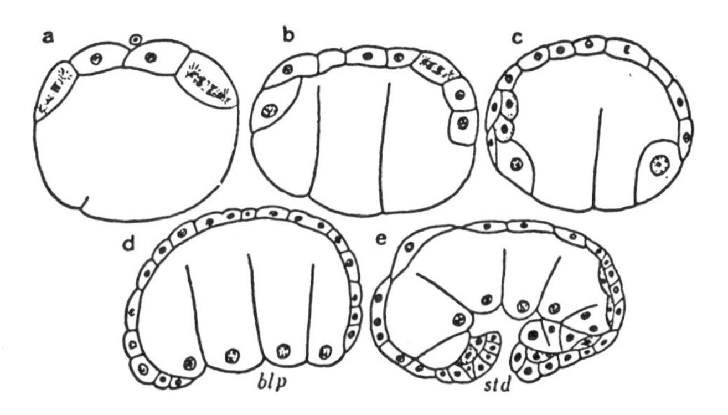

Fig. 11.14 Gastrulation by typical epiboly in *Crepidula fornicata* (Conklin)

At first a wide blastopore is seen as the result of entomere invagination, but later this is gradually reduced to a long, narrow longitudinal slit by the proliferation of the 2*q* and 3*q* cells, and in many species is finally closed altogether. The cells which participate in this process in *Crepidula* and *Ischnochiton* are those derived from 2*q* and 3*q*; in *Trochus* and *Planorbis* they are chiefly from 3*q*; while in *Fiona* they are said to be almost exclusively from 3*a* (Conklin, Heath, Robert, Holmes, Casteel).

At this period the posterior gastrular surface is occupied by cells of the 3*q* line, while the central part is composed of descendants of the 2*q* blastomeres.

The segregation of the ecto-mesoderm cells described above also takes place about the time of the reduction of the blastopore *(Fiona)*. In the Gastropoda a large cell which will form the excretory organ of the future larva can be seen on the right side in the posterior part of the embryo.

At the time when the archenteron becomes clearly defined, the cells composing it (in *Fiona,* for example) are the macromeres 5*A*, 5*B*, 5*C* and 4*D*, lying farthest inside; next to them are the 5*a*, 5*b* and 5*c* cells, and above these, $4c^2$, $4b^2$, $4a^2$ and

$4c^1$, $4b^1$, $4a^1$ lie directly below the ectoderm. In the posterior part of the body can be seen the enteroblasts E^1, E^2, e^1, e^2, which have arisen from $4d$.

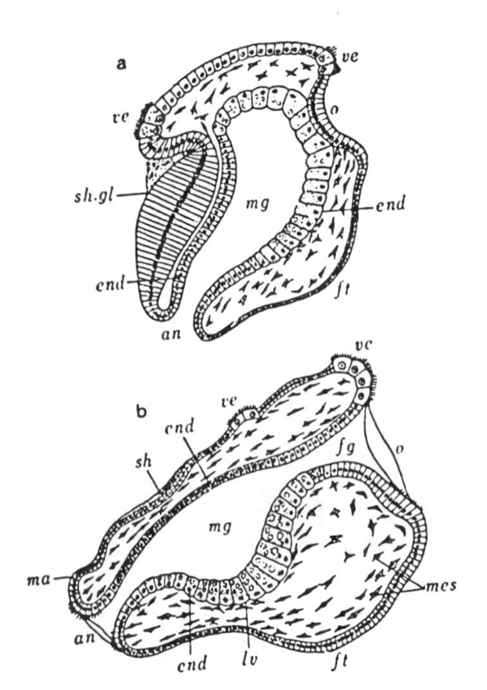

Fig. 11.15 Two sections of *Paludina vivipara* embryo (Tönniges)

Notice that the blastopore changes to become the anus and the mouth is newly formed, *an:* anus, *end:* endoderm, *fg:* fore-gut, *mg:* mid-gut, *o:* mouth, *sh:* shell, *sh.gl:* shell gland, *ve:* velum.

Accompanying the development of the archenteron and the contraction of the blastopore, the ectoderm around the latter gradually invaginates, forming the stomodaeum in this position. That is to say, the future mouth occupies the position of the blastopore, but among the Gastropoda there are some species in which the blastopore is closed and a new mouth opening is formed (*Crepidula, Trochus, Fiona, Nassa*). In others the blastopore remains open and changes over to a mouth (*Patella, Limnaea,* the Pteropoda and the Heteropoda). In this connection, *Paludina* (Fig. 11.15) forms a noteworthy exception, as has been observed and confirmed by many workers, in that the blastopore changes into the anus, and a new mouth is formed in a different position (see Korschelt u. Hieder 1936, p. 876).

(5) TROCHOPHORE AND VELIGER STAGES

With the exception of the Cephalopoda, molluscan larvae develop from the gastrula stage into a form which closely resembles the annelidan trochophore larva; this also is called a *trochophore*. In the molluscs the prototroch of the trochophore larva eventually develops prominently to form an organ of locomotion called the *velum,* a shell also forms and the larva takes on the characteristics of the *veliger stage.*[4] No drastic change in body form is involved in the shift from trochophore stage to veliger, and the boundary line drawn between the two stages varies from one investigator to another. The present description will treat the period lasting until the larval shell becomes a prominent organ as the trochophore stage. Moreover, in this section mainly the developmental processes common to all the species will be described, postponing the phenomena peculiar to the several species until the next section. The chief points in which the molluşcan trochophore differs from that of the annelids are in the development of the shell gland, the growth of the *foot,* and the formation of a *radula sac* (although this does not appear in pelecypod larvae).

---

[4] "Veliger" means "possessing a velum (sail)". Since the veliger larva generally has a well-developed shell, it is also called "shelled larva". Lang (1896) describes a veliger larva as "a Trochophora with Molluscan characteristics".

a. Shell Gland

Some time after the invagination of the entoderm, the cells of the opposite (dorsal) side of the body proliferate vigorously and invaginate toward the blastocoel (Fig. 11.13d), forming the shell gland. In this stage the embryo thus shows two invaginated parts, the shell gland dorsally and the archenteron on the ventral side. The shell gland invagination subsequently disappears, and the flattened tissue of the gland begins to secrete the larval shell. In *Amphibola* (Pulmonata) it has been observed (Farnie 1924) that the shell gland begins to secrete without invaginating, but in other gastropods and the pelecypods, invagination invariably precedes shell secretion. According to Raven (1952), in *Limnaea* the invaginated shell gland cells maintain very close histological contact with the epithelial cells of the midgut; he believes that the shell gland is induced by contact between the gut and the ectoderm, but he has found no experimental proof to support this hypothesis. Finally the shell gland is everted, causing a slight bulging of the body wall. As a result of this eversion, the hitherto practically solid embryo acquires a large body cavity.

The shell gland is formed of cells derived from the $2d$ blastomere. According to Holmes (1900), in *Planorbis* they are derivatives of $2d^{12}$ and $2d^{21}$.

b. Foot, Pedal Gland, Statocyst

One of the characteristics of the molluscan trochophore is that, as development proceeds, the anus moves toward the mouth; between the two — i.e., between the posterior side of the blastopore and the anus — the foot is formed. In *Patella* (Patten), *Fulgur* (McMurrich), *Trochus* (Robert) the foot first appears as a pair of protuberances, but in general it is first recognized as a single projection. There are some differences among species with respect to which cells form the foot: in *Unio* and *Crepidula* it develops from the $2q$ cells; in *Planorbis* and *Trochus*, from the $3q$ cells; in *Fiona* the derivatives of both these quartets are involved.

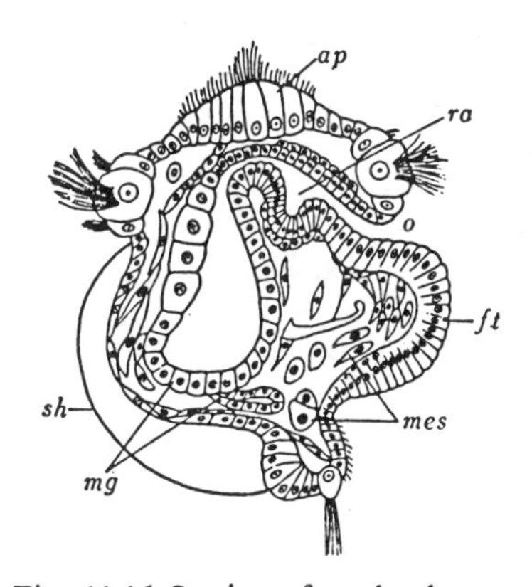

Fig. 11.16 Section of trochophore larva of Patella (Patten)

ap: apical plate, *ft:* foot, *mes:* mesoderm, *mg:* mid-gut, *o:* mouth, *ra:* radula sac, *sh:* shell.

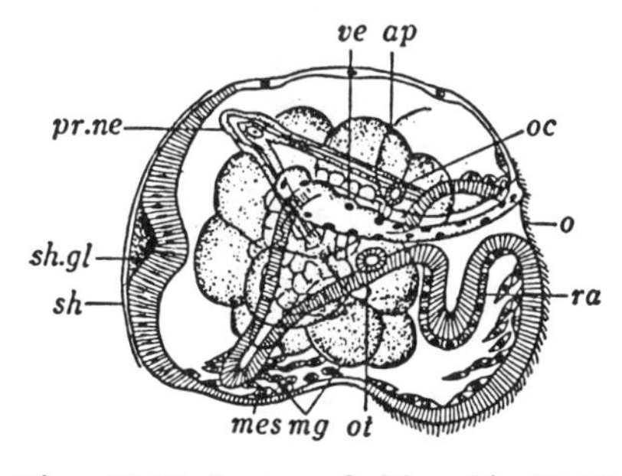

Fig. 11.17 Larva of *Planorbis* (Rabl)

ap: apical plate, *mes:* mesoderm, *mg:* mid-gut, *o:* mouth, *oc:* ocellus, *ot:* otocyst, *pr.ne:* protonephridium, *ra:* radula sac, *sh:* shell, *sh.gl:* shell gland, *ve:* velum.

In the Gastropoda the development of the foot lags somewhat behind that of the shell gland, but rather early in the late trochophore stage the *operculum* is secreted on the lower surface of the foot. The formation of an operculum is peculiar to the gastropods, and even when the adult form lacks an operculum, one develops in the larva. In *Paludina* the foot forms in front of the blastopore, posterior to the point where the future mouth will appear. In the Pelecypoda, the foot undergoes practically no development during the trochophore stage.

From the Amphineura to the Pelecypoda, all the Mollusca show an invagination of ectodermal cells to form a *pedal gland* on the posterior part of the foot (anterior in the Amphineura). On the two sides of the base of the foot there are also ectodermal invaginations which give rise to the *statocysts*. The statocysts later receive innervation from the brain ganglia; they develop earlier than these ganglia (Gastropoda) and near the statocysts the *foot ganglia* differentiate from ectoderm cells.

c. **Radula Sac**

The possession ofa *radula* is one of the most distinctive features of all the Mollusca except the Pelecypoda, the rudiment of the *radula sac* appearing immediately after the invagination of the stomodaeum and constituting one of the distinguishing characteristics of the trochophore stage (although it never appears during the whole life of the pelecypods). In general the rudiment of the radula sac develops from cells of the stomodaeum which grow and form a pocket extending toward the outside of the stomodaeum posteriorly (Fig. 11.16), but according to Wierzejski (1905), in *Physa* it appears as a pair of invaginations behind the mouth but unrelated to the stomodaeum.

d. **Velum**

As the trochophore larva approaches the veliger stage, its prototroch undergoes a striking growth; this expanded prototroch is then called the *velum*. In *Ischnochiton* and *Trochus*, the prototroch is formed by the trochoblast cells, the tip cells of the molluscan cross and others which are derived from $1q$ and $2q$ (with the exception of the D quadrant). In such Amphineura and primitive Gastropoda the prototroch, and, in its turn, the velum, are more radially symmetrical than those of the members of the other species. Moreover, the free-swimming period of these larvae is extremely short (several hours to several days), and as a consequence the velum is not highly developed. In the other Gastropoda and the Pelecypoda, the free-swimming period is prolonged, and the velum is well-developed. A typical velum is composed of two ciliated rings: one anterior to the mouth called the *preoral velum* and the other which develops posterior to the mouth, the *postoral velum*. In *Crepidula* and *Planorbis*, it is said that only $1a^{22}$, $1a^{21}$, $1b^{22}$, $1b^{21}$ among the prototroch cells take part in forming the preoral velum. In addition, the tip cells of the anterior arm and the lateral arms of the molluscan cross enter into the formation of the preoral velum in *Crepidula*, while the postoral velum is mainly composed of $2q$ cells, to which some $1q$ and $3q$ cells are added.

(6) **LARVAL EXCRETORY ORGAN**

Molluscan larvae have two main kinds of excretory organs. One of these is a *protonephridium* of the same sort as that found in the annelidan trochophore;

this appears during the larval period in the Pelecypoda, and among the Gastropoda, in the Pulmonata and freshwater Prosobranchia *(Paludina)*.

These protonephridia occur in each case as paired organs lying ventrally in the body on the two sides of the foot. In the pulmonate group Bassomatophora, this organ is formed of four cells; one end opens to the outside through an opening in the body wall, while the other end, consisting of a ciliated flame-cell *(solenocyte)*, connects with the primary body cavity. In the wall of the protonephridium one excretory cell with a very large nucleus can be seen (Fig. 11.18). In the Stylommatophora a larger number of cells participate in the formation of the protonephridium. Among the Pelecypoda, it is a straight organ composed of two or three cells, but the structure is similar to that in the Pulmonata (Fig. 11.20).

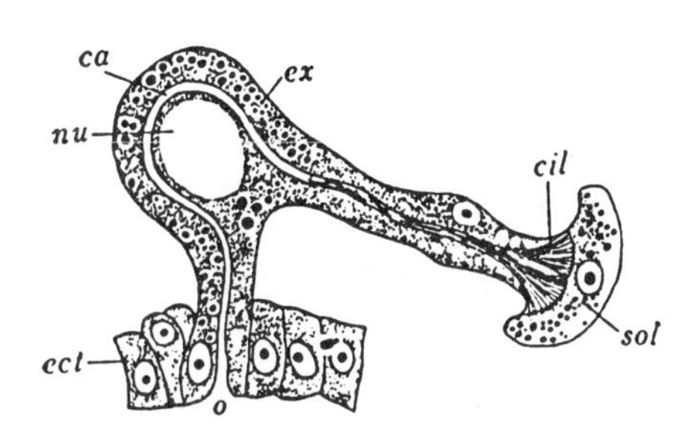

Fig. 11.18 Protonephridium of the Bassomatophora (Meisenheimer)

ca: protonephridial canal, *cil:* cilia, *ect:* ectoderm, *ex:* excreting cell, *nu:* nucleus of excreting cell, *o:* opening, *sol:* solenocyte.

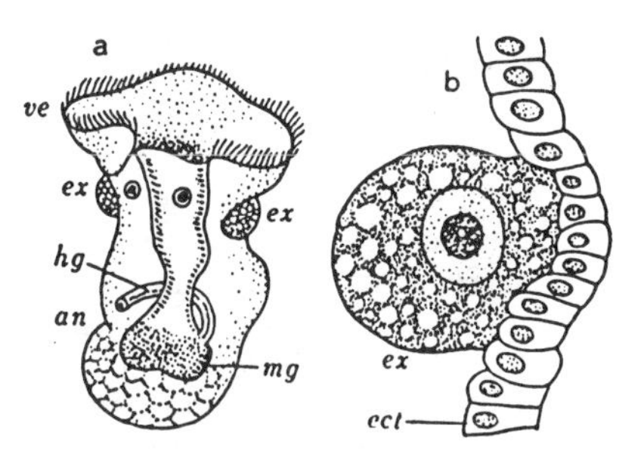

Fig. 11.19 Excretory cells of marine gastropod larvae (Portman)

a. larva of *Buccinum undatum*; b. enlargement of larval body wall showing excretory cell. *an:* anus, *ect:* ectoderm, *ex:* excretory cell, *hg:* hind-gut, *mg:* mid-gut, *ve:* velum.

Various different opinions have been expressed concerning the question of whether the protonephridium in these species is derived from the ectoderm or from the mesoderm. Concerning the Bassomatophora, Fol (1875, 1880), Wolfson (1880) and Meisenheimer (1899) maintain that it is of ectodermal origin, while Rabl (1879), Holmes (1909) and Wierzejski (1905) describe it as arising from the mesoderm. In the Stylomatophora, the observations of Meisenheimer (1898, 1899) seem to show clearly that its origin is ectodermal, but according to more recent studies it appears to be only the part near the epidermis which derives from the ectoderm, while the main inner portion is mesodermal (Fernando 1939, Carrick 1939). With respect to the pelecypod, protonephridium Meisenheimer (1901) describes it as differentiating from the ectoderm, but MacBride (1914) claims that it is probably of mesodermal origin. In the Amphineura, the marine Prosobranchia and Opisthobranchia (i.e., most of the Gastropoda) and the Scaphopoda, no protonephridium appears. Instead, in the marine Prosobranchia and Opisthobranchia, at each side of the foregut posterior to the velum, a pair of special organs consisting of several cells or one exceptionally large cell protrude from the body wall (Fig. 11.21). This organ is observed in the larvae of many of the gastropods which do not have protonephridia, and is considered to be derived from the ectoderm. It does not have an excretory tubule, but performs this function by receiving waste material from the free cells in the body cavity (Portmann 1930).

In addition to the above, in some gastropods, the so-called *nuchal cells* differentiate in the neck region. These organs are believed to have an excretory function (Wierzejski 1905; Bloch 1938); they are also called *cephalic kidneys*. Opinion is again divided as to whether these are of ectodermal or mesodermal origin. In the prosobranch *Fiona*, an "*anal kidney*" differentiates from the ectoderm above the anus, in the posterior part of the trochophore larva and the right side of the veliger.

(7) HEAD VESICLE

The part of the larva anterior to the ciliary ring of the velum is called the *head vesicle*. In this region the *apical organ* develops, as well as the ganglia and other nerves, the eyes and the other sense organs. In the Gastropoda, particularly the pulmonate Stylomatophora, it becomes extremely prominent.

The head vesicle is chiefly formed by the cells of the 1q, but in some cases the *2q* cells may be added to these. An *apical plate* occupies the anteriormost part. The apical plate is usually formed by the four apical cells $1a^{111}$—$1d^{111}$, although it is said that in *Planorbis* and *Physa*, two or three other cells of unknown origin also take part in this formation (Holmes 1900; Wierzejski 1905). A large number of cilia are produced on the apical plate. However, according to Casteel (1904), *Fiona* fails to form a distinct apical plate and no cilia appear. Apical cilia are similarly not seen in *Crassostrea*.

The lateral ectoderm of the apical region proliferates and gives rise to the rudiments of a pair of *brain ganglia*. In the pulmonates, to the left and right of the apical plate *cerebral plates* are formed and surrounded by apical cells (said to be derivatives of $1q^1$ in *Planorbis)*; the proliferation of these cerebral plates produces the brain ganglia. It therefore follows that the brain ganglia are of ectodermal origin, but Wierzejski (1905) states that mesoderm cells also take part in their formation. However, according to Raven (1952), some of the cells which are produced internally by the proliferation of the ectoderm develop into ganglia, while others become mesenchyme cells, corresponding exactly to the mesectoderm of vertebrate embryos, and as a result, cannot be invoked to settle the debate concerning the ectodermal or mesodermal nature of the ganglia.

In Pteropods such as *Vermetes, Helix* and *Limax,* the rudiments of the brain ganglia arise, not from a simple proliferation of ectoderm cells, but from an ectodermal invagination which appears as tubular tissue.

Among the Amphineura the rudiment of the central nervous system can be recognized as a thread-like tissue formed internally as the result of proliferation of the rosette cells (Hammarsten and Runnström 1925).

According to Meisenheimer, in *Dreissensia* the rudiment of the cerebral ganglia appears as a bilobate cerebral pit; from the bottom of this depression are separated off a pair of nerve cell clumps which become the ganglia and the commissure which connects them.

(8) DIGESTIVE SYSTEM

As has been described above, the rudiment of the digestive tract is formed by the invagination of the *4q* (except *4d*), the *5q* and the macromeres *(5A — 5D)*. As the result of the unequal proliferation of the body-wall cells, the blastopore, which was first in the posterior part of the body, moves in a ventral direction, becoming very small or entirely closed, and in its place a stomodaeum is newly formed by an ectodermal invagination. This makes a connection with the archenteron (midgut) and differentiates into the *foregut (oesophagus)*. In *Limnaea* some of the archenteron cells absorb the albumen of the egg case and become very large *albumen cells* (Raven 1952). The midgut differentiates into the *stomach, intestine* and *digestive diverticula* (also called *midgut glands*, or "*liver*"). The intestine is first recognized as a row of cells which do not form a tube (as in *Fiona*) or as an evagination of the archenteron which is tubular from the start *(Crepidula)*. The entoderm cells (entomeres) derived from the 4*d* blastomere take part in its formation.

The anus later opens at the place where the intestine comes into contact with the ectoderm. At this time, it is said that in *Limax* and *Dreissensia* a very slight ectodermal invagination — i.e., the *proctodaeum* — can be recognized (Meisenheimer 1898, 1901), but according to Wierzejski (1905) the hindgut arises from the primary mesoderm in *Physa*, and in *Limnaea* the hindgut is said to be formed by the growth of the midgut epidermis (Rabl 1879; Fol 1880; Bloch 1938; Raven 1952).

The digestive glands (digestive diverticula or "liver") arise from the left and right sides of the anterior part of the midgut — near the junction of oesophagus and midgut — as a pair of foliated organs having a lumen. The rudiment of the midgut gland usually appears before the re-opening of the blastopore, but its development is markedly retarded in embryos containing a large amount of yolk. Especially striking is the case of such forms as *Buccinum*, which obtain their energy for development from so-called *nurse eggs;* since the cells of the midgut in these species contain the yolk of the nurse eggs, the differentiation of the midgut is postponed until the veliger stage.

The pelecypods and plankton-eating gastropods posses a *crystalline style;* the *style sac* which secretes this differentiates fairly early in the veliger stage from the posterior part of the stomach as a blind tube.

(9) MESODERMAL BANDS

It was stated earlier that the 4*d* blastomere sinks inside the blastula and divides, the pair of daughter cells taking a bilaterally symmetrical position and giving rise to the mesoderm. These two cells are in contact with each other at the median plane (ME^1, ME^2); at first they lie on the dorsal side of the posterior part of the embryo, but as the blastopore shifts in a ventral anterior direction, they are carried to the ventral posterior region, and the cells which are derived from them unite to form bands. These are called the *mesodermal bands;* since the cells which gave rise to them lie at one end and produce them root and branch, these cells are called *telo-mesoblasts*.

This mode of *ento-mesoderm* formation is strikingly similar to that of the Annelida, but in the Mollusca the ento-mesoderm does not develop further until the beginning of metamorphosis, and before forming a coelomic cavity the mesodermal bands break up into individual cells and become the so-called *mesenchyme*. Moreover, the tissues of the larval musculature and other organs are chiefly formed by the ecto-mesoderm, which will be described in the next paragraph.

(10) ECTO-MESODERM AND LARVAL MUSCULAR SYSTEM

As described above, the mesoderm which has its origin in the ectoderm is called *ecto-mesoderm;* these cells are believed to become mesenchyme and form the *larval musculature* which degenerates at the time of metamorphosis.

The main components of the larval musculature — the *retractor muscles* of the velum, mantle, digestive tract and foot — are formed of spindle-shaped cells. In pelecypod veliger larvae, three sets of retractor muscles develop; each of these has an attachment point on the left or right shell near the hinge, while its other end ramifies on the velum, mantle, etc. (*Dreissensia;* Meisenheimer 1901). The action of these muscles draws the soft parts of the larval body inside the shells. In addition, an *anterior adductor muscle* develops early in the pelecypod veliger stage, and is used for closing the shells. This adductor muscle persists through metamorphosis and changes over into the adult organ.

In the gastropod veliger, taking *Fiona* as an example, a large retractor muscle appears, having its starting point in the posterior dorsal part of the body and attaching to the body wall near the digestive tract and oesophagus, while a further pair of muscles arise at the left and right sides, posteriorly, and each member ramifies to the foot. Besides these there can be seen highly branching muscles which extend from the neck region and the oesophagus to the velum (Casteel 1904).

Not all these parts of the larval musculature are thought to be of ecto-mesodermal origin. For example, it is said that at least the bilateral pair of pedal retractor muscles are derived from primary mesoderm (ento-mesoderm) cells.

(11) EXPERIMENTAL EVIDENCE RELATING TO CELL LINEAGE

From the many facts presented above which have been gathered from observation of normal development, it is possible to imagine that the various future organs which will form from the several blastomeres are determined in molluscs in the early cleavage stage; these facts have also been confirmed by a number of experiments. Beginning with the famous work of Wilson (1904) on the determination of the blastomeres in *Dentalium* and *Patella,* there are the experiments of Conklin (1902, 1912, 1917) on *Crepidula;* and on *Ilyanassa,* the early work of Crampton (1896) and the recent studies of Clement (1952). Still more recently Berg (1954) has published a study of the same sort on *Mytilus.* As has already been described, a polar lobe is formed in the first three cleavages of *Dentalium.* If the first polar lobe is removed in the trefoil stage, the trochopore developing from the remainder of the egg completely lacks the apical organs belonging to the anterior part of the body; furthermore, the post-trochal region — i.e., the portion of the body posterior to the prototroch (referred to temporarily as "the posterior part") fails to develop. If the polar lobe of the second cleavage is removed, the posterior part is small but in most cases a larva with apical organs is formed. Larvae lacking this polar lobe do not metamorphose, they fail to develop the foot, shell gland, shell, mantle and pedal ganglia, no mouth is seen and they are believed to lack a coelom and mesoderm cells. If the blastomeres are separated at the 2-cell stage, the *AB* blastomere develops into a trochophore with a small posterior part and lacking apical organs — in other words, the larva is like that formed when the first polar lobe is removed, as described above. On the other hand, from the *CD* blastomere a larva develops which is no different from the normal with respect to the size of these parts, and consequently has, for a half-larva, very large apical organs and posterior part. If the blastomeres are separated at the 4-cell stage, larvae lacking apical organs and posterior part develop from blastomeres *A*, *B* and *C*, while these structures are abnormally large in the larvae formed from isolated *D* blastomeres. Again, the larvae obtained from isolated 1*q* micromeres all lack a posterior part, and only in the larva produced from the 1*d* blastomere are apical cilia formed.

Wilson observed in the normal development of *Dentalium* that the material making up the polar lobe is chiefly shifted into the 2*d* blastomere (first somatoblast), and very probably a part of this is transferred to the 4*d* (second somatoblast) and 3*d* blastomeres; he further showed that the posterior part of the trochophore develops from these two somatoblasts. It is therefore possible to anticipate the same result as that seen in the experiments cited above: that the posterior part will not develop in blastomeres which do not contain polar lobe material; but it is very interesting that removal of the polar lobe leads to the loss of the apical organs which lie on the opposite side of the larva.

The egg of *Patella* belongs to the type which does not form a polar lobe; isolated 2-cell stage blastomeres both develop into larvae possessing apical organs. Also larvae with apical organs have been found to develop from each of the isolated 1*q* blastomeres. In other words, the basic material of the apical organs in *Patella* is believed to be divided equally among the four macromeres at the time of cleavage.

If the blastomeres of the *Patella* embryo are isolated at various cleavage stages thereafter, each of them follows the developmental course which it would have taken as a part of the whole embryo. For instance, a macromere isolated in the 8-cell stage forms a gastrula at one extremity of which are produced one or sometimes two secondary trochoblasts, while at the other end can be seen a group of entoderm cells bearing the delicate cilia characteristic of the normal embryo. From macromeres isolated in the 16-cell stage are formed gastrulae having the delicate cilia but lacking the secondary trochoblasts. Furthermore, if the $1q^2$ blastomeres, which are the primary trochoblasts, are isolated, they develop into four typical ciliated trochophore cells.

When Conklin (1912) isolated *Crepidula* blastomeres in the 2- and 4-cell stages and observed their later development, he found that the cleavage of each blastomere is the same as it would have been in the normal embryo. When the *D* blastomere is included, the 4*d* primordial mesoderm cell is formed; when the *D* blastomere is lacking, the 4*d* cell does not appear.

Like *Dentalium, Ilyanassa* forms a polar lobe at the time of the first three cleavages; Clement (1952) has removed the first polar lobe at the trefoil stage and followed the subsequent development. The main points in which this differs from the normal are that the size difference between the *D* blastomere and the other macromeres disappears, while the 4*d* cell, which is normally composed chiefly of clear cytoplasm, contains a large amount of yolk and presents the same appearance as 4*a*, 4*b* and 4*c*, and fails to produce the mesodermal bands. Moreover, in the normal embryo the cleavage of 3*D* precedes that of the other macromeres, but in lobeless embryos this difference is not seen.

The 1*d* blastomere also normally differs from the other 1*q* micromeres in such characteristics as size and time of cleavage, but these differences disappear on removal of the polar lobe. Among the molluscan cross tip cells ($2q^{11}$), the $2d^{11}$ cell is largest in normal embryos, but loss of the polar lobe removes this difference also. At the stage corresponding to the veliger, lobeless eggs develop enough to form abnormal partial larvae having a velum but lacking foot, shell and other organs; muscle tissue can be recognized, but Clement believes that this probably arises from ecto-mesoderm cells. Finally, he reports that embryos from which the *4D* blastomere is removed become small but perfect veligers.

2. LATER DEVELOPMENT

The preceding section has taken facts concerning the Gastropoda and Pelecypoda as the subject and described the developmental processes from egg to veliger stage in a general way; this section will treat various matters connected with the special characteristics of molluscan species in a more particular fashion, and describe later development from the veliger to the adult form.

(1) AMPHINEURA

a. Solenogastra or Aplacophora

Since there is very little detailed information concerning the development of these species and many points remain ambiguous, it was not discussed in the preceding section. Baba (1938, 1940, 1943, 1951) has published reports on a Japa-

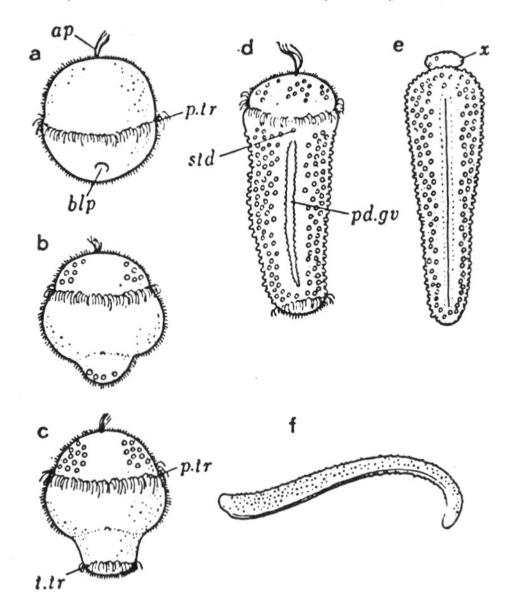

Fig. 11.20 Larval development in *Epimenia verrucosa* (Baba)

a. trochophore larva 20 hours after hatching, stomodaeum invaginating; b. trochophore, 24 hours; c. trochophore at 2 days. Telotroch appears at posterior end of larva; d. trochophore at 4 days (0.6 ~ 0.65 mm). Pedal groove appears; e. larva at late metamorphosis stage. Cells being sloughed off from degenerating head and tail regions; f. young adult, 19 ~ 33 days after hatching (3 mm). *ap:* apical cilium, *blp:* blastopore, *pd.gv:* pedal groove, *p.tr:* prototroch, *std:* stomodaeum, *t.tr:* telotroch, *x:* degenerating head region.

nese member of the Proneomeniidae, *Epimenia verrucosa* (Fig. 11.20). In this species the egg develops by spiral cleavage, produces apical cilia and an anterior prototroch at about the center of the embryo, and becomes a trochophore with the blastopore on its ventral side. Later the posterior end of the trochophore elongates, and a *pedal groove* is formed on the ventral side of this region. The tail extremity elongates still further, and a *terminal prototroch* is formed at its end. At the time of metamorphosis this tail region, including the prototroch, as well as most of the cells which made up the larva in the trochophore stage, are cast off and the adult worm-like form is assumed.

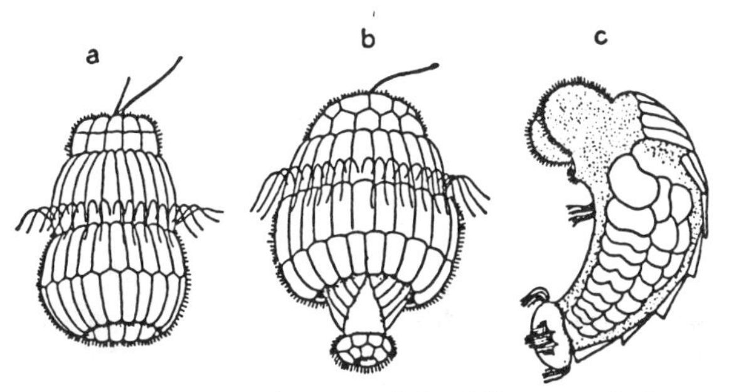

Fig. 11.21 Larvae and young adult of *Nematomenia banyulensis* (Pruvot)

a. larva at 36 hours (after hatching); b. larva at 100 hours; c. young worm just after metamorphosis (7 days).

The larva of *Nematomenia banyulensis* (Fig. 11.21a) resembles a trochophore with a conspicuous invagination at the vegetal side; the larval epidermis, which is formed of five rows of cells, is shed in metamorphosis, and the adult stage develops from the portion within the invagination and another part which extends posteriorly from this (Fig. 11.21b). At metamorphosis the part which changes over to become the adult body (Fig. 11.21c) develops lateral foliate plates and seven small calcareous plates along the back. These accessary structures are eventually exchanged for small calcareous spines (Pruvot, cited from Korschelt u. Heider 1936).

b. **Chitons (Loricata)**

The later development of the chitons has been studied by Heath (1899), Kowalevsky (1883), Hammarsten and Runnström (1925) and Okuda (1947).

The trochophore larvae of these species do not have protonephridia, but they do form the radula sac, pedal gland and other structures characteristic of molluscs, and pass through a free-swimming stage (Fig. 11.22a, b). However, in general this swimming period is very short, lasting several hours, or at most 2—3 days. At its end the larva begins to creep on the bottom with the foot which has developed on its ventral side during this period (Fig. 11.22c). Together with the degeneration of the apical cilia and prototroch, the posterior part of the body grows and assumes the adult form. A pair of *eyes* develop as temporary larval organs on the upper part of the foot posterior to the trochophore; these disappear with the completion of the mantle and shell.

One of the major distinguishing characteristics of adult chitons is the possession of a shell consisting of eight overlapping *plates;* the rudiments of these are also seen in the larvae. About the time when the larva changes to a bottom-crawling life, a transverse line first appears on its back, followed by six transverse folds which develop at successively posterior levels. A cuticle is secreted over the whole dorsal surface, while calcareous shells are secreted over the parts between the transverse folds. The eighth shell plate is formed considerably later than the others. The shell structure of these species is extremely complicated, the most conspicuous feature of the shells consisting in the fact that they are pierced at the sides by a number of cytoplasmic processes (the *aesthetes*) which arise from the mantle epithelium (Taki 1933). The ends of these processes perform a sensory function at the shell surface.

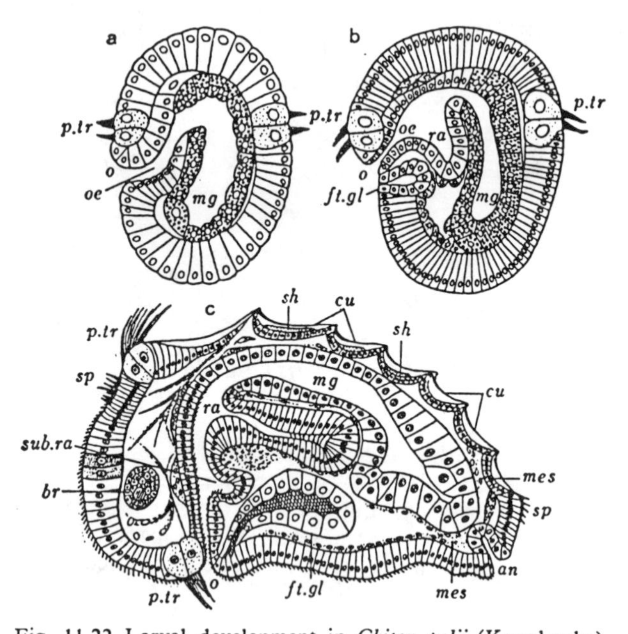

Fig. 11.22 Larval development in *Chiton polii* (Kowalevsky)

a,b. sections of trochophore larvae; c. section of larva which has assumed bottom-living habit. *an:* anus, *br:* cerebral ganglion, *cu:* chitin, *ft.gl:* foot gland, *mes:* mesoderm, *mg:* mid-gut, *o:* mouth, *oe:* oesophagus, *p.tr:* prototroch, *ra:* radula sac, *sh:* secreted calcareous shell, *sp:* spine, *sub.ra:* sub-radular organ.

The fact that the brain ganglia arise from four radially arranged rosette cells has been mentioned above; the paired longitudinal *pleural nerve cords* characteristic of the Amphineura, the pair of *pedal nerve cords* and the commissures which connect these transversely like the rungs of a ladder are believed to have the same origin as the brain ganglia (Korschelt u. Heider 1936). As a consequence this nervous system differs from that of the annelids in not showing metameric structure.

Furthermore, the head region is very poorly developed, without tentacles, etc., except for some species which have tentacle-like processes on the mantle in front of the mouth. The *labial palps*, which are formed around the mouth, and the *subradular organ* which is formed in the stomodaeum in front of the radula sac as an invagination of the stomodaeal epidermis (Fig. 11.22c), are also thought to be sensory organs.

(2) GASTROPODA

As important points in which the Gastropoda differ from other molluscs in their later development may be listed the growth of the head, the torsion of the posterior dorsal part of the body, the spiral growth of the visceral mass and shell, the secretion of the operculum on the dorsal posterior part of the foot. In particular the torsion of the body is a highly characteristic phenomenon; since it occurs relatively early in development, the following account will deal first with this subject.

a. Torsion

It has already been stated that the mouth (blastopore), which was located at the posterior end of the early gastrula, moves to a ventral position considerably behind the prototroch in the trochophore stage. An anus is first formed at the posterior end of the body, but as a result of the vigorous growth of the posterior dorsal region, the distance between mouth and anus, measured across the interposed foot, becomes relatively short. In other words, the digestive tract bends markedly toward the ventral side. This phenomenon is also seen in the Pelecypoda, where it is called *ventral flexion*. In the case of the Gastropoda, this growth of the posterior dorsal region is asymmetrical, the left side growing more markedly in most cases. As a result, the anus comes to be located at the right side of the mouth. The *mantle* develops as a fold of the epidermis; enveloped in this mantle, a *mantle cavity* is formed, and the anus comes to open into this cavity.

As a further result of this extremely asymmetrical growth, the anus moves past the right side of the mouth and finally takes a position above it and considerably to the left. In other words, it makes a turn of nearly 180° from its first position. This phenomenon is called *torsion*, and appears rather early in development — in the latter part of the trochophore stage. For example, according to Ino (1953), it occurs 45—46 hours after fertilization in the black abalone,

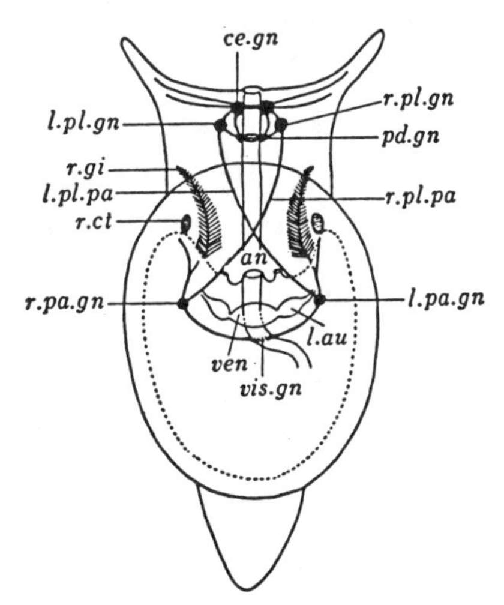

Fig. 11.23 Diagram of chiastoneury (Lang)

an: anus, *ce.gn:* cerebral ganglion, *l.au:* right (originally left) auricle, *l.pa.gn:* right (formerly left) parietal ganglion, *l.pl.gn:* left pleural ganglion, *l.pl.pa:* left pleuro-parietal connective, *pd.gn:* pedal ganglion, *r.ct:* left (formerly right) osphradium, *r.gi:* left (formerly right) gill, *r.pa.gn:* left (formerly right) parietal ganglion, *r.pl.gn:* right pleural ganglion, *r.pl.pa:* right pleuro-pariteal connective, *ven:* ventricle, *vis. gn:* visceral ganglion.

Haliotis discus. Furthermore, the process advances very rapidly, so that it is said to be completed in few hours in *Trochus* and within a few minutes in *Acmaea* (Robert 1902; Boutan 1899).

The mantle cavity, the shell and the visceral mass participate in this torsion, and accompanying them, the several organs which have some connection with the mantle (gills, heart and kidneys) also become asymmetrical. The organs of the right side usually develop better than those of the left side, or persist while the opposite member degenerates. Lang (1896) believed this to be due to the fact that in the torsion process, the organs of the left side are subjected to the full pressure of the visceral mass (i.e., develop under less favorable conditions). An especially striking result of this torsion is that the *pleuro-parietal connectives,* the pair of longitudinal nerves joining the (anterior) *pleural ganglia* with the (posterior) *parietal ganglia* cross each other, the right one above the oesophagus and the left one below it, forming a figure 8 (Fig. 11.23). This is known as *chiastoneury,* and constitutes an important characteristic of the Prosobranchia.

In most of the Opisthobranchia, this torsion secondarily undergoes a complete reversal at the time of metamorphosis; the mantle cavity becomes located in the posterior part of the body and the crossed nerves revert to their original condition. This phenomenon is called *detorsion.* However, a *gill* and *auricle* develop on one side only. Chiastoneury does not occur in the Pulmonata; this may be due to the fact that the ganglia concerned are located close to the head and are therefore not involved in the torsion process, or that the degree of torsion is slight (the anus, central point of the mantle organs, stops at the right side of the mouth).

The asymmetry, already extreme as the result of this torsion, becomes still further accentuated by unequal growth in a different direction from that of the torsion, causing the shell and visceral mass to develop in spiral fashion.

b. **Shell and Operculum**

The great majority of the Gastropoda are dextrally twisted (in a clockwise direction when viewed from above), although a few species show sinistral twisting, and in a few others both dextral and sinistral types are found. In the sinistral forms, the mantle cavity, anus and genital pore are also situated on the left side of the body, opposite to their location in the dextral animals. This dextral or sinistral organization can be traced back through the cleavage stages, and it is known to be an inheritable character that is already determined in the unfertilized egg (see Introduction, pp. [90—91]).

The shell of the veliger stage, i.e., the *larval shell,* is at first secreted by the shell gland, and later by the mantle; this shell is easily distinguishable by its color, shape and markings from the shell which is formed after metamorphosis. Even in species which do not have a shell in the adult stage, a larval shell is invariably secreted by the veliger shell gland; at metamorphosis this is enveloped in the mantle tissue and disappears.

In *Lamellaria* and *Cypraea,* beneath the shell first formed during the swimming larval period, another shell bearing pointed processes develops (this stage is called the "echinospira larva"). Among the Pyramidellidae the shells which are first formed are all sinistral, while the bodies of the animals are dextral. Shortly afterward dextral shells are secreted (Lebour 1937).

The *operculum* is secreted on the dorsal side of the foot as early as the late trochophore stage. The growth of the operculum also takes place in a spiral fashion but in the direction opposite to that of the dorsal shell, so that animals with dextral shells have sinistral opercula. An operculum is always formed in the larva, even in species in which none is found in the adult. There are some workers who claim that the dorsal shell and operculum of the gastropods are homologous with the two valves of the bivalves (viz., Fleischmann 1932), but the latter are both secreted by the shell gland on the dorsal part of the body and therefore correspond to the gastropod dorsal shell, while they have no homology with the operculum.

The shell of the adult gastropod is usually composed of an outer layer *(periostracum)* consisting mainly of *conchiolin,* a middle layer *(ostracum)* of which the chief component is calcium carbonate, and an inner layer *(hypostracum);* this three-layered structure closely resembles that of the pelecypod shell.

c. Nervous System

Compared with the other molluscan classes, the gastropods as a group have strikingly well-developed nervous systems, and already possess many ganglia in the larval stage. Chief among these are the *cerebral, pleural, pedal, parietal, visceral* and *buccal ganglia;* these develop from ectodermal thickenings which differentiate inward (Fig. 11.24).

It has already been stated that the central nervous system, like the apical plate, arises from the cells of 1*q*. The buccal ganglia are formed by proliferation of the cells making up the walls of the stomodaeum. In *Paludina* and *Bythinia* the pleural ganglia arise from the body wall posterior to the velum and toward the ventral side, while in *Littorina* they are formed by proliferation of the cells of the brain-ganglia.

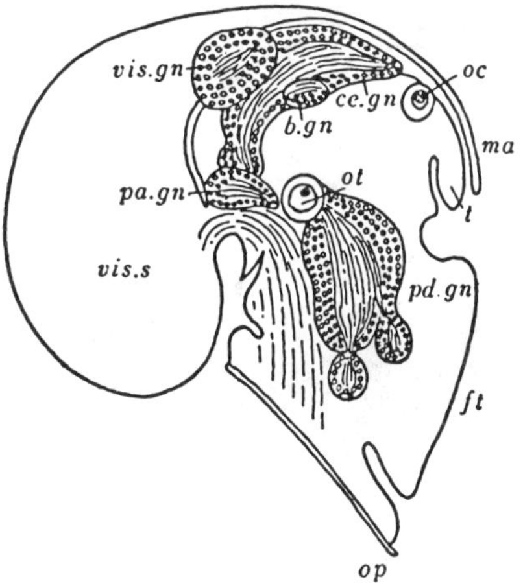

The relation between the nervous system and the phenomenon of torsion has been discussed above; it is an interesting fact that in *Acmaea, Trochus* and *Littorina,* at the time when torsion takes place, the ganglia which will participate in this process have not yet appeared (even though their anlagen may be present).

Fig. 11.24 Larval nervous system in *Littorina obtusa* (Delsman)

b.gn: buccal ganglion, *ce.gn:* cerebral ganglion, *ft:* foot, *ma:* mantle, *oc:* ocellus, *op:* operculum, *ot:* otocyst, *pa.gn:* parietal ganglion, *pd.gn:* pedal ganglion, *t:* tentacle, *vis.gn:* visceral ganglion, *vis.s:* visceral sac.

The various sense organs develop along with the development of the nervous system — in the veliger stage the head is already provided with *antennae* and *eyes,* while a pair of *statocysts* are well developed at the base of the foot. It has been observed in *Onchidella* that when the veliger metamorphoses, part of the apical plate becomes the rudiment of the *sensory processes* which form around the mouth (Fretter 1943).

(3) COELOM, PERICARDIUM, KIDNEY, HEART AND BLOOD VESSELS

It has been stated above that the mesodermal bands arise from a pair of teloblasts; the cells making up these bands separate and become mesenchyme, and then reaggregate into paired cell masses. Eventually a *coelomic cavity* appears in each of these, the two cavities develop in close contact with each other, and finally the intervening septum disappears, giving rise to a single large cavity (Fig. 11.25a-c). This is the *pericardium*. The part of the pericardium which originates in the right coelomic cavity is better developed than its counterpart on the left side.

The cells on the left and right sides of the ventral part of the pericardium proliferate, producing thickenings which eventually evaginate. These are the rudiments of the *kidneys* (Fig. 11.26), which are thus paired to begin with, but the one on the right side develops well, while (except among the primitive members of the Prosobranchia which have paired kidneys) the one on the left side degenerates, and the animal thus comes to have a single kidney. The evagination mentioned above develops into the kidney by forming with its walls a great number of folds and pouches. The mouth of the evagination becomes reduced in size, but a narrow connection between the pericardium and the kidney is maintained.

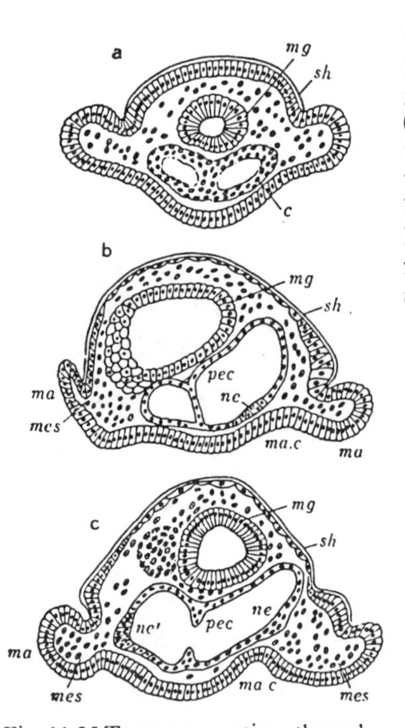

Fig. 11.25 Transverse sections through posterior region of *Paludina vivipara* larva, showing process of development of the pericardium (Tönniges)

c: coelom, *ma*: mantle, *ma.c*: mantle cavity, *mes*: mesoderm, *mg*: mid-gut, *ne*, *ne'*: nephridial primordia, *pec*: pericardium, *sh*: shell.

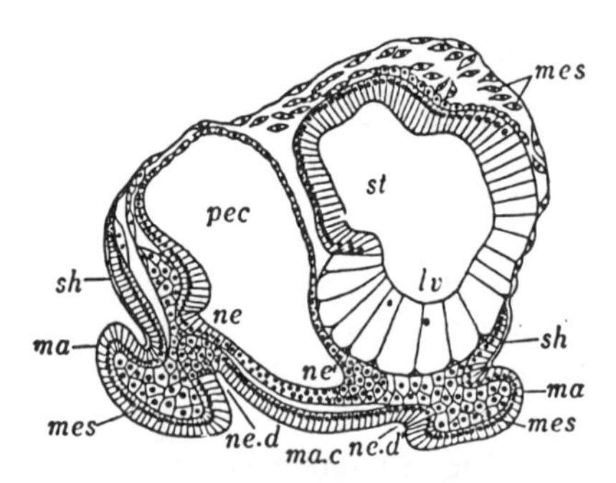

Fig. 11.26 Transverse section of *Paludina vivipara* larva, showing development of nephridium and its duct (Erlanger)

lv: liver, *ma*: mantle, *ma.c*: mantle cavity, *mes*: mesoderm, *ne*, *ne'*: nephridial primordia, *ne.d*, *ne.d'*: nephridial ducts, *pec*: pericardium, *sh*: shell, *st*: stomach.

The *urinary duct* connecting the kidney with the exterior, as well as the opening of this duct, are formed by an invagination of the wall of the mantle cavity opposite the kidney. Such invaginations actually occur on both left and right sides, the one on the right side developing and establishing contact with the right kidney while the one on the left side degenerates together with the left kidney rudiment. However, while the development of an ectodermally derived urinary duct of this kind is found in *Paludina* among the Prosobranchia and in many of the Pulmonata, the kidney opens directly into the mantle cavity in the other Prosobranchia, the Opisthobranchia and certain Pulmonata.

According to Erlanger's (1891) account of the developmental process followed by the heart, the cells of the pericardial wall above the kidney proliferate, giving rise to a thickening which sinks inward to form a groove. This groove is eventually closed over, except for its two ends, forming a tube which is suspended in the pericardium. This constitutes the *heart rudiment,* and its two ends communicate with the primary body cavity. This tubular *heart* is constricted at its middle into two parts; the anterior part becomes the rudiments of the *auricle* and *branchial vein,* while the wall of the posterior part thickens and forms the rudiments of the *ventricle* and the *aorta.*

The blood vessels can be seen as sinus-like spaces in the mesenchyme tissue even before the development of the heart. Moreover, these sinuses have occasionally been observed to pulsate before the heart is formed. The sinuses are surrounded by flattened cells; their cavities become more restricted as development proceeds and contribute to the formation of the blood vessels.

The description given here of the processes by which the heart and blood vessels develop in *Paludina* agree well with the results of observations on other gastropods.

(4) REPRODUCTIVE ORGANS

Observations made on the prosobranchs *Paludina* (Erlanger, Drummond, Otto and Tönniges) and *Littorina* (Delsman) indicate that the *gonads,* like the kidneys, develop by proliferation of the cells forming the wall of the pericardium. In *Paludina* the rudiment of the gonad arises near the left kidney (which will degenerate) and its tip fuses with the nephrostome of the kidney. As a result, the left urinary duct becomes the *genital duct*[5].

The reproductive organs of the Prosobranchia are thus considered to be of mesodermal origin, but opinion is divided with respect to the situation in the hermaphroditic Pulmonata, some workers believing them to be mesodermal, others claiming that they are ectodermal, while still others insist that one part is mesodermal and another part ectodermal (see Korschelt u. Heider 1936, p. 917).

(5) SCAPHOPODA

The development of the Scaphopoda has been studied by Lacaze-Duthiers (1851), Kowalevsky (1883) and Wilson (1904). The trochophore larva is characterized by having three rows of ciliated cells, but in general the mode of development of its various organ systems is closely similar to that of the Amphineura, Gastropoda and Pelecypoda (Fig. 11.27).

As in the Pelecypoda, the mantle first develops as paired left and right structures which later fuse ventrally. In consequence the shape of the shell also, seen from the dorsal side, at first resembles that of the gastropod trochophore,

[5] A general discussion of the problem of the genital duct in the Prosobranchia has been presented by Fretter (1941, 1946).

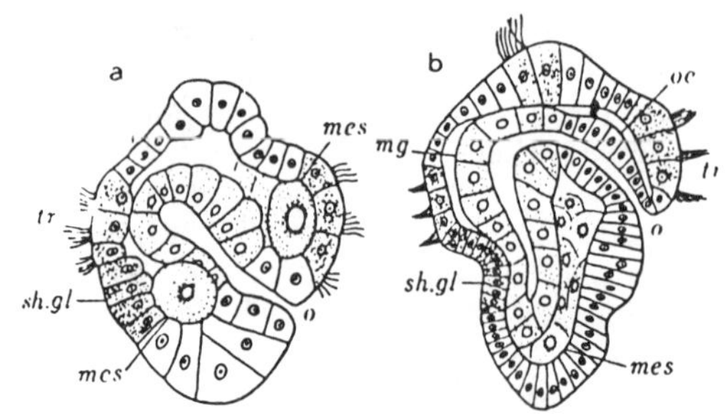

Fig. 11.27 Longitudinal sections of trochophore larvae of *Dentalium* (Kowalevsky)

a. larva at 14 hours after fertilization; b. larva at 34 hours. *mes:* mesoderm, *mg:* mid-gut, *o:* mouth, *oe:* oesophagus, *sh.gl:* shell gland, *tr:* trochal ring.

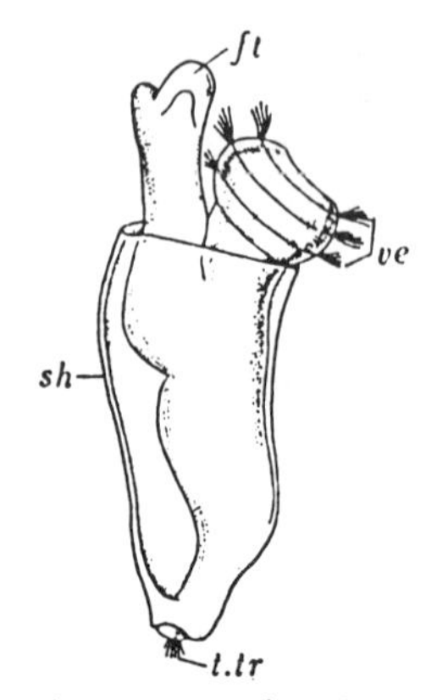

Fig. 11.28 Veliger larva of *Dentalium* (Wilson)

ft: trilobate foot, *sh:* shell, *t.tr:* telotrochal ring, *ve:* velum.

and from the ventral side, that of the pelecypod larva. Eventually the two edges fuse ventrally, forming a tube. The opening at the posterior end is small, and shell growth takes place at the anterior end. The foot is well developed, its division into three lobes forming one of the characteristic features of the scaphopod veliger larva (Fig. 11.28).

No protonephridia have been observed, but statocysts and pedal glands appear, and a radula sac is formed at a later stage.

(6) PELECYPODA

The fact has already been mentioned that the pelecypod trochophore differs from the larvae of the other molluscan classes in failing to form a radula sac. As in the gastropods, the shell is first secreted by the shell gland on the dorsal side of the embryo; eventually this differentiates into two (left and right) valves which make contact in a straight line in the region of the shell gland. By the time the shell valves develop sufficiently to cover the soft parts of the body (in *Crassostrea gigas,* about 18 hours after fertilization at 25° C), they exhibit a D-shape (Fig. 11.29a). The straight part of this D-shaped larva is the *hinge line; teeth* differentiate on the two opposing shell valves at this line. The larvae having shells of this shape are called "D-shaped larvae", and except for species in which the developmental morphology has been changed by the habit of rearing the young

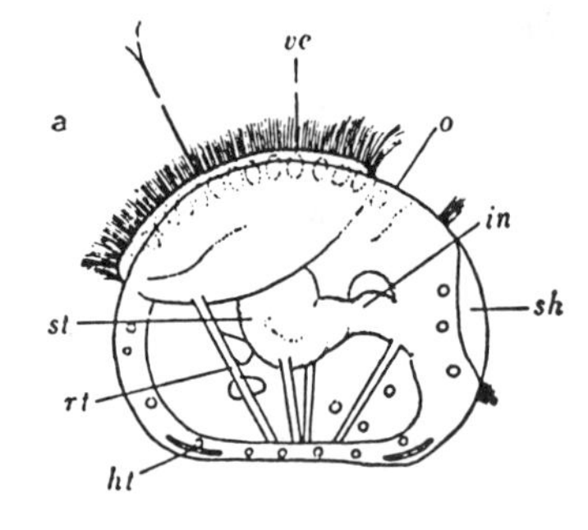

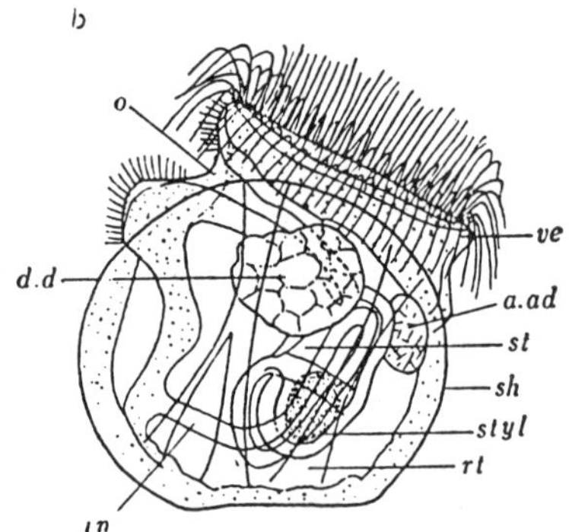

Fig. 11.29 D-shaped larva of the Pelecypoda

a. early D-shaped larva of *Pinctada maxima* (Wada); b. D-shaped larva of *Ostrea edulis* (Yonge). *a.ad:* anterior adductor, *d.d:* digestive sac, *ht:* hinge teeth, *in:* intestine, *o:* mouth, *rt:* retractor, *sh:* shell, *st:* stomach, *styl:* style sac, *ve:* velum.

in a broodpouch, a larva having such a shape can be found in practically all pelecypod species in the early veliger stage.

After this the shell shows rather more marked growth in the dorso-ventral direction (i.e., perpendicular to the hinge-line) than antero-posteriorly (parallel to the hinge-line), and becomes rounded, and eventually the *umbo* region begins to become prominent. When the larvae of *Crassostrea gigas, Pinctada martensii* and other bivalves are reared experimentally, an extremely high death rate is observed in this period. However, no particularly conspicuous change in the structure of the soft parts is associated with this stage, and the cause of such mortality is unknown.

The velum is well developed at this time, extending beyond the shell and moving vigorously during swimming and feeding activity (Fig. 11.29b). At times the larva may draw the velum inside the shell and close the valves, abruptly sinking to the bottom; a culture of such larvae is full of busy up-and-down movement.

In the central part of the velum apical cilia are usually developed. In the pearl oyster species, however, these are present only during the early D-shaped larval period, degenerating as development proceeds, while they are completely lacking in *Crassostrea* species.[6]

Muscles extending from the dorsal body wall (region of the hinge line) attach to the velum and digestive tract, and by their contraction the soft parts can be gathered completely inside the shell. An *anterior adductor muscle* also develops near the anterior part of the shell for the purpose of closing the valves. The development of the *posterior adductor muscle* occurs much later, this muscle first appearing late in the veliger stage.

With the opening of the anus early in the D-shaped larval stage the digestive tract becomes functional, and feeding activity begins; the paired rudiments of the *digestive diverticula* — the so-called "liver" — appear at this stage, and the left member of the pair develops rapidly. This "liver" takes on a conspicuous color which, however, varies according to the species of organism consumed as food (Wada 1942). The intestine, at first a straight tube, acquires loops as development proceeds; the stomach also develops, at its posterior end, a large blind sac, the *crystalline style sac* (Fig. 11.29b).

The nervous system of the adult pelecypod is extremely simple as compared with that of the gastropod, and this is also true of the larvae. These possess *cerebral ganglia* below the apical plate, and the rudiments of *pedal ganglia* appear in the region between mouth and anus, where the foot will later develop (Fig. 11.30). *Visceral ganglia* develop considerably later, near the pedal ganglia, as well as *parietal ganglia* lying close to the cerebral ganglia. In the higher Pelecypoda, the parietal ganglia afterwards fuse with the cerebral ganglia. In the vicinity of the pedal ganglia a pair of *statocysts* develop from the ectoderm late in the veliger stage. Below the velum a pair of black pigmeted *eye spots* or *pigmented spots* appear, forming a characteristic of the late veliger stage: whether or not these are sensory organs is a moot point, and it is also not clear whether the larvae in themselves exhibit phototaxis (see Cole 1938). Cole concludes that the pigmented spot of the *Ostrea edulis* larva has the structure of an eye. The statocysts persist in the adult animal, but the eye spots degenerate.

---

[6] The figure of Fujita (1922) is in error in this respect.

It has been stated above that a *protonephridium* of ectodermal origin develops near the tips of each of the pair of long, narrow mesodermal bands as one of the larval organs; the adult *kidney* is also formed adjacent to these.

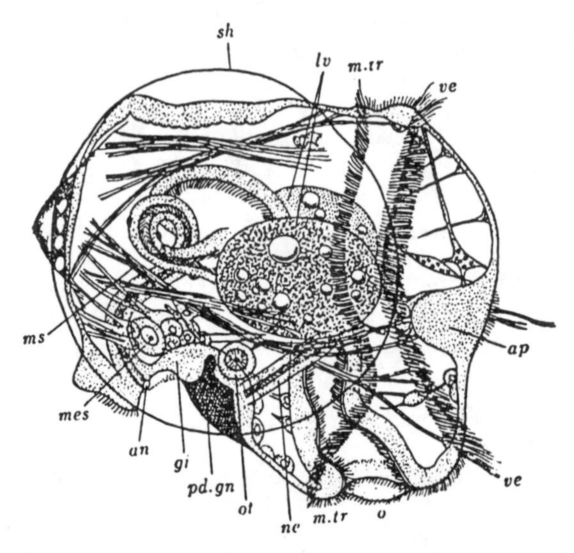

Fig. 11.30 Late veliger larva of *Teredo* (right side) (Hatschek)
an: anus, *ap:* apical plate, *gi:* gill rudiment, *lv:* liver, *mes:* mesoderm, *ms:* larval muscle, *m.tr:* metatrochal ring, *ne:* nephridium, *o:* mouth, *ot:* otocyst, *pd.gn:* rudiment of pedal ganglion, *sh:* shell, *ve:* velum.

The *foot* forms on the ventral side of the body, between the mouth and the anus; its development takes place in the latter part of the veliger stage, late in comparison with the Gastropoda. At the time when the veliger larva changes from a free-swimming to a bottom-dwelling mode of life, both velum and foot are well developed, and the larva sometimes crawls about with its foot, which is covered with slender cilia, and sometimes swims with its velum.

At the base of the foot the ectoderm invaginates, giving rise to the *pedal gland* or *byssus gland*. In most cases this is a single organ, but it may also occur as a pair of organs (viz. *Sphaerium*), and it is thought to be homologous with the mucus pedal gland of gastropods and chitons. When the larva takes up a benthic mode of life, it clings to the substratum with mucus or byssus threads secreted by this gland. In the common mussel and other species this gland is well developed in the adult form also; when the adult does not have such a gland (viz., certain Arcidae), it degenerates in a later developmental stage. The *cement gland* of the oysters is similar in nature to this gland.

The *mantle* appears first in the trochophore stage as an expansion of the ectoderm in the posterior part of the larva, and subsequently develops as lateral folds extending to its anterior end. The *shell,* which is secreted by the mantle, also grows along with the development of the mantle. The *ctenidium* appears in the late veliger stage (or sometimes at the time of metamorphosis) as a row of papillae, and develops rapidly after metamorphosis. Of the two (inner and outer) pairs of ctenidial lobes, the inner pair appears earlier than the outer pair.

It has been mentioned that the anterior adductor muscle develops in the early veliger stage; in the late veliger, spindle-shaped mesenchyme cells aggregate at a point anterior to the anus and form the *posterior adductor muscle.* Even in the species (Monomyaria) which have only one adductor muscle in the adult, paired anterior and posterior adductors are invariably formed in the larval period; later the anterior one degenerates and only the posterior adductor muscle develops (Fig. 11.31).

When the veliger changes to a bottom-dwelling mode of life, around the shell which is carried over from the free-swimming period — this may be called *larval shell, shell rudiment* or *prodissoconch*[7] — there is rapidly formed an *adult shell (dissoconch)* of a rather different nature.

As the veliger assumes the bottom-dwelling habit, most of its tissues become disassociated and are discarded or absorbed (Fig. 11.31). This process takes place with great rapidity; for example, in *Ostrea edulis* it is said to be completed within 48 hours after attachment. The distal part of the velum, however, is shown to persist and take part in the formation of the labial palps or their rudiments (Meisenheimer 1901; Cole 1938).

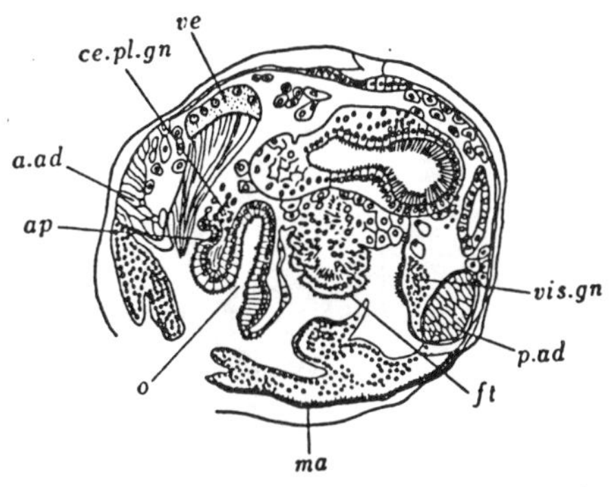

Fig. 11.31 Young oyster (*Ostrea edulis*) 24 hours after fixation. Foot and velum have degenerated (Cole)

a.ad: anterior adductor, *ap:* apical plate, *ce.pl.gn:* cerebro-pleural ganglion, *ft:* foot, *ma:* mantle, *o:* mouth, *p.ad:* posterior adductor, *ve:* velum, *vis.gn:* visceral ganglion.

In comparison with those of the Gastropoda, the mesodermal bands of the Pelecypoda are poorly developed in the early stages, but it has been observed

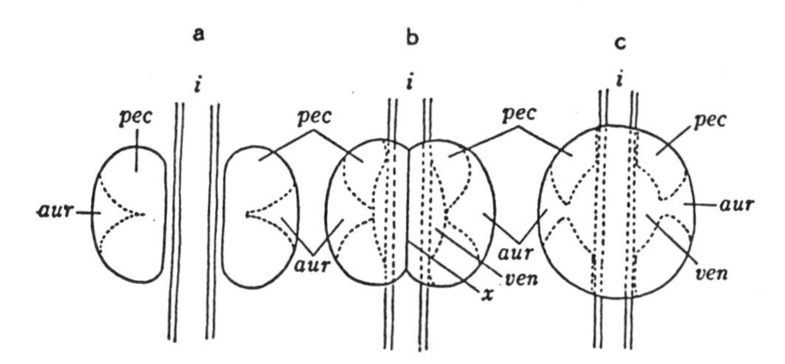

Fig. 11.32 Development of pericardium and heart in *Cyclas cornea* (Ziegler)

aur: auricle rudiment, *i:* intestine, *pec:* pericardium, *ven:* ventricle, *x:* line of fusion of right and left pericardial cavities.

[7] The larval shell consists of two different parts: that secreted by the shell gland in the late trochophore stage, and a part subsequently secreted by the mantle (Werner 1940; Watabe and Yuki 1952). The portion formed first is said to be composed of dahllite (apatite carbonate +apatite hydroxide) while the part laid down later is mainly composed of calcium carbonate (calcite), like the adult shell (Watabe 1956).

that the pericardium and its accessory organs have a mesodermal origin. According to the observations of Ziegler (1885) on *Cyclas*, the cells of the anterior dorsal parts of the left and right mesodermal bands form two vesicles, or coelomic cavities. These constitute the rudiment of the *pericardium*; the cavities presently increase in size on the two sides of the intestine, eventually coming in contact with each other and finally fusing (Fig. 11.32). The *ventricle* is formed between this cavity and the intestine, the inner wall of the cavity becoming the ventricular wall. Somewhat earlier than this, invaginations appear in the pair of cavities; these form the anlagen of the *auricles*, and make connection with the ventricles (Fig. 11.32).

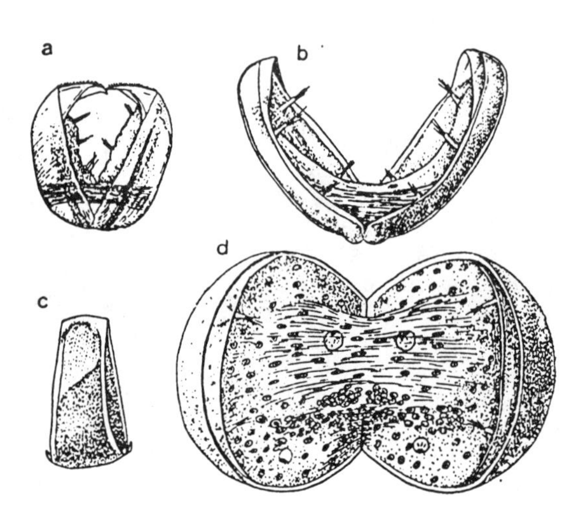

Fig. 11.33 Glochidium larvae (Lefevre u. Curtis)
a. *Symphynota cosata*; b. *Lampsilis* (*Proptera*) *alata*; c. *Lampsilis subrostrata*; d. *L. subrostrata* (ventral view).

The rudiment of the *kidney* is recognized adjacent to the pericardium as a filiform tissue formed by mesoderm cells and containing a lumen. This soon opens into the pericardium, while the other end extends toward the body wall and eventually forms an opening to the outside.

Although the pericardial rudiment can be seen in the veliger stage, the pulsation of the heart cannot be detected until after the bottom-dwelling habit has been assumed (*Ostrea edulis* — Fernando 1931).

Ordinarily the differentiation of the reproductive system lags far behind that of the other organs, but according to Woods (1931, 1932), it is possible to trace its rudiment back to a very early developmental stage in *Sphaerium corneum*. In any case, the reproductive organs maintain an intimate connection with the pericardium, heart and kidney, and are believed to be of mesodermal origin. When an independent *genital pore* is formed, it is usually close to the opening of the urinary duct; in other cases a common opening serves for both urinary and genital ducts, and in still other cases the reproductive cells are released into the kidney itself.

The species belonging to the fresh-water pelecypod family of the Unionidae all follow a developmental process which involves a larval stage called the *glochi-*

dium (Fig. 11.33), an extraordinary larva which lacks velum, foot and mouth. The eggs of these species develop for a certain period in brood-pouches formed by the water-tubes of the maternal gill laminae. The early development of *Unio,* which has been studied in detail by Lillie (1895), presents no fundamental differences from that of other pelecypod species. By the time the larva leaves the brood-pouch it is in the glochidium stage, and without going through a free-swimming period it becomes parasitic on the fins or other scaleless parts *(Anodonta, Symphynota)* or on the gills *(Unio, Quadrula, Lampsilis)* of various fishes.[8] After metamorphosis it leaves the host and takes up a bottom-dwelling mode of life.

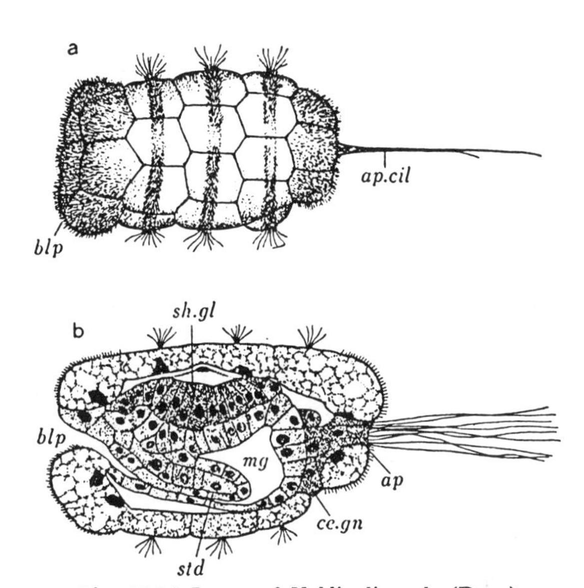

Fig. 11.34 Larva of *Yoldia limatula* (Drew)

a. trochophore larva 3 days after fertilization; b. longitudinal section of more advanced trochophore larva. *ap:* apical plate, *ap.cil:* apical cilial tuft, *blp:* blastopore, *ce.gn:* cerebral ganglion rudiment, *mg:* mid-gut, *sh.gl:* shell gland, *std:* stomodaeum.

In forms which are ectoparisitic on fishes the valves of the shell are provided with a pair of curved hooks on their ventral edges, while those which parasitize the gills have a blade-shaped protrusion in place of the hooks (Fig. 11.33). Sensory hairs are distributed here and there on the mantle, and there is one large adductor muscle in the center of the body. When a fish comes close to a glochidium, the larva uses this muscle to clap its valves together, and dancing upward, it secures a firm hold on the fish and becomes parasitic. In *Anodonta* and *Unio,* there is a long *larval thread* in the center of the larva, but this is not seen in other species (Lefevre and Curtis 1912). When the larva becomes parasitic, the cells of the inner part of the mantle develop large vacuoles and take on a tall columnar form; these feed on and absorb blood and tissue cells of the host fish.

---

[8] According to Lefevre and Curtis (1912), the glochidium larva of *Strophilus edentulus* metamorphoses without passing through a parasitic phase.

In the process of metamorphosis these special larval organs degenerate and are replaced by adult organs which do not differ from those of the other pelecypod species.

The members of the Protobranchiata (Nuculacea), which consists of *Yoldia, Nucula, Leda* and *Solenonaya,* also have larvae of a form which diverges widely from that of the other Pelecypoda. According to Drew (1899, 1901), who studied the development of these species, the swimming-stage larva is barrel-shaped and the outer body wall is composed almost entirely of the velum, which has three prototrochal rings; in its external appearance it resembles the larvae of the Solenogastra and Scaphopoda (Fig. 11.34). Enclosed within this body wall are not only the digestive system but also the shell gland and other ectodermal organs. The shell and foot develop thus internally; at metamorphosis the velum is cast off and the animal begins its life as a bottom-dweller.

REFERENCES

Allen, R. D. 1953: Fertilization and artificial activation in the egg of the surf-clam, *Spisula solidissima*. Biol. Bull., **105**: 219—239.

Ankel, W. E. 1929: Ueber die Bildung der Eikapsel bei *Nassa*-Arten. Verh. Dtsch. zoöl. Ges., 33, Zoöl. Anz., suppl. **4**: 219—230.

———— 1930: Nähreierbildung bei *Natica catena* (da Costa). Zoöl. Anz., **89**: 129—135.

———— 1937: Der feiner Bau des Kokons der Purpurschnecke *Nucella lapillus* (L.) und seine Bedeutung für das Laichleben. Verh. Dtsch. zoöl. Ges., 39, Zoöl. Anz., suppl. **10**: 77—86.

Baba, K. 1938: The later development of a solenogastre, *Epimenia verrucosa* (Nierstrasz). J. Dept. Agric. Kyushyu Imp. Univ., **6**: 21—40.

———— 1940: The early development of a solenogastre, *Epimenia verrucosa* (Nierstrasz). Annot. Zool. Jap., **19**: 107—113.

———— 1951: Outline of the development in *Epimenia verrucosa* (Nierstrasz). Rep. Nat. Resources Res. Inst., **19—21**: 38—46. (in Japanese)

Casteel, D. B. 1905: The cell-lineage and early larval development of *Fiona marina*, a nudibranch mollusk. Proc. Acad. Nat. Sci. Phila., **56**: 325—405.

Child, C. M. 1900: The early development of *Arenicola* and *Sternapsis*. Arch. Entw.-mech., **9**: 587—723.

Clement, A. C. 1952: Experimental studies on germinal localization in *Ilyanassa*. I. The role of the polar lobe in determination of the cleavage pattern and its influence in later development. J. Exp. Zool., **121**: 593—625.

Coe, W. R. 1943: Sexual differentiation in mollusks. I. Pelecypods. Quart. Rev. Biol., **18**: 154—164.

———— 1944: Sexual differentiation in mollusks. II. Gastropods, amphineurans, scaphopods and cephalopods. Quart. Rev. Biol., **19**: 85—97.

Cole, H. A. 1938: The fate of the larval organs in the metamorphosis of *Ostrea edulis*. J. Mar. Biol. Assoc., **22**: 469—484.

Conklin, E. G. 1897: The embryology of *Crepidula*. A contribution to the cell-lineage and early development of some marine gastropods. J. Morph., **13**: 1—226.

———— 1902: Karyokinesis and cytokinesis in the maturation, fertilization and cleavage of *Crepidula* and other Gastropoda. J. Acad. Nat. Sci. Phila., **21**: 1—121.

———— 1912: Experimental studies on nuclear and cell division in the eggs of *Crepidula*. J. Acad. Nat. Sci. Phila., **15**: 503—591.

———— 1917: Effects of centrifugal force on the structure and development of the eggs of *Crepidula*. J. Exp. Zool., **22**: 311—421.

———— 1924: Cellular Differentiation. In E. V. Cowdry (ed), General Cytology. Chicago.

Crampton, H. E. 1894: Reversal of cleavage in a sinistral gastropod. Ann. N. Y. Acad. Sci., **8**: 167—170.

———— 1896: Experimental studies on gastropod development. Arch. Entw.-mech., **3**: 1—18.

Crofts, D. R. 1937: The development of *Haliotis tuberculata*, with special reference to organogenesis during torsion. Phil. Trans. Roy. Soc. London, Ser. B, Biol. Sci., **228**: 552.
Dan, J. C. 1956: The acrosome reaction. Int. Rev. Cytol., **5**: 365.
——— & S. K. Wada 1955: Studies on the acrosome. IV. The acrosome reaction in some bivalve spermatozoa. Biol. Bull., **109**: 40—55.
Dauert, E. 1929: Keimblätterbildung von *Paludina vivipara*. Zoöl. Jahrb., Anat., **50**.
Delsman, H. C. 1914: Entwicklungsgeschichte von *Littorina obtusata*. Tijdschr. Nederl. Direkundige Ver., Ser. 2, **13**: 170—340.
Drew, G. A. 1899: Habits, anatomy and embryology of the Protobranchia. Anat. Anz., **15**.
——— 1901: Life history of *Nucula delphinodonta*. Quart. J. Micr. Sci., **44**.
Farnie, W. C. 1924: Development of *Amphibola crenata*. Quart. J. Micr. Sci., **68**.
Fernando, W. 1903: Origin and development of the pericardium and kidney in *Ostrea*. Proc. Roy. Soc. London, **107**.
Field, I. A. 1922: The biology and economic value of the sea mussel *Mytilus edulis*. Bull. Bur. Fish., Washington, **38**: 1—159.
Fleischmann, A. 1932: Vergleichende Betrachtungen über das Schalenwachstum der Weichtiere (Mollusca). II. Dechel und Hause der Schnecken. Zeitschr. wiss. Biol., Abt. A, **25**: 549—622.
Fretter, Vera 1941: The genital ducts of some British stenoglossan prosobranchs. J. Mar. Biol. Assoc., **25**: 173—211.
——— 1943: Studies on the functional morphology and embryology of *Onchidella celtica* (Forbes and Hanley) and their bearing on its relationships. J. Mar. Biol. Assoc., **25**: 685—720.
——— 1946: The genital ducts of *Theodoxis*, *Lamellaria* and *Trivia*, and a discussion on their evolution in the prosobranchs. J. Mar. Biol. Assoc., **26**: 312—351.
Fujita, T. 1904: Formation of the germ-layers in Gastropoda. J. Coll. Sci. Japan, **20**.
——— 1929: On the early development of the common Japanese oyster. Jap. J. Zool., **2**: 353—358.
Habe, T. 1943: Egg-sacs of marine molluscs. Venus, **13** (5—8), 199. (in Japanese)
——— 1953: Studies on the eggs and larvae of the Japanese gastropods (4). Publ. Seto Mar. Biol. Lab. **3**,: 161—167.
Hammarsten, O. D. & J. Runnström. 1925: Zur Embryologie von *Acanthochiton*. Zoöl. Jahrb., Anat., **47**.
Hatanaka, M., R. Itoh & F. Imai. 1943: The culture of *Venerupis* and *Meretrix*. Rep. Jap. Fisheries Assoc., **11** (616): 218. (in Japanese)
Hatchek, B. 1881: Entwicklungsgeschichte von *Teredo*. Arb. Zoöl. Inst. Wien, **3**.
Heath, H. 1899: Development of *Ischnochiton*. Zoöl. Jahrb., Anat., **12**.
Holmes, S. J. 1897: Secondary mesoblast in Mollusca. Science, **6**.
——— 1900: The early development of *Planorbis*. J. Morph., **16**.
Hyman, O. W. 1923, 1925: Spermic dimorphism and natural partial fertilization in *Fasciolaria tulipa*. J. Morph. **37**, **41**.
Ikeda, K. 1930: The fertilization cones of the land snail, *Eulota (Eulotella similaris stimpson* Pfeiffer.) Jap. J. Zool., **3**, 89—94.
Ino, S. 1950: Life history and method of culturing *Babylonia japonica* Reeve. In Okada (Ed.), Studies of Commercially Important Animals, **1**: 11—24. (in Japanese)
——— 1952: Biological aspects of Japanese abalone production. Report No. 5 of Bur. Fish. East. Dist. Res. Lab. 1—102. (in Japanese)
——— 1953: Biological aspect of Japanese abalone cultivation. Tokai Shobo, Tokyo. (in Japanese)
Iwata, K. S. 1951: Auto-activation of eggs of *Mactra veneriformis* in sea water. Annot. Zool. Jap., **24**: 187—193.
Korschelt und Heider, 1936: Vergleichende Entwicklungsgeschichte der Tiere. Jena.
Kowalevsky, A. 1883: Embryologie du Dentale. Ann. Mus. d'Hist. Nat. Marseille, 1.
——— 1883: Embryologénie du *Chiton*. Ann. Mus. d'Hist. Nat. Marseille, 1.
Kubota, T. 1954: The development of the egg of *Limnaea pervia* v. Martens. Sci. Rep. Kagoshima Univ., **3**: 61—73.
Lacaze-Duthiers, H. 1856—7: Organisation et dévelopement du Dentale. Ann. Sc. Nat., **6**, **7**.
Lamy, E. 1928: La ponte chez les gastéropodes prosobranches. Jour. Conchyliologie, **72**: 25—52, 80—126.
Lang, A. 1896: Textbook of Comparative Anatomy. Pt. 2. MacMillan, London
Lebour, M. V. 1936: Notes on the eggs and larvae of some Plymouth prosobranchs. J. Mar. Biol. Assoc., **20**: 547—565.

——— 1937: The eggs and larvae of the British prosobranchs with special reference to those living in the plankton. J. Mar. Biol. Assoc., **22**: 105—166.

——— 1938: The life history of *Kellia suborbicularis*. J. Mar. Biol. Assoc., **22**: 447—451.

——— 1945: The eggs and larvae of some prosobranchs from Bermuda. Proc. Zool. Soc. London, **114**: 462—489.

Lefevre, G. and W. C. Curtis 1912: Studies on the reproduction and artificial propagation of fresh-water mussels. Bull. Bur. Fisheries, **30**: 105—201.

Lillie, F. R. 1893: Preliminary account of the embryology of *Unio complanata*. J. Morph., **8**: 569—578.

——— 1895: The embryology of the Unionidae. J. Morph., **10**: 1—100.

MacBride, E. W. 1914: Textbook of Embryology. London.

MacGinitie, G. E. 1931: The egg-laying process of a gastropod, *Alectrion fossatus* Gould. Ann. Mag. Nat. Hist., Ser. 10, **8**: 258—261.

von Medem, G. 1943: Untersuchungen über die Ei- und Spermawirkstoffe bei marinen Mollusken. Zoöl. Jahrb., **61**: 1—44.

Meisenheimer, J. 1896—98: Entwicklungsgeschichte von *Limax maximus*. Zeitschr. wiss. Zoöl., **62** & **63**.

——— 1901: Entwicklungsgeschichte von *Dreissensia polymorpha* Pall. Zeitschr. wiss. Zoöl., **69**: 1—137.

Metcalf, M. M. 1893: Contributions to the embryology of *Chiton*. Stud. Biol. Lab. Johns Hopkins Univ., 5.

Miyawaki, M. 1953: Some observations on egg-laying in *Neptuna arthritica*. Zool. Mag., **62**: 199—201.

Miyazaki, I. 1935: Development of Japanese bivalves. I. J. Imp. Fish. Exp. Sta. Tokyo, **31**: 1—14.

——— 1936: Development of Japanese bivalves. II. J. Imp. Fish. Exp. Sta. Tokyo, **31**: 3 ' 50.

——— 1936: Brooding habit and larvae of bivalves. Plants & Animals, **4**: 1879—1886.

——— 1938: Brooding habit and larvae of bivalves. Plants and Animals, **6**: 1213—1218.

Morgan, T. H. 1927: Experimental Embryology. New York.

——— 1933: The formation of the antipolar lobe in *Ilyanassa*. J. Exp. Zool., **64**: 433—467

Okuda, S. 1947: Notes on the post-larval development of the giant chiton, *Cryptochiton steller* (Middendorff). J. Fac. Sci Hokkaido Imp. Univ., **9**: 267—275.

Ostergaard, J. M. 1950: Spawning and development of some Hawaaian Marine Gastropods. Pacific Science, **4**: 75—115.

Portmann, A. 1925: Der Einfluss der Nähreier auf die Larvenentwicklung von *Buccinum* und *Purpura*. Zeitschr. Morph. Oekol. Tiere, **3**: 526—541.

——— 1931: Entstehung der Nähreier bei *Purpura* durch atypische Befruchtung. Zeitschr. Zellf. mikr. Anat., **12**.

Purchon, R. D. 1941: On the biology and relationships of the lamellibranch *Xylophaga dorsalis* (Turton). J. Mar. Biol. Assoc., **25**: 1—39.

Rattenby, J. C. & W. E. Berg 1954: Embryonic segregation during early development of *Mytilus edulis*. J. Morph., **95**.

Raven, C. P. 1945: The development of the egg of *Limnaea stagnalis* L. from oviposition till first cleavage. Arch. néel. Zool., **7**: 91.

——— 1946: The development of the egg of *Limnaea stagnalis* L. from the first cleavage till the trochophore stage, with special reference to its 'chemical embryology'. Arch. neel. Zool., **7**: 353.

——— 1952: Morphogenesis in *Limnaea stagnalis* and its disturbance by lithium. J. Exp. Zool., **121**: 1—72.

Robert, A. 1902: Recherches sur le développement des Troques. Arch. Zool. Exp. Gen , **10**: 268—538.

Schleip, W. 1925: Die Furchung dispermer *Dentalium*-Eier. Arch. Entw.-mech., 106.

Smith, F. G. W. 1935: The development of *Patella vulgata*. Phil. Trans. Roy. Soc. London Ser. B Biol. Sci. No. 520, **225**: 95—125.

Staiger, H. 1950: Zur Determination der Nähreier bei Prosobranchiern. Rev. Suisse Zool. **57**: 495—503.

Stauffacher, H. 1894: Eibildung und Furchung bei *Cyclas cornea*. Zeitschr. Naturwiss., **28**.

Taki, 1933: Gastropoda, Pelecypoda. Iwanami Biol. Course, Iwanami Shoten, Tokyo.

Thorson, G. 1935: Studies on the egg capsules and development of Arctic marine prosobranchs. Meddel. om Grbnland., **100**: 1—71.

——— 1940: Studies on the egg masses and larval development of Gastropoda from the Iranian Gulf. Danish Sci. Investig. in Iran, pt. **11**: 159—238.

——— 1946: Reproduction and larval development of Danish marine bottom invertebrates. Kom. Danm. Fick. Havuders gesler, Meddel. Ser. Plankton, **4**(1): 1—523, 4(7): 1—276.

Wada, S. K. 1941: Artificial fertilization and development of *Pinctada maxima*. (Jameson) South Sea Sci., **4**: 202—208. (in Japanese)

——— 1942: The oysters of Palau. South Sea Sci., **5**: 69—73. (in Japanese)

——— 1942: Coral reef Mollusca. Ocean Science, **3**: 499—507.

——— 1947: Artificial fertilization in *Pinctada martensii* by means of ammoniacal sea water. Bull. Jap. Sci. Fish., **13**: 59.

——— 1947: Fertilization in *Mytilus edulis*, with special reference to the sperm acrosome reaction. Rep. Coll. Fish., Kagoshima Univ., **4**: 105—112.

——— 1953: Larviparous oysters from the tropical West Pacific. Rec. Oceanogr. Works Japan, N. S., **1**: 66—72.

——— 1954: Spawning in the tridacnid clams. Jap. J. Zool., **11**: 273—285.

——— & R. Wada 1953: On a new pearl oyster from the Pacific coast of Japan, with special reference to the cross-fertilization with another pearl oyster, *Pinctada martensii* (Dunker). S. Oceanogr. Soc. Jap., **8**: 127—138.

———, J. R. Collier & J. C. Dan 1956: Studies on the acrosome. V. An egg-membrane lysin from the acrosome of *Mytilus edulis* spermatozoa. Exp. Cell Res., **10**: 168—180.

Watabe, T. & R. Yuki 1952: On the mineral components of the shell from the larval to the young adult stage in *Pinctada martensii*. Zool. Mag., **61**: 118. (in Japanese)

——— 1956: X-ray analysis of the primary shell, adult shell and pearls of *Pinctada martensii* (Dunker). Sci., **26**: 359—360 (in Japanese)

Wierzejski, A. 1897: Über die Entwicklung der Mesoderm bei *Physa fontinalis*. Biol. Centralb., **17**.

——— 1905: Embryologie von *Physa fontinalis*. Zeitschr. wiss. Zoöl., 83.

Wilson, E. B. 1904: Experimental studies in germinal localization. (1) The germ region in the egg of *Dentalium*. (II) Experiments on the cleavage-mosaic in *Patella* and *Dentalium*. J. Exp. Zool., **1**: 197—268.

Woods, F. H. 1931—32: Keimbahn-determination and continuity of the germ-cells in *Sphaerium*. J. Morph., **51** and **53**.

Yamamoto, G. & U. Nishioka 1943: Artificial fertilization of some pelecypods. Zool. Mag., **55**: 372—373. (in Japanese)

Yoshida, H. 1953: Studies of the young forms of some useful shallow-water bivalves. J. Imp. Fish. Exp. Sta., Tokyo, **3**: 1—106. (in Japanese)

Yonge, C. M. 1926: The structure and physiology of the organs of feeding and digestion in *Ostrea edulis*. J. Mar. Biol. Assoc., **14**: 295—386.

Ziegler, H. H. 1885: Entwicklung von *Cyclas cornea*. Zeitschr. wiss. Zoöl., **41**.

II. CEPHALOPODA

INTRODUCTION

The Cephalopoda are the most highly evolved of the classes which make up the phylum Mollusca. The genera of this group which have served as the chief objects of embryological study may be classified as follows:

- Cephalopoda
  - Tetrabranchia - - - *Nautilus*
  - Dibranchia
    - Octopoda - - - *Argonauta*, *Octopus*
    - Decapoda
      - Myopsida - *Loligo*, *Sepioteuthis*, *Sepiella*, *Sepia*
      - Oegopsida - - *Ommastrephes*

The cephalopods are all marine animals, and heterosexual; their eggs are usually large, and heavily laden with yolk. The species found along the coast deposit attached egg masses, in which the eggs are largely non-buoyant, whereas species living in the open ocean often lay buoyant eggs. The development of the most common European squids, *Sepia officialis* and *Loligo vulgaris*, has been the object of study for many years: for example, the cleavage of *Sepia* was described by Vialleton in 1888. In Japan, also, the ease with which the eggs of *Sepia, Sepiella* and *Loligo* can be collected has made them objects of interest to many investigators; the artificial fertilization and early cleavages of *Ommastrephes sloani pacificus* have also been observed recently by Soeda (1952).

The developmental history of the Tetrabranchia is still not fully known, but it is believed to agree in general with the embryology of the other group. In this chapter the early development of *Ommastrephes sloani pacificus* will be described; details with respect to the larval forms and more complex organogenesis in the various species will be taken from appropriate sources.

1. EARLY DEVELOPMENT

(1) SPERMATOZOA

Among the cephalopods, the spermatozoa leave the body not as separate cells, but enclosed in small tubular sheaths, the *spermatophores* (Fig. 11.35). Sasaki (1929) has used specific differences in the size and structure of the spermatophores as diagnostic characters for classifying these animals. After the spermatozoa have been formed in the testis, they pass through a long, slender, twisted *sperm duct* into the *vesicula seminalis,* and are then apportioned into small lots which are each enclosed in a spermatophore. These are temporarily collected in bundles in a *spermatophore sac* (or *Needham's sac*) where they are stored until copulation takes place. At this time the male uses its *hectocotylized arms* to implant the spermatophores within the mantle cavity of the female, or on the buccal membrane surrounding the mouth. In some cases a *spermatophore receptacle* adapted to receive and attach the spermatophores develops ventrally on the inner edge of the buccal membrane in the females during the breeding season.

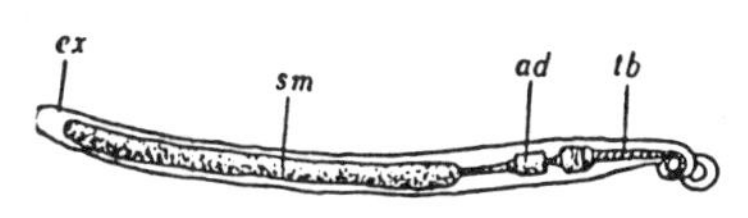

Fig. 11.35 Spermatophore of *Sepiella maindroni* (Ishikawa)
ad: adhesive body, *ex:* external sheath, *sm:* mass of spermatozoa, *tb:* emission duct.

(2) EGGS

In spring, *Sepiella maindroni* and *Sepia esculenta* come well into the bays along the coast of Japan from Tokyo toward the southwest, and lay their eggs on sea weeds and other objects in shallow water. Each of the eggs of *Sepiella* is surrounded by a jet-black, strong, rubber-like substance which sticks it to the substrate; the *Sepia* eggs are also attached separately by their white gelatinous coats, which are often covered with sand grains. These protective layers are secreted by the *nidamental glands* as the eggs are spawned. Both of these species spawn in the same localities, and at about the same time (although *Sepiella* may be slightly earlier); it is therefore quite usual to find these black and white eggs mixed together.

In early summer *Sepioteuthis lessoniana* comes close to the coast and spawns. The eggs of this species are laid in strings of about eight, surrounded by jelly. In *Loligo bleekeri*, the eggs are somewhat smaller than those of *Sepioteuthis,* and 50 to 60 of them in a row are enclosed in a jelly mass shaped like a finger; a number of these are deposited in a cluster. Since the sexually mature animals readily spawn in captivity, these eggs make a convenient material for embryological research.

The non-buoyant eggs of *Ommastrephes* are laid during the summer in Hokkaido waters. *Octopus vulgaris* attaches its smaller eggs one at a time during a long breeding season that lasts from spring to early autumn. These animals are easy to keep in an aquarium, where they often spawn. The embryos hatch,

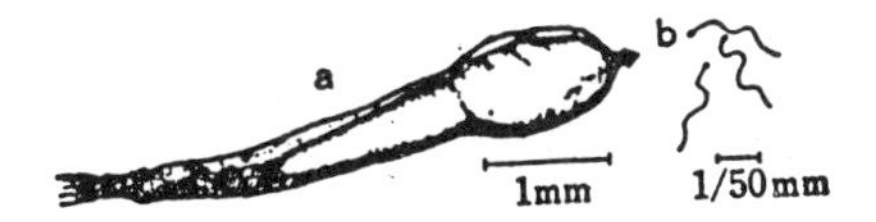

Fig. 11.36 a. Spermatophore of *Ommastrephes sloani pacificus*. b. spermatozoa of *O. sloani pacificus* taken from spermatophore receptacle. (Soeda)

and can frequently be seen swimming about in the tank, but it is difficult to rea the young animals. Among the *Argonauta*, small eggs like those of *Octopus* are attached inside the female's shell, where they are protected until they hatch.

(3) ARTIFICIAL FERTILIZATION IN *OMMASTREPHES*

The body of a sexually mature squid is enlarged so that the mantle reaches a length of about 25 cm. If the mantle of such a female is cut open along the mid-ventral line, the distended ovaries will be seen to be full of clear, amber-colored eggs, and the oviducts at the two sides will also be swollen and showing the amber color of the ripe eggs. Near the openings of the ducts ore the conspicuous, opaque-white *oviduct glands*, and the two centrally located nidamental glands are enlarged so that they nearly conceal the viscera.

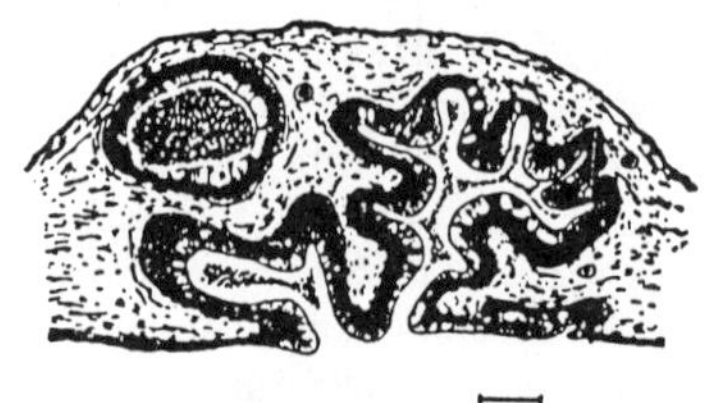

Fig. 11.37 Longitudinal section of part of seminal receptacle of *O. sloani pacificus* (Soeda)

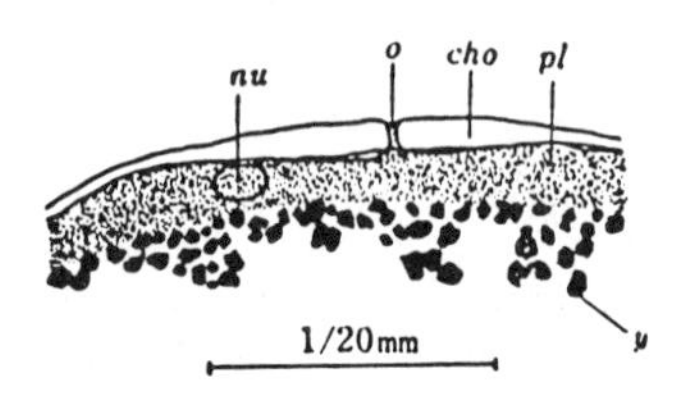

Fig. 11.38 Section through micropyle of unfertilized egg of *O. sloani pacificus* (Soeda)

cho: chorion *nu*: egg-pronucleus *o*: micropyle, *pl*: cytoplasm, *y*: yolk

To obtain eggs for a study of fertilization, the oviducts of a squid in this condition are gently pressed with a glass rod. Spermatozoa in the proper condition to fertilize these eggs may be squeezed out of the seminal receptacles, which can be recognized as white dots on the inner surface of the buccal membrane. There are about thirty of these seminal receptacles on the buccal membrane, but in the fresh material it is difficult to detect them with the naked eye. They have openings on the inner (mouth) side of the membrane, and branch internally; they are sometimes given the name of "fertilization sac". As these become filled with sperm they take on a white, opaque appearance, and form conspicuous protuberances on the outer side of the membrane. Spermatozoa contained in spermatophores which have been taken from the spermatophore sac of the male, or implanted on the buccal membrane or outer lip of the female (Fig. 11.36) are still not fully motile; Sasaki (1926), Isahaya (1933), Soeda (1952) and others have been unable to get satisfactory fertilization when they used sperm from such sources.

(4) UNFERTILIZED EGGS

The clear, amber-colored eggs taken from the oviducts of *Ommastrephes* are in the primary oocyte stage. There is a micropyle at the rounded end of the egg, and the chorion over this region shows a local thickening. The cytoplasm is separate from the yolk, lying at the egg surface in a layer which is slightly thicker at the animal than at the vegetal pole. Directly beneath the chorionic thickening this layer is somewhat depressed and displaced. The egg pronucleus lies in the layer of cytoplasm near the micropyle (Fig. 11.38).

(5) SYNGAMY

If these primary oocytes are inseminated with activated sperm from the seminal receptacles, fixed at intervals and sectioned, the head and midpiece of a spermatozoon can be found lying directly below the micropyle; nearby can be seen the metaphase spindle of the first maturation division (Fig. 11.39a). At first the spindle axis lies parallel to the egg surface, but it then turns to a perpendicular position and the first polar body is given off (Fig. 11.39c, e). This is followed

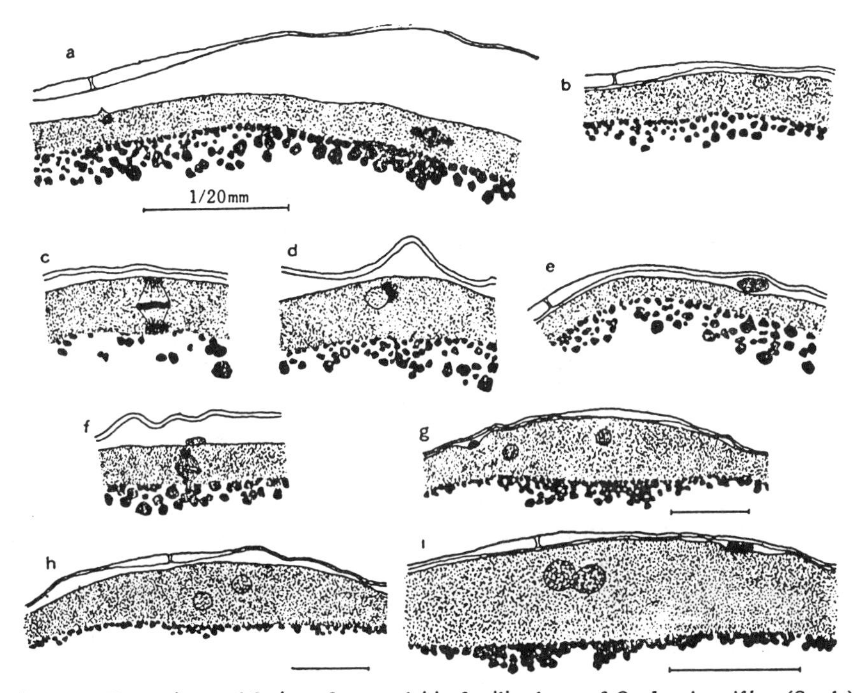

Fig. 11.39 Formation and fusion of pronuclei in fertilized egg of *O. sloani pacificus* (Soeda)

a. spermatozoon has entered through micropyle, first meiotic spindle at metaphase; b. sperm pronucleus, moving away from micropyle; c. rotation of first meiotic spindle; d. migration of sperm pronucleus; e. formation of first polar body; f. formation of second polar body; g,h. egg and sperm pronuclei approach each other; i. fusion of pronuclei near micropyle.

by the formation of the second polar body (Fig. 11.39f), which is somewhat smaller than the first. The chromosomes which are left in the cytoplasm immediately form the egg pronucleus; this begins to move through the cytoplasm to join the sperm pronucleus (Fig. 11.39g, h). In the meantime the sperm head and midpiece rotate around each other and continue to move about in the vicinity of the micropyle until the formation of both egg and sperm pronuclei has been completed. At this time the sperm pronucleus is practically identical with that of the egg in size and shape, and the two pronuclei finally approach each other and unite in syngamy (Fig. 11.39i).

(6) EARLY CLEAVAGE STAGE

At about 20 minutes after insemination the chorion can be seen to have expanded, with the micropyle clearly visible at the more rounded end. Finally it acquires the elongated shape of a hen's egg, the micropyle marking one pole of its long axis, while the egg itself is spherical (Fig. 11.40a). The polar bodies can be seen on the egg surface at the animal pole. About 45 minutes after fertili-

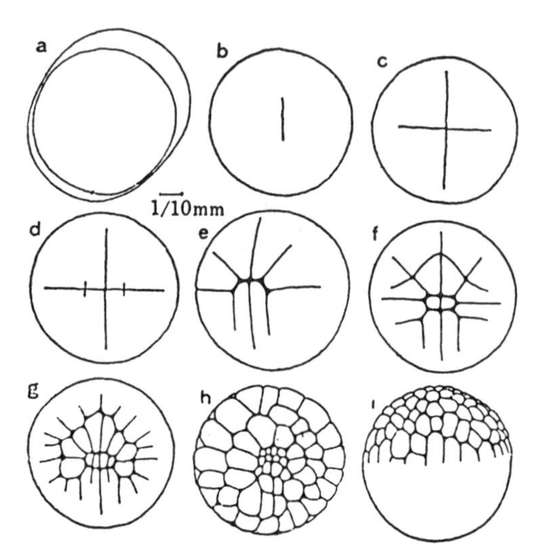

Fig. 11.40 Cleavage in *Ommastrephes sloani pacificus* (Soeda)

a. fertilized egg; b. 45 mins after fertilization; c. completion of second cleavage; d. 3 hours after fertilization; e. completion of third cleavage; f. 4 hours after fertilization; g. fifth cleavage (note 4 micromeres); h. about $6^1/_2$ hours after fertilization; i. about $8^1/_2$ hours after fertilization.

zation the first division begins with the appearance of a cleavage furrow passing through the animal pole (Fig. 11.40b). This gradually elongates so that it includes more than two-thirds of the egg diameter at one hour after fertilization. About an hour later the second cleavage furrow begins to appear at the same point as the first, but perpendicular to it; this elongates until it becomes as long as the first (Fig. 11.40c). At about three hours after fertilization the third division sets in, perpendicular to the second cleavage furrow (Fig. 11.40d). The points where these furrows begin are near the first cleavage plane, but whereas they run parallel to it on one side of the egg, they diverge from it by about 45° on the other (Fig. 11.40e). The fourth division is completed about four hours after fertilization; this results in the formation of two centrally located *micromeres* (Fig. 11.40f). After this, the rate of cleavage begins to show local differences, and the marked increase in the number of furrows makes it difficult to follow the segmentation process by simply observing the surface of the embryo. By six and a half hours after fertilization, 32 to 64 cells can be seen, and the number of central micromeres is clearly eight (Fig. 11.40h). During the next two hours, continued cleavage produces a single-layered *blastoderm,* but since these cells are confined to the animal hemisphere of the embryo, it should rather be called a *blastodisc* (Fig. 11.40i).

In a cross-section of this stage it is difficult to distinguish between the micromeres and macromeres; there is simply a single layer of blastomeres overlying the animal half of the egg, with as yet no formation of a yolk epithelium (Fig. 11.41a). About an hour and a half later (10.5 hrs after fertilization), a longitudinal section of the embryo shows a blastula stage which has a single layer of blasto-

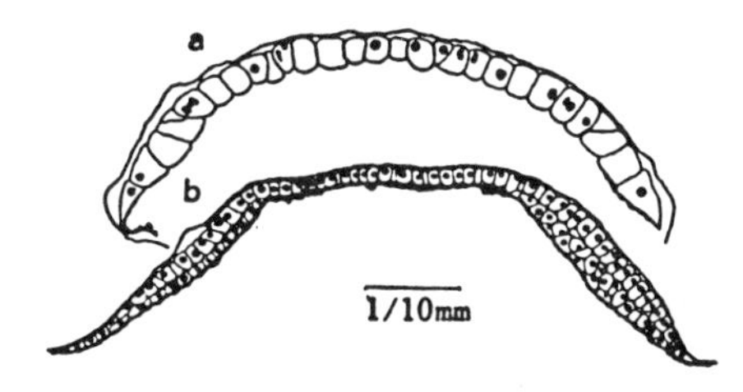

Fig. 11.41 Longitudinal sections of part of *Ommastrephes sloani pacificus* egg (Soeda)
a. about 9 hours after fertilization; b. about 10½ hours after fertilization.

meres at the center, encircled by a double-layered area (Fig. 11.41b). This description of the superficial cleavage by which the early embryo takes shape in the animal hemisphere of the egg should make it clear that the segmentation processs in the Cephalopoda is strikingly different from that found in the other molluscan classes.

(7) EMBRYONIC DEVELOPMENT

At one point on the edge of the blastodisc, very active cell proliferation takes place; this is the site of the future *anus.* Immediately beneath the blastoderm, endoderm is being formed as the result of this proliferation, at first in a crescent-shaped area, which gradually enlarges and becomes circular. The embryo thus acquires two layers, the *ectoderm* of the blastodisc overlying the newly formed *endoderm.*

The ectodermal cells of the anal side proliferate laterally, and extend toward the anterior in the space between the ecto- and endodermal layers, to give rise to the *mesoderm.* Next the endoderm thickens along the midline and forms a vesicle, the *archenteron,* from which will be differentiated a *stomach, intestine* and bilobate *liver.* The *anal invagination* forms near the vegetal pole of the blastodisc, and the *stomodaeal invagination* opposite it. The latter gives off the *buccal mass, radula sac* and *salivary glands,* and connects with the already formed stomach. The *proctodaeum* forms later, and makes connection with the intestine by a shallow invagination.

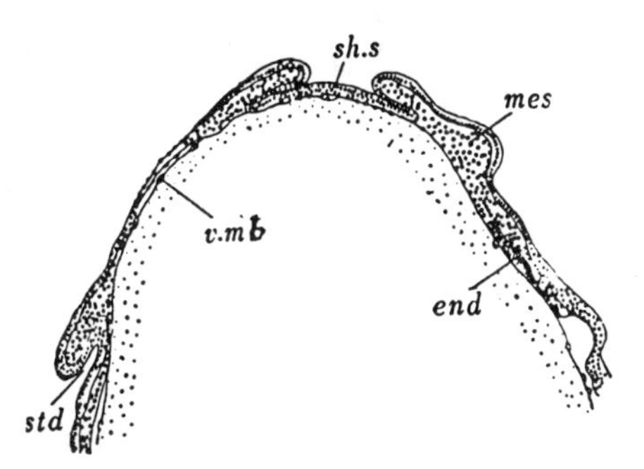

Fig. 11.42 Longitudinal section of part of *Loligo vulgaris* embryo at early blastoderm stage (Faussek)
end: endoderm, *mes:* mesoderm, *sh.s:* shell sac, *std:* stomodaeum, *v.mb:* vitelline membrane.

To the left and right of the intestine the mesoderm forms vesicles, the *body cavities.* From the inner walls of these coelomic cavities arise the *kidneys* and the *pericardium,* which in turn forms the *heart.* The left and right kidneys develop independently in their original positions. The *gonads* are formed from the walls of the coelomic cavities.

At the surface, in the meantime, the ectoderm in the central part of the blastodisc bulges to give rise to a heart-shaped protuberance, the rudiment of the *shell gland.* The periphery of this rudiment thickens and then bends inward, finally drawing together to produce a sac (Fig. 11.42 *sh.s*). This *shell sac,* within which the *shell* will be formed, does not correspond to the shell gland seen in the embryos of other molluscs, as was early pointed out by Lankester (1875).

In *Sepioteuthis, Loligo* and *Ommastrephes,* among others, the shell is a feather-shaped structure made of a transparent material, while that of *Sepia* is boat-shaped and made of calcium carbonate. The external shell of *Nautilus* and *Argonauta* does not make its appearance in the early part of embryonic development, since it arises as a secondarily derived structure. In some of the species of *Euprymna* and most species of *Octopus,* the shell sac is poorly developed, and fails to form any shell.

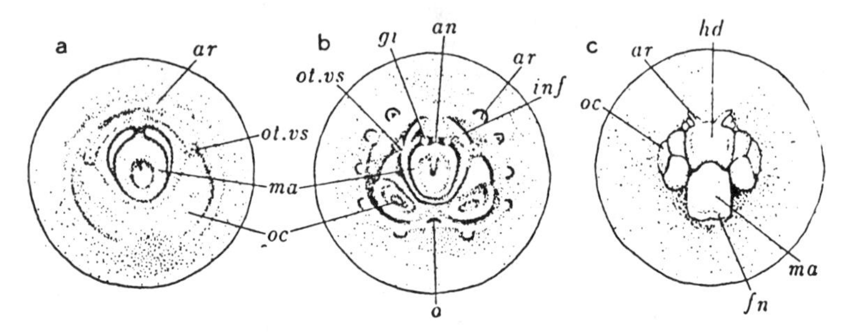

Fig. 11.43 Three successive stages of blastoderm development in *Sepiella maindroni* (Ishikawa)

a. embryo at 11 days after first cleavage (ca. 6 times natural size); b. more advanced stage (ca. 6 times),c. 21 days after first cleavage (ca. 4.5 times). *an:* anus, *ar:* arm, *fn:* fin rudiment, *gi:* gill, *hd:* head, *inf:* funnel, *ma:* mantle, *o:* mouth, *oc:* ocellus, *ot.vs:* otocyst.

Around the periphery of the shell sac, the rudiment of the *mantle* appears as another protuberance, and the ectoderm at the lateral and posterior edges of the blastodisc gives rise to outwardly directed processes, the *arm* rudiments (Figs. 11.43, 11.44a, c *ar*). As these arms elongate, they also move along the sides of the body toward the mouth, and their bases unite and finally surround it. Between the arm rudiments and the anus, a protuberance arises on each side of the latter. These are the *gill* rudiments, and the folds which form them are the *gill lobes* (Figs. 11.43, 11.44 *gi*). Two lobate projections which develop above the arms grow, and their edges turn inward and join together into a small tube which is the *funnel* (Fig. 11.44 *inf. a, inf.p*). In *Nautilus* these funnel folds simply increase in size and overlap each other, without fusing to form a tube.

The *head* is formed at the anterior side of the blastodisc (Fig. 11.43 *hd*), and two large lateral projections at its posterior corners represent the *eye stalks.* Invaginations at the ends of these become the *eyes* (Fig. 11.43, 11.44 *oc,* 11.47).

The *brain-, optic, visceral* and *pedal ganglia* all originate separately as streak-like areas of the ectoderm which gradually sink below the surface as they become differentiated (Figs. 11.44, 11.45). Two ectodermal masses at the sides of the optic nerves are the *white bodies,* which are believed to be vestiges of the lateral lobes of the brain-ganglia. The *cranial cartilage* surrounding the brain, which is characteristic of the cephalopods and not found in any other molluscan group, is formed by a modification of connective tissue, derived from the mesoderm near the bases of the arms.

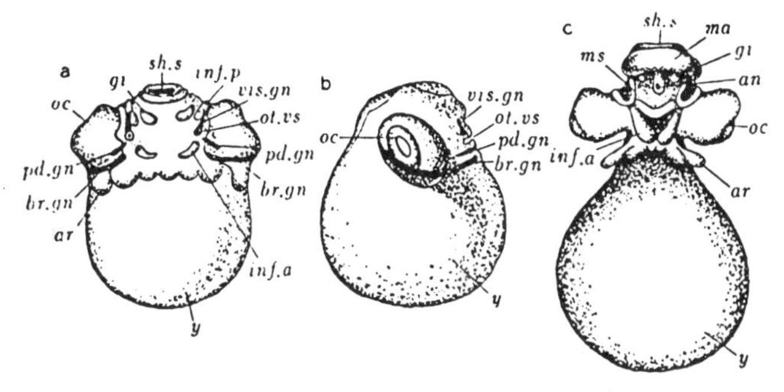

Fig. 11.44 Embryos of *Loligo vulgaris*

a. embryo viewed from posterior; b. embryo viewed from side (Faussek); more advanced stage (Korschelt). *an:* anus, *ar:* arm rudiment, *br.gn:* rudiment of cerebral ganglion, *gi:* gill rudiment, *inf.a:* rudiment of anterior funnel fold, *inf.p:* rudiment of posterior funnel fold, *ma:* mantle, *ms:* retractor muscle of funnel, *oc:* ocellus, *ot.vs:* otocyst, *pd.gn:* rudiment of pedal ganglion, *sh.s:* shell sac, *vis.gn:* rudiment of visceral ganglion, *y:* yolk.

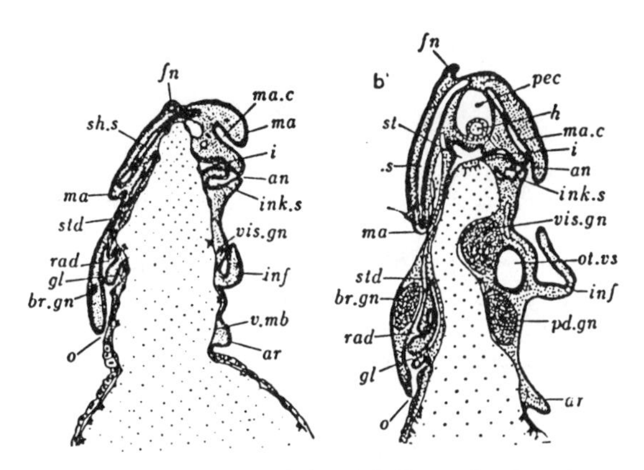

Fig. 11.45 Sagittal sections of *Loligo vulgaris* embryos to illustrate the formation of internal organs (Korschelt)

a. young embryo with incomplete alimentary canal; b. more advanced stage in which stomodaeum and mid-gut have joined. *an:* anus, *ar:* arm, *br.gn:* cerebral ganglion, *fn:* fin rudiment, *gl:* salivary gland, *h:* heart, *i:* intestine, *inf:* funnel, *ink.s:* ink sac, *ma:* mantle, *ma.c:* mantle cavity, *o:* mouth, *ot.vs:* otocyst, *pd.gn:* pedal ganglion, *pec:* pericardium, *rad:* radula sac, *sh.s:* shell sac, *st:* stomach, *std:* stomodaeum, *vis.gn:* visceral ganglion, *v.mb:* vitelline membrane.

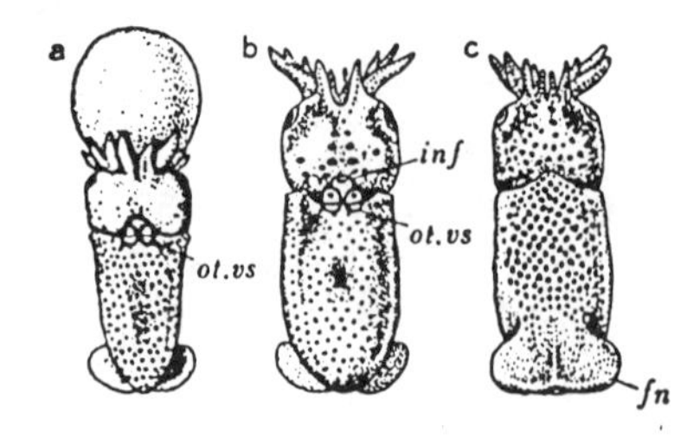

Fig. 11.46 Larvae of *Sepioteuthis lessoniana* x3.3 (Ishikawa)
a. ventral view, otocysts clearly visible beneath funnel; b. more advanced larva in which yolk has disappeared; c. dorsal view of same larva, in which fin is still limited to posterior end. *fn:* fin, *inf:* funnel, *ot.vs:* otocyst.

These developmental changes occur at the expense of the yolk, which is gradually reduced in amount. The *visceral sac* expands rapidly, and is soon definitely larger than the head. As the embryo acquires a shape practically identical with that of the adult, it secretes a hatching enzyme, dissolves its chorion, and swims forth to begin its free-living stage (Fig. 11.46).

2. LARVA

After hatching, cephalopod larvae pursue a direct course of development to the adult form, with practically no abrupt changes in their morphology. From this it follows that they by-pass the metamorphoses through the trochophore and veliger stages which characterize the other molluscs. As larvae, all the cephalopods lead a similar free-swimming life, including those like the octopus which will change to a sedentary habit after they are full-grown. Only after they have finally acquired the size and completed organ systems of the adult do they begin to carry on the various modes of life to which they are adapted.

(1) STRUCTURAL ACQUISITIONS

a. Fins

The fins which appeared as ectodermal projections from the posterior end of the embryonic body suddenly begin to grow, after hatching, and rapidly reach full size. Even the fins of *Sepia* and *Sepioteuthis,* which extend on both sides along the whole length of the mantle in the adult, are only small flaps on the outer end of the mantle when the larva hatches. These gradually spread toward the anterior, and finally the two sides together form an oval fin (Fig. 11.46). In general it can be said that the younger a larva is, the larger will its fin be in comparison with the size of its mantle.

b. Suckers

Although the embryo hatches with its suckers still extremely incomplete in structure, number and distribution, these organs become fully developed during the larval period. Among the squids the suckers are generally provided with a *horny ring* on the inner side of the pore, and are attached to the arms by prominent *stalks,* while those of the octopus species lack both. These features therefore serve as important criteria in cephalopod taxonomy.

(2) COMPLETION OF ORGANOGENESIS

That the cephalopods have outdistanced the other molluscan species to an extraordinary degree in their evolutionary development has been stressed repeatedly; their most conspicuous advance is seen in their nervous systems and sense organs.

a. Eyes

The eyes of these animals have a structure amazingly similar to that of the vertebrate eye, and their function is believed to be very highly developed (Fig. 11.47).

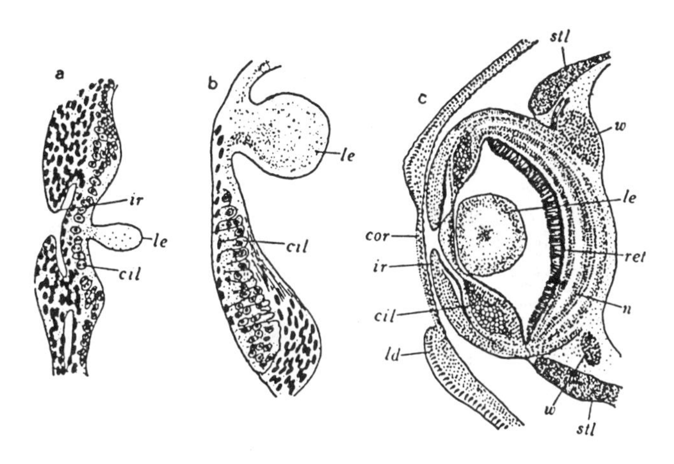

Fig. 11.47 Longitudinal sections through developing eyes of squid larvae (Faussek)

a. formation of lens in *Loligo vulgaris*; b. formation of lens in *Sepia officinalis*; c. almost completely developed eye of *S. officinalis* larva. *cil:* cells of ciliary body, *cor:* cornea, *ir:* iris, *ld:* rudiment of lower eye-lid, *le:* lens at early stage, *n:* retinal nerve cell, *ret:* retina, *stl:* eye stalk, *w:* rudiment of white body.

b. Statocysts

Although the *statocysts (otocysts)*, which like the eyes are formed as ectodermal invaginations at a very early stage in the development of the blastoderm, are at first only simple sacs, they occupy a prominent position at the center of the embryo and are clearly visible even from the outside (Figs. 11.46, 11.48). These organs have about the same structure in both squid and octopus embryos at the time of hatching, but during the larval period each species acquires particular characteristics of its own. For example, the internal morphology of octopus statocysts is extremely simple, with a single process close to the otolith, while there are five such processes in *Idiosepius*, six in *Euprymna*, nine in *Cranchia*, ten in *Watasenia*, eleven in *Sepia* and *Ommastrephes* and twelve in *Sepiella* and *Loligo*. Whereas the statocysts of a 5 mm larva of *Sepia* have, like the *Euprymna* larval statocyst, only six processes, they acquire the full number of eleven by the time the mantle length reaches 21 mm (Fig. 11.48d, e). As the animal grows the processes thicken, greatly reducing the space within the statocyst, so that it appears

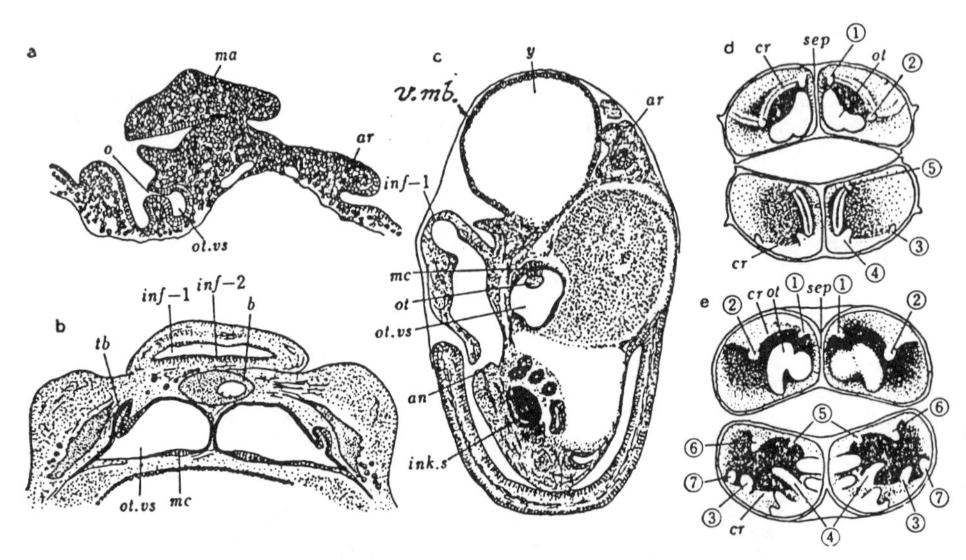

Fig. 11.48 Development of otocyst (Ishikawa)

a. longitudinal section of otocyst in *Sepia* larva, showing Kolliker's Passage; b. transverse section of otocyst of 2.5 mm *Loligo* larva x50; c. longitudinal section through otocyst of *Argonauta argo* larva x40; d. bisected otocyst of *Sepia* larva (mantle length 5 mm), sectioned just below funnel; e. bisected otocyst of *Sepia* larva (mantle length 21 mm). 1 ~ 7: numbers of otocyst processes. *an:* anus, *ar:* arm, *b:* blood vessel, *cr:* crista statica, *inf*-1: lumen of funnel, *inf*-2: funnel wall, *ink.s:* ink sac, *ma:* mantle, *mc:* macula statica, *o:* opening of invagination, *ot:* otolith, *ot.vs:* otocyst, *sep:* septum, *tb:* Kolliker's Passage, *v.mb:* vitelline membrane, *y:* yolk.

very different from that of the larval stage. In a comparative study of the development and structural characteristics of the cephalopod statocyst, the author (Ishikawa 1924, 1929) has found that the ontogeny of these organs reflects the evolution of the species.

REFERENCES

Behre, E. H. 1941: Observations on the later embryology of the squid, *Loligo brevis* Blainville. Anat. Rec., **81**: (4) suppl. 38.

Faussek, V. 1900: Untersuchungen über die Entwicklung der Cephalopoden. Mitt. Zoöl. Sta. Neapel, **14**.

Hirose, Y. 1956: Development of *Sepia subaculeata*. Formation of external structures and embryonic movements. Collected Studies, Dept. Ed., Kobe Univ., 12. (in Japanese)

Humoto, S. & Y. Hirose 1957: A study of hatching in squid eggs. Collect. of Works on Fish. Sci. (Amemiya Jubilee), Univ. Tokyo Press. (in Japanese)

Isahaya, T. & S. Kawakami 1933: An experimental study of breeding in *Ommastrephes sloani pacificus*. Sci. Papers Hokkaido Regional Fish. Res. Lab., 216. (in Japanese)

Ishikawa, M. 1924: On the phylogenetic position of the cephalopod genera of Japan based on the structure of statocysts. J. Coll. Agr. Imp. Univ. Tokyo, **7**.

——— 1929: On the statocysts of American cephalopod genera. J. Morph. Physiol., **48**.

——— 1932: More curious animals. Octopi and squids. Shinkosha, Tokyo. (in Japanese)

——— 1933: Zoology: Cephalopoda. Iwanami Study Courses. Iwanami Shoten, Tokyo. (in Japanese)

Korschelt, E. 1892: Beitrag zur Entwicklungsgeschichte der Cephalopoden. Festschr. Leuckart.

Lankester, E. R. 1875: Observation on the development of Cephalopoda. J. Micr. Sci., **15**.

MacBride, E. W. 1914: Textbook of Embryology. London.

Naef, A. 1921: Fauna e Flora del Golf di Napoli.

——— 1923: Die Cephalopoden. I. Fauna e Flora del Golf di Napoli.

——— 1928: Die Cephalopoden. II. Embryology. Fauna e Flora del Golf di Napoli, **35**.

Sasaki, M. 1921: On the life history of an economic cuttlefish of Japan, *Ommastrephes sloani pacificus*. Trans. Wagner Free Inst. IX, Phila.

——— 1929: Supplementary notes on the life history of an economic cuttlefish, *Ommastrephes sloani pacificus* (Steenstrupp). Jap. J. Zool., **2**.

——— 1929: A monograph of the dibranchian cephalopods of the Japanese and adjacent waters. J. Coll. Agr., Hokkaido Imp. Univ., **20**.

Soeda, J. 1949: Reproduction in *Ommastrephes sloani pacificus*. I. On the size and number of eggs. Monthly Rep. Hokkaido Fish. Sci. Inst., **6**. (in Japanese)

——— 1952: On the artificial fertilization and early cleavage of the *Ommastrephes* egg. Hokkaido Reg. Fish. Res. Rep., **5**. (in Japanese)

——— 1954: On the fertilization and shape of the *Ommastrephes* egg. Hokkaido Reg. Fish. Res. Rep., **11**. (in Japanese)

Watase, S. 1891: Studies on Cephalopods. I. Cleavage of the ovum. J. Morph., **4**.

Yamamoto, K. 1940: The effect of salinity on young squids. Bot & Zool., **8**.

——— 1941: Observations on the ecology of the *Sepia* larva and changes in the spots on its dorsal mantle. Bot. & Zool., **9**. (in Japanese)

——— 1942: Development of the *Sepia* egg. Bot. & Zool., **10**. (in Japanese)

——— 1946: On the eggs and larvae of *Ommastrephes* from the coastal waters of Korea. Jap. J. Malacology, **14** (5—8). (in Japanese)

——— 1949: Sexual dimorphism with respect to mantle length in *Doryteuthis bleekeri* and *Idiosepius paradoxa*. Jap. J. Malacology, **15** (5—8). (in Japanese).

Yasuda, J. 1951: Some observations on the ecology of *Sepia esculenta* Hoyle. Bull. Jap. Soc. Sci. Fish., **16** (8). (in Japanese)

TUNICATA

Chapter 12

INTRODUCTION

This report will deal exclusively with the development of the Ascidiacea, one of the two orders which make up the class Tunicata (Urochorda).

The tadpole larva of the Ascidiacea was discovered in 1841 by Milne-Edwards, and in 1847 by van Beneden; as the result of detailed studies by Kowalevsky (1867—1871) and Kupffer (1872), it became clear that the members of this group are highly evolved animals, possessing a notochord. The system currently in use for classifying the Ascidiacea is that of Huus (1937), who adopted a combination of Lahille's and Perrier's classifications; this system, which roughly divides the class into the two orders Enterogona and Pleurogona, does not draw a major line of distinction between the simple and compound species. Since, however, it is convenient from the embryological point of view to differentiate, following Savigny (1816), between the simple ascidians (Monoascidia) which reproduce only sexually, and the compound ascidians (Synascidia) which show both sexual and asexual reproduction, this article will follow the latter classification. The mode of sexual reproduction of simple and compound ascidians is essentially the same, although the conspicuously yolky eggs of some of the compound species follow a slightly different pattern, and the formation of colonies by budding is seen in this group.

The ascidians are objects of special interest because of their larval notochord, and since 1841 their embryology has been studied by a great many people, but such research in this country has been extremely scarce, limited to the author's work since 1939 on the embryology of simple ascidians and studies of the larval forms in both simple and compound species, the work of Oka since 1942 on metamorphosis and colony formation, as well as experimental morphological studies, and the studies of Usui and Watanabe on colony formation.

1. DEVELOPMENT OF SIMPLE ASCIDIANS

(1) REPRODUCTION

Ascidians are without exception hermaphroditic (Berrill 1950). The simple ascidians reproduce sexually; both *oviparous* and *viviparous* forms are found in this group. Among the former are *Ciona, Ascidia, Phallusia, Styela, Pyura, Molgula;* the latter include *Polycarpa, Boltenia, Molgula* and *Corella*. The Japanese species *Halocynthia roretzi* (Drasche) and *Chelyosoma siboja* are oviparous. The fertilized eggs of the viviparous forms develop in the cloacal cavity. With respect to the breeding season of simple ascidia, Berrill (1937) states that there is no special season in which the genera *Ascidia, Molgula, Ciona* and *Eugyra* become sexually mature, and although their breeding activity falls off somewhat during the cold months, they breed practically the year round. He reports that the Styelidae and Pyuridae are limited to the warm months, *Styela partita* reproducing in June to September, and *Pyura vittata* in summer. *Corella willmeriana* also spawns in summer (Child 1927), *Pelonaia corrugata* during two to four weeks in January and February, and *Pyura squamosa* in August (Millar 1951, 1954). In Japan *Halocynthia roretzi* spawns in Miyagi Prefecture from December to February, with a peak in January (Hirai, 1941), but spawning was observed in the middle of November, 1956, in Mutsu Bay, Aomori Prefecture (Hirai and Tsubata 1956). *Corrella japonica,* var. *asamusi,* also from Mutsu Bay, can be artificially fertilized in October; *Chelyosoma siboja* was observed to spawn there from December to January, 1954, and the development of *Styela clava* can be seen in Onagawa Bay, Miyagi Prefecture, during July. At Misaki in Kanagawa Prefecture, *Styela plicata* spawns in June (Hirai 1958)[1]. To be useful as material for embryological research, any given species should be plentiful and easy to collect. *Halocynthia roretzi* occurs in all parts of this country, and is especially numerous in the northeastern section (Oka, A. 1935); since the animals are cultivated on perpendicular rafts in Kesen Numa Bay, Miyagi Prefecture, they are particularly easy to obtain there in large numbers (Hirai 1939, 1940, 1941). *Halocynthia roretzi* and *Chelyosoma siboja* also occur very plentifully in the vicinity of the Asamushi Marine Biological Station of Tohoku University, where they are easily collected with a special ascidian collecting net.

(2) EARLY DEVELOPMENT

a. Methods of observation

The eggs of the Ascidiidae, Rhodosomatiidae, Cionidae, Deazonidae and Molgulidae are easy to fertilize artificially, while this is difficult with the Pyuridae and Styelidae (Berrill 1937; Grave 1937). *Halocynthia* may be sent from the culture grounds wrapped in sea weed dampened with sea water, and kept in a refrigerator for about a week. Artificial self-fertilization is possible with ripe individuals of this species (Hirai 1937, 1941), fertilized eggs can also be obtained following spontaneous spawning in a glass laboratory tank (Hirai and Tsubata 1956). *Corella japonica* eggs give a higher percentage of fertilization when the sperm of a different individual is used. If *Chelyosoma siboja* is cut open, fertilization is inhibited by sulfuric acid contained in the body fluid (Kobayashi 1938); for this reason the author has used naturally spawned eggs. Only a very few fertilized eggs are obtained with self-fertilization, but plentiful amounts of eggs, almost all of which are fertilized, can be secured by using eggs and sperm from different animals. Moreover, it is possible to induce shedding in an individual which has

[1] But more recent, unpublished observations (T. Hidaka, M. Kumé) indicate that the breeding season of this species at Misaki extends from April to November.

spawned two or three days previously by injecting about 3 ml of 1/2 M KCl into the reproductive tract. When this method is used with other species, immature eggs are shed along with the mature ones. In forms such as *Ciona intestinalis*, which have long genital ducts, the ripe gametes can be seen within the separate male and female ducts, and with care can be removed separately and later mixed at will to give artificial fertilization. When the male and female reproductive organs lie close to each other, as in *Styela clava* and *H. roretzi*, and especially in species with short genital ducts, it is impossible to separate eggs and sperm, since they become mixed as soon as the animal is cut open.

The excess spermatozoa should be removed after fertilization in order to observe the later development; as with all embryological materials, it is necessary to make the greatest effort to avoid contamination by other organic matter or reagents (Morgan 1945). As will be described below, the eggs of these animals are surrounded by complicated coverings which make observation difficult in many cases. Berrill (1932) suggests two methods for removing these coverings. One involves the use of the tadpole larva's hatching enzyme or commercially obtained proteolytic enzymes, and the other consists in digesting the membrane with the stomach fluid of decapod crustaceans *(Munid, Maia,* or *Homarus)*. Berg (1956) removed the chorion of the unfertilized *Ciona intestinalis* egg by digestion with a 3% solution of protease in sea water. The author uses sharp steel needles to tear off the covering.

From fertilization to the completion of tadpole metamorphosis, the larvae can be kept in a glass container and simply observed, but after this stage it is necessary to provide them with food. In order to study the later development of *Ciona intestinalis*, Berrill (1947) raised the metamorphosed animals on a diet consisting chiefly of a cultured diatom *(Nitzschia)*.

b. **Structure of the egg**

In the young oocyte of *Styela partita* (Conklin 1905), the nucleus occupies approximately the center of the cell, and the cytoplasm lacks yolk granules. At one side of the nucleus there is a spherical cytoplasmic body which stains deeply with eosin; Conklin described this as an "attraction sphere", but the appearance around it in later stages of yolk granules indicates that it is a *yolk nucleus*. These yolk granules are formed at the center of the egg cell, while pigment granules containing a yellow pigment collect around the periphery. The yolk granules present a slaty-grey appearance. The ascidian egg has a *chorion* (Fig. 12.1 *cho*), outside of which is attached a *follicle membrane* (Fig. 12.1 *fo*); the *perivitelline space* (Fig. 12.1 *perv*) between the chorion and the egg surface contains *test-cells* (Fig. 12.1 *ts*) and a colloidal substance. The membrane and other structures are formed along with the egg, and two modes of origin have been suggested (Berril 1950). According to one, they arise from rather undifferentiated cells among those which constitute the germinal epithelium, and differentiate into test-cells and follicle cells, the test cells becoming closely attached to the periphery of the oocyte, and the chorion being formed in the space between these and the outer layer of follicle cells. The other idea is that these structures arise from amoeboid cells of mesenchymal origin (Wanderzellen) which collect around the occytes. According to Berrill (1950), this sort of formation has been seen by Knaben (1936) in *Corella parallelogramma* and by Spek (1927) in *Clavelina lepadiformis*. Tucker (1942) was unable to confirm these observations in *Styela plicata*. Berrill (1950) has suggested that this failure of agreement may well be due to species differences. Hirai (1939) has observed the process to occur in *Halocynthia roretzi* according to the first scheme — with cells of the reproductive tissue giving rise to the structure in question.

In this species, the fully grown ovarian egg has five kinds of coverings. The outermost of these is the *basement membrane*, next to this is the *outer follicular layer*, then the *inner follicular layer*, the chorion and the test-cells deeply embedded in the periphery of the ovum (Fig. 12.2). The test--cells are thought by some to have a secretory function, and by others to be nutritive cells; Knaben (1936) has suggested that they may produce enzymes. In the egg of *H. roretzi*, selective staining of the test-cells in possible (Hirai 1949)[2]; with this method it can be shown that numerous long, very fine threads extend from the test-cells into the cytoplasm of ovarian eggs, and in the perivitelline space of ripe eggs a mucous secretion of these cells can be seen. It is thus possible to describe them as secretory cells of the perivitelline space.

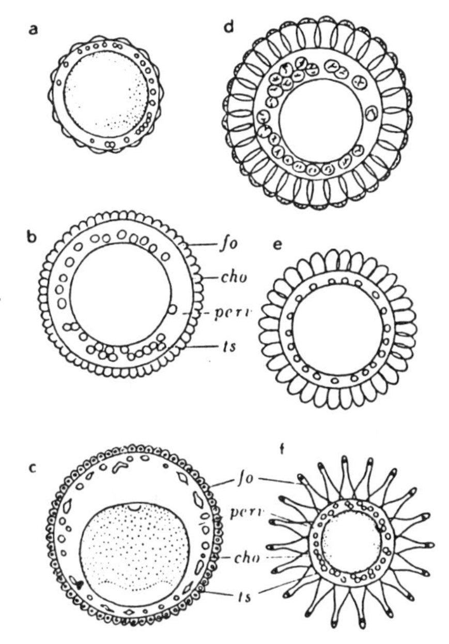

Fig. 12.1 Mature eggs of several simple ascidians

a. *Styelopsis grossularia:* viviparous form, with flattened follicle cells (Berrill 1950); b. *Chelyosoma siboja* (Hirai); c. *Halocynthia roretzi* (Drasche) (Hirai 1941); d. *Corella parallelogramma* (Knaben 1936); e. *C. japonica*, var. *asamusi* (Hirai); f. *Ciona intestinalis* (Berrill 1947), *cho:* chorion, *fo:* follicle cell, *perv:* perivitelline space, *ts:* test cell.

Fully grown ova of the simple ascidians are smaller than those of the compound species: the egg diameter among the Molgulidae is about 0.10—0.20 mm; of the Pyuridae and Styelidae, about 0.13—0.19 mm; of the Ascidiidae, 0.14—0.18 mm; of the Cionidae, 0.16—0.17 mm. The egg of *H. roretzi* is about 0.27 mm in diameter, or 0.30 mm including the envelope, but the perivitelline space is formed after the egg is spawned, practically doubling the diameter of the envelope. The dyameter of the *Chelyosama siboja* egg is about

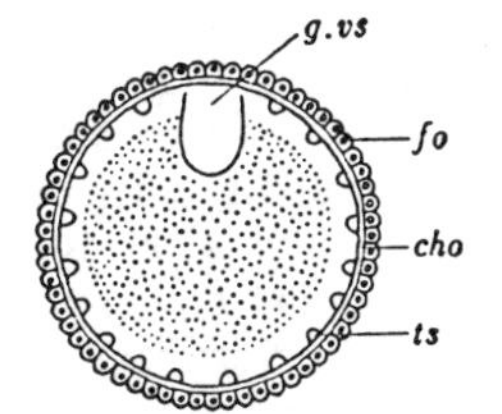

Fig. 12.2 Immature egg, immediately after ovulation (Hirai 1941)

cho: chorion, *fo:* follicle cell, *g.vs:* germinal vesicle, *ts:* test cell.

[2] The eggs were fixed by Zenker-formol solution, and the sections were stained with Nile Blue S or Toluidin Blue.

0.18 mm, or 0.30 mm including the membranes; the *Corella japonica* egg diameter is 0.15 mm, with the membranes, 0.26 mm.

The nearly ripe ascidian egg contains a large amount of granular yolk. The interior of the mature *Styela partita* oocyte consists of yolky cytoplasm, while around the periphery of the cell there is a homogeneous layer of *yellow protoplasm.* According to Conklin (1924), yellow or orange pigment granules and a great number of minute spherules in the superficial layer of the cytoplasm were identified as mitochondria by Duesberg (1915). A large germinal vesicle is seen at one pole of the cell. The spawned egg of *H. roretzi* is dark yellow and rather opaque; the large germinal vesicle, with a diameter about one-third that of the egg, lies excentrically, between the middle of the egg and one pole. In sections it can be seen that the yolk lies in two concentrically circular layers: a central part where the yolk granules are thickly crowded, and around it a region of lower density. At the periphery of the egg cell is a narrow clear zone, without yolk but containing very small granules. Having such a structure, these eggs can be thought of as homolecithal, but with a strong tendency toward the centrolecithal mode of yolk distribution. After maturation and fertilization, moreover, as will be described below, the cytoplasm forms layers perpendicular to the main egg axis, which later change to incline toward one side. The *Chelyosoma* egg is orange or orange-yellow; the egg of *Corella* is dark yellow. The form of the follicle cells differs from species to species, but they are always conspicuously vacuolated and display the characteristics of cells adapted for floating. The follicle cells of *Ascidiella aspersa, Corella parallelogramma* (Fig. 12.1d), and *Ciona intestinalis* (Fig. 12.1f) are large and especially buoyant. Those of *H. roretzi* (Fig. 12.1c) and *Ch. siboja* (Fig. 12.1b) are smaller, but *Corella japonica* (Fig. 12.1e) has large follicle cells. The naturally spawned eggs of *Halocynthia* and *Chelyosoma* are collected with a plankton net at a rather deep level in the sea.

c. Maturation and fertilization

The egg of *Ciona intestinalis* begins the maturation divisions within the ovary, and then immediately moves into the oviduct, where it remains for 24 hours or longer. In the oviduct meiosis proceeds as far as the metaphase of the first maturation division, but the first polar body is not given off until after fertilization. Spawning takes place in the morning. The mature spermatozoa are similarly stored in the sperm duct (Morgan 1945). The ovarian eggs of *Styela partita, S. clava, H. roretzi* and *Co. japonica* will mature in sea water. If the ovary of *Halocynthia* is cut and the oocytes removed to sea water, they are seen to be immature, having a large germinal vesicle, test cells embedded in the cytoplasm, and the follicle cells and chorion not yet elevated from the oocyte surface (Fig. 12.2). About two hours later a perivitelline space begins to be formed between egg surface and chorion, the test cells gradually move toward the egg surface and separate from it to become free in the perivitelline space. Finally the egg is seen to be suspended in the lower part of the distended vesicular envelope (Fig. 12.1c). The elevation of the membrane has no connection with fertilization. About one hour after removal to sea water, the germinal vesicle begins to diminish in size, and after about two hours the nuclear material appears as a small bright spot at the animal pole side of the egg. In the meantime the clear peripheral layer of the egg gradually darkens, and a bright layer becomes visible at the vegetal side. The movement of this cytoplasmic layer result from the cytoplasmic streaming which accompanies maturation and fertilization; since it has been observed by Conklin (1905), Berrill (1929 — cited by Berrill 1950, 1948), Hirai (1941), Millar (1951) and Sebastian (1953 cited by Millar 1954) in various ascidian species, it can be considered to be a phenomenon characteristic of ascidians in general.

According to Conklin (1905), the egg of *Styela partita* matures after being spawned and the nucleoplasm of the germinal vesicle moves toward the animal pole, where it forms a clear cap over this region. In its center lie the chromosomes, together with very fine granules. The spindle of the first maturation division is formed, but development proceeds no further unless fertilization takes place. In other words, fertilization occurs at the metaphase of the first maturation division. In most cases *Ciona intestinalis* is self-sterile, *Styela partita* is sometimes self-fertile, and *Molgula manhattensis* shows both types of fertilization (Morgan 1942). *Halocynthia roretzi* and *Chelyosoma siboja* are sometimes self-fertile in very low percentages. In *Styela partita*, the spermatozoa pass through the chorion and enter the egg from the vegetal side. Immediately after sperm entry the first polar body is formed, followed by the second. Accompanying maturation and fertilization, a pronounced cytoplasmic streaming takes place: the *clear cytoplasm*, which has been capping the animal pole area, and the peripheral layer of yellow granular material descend together to the vegetal side (Fig. 12.3b), where they form a cap that includes the sperm head. This downward streaming is so violent that it is said to carry the test-cells along with it so that they accumulate at the vegetal region in some cases. A small mass of hyaline cytoplasm is left surrounding the maturation spindle at the animal pole, and the *gray yolk* granules gather around it. The two kinds of protoplasm which descended to the vegetal side become arranged so that the clear cytoplasm spreads out over the plasm containing the yellow pigment granules. The sperm head proceeds from its point of entrance through the clear area until it reaches the yolk region; there it changes its course and, now in the form of the *sperm pronucleus*, moves toward the posterior (see below) side of the egg. Most of the yellow plasm shifts with it, forming a *yellow crescent* which extends to left and right below the equator, including half the egg circumference (Fig. 12.3 *ye.cr.*). The clear cytoplasm also moves posteriorly and takes a position at the egg surface above the yellow crescent (Fig. 12.3 *cl.pl.*), and the gray yolky plasm moves anteriorly downward, finally reaching the vegetal region (Fig. 12.3 *gr.y.*).

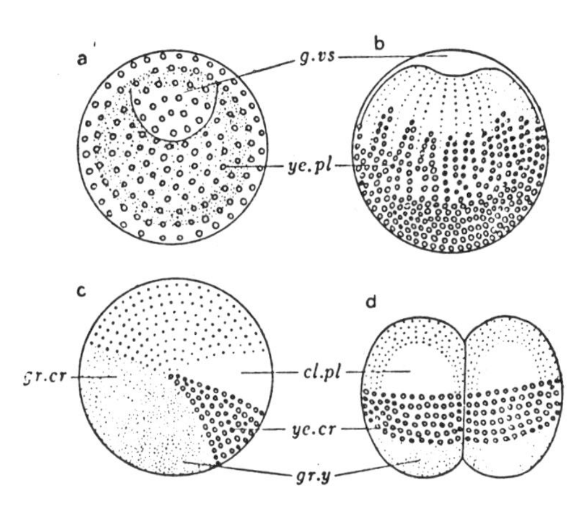

Fig. 12.3 Successive stages in the egg of *Styela partita* (Conklin 1927)

a. unfertilized egg; b. immediately after sperm entrance, showing cytoplasmic streaming; c. left side view of egg immediately before first cleavage; d. posterior view of 2-cell stage. *cl.pl:* clear cytoplasm, *gr.cr:* grey crescent, *ye.pl:* yellow cytoplasm.

After extrusion of the second polar body, the remaining nuclear material, as the *egg pronucleus*, moves downward to unite with the sperm pronucleus, the two pronuclei coming into contact in the vegetal hemisphere. The synkaryon thus formed then shifts to the center of the egg. Before syngamy a well-defined aster is formed at one side of the sperm pronucleus; before the synkaryon divides, this aster cleaves into two, which come to be located at the two ends of the first cleavage spindle.

d. Cleavage and Blastula

In *Styela partita* a yellow layer and a clear layer thus appear in the posterior part of the egg at fertilization; in addition other parts of the cytoplasm also form layers so that five different regions can be distinguished before the first cleavage (Fig. 12.3). These are the posterior *yellow crescent* and *clear protoplasm*, a *gray crescent* in the anterior part, the gray yolk at the vegetal side and a layer of *clear yolk* at the animal side (Conklin 1905). Similar layers resulting from cytoplasmic streaming can also be observed in other species. In *Halocynthia roretzi* the layer which corresponds to the yellow crescent of *S. partita* is hyaline (Hirai 1941), in *Chelyosoma siboja* it is opaque white (unpublished), and in *Pyura microcosmus* it is grayish-brown (Millar 1954); in none of these other species, however, is its formation as conspicuous as that of *Styela partita*. As the result of such layer formation, the uncleaved egg of ascidians exhibits bilateral symmetry, the polarity of which coincides with the larval polarity. Sections of the unfertilized egg of *Ascidiella scabra* show

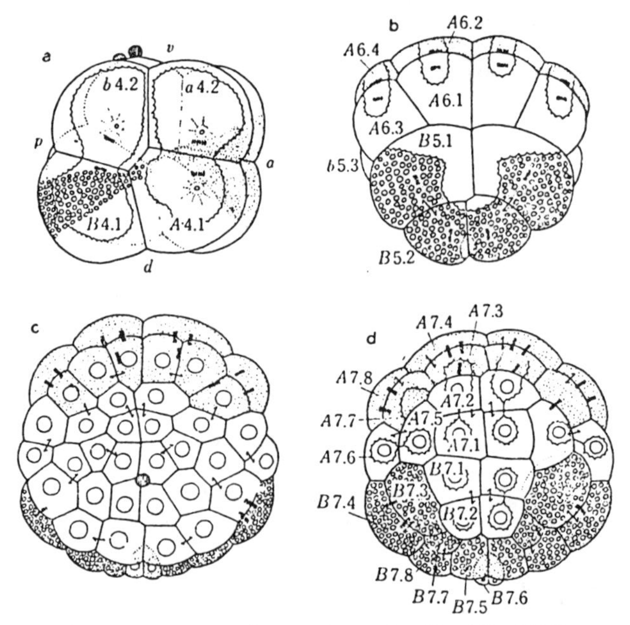

Fig. 12.4 Cleavage in *Styela partita* (Conklin 1905)

a. right side view of 8-cell stage; *a:* anterior. *p:* posterior. *d:* dorsal. *v:* ventral. b. dorsal view of transition stage from 20- to 24-cell stage. Cytoplasm derived from grey crescent is contained in *A* 6.2- and *A* 6.4-blastomeres; c,d. ventral and dorsal views of 64-cell stage. *A* 7.4 and *A* 7.8 are presumptive neural plate areas. *A* 7.3 and *A* 7.7 are presumptive notochord areas. *B* 7.4 and *B* 7.8 are presumptive muscle areas. Other cells containing yellow cytoplasm, together with *A* 7.6, develop into mesenchyme. Remaining vegetal region develops into endoderm. Extra-crescent blastomeres of ventral side will form ectoderm.

that the distribution of the cytoplasm is already bilaterally symmetrical before fertilization (Dalcq 1938).

In *Ciona*, peroxidase activity (Ries 1937, cited by Brachet 1950) can be recognized only in a crescent in the unfertilized egg. This crescent is identical with Conklin's yellow crescent, which contains the material destined to develop into muscle (myoplasm). According to Berg (1956), the cytochrome oxidase complex is located in the mitochondria. Working with a microspectrophotometric method on *Ciona intestinalis*, he found the cytochrome oxidase activity of homogenates of the posterior blastomeres to be 2.7 times that of the anterior blastomeres in the 4-cell stage, indicating a localization of the mitochondria in the posterior blastomeres. Ries (1937, 1939), Reverberi and Pitotti (1939), Urbani and Mistruzzi (1947), using the indophenol reaction, demonstrated a localization of indophenol blue oxidase (presumably cytochrome oxidase) in the posterior blastomeres of various ascidian embryos (cited by Berg 1956).

In *Styela*, most of the clear plasm as well as some of the yellow plasm accompanies the synkaryon as it moves toward the center of the egg. At the surface a thin band of clear plasm remains above the yellow crescent; the egg in this condition begins the first cleavage. This first cleavage is perpendicular and divides the egg equally, bisecting the yellow crescent and the other layers, and indicating the plane of bilateral symmetry of the embryo (Fig. 12.3d). At the end of the first cleavage the nuclei, surrounded by hyaline cytoplasm, move toward the animal pole, pushing aside the yolk which fills this region. The second cleavage is also perpendicular, and at right angles to the first. In the four-cell stage the anterior blastomeres, containing the gray crescent, are somewhat larger than the posterior two which include the yellow crescent material. The third cleavage, leading to the eight-cell stage, is horizontal (Fig. 12.4a), the four blastomeres of the animal side all being smaller than the vegetal four, and consisting of hyaline cytoplasm. The planes dividing the blastomeres into upper and lower cells in this stage tend to slant downward toward the anterior side, so that they fail to meet the second cleavage plane perpendicularly; this result in an H-shaped pattern when seen from the side. The animal pole side of the eight-cell stage corresponds to the antero-ventral part of the larva, and the vegetal side to the larval postero-dorsal region.

The fourth cleavage produces a 16-cell stage consisting of two tiers of eight cells each. In the anterior two of the four cells on the animal pole side at this division, the cleavage spindles converge anteriorly, while the spindles of the posterior two cells diverge anteriorly. As a result, the cleavage planes of the anterior two cells intersect the first cleavage plane, while those of the posterior cells intersect the second. This performance is exactly reversed at the vegetal side (Fig. 12.4b), the anterior two planes intersecting the second cleavage planes, and the posterior two meeting the first. At this cleavage the two smallest blastomeres are formed in the posterior part of the vegetal region (*B* 5.2); these blastomeres consist mainly of yellow cytoplasm. The two blastomeres lying directly anterior to these contain the rest of the yellow plasm, as well as a large amount of yolk. The four blastomeres located in the anterior part of the vegetal side contain gray yolk, but hyaline plasm can also be seen on the anterior edge of each.

The next (fifth) cleavage gives rise to the 32-cell stage. The two smallest blastomeres at the vegetal pole containing yellow plasm divide into four cells which form a straight line. The two blastomeres which lie directly in front of these cleave so as to separate their yellow plasm and yolk, each forming a yellow and a gray blastomere. The yellow cells lie peripheral to the gray ones, joining the posterior line of yellow blastomeres and bringing their number to six. The four gray yolk blastomeres at the anterior part of the vegetal region divide across their antero-posterior axis, forming anterior and posterior rows of four cells each, the anterior row lying below the equator. The cells of the animal hemisphere divide in the same way, each blastomere containing clear cytoplasm and a small amount of yolk. These cells of the animal half in the 32-cell stage are columnar, while those of the vegetal hemisphere are wide and flattened. These polar differences in the shape of the blastomeres, however, are later reversed completely, the blastomeres of the vegetal side becoming columnar.

In this stage the blastomeres become separated from each other at the center of the embryo, giving rise to a blastocoel; the embryo thus becomes a blastula. The cells making up the animal hemisphere of the blastula will become the ectoderm; the two of these lying most anteriorly on the equator correspond to the anterior part of the *neural plate*. Among the cells of the vegetal hemisphere, the gray cells will form the endoderm, while the cells lying most anteriorly and containing clear plasm and gray yolk will give rise to the *dorsal* or *anterior lip of the blastopore*. These dorsal lip cells were given the name of *chorda-neural cells* by Conklin because they eventually separate into one group of hyaline cells and another group of gray cells; the former make up the neural plate, while the latter form the notochord. The row of yellow cells along the posterior part of the vegetal side will become mesoderm and give rise to the larval muscles and reproductive organs. In other words, the 32-cell stage consists of 14 ectoderm cells, 2 neural plate cells, 4 chorda-neural plate cells, 6 endoderm and 6 mesoderm cells. The broad, flat cells in the anterior part of the gray cell group will later form the larval *pharynx*, and the invaginated cells in the center of the yellow cell row will produce the *tail endoderm*. Thus it is known that as development proceeds, the cytoplasmic layers of the uncleaved egg become groups of cells which incorporate the material of those layers, and these eventually form the various organs. Projecting this result back on the uncleaved egg, Conklin concluded that five kinds of cytoplasm can be distinguished in the uncleaved egg of *Styela partita*, of which the deep yellow plasm will later enter the larval tail muscles, the light yellow plasm will become mesenchyme, the light gray plasm will form notochord and neural plate, the slate gray substance will become endoderm and the clear plasm will become ectoderm. The scheme for the distribution of embryonic organ rudiments proposed by Conklin (1905) is a fundamental achievement, performed by the method of tracing cell lineage with the cytoplasmic layers as landmarks. Following his work, according to Berrill (1950), investigation into the presumptive location of organ systems has been performed by Tung (1932), Cohen and Berrill (1936) and Vandebroek (1937 cited by Dalcq 1938), who used local vital staining, and Ortolani (1952), using adhering carbon particles. According to the results which Vandebroek obtained by locally staining 8-cell stage embryos of *Ascidiella scabra* (Fig. 12.5), most of the material of the four animal hemisphere blastomeres is ectoderm primordium *(ect.pl)*, while the remaining narrow wedge-shaped strip extending from the front around the two sides above the equator is the neural rudiment *(n.pl)*, with the wide anterior part representing the brain. In the four blastomeres of the vegetal hemisphere are found the rudiments of the notchord *(ch.pl)*, the endoderm and the mesoderm *(ms.pl)*. Of these, the notochord material lies below the neural rudiment, in the parts of the two anterior vegetal blastomeres which lie just under the equator; in the yellow crescent regions of the two posterior

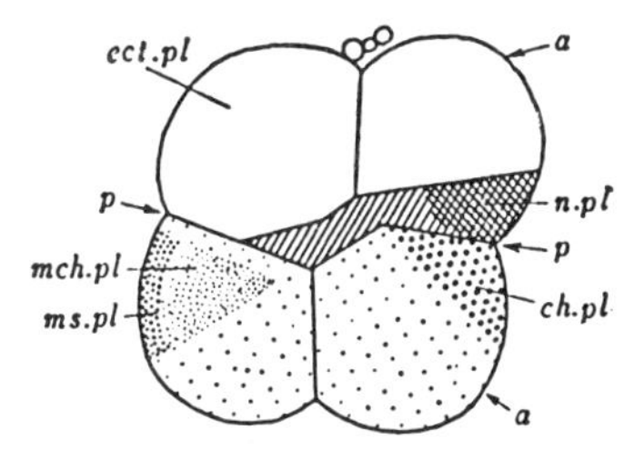

Fig. 12.5 Map of presumptive organ rudiments at 8-cell stage of *Ascidiella scabra* (Vandebroek 1937)

a: anterior, *ch.pl:* notochord rudiment, *ect.pl:* ectoderm rudiment, *mch.pl:* mesenchyme rudiment, *ms.pl:* muscle rudiment, *n.pl:* nerve rudiment (cross-hatched area in anterior part of nerve rudiment is presumptive brain) *p:* posterior.

blastomeres are the rudiments of the muscles and the mesenchyme *(mch.pl).* The remainder of the vegetal side consists of endoderm. It is found that the germ layer movements by which gastrulation is accomplished shift the material from the anterior equatorial region (*p* at right side of Fig. 12.5) through the blastopore toward the posterior side (*p* at left side of Fig. 12.5), consequently bringing together the upper and lower blastomere surfaces marked *a* in the figure, to form the anterior side of the embryo.

e. **Gastrula and Young Tadpole**

From the 32-cell stage the embryo proceeds without intermission to the 64-cell stage (Fig. 12.4c,d). The animal hemisphere (ventral side) of the embryo in this stage consists of 26 ectoderm cells and an arch of 4 neural plate cells, to the ends of which are added 2 ectodermal cells to make a total of 6 neural cells. In the vegetal hemisphere the chorda-neural cells divide into an anterior and a posterior row; the anterior 4 (*A* 7.4, *A* 7.8 group) become neural cells and join the 6 neural cells described above in forming the *neural plate.* The 4 cells of the posterior row (*A* 7.3, *A* 7.7 group) become the *notochord rudiment.*

Most of the gray cells divide antero-posteriorly, except for the two at the lateral extremes, which divide into inner and outer cells; of these the outer ones (*A* 7.6 group) have fewer yolk granules than their sisters and become the *anterior mesenchyme cells* which will form the anteriormost part of the mesoderm. The 6 yellow cells cleave into 12; 4 of these (*B* 7.4, *B* 7.8 group) will become *muscle cells,* while the other 8 (*B* 7.3, *B* 7.7, *B* 7.5, *B* 7.6 group) will become mesenchyme, with the addition of the *A* 7.6 pair mentioned above.

In this stage the endoderm cells commence to invaginate into the small blastocoel, beginning the gastrulation process. As the cells of the vegetal hemisphere change to a columnar shape, the surface of the vegetal region shrinks and at the same time the cells of the animal hemisphere become flattened, causing this surface to expand. In the anterior part of the vegetal region an arch of eight chorda cells curves toward the posterior, its ends meeting the forward-pointing ends of another curved row of 12 mesenchyme cells so that the 20 cells together form a circle, within which lie the endoderm cells. Underneath the mesenchymal arch are six muscle cells, and outside of the chorda cells is a row of eight nerve cells.

Next the neural cells of the dorsal lip extend posteriorly so that they cover the chorda and endoderm cells, and the chorda cells at their lower level accompany the nerve cell layer in elongating posteriorward (Fig. 12.6a). Together with this movement the ectodermal cells adjacent to the mesoderm proliferate and keep

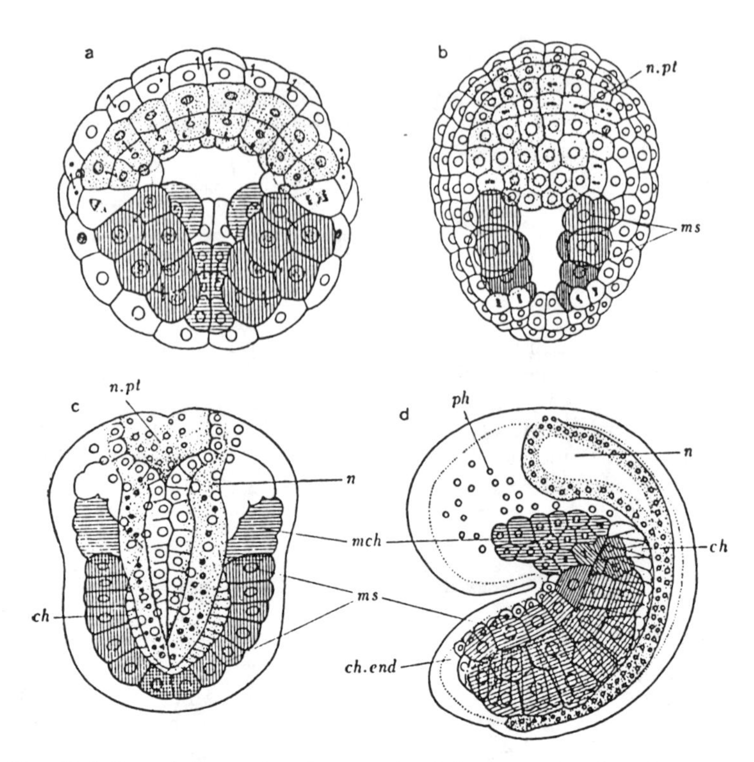

Fig. 12.6 Gastrulation and early gastrula of *Styela partita* (Conklin 1905)
a. very early gastrula (180-cell stage), shawing T- shaped blastopore; b. later stage; dorsal lip grows backward, laterally-lying muscle cells draw closer to median line and blastopore narrows vertically in posterior half of embryo; c. dorsal view of voung embryo at formation of neural tube: d. left side view of young embryo, showing orientation of 3 rows of muscle cells and other organ rudiments. *ch:* notochord, *mch:* mesenchyme, *ms:* muscle cell (hatched), *n:* neural tube, *n.pt:* neural plate (stippled), *ph:* pharyngeal region

pace with the extension of the dorsal lip region. This causes the margin of the blastopore to assume a T-shape, as seen from the vegetal side (Fig. 12.6b), with the wide part of the opening toward the anterior, and posteriorly a narrow slit between the left and right rows of mesodermal cells. About the time the dorsal lip has advanced posteriorly to enclose some three-fourths of the embryo, the *ventral* or *posterior lip* suddenly elongates, causing this posterior part of the blastopore to turn upward so that it is practically perpendicular (Fig. 12.7a) and changes its position from posterior to dorsal. The mass of mesenchymal cells eventually divides into two, an anterior mass lying above the column of muscle cells, and a posterior mass located below it. As the gastrulation process reaches completion, the posterior end of the embryo elongates in the beginning of *tail* formation, and the embryo is differentiated into *trunk* and tail regions. The muscle cells, which were arranged in perpendicular rows during the gastrula stage, return to a horizontal or antero-posterior alignment in this stage of tail-formation, eventually making up two rows, one above the other, on each side of the embryo. In *Styela partita* there are six muscle cells in each row at this stage; each cell acquires a slender, elongated lozenge shape and *myofibrils* differentiate within the cytoplasm. The

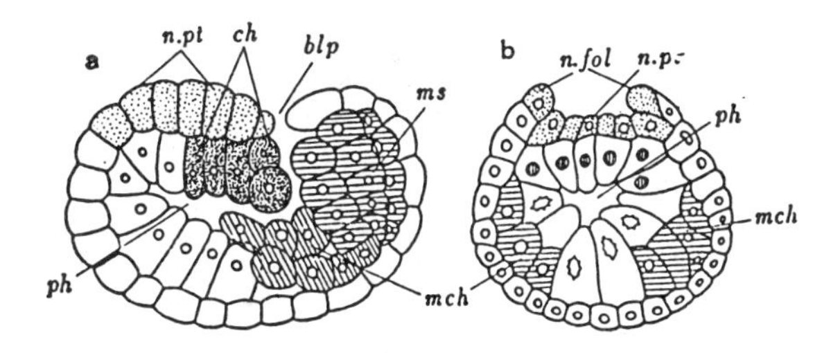

Fig. 12.7 Gastrula of *S. partita* (Conklin 1905)

a. sagittal section; b. median transverse section. *blp:* blastopore, *ch:* notochord, *mch:* mesenchyme cell, *ms:* muscle cell arranged in vertical row, *n. fol:* neural fold, *n.pt:* neural plate, *ph:* pharynx.

anterior mesenchyme lies in front of the muscle cells and enters the trunk region of the embryo; i.e., these cells are located on the two sides of the *pharyngeal wall,* interposed between this structure and the ectoderm (Fig. 12.7b).

The time when the blastopore of the gastrula can be seen on the dorsal side marks the beginning of *neural fold* formation. The neural folds arise from the two sides of the neural plate; in the region posterior to the blastopore these folds meet at the midline, but are widely separated anteriorly. Gradually, however, they draw together, closing the gap from the posterior toward the anterior. In this process the blastopore becomes covered by the neural folds, forming the *neurenteric canal*; the folds next cover the neural plate to form the *neural tube,* and this is finally covered by the ectoderm (Fig. 12.6c,d). In the wide, thick, anterior part of the neural tube a *brain-vesicle (sensory vesicle)* differentiates, while the slender posterior part forms a *spinal cord.* The anteriormost end of the brain vesicle remains open for a short time as the *neuropore.* As will be described below, the neuropore in simple ascidians is said to close temporarily in the tadpole larval stage, to open again at metamorphosis. The chorda cells, which were first aligned in a plate, become arranged to form a groove as the tail elongates; several cells can be seen in a cross-section of this structure. These cells next redistribute themselves into a single line, and as the tail grows longer they also elongate. If a local vital stain is applied to the two sides of the row of chorda cells in the *H. roretzi* embryo, the stained part is found on the ventral side of the cylindrical notochord after the completion of tail formation. The posterior part of the endoerm lies along the under side of the notochord, where it forms a ventral cord *(sub-chordal endodermal cord),* but after a short time these cells become dispersed. In the dorsal part of the larva, the rudiments of the *atrial,* or *peribranchial, cavities* appear in the Pleurogona as a single ectodermal invagination bifurcating ventrally, and in the Enterogona as a pair of lateral and independent ectodermal sacs, which later fuse mid-dorsally to form the atrial siphon (Berrill 1928, 1950). The walls of these cavities unite with the pharyngeal wall and differentiate into the *gill-slits.* A blind tube arises from the posterior wall of the pharynx to form the rudiment of the *digestive tract.* Underlying the ventral wall of the pharynx is a wide band of columnar cells which develops into the *endostyle* (Fig. 12.10 *endst).* In the brain-vesicle there appears an *eye (ocellus),* having *pigment granules* and a *lens* (Fig. 12.8d), although this may be vestigial or lacking in some species. A *balancing (static) organ* is formed, which also contains a spherical pigmented concretion. A thickening in the posterior wall of the brain vesicle develops into the *visceral ganglion*. The anterior part of the trunk becomes considerably distended and is called the *chin*. The front end of this structure develops typically into three processes, the *adhesive papillae*, consisting of ectodermal secretory cells; one of these is situated on the midline and the others to the left and right of it in the simple ascidians (Fig. 12.8a, b, c).

Gastrulation in the simple ascidians thus takes place by the combined processes of invagination and epiboly; to these factors, according to Dalcq (1938), Vandebroek's (1937) localized vital staining experiments on *Ascidiella* led him to add "convergence" and "elongation". The author, moreover, using the same vital staining technique on *H. roretzi*, has observed during gastrulation complicated movements which can be described as convergence and elongation, as well as "divergence".

(3) LARVA

Once gastrula formation is complete, the tail of the embryo elongates so that it is wrapped around its trunk within the chorion. It then hatches out of this membrane and begins a period of vigorous swimming (Fig. 12.8). The size at this time differs from species to species: the total length of the larva in *Ciona intestinalis* is about 0.7 mm (Berrill 1950); in *H. roretzi*, 1.5 mm ; in *Styela clava*, 1.0 mm; in *Corella japonica*, 0.7 mm; in *Chelyosoma siboja*, 1.0 mm; in *Ascidia zara*, 0.5 mm. The length of the swimming period also varies with the individual as well as with the species: among those with a short swimming stage are the viviparous species of *Molgula*, in which it lasts for a few minutes to two hours; in the oviparous species the larvae swim for 12 hours or longer (Berrill 1950). *C. intestinalis* (Berrill 1947) and *H. roretzi* (Hirai 1941) larvae swim for at least 12 hours, and in *C. japonica* this stage lasts at least five hours, and may continue for more than two days.

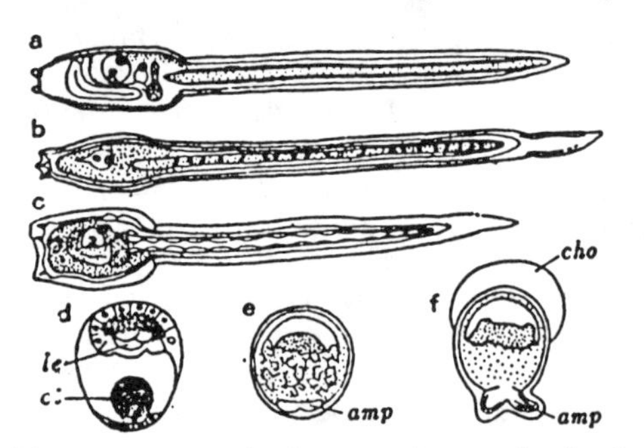

Fig. 12.8 Tadpole larvae of several simple ascidians

a. *Ciona intestinalis* (Berrill 1947); b. *Halocynthia roretzi* (Hirai 1951); c. *Corella japonica* (Hirai); d. brain-vesicle of *C. intestinalis* tadpole (Berrill); e. anural embryo of *Pelonaia corrugata* (Millar 1954); f. hatching of larva of *P. corrugata* (Millar). *amp:* ampulla at anterior end of body, *cho:* chorion, *le:* lens, *ot:* otolith.

The completely formed ascidian larva is called a *tadpole larva*, and displays a characteristic structure which is found only in the offspring resulting from the sexual type of reproduction. The shape differs from one species to another, but except for a few tailless *(anural)* larvae, it is composed of a trunk and a tail. (Fig. 12.8a, b, c). The surface of the epithelium is covered with a gelatinous cuticle, the *tunica (tunic, test)*. At the anterior end of the trunk there are typically found three adhesive papillae, *branchial* and *atrical apertures*, brain vesicle, visceral ganglion and the *rudiments of the adult organs*. Within the brain vesicle is an eye formed of pigment granules and a lens, as well as a static or balancing organ *(otocyst)*, consisting of a pigmented *otolith* and a *supporting cell*. In the tail region is the central notochord, which is composed of large cells, with the neural tube lying dorsal to it, and the two rows of large muscle cells at its left and right sides (Figs. 12.8 and 12.10a).

The differentiation of these larval organs, with the exception of the adult organ rudiments, is in every case completed during the larval period, and they all regress or undergo modification during the subsequent metamorphosis. The organ formation by means of budding which occurs in the compound ascidians omits such special larval organogenesis. Then if, on the one hand, the larval form is compared with the post-metamorphosis structure by mapping the presumptive embryonic organ rudiments in the uncleaved egg, and on the other hand, the organogenesis accompanying budding is compared with the larval structure, it becomes apparent that the adhesive palps, brain-vesicle, sensory organ, the visceral ganglion and notochord of the trunk, and the tail region with its muscles, nerves and fin are all larval structures peculiar to the tadpole stage. This larva, with its special organs centering around the notochord, may be described as the focal point of the significance attached to the Protochordata.

A broad comparison of tunicate tadpole larval forms reveals certain differences in the structure and arrangement of the larval organs and the degree of differentiation of the adult organ rudiments. With respect to these points of difference, three types of tunicate larvae can be distinguished: I, Cynthia Type; II, Amaroucium Type; and III, Botrylloides Type (Hirai 1951). The larvae of the simple ascidians are included in Subtype 1 (Simple Type) of Type I. This group is characterized by a triangular arrangement of the stalkless adhesive papillae, a failure to differentiate the trunk epidermal structures found in the compound ascidians, a horizontal and practically linear arrangement of the brain-vesicle and branchial and atrial apertures, a horizontal visceral ganglion, vertical fin and inconspicuous adult organ rudiments. Finally, the over-all shape of the larva is long and slender.

The above description applies to the larval form in general, but *Molgula, Engyra,* and other genera have larvae of a tailless type. In *Molgula,* the early stages of such a tailless larva do not differ at all from the developmental formula of the larvae with tails, but the chorda cells fail to increase in size, and no tail region forms. Among nine free-living species of *Molgula* which fail to attach themselves in sand or mud, eight have tailless, and one has tailed larvae, while seven of nine fixed species have tailed, and only two tailless, larvae. Furthermore, the larvae of *Molgula* lack eyes, and both types of sensory organs are missing in the tailless larvae (Berrill 1928, 1950). Eyeless larvae also occur in some other groups, and tailless larvae are found in *Pelonaia corrugata* (Millar 1954) (Fig. 12.8e, f). Eyes and balancing organs similarly fail to differentiate in this species. It appears that no thorough study has yet been made of the embryological details.

(4) METAMORPHOSIS

In the simple ascidian *Ciona intestinalis,* the duration of swimming is generally 12 hours, although it may last as long as 36 hours (Berrill 1950). In *Halocynthia roretzi,* metamorphosis generally occurs about 12 hours after hatching (Hirai 1941), but many larvae continue to swim without metamorphosing for several days at 10°C. In the compound ascidian *Botryllus schlosseri,* the swimming period is usually one to three hours, although it may be as short as 30 minutes or as long as 27 hours (Grave 1924). As these examples show, the duration of swimming is characteristic of the species, but is far fom being a constant trait.

The onset of metamorphosis is indicated by the resorption of the tail. Oka (1958), in an experimental study of the compound ascidian *Perophora japonica*, finds that metamorphosis can be induced by exposing the larvae to narcotics, distilled water, or solutions of vital stains such as methylene blue (0.004% for 5 min), neutral red (0.004% for 30 min.) or nile blue (0.002% for 10 min.). He is also able to cause the initiation of metamorphosis by shaking the larvae in a test-tube.

The author has found (Hirai 1961) that metamorphosis is induced in *H. roretzi, Ch. siboja* and some compound ascidians by vital staining with nile blue sulfate, neutral red and toluidin blue, nile blue sulfate being the most effective. When larvae of *H. roretzi* are treated, a day after hatching, with this dye (1 drop of 1% nile blue sulfate in 10 cc sea water) for 3 to 5 minutes and then returned to sea water, the resorption of the tail begins within a few minutes. The tails of more than 90% of the larvae are absorbed within one and a half hours. This method causes the onset of metamorphosis even in larvae which would otherwise continue to swim for four days.

By this method it is found that the larva of *H. roretzi* shows no potency for metamorphosis at the time of hatching; after swimming for about six hours it begins to acquire such potency, and is completely differentiated and ready to metamorphose 12 hours after hatching. In *Ch. siboja* exposure for 3 minutes to a more dilute solution of nile blue sulfate (1 drop of 1% dye in 40 cc sea water) is equally effective in initiating metamorphosis, and even without such treatment the larvae are able to metamorphose soon after hatching or even before (Hirai, unpublished). Such differences in the relation between degree of differentiation and time of hatching underlie the observed variation in the duration of the larval swimming period.

a. **Metamorphosis and differentiation of adult organs**

When its free-swimming period is over, the tadpole larva secretes an adhesive substance from its adhesive papillae and attaches to some object; its movements become sluggish and it begins the process of metamorphosis. The resulting changes consist in the disappearance of the larval tail and adhesive papillae, accompanied by the development of *attaching organs* or *ampullae*; the brain-vesicle is transformed into the *cerebral ganglion* and its accessary organs, and the special larval character is completely changed. At the same time the adult organ rudiments proceed to differentiate in a way which transforms the motile larva into a fixed, sessile adult. Externally the process of metamorphosis alters the larval structure abruptly, within a short time, as the larval body axis rotates to produce a new adult polarity and the specialization of the adult organ rudiments advances step by step. External observation of the changes in *C. japonica* (Hirai, unpublished) (Fig. 12.9) shows that the larva may attach to the bottom of the glass container or simply become quiescent without attaching. The most conspicuous phenomenon as metamorphosis begins is the shrivelling of the tail, which is drawn into the posterior part of the trunk by a folding-up process starting at its basal end. After about 30 minutes the length of the tail is reduced to half, and all but a small part of it is within the body within about an hour. Another hour is required for it to vanish completely. Inside the body it bends three or four times and lies folded up in the posterior region (Fig. 12.9c, *tl*). While the tail is shrinking, the epidermis in the neighborhood of the adhesive palps thickens and expands, giving rise to three ampullae (Fig. 12.9 *amp*). No change in the arrangement of the various organs can be observed until resorption of the tail is complete, but from about five hours

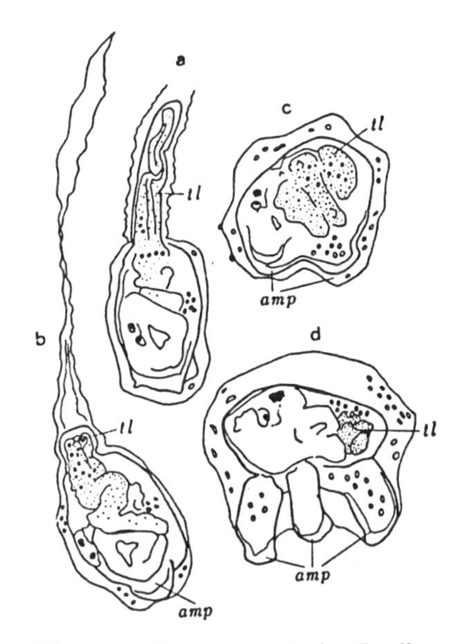

Fig. 12.9 Metamorphosis in *Corella japonica* (Hirai)

a. 30 minutes after beginning of metamorphosis; b. 70 minutes after beginning of metamorphosis; c. 2½ hours, d. 20 hours. *amp:* ampulla, *tl:* tail

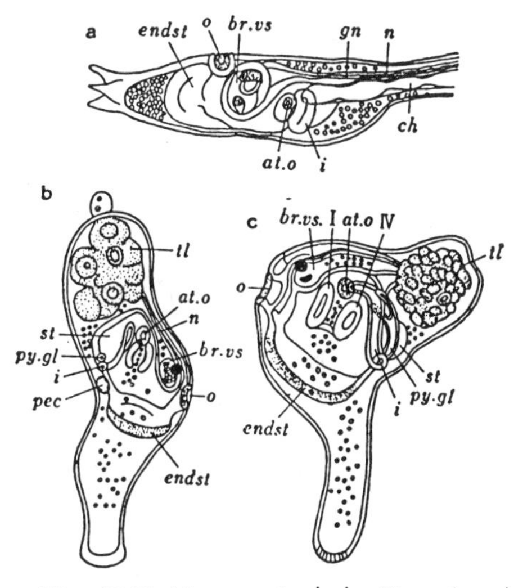

Fig. 12.10 Metamorphosis in *Ciona intestinalis* I (Willey 1893)

a. tadpole larva; b,c. successive stages in metamorphosis. I,IV: first and fourth gill slits. *at.o:* atrial canal, *br.vs:* brain vesicle, *ch:* notochord, *endst:* endostyle, *gn:* somatic ganglion, *i:* intestine, *n:* neural tube, *o:* mouth, *pec:* pericardium, *py.gl:* pyrolic gland, *st:* stomach, *tl:* tail.

after the beginning of metamorphosis a rotation in the body axis takes place. The sensory organ, which is located dorsally in the larva, begins to turn toward the posterior part of the larval axis, and finally makes a revolution of about 90° and comes to lie in a position posterior in relation to the adhesive papillae, but at the anterior end of the post-metamorphosis body axis (Fig. 12.9a-d). Simultaneously the tail material, which had lain posteriorly in the larva, moves around by 90° to the ventral side. This external rotation is easily observed in the living state; accompanying it a corresponding rotation of the internal organs is taking place, by which the positions of the larval structures are shifted to the adult organ arrangement.

In *Ciona intestinalis,* the early developmental stages have been described by Conklin (1905), and the later development by A. Willey (1893, 1894) and Berrill (1947). Willey has also observed the later stages in *Ascidia mentula*; these reports contain the basic data covering the subject. About 24 hours after fertilization the embryo of *Ciona* reaches the tadpole stage, and swims for 12 hours or more. When metamorphosis begins (Fig. 12.10, 12.11), the tail epidermis contracts, causing the notochord, nerve cord and muscles to be bent together and brought inside the trunk, where they fragment and are consumed by wandering phagocytic cells (Fig. 12.10 *tl*); the tail epidermis invaginates to form pockets. While the tail is thus being resorbed, the front end of the trunk elongates, and forms the center around which the anterior part of the body grows (Figs. 12.10a, b, c, and 12.11a, b) so that it excutes a turn of 90° posteriorly, or toward the former tail position.

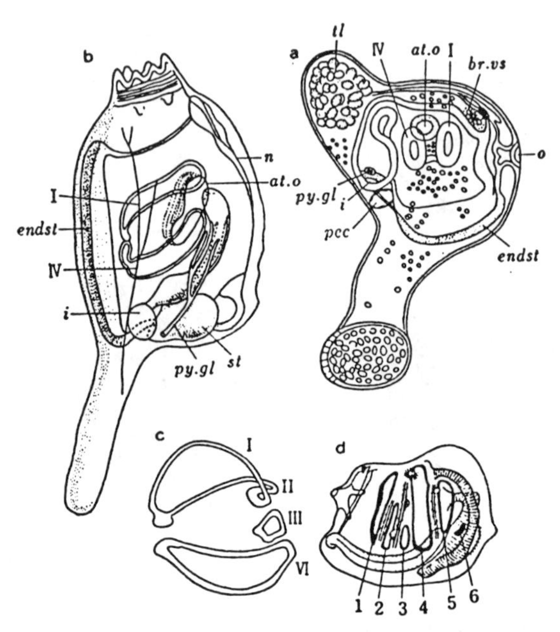

Fig. 12.11 Metamorphosis in *C. intestinalis* II (Willey), and *Phallusia scabroides* (Korschelt u. Heider)

a,b. later stages in metamorphosis of *C. intestinalis* larva; c. differentiation of gill slits in larva of *C. intestinalis*; d. differentiation of gill slits in larva of *Phallusia scabroides*. Abbreviations as above.

The rudiment of the endostyle lies in the anterior part of the endodermal organ rudiments in the larva (Fig. 12.10 *endst*), so that its long axis runs dorsoventrally. As the result of the body rotation, the endostyle is gradually shifted posteriorly from its larval position, finally turning through 90°. The mouth, which was first situated at the anteriormost end of the endostyle, is thus moved posteriorly, as the endostyle turns, to its definitive adult location. Since the position of the mouth marks the anterior end of the body, this means that the adult body axis is exactly opposite to that of the larva; the endostyle thus lies posterior to the mouth and elongates in a direction parallel to the long axis of the body. While these externally visible changes are taking place, the differentiation of the adult organs is going forward inside the body.

The ectoderm of the dorsal side of the larva forms either a single or paired atrial invaginations which fuse with the pharynx to produce the gill-slits. In *Ciona intestinalis* a pair of lateral atrial invaginations are formed, and each of these gives rise simultaneously to two gill-slits (Fig. 12.10 I, IV). According to Willey (1893), however, the formation of these gill-slits does not involve the formation of two corresponding outgrowths of the pharynx wall. First an atrial invagination is formed on each side, and then the wall of the pharynx adjacent to its inner end also invaginates. Both atrial and pharyngeal pockets continue to deepen until the inner part of the atrial invagination makes contact with two angles on the rim of the pharyngeal invagination: the fused walls at these points then break through to form two gill-slits. At first there is only one opening on each side, but the tissues of the atrial and pharyngeal invaginations give rise to a structure called the *tongue-bar* which divides this opening into the two primary gill-slits (Figs. 12.10, 12.11). These two gill-slits are gradually caused to elongate laterally by the growth of the pharynx, as it develops into the *branchial sac* and the *atrial cavity,* and finally their two ends at the endostyle side bend toward each other. These bent portions eventually separate off, forming two new gill slits in the space between the first two (Fig. 12.11b, c). This gives a total of four pairs; later two more pairs are formed posterior to these but having no connection with them. Since the first four of these six gill-slits arise from a common rudiment, Willey gave them the collective name of *1st gill-slits*; the two pairs which are formed later he called the *2nd* and *3rd gill-slits,* respectively, and considered these three "pairs" to be the basic rudiments of gill-slit formation. The completed gill-slits he named the *first, second, third ...... sixth primary stigmata.*

After this the six primary stigmata of each side become laterally extended, and then small, tongue-shaped processes are formed at various points along their anterior or posterior margins. These elongate across the slits until they fuse with the opposite wall, thereby dividing each slit into a number of small ones. Willey calls the gill-slits of this period the *secondary stigmata.* In the next stage a change takes place in the direction of these stigmata. The individual stigmata undergo a bow-like curvature, and then divide in the hollow of the bow, with the result that the new stigmata thus produced have approximately a vertical instead of a horizontal long axis. These are at first arranged in six horizontal rows, corresponding to the six larval primary stigmata; in the adult, however, there are many more such rows, which, according to Willey, arise by further transverse division of the secondary stigmata.

When metamorphosis is about complete in *Ciona intestinalis,* the heart and gill-slits become active; this is the stage in which each of the paired gill-slit primordia first divides into a left and right part, and the atrial cavity still has a pair of openings, one on each side of the body. Berrill (1947) has named this the *first ascidian stage.* Next the first to fourth primary stigmata differentiate; then, following the formation of primary stigmata V and VI, each gill-slit divides into 2, 4, 8 . . . . As each of these gill-slits turns to a vertical position, the paired atrial openings move toward each other on the dorsal side, and finally unite to form the single *atrical aperture.* Berrill calls this the *second ascidian stage.* While the gill-slits in *Ciona* thus arise from a common original gill-slit, in other genera they may be formed independently of each other (although Willey doubted this), as in *Phallusia scabroides* (van Beneden & Julin 1887; cited by Korschelt und Heider 1936) (Fig. 12.11d). In this type of gill-slit formation, the slit shown as No. 4 in Figure 12.11 is formed first, then Nos. 1, 5 and 2, and finally Nos. 3 and 6.

From the posterior part of the pharynx, an *intestine* takes shape, in the basal part of which a spherical sac develops into a *stomach.* The *pyloric gland* in *Ciona* differentiates as a *coecum* in the region where the stomach adjoins the intestine

(Figs. 12.10b, c *py.gl*; 12.11a, b *py.gl*). It is a simple hollow process, which probably develops before fixation but is only recognizable afterward. This rudiment grows and elongates; its tip first bifurcates and then redivides many times to produce the complicated structure of the pyloric gland, which surrounds the digestive tract from the stomach nearly to the anal region with its finger-shaped branches.

According to Kühn (1893, cited by MacBride), the heart has an endodermal origin, appearing as a thickening of two cellular layers on the ventral wall of the pharynx in the tadpole larval stage of *Ciona intestinalis.* In the latter part of this stage the outer of these layers becomes a round clump of cells with a cavity in its center which is at first continuous with the pharynx lumen, but soon separates from it to form the *pericardial sac* (Figs. 12.10b and 12.11a *pec*). Eventually the dorsal wall of this structure invaginates to give rise to a fold, the inner part of which becomes the heart chamber, connecting with the primary body cavity of blastocoelic origin. After metamorphosis the pericardial sac enlarges and its outer wall becomes thin, while myofibrils differentiate in the wall of the heart, and it begins to perform peristaltic movements. In this fashion the *pericardium* and *heart* are formed.

According to the observations of Willey (1893) on this same *Ciona intestinalis,* the pericardial sac of ascidians differentiates from the endoderm of the branchial sac, and the heart lacks a lining layer of endothelium; in this he agrees with the observations of van Beneden and Julin (1887). On the other hand, Damas (1899), working with *Ciona,* Sélys-Longchamps (1938) studying *Clavelina,* and Brien and Blanjean (1939) studying the compound ascidian *Morchellium* (all cited by Berrill 1950) maintain that the heart develops from a mesodermal cell-plate which lies between the endoderm and ectoderm. When the larva of *Ciona* completes its metamorphosis and enters the first ascidian stage, two lateral swellings arise on the wall of the pharynx where the pericardium is being formed, and elongate posteriorly. These are the rudiments of the *epicardial tubes (epicardia, perivisceral sac).* According to Damas (1899) the left epicardium is larger than the right one; both tubes grow rapidly to form thin-walled vesicles which surround the heart, intestine, stomach and gonads, while the primary coelom is compressed into blood sinuses. As a result the epicardial cavities become the *body cavity,* and the internal organs are suspended from the body wall by a mesentery formed from the epicardial epithelium. The pair of openings connecting the epicardia with the pharynx become narrower, but persist for a rather long time as slits. The structure of the epicardia varies from species to species; as will be described below, in some of the compound ascidians it provides the tissue from which the new organs are formed during budding. In *Ciona,* however, its structure is completely typical of the simple ascidians, and it has no capacity for bud formation.

Willey (1894) has observed the differentiation of the neuropore and sensory organs in *Ascidia mentula.* At the stage when the larval tail is elongating whithin the chorion and the body is just beginning to bend, the neuropore closes. With this the neural tube is brought under the epithelium and becomes completely closed, the anterior portion forming a brain vesicle with a spacious cavity. The first differentiation of the sensory organs is recognizable as pigment granules which appear in several of the cells constituting the dorsal wall of the brain-vesicle (Fig. 12.12a). The stage in which these sensory organs — i.e., the eye and static organ — take shape is not uniform; in some individuals the 4—5 pigment granules of the eye rudiment differentiate first, while in other cases eye and static organ may be formed simultaneously. Both of these structures begin as a few pigment granules, but in the eye rudiment the granules are scattered among several cells, while those which make up the static organ rudiment are somewhat larger and concentrated in a single cell. The eye-forming granules decrease in size as they increase in number, and congregate at the inner borders of the cells. The pigment granules of the statocyst rudiment fuse together without proliferating. At first this rudiment is directly in front of the eye, but a shift in its position takes place, which is said to be caused by local inequalities in the growth of the brain-vesicle wall: the cells making up the part where the two rudiments are at first in contact form a thin, structureless membrane which expands, so that eye and statocyst are separated from each other (Fig. 12.12b, c, d). After this the cells of the eye rudiment draw together and become located at the right posterior side of the brain vesicle, and the statocyst rudiment moves to its ventral wall. After the sensory organs take their prescribed positions, the rudiment of the *stomodaeum* is formed by an ectodermal invagination; this grows inward and the *mouth* opens where it makes contact with the rudiment of the branchial sac. Kowalevsky (1866, 1871) states that the brain-vesicle opens into the stomodaeum in this stage, but Willey denies that this is so.

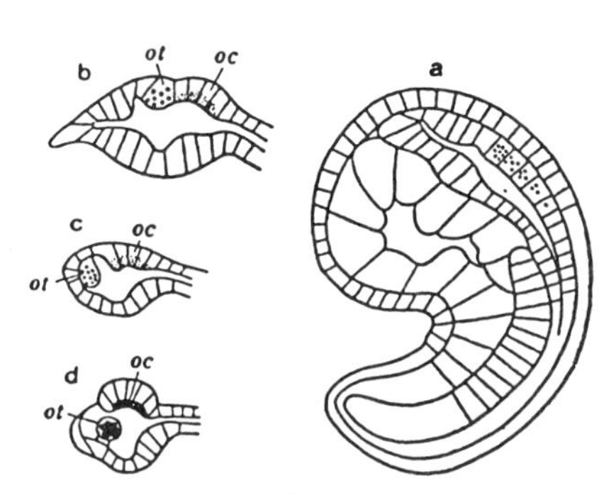

Fig. 12.12 Formation of sensory organs in *Ascidia mentula* (Willey 1893)

a, b ~ d. process of formation of sensory organs *oc:* eye rudiment, *ot:* otocyst rudiment.

The differentiation of the nervous system, which accompanies the formation of the *subneural gland* in *Ciona intestinalis*, is described by Willey (1894) as divided into five stages.

Stage I

While the larva, still enclosed by the chorion, is making convulsive movements with its tail, a small cavity is formed in the left side wall of the anterior part of the brain-vesicle; the posterior part of this cavity connects with the cavity of the brain vesicle (Fig. 12.13a, b). The anterior end of the cavity is a blind tube, the rudiment of the *neuro-hypophysial canal.*

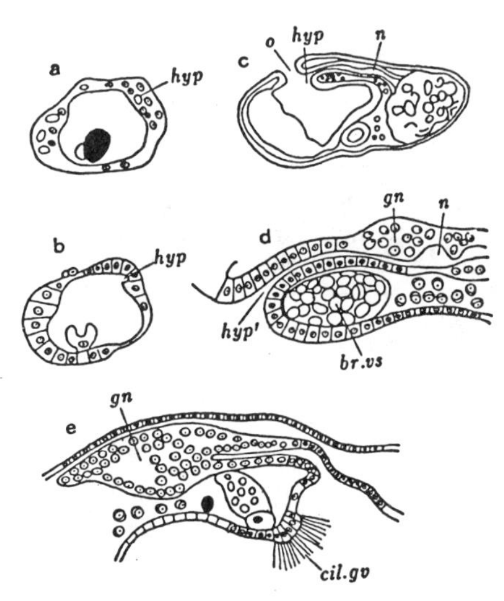

Fig. 12.13 Formation of nervous system in *C. intestinalis* (Willey 1893)

a. formation of neuro-hypophysial canal in side wall of brain vesicle; b. opening of neuro-hypophysial canal into brain vesicle; c. sketch showing position of neuro-hypophysial canal; d. differentiation of ganglion in dorsal wall of neuro-hypophysial canal; e. ganglion at more advanced stage. *br.vs:* vestige of degenerating larval brain vesicle, *cil.gv:* rudiment of ciliated groove, *gn:* ganglion rudiment, *hyp:* neuro-hypophysial canal, *hyp':* opening of neurohypophysial canal into gill sac, *n:* neural tube, *o:* mouth.

Stage II

About the time when the larva attaches, the neuro-hypophysial canal rudiment becomes prominent, and begins to separate from the brain-vesicle. At this time the blind tube approaches the midline of the brain-vesicle.

Stage III

In this stage the neural tube forms an opening at the base of the stomodaeal rudiment, while the blind tube of the neuro-hypophysial canal rudiment separates from the brain-vesicle and develops an opening into the stomodaeum. About this time the brain-vesicle is in the process of disappearing, and the tissue is breaking down. The opening of this connection between the hypophysial canal and the neuropore means that the temporarily closed neuropore reopens after the formation of the stomodaeum. The sagittal section of an attached larva in Figure 12.13c shows the relation between the nervous system and the other organs; the hypophysial canal can be seen to open into the stomodaeum. Figure d, an enlargement of the hypophysial canal region, shows the canal to be surrounded by columnar cells. Near the posterior end its cavity becomes irregular, and the cellular wall of its upper part is composed of numerous cell masses; this marks the beginning of adult ganglion formation.

Stage IV

During this stage, while the first four primary gill slits are being formed, the ganglion grows and increases in size (Fig. 12.13d). The lumen of the larval neural tube is being obliterated, the ganglion acquires a large, solid structure, and anterior to it is a funnel-shaped passage leading to the stomodaeal part of the branchial sac *(hyp)*. Below this opening, a ciliated process appears on the dorsal side of the branchial sac; this is the rudiment of the adult *ciliated groove* (Fig. 12.13e, *cil.gv)*.

Stage V

"In the younger immature adults, anteriorly the duct of the hypophysis expands into a large funnel-shaped dilatation, which in its turn opens into the branchial sac at the end of a papilliform prominence, which projects boldly into the branchial chamber." (Willey, 1894, p. 305) In this stage, "the hypophysis and the ganglion, which have been gradually differentiating themselves from the common neuro-hypophysial tube, have at last separated entirely from one another. In the posterior region of the hypophysial tube, which lies closely applied to, but at the same time distinct from, the ganglion, glandular tissue is seen to be developing from its ventral wall" (loc. cit. p.306). The hypophysis rudiment includes a lumen which is reduced in the anterior part. The cerebral ganglion arises by proliferation and constriction from the dorsal wall of the neuro-hypophysial tube. "The anterior portion of the hypophysis, including the funnel-shaped dilatation and the duct, is derived from a secondary evagination from the stomodaeal region of the branchial sac." (loc. cit. p. 306)

The outer surface of the larval epithelium is covered with a hyaline layer, the tunic or test, which may be either hard or gelatinous, while the tail is provided with a *fin*; certain cells, the *test cells*, become visible within the test. This hyaline layer contains cellulose, and is secreted by the ectoderm during the stage of tail formation. According to Kowalevsky (1892), in *Phallusia mammillata* the test cells pass through the ectoderm to reach the hyaline layer; Brien (1930, cited by Berrill 1950) made the same observation, and considers that mesoderm cells play an essential role in the growth and vitality of the test. Among the compound ascidians also, mesenchyme cells have been observed to move slowly through the epithelium into the test in metamorphosing *Amaroucium constellatum* larvae (Scott 1952).

b. Time required for development

In *Ciona intestinalis,* both eggs and sperm mature within their respective ducts, and fertilization takes place externally, after they have been shed. The fertilized egg begins to cleave within about an hour, and reaches the tadpole larval stage 25 hours after fertilization, at 16°C. The swimming stage lasts for about 12 hours, although it may be as short as 6, or as long as 36 hours. Post-metamorphosis larvae can be raised to maturity in the laboratory with diatoms as food. The egg is green or yellowish-green, diameter 0.16 mm; artificial fertilization is possible (Morgan 1945; Berrill 1947, 1950). The eggs of *Pyura microcosmus* are reddish gray, their diameter is 0.2 mm, and they can be artificially fertilized. After the egg is fertilized, the germinal vesicle disappears, maturation takes place, anda grayish-brown crescent is formed. Cleavage begins an hour and 20 minutes after fertilization (18.0—20.0°C), the gastrula stage is attained after about 6 hours, and the tadpole stage after 22 hours. Within 24 hours after hatching, metamorphosis begins. The gill-slits are differentiated and the heart becomes active about 11 days after fertilization (Millar 1954). In *Pelonaia corrugata,* which forms a tailless larva, artificial fertilization in possible and a few can be raised. Four to six days are required for the completion of gastrulation, and 20 days for the differentiation of the gill-slits (Millar 1954).

As soon as the gonads of the Japanese tunicate *Halocynthia roretzi* are extirpated, the eggs and sperm become mixed. The eggs are yellow-gray, 0.27 mm in diameter; as has been described above, the germinal vesicles are intact in artificially obtained eggs and about two hours are required for them to complete the maturation divisions. The percentage of fertilization varies with the individual

Fig. 12.14 Photograph showing spawning of *Halocynthia roretzi* in laboratory. White dots in photograph are eggs (Hirai)

when the gametes are obtained from extirpated gonads, but this species can be made to spawn naturally. Ten individuals, which were collected from Mutsu Bay on October 10, 1955, and kept in a glass container in the laboratory, were observed to shed eggs and sperm 16 times between this date and December 16. The gametes are expelled by abrupt contractions of the body wall and atrial aperture (Fig. 12.14). Most of the spawning occurred during the hour between eleven and twelve in the morning, at two- to three-day intervals during the beginning of the period, but nearly every day toward its end. A single individual ejected eggs about nine to twelve times in one day, and the number of eggs shed by one individual was estimated at 10,000 to 12,000.

The 2-cell stage in this species is reached about one hour and 25 minutes after spawning, and the 4-cell stage at two hours and 35 minutes (water temperature 11° C). The larvae begin to hatch after about 50 hours (9—11° C); about 13 hours later individuals can be found which are beginning to metamorphose, and gill-slit differentiation takes place after ten days (Hirai and Tsubata 1957).

In observations made on *Halocynthia* gonads dissected out into sea water (Hirai 1941), it was found that maturation takes place only after the eggs come in contact with the sea water, although eggs collected directly from the atrial aperture during natural spawning showed the germinal vesicles broken down and incompletely elevated chorions. This indicates that the eggs normally begin the division in the ovary, or more probably in the oviduct, but that immature eggs are able to mature in sea water.

In *Chelyosoma siboja*, as in *Halocynthia*, developing eggs can be obtained from natural spawning. The eggs are grayish-yellow or orange, 0.18 mm in diameter, and the germinal vesicle is already broken down immediately after shedding. The first cleavage occurs at about two hours, and the second 30 minutes later; the larva hatches after about 46 hours (13° C) and the free-swimming period generally lasts for 12 hours or more, although the larvae sometimes metamorphose soon after, or even before, hatching. Eggs of *Corella japonica* var. *asamusi* obtained from extirpated ovaries can be fertilized artificially; they are 0.15 mm in diameter, light ochre in color, and form a white crescent after fertilization. The time required for the first cleavage is about one hour and 20 minutes, and 30 minutes more to reach the 4-cell stage. The larvae hatch after about 20 hours, and some individuals begin metamorphosis five hours later (water temperature about 18° C), but other larvae may be found swimming for two days (Hirai, unpublished).

2. DEVELOPMENT OF COMPOUND ASCIDIANS

(1) REPRODUCTION

The compound ascidians reproduce both sexually and asexually. Sexual reproduction is almost exclusively of the viviparous type, with the sole exception of the oviparous genus *Diazona* (Berrill 1948). In the viviparous forms, the eggs are fertilized within the parental body, and development takes place in the atrium or in a specially differentiated pouch-shaped *brood-chamber*. The species of *Clavelina* reproduce asexually by budding during the winter, and these new individuals resulting from the budding reproduce sexually the following summer. Such genera as *Botryllus, Botrylloides, Polycitor* and *Perophora* carry out sexual and asexual reproduction simultaneous. In *Perophora formosana* (Oka), which occurs near Misaki, tadpole larvae are to be found in August, and at the same time stolon-budding is going on. Within the summer buds of *Polycitor proliferus,* formed by transverse fission, can be seen sexually produced larvae (Oka, H. and M. Usui 1944). According to Berrill (1937), the sexual breeding seasons of various compound ascidia are as follows: *Perophora viridis* — August, September; *P. annectans* — July to September; *P. bermudensis* — September, October; *Ecteinascidia turbinata* — June to August; *Botryllus schlosseri* — June to September; *Botrylloides niger* — June to August; *Clavelina picta* — July to September; *Distaplia clava* — summer; *D. bermudensis* — June to September; *Aplidium* — summer. The breeding seasons of Japanese compound ascidians have not yet been thoroughly determined, but it is known that *Botryllus primigenus* breeds from July to September (Watanabe 1953), *Polyciter proliferus* in July (Oka, H. 1942) and *Distaplia imaii,* in July (Hirai 1954). In addition to these, the larvae of *Botryllus, Botrylloides, Amaroucium,* etc., can be seen during the summer in the neighborhood of Asamushi. Unfortunately very little embryological investigation has been done on the compound ascidia in this country, and the details of their development are largely unknown. According to the study of Oka and Watanabe (1957), *Polycitor proliferus* and *Botryllus primigenus* are plentiful near the Shimoda Marine Biological Station, and can be reared in boxes set in the bay. *Perophora formosana* can be found along the shore at the Misaki Marine Biological Station.

(2) EARLY DEVELOPMENT

a. Methods of observation

Diazona violacea is an oviparous species, and can be observed by the methods employed for studying the simple ascidians, but its short breeding season and the fact that it not easy to keep the colonies in the laboratory make it difficult to study the early stages (Berrill 1953). The other species are viviparous and hard to observe in most cases.

Berrill (1935) has described the following methods for making embryological observation. The eggs of *Perophora* and *Clavelina lepadiformis* can be removed from the body for observation and will develop like those of the oviparous species. Since all the eggs of a given colony of *Botryllus* are in the same stage of development, the removal of a few at a time for study will give a continuous picture of the successive stages. In the case of *Diplosoma,* a piece of the colony can be cut off and placed on a glass plate, where it will become attached. If the plate is then turned upside down, the embryos developing in the common cloaca can be seen from the underside.

Two methods are described for handling developing embryos of *Archidistoma, Didemnum, Trididemnum, Morchellium, Distaplia, Distomus* and *Stolonica.* One of these consists in constructing an artificial brood-chember by tying bolting silk over the wide upper end of a funnel-shaped glass container. The embryos are introduced through the small lower opening, to which a T-shaped piece of glass tubing is attached. If air is bubbled through the tubing, water currents will be set up which carry oxygen to the embryos and remove waste materials and debris so that the embryos develop well. The other method is to introduce the embryos of other species through the atrial aperture into the common atrium of a piece of *Diplosoma* attached to glass as described above, and make them develop there, but with this method there is some danger that the introduced embryos will be ejected by the contractions of the cloaca. In general it is possible to remove embryos to observe them during development, but difficult to rear them so as to follow the movement of the blastomeres and germ layers.

The mode of development of compound ascidians in which the eggs contain moderate amounts of yolk is similar to that of the simple ascidians and will be omitted here; the following description will apply to species with conspicuously yolky eggs.

The older studies of this kind include those by Dawydoff (1899—1901) on *Distaplia* and by Maurice and Schulgin (1884) on *Amaroucium proliferum;* recently F. M. Scott (1945, 1946 1952) has published detailed observations on the development of *Amaroucium constellatum.* This species is common on the east coast of the United States, where is it known as 'sea pork'; its eggs are fertilized internally and develop within "brood spaces". The youngest embryos are found in the post-abdominal region, the late embryonic stages in the lower abdominal region, the tadpole larvae are in the thoracic region, and the fully developed larvae swim away through the mouth opening. The sexual reproduction of this species reaches its peak in July and August. Eggs and embryos are obtained by dissecting them out of the individuals that make up the colony. Embryos of the late gastrula stage will remain alive for some time after being dissected out, and their development can be followed, but early gastrulae disintegrate soon after being removed, so that observation must be done rapidly. Scott used this method to some extent to study living material, but most of her studies were made on embryos fixed in Bouin's fluid, preserved in 70% alcohol, and observed in hollow slides. The picric acid of the fixative stains the yolk a pale yellow, while the yolk-free areas close to the nuclei remain clear so that it is easy to differentiate among the blastomeres, and since the mitotic spindles are visible, it is possible to determine the direction of cleavage.

b. **Structure of the egg**

The structure of the compound ascidian egg is of course essentially similar to that of the simple ascidians; the main difference between the two types is rather concerned with whether they develop outside the parental body or within a brood-chamber. For example, the eggs of the oviparous compound species, *Diazona,* have rather large follicle cells which make them buoyant, in contrast to those of the viviparous simple ascidian *Styelopsis grossularia,* which have flat follicle cells; compound ascidian eggs are similar to the latter. The eggs are larger than those of the simple ascidians; the diameters as given by Berrill (1935) are *Botrylloides leachi* — 0.26 mm; *Botryllus gigas* — 0.45 mm; *Stolonica socialis* — 0.27 mm; *Distomus variolosus* — 0.59 mm; *Perophora listeri* — 0.24 mm, *Clavelina lepadiformis* — 0.26 mm. According to Scott (1945), the diameter of the *Amaroucium constellatum* egg is 0.25 mm, and it contains more yolk than that of any other ascidian of which the embryology has been studied.

c. **Early development (Scott 1945)**

Within the peripheral layer of the *Amaroucium constellatum* egg are embedded test cells; these move out into the space between egg surface and chorion after fertilization when the perivitelline space is formed. The first polar body is given off after fertilization. The meridional first cleavage plane divides the egg into a slightly smaller right, and larger left blastomere, which correspond respectively to the right and left halves of the embryo (Fig. 12.15a).

In all the sections of *Amaroucium* eggs examined, the polar bodies lie eccentrically with respect to the main mass of yolk. By assuming that the germinal vesicle always lies to the right of the apex of the elliptical egg and also that the sperm penetrates on the same side, it is possible to explain the lateral eccentricity of the zygote nucleus to the right of the median axis; this, in turn explains the constant inequality in size of the first two blastomeres.

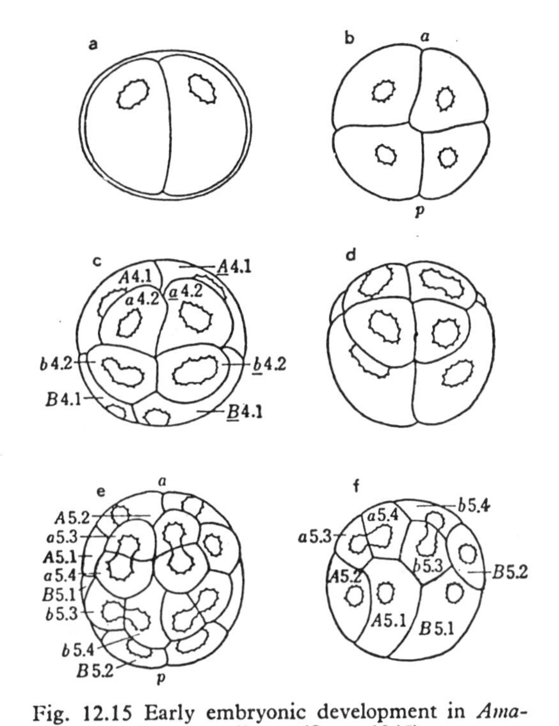

Fig. 12.15 Early embryonic development in *Amaroucium constellatum* (Scott 1945)

a. 2-cell stage; b. 4-cell stage (animal side); c. 8-cell stage (animal side); d. 8-cell stage (anterior view); e. 16-cell stage (animal side); f. 16-cell stage (from right side).

"The second division is also meridional and at right angles to the first, thus dividing the egg into two anterior and two posterior blastomeres. The posterior blastomeres are smaller than the anterior two. Beginning with the smallest of the four they fall into this order; the right posterior, the right anterior, the left posterior and the left anterior." (Scott 1945; p. 128)[3]. In each of these blastomeres the nucleus lies at the animal pole side of the cell. The third cleavage, which is horizontal, divides the egg into four micromeres at the animal pole and four yolk-filled macromeres at the vegetal pole (Fig. 12.15c, d). According to Conklin's system of indicating the various blastomeres of the 8-cell stage, the left anterior macromere is $\underline{A}$ 4.1, the right anterior macromere is *A* 4.1, and the corresponding micromeres are $\underline{a}$ 4.2 and *a* 4.2. The left posterior macromere is $\underline{B}$ 4.1, the right posterior macromere is *B* 4.1, and their corresponding micromeres are $\underline{b}$ 4.2 and *b* 4.2 (Fig. 12.15c, d). At the fourth cleavage the micromeres cleave practically equally, while both the anterior and posterior pairs of macromeres undergo unequal cleavage. The micromeres produced by this cleavage all tend to accumulate toward the animal pole side, and the yolky macromeres occupy the vegetal side of the embryo. The prospective fate of each of these 16 blastomeres is as follows: $\underline{A}$ 5.2 and *A* 5.2 will become notochord and nerve; *A* 5.1 and $\underline{A}$ 5.1 will

[3] The following description is also cited directly, with some omissions and slight changes in terminology, from Scott's paper.

36*

form the anterior half of the endoderm. $\underline{B}$ 5.2 and *B* 5.2 will form mesoderm, and $\underline{B}$ 5.1 and *B* 5.1 will provide the posterior half of the endoderm. The micromeres, indicated by small letters, form the ectoderm which will cover the whole embryo (Fig. 12.15e, f).

After the 16-cell stage a discrepancy in cleavage time appears. The mode of cleavage outlined above can be regarded as similar to that of *Styela,* but the extremely large amount of yolk in the eggs of this species interferes with the formation of a blastocoel. Furthermore, the mesoderm cells are distributed nearer to the animal pole than they are in an embryo like that of *Styela,* in which they are larger than the endoderm cells. There is thus no difference from *Styela* with respect to the mutual relations of the germ layers, but some divergence can be found in the positions in which they are arranged. For example, the vegetal pole occupied by the macromeres will be the dorsal side of the embryo; in its anterior part are the two chorda-neural cells, and in its posterior part, the two mesodermal cells. The animal pole side, on the other hand, consists of ectodermal cells and will become the ventral part of the embryo. As the result of the unequal first cleavage, all the blastomeres of the right side are smaller than those of the left side, but the arrangement of the two sides is bilaterally symmetrical.

Before the fifth cleavage, the *a* cells spread laterally, and the *b* 5.4 cells approach the *a* cells, while the *b* 5.3 cells come to lie on top of the mesoderm. At this time the embryo is in the 22-cell stage (Fig. 12.16a, b). The mesodermal blastomeres cleave vertically to form four cells arranged in an arch. The posterior macromeres also cleave vertically, to form the lateral mesoderm toward the animal pole and elongated cells lying centrally. The anterior macromeres divide into micromeres lying at the two sides of the chorda-neural cells and centrally located macromeres. The chorda-neural blastomeres cleave later, to form a transverse row of cells. At the 22-cell stage there are three mesodermal cells arranged on each side (*B* 6.2, *B* 6.3, *B* 6.4 and the corresponding $\underline{B}$ set). When the fifth cleavage cycle divides the embryo into 32 cells, the *B* 7.4 blastomere formed dorsally from *B* 6.2, and *B* 7.8 formed by *B* 6.4 become muscle cells, while the ventral blastomeres produced by these divisions, together with *B* 6.3, become mesenchyme. The chorda-neural cells separate into four chorda and four neural cells, the chorda cells being arranged at the vegetal side and the neural cells at the animal pole side (Fig. 12.16c). The *B* 6.3 blastomere of the 22-cell stage will be located at the center of the future ventral lip of the blastopore.

d. **Formation of gastrula and larval stages**

Gastrulation in *Amaroucium* is slightly different from that of forms like *Styela.* The margins of its blastopore are established at the sixth cleavage. The dorsal lip consists of four chordal cells; the ventral lip is formed by mesodermal cells. Enclosed by the blastoporal lips are the large vegetative macromeres (Fig. 12.16c). The cells of the animal pole side surround the embryo by the process of epiboly, giving rise to the ectoderm. In this gastrulation process, the large amount of yolk in the endodermal macromeres prevents them from invaginating; as a result, they do not participate in such movements of the germ layers as were seen in the early gastrula of *Styela.* These are chiefly begun by the mesodermal crescent cells, which move over the surface of the endoderm cells toward the vegetal pole, dividing as they go; furthermore, the mesodermal cells of the right side are clearly more active than those of the left side. The lateral lips of the blastopore converge toward the median line from the posterior side of the embryo, but the greater activity of the right side causes the circular blastopore to become irregularly ovoid, its right edge defining a horizontal line (Fig. 12.16e).

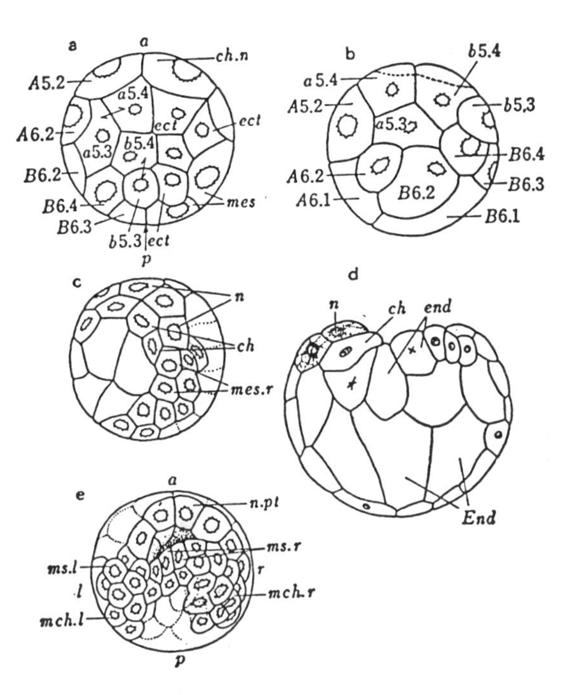

Fig. 12.16 Cleavage and gastrulation in *Amaroucium constellatum* (Scott)

a. 22-cell stage (animal side); b. 22-cell stage (right side); c. embryo at gastrulation (right side); d. section of central region of gastrula through blastopore; e. dorsal view of late gastrula. ***a:*** anterior, ***ch:*** notochord cell (chordal cell), ***ch.n:*** chorda-neural cell, ***ect:*** ectoderm, ***End***: large endodermal cell, ***end:*** small endodermal cell, ***mch.l:*** left mesenchyme cell, ***mch.r:*** right mesenchyme cell, ***mes:*** mesodermal cell, ***mes.r:*** right mesodermal cell, ***ms.l:*** left muscle cell, ***ms.r:*** right muscle cell, ***n:*** nerve cell, ***n.pt:*** neural plate, ***p:*** posterior.

As the blastopore becomes smaller, changes occur in the endodermal cell mass. The macromeres divide into a number of polyhedral cells, those immediately enclosed by the blastopore being smaller than the yolk-containing internal cells. As the size of the blastopore diminishes, these smaller cells change to a pyramidal shape, with their apices converging toward the closing blastopore. A cleft-like depression is formed here, but since it does not invaginate to form an archenteron, Scott cells this a "pseudo-invagination". The broad bases of these pyramidal cells rest on the large endodermal cells in the interior of the embryo (Fig. 12.16d).

The chordal cells adjacent to the endoderm invaginate and come to lie under the rudiment of the neural plate and dorsal to the endoderm. In contrast to the regular T-shape of the blastopore region in *Styela*, the *Amaroucium* blastopore forms an extremely irregular T, with the right side of its horizontal bar longer than the left side. The convergence of the lateral margins toward the median axial plane is asymmetrical, and the right margin fuses with the left well to the left of the median line. The ensuing closure of this blastopore also starts as the right side and proceeds toward the left; as a result the neural plate slants to the left by 90° in comparison with its position in *Styela.* This asymmetrical closure of the blastopore shifts the muscle cells to positions dorsal and ventral to the notochord, the neural tube to its left side and the endodermal cord to its right.

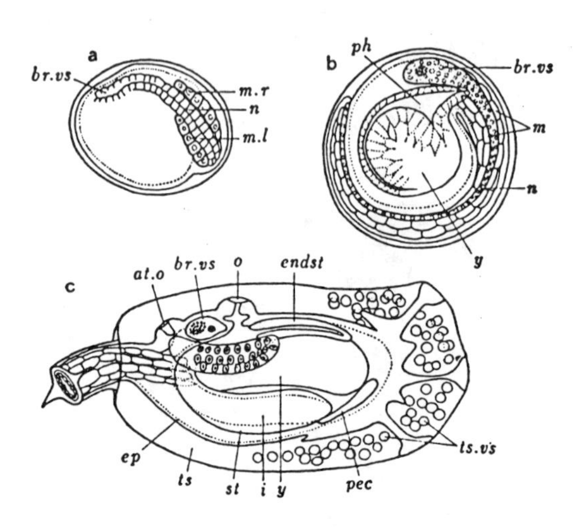

Fig. 12.17 Formation of larva of *Amaroucium constellatum* (Scott 1946)

a. before closure of neural fold; b. early larva; c. trunk and part of tail of well-developed tadpole larva. *at.o:* atrial opening, *br.vs:* brain vesicle, *endst:* endostyle, *ep:* epithelium, *i:* intestine, *m:* muscle, *m.l* left muscle, *m.r:* right muscle, *n:* neural tube, *o:* branchial opening, *per:* pericardium, *ph:* pharynx, *st:* stomach, *ts:* test, *ts.vs:* test vesicle, *y:* yolk mass.

When gastrulation is complete, the embryo is approximately spherical, with a shallow depression in its posterior ventral region where the tail bud is formed (Scott 1946). This tail bud is short and rounded, and curves around the ventral side of the embryo. Through the thin epidermis the large muscle cells can be seen lying dorsal and ventral to the notochord. The neural plate is being closed over progressively toward the anterior end, where the cavity of the brain vesicle rudiment is located; posteriorly the neural tube can be seen running parallel to the notochord on its left side (Fig. 12.17a).

After this the embryo increases in size and comes to exhibit the shape of a tadpole larva. The trunk elongates slightly in the antero-posterior axis, its anterior part curving ventrally. As the tail elongates it encircles the body meridionally. The neural folds are closed, and black pigment granules can be seen in the brain vesicle; the muscle cells lying dorsal and ventral to the notochord become more conspicuously visible. On the dorsal surface of the body, at the two sides posterior to the brain vesicle a pair of small ectodermal invaginations form the rudiments of the atrial chambers. The embryo is surrounded by a thin test rudiment (Fig. 12.17b).

The pharyngeal cavity appears as a space between the large and small endodermal cells when these become arranged in two layers. This cavity first arises below the brain vesicle, and gradually extends posteriorly; below the notochord rudiment it turns upward and gives rise to a projecting region. An invagination in the ventral part of this projection forms the stomach and intestinal rudiments on the right side of the body.

The trunk of the completely formed tadpole larva measures about 0.60 mm in length and 0.27 mm in depth, and has three rows of gill slits on each side (Fig. 12.17c). The endostyle is visible in the dorsal part of the pharynx, and in the ventral wall is a conspicuous mass of yolk. Below this yolk mass the large, transparent pericardium lies near the anterior part of the digestive tract. In the right ventral part of the body the stomach fits into a depression in the lower side of the yolk mass, and along its left side the intestine extends toward the left atrium. The basal part of the tail lies in the posterior third of the larval trunk. At the anterior end of the body, the adhesive papillae project into the tunic in a vertical row slightly to the right of the median plane. Near them the *test vesicles* lie scattered within the tunic, most of them connected with the epithelium by a slender thread. At the junction of trunk and tail the tunic is tucked into a pocket, while it forms wide horizontal fins along the tail. The tail muscles lie dorsal and ventral to the notochord, and the neural tube at its left side.

(3) LARVAL STAGE

Compared with the tadpole larva of the simple ascidians, the compound ascidian tadpole is generally larger, with a larger trunk and slenderer tail, and the rudiments of the adult organs are usually more highly developed (MacBride 1914). To cite the total length of some compound ascidian tadpoles: *Distaplia imaii* — 3.50 mm; *Cyathocormus mirabilis* — 4.00 mm; *Sidneioides snamoti* — 1.70 mm; *Didemnum (Didemnum) misakiense* — 2.00 mm; *Leptoclinum mitsukurii* — 1.60 mm; *Botrylloides* sp. — 2.30 mm; *Botryllus schlosseri* — 1.50 mm; *Ecteinascidia turbinata* — 5.00 mm; *Amaroucium constellatum* — 2.25 mm; *Polycitor proliferus* — 2.50 mm.

Among the three larval types mentioned previously, the larvae of the compound ascidians belong to Subtype 2 (Compound Type) of Type I (*Cynthia* type), Type II (*Amaroucium* type) and Type III (*Botrylloides* type) (Hirai 1951). In Type I Subtype 2 (Fig. 12.18b), which includes *Perophora formosana* and the rest of the genus *Perophora,* the genus *Ecteinascidia, Clavelina lepadiformis, Distaplia imaii* and the rest of the genus, and *Cyathocormus mirabilis,* the trunk is large; the adhesive papillae show various shapes and are arranged in a triangle; the arrangement of the sensory vesicle, branchial and atrial apertures is linear and parallel to the cranio-caudal axis; the trunk-part of the nerve cord is either parallel or perpendicular to the cranio-caudal axis; the rudiments of the adult organ are visible.

Type II (Fig. 12.17c) includes the genus *Amarouciоum, Sidneioides snamoti, Sigillinaria clavata, Polycitor proliferus* (Fig. 12.19), the genus *Didemnum, Diplosoma mitsukurii;* the trunk is large; the stalked adhesive papillae are arranged in a dorso-ventral line; epithelial process and vesicles are present; the sensory vesicle, branchial and atrial apertures lie in a straight line parallel to the cranio-caudal axis; the rudiments of the adult organs are distinctly visible. The trunk portion of the nerve cord extends dorso-ventrally, and the tail fin is laterally flatened.

Type III (Fig. 12.18a) includes the genera *Botryllus* and *Botrylloides, Stylopsis grossularia, Stolonica socialis.* The adhesive papillae in this group are stalkless, and arranged in a triangle; the epithelial organs (ampullae) are arranged in a ring anteriorly; the sensory vesicle is sunk deep in the trunk region and does not maintain a linear relation with the branchial and atrial apertures; the rudiments of the adult organs are visible; the nerve ganglion lies parallel to the cranio-caudal axis and the tail fin is dorso-ventral.

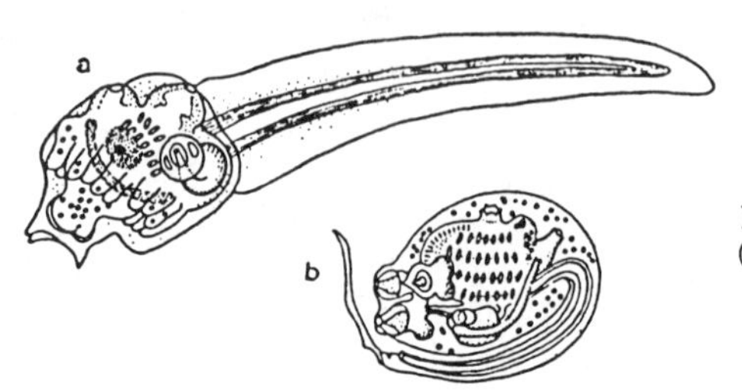

Fig. 12.18 a. Larva of *Botrylloides* sp. (Hirai 1951). b. Larva of *Distaplia imaii* (Hirai 1952).

The larvae of the compound ascidians can thus be described as more complicated in structure and less active than those of the simple ascidians. The tadpoles can be obtained during the breeding season by simply leaving part of a colony in a container of sea water; or if the colony is torn with a needle, the larvae will swim out. According to H. Oka (1943), if a colony of *Polycitor poliferus* is cut into several pieces and placed in sea water, the larvae which are released show a characteristic type of activity in which swimming periods alternate with periods of rest. The swimming stage of this species is extremely short, lasting 15 minutes at the longest, while some individuals attach and begin to metamorphose within three minutes. Grave (1924) reports that *Botryllus schlosseri* breeds in summer; the swimming stage may last as long as 27 hours, or end at 30 minutes, although the greatest number of larvae swim for one to three hours.

(4) METAMORPHOSIS

The larvae of the compound ascidians begin their metamorphosis after a swimming period which is generally shorter than those of the simple ascidians. In *Polycitor proliferus* (Oka, H. 1943), the adhesive papillae have completely disappeared within two hours after the beginning of their regression at the time of

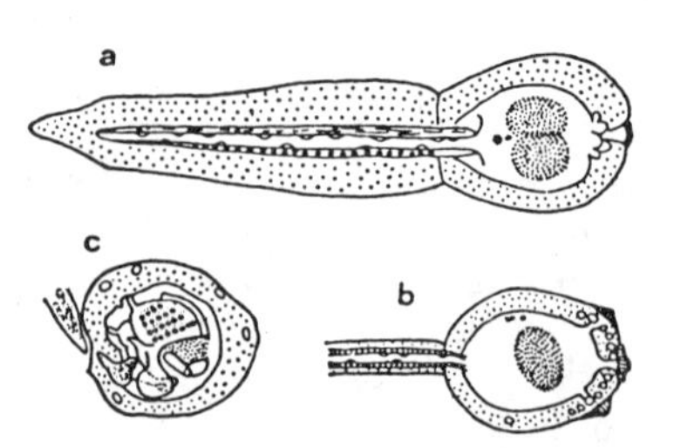

Fig. 12.19 Larva of *Polycitor proliferus* and its metamorphosis (H. Oka 1943)

a. dorsal view of larva; b. lateral view; c. larva 1 day after metamorphosis

larval fixation. The tail also begins to degenerate at this time, shortening and simultaneously becoming irregularly coiled up; eventually a constriction takes place at the base of the tail, cutting it off and isolating it. The part anterior to the point of section is absorbed into the trunk, and 24 hours later a young animal with the adult structure has been formed (Fig. 12.19).

The process of metamorphosis includes the two phenomena of regression of the larval structures and differentiation of the adult organs; in parallel with these can be seen the regression of the tail and the rotation of the adult organ axis. As the result of this rotation, the larval arrangement of the adult organ rudiments is changed to the adult organ arrangement, but the degree of rotation differs from species to species.

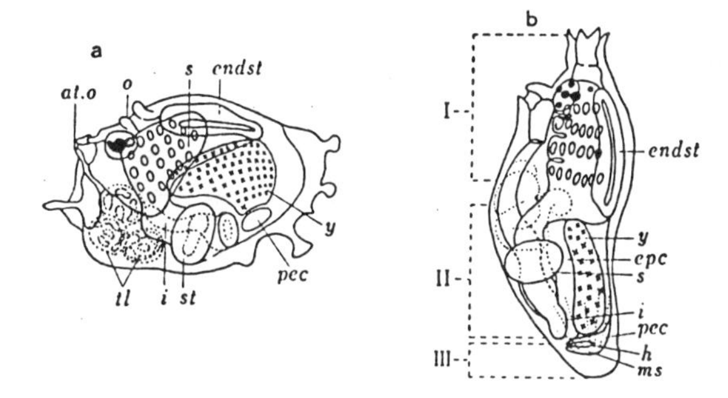

Fig. 12.20 Metamorphosis in *Amaroucium constellatum* (Scott 1952)

a. larva after resorption of tail; b. larva after completing rotation. I: thorax, II: abdomen, III: post-abdomen.
at.o: atrial opening, *endst:* endostyle, *epc:* epicardium, *h:* heart, *i:* intestine, *ms:* muscle, *o:* branchial opening, *pec:* pericardium, *s:* gill sac, *st:* stomach, *tl:* resorbed tail, *y:* yolk mass.

According to Scott (1952), metamorphosis of the *Amaroucium constellatum* larva begins when it becomes attached, after swimming for a period lasting from a few to 30 minutes at the most. When the larva comes to rest and attaches, the contents of the adhesive papillae are secreted from their tips; the tail stops moving and within two minutes its tissues begin to shrink, moving with a flowing motion into the hyaline part of the trunk where they are drawn into a compact mass. This resorption is complete within ten minutes. The empty tail envelope remains attached for two or three days and eventually falls off. In this species the regression of the tail does not necessarily occur simultaneusly with the metamorphosis of the trunk; in many cases the tail may continue to hang onto the body for several days without being resorbed.

Within one to eight minutes after the adhesive substance is released from the adhesive papillae, the rotation of the body axis begins. In the larva, the position of the adhesive papillae marks the anterior end, and the base of the tail, the posterior end of the antero-posterior axis; the siphons are on the dorsal side and the digestive tract lies ventrally. In the adult body, on the other hand, the antero-posterior axis extends from the siphons to the attached region; the ganglion marks the dorsal side, and the endostyle lies ventrally. At the time of metamorphosis the larval polarity shifts to that of the adult in such a way that the larval dorso-ventral axis practically corresponds to the antero-posterior axis of the adult. The pericardium is at almost the anterior end of the larva, while in the adult it is located in the post-abdomen.

The process of rotation begins with a series of contractions. One strong contraction can be seen to move from the anterior, endostyle end around the ventral side to the base of the tail. This pushes the digestive tract against the mass of caudal material, and the siphons, endostyle and yolk mass, which in the larva formed a straight antero-posterior line, move to a line which coincides with the larval dorso-ventral axis, but will constitute the adult antero-posterior axis. The adhesive palps are also pulled strongly away from the attached surface region, toward the ventral side. A reverse movement toward the dorsal side follows this ventrally directed contraction, but it is less effective. These two movements alternate rhythmically, with a period varying between six and twelve minutes, for about two hours; by the time they stop the adult arrangement of organs is established. During this period the whole body has turned by about 70°, the endostyle by 90—100°, and the pericardium and yolk mass by at least 90° (Fig. 12.20). While these contracting movements are going on, the entire body shrinks and gives the appearance of a hard lump, but once the rotation of the body axis has been achieved, the animal expands again and begins to recover its transparency. In the larva of *Clavelina* (Seelinger 1885, in Korschelt und Heider 1936) a deep depression is formed between the adhesive palps and the branchial aperture. When the larva attaches, this depression turns inside out and widens, causing the branchial aperture to become located on the opposite side from the adhesive palps.

(5) BUDDING AND COLONY FORMATION

In addition to sexual reproduction, the compound ascidians also reproduce by *budding* and form colonies. The manner of budding is not identical in all species, but tends to be similar among the members of any given taxonomic group. In general, the epidermis forms the contour of the new individual *(zooid)* that will result from the budding; the various organs arise from the tissues of the epicardium, *mesenchymal septum,* atrium and blood.

a. In the family Diazonidae and the sub-family Polycitorinae, budding takes place by the strobilation of the zooids. The body in these groups is composed of a *thorax* and an *abdomen*. At the time of strobilation the thorax regresses, and trophocytes migrate into the abdomen, where they accumulate; it is then cut transversely to given rise to *buds* by one or more constrictions of the epidermis.

In *Diazona violacea,* studied by Berrill (1948), three to eight buds are formed between late autumn and early winter by such a strobilation process; about the end or the winter or the beginning of spring the buds develop into new zooids. During late spring and early summer these grow and reach maturity, and for a period of six to eight weeks carry on sexual reproduction of the oviparous type. Several weeks after the period of sexual reproduction ends, they again begin to reproduce, this time asexually (Fig. 12.21).

As this process starts, shrinkage appears in the thorax, and simultaneously *trophocytes,* which are believed to arise from blood cells, become active. These nutritive cells migrate to the posterior part of the body and accumulate ventrally and in the oesophageal region, causing a distension of those regions. The thorax undergoes autolysis, temporarily becoming a *ghost* and finally disappearing altogether. The trophocytes supply nourishment but do not take part in forming the bud or the new organs. Next, the epidermis constricts so as to cut deeply into the underlying tissues. This usually divides the anterior part of the abdomen into four or five spherical buds. In some cases, the anterior part of the next-to-the-last of these may be already beginning to regenerate, while the posterior part is only constricting, but this also eventually separates off. Sometimes a bud is formed by the oesophageal region but not by the part directly behind it, so that the bud is connected with the posteriormost region only by the ghost tissue (Fig. 12.21b). The bud formed by the posterior end is the largest. The contraction of the epidermis appears to be the first step in bud formation; that this is not simply an apparent local contraction is shown by the fact that sections show the epidermal cells in the constricted part to be unusually large and have every appearance of extensive growth toward the interior, indicating that the internal underlying tissues are passively pinched off by these cells. The constricted bud is practically spherical. This appears to be due to the freedom of movement of the large trophocytes.

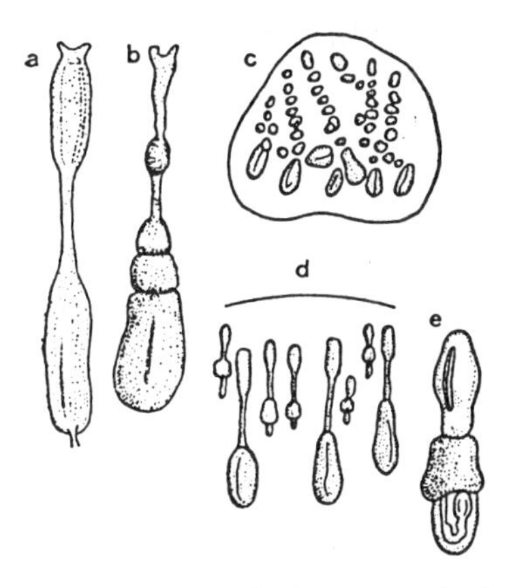

Fig. 12.21 Strobilation of zooid in *Diazona* (Berrill 1948, 1952)
a. individual degenerating before strobilation; b. one type of strobilation; c. representative case of strobilation in January; d. development of buds in March: buds from oesophageal region regenerate both anteriorly, and posteriorly; large bud from posterior part regenerates only anteriorly; e. buds from oesophageal region form anterior thorax containing endostyle and posterior abdomen including digestive tract.

Next these buds set about reconstituting the organs which they have lost. The surfaces of both the anterior and posterior parts of the bud are covered by a new epidermis which appeared as the result of the constriction. The atrium, pericardium, ganglion and part of the intestine of the new thorax are differentiated from the tissue of the anterior part of the epicardium, and the newly formed intestine establishes connection with the surviving part of the old intestine. The constriction which cut off the bud severed the digestive tract in its posterior part at two places; these two cut ends unite to form the *intestinal loop,* which elongates posteriorly, keeping pace with the epidermal outgrowth, and finally differentiates into the *stomach, post-stomach* and *mid-intestine.* The new pericardium and heart are formed from the posterior end of the epicardium. The lost anterior part is formed from the posteriormost large bud.

A study of strobilation in the Japanese compound ascidian *Polycitor proliferus* has been made by H. Oka (1942, 1944). This species can be studied by causing part of a colony to adhere to a glass slide. In an especially interesting study H. Oka and Usui (1944) removed one individual from a colony to a glass slide and observed its fission and the development of the buds. In this species the post-abdominal region is short in the spring, but by July it has lengthened and sexual reproduction is going on. As bud-formation approaches, the zooid regresses, and a constriction between the oesophagus and stomach divides it into thorasic and abdominal sections (Fig. 12.22c). The thorax includes the branchial sac, oesophagus, a part of the epicardium and the end of the rectum; in the abdomen are the stomach, mid-intestine, heart, a part of the epicardium and the proximal part of the rectum. Next, the abdomen is separated by deep constrictions into as many as six buds; these later migrate to the surface of the colony to complete their development, which requires two days. The isolated thorax regenerates the lost abdominal portions. The new zooids formed from the buds retain the polarity of the parental organ arrangement.

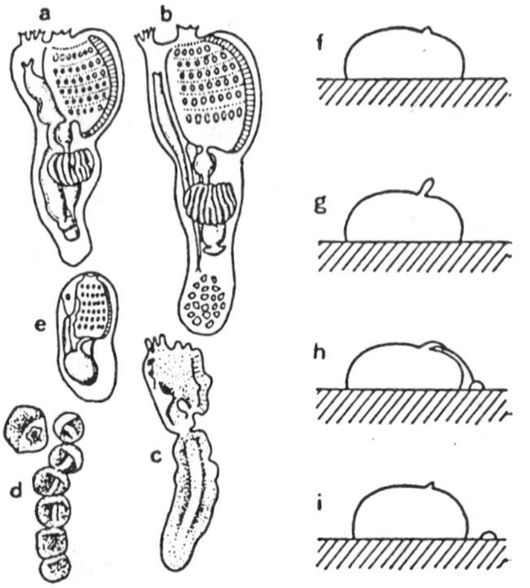

Fig. 12.22 Budding by strobilation (a ~ e) and colonial budding (f ~ i) in *Polycitor proliferus* (H. Oka 1942; H. Oka and M. Usui 1944)
a. zooid in April; b. zooid in July; c,d. successive stages in strobilation; e. new individual resulting from bud; f ~ i. order of colonial budding.

In addition to such asexual reproduction by means of zooid budding, this species also propagates by *division of the colony* as well as by colonial budding. In the first of these methods, pseudopodia-like processes are extended from the tunic, and the colony slowly moves after them. When the processes extend in different directions, the colony is pulled apart into as many pieces. Under favorable conditions a colony may increase by this method to twelve or more units within one month.

The phenomenon of *colonial budding* was observed in fixed material by A. Oka (1933) and by H. Oka (1944) in the living state. In this method of reproduction some zooids, together with some test substance, escape from the surface of the colony and slip out of it altogether. From the first only a thin thread extends between the mother colony and the bud, and eventually the latter slides gently down and forms a new, separate colony. The individual animals included in the bud are small, their branchial and atrial apertures are both closed and the atrium is in a contracted state (Fig. 12.22f-i).

Among the Polyclinidae, the zooid structure includes a post-abdomen, and these species undergo strobilation. The epicardial tissue takes the lead in forming the new individuals from the buds.

b. The zooids of the family Didemnidae, which have a thorax and a short abdomen, undergo a type of budding *(pyloric budding)* in which one or two swellings of the epidermis, including within them epicardial tissue, form in the central or in the oesophageal part of the body. Although the epicardium in the didemnid species is reduced and rudimentary in form, consisting of a pair of small sacs located in the oesophageal region, it plays an important role as the budding tissue of this group. A bud formed in the anterior part of the oesophageal region gives rise to an abdomen, or posterior region; one formed near the posterior part of the oesophagus will give rise to a thorax, so that two new zooids result from this type of budding. Sometimes only one bud appears; in such a case only a thorax is formed, and the thoracic region of the original zooid retrogresses. In other cases two buds appear, each of which again gives rise to a bud which in turn forms a complete new individual (Salfi 1933, cited in Berrill 1950) (Fig. 12.23a, b). Della Valle (1883, cited in Korschelt and Heider 1900) reports that in some cases the anterior bud of *Didemnum* forms an abdomen, the oesophagus and intestine of which connect with those of the parent zooid, while the posterior bud forms a thorax, with oesophagus and rectum which also join those of the parent. As a result, the organ systems formed by the two buds and the original zooid are connected together and may remain so. He also describes cases in which a thorax formed by a single bud comes to have two abdomens, or two thoraces may have a single abdomen in common. Among the species belonging to the genus *Diplosoma,* the tadpole larva has two atria; one of these is believed to arise by budding. The original atrium has a sensory vesicle, but the one derived by budding lacks such a structure. In some species, larvae can sometimes be found with three atria.

c. In the subfamily Claveliniae of the family Clavelinidae, the zooid consists of thorax and abdomen, and the buds make their appearance from swellings *(ampullae)* at the tips of the *stolons* (Fig. 12.23c, d). The colony of *Clavelina lepadiformis* is covered with a common tunic, but the zooids are entirely isolated from each other except for the fact that they are embedded in a common test. According to Berrill (1950), the adult zooids of this species retrogress, the trophocytes migrate into the stolons and ampullae, and the zooids are sloughed off at the end of the summer. The ampullae are thus isolated, and these survive into the winter. The low temperature inhibits their development so that their budding occurs in spring; two or three months later the new individuals begin to reproduce sexually, and continue until the end of the summer, when the retrogression of the zooids takes place.

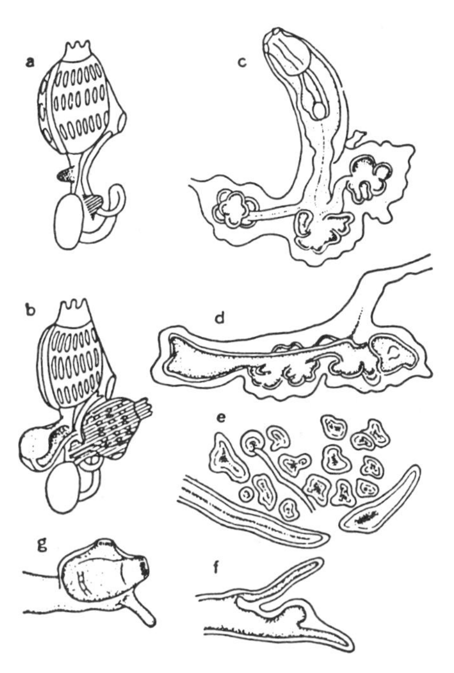

Fig. 12.23 Various types of budding
a,b. budding in *Trididemnum* (Salfi, Berrill 1950); c,d. before and after bud formation in *Clavelina lepadiformis* (Huxley, Berrill 1950); f,g. budding near end of stolon in *Ecteinascidia tortugensis* (Plough Jones, Berrill and 1950).

The ampullae consist of epidermis, the mesenchyme of the *septum* and an accumulation of trophocytes; the epicardium is a sac-like structure which separates from the atrium and does not extend as far as the stolon. Budding and the formation of the new zooids in this species depend on the septal mesenchyme. This is true also of *Pycnoclavella* (Fig. 12.23e), *Chondrostachys* and other species.

On the basis of the study of Brien and Brien-Gavage (1927), Berrill (1950) rejects the observations of Seeliger (1885, 1893, cited by Korschelt und Heider 1936). Seeliger maintains that the septum of the stolon is pericardial rather than mesenchymatous, and the bud-forming ampullae of the stolons simply nutritive chambers; that buds arise from the stolon, and are lined up in order of development from its base to its tip.

The members of the Perophoridae lack an abdomen, and form buds from the stolon (*stolonic germination*). *Perophora* and *Eckteinascidia* (Fig. 12.23f, g) follow this method. Unlike other groups, in which the process of budding is affected by the condition of the parent zooid, these species form buds while the original zooid maintains its usual activity. Their stolons are not only wide, but also rather long. Near the growing tip of the stolon a swelling develops in the epidermis of the upper surface, and also in the mesenchyme of the septum contained within it. The epidermis initiates the process of budding, and the mesenchyme of the septum is the formative tissue. Since there is no epicardium in these species, as in *Clavelina* it takes no part in the regeneration process.

d. In the subfamily Botryllinae of the family Styelidae, peribranchial and vascular budding take place. In *peribranchial budding*, all the layers of the body wall are involved; depending on the species, either a bud or a stolon may be formed, but if a stolon develops, it tip becomes a bud. The formation of the new zooid depends on the epithelial tissue of the *atrial* or *peribranchial lining*.

In *Botryllus schlosseri*, the first colony to result from sexual reproduction consists of the zooid which has developed from the egg *(oozooid)*, with a *blastozooid* on each side of it; these are formed as buds from the oozooid. The original zooid has protostigmata; the two blastozooids are inactive, have four rows of gill-slits, and each in turn produces one bud on its left side and two on its right side. About 12 days after the tadpole larva attaches, the first bud becomes active, and this is followed by the dissolution and resorption of the original zooid (Berrill 1950). According to Berrill (1941), the buds formed by sexually mature zooids of *Botryllus* arise as thickened discs on the atrial epithelium directly in front of the gonads (Fig. 12.24a, b). The adult epithelium gives rise to only the external body wall of the bud, taking no part in the formation of the other organs. The earliest detectable rudiment of the bud consists of the disc, made up of eight atrial epithelial cells which are flat at first, but later become columnar. The rudiment also changes in shape from an arch to a hemisphere, and finally forms a closed sphere that is pinched off and separates from the main zooid, except for an epidermal stalk. The internal organs eventually differentiate from this hollow sphere. The reproductive cells are already differentiated in this stage, originating in the wall of the sac and separating to the outside. An epithelial projection which arises from the anterior end of the bud establishes connection with the circulatory system and forms a stolon. Next, two folds appear at the anterior end; from these anterior folds two longitudinal partitions cut inward and extend posteriorly, forming a median and two lateral chambers. The central one becomes the pharyngeal

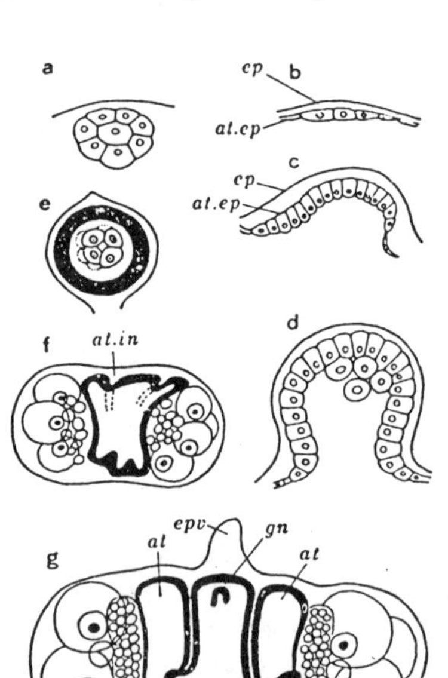

Fig. 12.24 Formation of the bud and its development in *Botryllus* (Berrill 1941)

a,b. superficial view and transverse section of disc-like thickening at early stage of bud formation; c. arched bud; d. bud developing into pouch-like form; e. lateral view of pouch-shaped bud, containing germ cells; f. early stage of formation of atrium and intestine; g. pharynx and atrium complete; heart, nerve vesicle and intestine are being formed. *at:* atrial cavity, *at.ep:* atrial epithelium, *at.in:* atrial invagination, *ep:* epidermis, *epv:* epidermal vessel, *gn:* ganglion rudiment, *h:* heart rudiment, *st:* stomach rudiment, *stl:* stalk.

chamber and the lateral ones form paired atrial chambers (Fig. 12.24 f, g *at.in*). At the same time a posterior evagination gives rise to the rudiments of the stomach and intestine, and two other evaginations are formed from the pharyngeal chamber. The anterior of these becomes the neural mass, and the other, on the left posterior side of the pharynx, forms the heart rudiment (Fig. 12.24 g, *gn.h*). The cells which were separated from the wall of the spherical bud give rise to the gonad rudiment.

A study of budding in a Japanese species of this genus, *Botryllus primigenus*, has been made by H. Watanabe (1953). The mode of budding is the same as that observed by Berrill (1941): first a disc-like thickening is formed (about 24μ, first stage), which grows (about 30μ, second stage), changes from a hemisphere to a closed sphere and separates from the parent zooid (about 48μ, third stage), and differentiates pharyngeal and atrial chamber rudiments (about 90—100μ, fourth stage). Heart neural mass and intestinal rudiments are formed (about 100—120μ, fiftrh stage); as the ows of definitive stigmata begin to open, the first stage of the next bud is being formed (about 300—360μ, sixth stage). The stigmata open completely (about 300—500 μ, seventh stage), the heart begins to function and the zooid shows its greatest activity; this state lasts four days (about 1.0—2.4 mm, ninth stage). During the period of sexual reproduction, from July to September, the gametes become mature (about 2.4 mm, tenth stage). Ninth and tenth stage zooids begin to degenerate as soon as the bud of the next generation has completed its differentiation, and disappear into the common circulatory system (840—312—0μ, eleventh stage). This life cycle, from the first to the eleventh stage, takes about 12 days.

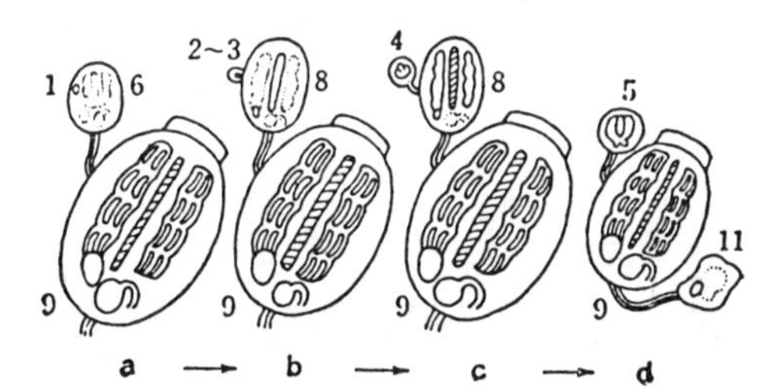

Fig. 12.25 Four successive phases (a ~ d) of a unit in a colony of *Botryllus primigenus* (Watanabe 1953)

Numbers (1, 2, . . . , 11) show stages in development of the bud. Stage 9 zooid in phases a, b and c degenerates to Stage 11 in phase d; new bud (Stage 1) develops inside Stage 6 bud in phase a, etc.

In going through this series of stages, each individual zooid is accompanied by two buds; this is considered as one *unit*, and each unit passes through four *asexual reproductive phases* (a, b, c, d) in its developmental progression (Fig. 12.25). On the first day (Phase a), the ninth stage zooid has a sixth stage bud, on which in turn is a first stage bud — i.e., the unit consists of Stages 1-6-9. On the second day (Phase b), the unit is composed of Stages (2-3)-8-9. Phase c, occupying the third day, consists of Stages 4-8-9-, and in Phase d, on the fourth day, the unit is composed of Stages 5-9-11. On the fifth day the eleventh stage zooid has disappeared, the fifth stage bud has advanced to the sixth stage, and a first stage bud is forming on it, so that the unit again consists of Stages 1-6-9 and the cycle reverts again to Phase a. These four phases are repeated in regular sequence.

Until recently, this type of ectodermal budding, originating in the epithelium of the atrial cavity *(palleal,* or *peribranchial, budding)* was considered to be the only method by which asexual reproduction takes place in the order Stolidobranchiata, which includes the genus *Botryllus.* Oka and Watanabe (1957), however, have just found that this genus also shows another type of budding, originating in the vascular system *(mesoblastic-vascular budding)*: in *Botryllus primigenus,* buds are formed by lymphocytes, which aggregate in the bases of the ampullae. This new type has been given the name of *vascular budding.* In *Botrylloides violaceum* Oka, vascular budding takes place only when a small piece of a colony devoid of zooids is isolated. As in *Botryllus,* these new buds are formed from aggregations of lymphocytes on the walls of the blood vessels in the test (Oka and Watanabe 1959).

REFERENCES

Berrill, N. J. 1928: The identification and validity of certain species of Ascidians. J. Mar. Biol Assoc. **15**: 159—175.

———— 1932: Mosaic development of Ascidian egg. Biol. Bull., **63**: 381—386.

———— 1935: Studies in Tunicate development. Part III. Differential retardation and acceleration. Phil. Trans. Roy. Soc. London, B. **225**: 255—326.

———— 1941a: The development of the bud in *Botryllus.* Biol. Bull., **80**: 169—184.

———— 1941b: Size and morphogenesis in the bud of *Botryllus.* Biol. Bull., **80**: 185—193.

———— 1947a: The structure, tadpole and budding of ascidian *Pycnoclavella aurilucens* Garstang. J. Mar. Biol. Assoc., **27**: 245—251.

———— 1947b: The development and growth of *Ciona.* J. Mar. Biol. Assoc., **26**: 616—625.

———— 1948a: Structure, tadpole and bud formation in ascidian Archidistoma. J. Mar. Biol. Assoc., **27**: 380—388.

———— 1948b: The development, morphology and budding of ascidian *Diazona.* J. Mar. Biol. Assoc., **27**: 389—399.

———— 1950: The Tunicata. Ray Society, London.

Child, C. M. 1927: Developmental modification and elimination of the larval stage in the Ascidian, *Corella willmeriana.* J. Morph., **44**: 467—514.

Cohen, A. and N. J. Berrill 1936: The early development of Ascidian eggs. Biol. Bull., **70**: 78—88.

Conklin, E. G. 1905: Mosaic development in ascidian eggs. J. Exp. Zoöl., **2**: 145—223.

Cowdry, E. V. 1924: General Cytology. Chicago.

Dalcq, A. M. 1938: Form and causality in early development. Cambridge.

Galtsoff, P. S. and others. 1937: Culture methods for Invertebrate Animals. New York.

Garstang, S. L. and W. Garstang 1928: On the development of *Botrylloides.* Quart. Jour. Micr. Sci., **72**: 1—49.

Grave, C. 1921: *Amaroucium constellatum* (Verrill). II. The structure and organization of the tadpole larva. J. Morph., **36**: 71—101.

———— 1924: *Botryllus schlosseri* (Pallas): The behavior and morphology of the free-swimming larva. J. Morph., **38**: 207—247.

Hirai, E. 1939: A report of observation made of the egg-follicle formation in the case of *Cynthia roretzi* Drasche. Sci. Rep. Tohoku Imp. Univ., Biology, **14**: 305—318.

———— 1940: On the recent condition of *Cynthia roretzi* Drasche culturing. Zool. Mag., **52**: 467—471. (in Japanese)

——— 1941a: The early development of *Cynthia roretzi* Drasche. Sci. Rep. Tohoku Imp. Univ., Biology, **16**: 217—232.

——— 1941b: An outline of the development of *Cynthia roretzi* Drasche. Sci. Rep. Tohoku Imp. Univ., Biology, **16**: 257—261.

——— 1949: Concerning the test cells and their secretory substance in *Cynthia roretzi* Drasche. Zool. Mag., **58**: 205—206. (in Japanese)

Hirai, E. 1951: A comparative study on the structures of tadpoles of ascidians. Sci. Rep. Tohoku Univ., Biology, **19**: 79—87.

——— 1961: Accelerating effect on metamorphosis of vital staining of tadpole larva of an ascidian, *Halocynthia roretzi* (v. Drasche). Bull. Mar. Biol. Sta. Asamushi, **10**: 189—192.

Hirai, E. & B. Tsubata 1956: On the spawning of an ascidian, *Halocynthia roretzi* (Drasche). Bull. Mar. Biol. sta. Asamushi. **8**: 1—4.

Huus, J. 1937—1940: *Ascidiaceae*: in Kükenthal und Krumbach, Handbuch der Zoologie, Bd. V. zweite Hälfte, Tunicata. : 545—672.

Knaben, N. 1936: Über Entwicklung und Funktion der Testazellen bei *Corella parallelogramma* Mull. Bergen Mus. Arb. Nat. Rek., 1936: 5—33.

Kobayashi, S. 1938: Studies on the body fluid of an ascidian, *Chelyosoma siboja* Oka with special reference to its blood system. Sci. Rep. Tohoku Imp. Univ., Biology, **13**: 25—35.

Korschelt u. Heider 1900: Text Book of Embryology. IV. Invertebrata. London.

——— 1936: Vergleichende Entwicklungsgeschite der Tiere. II. Jena.

MacBride, E. W. 1914: Text Book of Embryology. London.

Millar, R. H. 1951: The development and early stages of the ascidian *Pyura squamulosa* (Alder). J. Mar. Biol. Assoc., **30**: 27—31.

Millar, R. H. 1954: The development of the ascidian *Pyura microcosmus* (Savigny). J. Mar. Biol. Assoc., **33**: 403—407.

——— 1954: The breeding and development of the ascidian *Pelonaia corrugata* Forbes and Goodsir. J. Mar. Biol. Assoc., **33**: 681—687.

Morgan, T. H. 1942a: Cross- and self fertilization in the ascidian *Styela*. Biol. Bull., **82**: 161—171.

——— 1942b: Cross- and self-fertilization in the ascidian *Molgula manhattensis*. Biol. Bull., **82**: 172—177.

——— 1945: The conditions that lead to normal or abnormal development of *Ciona*. Biol. Bull. **88**: 50—62.

Oka, A. 1935: Report of the biological survey of Mutsu Bay. 28. Ascidiae simplices. Sci. Rep. Tohoku Imp. Univ., Biology, **10**: 427—466.

Oka, H. 1942: On a new species of *Polycitor* from Japan with some remarks on its mode of budding (Ascidiae compositae). Annot. Zool. Japon., **21**: 155—162.

——— 1943: Metamorphosis of *Polycitor mutabilis* (Ascidiae compositae). Annot. Zool. Japon., **22**: 54—58.

Oka, H. & M. Usui 1944: On the growth and propagation of the colonies in *Polycitor mutabilis* (Ascidiae compositae). Sci. Rep. Tokyo Bunrika Daigaku, Sec. B, **7**: 23—53.

Oka, H. and H. Watanabe 1957: Vascular budding, a new type of budding in *Botryllus*. Biol. Bull., **112**: 225—240.

Ortolani, G. 1952: Risultati sulla localizzazione dei territori presuntivi degli organi nel germe di Ascidie allo stadio VIII—XVI, determinata con le marche al carbone. Pubbl. Staz. Zool. Napoli, **23**: 271—283.

Seeliger-Hartmeyer 1893—1911: Bronn's Klassen und Ordnungen der Thier-Reichs, **2** und **3**.

Scott, F. M. 1945: The developmental history of *Amaroecium constellatum*. I. Early embryonic development. Biol. Bull., **88**: 126—138.

——— 1946: The developmental history of *Amaroecium constellatum*. II. Organogenesis of the larval action system. Biol. Bull., **91**: 66—80.

——— 1952: The developmental history of *Amaroecium constellatum*. III. Metamorphosis. Biol. Bull., **103**: 226—241.

Watanabe, H 1953: Studies on the regulation in fused colonies in *Botryllus primigenus* (Ascidiae compositae). Sci. Rep. Tokyo Bunrika Daigaku, Sect B, **7**: 183—198.

Willey, A. 1893: Studies on the Protochordata I. On the origin of the branchial stigmata, preoral lobe, endostyle, atrial cavities, etc., in *Ciona intestinalis* L. with remarks on *Clavelina lepadiformis*. Quart. Jour. Micr. Sci. **34**: 317—360.

——— 1893: Studies on the protochordata. II. The development of the neuro-hypophysial system in *Ciona intestinalis* and *Clavelina lepadiformis*, with an account of the origin of the sense-organs in *Ascidia mentula*. Quart. Jour. Micr. Sci. **35**: 295—316.

AUTHOR INDEX

C

D

E

F

O

P

R

SPECIES INDEX

D

E

F

G

H

I

J

K

L

M

N

O

P

T

U

V

W

X

Y

Z

GENERAL INDEX

D

E

F

G

T

V

W

Y

Z

Izdavač
IZDAVAČKO PREDUZEĆE „NOLIT“, BEOGRAD, TERAZIJE 27/II

*

Štampa
GRAFIČKO PREDUZEĆE „PROSVETA“, BEOGRAD
ĐURE ĐAKOVIĆA 21